Khanna's Multi-Choice Questions & Answers in

ENVIRONMENTAL ENGINEERING

OBJECTIVE TYPE QUESTIONS & ANSWERS BY THE SAME AUTHOR

- Khanna's Multi-Choice Questions & Answers in Chemical Engineering (ISBN: 9789386173201)
- Khanna's Multi-Choice Questions and Answers in Metallurgical Engineering (ISBN: 9789386173119)
- Khanna's Objective Type Questions & Answers in Chemical Engineering (ISBN: 9789382609773)

Khanna's Multi-Choice Questions & Answers in

ENVIRONMENTAL ENGINEERING

Om Prakash Gupta

B. Tech. (Chemical Engg.)

Ex DGM, HRD

Bokaro Steel Plant

Steel Authority of India Limited

(SAIL)

KHANNA BOOK PUBLISHING CO. (P) LTD.

Publisher of Engineering and Computer Books

4C/4344, Ansari Road, Darya Ganj, New Delhi-110002

Phone : 011-23244447-48 **Mobile:** +91-9910909320

E-mail : contact@khannabooks.com

Website : www.khannabooks.com

Dear Readers,

To prevent the piracy, this book is secured with HIGH SECURITY HOLOGRAM on the front title cover. In case you don't find the hologram on the front cover title, please write us to at *contact@khannabooks.com* or whatsapp us at +91-99109 09320 and avail special gift voucher for yourself.

Specimen of Hologram on front Cover title:

Moreover, there is a SPECIAL DISCOUNT COUPON for you with EVERY HOLOGRAM.

How to avail this SPECIAL DISCOUNT:

Step 1: Scratch the hologram
Step 2: Under the scratch area, your "coupon code" is available
Step 3: Logon to *www.khannabooks.com*
Step 4: Use your "coupon code" in the shopping cart and get your copy at a special discount
Step 5: Enjoy your reading!

Thanks,
Khanna Publishing House

Price: ₹ 450.00

Khanna's Multi-Choice Questions & Answer In
ENVIRONMENTAL ENGINEERING
O.P. Gupta

ISBN: 978-93-91505-07-3

Edition: 2021

Published by:

Khanna Book Publishing Co. (P) Ltd.
4C/4344, Ansari Road, Darya Ganj, New Delhi-110 002
Phone: 011-23244447-48 **Mobile:** +91-9910909320
E-mail: contact@khannabooks.com

Printed in India by: SPS Enterprises, Delhi

Khanna's Multi-Choice Questions & Answers in

ENVIRONMENTAL ENGINEERING

Om Prakash Gupta
B. Tech. (Chemical Engg.)
Ex DGM, HRD
Bokaro Steel Plant
Steel Authority of India Limited
(SAIL)

KHANNA BOOK PUBLISHING CO. (P) LTD.

Publisher of Engineering and Computer Books

4C/4344, Ansari Road, Darya Ganj, New Delhi-110002

Phone : 011-23244447-48 **Mobile:** +91-9910909320

E-mail : contact@khannabooks.com

Website : www.khannabooks.com

Dear Readers,

To prevent the piracy, this book is secured with HIGH SECURITY HOLOGRAM on the front title cover. In case you don't find the hologram on the front cover title, please write us to at *contact@khannabooks.com* or whatsapp us at +91-99109 09320 and avail special gift voucher for yourself.

Specimen of Hologram on front Cover title:

Moreover, there is a SPECIAL DISCOUNT COUPON for you with EVERY HOLOGRAM.

How to avail this SPECIAL DISCOUNT:

Step 1: Scratch the hologram
Step 2: Under the scratch area, your "coupon code" is available
Step 3: Logon to *www.khannabooks.com*
Step 4: Use your "coupon code" in the shopping cart and get your copy at a special discount
Step 5: Enjoy your reading!

Thanks,
Khanna Publishing House

Price: ₹ 450.00

Khanna's Multi-Choice Questions & Answer In
ENVIRONMENTAL ENGINEERING

O.P. Gupta

ISBN: 978-93-91505-07-3

Edition: 2021

Published by:

Khanna Book Publishing Co. (P) Ltd.
4C/4344, Ansari Road, Darya Ganj, New Delhi-110 002
Phone: 011-23244447-48 **Mobile:** +91-9910909320
E-mail: contact@khannabooks.com

Printed in India by: SPS Enterprises, Delhi

PREFACE

Environmental Engineering is an integral part of the degree, postgraduate & diploma curriculum of various branches of engineering. Besides, it is also a compulsory paper for various associate membership examinations conducted by professional bodies like Institution of Engineers (AMIE), Indian Institute of Metals (AMIIM), Indian Institute of Chemical Engineers (AMIIChE) etc. This book has been prepared keeping in view the syllabus of these examinations.

This book contains more than 1600 Multiple Choice Questions & Answers in Section-A followed by short Questions & Answers in Section-B on the following topics:

1. Environmental Pollution
2. Air Pollution
3. Water Pollution
4. Waste Water Treatment
5. Marine Pollution
6. Noise Pollution
7. Soil/Land Pollution
8. Solid Waste Pollution
9. Radioactive/Nuclear Pollution
10. Environmental Studies
11. Ozone Layer Depletion and their effects
12. Climate Change
13. ISO-14000
14. Smoking & its Health Effects
15. Electrical Hazards

This book is targeted to benefit the following:

- Diploma in engineering students
- Degree in engineering students
- Postgraduate in engineering students
- AMIE, AMIIM, AMIICHE Students.
- M.Sc. & B.Sc. students
- Faculty members and teaching staff
- Practicing engineers/professionals/consultants

Subject matter has been presented in simple, lucid and easy to understand language. I hope that the book in its present form will be useful for the students.

Constructive suggestion for the improvement of the book are welcome.

15th April, 2021

Om Prakash Gupta
Ranchi

TABLE OF CONTENTS

Section A: Short Questions & Answers

1. Environmental Pollution 1.1-1.28
2. Air Pollution 2.1-2.46
3. Water Pollution 3.1-3.74
4. Waste Water Treatment 4.1-4.33
5. Marine Pollution 5.1-5.13
6. Noise Pollution 6.1-6.17
7. Soil/Land Pollution 7.1-7.17
8. Solid Waste Pollution 8.1-8.109
9. Radioactive/Nuclear Pollution 9.1-9.17
10. Environmental Studies 10.1-10.20
11. Ozone Layer Depletion and its Health Effects 11.1-11.26
12. Climate Change 12.1-12.4
13. ISO-14000 .... 13.1-13.2
14. Smoking and its Health Effects 14.1-14.9
15. Electrical Hazards 15.1-15.4

Section B: Multiple Choice Questions and Answers

16. Multiple Choice Question and Answers on Environmental Engineering 16.1-16.135

    Immediately Dangerous to Life or Health (IDLH) Concentrations of Various Substances/Chemicals (as per NIOSH as on 03/01/1995) 16.136-16.144

Section A: Short Questions & Answers

Chapter 1 : Environmental Pollution
Chapter 2 : Air Pollution
Chapter 3 : Water Pollution
Chapter 4 : Waste Water Treatment
Chapter 5 : Marine Pollution
Chapter 6 : Noise Pollution
Chapter 7 : Soil/Land Pollution
Chapter 8 : Solid Waste Pollution
Chapter 9 : Radioactive/Nuclear Pollution
Chapter 10 : Environmental Studies
Chapter 11 : Ozone Layer Depletion and its Health Effects
Chapter 12 : Climate Change
Chapter 13 : ISO-14000
Chapter 14 : Smoking and its Health Effects
Chapter 15 : Electrical Hazards

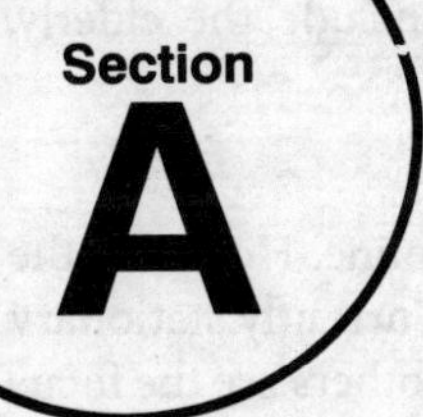

Section A

1 Environmental Pollution

SHORT QUESTIONS AND ANSWERS ON ENVIRONMENTAL POLLUTION

Q.1. Define air pollution.

Ans. Air pollution is defined as the presence of any solids, liquid or gaseous matter present in the atmosphere in such concentrations that may or tend to be injurious to human beings or other living creatures, plants, vegetation.

Q.2. What is primary and secondary air pollutant?

Ans. The pollutant emitted directly from the sources, either from the natural hazardous events like dust storms volcanoes or by human activities are called primary pollutant.

The primary pollutant often reacts with one another, water vapour or with other atmospheric air to form a new pollutant is called secondary pollutant.

Q.3. What are the sources and effects of the particulate matter pollutant?

Ans. Particulate matter originates from many different stationary and mobile sources as well as from natural sources. Fine particles result from the fuel combustion of motor vehicles, power generation, and industrial facilities, as well as from residential fire places and wood stoves.

Sources of coarse particles include vehicles travelling on unpaved roads, materials handling, crushing and grinding operations, and wind blown dust. Some particles are emitted directly from their sources, such as smoke stacks and cars. In other cases, gases such as SO_2, NO_x, and VOCs interact with other compounds in the air to form fine particles.

Health effects:

- Some particles are large or dark enough to be seen as soot or smoke. Others are so small they can hardly be detected with an electron microscope.
- PM10 refers to all particles less than or equal to 10 micrometers in diameter. Ten micrometers are about one-seventh the diameter of human hair.
- Particulate matter includes both fine and coarse particles. When inhaled, these particles can accumulate in the respiratory system and are closely associated with numerous health effects. Exposure to coarse particles is primarily associated with aggravation of respiratory conditions, such as asthma.
- Fine particles are also closely associated with such effects as premature mortality of elderly persons, increased hospital admissions and emergency room visits for heart and lung patients, increased respiratory symptoms and disease among the chronically ill, and decreased lung function in almost all population groups.

- However, those groups that appear to be at greatest risk to such effects include the elderly, individuals with cardio pulmonary disease such as asthma, and children.

Q.4. What are sources and effects of lead?

Ans. Historically atmospheric lead came primarily from the combustion of leaded gasoline. However, the use of unleaded gasoline has reduced mobile source lead emissions by over 90%. Currently stationary sources, such as lead smelters, battery manufacturers, iron and steel producers and others are the focus of regulatory action.

Health effects of lead:

- Lead (Pb) forms stable compounds, which persist and accumulate both in the environment and in the human body.
- Lead enters the human body through ingestion and inhalation with consequent absorption into the blood stream and distribution to all body tissues.
- Clinical, epidemiological and toxicological studies have demonstrated that exposure to lead adversely affects human health. Low-level lead exposure has been found to interfere with specific enzyme systems and blood production. Kidney and neurological cell damage has also been associated with lead exposure.
- Animal studies have demonstrated that lead can contribute to reduced fertility and birth defects.
- Children are the most sensitive to many of lead's adverse effects. Recent studies show that lead also may be an important factor in high blood pressure and subsequent heart disease. Other serious potential effects from lead exposure are behavioural.

Q.5. Describe the sources and effects of sulfur dioxide (SO_2).

Ans. Sources of SO_2:

- Sulfur dioxide is an atmospheric pollutant that results from combustion processes (mainly burning of fossil fuels containing sulfur compounds), petroleum refining, sulfuric acid manufacturing and smelting of ores.
- Fuel combustion, largely from coal-fired power plants, accounts for most of the total SO_2 emissions.
- The reduction of sulfur dioxide pollution levels can generally be achieved by using low sulfur content fuels or chemical sulfur removal systems.

Health effects of SO_2:

- The effects of SO_2 on health are irritation and inflammation of tissue that it directly contacts.
- Inhalation of SO_2 causes bronchial constriction resulting in an increased resistance to air flow, reduction of air volume and a marked increase in the respiratory and heart rate.
- High concentrations of SO_2 can result in temporary breathing impairment for asthmatic children and adults who are active outdoors.
- Short-term exposures of asthmatic individuals to elevated SO_2 levels during moderate exertion may result in breathing difficulties that can be accompanied by such symptoms as wheezing, chest tightness, or shortness of breath.
- Other effects that often have been associated with longer-term exposures to high concentrations of SO_2, in conjunction with high levels of PM, include respiratory illness, alterations in the lungs' defenses, and aggravation of existing cardiovascular disease. The population groups most susceptible to these conditions include individuals with cardiovascular or chronic lung disease, as well as children and the elderly.
- SO_2 can exacerbate pre-existing respiratory diseases (asthma, bronchitis, emphysema).

- The enhancement (synergism) by particulate matter of the toxic response to sulfur dioxide has been respiratory system's mucous secretions, which reduces the system's ability to remove particulates via mucociliary clearance. This can result in an increase incidence of respiratory infection.

Q.6. Describe the sources and effects of carbon monoxide (CO).

Ans. Sources of the (CO):

- Carbon monoxide (CO) is a colorless, odorless and, at high levels, a poisonous gas, formed when carbon in fuel is not burned completely.
- It is a component of motor vehicle exhaust, which contributes about 60 percent of all CO emissions nationwide.
- Non-road vehicles account for the remaining CO emissions from the transportation sources category.
- High concentrations of CO generally occur in areas with heavy traffic congestion. In cities, as much as 95 percent of all CO emissions may come from automobile exhaust.

Effects of (CO):

- The toxic effects of high concentrations of CO on the body are well known. Carbon monoxide is absorbed by the lungs and reacts with hemoglobin (the oxygen carrying molecule in the blood) to form carboxyhemoglobin (COHb).
- This reaction reduces the oxygen carrying capacity of blood because the affinity of hemoglobin for CO is over 240 times that for oxygen.
- The higher the percentage of hemoglobin bound up in the form of carboxyhemoglobin, the more serious is the health effect.
- The level of COHb in the blood is directly related to the CO concentration of the inhaled air.
- For any given ambient air CO concentration, the COHb level in the blood will reach an equilibrium concentration after a sufficient time period.
- This COHb equilibrium level will be maintained in the blood as long as CO ambient air level remains unchanged.

Q.7. Describe the sources and effects of nitrogen dioxide (NO_2).

Ans. Sources of NO_2:

- The major sources of manmade NO_x emissions are high temperature combustion processes, burning of wood, from automobiles and power plants.
- Home heaters and gas stoves also produce substantial amounts of NO_2 in indoor settings.

Health effects of NO_2:

- Nitrogen dioxide (NO_2) is a reddish brown, highly reactive gas that is formed in the ambient air through the oxidation of nitric oxide (NO).
- Nitrogen oxides (NO_x), the term used to describe the sum of NO, NO_2 and other oxides of nitrogen, play a major role in the formation of ozone, particulate matter, and acid rain.
- Short-term exposures (*i.e.*, less than 3 hours) to low levels of nitrogen dioxide (NO_2) may lead to changes in airway responsiveness and lung function in individuals with pre-existing respiratory difficulties and increases in respiratory illness in children (5–12 years old).
- Long-term exposures to NO_2 may lead to increased susceptibility to respiratory infection and may cause permanent alterations in the lung. Nitrogen oxides react in the air to form ground level ozone and fine particle pollution, which is associated with adverse health effects.

Q.8. Describe the sources and effects of ozone (O_3).

Ans. Sources of ozone:

- Ozone is not emitted directly into the air but is formed by the reaction of VOCs and NOx in the presence of heat and sunlight.
- Ground-level ozone forms readily in the atmosphere, usually during hot summer weather.
- VOCs are emitted from a variety of sources, including motor vehicles, chemical plants, refineries, factories, consumer and commercial products, and other industrial sources.

Health effects of ozone:

- Short-term (1–3 hours) and prolonged (6–8 hours) exposures to ambient O_3 have been linked to a number of health effects of concern.
- For example, increased hospital admissions and emergency room visits for respiratory causes have been associated with ambient O_3 exposures.
- Exposures to O_3 can make people more susceptible to respiratory infection, result in lung inflammation, and aggravate pre-existing respiratory diseases such as asthma.

Q.9. Describe the sources and effects of toxic air pollutants.

Ans. Sources: Toxic or hazardous air pollutants are those pollutants that cause or may cause cancer or other serious health effects, such as reproductive disorders, birth defects, or adverse environmental and ecological effects.

- Examples of toxic air pollutants include benzene, found in gasoline; perchloroethylene, emitted from some dry cleaning facilities; and Methylene chloride, used as a solvent by a number of industries.
- Most air toxics originate from man-made sources, including mobile sources (*i.e.*, cars, trucks, and construction equipment) and stationary sources (*i.e.*, factories, refineries, power plants), as well as indoor sources (*i.e.*, building materials and activities such as cleaning).
- Some air toxics are also released from natural sources such as volcanic eruptions and forest fires.

Effects:

- People exposed to toxic air pollutants at sufficient concentrations may experience various health effects including damage to the immune system, as well as neurological, reproductive (*i.e.*, reduced fertility), developmental, respiratory and other health problems.
- Many of these toxins may increase the risk of developing cancer or experiencing other serious health effects.
- In addition to exposure from breathing air toxics, risks also are associated with the deposition of toxic pollutants onto soils or surface waters, where they are taken up by plants and ingested by animals and eventually magnified up through the food chain.
- Like humans, animals may experience health problems due to air toxics exposure.

Q.10. Describe the environmental effects of air pollution.

Ans. Various effects are discussed below:

(*i*) **Ecosystem:** Long before pollutant effects become manifest in human health, they are seen in the environment.

- When sulfur dioxide (SO_2) and nitrogen oxides (NO_x) are transformed into acids in the Atmosphere, the resulting precipitation, rain, snow, or fog) is deposited onto vegetation and into the lakes and soil.

(*ii*) **Damage to vegetation:**

- **Usually occurs in the structural mechanisms of the leaf since it is the leaf that provides the mechanics for growth.**
- **A leaf is typically divided into three parts: (1) the epidermis (outer protective layer), (2) the mesophyll (the spongy center section), and (3) the veins (a dense network of tubes that carries nutrients throughout the plant).**
- **When plants are exposed to SO_2, O_3, NO_2, and acid rain, all three structures are inextricably damaged.**

(*iii*) **Property damage:**

- **Property damage is another significant by-product of air pollution. Research has shown that acid rain can destroy irreplaceable cultural treasures such as buildings and outdoor art.**
- **In addition, acid rain cracks and splits rubber products, nylon, polymer plastics, paint dyes, and other textiles. Acidification not only soils buildings, clothing, and automobiles, but also deteriorates the metal and stone in our buildings and monuments.**

(*iv*) **Quality of life:**

- **Quality of life is another matter that is also diminished by particulate matter and ozone formation.**
- **Pristine vistas can be significantly altered by air pollution transported several hundred miles away. For example, pollutants can substantially affect outdoor sports and recreation.**
- **One survey of sixty-one native trout streams in southwestern Pennsylvania found that 26% of the streams studied did not support populations of brook trout due to acidic conditions Likewise the cosmetic appearance of our homes and cars require significant expenditures of time and money to remain free of the damaging effects of air pollution.**

Q.11. How are the concentration of air pollutants expressed?

Ans. **Concentrations of air pollutants are commonly expressed as the mass of pollutant per unit volume of air, as mg/m³, μg/m³, ng/m³.**

Concentration of gaseous pollutants may also be expressed as volume of pollutant per million volumes of the air plus pollutant mixture (ppm) where 1 ppm= 0.0001 % by volume. It is sometimes necessary to convert from volumetric units to mass per unit volume and vice versa.

The relationship between ppm and mg/m³ depends on the gas density, which in turn depends on Temperature, Pressure, and Molecular weight of the pollutant. The following expression can be used to convert of between ppm and mg/m³ at any temperature or pressure.

$$\text{mg/m}^3 = \frac{(273 \times \text{ppm} \times \text{molecular weight} \times \text{pressure})}{22.4 \times \text{temperature}}$$

Multiply the calculated value of mg/m³ by 1000 to obtain μg/m³. The constant 22.4 is the volume in liter occupied by 1 mole of an ideal gas at standard concentration (0°C and 1 atm.). One mole of any substance is a quantity of that substance whose mass in grams numerically equals its molecular weight

Q.12. What are various particulate pollutants found in atmosphere?

Ans. **Particulate matter suspended in air are dust and soot released from the industrial chimneys. Their size ranges from 0.001 to 500 μm in diameter. Particles less than 10 μm float and move freely with the air current. Particles which are more than 10 μm in diameter settle down. Particles less than 0.02 μm form persistent aerosols. Major source of SPM (suspended particulate matter) are vehicles, power plants, construction activities, oil refinery, railway yard, market place, industries, etc.**

- **Fly ash:** Fly ash is ejected mostly by thermal power plants as by products of coal burning operations. Fly ash pollutes air and water and may cause heavy metal pollution in water bodies. Fly ash affects vegetation as a result of its direct deposition on leaf surfaces or indirectly through its deposition on soil. Fly ash is now being used for making bricks and as a land fill material.
- **Lead and other metals particles:** Tetraethyl lead (TEL) is used as an anti-knock agent in petrol for smooth and easy running of vehicles. The lead particles coming out from the exhaust pipes of vehicles is mixed with air. If inhaled it produces injurious effects on kidney and liver and interferes with development of red blood cells. Lead mixed with water and food can create cumulative poisoning. It has long term effects on children as it lowers intelligence.

Oxides of iron, aluminum, manganese, magnesium, zinc and other metals have adverse effect due to deposition of dust on plants during mining operations and metallurgical processes. They create physiological, biochemical and developmental disorders in plants and also contribute towards reproductive failure in plants.

Q.13. What are the sources and effects of various gaseous pollutants in the atmosphere?

Ans. Power plants, industries, different types of vehicles – both private and commercial use petrol, diesel as fuel and release gaseous pollutants like carbon dioxide, oxides of nitrogen and Sulphur dioxide along with particulate matter in the form of smoke. All of these have harmful effects on plants and humans. Table below lists some of these pollutants, their sources and harmful effects.

Table 1.1: Gaseous air pollutants: their sources and effects

Pollutants	Sources	Harmful effects
Carbon compound (CO and CO_2)	Automobile exhaust, burning of wood and coal	• Respiratory problems • Green house effect
Sulphur compounds (SO_2 and H_2S)	Power plants and refineries, volcanic eruptions	• Respiratory problems in humans • Loss of chlorophyll in plants (chlorosis) • Acid rain
Nitrogen Compounds (NO and N_2O)	Motor vehicle exhaust, atmospheric reaction	• Irritation in eyes and lungs • Low productivity in plants • Acid rain damages materials
Hydrocarbons (benzene, ethylene)	Automobiles and petroleum industries	• Respiratory problem • Cancer causing properties
SPM (Suspended Particulate Matter) (Any solid and liquid particles suspended in the air e.g., flush, dust, lead)	Thermal power plants, construction activities, metallurgical processes and automobiles	• Poor visibility, breathing problems • Lead interferes with the development of red blood diseases and cancer. • Smog (smoke and fog) formation leads to poor visibility and aggravates asthma in patients
Fibers (Cotton, wool)	Textiles and carpet weaving industries	Lung disorders

Q.14. What are the common causes of indoor air quality (IAQ) problems?

Ans. IAQ problems result from interactions between building materials and furnishing, activities within the building, climate, and building occupants. IAQ problems may arise from one or more of the following causes:

Indoor environment: Inadequate temperature, humidity, lighting, excessive noise

Indoor air contaminants: Chemicals, dusts, moulds or fungi, bacteria, gases, vapours, odors

Insufficient outdoor air intake

Indoor air quality has become an important health and safety concern. Common issues associated with IAQ include:

Improper or inadequately maintained heating and ventilation systems. Contamination by construction materials, glues, fiberglass, particle boards, paints, chemicals, etc. Increase in number of building occupants and time spent indoors

Q.15. What are sources of some indoor air contaminants?

Ans. Following are some examples of common indoor air contaminants and their main sources:

Carbon dioxide (CO_2), tobacco smoke, perfume, body odors – from building occupants

Dust, fiberglass, asbestos, gases, including formaldehyde – from building materials

Toxic vapors, volatile organic compounds (VOCs) – from workplace cleansers, solvents, pesticides, disinfectants, glues

Gases, vapors, odors – off-gas emissions from furniture, carpets, and paints

Dust mites – from carpets, fabric, foam , chair cushions

Microbial contaminants, fungi, moulds, bacteria, – from damp areas, stagnant water and condensate pans

Ozone – from photocopiers, electric motors, electrostatic air cleaners

Q.16. What are the sources, effects and control measures of indoor air pollution?

Ans. Poor ventilation due to faulty design of buildings leads to pollution of the confined space. Paints, carpets, furniture, etc. in rooms may give out volatile organic compounds (VOCs). Use of disinfectants, fumigants, etc. may release hazardous gases. In hospitals, pathogens present in waste remain in the air in the form of spores. This can result in hospital acquired infections and is an occupational health hazard. In congested areas, slums and rural areas burning of firewood and biomass results in lot of smoke. These all are sources of indoor air pollution.

Children and ladies exposed to smoke may suffer from acute respiratory problems which include running nose, cough, sore throat, lung infection, asthma, difficulty in breathing, noisy respiration and wheezing.

Control of indoor air pollution: Use of wood and dung cakes should be replaced by cleaner fuels such as biogas, kerosene or electricity. But supply of electricity is limited. Similarly kerosene is also limited. Improved stoves for looking like smokeless chullahs have high thermal efficiency and reduced emission of pollutants including smoke. The house designs should incorporate a well ventilated kitchen. Use of biogas and CNG (Compressed Natural Gas) need to be encouraged.

Those species of trees such as baval (*Acacia nilotica*) which are least smoky should be planted and used. Charcoal is a comparatively cleaner fuel. Indoor pollution due to decay of exposed kitchen waste can be reduced by covering the waste properly. Segregation of waste, pretreatment at source, sterilization of rooms will help in checking indoor air pollution.

Q.17. Describe some methods for prevention and control of industrial pollution.

Ans. Industrial pollution can be reduced by:

(*a*) use of cleaner fuels such as natural gas in power plants, fertilizer plants etc. which is cheaper in addition to being environmentally friendly.

(*b*) employing environment friendly industrial processes so that emission of pollutants and hazardous waste is minimized.

(*c*) installing devices which reduce release of pollutants. Devices like bag filters, electrostatic precipitators, inertial collectors, scrubbers, gravel bed filters or dry scrubbers etc. They are described below:

(*i*) **Bag filters:** Filters remove particulate matter from the gas stream. The medium of a filter may be made of fibrous materials like cloth, granular material like sand, a rigid material like screen, or any mat like felt pad. Baghouse filtration system is the most common one and is made of cotton or synthetic fibers (for low temperatures) or glass cloth fabrics (for higher temperature up to 290°C).

(*ii*) **Electrostatic precipitators (ESP):** The emanating dust is charged with ions and the ionized particulate matter is collected on an oppositely charged surface. The particles are removed from the collection surface by occasional shaking or by rapping the surface. ESPs are used in boilers, furnaces, and many other units of thermal power plants, cement factories, steel plants, etc.

(*iii*) **Inertial collectors:** It works on the principle that inertia of SPM in a gas is higher than its solvent and as inertia is a function of the mass of the particulate matter this device collects heavier particles more efficiently. 'Cyclone' is a common inertial collector used in gas cleaning plants.

(*iv*) **Scrubbers:** Scrubbers are wet collectors. They remove aerosols from a stream of gas either by collecting wet particles on a surface followed by their removal, or else the particles are wetted by a scrubbing liquid. The particles get trapped as they travel from supporting gaseous medium across the interface to the liquid scrubbing medium.

Gaseous pollutants can be removed by absorption in a liquid using a wet scrubber and depends on the type of the gas to be removed e.g., for removal of Sulphur dioxide alkaline solution is needed as it dissolves Sulphur dioxide. Gaseous pollutants may be adsorbed on an activated solid surface like silica gel, alumina, carbon, etc. Silica gel can remove water vapor. Condensation allows the recovery of many by products in coal and petroleum processing industries from their liquid effluents.

Apart from the use of above mentioned devices, other control measures are-

- increasing the height of chimneys.
- closing industries which pollute the environment.
- shifting of polluting industries away from cities and heavily populated areas.
- development and maintenance of green belt of adequate width.

Q.18. Discuss some means adopted for control of vehicular pollution.

Ans.
- The emission standards for automobiles have been set which if followed will reduce the pollution. Standards have been set for the installation of catalytic converters which reduce vehicular emission.
- In cities, motor vehicles need to obtain Pollution Under Control (PUC) certificate at regular intervals. This ensures that levels of pollutants emitted from vehicle exhaust are not beyond the prescribed legal limits.
- The price of diesel is lower than petrol which promotes use of diesel. To reduce emission of Sulphur dioxide, Sulphur content in diesel has been reduced to 0.05%.
- Earlier lead in the form of tetraethyl lead was added in the petrol to raise octane level for smooth running of engines. Addition of lead in petrol has been banned to prevent emission of lead particles with the vehicular emission.

Q.19. What is acid rain?

Ans. "Acid rain" is a broad term used to describe several ways that acids fall out of the atmosphere. A more precise term is acid deposition, which has two parts: wet (rain, fog, and snow) and dry (particles and gases).

Acid rain's effects include:

- harming fish and other organisms living in lakes and streams harming to a variety of plants and animals on land and damaging human health.
- reducing how clearly, we see through the air; damaging materials like those found in statues and buildings.

The specific effects and their severity depends on several factors, including soil and surface water chemistry, the amount of air pollution that creates acid rain, and the specific species involved.

Q.20. What are the effects of agricultural burning?

Ans. Burning agricultural waste is a source of particulate matter and other pollutants. It can greatly affect the regional air quality, visibility and ground level ozone potential.

Q.21. What are the sources of carbon monoxide (CO) at home?

Ans. Sources of CO include clogged chimney, wood stove, wood and gas fireplace, automobile garage, gas and water heater, gas appliances, gas or kerosene heater, cigarette smoke.

Q.22. What are the health effects of carbon monoxide?

Ans. When we breathe air containing carbon monoxide, it is absorbed through the bloodstream where it displaces oxygen and bonds with the hemoglobin in our blood. Carbon monoxide has a greater affinity to hemoglobin than oxygen; CO bonds to hemoglobin about 240 times better than oxygen. Without oxygen, vital organs, heart and brain become deprived of oxygen and will begin to deteriorate. To compensate, heart rate increases, breathing may become difficult and in the most serious circumstances cardiac trauma, brain damage, coma and even death will result.

Q.23. What are CFC's?

Ans. Chlorofluorocarbons (CFCs) are nontoxic, nonflammable chemicals containing atoms of carbon, chlorine, and fluorine. They are used in the manufacture of aerosol sprays, blowing agents for foams and packing materials, as solvents, and as refrigerants. CFCs react with sunlight in the earth's stratosphere to break down the protective ozone layer, a layer of gas that shields the earth's surface from damaging ultra violet (UV-B)rays.

Q.24. What are ozone and the ozone layer?

Ans. Ozone (O_3) is a molecule made up of three oxygen atoms. The oxygen we breathe (O_2) is similar but has only two oxygen atoms. High up in a region of the upper atmosphere known as the stratosphere, light rays can break down breathable oxygen into two individual oxygen atoms. Single oxygen atoms are quite reactive, and ozone is formed when one of these lone oxygen bump into and combine with O_2. But ozone isn't very stable either. If a high-energy light ray, in particular ultraviolet-B (UV-B) rays, strikes an ozone molecule, it too will break down, back to the lone oxygen and O_2. Other molecules naturally found in the stratosphere, such as nitrous oxide, can also react with O_3 and break it down. Over time, as the atmosphere formed, the reactions that create and break down ozone came into a dynamic equilibrium. The result was a small residual amount of ozone concentrated in a band between 20 and 30 kms. high in the stratosphere. This is the band popularly referred to as the ozone layer. But the equilibrium is delicate, and ozone is rare even in the ozone layer. For every ten million molecules of air, two million are breathable O_2, and only three are ozone. Yet this small amount of ozone is enough to prevent most UV-B radiation from reaching the surface of the earth.

Q.25. What is ozone hole and its effects?

Ans. The stratosphere has an ozone layer which protects the earth's surface from excessive ultraviolet (UV) radiation from the Sun. Chlorine from chemicals such as chlorofluorocarbons (CFCs) used for refrigeration, air conditioning, fire extinguishers, cleaning solvents, aerosols (spray cans of perfumes, medicine, insecticide) cause damage to ozone layer as the chlorine contained in the CFCs on reaching the ozone (O_3) layer split the ozone molecules to form oxygen (O_2). Amount of ozone, thus gets reduced and cannot prevent the entry of UV radiation. There has been a reduction of ozone umbrella or shield over the Arctic and Antarctic regions. This is known as **ozone hole**. This permits passage of UV radiation on earth's atmosphere which causes sunburn, cataract in eyes leading to blindness, skin cancer, reduced productivity of forests, etc. Under the "Montreal Protocol" amended in 1990 it was decided to completely phase out the use of CFCs to prevent damage of ozone layer.

Q.26. How is increased surface UV-B radiation harmful?

Ans. The harmfulness of UV-B stems from the high energy of these light rays, which allows them to penetrate deeply into water, leaves, and skin. Because of this, increased UV-B radiation can harm the metabolism of cells and even damage genetic material. Although living organisms have always been exposed to some UV-B, cellular repair mechanisms evolved to protect against its damaging effects. The problem with increased UV-B is that it causes more damage than the natural protection can cope with. Increased UV-B radiation leads to increased incidence of such problems as skin cancer, eye damage and cataracts, and possible inhibition of immune system function in humans as well as other animals. Plants also suffer under increased UV-B, and their vulnerability could result in reduced crop yields, damage to forest ecosystems, and decreased populations of phytoplankton in the world's oceans.

Q.27. What has been done to protect the ozone layer?

Ans. Through extensive research, scientists identified the human-produced chemicals that are responsible for the destruction of stratospheric ozone. As evidence emerged on the extent of the threat to the ozone layer, the international community agreed to control ozone-depleting substances and schedule a timetable for completely phasing them out. This agreement is known as the Montreal Protocol and is a monumental achievement in international cooperation and environmental protection. Furthermore, the protocol provides for an on-going process so that, as the scientific understanding of ozone depletion improved, the phasing out process of ozone depleting substances (ODS) could be accelerated. The agreement also provides a powerful precedent for similar international efforts to deal with global warming and loss of biodiversity.

The Environmental Protection Agency enforces the terms of the Montreal Protocol. The treaty provisions are given the force of law and accordingly, chlorofluorocarbon, carbon tetrachloride, methyl chloroform and methyl bromide production ended and all hydrochlorofluorocarbons (HCFCs) will be phased out by 2030.

Q.28. Is ozone depletion related to global warming?

Ans. No, Ozone depletion and global warming are separate problems, though some agents contribute to both. Chlorofluorocarbons (CFCs) are the principle cause of ozone deletion, but they also happen to be potent heat-trapping gases. Still, CFCs are responsible for less than 10 percent of total atmospheric warming, far less than the 60 percent contribution of carbon dioxide. Thus, attention paid to CFCs has been on their ozone depletion role. This will change as CFCs are phased out and replaced by hydrochlorofluorocarbons (HCFCs) and hydrofluorocarbons (HFCs such as R-134a). These chemicals have little or no effect on the ozone layer but are strong heat-trapping gases. As their concentration in the atmosphere is already rising, the likely net effect in the future is that reductions in the CFC-related contribution to global warming will be offset by the presence of HCFCs and HFCs.

Q.29. What is global warming and greenhouse effect?

Ans. Atmospheric gases like carbon dioxide, methane, nitrous oxide, water vapor, and chlorofluorocarbons are capable of trapping the out-going infrared radiation from the earth. Infra-red radiations trapped by the earth's surface cannot pass through these gases and increase thermal energy or heat in the atmosphere. Thus, the temperature of the global atmosphere is increased. As this phenomenon of increase in temperature is observed in **green** houses, in the botanical gardens these gases are known as green house gases and the heating effect is known as green house effect. If greenhouse gases are not checked, by the turn of the century the temperature may rise by 3°C. This will melt the polar ice caps and increase the sea level leading to coastal flooding, loss of coastal areas and ecosystems like swamps and marshes, etc.

Q.30. What is biodiesel?

Ans. Biodiesel is a clean burning mono-alkyl ester-based oxygenated fuel, made from vegetable oil, cooking grease, or animal fats. Currently, the most common source is virgin rapeseed oil, with biodiesel also being processed from recycled frying oils. Biodiesel is a renewable fuel that can be domestically produced from agricultural resources, simple to use, biodegradable, nontoxic and free of sulfur and aromatic compounds and meets the sulfur standards. Neat biodiesel fuel (100% biodiesel) has been designated as an alternative fuel for automobiles.

Q.31. What is diesel particulate matter?

Ans. Diesel particulate matter is part of a complex mixture that makes up diesel exhaust. Diesel exhaust is commonly found throughout the environment. Diesel exhaust is composed of two phases, either gas or particle and both phases contribute to the health risk. The gas phase is composed of many of the urban hazardous air pollutants, such as acetaldehyde, acrolein, benzene, 1,3-butadiene, formaldehyde and polycyclic aromatic hydrocarbons (PAHs). The particle phase also has many different types of particles that can be classified by size or composition. The size of diesel particulates that are of greatest health concern are those that are in the categories of fine, and ultra fine particles. The composition of these fine and ultra fine particles maybe composed of elemental carbon with adsorbed compounds such as organic compounds, sulfate, nitrate, metals and other trace elements. Diesel exhaust is emitted from a broad range of diesel engines; the on road diesel engines of trucks, buses and cars and the off road diesel engines that include locomotives, marine vessels and heavy duty equipment.

Q.31. How do we get exposed to diesel particulate matter?

Ans. The most common exposure pathway is breathing the air that contains the diesel particulate matter. The fine and ultra-fine particles are respirable which means that they can devoid many of the human respiratory system defense mechanisms and enter deeply into the lung. Exposure to diesel particulate matter comes from both on road and off road engine exhaust that is either directly emitted from the engines or aged through lingering in the atmosphere.

Q.32. How can diesel particulate matter affect human health?

Ans. Diesel exhaust causes health effects from both short term or acute exposures and also long term chronic exposures, such as repeated occupational exposures. The type and severity of health effects depends upon several factors including the amount of chemical we are exposed to and the length of time we are exposed. Individuals also react differently to different levels of exposure. There is limited information on exposure to just diesel particulate matter but there is enough evidence to indicate that inhalation exposure to diesel exhaust causes acute and chronic health effects. Acute exposure to diesel exhaust may cause irritation to the eyes, nose, throat and lungs, and some neurological effects such as light headedness. Acute exposure may also elicit a cough or nausea as well as exacerbate asthma. Chronic exposure in experimental animal inhalation studies have shown a range of dose dependent lung inflammation and cellular changes in the lung and there are also diesel exhaust immunological effects. Based upon human and laboratory studies, there is considerable evidence that diesel exhaust is a likely carcinogen. Human epidemiological studies demonstrate an association between diesel exhaust exposure and increased lung cancer rates in occupational settings.

Q.33. Define atmospheric stability and instability.

Ans. Atmospheric stability is the resistance of the atmosphere to vertical motion. Temperatures normally increase as we get closer to the earth's surface. This is due in part to the greater molecular activity of denser, more compressed air at lower altitudes. These conditions change throughout a 24-hour period, as the day time solar heating and night time heat loss to and through the atmosphere tend to modify the temperature distributions.

Weather is strongly affected by how stable or unstable the atmosphere is. Stable air means that the weather is likely to be calm. It may rain or snow slowly and steadily, it may be sunny, but the weather will not change quickly. Unstable air means that the weather might change quickly with very little warning. Unstable air leads to sudden thunder storms.

Q.34. Describe the sources, effects and control of water pollution.

Ans. Addition or presence of undesirable substances in water is called **water pollution**. Water pollution is one of the most serious environmental problems. Water pollution is caused by a variety of human activities such as industrial, agricultural and domestic. Agricultural run off laden with excess fertilizers and pesticides, industrial effluents with toxic substances and sewage water with human and animal wastes pollute our water thoroughly. Natural sources of pollution of water are soil erosion, leaching of minerals from rocks and decaying of organic matter. Rivers, lakes, seas, oceans, estuaries and ground water sources may be polluted by point or non-point sources.

When pollutants are discharged from a specific location such as a drain pipe carrying industrial effluents discharged directly into a water body it represents **point source pollution.** In contrast **non-point sources** include discharge of pollutants from diffused sources or from a larger area such as run off from agricultural fields, grazing lands, construction sites, abandoned mines and pits, roads and streets.

Sources of water pollution:

Water pollution is the major source of water born diseases and other health problems.

Sediments brought by runoff water from agricultural fields and discharge of untreated or partially treated sewage and industrial effluents, disposal of fly ash or solid waste into or close to a water body cause severe problems of water pollution. Increased turbidity of water because of sediments reduces penetration of light in water that reduces photosynthesis by aquatic plants.

Pollution due to pesticides

- Pesticides like DDT and others used in agriculture may contaminate water bodies. Aquatic organisms take up pesticides from water get into the food chain (aquatic in this case) and move up the food chain. At higher trophic level they get concentrated and may reach the upper end of the food chain.

Pollution due to inorganic chemicals

- Metals like lead, zinc, arsenic, copper, mercury and cadmium in industrial waste waters adversely affect humans and other animals. Arsenic pollution of ground water has been reported from West Bengal, Orissa, Bihar, Western U.P. Bangladesh is the worst affected from arsenic pollution of its drinking water sources. Consumption of such arsenic polluted water leads to accumulation of arsenic in the body parts like blood, nails and hairs causing skin lesions, rough skin, dry and thickening of skin and ultimately skin cancer.

Effects of water pollution

- Pollution of water bodies by mercury causes **Minamata disease** in humans and **dropsy** in fishes. Lead causes **dyslexia,** cadmium poisoning causes **Itai-Itai disease** etc.

- Oil pollution of sea occurs from leakage from ships, oil tankers, rigs and pipelines. Accidents of oil tankers spill large quantity of oil in seas which kills marine birds and adversely affects other marine life and beaches.

Thermal pollution of water

Power plants- thermal and nuclear, chemical and other industries use lot of water for cooling purposes and the used hot water is discharged into rivers, streams or oceans. The waste heat from the boilers and heating processes increases the temperature of the cooling water. Discharge of hot water may increase the temperature of the receiving water by 10°C above the ambient water temperature.

This is called **thermal pollution of water.** Increase in water temperature decreases dissolved oxygen in water which adversely affects aquatic life. Unlike terrestrial ecosystems, the temperature of water bodies remains steady and does not change very much. Accordingly, aquatic organisms are adopted to a uniform steady temperature of environment and any fluctuation in water temperature severely affects aquatic plants and animals. Hence discharge of hot water from power plants adversely affects aquatic organisms. Aquatic plants and animals in the warm tropical water live dangerously close to their upper limit of temperature, particularly during the warm summer months. It requires only a slight deviation from this limit to cause a thermal stress to these organisms.

Discharge of hot water in water body affects feeding in fishes, increases their metabolism and affects their growth. Their swimming efficiency declines. Running away from predators or chasing prey becomes difficult. Their resistance to diseases and parasites decreases.

Due to thermal pollution biological diversity is reduced. One of the best methods of reducing thermal pollution is to store the hot water in cooling ponds, allow the water to cool before releasing into any receiving water body.

Ground water pollution: Lot of people around the world depend on ground water for drinking, domestic, industrial and agricultural uses. Generally groundwater is a clean source of water. However, human activities such as improper sewage disposal, dumping of farm yard manures and agricultural chemicals, industrial effluents are causing pollution of ground water.

Eutrophication

- 'Eu' means well or healthy and 'trophy' means nutrition. The enrichment of water bodies with nutrients causes eutrophication of the water body. Discharge of domestic waste, agricultural surface runoff, land drainage and industrial effluents in a water body leads to rapid nutrients enrichment in a water body. The excessive nutrient enrichment in a water body encourages the growth of algae duckweed, water hyacinth, phytoplankton and other aquatic plants. The biological demand for oxygen (BOD) increases with the increase in aquatic organisms. As more plants grow and die, the dead and decaying plants and organic matter acted upon by heterotrophic protozoans and bacteria, deplete the water of dissolved oxygen (DO). Decrease in DO result in sudden death of large population of fish and other aquatic organisms including plants, releasing offensive smell and makes the water unfit for human use. The sudden and explosive growth of phytoplankton and algae impart green color to the water is known as water bloom, or "algal blooms". These phytoplankton release toxic substances in water that causes sudden death of large population of fishes. This phenomenon of nutrient enrichment of a water body is called **eutrophication**. Human activities are mainly responsible for the eutrophication of a growing number of lakes and water bodies in the country

Methods for control of water pollution

Waste water from domestic or industrial sources or from garbage dumps is generally known as **sewage.** It may also contain rain water and surface runoff. The sewage water can be treated to make it safe for disposal into water bodies like rivers, lakes etc. The treatment involves three stages: primary, secondary and tertiary. This includes 1. sedimentation, 2. coagulation/flocculation, 3. filtration, 4. disinfection, 5. softening and 6. aeration. The first three steps are involved in primary treatment which remove suspended particulate matter. Secondary treatment removes organic solids, left out after primary treatment, through their microbial decomposition. Effluents after secondary treatment may be clean but contain large amounts of nitrogen, in form of ammonia, nitrates and phosphorous which can cause problem of eutrophication upon their discharge into a receiving water body such as river, lake or pond. The tertiary treatment is meant to remove nutrients, disinfect for removing pathogenic bacteria, and aeration removes hydrogen sulphide and reduce the amount of carbon dioxide and make water healthy and fit for aquatic organisms. This treatment of waste water or sewage is carried out in effluent treatment plants especially built for this purpose. The residue obtained from primary treatment is known as sludge.

The following measures can be adopted to control water pollution:

(*a*) The water requirement should be minimized by altering the techniques involved.

(*b*) Water should be reused with or without treatment.

(*c*) Recycling of water after treatment should be practiced to the maximum extent possible.

(*d*) The quantity of waste water discharge should be minimized.

Water recycling: With increasing population the requirement for water is increasing rapidly. However, the availability of water is limited but an ever-increasing water withdrawal from different sources such as rivers, lakes and ground water is depleting these sources and deteriorating their water quality. Therefore, it is essential to utilize the available water with maximum economy. This involves recycling of waste water for certain uses with or without treatment. Recycling refers to the use of waste-water by the original user prior to the discharge either to a treatment system or to a receiving water body. Thus the waste water is recovered and repetitively recycled with or without treatment by the same user.

Q.35. Describe the sources, effects and control of soil pollution.

Ans. Addition of substances which adversely affect the quality of soil or its fertility is known **as soil pollution.** Generally polluted water also pollute soil. Solid waste is a mixture of plastics, cloth, glass, metal and organic matter, sewage, sewage sludge, building debris, waste water generated from households, commercial and industries establishments add to soil pollution. Fly ash, iron and steel slag, medical and industrial wastes disposed on land are important sources of soil pollution. In addition, fertilizers and pesticides from agricultural use which reach soil as run-off and land filling by municipal waste are growing cause of soil pollution. Acid rain and dry deposition of pollutants on land surface also contribute to soil pollution.

Sources and effects of soil pollution

(*a*) **Plastic bags:** Plastic bags made from low density polyethylene (LDPE), is virtually indestructible, create colossal environmental hazard. The discarded bags block drains and sewage systems. Left over food, vegetable waste etc. on which cows and dogs feed may die due to the choking by plastic bags. Plastic is non-biodegradable and burning of plastic in garbage dumps release highly toxic and poisonous gases like carbon monoxide, carbon dioxide, phosgene, dioxane and other poisonous chlorinated compounds.

(*b*) **Industrial sources:** It includes fly ash, chemical residues, metallic and nuclear wastes. Large number of industrial chemicals, dyes, acids, etc. find their way into the soil and are known to create many health hazards including cancer.

(*c*) **Agricultural sources:** Agricultural chemicals especially fertilizers and pesticides pollute the soil. Fertilizers in the runoff water from these fields can cause eutrophication in water bodies. Pesticides are highly toxic chemicals which affect humans and other animals adversely causing respiratory problems, cancer and death.

Control of soil pollution: Indiscriminate disposal of solid waste should be avoided. To control soil pollution, it is essential to stop the use of plastic bags and instead use bags of degradable materials like paper and cloth. Sewage should be treated properly before using as fertilizer and as landfills. The organic matter from domestic, agricultural and other waste should be segregated and subjected to vermicomposting which generates useful manure as a byproduct. The industrial wastes prior to disposal should be properly treated for removing hazardous materials. Biomedical waste should be separately collected and incinerated in proper incinerators.

Q.36. Describe sources and hazards of radiation pollution.

Ans. Radiation is a form of energy travelling through space. The radiation emanating from the decay of radioactive nuclides are a major sources of radiation pollution. Radiations can be categorized into two groups namely the non-ionizing radiations and the ionizing radiations.

Non-ionizing radiations are constituted by the electromagnetic waves at the longer wavelength of the spectrum ranging from near infra-red rays to radio waves. These waves have energies enough to excite the atoms and molecules of the medium through which they pass, causing them to vibrate faster but not strong enough to ionize them. In a microwave oven the radiation causes water molecules in the cooking medium to vibrate faster and thus raising its temperature.

Ionizing radiations cause ionization of atoms and molecules of the medium through which they pass. Electromagnetic radiations such as short wavelength ultra violet radiations (UV), X-rays and gamma rays and energetic particles produced in nuclear processes, electrically charged particles like alpha and beta particles produced in radioactive decay and neutrons produced in nuclear fission, are highly damaging to living organisms. Electrically charged particles produced in the nuclear processes can have sufficient energy to knock electrons out of the atoms or molecules of the medium, thereby producing *ions.* The ions produced in water molecules, for example, can induce reactions that can break bonds in proteins and other important molecules. An example of this would be when a gamma ray passes through a cell, the water molecules near the DNA might be ionized and the ions might react with the DNA causing it to break. They can also cause chemical changes by breaking the chemical bonds, which can damage living tissues. The **ionizing radiations** cause damage to biological systems and are, therefore, pollutants.

Radiation damage

The biological damage resulting from ionizing radiations is generally termed as **radiation damage.** Large amounts of radiation can kill cells that can dramatically affect the exposed organism as well as possibly its offspring. Affected cells can mutate and result in cancer. A large enough dose of radiation can kill the organism. Radiation damage can be divided into two types: (*a*) **somatic damage** (also called *radiation sickness*) and (*b*) **genetic damage.**

Somatic damage refers to damage to cells that are not associated with reproduction. Effects of somatic radiation damage include reddening of the skin, loss of hair, ulceration, fibrosis of the lungs, the formation of holes in tissue, a reduction of white blood cells, and the induction of cataract in the eyes. This damage can also result in cancer and death. Genetic damage refers to damage to cells associated with reproduction. This damage can subsequently cause genetic damage from gene mutation resulting in abnormalities. Genetic damages are passed on to next generation.

Radiation dose

The biological damage caused by the radiation is determined by the intensity of radiation and duration of the exposure. It depends on the amount of energy deposited by the radiation in the biological system.

In studying the effects of radiation exposure in *humans*, it is important to realize that the biological damage caused by a particle depends not only on the total energy deposited but also on the rate of energy loss per unit distance traversed by the particle. For example, alpha particles do much more damage per unit energy deposited than do electrons.

Radiation effects and radiation doses

A traditional unit of human-equivalent dose is the ***rem***, which stands for ***radiation equivalent in man***. At low doses, such as what we receive every day from background radiation (< 1 *m rem*), the cells repair the damage rapidly. **At higher doses (up to 100 *rem*), the cells might not be able to repair the damage, and the cells may either be changed permanently or die. Cells changed permanently may go on to produce abnormal cells when they divide and may become cancerous.**

At even higher doses, the cells cannot be replaced fast enough and tissues fail to function. An example of this would be "radiation sickness." This is a condition that results after high doses is given to the whole body (>100 *rem*). Nuclear explosions and accidents in nuclear reactors are a serious source of radiation hazard.

Radiation pollution is the increase in over the natural background radiation. There are many sources of radiation pollution such as nuclear wastes from nuclear power plants, mining and processing of nuclear material etc. The worse case of nuclear pollution was the Chernobyl disaster in Russia occurred in 1986 but the effects still linger today. The effects of atomic explosions in Nagasaki and Hiroshima are still not forgotten. The nuclear reactor accident at Chernobyl in 1986 led to deaths of many reactor personnel and a very large release of radionuclide to the environment causing a long term radiation damage to the people living in the neighbouring regions.

Accidents at nuclear power plants

Nuclear fission in the reactor core produces lot of heat which if not controlled can lead to a meltdown of fuel rods in the reactor core. If a meltdown happens by accident, it will release large quantities of highly dangerous radioactive materials in the environment with disastrous consequences to the humans, animals and plants. To prevent this type of accidents and reactor blow up, the reactors are designed to have a number of safety features. In spite of these safety measures two disasters in the nuclear power plants are noteworthy- namely at 'Three Mile Island' in Middletown (U.S.A.) in 1979, and at Chernobyl (U.S.S.R.) in 1986. In both these cases a series of mishaps and errors resulted in over heating of the reactor core and lot of radiation was released into the environment. The leakage from Three Mile Island reactor was apparently low and no one was injured immediately. However, in case of Chernobyl the leakage was very heavy causing death of some workers and radiation spread over large areas scattered all over Europe. People of the city had to be evacuated to safer places and the plant had to be closed down. These two disasters are a reminder that nuclear power reactors require a constant upgradation of safety measures. Accidents with nuclear submarines also points to the same.

Q.37. What is noise and its various types?

Ans. Noise is unwanted sound. Sound is a form of energy which is emitted by a vibrating body and on reaching the ear causes the sensation of hearing through nerves. Sounds produced by all vibrating bodies are not audible. The frequency limits of human audibility are from 20 Hz to 20,000 Hz.

A noise problem generally consists of three inter-related elements- the source, the receiver and the transmission path. This transmission path is usually the atmosphere through which the sound is propagated, but can include the structural materials of any building containing the receiver.

Noise may be continuous or intermittent. Noise may be of high frequency or of low frequency which is undesired for a normal hearing. For example, the typical cry of a child produces sound, which is mostly unfavorable to normal hearing. Since it is unwanted sound, we call it noise.

Sounds of frequencies less than 20 Hz are called infrasonic and greater than 20,0000 Hz are called ultrasonic. Since noise is also a sound, the terms noise and sound are synonymously used.

Q.38. How noise is computed?

Ans. The intensity of sound is measured in sound pressure levels (SPL) and common unit of measurement is decibel, dB. The community (ambient) noise levels are measured in the *A* - weighted SPL, abbreviated as dB(A). This scale resembles the audible response of human ear. Sounds of frequencies from 800 to 3000 Hz are covered by the A - weighted scale. If the sound pressure level, L_1 in dB is measured at r_1 meters, then the sound pressure level, L_2 in dB at r_2 meters is given by,

$$L_2 = L_1 - 20\log_{10}\left(\frac{r_2}{r_1}\right)$$

If the sound levels are measured in terms of pressure, then, sound pressure level, L_P is given by,

$$L_P = 20\log_{10}\left(\frac{P}{P_0}\right)\text{dB(A)}$$

The L_P is measured against a standard reference pressure, $P_0 = 2 \times 10^{-5}$ N/m^2 which is equivalent to zero decibels. The sound pressure is the pressure exerted at a point due to a sound producing source.

Day-night equivalent noise levels (Ldn): The day night equivalent noise levels of a community can be expressed as:

$$Ldn\text{, dB(A)} = 10\times\log_{10}\left[\frac{15}{24}(10^{\text{Ld}/10})+\frac{9}{24}(10^{\text{Ln}+10/10})\right]$$

where, Ld = day-equivalent noise levels (from 6AM–9 PM), dB (*A*)

Ln = night equivalent noise levels (from 9 PM–6 AM), dB (*A*)

The day hours in respect to assessment of noise levels, is fixed from 6 AM–9 PM (*i.e.*, 15 hrs) and night hours from 9 PM–6 AM (*i.e.*, 9 hrs). A sound level of 10 dB is added to *Ln* due to the low ambient sound levels during night for assessing the *Ldn* values.

Addition of sound levels: The effective sound levels from two or more sources cannot be simply added algebraically. For example, the effective sound level from two air conditioners 60 dB(A) each, say is not 60 + 60 = 120 dB(A) but 60 + 3 = 63 dB(A). (See table 1). Similarly, the effective sound level of 57 dB, 63 dB, 63 dB, 66 dB and 69 dB is 72 dB.

Table 1.2: Addition of sound levels, L_1 and L_2 ($L_1 > L_2$)

$L_1 - L_2$, dB	Add to L_1
0 or 1	3 dB
2 or 3	2 dB
4 or 8	1 dB
9 or more	0 dB

Q.39. Describe some typical sources of noise.

Ans. The sources of noise may vary according to daily activities. They sources may be domestic (movement of utensils, cutting and peeling of fruits/vegetables etc.) natural (shores, birds/animal shouts, wind movement, sea tide movement, water falls etc.), commercial (vendor shouts), automobiles, aero planes, marriages, laboratory, machinery etc.), industrial (generator sets, boilers, plant operations, trolley movement, transport vehicles, pumps, motors etc.). Road traffic is identified as the major source of noise pollution while at home or outdoors or at work.

Typical noise levels of some point sources are given below.

Source	Noise level dB(A)	Source	Noise level, dB(A)
Air compressors	95-104	Quiet garden	30
110 KVA diesel generator	95	Ticking clock	30
Lathe Machine	87	Computer rooms	55-60
Milling machine	112	Type institute	60
Oxy-acetylene cutting	96	Printing press	80
Pulveriser	92	Sports car	80-95
Riveting	95	Trains	96
Power operated portable saw	108	Trucks	90-100
Steam turbine (12,500 kW)	91	Car horns	90-105
Pneumatic Chiseling	118	Jet takeoff	120

Q.40. How aircraft noise is expressed?

Ans. The noise of aircraft is described in terms of Perceived Noise Levels (PNL), a scale of noisiness, expressed in pNdB. There is no simple relationship between the dB(A) value and pNdB value for all noises. However, the pNdB value for a noise is approximately 13 units greater than the dB(A) value for the noise.

A further refinement resulting from the study of aircraft noise is the Effective Noise Level, a scale of noisiness of a time-varying event, expressed in EPNdB. It is used to describe the noise of a single aircraft activity. In order to describe the noise exposure associated with an airport, the EPNdB values are supplemented with such information as the number of flights of each aircraft type, the flight paths that the aircraft use and the time of day at which the operations occur. The result is presented in terms as Noise Exposure Forecast (NEF) contours, which represent the long-term average noise exposure in communities around airports.

Q.41. Describe in brief the various impacts of noise

Ans. Noise induces a severe impact on humans and on living organisms. Some of the adverse effects of noise are summarized below.

- **Annoyance:** It creates annoyance to the receptors due to sound level fluctuations. The aperiodic sound due to its irregular occurrences causes displeasure to hearing and causes annoyance.
- **Physiological effects:** The physiological features like breathing amplitude, blood pressure, heart-beat rate, pulse rate, blood cholesterol are effected.
- **Loss of hearing:** Long exposure to high sound levels cause loss of hearing. This is mostly unnoticed, but has an adverse impact on hearing function.
- **Human performance:** The working performance of workers/human will be affected as they'll be losing their concentration.
- **Nervous system:** It causes pain, ringing in the ears, feeling of tiredness, thereby effecting the functioning of human system.
- **Sleeplessness:** It affects the sleeping there by inducing the people to become restless and loose concentration and presence of mind during their activities
- **Damage to material:** The buildings and materials may get damaged by exposure to infrasonic/ ultrasonic waves and even get collapsed.

Q.42. What problems does noise pollution cause for animals?

Ans. Wild animals rely on their hearing for detecting predators, finding mates, establishing territory, and recognizing warning alerts. Unnaturally high levels of noise can damage their hearing and can also mask more subtle sounds that they need to hear in order to survive and reproduce. They may also react with a fight-or-flight response to artificial sounds such as aircraft noise, thereby using up valuable energy reserves to flee from a non-existent predator. If noise in an area becomes too intrusive, animals may shift to a new territory or alter their migration patterns, which can create new complications for their mating and survival.

Q.43. What problems does noise pollution cause for people as per WHO?

Ans. The World Health Organization (WHO) cites seven categories for the ways noise adversely affects human health:

- Noise-induced hearing impairment
- Interference with speech communication
- Sleep disturbances
- Cardiovascular and physiological effects
- Mental health effects
- Effects on performance of tasks
- Annoyance and effects on behaviour

Q.44. How many decibels can the human ear handle?

Ans. Immediate and irreversible nerve damage can be caused by sounds at 140 dB or higher (120 dB in young children). However, damage also occurs at lower sound levels, and this harm accumulates over time. Any sound above 85 dB can cause wear and tear on ears that reduces hearing acuity over time.

Q.45. What is the loudest sound possible?

Ans. Sound is normally carried in air as a pressure wave. When the pressure of a sound wave becomes as high as the air pressure itself, the sound becomes a shock wave. Normal air pressure at sea level is 101,325 pascals (Pa), which is equivalent to 194 decibels (dB). So 194 dB is the loudest sound possible in air at sea level; beyond that point it becomes a shock wave. (Sound waves that are transmitted through water or other substances would have different limits.)

Q.46. What are the physiological effects of noise on human health?

Ans. Noise has direct physiological effects such as hearing damage (including hearing loss and tinnitus, or ringing in the ears), as well as cardiovascular and hormonal disturbances. Indirect effects include sleep loss, interference with concentration and learning, mood changes and aggression, and social isolation.

Q.47. How does noise affect babies and children?

Ans. Because the ear canal of a young child is smaller than an adult's, sound pressure is up to 20 dB greater than that in an adult ear. In addition to the threat to a child's hearing, noise causes physiological and mental stress, and significantly impacts learning and cognitive development. Background noise also interferes with speech perception and language acquisition.

Q.48. What is "white noise"?

Ans. White noise is a sound similar to radio static, or the sound a fan makes, that is often used to mask unpleasant sounds. Some people find it helpful for sleeping, and it can be a soothing sound for babies.

Q.49. Describe in detail the control methods of noise pollution

Ans. Noise generation is associated with most of our daily activities. A healthy human ear responds to a very wide range of SPL from - the threshold of hearing at zero dB, to uncomfortable at 100-120 dB and painful at 130-140 dB. Due to the various adverse impacts of noise on humans and environment, noise should be controlled. The technique or the combination of techniques to be employed for noise control depend upon the extent of the noise reduction required, nature of the equipment used and the economy aspects of the available techniques.

Reduction in the noise exposure time or isolation of species from the sources form part of the noise control techniques besides providing personal ear protection, engineered control for noise reduction at source and/or diversion in the trajectory of sound waves.

The techniques employed for noise control can be broadly classified as:

- Noise control at source
- Control in the transmission path
- Using protective equipment.

Noise control at source: The noise pollution can be controlled at the source of generation itself by employing techniques like:

- **Reducing the noise levels from domestic sectors:** The domestic noise coming from radio, tape recorders, television sets, mixers, washing machines, cooking operations can be minimized by their selective and judicious operation. By usage of carpets or any absorbing material, the noise generated from felling of items in house can be minimized.
- **Maintenance of automobiles:** Regular servicing and tuning of vehicles will reduce the noise levels. Fixing of silencers to automobiles, two wheelers etc., will reduce the noise levels.
- **Control over vibrations:** The vibrations of materials may be controlled using proper foundations, rubber padding etc. to reduce the noise levels caused by vibrations.
- **Low voice speaking:** Speaking at low voices enough for communication reduces the excess noise levels.
- **Prohibition on usage of loud speakers:** By not permitting the usage of loudspeakers in the habitant zones except for important meetings / functions
- **Selection of machinery:** Optimum selection of machinery tools or equipment reduces excess noise levels. For example, selection of chairs, or selection of certain machinery/equipment which generate less noise (Sound) due to its superior technology etc. is also an important factor in noise minimization strategy.
- **Maintenance of machines:** Proper lubrication and maintenance of machines, vehicles etc. will reduce noise levels. For example, it is a common experience that, many parts of a vehicle will become loose while on a rugged path of journey. If these loose parts are not properly fitted, they will generate noise and cause annoyance to the driver/passenger. Similarly is the case of machines. Proper handling and regular maintenance is essential not only for noise control but also to improve the life of machine.

Noise control in the transmission path: The change in the transmission path will increase the length of travel for the wave and get absorbed/refracted/radiated in the surrounding environment. The available techniques are briefly discussed below.

Installation of barriers: Installation of barriers between noise source and receiver can attenuate the noise levels. For a barrier to be effective, its lateral width should extend beyond the line-of-sight at least as much as the height. The barrier may be either close to the source or receiver, to increase the traverse length for the sound wave. The presence of the barrier itself can reflect sound back towards the source. At very large distances, the barrier becomes less effective because of the possibility of refractive atmospheric effects

Design of building: The design of the building incorporating the use of suitable noise absorbing material for wall/door/window/ceiling will reduce the noise levels.

- **Installation of panels or enclosures:** A sound source may be enclosed within a paneled structure such as room as a means of reducing the noise levels at the receiver. The actual difference between the sound pressure levels inside and outside an enclosure depends not only on the transmission loss of the enclosure panels but also on the acoustic absorption within the enclosure and the details of the panel penetrations which may include windows or doors.

  The product of *frequency of interest and surface weight of the absorbing material* is the key parameter in noise reduction through transmission loss. With conventional construction practices, the high-frequency transmission loss of a panel becomes limited to around 40 dB, owing to the transmission of sound through flanking paths other than the panel itself. Examples of such flanking are structural connections or ducts joining the two spaces on either side of the panel of interest.
- **Green belt development:** Green belt development can attenuate the sound levels. The degree of attenuation varies with species of green belt. The statutory regulations direct the industry to develop greenbelt four times the built-up area for attenuation of various atmospheric pollutants, including noise.

Using protection equipments: Protective equipment usage is the *ultimate* step in noise control technology, *i.e.*, after noise reduction at source and/or after the diversion or engineered control of transmission path of noise. The first step in the technique of using protective equipment is to gauge the intensity of the problem, identification of the sufferer and his exposure to the noise levels.

The usage of protective equipment and the worker's exposure to the high noise levels can be minimized by:

- **Job rotation:** By rotating the job between the workers working at a particular noise source or isolating a person, the adverse impacts can be reduced.
- **Exposure reduction:** Regulations prescribe that, noise level of 90 dB (A) for more than 8 hr continuous exposure is prohibited. Persons who are working under such conditions will be exposed to occupational health hazards. The schedule of the workers should be planned in such a way that, they should not be over exposed to the high noise levels.
- **Hearing protection:** Equipment like ear muffs, ear plugs etc. are the commonly used devices for hearing protection. Attenuation provided by ear-muffs vary widely in respect to their size, shape, seal material etc. An average noise attenuation up to 32 dB can be achieved using earmuffs.

Q.50. What are various regulatory guidelines for noise pollution.

Ans. Statutory Regulatory guidelines are prescribed both for the ambient noise levels and for workspace environment noise levels. Indian Factories Act, 1948 prescribes the protection of workers against high noise levels (noise level > 90 dB (A)). The State Pollution Control Board and Inspector of Factories have powers to administer the control of noise pollution.

Table 1.3: Indian Standards for ambient noise levels

Area	Noise Limits, Leq, dB (A)	
	Day Time	Night Time
Silence zone	50	45
Residential area	55	45
Commercial area	65	55
Industrial area	75	65

1. Ministry of Environment and Forest (MOEF) Guidelines vide Environment (Protection) Act, 1986 third amendment rules, dated 26/12/89.

2. Day time from (600 hrs to 2100 hrs, IST)
3. Night time from (2100 hrs to 600 hrs IST)
4. *Silence zone*: Up to 100m around hospitals, educational institutions and courts. The zones are to be declared by competent authority. Use of vehicle horns, loud speakers and bursting of crackers shall be banned in these zones.

Table 1.4: Damage risk criteria for hearing loss – Occupational Safety and Health Administration (OSHA) regulations

Maximum allowable duration per day in hour	Sound pressure level, dB (A)
8	90
6	92
4	95
3	97
2	100
1.5	102
1	105
0.75	107
0.5	110
0.25	115

No exposure in excess of 115 dB(A) is permitted.

Q.51. Describe the application of noise level studies in Environmental Impact Assessment (EIA) studies.

Ans. The **EIA** study are carried out to evaluate and assess the impacts of any proposed (or existing) activity on the environment. Noise is one of the environmental attributes, on which the likely impacts due to the proposed (or existing) activity need to studied.

The likely steps to be carried out while conducting noise level studies for an EIA project are summarized below. The sequential steps involved will be same for both the activities.

- The likely activities that generate noise from the proposed activity are to be identified.
- The typical sound (noise) levels of the noise generating sources are to be assessed either from literature or from a similar source
- The likely exposure time of a worker at a noise generating source is to be assessed from the plant / utility records.
- The workspace environment noise levels are to be checked with OSHA standards. If the noise exposure levels are higher, suitable noise control measures like personal protective equipment, installation of barriers, enclosures etc., need to be suggested
- The EIA will usually be carried out in an impact circle of radius 3 Km to 25 Km or even more depending on the objective and the likely activities of the proposed project. The representative baseline (or back ground) status of the ambient noise levels need to be collected by monitoring at various stations in the study zone.
- The ambient noise levels are to be analyzed for the prescribed parameters like, Leq, Ldn etc., and compared with the ambient noise level standards for the study region. If these values are higher than the prescribed limits, the likely causes for the high values need to be assessed.
- The likely impact of the noise levels from the proposed activity on the local environment keeping in view the baseline status of noise levels needs to be predicted.

- If the predicted impact is adverse, suitable measures for attenuating the noise levels like, green belt development, in-plant control measures etc., need to be suggested.
- The objective of the EIA study is to make sure that, the local environment, say noise, will **not** get affected by the noise levels emanated from the proposed activity. If the ambient noise levels are high, then control measures be suggested to the project proponent to ensure that, ambient noise levels will not increase due to the proposed activity.

Q.52. Describe the sources, effects and control of noise pollution.

Ans. Noise is one of the most pervasive pollutant. A musical clock may be nice to listen during the day, but may be an irritant during sleep at night. Noise by definition is "sound without value" or "any noise that is unwanted by the recipient". Noise in industries such as stone cutting and crushing, steel forgings , loudspeakers, shouting by hawkers selling their wares, movement of heavy transport vehicles, railways and airports leads to irritation and an increased blood pressure, loss of temper, decrease in work efficiency, loss of hearing which may be first temporary but can become permanent if the noise stress continues. It is therefore of utmost importance that excessive noise is controlled. Noise level is measured in **terms** of decibels (dB). W.H.O. (World Health Organization) has prescribed optimum noise level as 45 dB by day and 35 dB by night. Any noise above 80 dB is hazardous.

The table below gives the noise intensity in some of the common activities.

Source	Intensity(dB)	Source	Intensity(dB)
Quiet conversation	20-30	Radio music	50-60
Loud conversation	60	Traffic noise	60-90
Lawn mower	60-80	Heavy truck	90-100
Aircraft noise	90-120	Space vehicle	140-179
Beat music	120	Jet engine	140
Motor cycle	105		

Sources of noise pollution: Noise pollution is a growing problem. All human activities contribute to noise pollution to varying extent. Sources of noise pollution are many and may be located indoors or outdoors.

(*i*) **Indoor sources** include noise produced by radio, television, generators, electric fans, air coolers, air conditioners, different home appliances, and family conflict. Noise pollution is more in cities due to a higher concentration of population and industries and activities such as transportation. Noise like other pollutants is a byproduct of industrialization, urbanization and modern civilization.

(*ii*) **Outdoor sources** of noise pollution include indiscriminate use of loudspeakers, industrial activities, automobiles, rail traffic, aero planes and activities such as those at market place, religious, social, and cultural functions, sports and political rallies. In rural areas farm machines, pump sets are main sources of noise pollution. During festivals, marriage and many other occasions, use of fire crackers contribute to noise pollution.

Effects of noise pollution: Noise pollution is highly annoying and irritating. Noise disturbs sleep, causes hypertension (high blood pressure), emotional problems such as aggression, mental depression and annoyance. Noise pollution adversely affects efficiency and performance of individuals.

Prevention and control of noise pollution:

Following steps can be taken to control or minimize noise pollution:

- Road traffic noise can be reduced by better designing and proper maintenance of vehicles.

- Noise abatement measures include creating noise mounds, noise attenuation walls and well maintained roads and smooth surfacing of roads.
- Retrofitting of locomotives, continuously welded rail track, use of electric locomotives or deployment of quieter rolling stock will reduce noises emanating from trains.
- Air traffic noise can be reduced by appropriate insulation and introduction of noise regulations for take off and landing of aircrafts at the airport.
- Industrial noises can be reduced by sound proofing equipment like generators and areas producing lot of noise.
- Power tools, very loud music and land movers, public functions using loudspeakers, etc should not be permitted at night. Use of horns, alarms, refrigeration units, etc. is to be restricted. Use of fire crackers which are noisy and cause air pollution should be restricted.
- A green belt of trees is an efficient noise absorber.

Q.53. What are the sources, effects and control measures of thermal pollution of water?

Ans. The discharge of thermal energy (e.g., heated water) into natural waters like streams, rivers etc., is called as thermal pollution. The addition of heat to water reduces the ability of water to hold dissolved gases like oxygen which is vital for aquatic life. If the temperature of water exceeds 35°C the dissolved oxygen content in water is too low to support aquatic life.

Sources of thermal pollution: The following sources contribute to thermal pollution (1) Nuclear Power plants (2) Coal fired power plant (3) Industrial effluent (4) Domestic sewage (5) Hydroelectric power

(1) **Nuclear power plants:** Nuclear power plants discharge a lot of unutilized heat and trace of toxic radio nuclides into nearby water streams, thereby increasing the temperature of water bodies which affects the aquatic flora and fauna.

(2) **Coal fired power plants:** The condensers coils of these plants are cooled by water from nearby lake or river and discharge hot water back affecting the aquatic life.

(3) **Industrial effluents:** Similar effect is found in industrial effluent which uses water as coolant.

(4) **Domestic sewage:** The organic matter present in sewage use dissolved oxygen for oxidation, which emits foul gas and heat. The heat has profound influence on water bodies.

(5) **Hydroelectric power plant:** The generation of Hydro Electric power, sometimes result in negative thermal loading in water systems.

Thermal power plant pollution: A coal fired power plant releases hot water from condenser cooling used in turbine to water stream.

Harmful effects of thermal pollution of water are as follows:

- Reduction in dissolved O_2.
- Change in water properties.
- Increase in toxicity owing to rise in temperature.
- Interference with biological activity.
- Interference with reproduction.
- Variations in reproductive rate.
- Change in metabolic rate.
- Increased vulnerability to disease.
- Invasion of destructive organism.
- Destruction of organism in cold water.
- Undesirable algae production.
- Growth of blue green algae.
- Increasing the demand of O_2.
- Bio-chemical oxygen demand.
- Direct mortality of aquatic creatures.
- Affecting distribution of organism.

The parameters and instruments to measure thermal pollution includes electronic temperature meters, dissolved oxygen meters and lux meters.

Control of thermal pollution: To reduce the magnitude of thermal pollution the outlet water is made to give out some heat to the environment and then may be discharged into water course.

The following methods can be adopted to control the high-water temperature caused by thermal discharges:

- **Cooling towers** are used to transfer some of the heat from cooling water in the surrounding atmosphere by the process of evaporation.
- **Cooling ponds:** Heated effluents on the surface of water in cooling ponds maximize dissipation of heat to the atmosphere and minimize the water area and volume. By forming a warm water wedge over the cold receiving water, rapid heat loss to the atmosphere is encouraged.
- **Artificial lake:** The heated effluent can be discharged in to lake at one end and the water for cooling ponds may be withdrawn from the other end. The heat is dissipated through evaporation.
- **Prevention of thermal pollution:** The problem associated with thermal pollution can be alleviated by the following methods:
  - — Plant is suitably sited with river system.
  - — Channeling thermal effluent.
  - — Using adequate cooling of the water and ponds.
  - — Avoiding interference of hot water mass with fish migration.
  - — Adopting heat transfer mechanism.
  - — By improving efficiency of electric power generating plants.

Q.54. Sum up the facts of environmental pollution.

Ans.
- Nature's components such as air, water, soil, forest and fisheries are resources exploited by humans and their pollution are undesirable by-product of urbanization and industrialization. The agents directly or indirectly responsible for the pollution of the environment are known as pollutants.
- There are six types of pollutions: Air pollution, water pollution, noise pollution, soil pollution, thermal pollution, radiation pollution etc.
- Air pollution is a result of industrial and certain domestic activity. Air pollutants are of two types (1) suspended particulate matter, and (2) gases like CO, CO_2, NO_x etc. Use of cleaner fuels such as biogas, CNG and electricity prevent air pollution. Segregation of waste, pretreatment at source, sterilization of rooms will help in checking indoor pollution.
- Prevention and control of industrial pollution can be reduced by using cleaner fuels, filters, electrostatic precipitators, inertial collectors, scrubbers etc.
- Use of chlorofluorocarbons cause damage of ozone layer which has resulted in its thinning over the Arctic and Antarctic regions, is known as ozone hole. Increase in global temperature or heating effect by greenhouse gases (CO_2, methane) is known as green house effect.
- Noise like other pollution is a byproduct of industrialization, urbanization and modern civilization. Indoor sources include noise produced by radio, television and outdoor source includes indiscriminate use of loudspeakers, industrial activities, automobile, rail traffic and aero planes etc.
- Addition of undesirable substances in water is called water pollution. Natural sources of water pollution are soil erosion, leaching of minerals from rocks and decaying of organic matter. Power plants and various industries uses lot of water for cooling purposes and hot water is discharged

into rivers, streams or oceans. This waste heat increases the temperature of the cooling water upto 10°C. This is called thermal pollution.

- Improper sewage disposal, dumping of farm yard manures and agricultural chemicals, industrial effluents are causing pollution of ground water. Nutrient enrichment of a water body is called eutrophication. Waste water from domestic or industry or garbage dump is generally known as sewage.
- Addition of substances which adversely affect the quality of soil or fertility is known as soil pollution. Sources of soil pollution are plastic bags, industrial sources, agricultural sources etc.
- Radiation is a form of energy travelling through space. Radiation can be grouped into non-ionizing radiation and the ionizing radiations.

❑❑❑

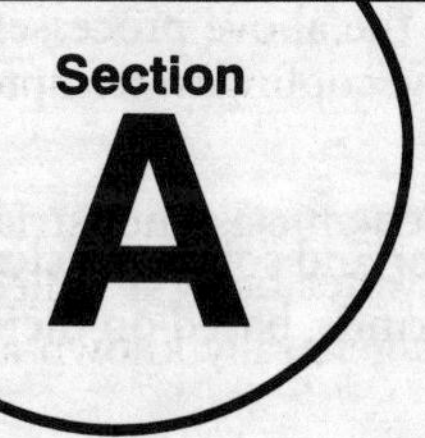

2

Air Pollution

SHORT QUESTIONS AND ANSWERS ON AIR POLLUTION

Q.1. What is atmosphere?

Ans. Atmosphere is a gaseous layer surrounding the earth. In other words, our earth is surrounded by a thin layer of gases, called atmosphere. This thin layer has influences on various processes that take place on earth. It contains a mixture of gases with some impurities. The atmosphere is special because it contains life-sustaining oxygen in large quantities (about 21% by volume).

Atmosphere constantly exchange energy and matter with other components of the earth – lithosphere, hydrosphere and biosphere.

Atmosphere extends from a few meters below the earth's surfaces or water's surface to a height of about 60,000 km. However, about 90% of the atmosphere is within 70 km from the ground.

Q.2. What is the importance of earth's atmosphere?

Ans. Atmosphere is very important to sustain life on earth. It contains life-supporting oxygen in sufficiently large quantities. Constant concentration of oxygen is maintained through oxygen cycle. The oxygen cycle does not take place in isolation but It takes place along with other bio-geo chemical cycles. These cycles connect the atmosphere with hydrosphere, lithosphere and biosphere.

Atmosphere is the transparent layer through which, life-sustaining solar radiation passes and reaches the earth's surface or into the water. Solar radiation is the only source to supply energy for photosynthesis to take place on earth, which thereby supports all other life. When solar radiation passes through the atmosphere harmful ultra violet radiation is "absorbed" by ozone layer, which is a part of our atmosphere. Ozone layer prevents about 95% of harmful ultra-violet radiation from reaching the earth's surface.

Solar radiation when it passes through the troposphere, the lowest layer of a atmosphere, part of the radiation, may be reflected back by the clouds, scattered by atmospheric constituents, thereby reducing visibility, absorbed by atmospheric constituents and the remainder reach the earth's surface (land or water).

Heated earth emits the energy in the form of infra-red radiation during night time and this radiation is absorbed by carbon dioxide, water and few other gases. This process results in "greenhouse effect". Thus, the atmosphere is kept warm during night, otherwise it will become so cool and intolerable for living organisms. Earth is not heated by solar radiation uniformly due to its inclination. As a result, different weather patterns exist over the earth. In order to compensate these differences, air sets in motion resulting in winds and circulation of air. These wind currents are of global scale as well as of local scale. They are responsible for cyclones, dust storms, tornadoes etc. These wind currents also influence water currents in the oceans; they in turn affect the wind currents. All the above processes make the atmosphere, a dynamic atmosphere.

Q.3. What is the composition of atmosphere?

Ans. The composition of the atmosphere are gases present in large amounts, water vapor and solid particles in considerably less amounts. Gases that in the atmosphere are divided into two kinds, based on their concentration, viz., ***constant gases*** and ***variable gases.***

Constant gases are the ones, whose concentrations do not change over time, and their concentrations almost remain same. But variable gases are present in different concentrations at different places and times.

Nitrogen and oxygen are the two major **constant gases** that make up 99 percent of the air. Both are important to sustain life on earth. Nitrogen constitutes 78.09% and oxygen 20.94 percent by volume thus, making the bulk of the atmosphere. The remaining 0.97% is constituted by nitrous oxide and ***inert gases*** such as, argon (0.94 %), helium, krypton, xenon and also by ***variable gases*** – carbon dioxide, water vapor, ozone etc.

Constituents of atmosphere (average percentage by volume in dry air):

- Nitrogen - 78.09
- Oxygen - 20.94
- Argon - 0.94
- Carbon dioxide - 0.03
- Neon - 0.002
- Helium - 0.0005
- Methane - 0.0002
- Krypton - 0.0001
- Hydrogen - 0.00005
- Nitrous oxide (N_2O) - 0.00005

Q.4. What is the importance of atmospheric gases?

Ans. Nitrogen present in the atmosphere is not that important. It is non-poisonous but indirectly, it is very important, as it is converted into useful forms for life by certain microorganisms. Nitrogen is very important in the formation of amino acids, which are building blocks of proteins, and also in the formation of nucleotides, which are part of the genetic materials, viz., DNA and RNA. This is the best examples that the natural processes are in perfect harmony with each other.

It is needless to stress the importance of oxygen. Every living organism, including humans, need oxygen for respiration. Respiration is the process through which the chemical energy (food) is converted into usable form of energy by the living cells. Inert gases, as such do not play any role in the environment. However, they are used for commercial purposes such as in neon lights.

Q.5. What is the role of carbon dioxide in atmosphere?

Ans. Carbon dioxide is emitted by all living organisms as an end product of respiration.

This carbon dioxide, in turn is used by producer organisms (green plants and certain micro organisms) for the synthesis of food. This process is called, ***photosynthesis.*** This is another example how natural process are inter-related and inter-dependent. As a large amount of carbon dioxide is utilized by plants, on average, it comprises only 0.04 percent of dry air. However, it plays significant role in keeping the atmosphere at temperatures that permit life. The concentration of carbon dioxide, nowadays, is increasing due to human activities. Today, man burns large quantities of fossil fuels for various purposes. As a result, huge amounts of carbon dioxide are emitted into the atmosphere. Its ever-increasing concentration has already resulted in ***global warming*** and in some places, melting of ice.

Water vapor, though present in small quantities, plays a crucial role. It is responsible for cloud formation in the atmosphere and precipitation. Its concentration varies over time at a given place and at different places. It is an important component of the atmosphere in determining the weather of a place at a given time. It is responsible for fog formation during early morning hours in winter, for feeling sultry near coastal regions, during summer. Water vapor also absorbs outgoing radiation from earth as CO_2

does and it has a crucial role in greenhouse effect. Thus, both CO_2 and H_2O along with ozone (in troposphere), methane and N_2O are called, ***greenhouse gases.***

Q.6. What is the role of other constituents of atmosphere?

Ans. In addition, water present in clouds, can reflect and absorb the part of the incoming solar radiation. It also determines the weather of a place at a given time to a large extent. Water vapor is also transported along with the wind and its circulation. As a result, it is distributed to the parts away from sea. Nevertheless, the places that are far away from sea will have less moisture content than the places near the sea.

Other variable gases are also present in the atmosphere but in minute quantities. They are hydrogen, helium, sulfur dioxide, methane, carbon monoxide and oxides of nitrogen. Some of them are air pollutants, emitted by human activities. Solid particles are present in minute quantities in the atmosphere. They might have originated either naturally or by human activities or both. In acceptable quantities, they are beneficial as they initiate and take part in cloud formation. The solid particles absorb and scatter the solar radiation. When their concentration goes high, they reduce the visibility of the atmosphere and are deleterious to human health. They also affect other living organisms in many ways.

Q.7. Discuss the structure of atmosphere and layers of atmosphere based on composition of constituents?

Ans. The earth is surrounded by a thin layer of air, called, atmosphere. The atmosphere, from the surface of earth extends up to 60,000 km. Such a thickness is called a thin layer. Because most of the mass of the atmosphere is found near the planetary surface. It is near the earths surface from surface to about 70 km. This is due to the earth's gravity, which pulls the atmospheric constituents, towards its center. Asa result, the most of the atmosphere are within this thickness. It is this thin layer, in which most of the atmospheric processes take place.

According to the concentration of the gases, atmosphere is divided into:

Homosphere: the lower region, extending from the surface of the earth to a height of 80 km above the earth. In this layer, gases are more or less uniform in their chemical composition.

Heterosphere: it starts from the upper portion of homosphere and extends to height of 60,000 km above the earth. In this layer, chemical composition changes with height. Concentration of gases keep decreasing as one goes up. Inter molecular distance increases with height and hence, concentration decreases.

Q.8. How do we classify layers of atmosphere based on temperature variation?

Ans. Atmosphere is divided into four distinct layers according to their temperature characteristics:

1. **Troposphere:** It is the bottom layer of the atmosphere. It contains 70% mass of the atmosphere. It extends to an average height of 12 km. However, its thickness varies with latitude: over the poles, only about 8 km; above the equator, it is about 16 km. A very important feature of this layer is that temperature decreases with height in this layer. The rate of decrease of temperature with altitude is called, ***lapse rate.*** Average lapse rate in troposphere is – 6.4 °C / km. Troposphere ends at **tropopause.** Tropopause is just like a lid over the troposphere, where temperature stops decreasing with height.

2. **Stratosphere:** It lies just above the tropopause. It extends to a height of 50 km from earth's surface. ***Ozonosphere***, a very important layer is found within this stratosphere. Ozone present in ozonosphere prevents the harmful ultra-violet rays from reaching the earth, thereby protecting the life. Thus, ozonosphere acts as a protective umbrella.

   Stratosphere is a calm layer consisting of relatively clean air. Water vapor in this layer is almost absent, and hence clouds will not form in this layer. In this layer, temperature increases with height;

just opposite to that in troposphere. This temperature increase with height prevents the vertical winds. Only horizontal winds are seen. These horizontal winds flow almost always parallel to the earth's surface. Absence of vertical winds and the existence of horizontal winds parallel to earth's surface result in relatively calm atmosphere (absence of turbulence). This ensures smooth travel for fights. Absence of clouds also provide with good visibility for pilots. All the above features led to the operation of jet air planes in this layer. The flying of jet air planes was partly responsible for the destruction of ozone. Above stratosphere, temperature neither decreases nor increases with height up to some level. This small layer is called ***stratopause***

3. **Mesosphere:** It starts from the edge of the stratopause, just at an approximate height of 52 km from earth's surface. It extends to height of 80 km from the ground. In this layer, temperature decreases with height as in troposphere. This layer, as such does not have any impact on life. But, it gains importance as it plays crucial role in radio communication.

   Sunlight passing through this layer, converts the individual molecules to individual charged ions *i.e.*, ionization. Ionized particles are concentrated as a zone called the **D**-layer. This D-layer reflects radio waves sent from earth. But this D-layer blocks the communications between earth and astronauts. In this layer, during summer, at night times, a spectacular display of wispy clouds can be seen sometimes over high latitudes. It is presumed that meteoric dust particles coated with ice crystals reflect the sunlight resulting in wispy clouds. Just above the mesosphere lies, ***mesopause***, in which temperature neither decreases nor increases.

4. **Thermosphere:** It is found approximately above 80 km from earth's surface. It extends to the edge of space at about 60,000 km from earth's surface. Temperature keeps rising with altitude in this layer. It is likely to reach 900 °C at an altitude of 350 km. However, these high temperatures are not felt as in lower layers. The air molecules are so far apart in this layer. As a result, these temperatures really apply to individual molecules only. In this layer too, ionization of molecules take place. It results in individual charged ions. This process produces two ionized belts, viz., **E**- and **F**-layers. These layers also reflect radio waves and have influence over radio communications. In the upper thermosphere, further concentration of ions are seen that comprise the Van Allen radiation belts. This layer is sometimes called, ***magnetosphere.*** It is thus called as earth's magnetic field which has more influence over the movement of particles rather than earth's gravitational field. Thermosphere as such has no definable upper boundary and gradually blends with space.

Q.9. Discuss the photochemical reactions occurring in the ozone layer?

Ans. The existence of ozone layer was discovered in 1913. It is located in the lower part of the stratosphere between the altitudes of 15 and 35 km. Concentration of ozone at its maximum is 10 ppm at an altitude of 25-30 km. Between the altitudes of 15 and 40 km, its concentration ranges from 2 to 8 ppm. Concentration of ozone is maintained at their levels in the above altitudes by **"ozone-oxygen" cycle.**

When ultra-violet radiation strikes the oxygen molecule, O_2, it splits the molecule into two individual oxygen atoms (O and O). The atomic oxygen thus produced will combine with un-split oxygen molecule (O_2) to produce ozone molecule (O_3) once again. In this reaction a third body, M plays a crucial role in absorbing the excess energy liberated. This M may be a N_2 or another O_2. Thus, formed ozone will be split by striking ultra-violet radiation at the wavelength around 308 nm into a molecule of O_2 and an atom of oxygen. By this ozone-oxygen cycle, the concentration of ozone is maintained. The actual concentration of ozone is determined by the rate of its formation and its destruction.

The amount of ozone present in the stratosphere is measured by simple spectrophotometer from the ground. This instrument was developed by the British meteorologist, G.M.B. Dobson who explored the properties of ozone layer in detail. Total amount of ozone in a column overhead is measured and expressed in "***Dobson units***"

Q.10. Discuss the phenomenon of ozone layer depletion and its effects?

Ans. Today, due to human activities, the ozone layer is becoming thin. This thinning is called, ozone depletion. At the zones, where thinning is too severe, they are termed as **"Ozone holes"**: The ozone hole is defined as the area having less than 220 Dobson units (DU) of ozone in the overhead column (*i.e.*, between the ground and space). Ozone hole, was observed over Antarctic region. Chloro-fluoro carbons (CFC) used in refrigerators, air conditioners, propellants etc. and oxides of nitrogen emitted by air crafts flying near stratosphere are found to be the causes for ozone layer depletion.

Ozone layer is depleted by free radical catalysts – nitric oxide (NO), hydroxyl (OH), atomic chlorine (Cl), and atomic bromine (Br). Halogens have the ability to catalyze ozone breakdown.

Chlorine atom or any other halogen atom released from CFC or BFC by striking ultraviolet radiation catalyzes ozone breakdown. Although these species occur naturally, large amounts are released by human activities through the use of CFCs (Chloro-fluoro carbons) and Bromo fluorocarbons (BFC).

Ozone layer depletion by CFCs:

CFCs and BFCs are stable compounds and live long in the atmosphere. As they live longer, they are able to rise to the stratosphere. Cl and Br radicals are liberated from these compounds by the action of ultraviolet radiation. These radicals initiate and catalyze breaking the ozone molecules. One single radical is capable of breaking down over 100,000 ozone molecules. Ozone concentration is decreasing at the rate of 4% per decade over northern hemisphere. CFCs have long life time – 50 to 100 years. As they remain for such a long duration, they deplete ozone layer continuously. Moreover, this depletion rate keeps increasing as more and more CFCs are released.

After realizing the seriousness of this problem, countries have come forward to ban completely or phase out and the ban the use of CFCs. Ozone hole was discovered in the year 1985. After this, countries came forward for an international treaty, the Montreal Protocol for complete phase-out of CFCs by 1996. This effort has yielded positive result. Scientists announced on August 2, 2003 that depletion of ozone layer had slowed down due to the ban on CFCs.

Scientists have developed HCFC to replace CFC for the same purpose. These HCFCs are short-lived in the atmosphere to reach the stratosphere and damage ozone layer.

Ozone layer depletion by nitric oxide:

When a nitric oxide (NO) molecule combines with O_3, it is oxidized to nitrogen dioxide (NO_2). This NO_2 now combines with another O_3 molecule to become NO_3 and O_2. Both NO_2 and NO_3 may combine to form N_2O_5. Even if atomic oxygen is available, it readily combines with NO_2 to yield NO_3. Thus, O_3 is completely utilized for the above reactions and thereby depleted.

International community, after realizing its seriousness, has agreed to withdraw the operation of jet airplanes that emit NO in stratosphere

Q.11. What is meteorology?

Ans. Physical processes taking place in the atmosphere, especially in troposphere, which are responsible for weather and climate of a place are studied as a branch of science called, **meteorology**. Meteorology is the science of atmosphere and deals with basic physical principles as they apply to atmospheric phenomena. When it is said atmosphere, it may represent atmosphere of not only earth but also of other planets. The study of "science of earth's atmosphere" is called metrology.

The atmosphere behaves like a mega heat engine, in which many processes like movement of air, storms, cyclones, formation of clouds and precipitation etc. take place. Major driving force for this "engine" is solar radiation. Incoming solar radiation (called, ***insolation***) supply all the energy for all these processes. When radiation strikes the earth, earth is heated up. This heat energy then transferred to the overlying air through conduction and convection. Now air starts moving horizontally as well as vertically. This causes all the weather and insolation is the only force for all these, weathers.

Q.12. What is incoming solar radiation-insolation and its importance?

Ans. Insolation is sun's radiant energy that strikes the earth. It is short form of "incoming solar radiation". Radiation emanating from the sun consists of a range of wave lengths, known as ***solar spectrum***. Each wavelength contains its own energy. Solar spectrum includes X-rays, gamma rays, ultraviolet rays, visible light, infrared rays, more than 95% of the energy received by earth.

Most of the ultraviolet radiation is prevented by existing ozone layer in the upper atmosphere. Part of the long waves is absorbed in the troposphere. The radiation which ultimately reaches the earth mainly comprises visible light. This visible light is composed of seven colors that make the rainbow after the rainfall at some times. These seven colors are: violet, indigo, blue, green, yellow, orange and red (VIBGYOR).

Out of the total energy emitted by sun, only two-billionth of it is intercepted by the earth. Only this small amount of energy is responsible for all the physical processes taking place in the atmosphere and all the biological processes on earth. The average amount of energy striking the outer atmosphere per unit area is known as the ***solar constant***. It is estimated that a mean value of 1.940 langleys per minute of energy strikes the outer atmosphere. One Langley equals to one gram-calorie per centimeter. In other words, 1.940 g cal per of energy reaches one-centimeter area of outer atmosphere per minute (*i.e.*, solar constant = 1.940 g cal/cm/min).

In addition to the radiation, sun emits energy as "***solar wind***". It is a stream of charged particles (*i.e.*, plasma) ejected from upper atmosphere of the sun. It also consists of magnetic fields. The effects of solar wind are not well understood. However, it is known that it interferes with radio communication, knock out power grids on earth, and disturbs earth's magnetic field.

Q.13. Discuss solar radiation balance, its effects on earth and its atmosphere.

Ans. As far as earth is concerned, radiation means both incoming solar radiation and outgoing radiation emitted by earth's surface. Before radiation reaching the earth's surface, it travels through the atmosphere. When it passes the atmosphere, some of the radiation energy is reflected by clouds, and some are scattered and absorbed by gases and particles in the atmosphere. The remaining radiation reaches the earth; even some of the radiation reaching the earth's surface or water surface is reflected back.

The amount of radiation reflected depends upon the surface characteristics. The reflected solar radiation is called, ***albedo***. Albedo of a water body is different from that of a land surface; albedo of a forest is different from that of a desert and so on. That is, albedo values vary with the type of the surfaces on which the radiation strikes. The scattered radiation by atmospheric constituents, may reach the earth and this is called ***diffuse radiation***. On a cloudy day, the light we receive is not direct radiation but diffused radiation passing through the clouds.

Out of the total energy received by outer atmosphere at a given location, some are reflected, absorbed, and/or scattered. Some units finally reach the earth's surface directly or through diffuse radiation. The earth re-emits either all or part of the received radiation. All the above details constitute ***radiation balance***.

On an average, out of all the energy received at the edge of the outer atmosphere, only 31% directly reaches the earth's surface (or water surface). This energy is called ***direct radiation***. Almost 30% is reflected by the atmosphere to the outer space. Clouds reflect approximately 25% and the dust particles, about 5%. 17% of the energy is absorbed by clouds and other atmospheric constituents (dust particles and gaseous molecules). Some of the particles and gases scatter the radiation in the atmosphere. Some of the scattered radiation eventually reaches the earth' surface. It is called ***diffuse radiation***. Diffuse radiation reaching the earth's surface accounts for about 22%. Thus, earth receives on average, 53% of the solar energy both directly and through diffuse radiation.

If all the energy received by the earth's surface is retained, then earth will become very hot and inhabitable place. Surface of the earth also reflect some amount of energy. The earth's surface is heated up by the energy received by it. Of course, the remaining energy available after reflection heats up the surface. Some of this heat energy is transmitted to overlying air through conduction and this in turn will initiate the convection in the air above the earth's surface. As the earth's surface is heated up, it will become hot and hold the thermal energy. The earth at this point starts emitting energy in the form of long wave radiation. This is called ***outgoing radiation***.

Q.14. What is green house effect and its importance?

Ans. Part of the outgoing radiation is absorbed by the atmosphere and retained as heat energy. The remaining energy escapes into outer space. Carbon dioxide, water vapor, and few other gases in the atmosphere are capable of absorbing the outgoing radiation. (called, ***greenhouse gases***); higher the concentration of these gases, more the amount of long wave radiation is absorbed. It is a natural phenomenon occurring in the atmosphere. It helps preventing the earth from drastic cooling of the earth and its atmosphere. If all the heat energy from the earth's surface is lost to the space, then earth will become too cold. These gases act like a blanket and provide the earth with "blanketing effect". This process is called ***Greenhouse effect***.

However, the intensity of this effect may vary from place to place depending upon the concentration of these gases in the atmosphere. For example, over desert atmosphere, the amount of water vapor is far less as it is a dry place. Carbon dioxide concentration is also very less as life is sparsely distributed and a smaller number of living organisms exist here. Hence, more radiation will escape into the space and very little amount of radiation is retained. Consequently, the desert atmosphere will be cool during night.

Q.15. What is weather? What causes it and how it is different from climate?

Ans. Weather is the conditions of the atmosphere experienced by humans at any given moment. Weather is used to refer to the atmospheric conditions that exist for a short duration (hours or days). These conditions fluctuate and vary temporally over a given place. Often people confuse this with the term, ***climate***. Climate refers to the average atmospheric conditions over a long period of time (month or season) of a place.

At times, we feel our atmosphere is very hot / cold, sultry and humid/dry, rainy, stormy etc. All these experiences are due to atmospheric conditions such as, temperature, humidity, wind speed and its direction, rainfall, etc. These conditions vary both spatially and temporally. All the weather phenomena occur in lower atmosphere *i.e.*, troposphere. Our weather is seldom affected by conditions that exist above the troposphere. Troposphere is the layer in which we all live; movement of air occurs both vertically and horizontally. The temperature decreases with height in troposphere. Rate of decrease of temperature with height is called ***lapse rate***. The average lapse rate in troposphere is – 6.4 °C/km. However, it may vary at different conditions.

Q.16. Discuss about the horizontal temperature distribution in the atmosphere?

Ans. Temperature differs horizontally at different locations. Temperature of a place also varies at different times of a day and at different months and seasons of a year. These variations are due to several factors. Incoming solar radiation is the major factor which determines the temperature of a place. The angle of the sun's rays and length of daylight determine the amount of insolation received in a certain place. The angle of sun's rays and length of daylight are determined by latitudinal location of a place and time of the day.

As the earth revolves around the sun, one half of the globe is exposed to sun's rays and other half remains dark at any given time. The illuminated surface of the earth keeps changing as the earth rotates itself on its own axis. This results in occurrence of day and night time alternately over a given place. However, entire half of the globe does not receive equal amount of sun's radiation. In addition, as the earth keeps rotating, the amount of radiation received at a given location varies over time during day time.

The sun's rays strike almost perpendicular near the equator and at angles over high latitudes. This means farther away from equator more will be the angle of striking of sun's rays over a place. Moreover, the earth's axis is tilted at angle of 66½ degrees to the plane of the ecliptic. This tilt is always in the same direction throughout its orbit around the sun. The tilting of the earth in the same direction is called ***parallelism***. This parallelism is the key responsible factor for seasonal variation over the earth during a year.

Northern hemisphere is tilted to a maximum extent towards the sun around June 22. Months closer to June are summer months in northern hemisphere. During this period, northern hemisphere receives greater amount of sun's energy than southern hemisphere does. Around December 22, the southern hemisphere is tilted to the maximum extent towards the sun. As a result, just opposite occurs during this period. That is southern hemisphere will receive greater amounts of solar energy than the northern hemisphere does. This variation determines the length of the day time over a place. That is the length of the day time and is a function of latitude and month of the year.

In general, the places near equator (lower latitudes) receive more amounts of sun's energy than the places over higher latitudes. In addition to this variation, morning hours and evening hours receive less amounts of sun's energy when compared to the sun's energy received during noon hours. During noon, the sun's rays strike vertically overhead. But during morning and evening hours, the sun's rays strike at angles. These variations cause the same amount of radiation fall over a small area during noon and over a relatively larger area during morning and afternoon hours.

As the earth's surface receives the energy from the sun, it gets heated up. But this heating up also differs as the type of the surface differs. Different surfaces get heated up differently. For example, rocky surfaces are heated up rapidly while water surfaces take long time to get heated up. Similarly, the rocky surfaces loose the energy rapidly while water surfaces loose the energy slowly. Other surfaces too behave differently depending on their thermal properties. Over a certain place many different types of surfaces may exist. All the anomalies discussed above, determine the weather of a given place at a given time.

Heat energy received by a surface is transferred both downward and upward. The overlying air molecules are heated up due to conduction. Thus, heated air become less dense and they tend to move upward. As the heated air moves upward, the air above this air (relatively cool air) tend to sink downward. In this way air starts moving up and down and thereby, heat energy is transferred by air currents. This process is called, convection. Through the processes of conduction and convection, air either is heated up or cooled down. All these determine the temperature of overlying air-atmosphere. The heating up of air is different as the underlying surfaces are different. For example, air over water surfaces will be cooler than the air over land surfaces of a given place. Consequently, air over the water will be denser than the air over the land surface. This will result in movement of air from over water surface towards over the land surfaces.

During night time, opposite will occur – air from over the land move towards water. This movement of air is called ***sea-land breeze***.

As the different surfaces of the earth are heated up differently due to variations in surface characteristics, the air above these places will have different temperatures accordingly. This variation will tend to move horizontally. Horizontal movement of air is termed as ***wind***. Thus, winds are generated. In the same way, the differences in air temperature over different latitudes cause the movement of air in large scale. This large scale movement of air is called ***global circulation***.

Temperature of the atmosphere over a place changes during day and this is called the ***diurnal cycle***. It also changes during a year and this change is called ***annual cycle***. Temperature reaches at its maximum at around 3 p.m. of the day. Earth receives the sun's energy from dawn till dusk. Maximum energy is received at noon and as a result earth is heated up to a maximum extent. Then this heat energy is transferred to overlying air. There is a time gap between these two events: heating up of the earth's surface and transfer of the heat to the overlying air. Thus, the maximum temperature is attained around 3 p.m. After sunset, earth starts losing its heat energy through outgoing infrared radiation. As it loses heat gradually, the minimum temperature is attained around 4 a.m.

Q.17. What is the role of humidity in the atmosphere?

Ans. Humidity is the amount of water vapor in the air. Water vapor present in the atmosphere plays important roles in determining weather of a place. When high amount of water vapor is present in the atmosphere, we will feel discomfort. This discomfort is called sultry. Atmosphere will be dry when water vapor is less. This also causes discomfort. The atmosphere of the places near the water bodies will have high moisture content. Water vapor enters the atmosphere from water bodies through evaporation. Subsoil water enters the atmosphere both through evaporation and evapotranspiration by plants.

Water vapor present in the atmosphere may form clouds as they are lifted to higher altitudes.

Clouds get saturated as they gain more and more water vapor. At certain point of time after saturation, the water may fall from the atmosphere through the process called ***precipitation***. The falling of water either in liquid form or in solid form from the atmosphere is called precipitation. Thus atmosphere becomes part of the hydrological cycle. At any given point of time, atmosphere holds water vapor. The amount of water vapor held by atmosphere varies over time during a day and over a year. Air is a mixture of gases present in the atmosphere. As such air does not hold the moisture; but the space available among the gaseous molecules hold moisture. The water vapor holding capacity of air varies according to the temperature and pressure of air.

Phases of water in atmosphere

Water vapor present in the atmosphere plays significant roles in determining weather and climate. Water vapor in the atmosphere is also a part of water cycle. Water may be present in the atmosphere in all of its three physical forms: solid, liquid and gas. It may change its form at any point of time in the atmosphere by exchanging latent heat. The change of state from liquid to vapor is called ***evaporation***. It results when molecules escape from any water surface.

Evaporation takes place by utilizing vast amounts of energy. In this way, a large amount of energy in the form of latent heat is transferred from earth's surface to the atmosphere. The rate of evaporation is determined by three factors: vapor pressure, temperature and air movement. Evaporation increases as the saturation vapor pressure at the water surface becomes greater than the actual vapor pressure of the adjacent air. Hence, evaporation takes place at faster rate into dry air than into air with high relative humidity.

When temperature of water increases with other factors being equal, the rate of evaporation also increases. That means, whenever temperature of water is greater than that of air, evaporation always takes place. Wind movement over the surface of water replaces moist air by relatively dry air. This also increases evaporation.

Evapotranspiration: Loss of water vapour by terrestrial plants through their leaves and stems is called ***transpiration***. The combined losses of water vapour by evaporation and transpiration from a given area are called ***evapotranspiration***. Thus, the atmosphere over thick vegetation like forest will always have greater humidity than that over non-vegetated or less vegetated atmospheres. The amount of solar energy received by a place is an important factor for evapotranspiration. The intensity of solar radiation is determined by the latitude of a place. Mean annual rates of evapotranspiration are greatest between 20° N and 20° S where the highest amount of net radiation is received. Greater evapotranspiration is accounted over southern hemisphere than that over northern hemisphere as southern hemisphere consists of larger proportion of oceanic surface.

When water changes its state from vapor to liquid, it is called ***condensation***. Condensation is a very important process for the formation of clouds. As the moist air comes into contact with cool surfaces, it gets cooled to the dew point temperature and a part of the vapor condenses into liquid form on that cool surface. Condensed water over the surface is called ***dew***. During the condensation process, the heat is released as the heat is assimilated during vaporization. This released heat is called ***latent heat of condensation***. Liberation of the latent heat of condensation may slow down the cooling process. This latent heat released during condensation is one of the energy sources for atmospheric processes. Condensation also occurs from cooling in the free air. However, it requires the presence of very small particles called ***condensation nuclei***. Some of these particles are hygroscopic and hence have greater affinity towards water. As a result, condensation will occur on the surface of these hygroscopic particles even above the dew point. On the surface of other particles, water vapor may condense upon cooling *i.e.*, at dew point. After sufficient amount of vapor is thus condensed, they become visible as ***haze***. When more amount of vapor condenses, condensed water droplets increase in number and become dense to form ***fog*** or ***cloud***. Since the particles play vital role in the process of condensation and the formation of clouds, they are called ***cloud condensation nuclei***.

Water may contain foreign materials in dissolved form and/or in suspended form. It may also be covered by foreign substances. Presence of foreign substances in water will lower the freezing point below 0 °C. Amount of water involved and electrical charges also lower the freezing point.

Q.18. What is precipitation and its various forms?

Ans. ***Precipitation*** is water falling on earth in either liquid or solid form. Precipitation occurs in the form of ***rain***, ***drizzle***, ***snow***, ***hail*** and/or their modifications. Precipitation is an important part of the hydrological cycle which brings water over the continental regions.

The precipitated water either runs as stream or infiltrate into the ground to become groundwater. Some part of the precipitated water is stored in natural and/or artificial reservoirs such as dams, lakes, ponds and tanks. Water vapor condensed over condensation nuclei form clouds in the atmosphere and these clouds give out the water as precipitation. The formation of clouds and the process of precipitation are quite complex.

Rain: It is precipitation with large liquid water droplets. Rain drops range from less than 1 mm to 5 mm in diameter.

Drizzle: It is precipitation consisting of uniformly minute droplets of water (less than 0.5 mm in diameter).

Snow: It is precipitation consisting of large ice crystals called snow-flakes. It is formed by the ice-crystal process. These crystals do not have time to melt before they reach the ground.

Hail: It is precipitation in the form of ice pellets called hail stones. The process of formation of hail stone is complex one.

Q.19. How does atmospheric pressure vary with altitude?

Ans. Pressure is "*the force per unit area acting on a surface*". The SI unit for pressure is the *pascal*. One pascal is equivalent to 1 N/m^2. The atmospheric pressure is expressed in millibars. 1 millibar is equivalent to 100 N/m^2 (= 100 pascal).

Atmosphere is held by earth by its gravitational pull. The combined mass of all air molecules lying above the earth exert a force on earth. So, atmospheric pressure is the force exerted by the column of air per unit area above the earth's surface. The amount of air in a column over a unit area will differ from place to place as there is variation in its altitude. Over top of a hill, the amount of air in a column will be less than the amount of air in a column over a sea surface. Hence, the atmospheric pressure will be less over the hill than over sea surface or a land surface at low lying areas. As the height of a place increases from the mean sea level, the pressure decreases. As we go deep under the earth, the pressure will be more than that over land. Under the sea, the water column along with overlying air will exert pressure which will be unbearable.

The atmospheric pressure over mean sea level under normal conditions is 1013 mb. The atmospheric pressure over the top of the Mount Everest is 320 mb. In addition, cold air will generally be denser than warm air. As a result, a column of warm air will exert less pressure than that by a column of cold air. The differential heating of surfaces (due to differences in surface characteristics), the air lying over these surfaces will be heated differently and hence will exert pressure differently. This will cause anomalies of pressure values in horizontal atmospheres. Consequently, air from high pressure region tends to move towards the low-pressure region. Thus, winds are generated. The strength of the winds is determined by the difference in pressure values. Greater the difference in pressure values, stronger will be the speed of the wind.

Q.20. What is wind and its importance?

Ans. The movement of air is called wind. Winds follow same pattern globally. This pattern is called ***global atmospheric circulation***. In the past the sailors took the advantage of the wind for the movement of ships and boats. The sea routes were chosen according to the winds blown and their direction.

Earth receives unequal amount of radiation and hence is heated differently at different latitudes. Earth also rotates on its axis. These two factors cause the formation of winds and result in global circulation. Through this the surplus energy received over low latitudes are transported as kinetic energy in the form of winds towards high latitudes. Different types of surfaces of the earth are heated up differently as their thermal properties differ. Even if places are close to each other in their latitudinal position, differences in their surface characteristics make them have different thermal properties. Consequently, the pressures over these different places also differ. This will create horizontal pressure variation over an area of few meters to several kilometers. This variation results in a gradient of pressure anomalies over the entire area. This horizontal pressure gradient is responsible for the movement of air, wind. This pressure gradient drives the air to move, and hence this force is called ***pressure gradient force***.

Winds may blow gently or violently or with moderate speeds. Gentle winds bring relief and comfort from heat of the day. Violent winds may sometimes bring devastating effects. Storms, cyclones, tornado etc. are such violent winds that bring devastating effects. In addition, winds carry moisture, heat, pollutants etc. from one place to other.

Q.21. What is geostrophic wind?

Ans. Due to pressure gradient force, air moves from high pressure region to low pressure region. When a surface is heated up, the overlying air is heated up and become light. The light air tends to ascend. As the warm air rises, the pressure over that surface becomes low. To compensate this low-pressure air will start moving from a high-pressure region towards this low-pressure region. When air moves from high pressure region, the overlying air at that point will tend to sink. The air ascended will move towards the region where from air is sinking. This completes the circulation of air between two different places vertically. Consider a large area, and the pressure keeps varying over this area as we move away from one place farther and farther. That means pressure keeps changing as we move away from a place to another place little by little. The line connecting areas with equal pressure is called ***isobar***. When pressure changes over a large area, that area may be occupied by several isobars, each isobar denoting a particular pressure value. Sometimes, these isobars may be parallel or circular over a large area of consideration. Once the air start moving, other forces come into action: ***Coriolis force*** and ***frictional force***.

Coriolis force: It is a deflective force. Any moving object, stream of water or air over the surface of spinning planet, will be deflected due to the rotation of earth. This is called Coriolis force. Coriolis force will deflect moving bodies towards their right in northern hemisphere and towards their left in southern hemisphere. Due to this Coriolis force, the moving air over the earth's surface will be deflected accordingly. However, the effect of Coriolis force is nil over equator and its effect increases toward the poles. It acts at an angle of 90° to the horizontal direction of the wind and is directly proportional to horizontal wind speed. When pressure gradient initiates the air to flow, the Coriolis force deflects it to flow parallel to the isobars. The winds must flow from an isobar with low pressure value towards the isobar with high pressure value. Instead of this perpendicular flow, the wind flows parallel to the isobars as a result of Coriolis force. The wind thus flowing is called ***geostrophic wind***.

Q.22. What is gradient wind?

Ans. If the air moves along curved isobars, a net centripetal acceleration tends to pull it toward the center of curvature. This will result in a rotating motion of air relative to earth's surface. Such wind is called ***gradient wind***. If this gradient wind circulates in counter clock direction, then it is termed as ***cyclone*** in northern hemisphere while it is called ***anticyclone*** in southern hemisphere. Similarly, the clockwise motion of air is called ***cyclone*** in southern hemisphere while it is called ***anticyclone*** in northern hemisphere. At the place under the low pressure region, air flowing toward this region tends to converge and ascend and hence, this place is called ***area of convergence***. Likewise, under the high pressure region, air flows away from it as the air sinks from above and hence, this region is called ***area of divergence***.

In cyclonic circulation, surface air moves towards the center, the low-pressure cell, and converges. Thus, converged air ascends vertically in the center of the low-pressure cell.

The reverse happens in the center of an anticyclone – as the surface air diverges from the center of the anticyclone (high pressure cell), the air sinks from high in the center of the anticyclone (high pressure cell). In these ways, the cyclones are associated with rising air at their centers while, anticyclones are associated with descending air in their centers.

Q.23. What creates turbulence in wind?

Ans. Another factor may come into force when air starts moving due to friction exhibited by surface. The earth's surface may consist smooth surfaces like water, or highly rough surfaces like rocky terrain, urban buildings etc. When air flows over the smooth water surface, the friction encountered is almost zero. But when wind flows over a rough terrain like a land with rocks or a land with urban buildings of various heights, it encounters friction. The magnitude of friction differs as the surface features differ. The friction lowers the speed as well changes the direction of the wind. It creates ***turbulence*** in winds. Turbulence is associated with lulls, gusts and eddies. Turbulence also increases with the increase of the wind speed.

However, the surface relief does not pose any friction at the height of about 500 m. At the height of about 1000 m, the winds are no more under the influence of friction. As a result, the winds flowing at this height will be approximately equal to the theoretical geostrophic wind and/or gradient wind.

Q.24. What are local winds like sea land breeze and mountain valley winds?

Ans. Apart from the atmospheric wind circulation at global level, the local winds are generated by differences in surface characteristics: e.g., ***sea-land breeze***, ***mountain-valley winds***, and other local winds.

Sea-land breeze: Some land areas are very close to large water bodies. Lands near sea, large reservoir, lakes and wide rivers are the best examples. During day time the incoming solar radiation heats both the land surface and water surface. However, the land surfaces are heated up rapidly while water surfaces are not. The rapidly heated land surface in turn, heats up the overlying air in relatively short duration. Heated air becomes light and ascends high creating a low-pressure region over the land surface. At the same time, the air above the water remains relatively cooler and denser to create a high-pressure region over the water. This anomaly results in pressure gradient between the land and water surfaces. This pressure gradient force will start driving the air lying above water to move towards the land. This wind is called ***sea breeze***.

During night, the earth loses it heat energy rapidly and gets cooled down. The water does not cool down so fast but remain relatively warmer during night. Relatively warmer water surface transfer its heat energy to the overlying air and thereby air becomes warmer and light to ascend high. This will create a low pressure over the water. At the same time, the overlying air at the land will be relatively cooler and denser creating a high-pressure region. This anomaly results in pressure gradient (just opposite of what exists during day time) between the land and water. Now the air from land will tend to move towards the water and this wind is called ***land breeze***.

Mountain-valley winds: On mountain sides, under a clear night sky the lands situated at high altitude radiate heat and get cooled down. This in turn cools the air lying over those lands. Now the cool and dense air tends to slide down the mountain slopes into valleys and low lands. This air flow is called ***mountain breeze***. In the morning hours, temperature inversions may occur over the low lying areas. As a result, the valley bottoms are colder than the hillsides. In general, inversions over a land area prevent the vertical motion of air. But the mountain breeze reached the valley, may gain considerable velocity and generate enough turbulent mixing to break the inversion. On warm sunny days, the slopes of mountain will be heated up and thereby the overlying air. This will generate an upslope flow of air towards the top of the mountain.

The flow of air from slope of the mountains to the top is called ***valley breeze***. When warm air moves up, it is replaced by cooler air from above the valley. This results in moderation of surface temperature.

Other local winds: Under the influence of gravity, the cold and dense air moves downward along the steep slopes. These winds are called ***gravity winds*** or ***katabatic winds***. These winds are prominent under calm, clear conditions where the edges of highlands plunge sharply toward low lying terrain. These winds are fed by large quantities of extremely cold air that collect over the high land. Katabatic winds sometimes may reach destructive intensities as of a waterfall from a high land. This kind of winds is experienced in the Rhône River valley of southern France during winter. Icy, high velocity winds drain from the snowy mountains to the valleys in this area. These winds are called ***mistral winds***.

Another type of local wind is generated when the air is forced to pass across the mountainous terrain. As the air move upward, the moisture present originally is almost wrings out. After reaching the top, the winds cross over the mountain and slide down on the other side of the hill along its slopes. As the air has already exhausted its moisture, only dry and relatively warm air flows down with high speeds. Sometimes the speeds reach to 120 km per hour.

Q.25. How atmospheric dispersion of air pollutants are achieved?

Ans. The atmospheric processes determine the concentration of air pollutants emitted into it.

Atmospheric dispersion is the ability of the atmosphere to disperse the air pollutants and dilute them over a given period of time at a given geographical area. Pollutants are released by various sources either continuously or irregularly and either in large quantities or in small quantities. When they are present in atmosphere in such concentrations and for such durations, may cause adverse effects on humans, plants, animals and materials. Hence, the dispersion of these pollutants is essential and desirable. Prevailing winds carry and transport the pollutants horizontally away from the source and its vicinity. While they are transported, they may diffuse laterally as well as vertically. This makes the pollutants concentration to become less and less. That is the pollutants are diluted to such level so as not to cause adverse effects. The diluted pollutants may be removed from atmosphere by natural atmospheric cleansing mechanisms. Two factors play important roles in dispersion: one is wind speed and its direction, another is the atmospheric stability.

The prevailing winds will transport the pollutants horizontally. Greater the wind speed, greater the distance the pollutants are transported. Speed of the wind is determined by the strength of the pressure gradient force. As the pressure gradient force varies temporally and seasonally as well as spatially, the speed of the wind also varies. As the location of high pressure and low-pressure regions change over time, the direction of the wind also changes. In meteorology, the direction of wind is the direction from which the wind blows. Presence of physical barriers will change the direction of the flow of the winds.

Over a geographical area, the surface winds may blow at any speed and from any direction in a given time. When we consider a long period of time, say, a month or a season, over that place, we can get the information about the wind speed and direction that prevailed during that time period. This information is presented in a diagram called ***wind rose***. Which is helpful in locating new industries at a location so as to avoid or minimize the effects of pollutants on residents of a densely populated area.

Q.26. How does the atmospheric stability affects the pollutants dispersion?

Ans. The atmospheric stability is an important factor in determining the diffusion of the pollutants. It is the tendency of the atmosphere to resist or enhance vertical motion. It is determined by vertical temperature change in the atmosphere and the speed of the wind.

There are basically three types of atmospheric stability viz. unstable condition, stable condition and neutral condition.

Lapse rate: Before understanding the stability conditions of the atmosphere, we should be familiar with lapse rate. Lapse rate is defined as rate of decrease of an atmospheric variable with height. Temperature is a variable decreasing with the height of the atmosphere. It is an important property of troposphere. However, this rate may change diurnally, monthly and seasonally at a given location.

Environmental lapse rate/Ambient lapse rate (ELR): It is the actual temperature at each level vertically. It is the rate of decrease of temperature in the actual atmosphere vertically.

Adiabatic lapse rate: When a parcel of air moves vertically, its temperature will decrease without exchange of heat (adiabatically) to its surrounding air as it rises up. The rate of decrease of temperature of a rising parcel of air vertically is called adiabatic lapse rate.

Dry adiabatic lapse rate (DALR): when a dry parcel of air (without moisture content) moves upward, its temperature decreases without exchange of heat to the surrounding atmosphere (adiabatic). The rate of decrease of temperature of a rising dry parcel of air is called dry adiabatic lapse rate. The rate of decrease of temperature of a rising dry parcel of air is calculated and determined as – 0.98 °C per 100 m (= – 9.8 °C per km). For practical purposes, it can be approximated to be – 1 °C per 100 m (= – 10 °C per km).

Saturated adiabatic lapse rate/wet adiabatic lapse rate (SALR): When a rising parcel of air is saturated with water vapor, the temperature will decrease and at the same time water vapor will condense. The condensation of water vapor will release latent heat and this will heat the air parcel. Consequently, the lapse rate of the rising parcel air will be different from that of DALR. The saturation lapse rate is determined as – 0.44 °C per 100 m (= – 4.4 °C per km) at 20 °C of atmospheric temperature. The rising parcel of air may not be saturated with water vapor all the time, but some amount of water vapor will be present in it. This will alter the lapse rate accordingly.

Q.27. What are the stable, unstable and neutral condition in the atmosphere?

Ans. Stable condition in the atmosphere: Stability of the atmosphere is its resistance to the vertical movement of air parcel. It does not allow the air parcel to move either upward or downward but to allow the parcel to remain at the same position. Even though a parcel of air receives some upward force, it will return to its original position from where it was released. When the environmental lapse rate is less than the dry adiabatic lapse rate (DALR), the condition is said to be "sub adiabatic" or "stable atmosphere". Under this condition, when a parcel of air rises up, it will cool down adiabatically but the surrounding atmosphere at this level will be relatively warmer than the air parcel. As a result, the risen parcel of air will tend to sink until it reaches its original level where from it was released.

Unstable atmosphere: When environmental lapse rate is greater than the dry adiabatic lapse rate, the environmental temperature decreases faster with height and this condition is called "super adiabatic" or unstable condition. Under this condition, when the parcel of air moves up it will cool down adiabatically but the surrounding atmosphere will be relatively cooler. This will cause the risen parcel of air to further move up. This condition favors the pollutant dispersion.

Neutral atmosphere: When environmental lapse rate and dry adiabatic lapse rate are equal, then the atmosphere is said to be neutral. When air parcel rises up, it will cool down adiabatically. If the temperature attained after cooling by the air parcel is higher than the surrounding air, then the parcel will not rise. However, if the parcel is forced to move up, it will rise.

Q.28. What is inversion? How does it affect the pollutants dispersion?

Ans. Inversion: When temperature increases with height, the environmental lapse rate becomes just opposite. This condition is extremely sub adiabatic. Under this condition, vertical movement of air parcel is prohibited as the rising air parcel will become cooler than surrounding warmer atmosphere.

- Inversions occur during night or early morning hours.
- Inversions are more frequent during the fall than during other seasons.
- When inversions accompany low wind speeds (10 km/hr.) – limited horizontal and vertical dispersion occurs.

Sometimes, vertical temperature gradients are grouped to result in combination of different stability conditions.

Ground based inversion with elevated neutral or unstable condition aloft: At times, inversion layer exists near the ground and either neutral or unstable condition exists above it. Under these circumstances, if the air parcel is released within the inversion layer, it will tend to sink and if it is released above the inversion layer, it will behave accordingly (rise further).

Neutral or unstable near the ground with inversion aloft: sometimes, the neutral or unstable condition exists near the ground and inversion above it. Under these circumstances, if the air parcel is released within the neutral or unstable layer, the parcel will tend to move upward until it reaches the bottom of the elevated inversion layer. If air parcel is released above the inversion layer, it will remain at the same level.

Q.29. What is the role of atmospheric stability in dispersion of air pollutants?

Ans. The stability of the atmosphere determines the strength of dispersion of the pollutants vertically. Stable atmosphere and inversion inhibit the vertical dispersion of the pollutants while unstable and neutral conditions favor the vertical dispersion. The wind speeds and existing stability determine the overall dispersion both vertically and horizontally. Greater the wind speed, farther the pollutants will be transported horizontally. If low wind speed accompanies stable atmosphere or inversion, less/no dispersion of pollutants will occur. Under these circumstances, the concentration of the pollutants will build up, if they are released continuously. It will result in high pollution potential.

Q.30. What are different types of plume behavior of released air pollutants?

Ans. The plume of the released air pollutants from chimney will be dispersed differently. Various types of plume behavior are discussed below:

Fanning: This plume has a large spread horizontally and very little vertically. This typically occurs at night in a very stable boundary layer with strong surface inversion and weak variable winds. The inversion does not permit the vertical movement of the plume (air parcel) and hence, there is no dispersion or little dispersion vertically.

Fumigation: When the plume material gets rapidly brought down to the ground level due to downward mixing, it is called fumigation. This situation occurs shortly after sunrise due to surface heating and is slowly replaced by an unstable layer that grows up to the top of the plume. This condition is usually short-lived but results in the highest ground level concentrations.

Looping: It occurs in very unstable and convective conditions during midday and afternoon. Large convection eddies take the plume material in successively upward and downward motions. The atmosphere is unstable which permits and encourages the vertical movement of plume to a greater extent. During this atmospheric condition, the large convective eddies are formed. These eddies take the plume up and down vertically as it disperses vertically.

Coning: This is when the plume looks like a cone in both the horizontal and vertical scale. This usually occurs under cloudy and windy conditions.

Lofting: The plume stays above the surface inversion. This occurs shortly after transition from unstable to stable conditions near sunset. The plume can be thin or become quite thick.

Depending on the height of the stack and the rate of deepening of the inversion layer the lofting condition may be very transitory or it may persist for several hours.

Trapping: Plumes released in unstable atmosphere disperse their material uniformly throughout the air. Trapping can lead to very high ground level concentrations of the pollutants when the inversion layer is low and there are weak winds.

Q.31. What is mixing height and its effect on air pollutants dispersion?

Ans. Mixing height is the vertical distance above the earth's surface at a given location and at a given time for the mixing of air pollutants. In other words, mixing height is the thickness of the atmospheric layer measured from the surface upward, through which pollutants are presumed to mix by virtue of convection caused by day time heating of the surface. Mixing height vary diurnally, daily, monthly and seasonally. They also vary depending upon the topographic features.

The extent of mixing height determines the strength of dispersion of pollutants vertically. However, low mixing heights with strong winds have the same effect on pollutants transport as light winds with larger mixing heights. In air pollution potential forecasting, the ventilation index (the product of mixing height and average wind speed) is generally employed. Ventilation index determines the air pollution potential of a given place. High air pollution potential is the occurrence of high concentration of pollutants due to low dispersion of the pollutants. Mixing heights are not measured directly but are determined from the vertical temperature distribution with assumption that in thoroughly mixed unsaturated atmosphere the temperature lapse rate is "dry adiabatic".

The morning mixing height is the height above ground level where dry adiabatic extension of the morning minimum surface temperature plus 5 °C (this addition is done for urban areas) intersects the vertical temperature profile near sunrise that morning. 5 °C is added to incorporate urban heat island effects.

Afternoon mixing height is computed in the same way from the afternoon maximum surface temperature. While calculating the afternoon mixing height, there is no need for adding any extra value with the maximum temperature value.

Q.32. What is the role of stack height in air pollutants dispersion?

Ans. Stack height is the height of the stack of a pollutant source. The pollutant source is usually an industry or industrial complex. The stack is the physical provision through which the air pollutants are emitted/released into the atmosphere. It is also called ***chimney***. Earlier, people thought greater the height of the stack, the farther the pollutants will be transported and diluted before reaching the ground level. However, in Western Europe and North America, the tall stacks created yet another problem – acid rain over the downwind localities. The taller stacks transported the pollutants to farther distances but during transportation, they underwent chemical transformation to form acid rain in the places/countries located in the downwind.

Stack height is an important parameter in calculating the downwind concentration of the pollutants emitted. No doubt, tall stacks will transport the pollutants to longer distances than the short stacks. While stack height is considered, we should bear in mind that there are two stack heights of the same stack. One is "physical stack height" and another is "effective stack height".

Physical stack height: It is the height of the stack measured from the surface of the ground till the top edge of the stack. It is denoted by $\boldsymbol{h}$.

Effective stack height: It is the physical height plus plume rise. When pollutants(plume) are emitted from the stack they go upward to a certain height due to forced ejection and being warmer than the atmosphere. The height up to which plume rises above the top edge of the stack is called plume rise. After rising to this level, the plume will take a bend due to prevailing wind and transported horizontally in the downwind direction. The plume rise is symbolized by $\boldsymbol{h_r}$.

Now, $h + h_r$ will yield $\boldsymbol{H}$, the effective stack height. $H = h + h_r$

For calculating plume rise several equations are available.

Q.33. What is Gaussian plume dispersion model?

Ans. It is a model used to determine the concentration of the pollutant(s) at any point in the downwind direction. It is based on normal distribution (Gaussian distribution) of the observations in a data set. Here concentration values are the observations of the normal distribution. It is assumed that the centerline of the distribution will have the maximum concentration while areas away from centerline will have minimum concentration.

Concentration will decrease as one goes away from the centerline. It is given by:

$$C(x, y, z) = \frac{Q}{2\pi u\sigma_y\sigma_z} \times \left[\exp-\left(\frac{y^2}{2\sigma_y^2}\right)\right]\left\{\exp\left(\frac{-(z-H)^2}{2\sigma_z^2}\right)+\exp\left(\frac{-(z+H)^2}{2\sigma_z^2}\right)\right\}$$

Where, C = ground level concentration at some distance × downwind (g/m³)

Q = average emission rate (g/sec)

u = mean wind speed (m/sec)

H = effective stack height (m)

σ_y = standard deviation of wind direction in the lateral (m)

σ_z = standard deviation of wind direction in the vertical (m)

y = off-centerline distance (m)

e = natural log equal to 2.71828

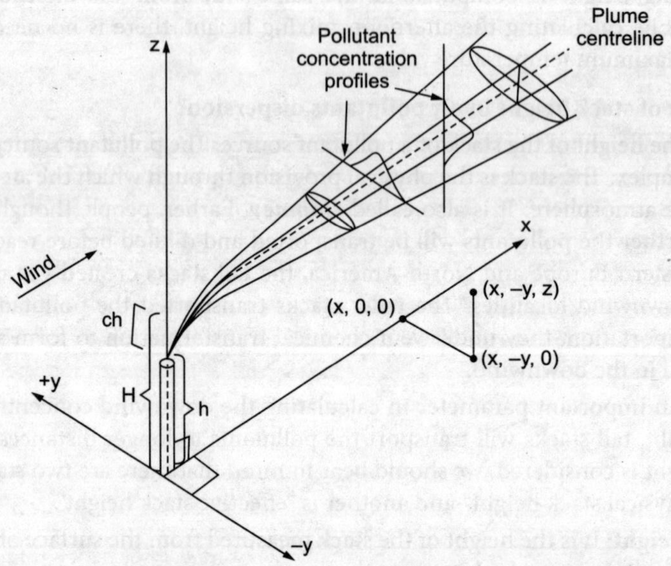

Fig. 2.1: Gaussian plume dispersion model

This equation is the general equation. It can be modified to suit local conditions. Further this equation can be modified according to the requirement. For example, to compute ground level concentration alone, the equation will become simpler than this. In this model, the dispersion coefficients, σ_y and σ_z are used. These are the standard deviation values in lateral and vertical directions respectively. These values cannot be determined directly. Pasquill and Gifford have developed the curves called P-G curves for determining these σ_y and σ_z values.

Q.34. What is a wind rose? Discuss its usefulness.

Ans. Wind rose is a diagrammatic representation of wind data over a period of time in a given place. It represents the speed and direction of the wind blown. It is a frequency distribution of the wind classes grouped based on wind speed and direction. Here wind direction is the direction from which the wind is blowing.

Wind rose diagrams are useful in knowing the predominant wind direction and the strength of the wind over a place. Wind roses will differ monthly and seasonally. They may also vary between day time and night time. It represents climatology of a location with reference to wind speed and wind direction. Using the wind data of a place over a period of time, we can prepare the following wind roses:

- Day time wind roses
- Night time wind roses
- Monthly wind roses
- Seasonal wind roses
- Wind roses under different stability conditions

From the wind rose, we may infer the downwind locations that will be affected by the air pollutants. We can also decide the location of any proposed industry in such a way that the highly populated area is not affected.

Wind rose diagrams will be useful while predicting the pollutant concentrations using Gaussian plume dispersion equations. Wind roses are used to determine the average wind speed for the winds blowing from a given direction. This average value is used in Gaussian plume dispersion model. The downwind direction is also determined from wind rose diagram; for each direction, corresponding downwind direction will be chosen and pollutant concentration will be predicted in that downwind direction. In this way, pollution concentrations can be determined for all the wind directions and corresponding downwind directions.

Q.35. What is pollution and air pollution?

Ans. "Pollution is an undesirable change in the physical, chemical, or biological characteristics of our land, air or water that may or will harmfully affect human life or that of desirable species". This definition includes all kinds of pollution.

Air pollution is defined as "the presence in the outdoor atmosphere of one or more contaminants such as dust, fume, gas, mist, mist, odor, smoke or vapor in quantities, of characteristics, and of duration such as to be injurious to human, plant or animal life or to property, or which unreasonably interferes with the comfortable enjoyment of life and property".

Q.36. What are the sources of air pollutants?

Ans. Air pollutants are released by both natural and anthropogenic(manmade) sources. Natural sources include volcanoes, sea salt spray, forest fires, dust storms, etc. Major anthropogenic sources are transportation, power generation by combustion of fossil fuels, refuse burning, industrial and domestic fuel burning and industrial processes.

Though the amounts of pollutants released by natural sources are greater than that by anthropogenic sources, pollution by anthropogenic sources is of great concern. Anthropogenic sources emit pollutants in large quantities within a confined location, for example, in an urban area, an industrial town, which result in high concentrations to cause deleterious effects. In addition, anthropogenic sources emit the substances which were not originally present in natural atmosphere. These substances are called "foreign substances".

The air pollutant sources are categorized as "**Stationary sources**" and "**Mobile sources**". The stationary sources are the ones that stay permanently in their places and emit the air pollutants regularly and/or continuously. Examples of **stationary** sources include: industries and power plants. Mobile sources are the ones that keep moving from one place to other and during their movement they emit air pollutants into the atmosphere. All the vehicles ply on roads – two wheelers, three wheelers, cars, buses, lorries, trucks etc., diesel powered trains, air planes, ships and boats etc. are the mobile sources that emit large quantities of air pollutants.

Q.37. What are various ways of classification of air pollutants?

Ans. Air pollutants are classified in following ways:

(*a*) based on their physical state,
(*b*) based on their chemical properties and
(*c*) based on their origin

Q.38. Discuss the classification of air pollutants based on their physical state?

Ans. Based on physical state, the air pollutants are classified as:

1. **Particulates:** They include both solid and liquid particles. Small solid particles and liquid droplets are collectively known as particulates. They are present in atmosphere in fairly large amounts and pose a serious air pollution problem. Particulate pollutants are classified according to their particle size and nature into fumes, dust, ash, carbon smoke, lead, asbestos, mist, spray, oil, grease etc.

   (*a*) **Suspended Particulate Matter:** Suspended Particulate Matter (SPM) are defined as "all solid and liquid particles in the air that are small enough not to settle out on to the earth's surface under the influence of gravity".

(*b*) **Respirable Particulate Matter (RPM) or PM10:** Particles of size less than 10 µm can enter into the lungs. Particulates of this size and below cannot be prevented by the filtering mechanism available in our respiratory tract and hence their entry into the lungs. Special attention is given to RPM due to its potential harm to human respiratory system.

(*c*) **Dust:** It is made up of solid particles predominantly larger than those found in colloids and capable of temporary suspension in air or other gases. They do not tend to flocculate except under electrostatic forces. They do not diffuse but settle under influence of gravity. Though the dust can also be included under particulates, it is distinguished from the particulates based on its ability to settle down by gravity.

(*d*) **Aerosol:** It is a suspension of solid and/or liquid particles having a negligible falling velocity. The size of the particle in an aerosol frequently exceeds the normal colloidal limits (approximately 1 nm to 1 µm).

(*e*) **Smoke:** Fine, solid particles resulting from the incomplete combustion of organic substances such as wood, coal, tobacco. It consists mainly of carbon and other combustible materials. Smoke particles range from 0.5 to 1 µm in diameter.

(*f*) **Fume:** Fumes are fine, solid particles (often metallic oxides such as zinc and lead oxides) formed by condensation of vapors of solid materials. Fumes may be from sublimation, distillation, calcinations, or molten metal processes and they range in size from 0.03 to 0.3 µm. Fumes flocculate and coalesce, then settle out.

(*g*) **Fly ash:** It consists of finely divided, non-combustible particles contained in the gases arising from combustion of coal. Inherent in coal, these mineral or metallic substances are released when the organic portion of coal is burned. Fly ash shares characteristics dust, smoke and fumes. Like dust, fly ash particles in size, range from 1.0 to 1000 µm; like smoke, it results from burning; and like fumes, it consists of inorganic metallic or mineral substances.

(*h*) **Mist:** It consists of liquid particles or droplets formed by the condensation of a vapor, the dispersion of a liquid, or the enactment of a chemical reaction. Mists are usually less than 10 µm in diameter. If mist concentration is high enough to obscure visibility, then it is called a fog.

(*i*) **Spray:** It consists of liquid particles formed by the atomization of parent liquids, such as pesticides and herbicides. Spray particles range in size from 10 to 1000 µm.

2. **Gaseous contaminants:** Pollutants that occur in gaseous state are called gaseous contaminants. These pollutants are gaseous in nature at normal temperature and pressure.

Gaseous pollutants are formless fluids that occupy the space into which they are released. They behave like air and do not settle out of the atmosphere. They include a variety of organic and inorganic gaseous materials.

(*a*) **Inorganic gases:** These include noxious gaseous pollutants like oxides of nitrogen (NO_x), oxides of sulfur (SO_x), hydrogen sulfide (H_2S), ammonia, chlorine, hydrogen fluoride, hydrogen chloride, oxides of phosphorous, hydrogen cyanide, bromine and mercaptans etc.

(*b*) **Organic gases:** These pollutants include hydrocarbons (C_xH_y) such as CH_4, C_3H_8, C_2H_2, C_2H_4, C_6H_6, C_8H_{18} and other compounds such as formaldehyde, acetone vapours, alcohols, organic acids, methyl isocyanate, chlorinated hydrocarbons etc.

Q.39. Discuss the classification of air pollutants based on chemical properties.

Ans. Based on their chemical properties, the air pollutants are classified as follows.

These may include both particulates and gases.

- **Sulfur-containing compounds:** All air pollutants that contain sulfur atom in their molecular structure is called "sulfur-containing compounds". Examples are: SO_2, SO_3, H_2S, mercaptans etc.
- **Nitrogen-containing compounds:** All the air pollutants that contain nitrogen atom in their molecular structure are called "nitrogen-containing compounds". Examples are: NO, NO_2, NH_3 etc.
- **Carbon-containing compounds:** All the air pollutants that contain carbon atom in their molecular structure are called "carbon-containing compounds". Examples are: CO, CO_2 etc.
- **Halogen compounds:** They include, HF, HCl etc.
- **Radioactive compounds:** The substances that are radioactive and airborne are called radioactive compounds.

Q.40. Discuss the classification of air pollutants based on origin.

Ans. Based on their origin, the air pollutants are classified as:

- **Primary air pollutant:** A pollutant emitted into the atmosphere from an identifiable source. (e.g., CO, SO_2, H_2S, CO_2 etc.)
- **Secondary air pollutant:** A pollutant formed by chemical reaction in the atmosphere (e.g., photochemical smog, NO_2, O_3 etc.)

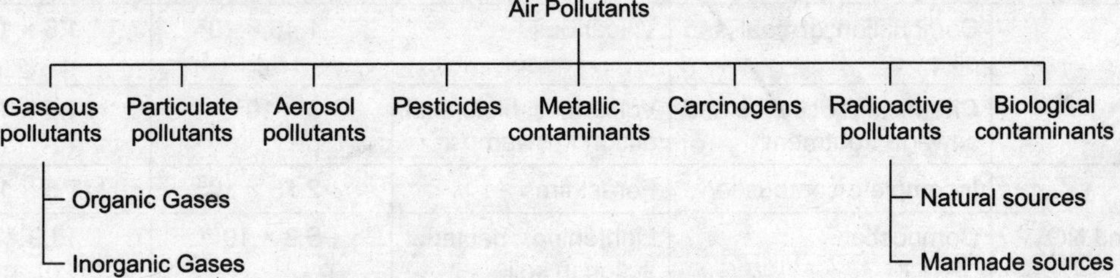

Fig. 2.2: Classification of Air Pollutants

Q.41. What are the natural and anthropogenic sources of particulates?

Ans. Several natural processes such as volcanic eruptions, dust storms, wind-driven dust from soil, sea spray etc. release huge quantities of particulates (1000-2000 million tons each year) into the atmosphere. The anthropogenic sources include power plants, smelters, mining operations, industrial processes, etc.

It has been estimated that combustion, industrial processes and miscellaneous sources contribute almost about 1000 million tons per year. Metal oxides comprise a major class of inorganic particulates in the atmosphere. They are produced when metal-containing fuels are burnt.

Formation of H_2SO_4 droplets generates aerosol mists in the atmosphere. If basic substances such as NH_3 or CaO are present in the atmosphere, the H_2SO_4 will react to form salt particles such as $(NH_4)_2SO_4$ and $CaSO_4$. These salts may be present in dissolved form (mist) or they may get dried by sunlight to appear as solid salt particles. Atmospheric particulates may comprise a variety of organic and inorganic chemical compounds and radionuclide.

Table 2.1: Sources of particulate emissions

Source	Examples
Combustion	Fuel burning (coal, wood, fuel, oil) Incineration Others – open fires, etc.

Materials handling and processing	Loading and unloading Crushing and grinding Cutting and forming
Earth moving operations	Construction (road, buildings, dams, etc.) Mining Agriculture
Miscellaneous	House cleaning Mud road cleaning Engine exhaust, etc.

Q.42. Discuss the sources of gaseous pollutants.

Ans. Gaseous pollutants are released by a wide variety of sources. They include both natural and man-made(anthropogenic) sources. Volcanic eruptions, biological processes, lightning, forest fires and oceans are major natural sources. Major man-made sources are combustion of fossil fuels, chemical processes in industries, incineration and sewage treatment processes.

Table 2.2: Summary of sources and contributions of gaseous pollutants

Gases	Man-made sources	Natural sources	Estimated annual emissions (tons)	
			Man-made	Natural
SO_2	Combustion of coal and oil	Volcanoes	1.46×10^6	1.5×10^6
H_2S	Chemical processes and sewage treatment	Volcanoes, biological action in swamps	3×10^6	1.0×10^4
CO	Incomplete combustion	Forest fires	2.75×10^8	7.5×10^7
NO and NO_2	Combustion	Lightening, bacterial action in soil	5.3×10^7	10.9×10^8
NH_3	Sewage treatment	Biological decay	4.0×10^5	1.16×10^9
N_2O	—	Biological action	—	5.9×10^8
Hydrocarbons	Incomplete combustion, chemical process	Biological processes	8.8×10^7	4.8×10^8
CO_2	Combustion	Biological processes, oceans	1.4×10^{10}	10^{12}

Q.43. Discuss the effects of air pollutants on human health?

Ans. Human beings breathe air continuously throughout our life. While breathing, we inhale the surrounding air and its constituents. The air may contain both particulates and gaseous contaminants.

Human beings depend on atmosphere for our metabolic requirements and for our comfort. We inhale and exhale air almost ceaselessly throughout our life. This process is called breathing; biologically, called respiration. An average man breathes 22,000 times a day and takes in about 16 kg of air daily. We use oxygen for metabolism and exhale carbon dioxide into the atmosphere. During inhalation, we also take other impurities present in air along with. The impurities may be of natural origin like pollens, natural allergens, dust particles etc. or may be introduced by pollution. The impurities entering into the human lungs may affect in various ways. All the impurities entering do not necessarily cause harm however. The health effects caused by the inhaled pollutants depend upon following factors:

- Nature of the pollutants
- Concentration of the pollutants

- Duration and frequency of exposure
- State of health of the receptor
- Age group of the receptor

The chemical nature of the pollutants determines the type and extent of the health problem posed. The physical property such as size and shape of the particle also is important factor in causing the effect. Concentration of the pollutants determines the magnitude of the harm. Generally, low concentrations may produce little or no effect. Duration and frequency of exposure by the individual and concentration are important factors that determine the effect. The effects may be acute or chronic in nature.

The young children, old persons, people already with certain illness/sickness and infirmity are highly susceptible to the effects of air pollution.

Q.44. What are various routes of entry of air pollutants in human beings?

Ans. Air pollutants may enter human body through:

- Eyes
- Nose (also orally while opening the mouth)
- Skin (this route is not important except in case of fumes and vapours)

Of the above three routes, air pollutants enter predominantly in large quantities through nose and reach the throat, larynx, trachea-branchial tree and finally lungs. From lungs they may enter the blood stream and reach other tissues. Some irritants may even reach the mucosa of the digestive tract.

Q.45. What are the problems and health effects caused by air pollutants?

Ans. Air pollutants pose a variety of problems and cause a number of effects as given below:

- Eye irritation.
- Nose and throat irritation.
- Irritation of the respiratory tract.
- Nuisance by odor from gases like H_2S, NH_3 and mercaptans even at low concentrations.
- Increase in mortality and morbidity rates.
- A variety of particulates particularly pollens are known to initiate asthmatic attack.
- Chronic diseases such as bronchitis and asthma may get aggravated by high concentration of SO_2, NO_x, photochemical smog and particulates.
- Carbon monoxide readily combines with hemoglobin (Hb). It has 240 times greater affinity to Hb than that of oxygen. It causes anoxic conditions, and may result in death when an exposure to > 1000 ppm for one hour is encountered. At lower concentrations, it increases stress, headache, drowsiness, fatigue etc.
- Hydrogen fluoride will cause fluorosis.
- Certain pollutants are carcinogenic in nature; exposure to which will result in cancer.
- Dust particles may cause respiratory diseases (silicosis, asbestosis, etc.).
- Heavy metals in the form of particles or vapour entering the lungs are poisonous.

Table 2.3: Health effects caused by particulate matter

Concentration, $\mu g\ m^{-3}$	Accompanied by	Time	Effects
750	715 $\mu g\ m^{-3}$ SO_2	24 hr average	Considerable increase in illness
300	630 $\mu g\ m^{-3}$ SO_2	24 hr average	Acute worsening of chronic bronchitis patients
200	250 $\mu g\ m^{-3}$ SO_2	24 hr average	Increased absence of industrial workers

Concentration, $\mu g\ m^{-3}$	Accompanied by	Time	Effects
100-130	120 mg m^{-3} SO_2	Annual mean	Children likely to experience increased incidence of respiratory diseases
100	Sulfation rate above 30 mg/cm^2/mo	Annual geometric mean	Increased death rate for those over 50 likely
80-100	Sulfation rate above 30 mg/cm^2/mo	2-yr geometric mean	Increased death rate for those 50 to 69 years

Q.46. What are various respiratory defense mechanisms in human beings against air pollutants?

Ans. The fine particles with size less than 3 μm can penetrate into the deep air passages of lungs and cause damage to lungs. Once, they reach the lungs, they are not removed. Hence, they remain lodged in lungs. They cause severe breathing trouble by physical blockage and irritation of the lung capillaries. Black-lung disease, pulmonary fibrosis and emphysema are examples of lung damage.

Human beings possess certain defense mechanisms to prevent the entry of foreign particles into their respiratory tract. Coarse particles are trapped in the nose and throat and eliminated. When fine particles enter, the defense mechanism fails to prevent them. Consequently, the entered particles cause adverse effects. The effects become multiplied when other harmful gases like SO_2 are also present with particles in the atmosphere.

- Particulates, in general, range in size from 0.0002 μm to 500 μm.
- Larger particulates (> 3 μm) are trapped in nose and throat from which they are easily eliminated.
- Fine particulates (< 3 μm) can penetrate into deep air passages and cause lung damages.
- The lodged particles in the lungs cause severe breathing trouble by physical blockage and to the lung capillaries.

Table 2.4: Respiratory defense mechanisms in human beings

Particle size	Description	Defense mechanism
Over 10 μm	Coarse dust, fly ash (visible to naked eye)	Hairs at the front of the nose remove all particles over 10 mm
2-10 μm	Fumes, dust, smoke	Movement of cilia sweeps mucus upward, carrying particles from windpipe to mouth, where they can be swallowed
Less than 2 μm	Aerosols, fumes	Lymphocytes and phagocytes in the lung attack some submicron particles

- Respiratory defense mechanism depends largely on the size of the particles entering into the system and the depth of their penetration. Approximately, 40% of the particles between 1 and 2 μm in size are retained in the bronchioles and alveoli. Particles ranging in size from 0.25 to 1 μm show a decrease in retention, because many particles in this range are breathed in and out again. However, particles below 0.25 μm show another increase in retention due to Brownian motion, which results in impingement.

Q.47. What is the health effects of exposure to carbon monoxide?

Ans. Carbon monoxide (CO) combines with hemoglobin (Hb) to form carboxyhemoglobin (COHb). Due to its stronger affinity (240 times than that of oxygen) towards Hb, it readily combines with Hb even at very low concentrations. As a result, hemoglobin to combine with oxygen is less available or not available leading to anoxic conditions. The effects due to the formation of COHb are presented in table below.

Table 2.5: Health effects of COHb

COHb level in 1%	Effects
< 1.0	No apparent effect
1.0–2.0	Some evidence of effect on behavioural response
2.0–5.0	Effects on central nervous system, impairment of time interval discrimination, visual acuity, brightness discrimination and certain other psychomotor functions
> 5.0	Cardiac and pulmonary functional changes
10.0–80.0	Headaches, fatigue, drowsiness, coma, respiratory failure, death

CO poisoning will have initial effect of loss of awareness and judgment. It is responsible for many automobile accidents as the drivers are exposed to CO while driving. Traffic policemen are severely affected as they are exposed to CO for long duration.

Q.48. What are the health effects of exposure to sulfur dioxide?

Ans. Sulfur dioxide is an irritant gas. It affects the mucous membrane when inhaled. Under certain conditions, some of the SO_2 is oxidized to SO_3 in the atmosphere. Each of these gases, in the presence of water vapor forms sulfurous and sulfuric acid respectively. These acid vapors, when inhaled cause corrosive action on mucous membranes. SO_3 is a strong irritant and causes severe bronchospasms even at relatively low concentrations. SO_2 may cause severe problem when it is accompanied by smoke. In 1952, a major disaster occurred with a death toll of 4000 in London, when SO_2 accompanied with smoke – a condition is called smog. The combination resulted due to temperature inversion that existed in atmosphere.

Q.49. What are the health effects of exposure to oxides of nitrogen(NO_x)?

Ans. Of the seven oxides of nitrogen, only two are known to harm human health. They are nitric oxide (NO) and nitrogen dioxide (NO_2). Nitric oxide is less toxic than nitrogen dioxide. NO_2 causes eye and nasal irritation at a concentration of about 15 ppm and pulmonary discomfort at a concentration of about 25 ppm after brief exposure. Like CO, NO_2 combines with hemoglobin and thereby causes anoxic condition.

Mostly, NO_2 present in the atmosphere is always less than CO concentration and hence the effects on hemoglobin are not that severe. However, NO_2 pose other serious problems and they are presented in table below.

Table 2.6: Effects of NO_2 on human health

NO_2 concentration, ppm	Duration of exposure	Effects
50-100	Up to 1 hour	Inflammation of lung tissue for 6-8 weeks
150-200	—	Bronchiolitis figrosa obligerans – fatal result within 3-5 weeks
500 or more	2-10 days	Death

Q.50. What are the health effects of exposure to oxidants like ozone(O_3) and peroxyacetylnitrate (PAN)?

Ans. Ozone has an irritation effect in the respiratory tract and in lungs. It reaches much deeper into the lungs than SO_x. Both ozone and PAN are formed in the atmosphere through photochemical reactions.

Both of them cause irritation to eyes and to respiratory tract. Exposure to 50 ppm of ozone for several hours will cause mortality due to pulmonary edema, a condition characterized by accumulation of fluid in lungs. At lower levels, ozone brings about non-lethal accumulation of fluid in the lungs and damage to lung capillaries. Young children are more susceptible to ozone exposure than adults.

Both ozone and PAN generate free radicals. The free radicals attack the sulfhydryl groups (–SH) on enzymes. The –SH groups are oxidized by these free radicals and are acetylated by PAN resulting in activation of the enzymes. The affected enzymes include isocitric dehydrogenase, malic dehydrogenase and glucose-6-phosphate dehydrogenase. These enzymes catalyze some reactions of citric acid cycle and glycolysis. Thus, the energy yielding biochemical reactions are affected which will deprive energy in cells.

Q.51. Discuss the health effects of few other gaseous air pollutants?

Ans. Health effects posed by few more gaseous air pollutants are described below briefly:

Effects of H_2S and Mercaptans exposure: H_2S is known for its foul rotten egg like odor. Exposures to H_2S for short periods can result in fatigue in the sense of smell. Methyl mercaptans and ethyl mercaptans are known for their strong odors. Other than odor nuisance, they do not pose any health problems. In fact, mercaptans are added in LPG to detect any leakage of the gas.

Effects of Fluorides exposure: Some fluorides are irritants. Hydrogen fluoride in the air has corrosive effect. Prolonged exposure to fluorides may cause flourosis.

Effects of Hydrocarbon vapors exposure: Hydrocarbon vapors are known to cause eye irritation and respiratory tract irritation.

Effects of exposure to carcinogens: Carcinogenic substances like poly cyclic aromatic hydrocarbons(PAHs), lead etc. are known to cause cancer. PAHs are released mainly from diesel driven vehicles. At present, in India, we use lead-free petrol; hence there is no possibility of lead from automobiles. However, lead can be emitted by other sources like industrial operation.

Q.52. List the air pollutants which adversely affects the plants/vegetation?

Ans. The following air pollutants are known to cause adverse effects on plants:

- sulfur dioxide
- fluoride compounds
- ozone
- chlorine
- hydrogen chloride
- oxides of nitrogen
- ammonia
- hydrogen sulfide
- hydrogen cyanide
- mercury
- ethylene
- PAN
- herbicides
- smog

The above pollutants interfere with plant growth and the process of photosynthesis. Smog and particulates settle on the leaves and coat and clog the stomata. Coating of leaves reduces the amount of light reaching the leaves while clogging of stomata reduces the intake of CO_2. Thus, the particles by coating and clogging will interfere in the photosynthesis.

Q.53. What are the different ways in which damage to leaves may occur?

Ans. Leaves may be damaged in several ways as described below. These damages may occur due to infection, nutrition deficiency, weather or exposure to some chemicals:

- **Necrosis:** Death or collapse of the tissue is called necrosis
- **Chlorosis:** It is the loss of green pigment (chlorophyll). The leaf becomes either pale green or yellow in color due to loss of chlorophyll.

- **Abscission:** Dropping of leaves is called abscission.
- **Epinasty:** It is a downward curvature of the leaf due to high growth on the upper surface.

Sometimes, external factors described above may retard the growth of the plant or reduce the yield of the plant. The effect caused by air pollutants may be of acute or chronic in nature. The effects of air pollutants are summarized in table below.

Table 2.7: Effects of various air pollutants on plants

Pollutant	Dose	Effects
SO_2	Mild	Interveinal chlorotic bleaching of leaves
	Severe	Necrosis in interveinal areas and skeletonized leaves
Ozone	Mild	Flecks on upper surfaces, premature aging and suppressed growth
	Severe	Collapse of leaf, necrosis and bleaching
Fluorides	Cumulative effect	Necrosis at leaf tip
NO_2	Mild	Suppressed growth, leaf bleaching
Ethylene	Mild	Epinasty, leaf abscission
PAN	Mild	Bronzing of lower leaf surface (upper surface normal), suppressed growth. Young leaves are more susceptible

Q.54. Discuss in detail the effects of some important air pollutants on plants/trees/vegetation.

Ans. The effects of various air pollutants are discussed below.

Sulfur dioxide: It produces two types of injury on leaves: acute and chronic, depending on the concentration and period of exposure. The acute injury is characterized by the killing of marginal or interveinal areas of the leaf. Chronic injury is caused by the slow, long-continued absorption of sublethal amounts of gas or by absorption of an amount of gas less than that necessary to cause acute injury.

SO_2 is toxic to the plants in concentrations above 0.1 to 0.2 ppm. Below 0.4 ppm, it tends to be oxidized in the cells and slightly interfere with photosynthesis. Chronic effects are exhibited in such small concentrations. Above 0.4 ppm acute injury occurs. However, the mechanism by which the effects occur is not fully understood yet.

Hydrogen fluoride: It behaves somewhat similar to SO_2 in many plants. It causes lesions and interferes with photosynthesis. Plants will recover slower from effects of fluorides than that of SO_2. Forage may be rendered unsafe for animal consumption, if the plant has absorbed more than 50 ppm of fluoride.

Ozone: Lesions occur and these lesions are confined to the upper surface. The effects are noticed in exposures for few hours at about 0.2 ppm.

Oxides of nitrogen: They cause the appearance of brown margins and brownish black spots on the leaves at the concentrations of about 25 ppm.

Photochemical smog: Injury occurs from two sources: one due to gases of smog and second due to deposition of fog droplets on leaves. The smog causes leaf lesions which are quite different from those produced by other pollutants. It is also believed to cause some "invisible injury".

Plants respond to the pollutants differently as their sensitivity differs due to several factors:

- **Genetic factors:** The genetic make-up of the plant is different in different species. Some plants are more susceptible while others are capable of withstanding to a certain extent.

- **Climatic factors:** Climatic factors such as duration of light, wavelength of the light, light intensity, temperature and humidity cause the plants to respond differently to air pollutants.
- **Miscellaneous factors:** Soil, water and fertility also affect the sensitivity of the plants.

Q.55. Discuss in detail the effects of various air pollutants on animals.

Ans. Information on effects of air pollution on animals comes mainly from two sources: from the animals affected during major air pollution episodes and from the laboratory research conducted with test animals which include mice, rabbits, rats, guinea pigs and monkeys.

The effects on animals due to air pollutants are not mainly through the inhalation as with human beings but are largely by ingestion of contaminated vegetation. Air borne pollutants both settle on the parts of the plants and form a coating over them. Some pollutants enter and interfere with metabolic processes of the plants. Either way, the plants are contaminated with the pollutants. When this contaminated forage is ingested by animals they enter into body of animals to cause deleterious effects. The effects of various air pollutants are described below.

Fluorides: Cattle and sheep are the most susceptible animals to fluorine. Horses and poultry are the most resistant to fluorine poisoning. Large concentration causes acute effects while low concentration causes chronic effects due to prolonged exposure.

Acute fluoride poisoning: Symptoms include lack of appetite, rapid loss of weight, decline in health and vigor, lameness, periodic diarrhea, muscular weakness and death. Skeletal fluorosis also results.

Chronic fluoride poisoning: Fluoride is accumulated, if the animals are exposed continuously in sub-acute doses. It also poisons protoplasm. It has high affinity with calcium and hence, interferes with normal calcification.

Though animals have more resistance than humans to dental mottling, cattle and sheep are affected mostly. Teeth in the process of formation are easily affected. Hence, tooth symptoms are noted to indicate chronic fluorosis. Bone lesions also result due to chronic fluorine poisoning. Overgrowth of bones is observed in legs, jaws and ribs which will result in lameness. Advanced stages of fluorosis exhibit the symptoms such as lack of appetite, general ill-health, lowered fertility, reduced milk production and growth retardation. Other symptoms include mottling, staining and wearing of the teeth, stiffness and lethargy.

Arsenic: It occurs as an impurity in many ores and coal. Livestock near various industrial processes are reported to be poisoned by arsenic. Arsenic in dusts spreading over the plants cause adverse effects on animals when ingested.

Acute poisoning by arsenic: The symptoms include severe salivation, thirst, vomiting, uneasiness, feeble and irregular pulse and respiration. The animal stamps, lies and gets up. Occurrence of diarrhea with garlic odor and sometimes with blood. The ears will become cold and body will tremble. This will lead to abnormal temperature and convulsion. Finally, animal will die within few hours or days.

Chronic poisoning by arsenic: It causes depressing effects on the central nervous system. Animal will become dull and exhibit lack of appetite. This will result in weight loss. Chronic diarrhea may occur continuously. Skin may become thick. Animal will become anemic. Abortion may also occur. Animals may become sterile in course of time. Chronic poisoning eventually leads to paralysis and death.

Lead: Industrial processes such as smelters, coke ovens and coal combustion processes emit lead into the atmosphere. The emitted lead settles over the plants and lead poisoning of animal results through ingestion of forage.

Acute poisoning by lead: The symptoms include prostration, staggering and inability to rise. Pulse becomes fast but weak. Some animals may fall suddenly. Stiffness may occur in legs and the animal may develop convulsions. Complete loss of appetite will result. Digestive tract will be paralyzed and there will be diarrhea. Grinding of the teeth and rapid chewing of cud are other symptoms. Eventually, the animal may die.

Chronic poisoning by lead: Muscles of larynx will be paralyzed and there will be difficulty in breathing. Convulsions may start occurring due to paralysis.

Q.56. What are the types of effects of air pollution on materials?

Ans. A wide variety of effects of air pollution on materials occurs which includes corrosion of metals, soiling and eroding of building surfaces, fading of dyed material, rubber cracking, etc.

Air pollutants damage the materials by the following mechanisms:

Abrasion: When sufficiently large solid particles travel at high speeds with wind, they abrade the surfaces they come into contact with. Large sharp-edged particles may get embedded on fabrics to accelerate wear.

Deposition and removal: As such the solid/ liquid particles deposited on surface do not damage materials except spoiling its appearance. But when these particles are removed from the surface by cleaning or washing, they may cause some deterioration. The deterioration may be unnoticed initially, but repeated removal of the particles after deposition will deteriorate the materials to a considerable extent.

Direct chemical attack: Some air pollutants react chemically with the exposed materials. Sulfur dioxide will react with marble and deteriorate. Hydrogen sulfide will tarnish the surface of silver. Acid mists etch metallic surfaces.

Indirect chemical attack: Some chemicals may not attack as such. But they may get adsorbed/absorbed by material. Thus lodged chemicals will undergo changes chemically. The new compound may damage the material. For example, SO_2 does not damage leather but when it is absorbed by leather, it becomes sulfuric acid which damages the leather.

Corrosion: The metals exposed to atmosphere are corroded by existing atmospheric conditions. The corrosion may be facilitated or enhanced by the presence of some air pollutants.

Damage by acid rain: Acid rain causes extensive damage to building and sculptural materials of marble, limestone, slate, mortar etc. These materials become pitted and weakened mechanically as the soluble sulfates are leached out by the rainwater.

Q.57. What are various economic losses caused due to air pollution?

Ans. Economic loss of a region, state and/or country occurs due to the following effects of air pollution:

Economic loss due to effects on human health: When air pollution cause effects on human health, the victims spend money on medical care. Many people are affected and end up with cancer of various types. Apart from the mental agony they undergo, they spend huge amounts of money for the treatment. People may not be aware that the illness was caused by exposure to air pollution. People with chronic illness due to exposure to air pollutants will be spending throughout their lifetime on medical care. If we calculate this expenditure for the entire exposed community, we will realize that this expenditure affects the GDP of a nation.

Economic loss due to effects on plants: When crops and other vegetation are affected by air pollution, yield is reduced substantially. It will cause huge loss for the farm owners. This will reflect in country's economy. In addition, the contaminated plants, when consumed by humans and/or animals, cause deleterious effects on humans/animals. This will further add to the economic loss.

Economic loss due to effects on animals: When animals are affected, especially farm animals, the productivity goes down. The affected animals are also provided with some medical care. All these will lead to heavy expenses and thereby economic loss.

Economic loss due to effects on materials: When the materials are affected, we will have to incur the replacement cost, repair cost, cleaning cost etc.

Q.58. What do you understand by global effects of air pollution?

Ans. Air pollution affects humans, plants, animals and materials at local level. But it also has effects at regional and global level which includes global warming, acid rain, photochemical smog and ozone layer depletion by CFCs.

The air pollution effects are felt locally due to noticeable effects on human health, animals, plants, materials and on visibility. When these pollutants are transported and dispersed in the atmosphere, their effects may be felt regionally and even globally. Some of the dispersed pollutants may not be harmful locally but may create problems globally. For example, the CFCs emitted do not cause any health effects but when it reaches the lower atmosphere it is a greenhouse gas; when it reaches the stratosphere, it depletes the ozone. At global level, "global warming", "acid rain" and "ozone layer depletion" are the environmental problems we face today.

Q.59. What is green house and greenhouse effect?

Ans. Greenhouse means a building made of glass, with heat and humidity regulated for growing plants. Solar radiation passes through the glass and the inside environment is heated up. The heat inside the house is contained within as the heated air does not leave the house. Consequently, the glass house will be warmer than the outside atmosphere.

In real atmosphere, the heat energy is contained by certain gases and hence, called ***Greenhouse effect***. During day time, solar energy passes through the atmosphere and reaches the surface of the earth. During its passage, parts of the incoming solar radiation are scattered, reflected, or absorbed. The remaining radiation reaches the earth's surface and thereby the earth's surface is heated up. The energy received by the earth's surface throughout the day time heats up the earth's surface continuously. As the earth's surface is heated up, it will become hot and hold the thermal energy. The earth at this point, starts emitting energy in the form of long wave radiation. This is called ***outgoing radiation***. Part of this outgoing radiation is absorbed by the atmosphere and retained as heat energy. The remaining energy escapes into the outer spaces. Carbon dioxide, water vapor, and few other gases in the atmosphere are capable of absorbing this outgoing radiation. The concentration of these gases (called, ***greenhouse gases***) determine the amount of radiation absorbed. More the concentration of these gases, more the amount of long wave radiation absorbed. It is a natural phenomenon occurring in the atmosphere. It helps preventing the earth from drastic cooling of the earth and its atmosphere. If all the heat energy from the earth's surface is lost to the space, then earth will become too cold. These gases act like a blanket and provide the earth with "blanketing effect". This process is called ***Greenhouse effect***. In nutshell, atmosphere, like glass absorbs some of the long wave radiation emitted by earth and radiates the energy back to the earth. In this way temperature of the earth is maintained.

Actually, the atmosphere does not behave like a real green house. The primary reason is that enclosed structure by glass keeps the air from being carried away by convection to the outside air. This retains the warmth inside the real green house. In contrast, the heated molecules of green house gases by absorbing the long wave radiation spread the heat to its surrounding by convection.

Q.60. What are various green house gases and their characteristics?

Ans. Green house gases are those gases present in atmosphere (e.g., Water vapor, carbon dioxide, methane, nitrous oxide etc.) which trap the heat going out of the atmosphere and thus keep earth warm.

Actually for the Earth, **water vapour** is the most important of all greenhouse gases. Water vapor absorbs IR(Infra Red) in the region 2.5 to 3.5 μm, 5 to 7 μm as well as over a broad range above 13 μm. The mixing ratio of water vapor is highly variable in time and space, but the global relative humidity is constant at about 1% and there are no anthropogenic activities that directly cause it to increase. Nevertheless, gaseous water is involved in feedback processes: positive in that increased global warming means increased evaporation from ocean and land surfaces leading to higher ratios of atmospheric water, thereby enhancing warming: negative in that the troposphere becomes cloudier leading to increased reflection and absorption in the atmosphere so that the solar flux at the solid/liquid surface of the Earth is reduced. In 1995, greenhouse warming associated with water vapor was estimated to be 110 Wm^{-2}.

Carbon dioxide is a major contributor to greenhouse effect. Carbon dioxide absorbs in the range of 14 to 19 μm, and completely blocks the radiative flux between 15 and 16 μm. It also absorbs at wavelengths between 4 and 4.5 μm. Water vapor and carbon dioxide are the two most important greenhouse gases. Together, they absorb much of the radiation in the thermal infrared region below 7.5 μm and above 13 μm. Between these two wave lengths, very little IR radiation is absorbed by these gases, and this part of the spectrum is regarded as window, transparent to the radiation of interest. Other gases which absorb in that region, however, partially close the window and can have a major effect on heat retention in the region of the earth.

Methane, another greenhouse gas, absorbs radiation in the wavelength ranges from 3 to 4 μm and 7 to 8.4 μm. According to 1995 estimates, methane contributed 1.7 Wm^{-2} to greenhouse warming. Ozone absorbs between 9 to 10 μm in the IR region and it therefore acts as a highly efficient greenhouse gas. In the troposphere, ozone concentrations are highly variable, partially because it has a short lifetime. Greater production of NO_x by fossil fuels burning and a forest and grassland fire has resulted in net lower tropospheric concentration of ozone by about 1.6% in the northern hemisphere. The overall contribution of ozone to warming has been estimated to be 1.3 Wm^{-2} in 1995.

Nitrous oxide absorbs IR radiation in the ranges of 3 to 5 μm and 7.5 to 9 μm. According to 1995 estimates its concentration was about 312 ppbv and it is increasing at a rate of 0.3% per year. It has approximately the same effect on warming as by ozone. There are no important troposphere sinks for nitrous oxide gas. So it is lost only by slow leakage into the stratosphere where it undergoes photolytic degradation. It, therefore, has a substantial troposphere residence time estimated to be about 120 years.

In addition to their role as agents for the catalytic decomposition of stratospheric ozone, chlorofluorocarbons (CFCs) are also important greenhouse gases. They absorb in the range 8 to 12 μm with each CFC having specific absorption bands in this region. Thus CFC-11 absorbs at 9.5 μm and 11.5 μm, and CFC-12 at 9.5 μm and 11.0 μm.

The **Hydrochlorofluorocarbons (HCFCs)** also attenuate radiation within the same range, but their residence time in the troposphere is much shorter than that of the CFCs. The rate of increase of the CFCs has declined by a factor greater than two in the past decade, but HCFC concentrations are increasing at a much higher rate. Three fully fluorinated gases of industrial origin have recently been recognized as potential contributors to global warming. They are present in trace amounts, but have lifetime of thousands or ten of thousands of years. Tetrafluoromethane (CF_4) and hexachloroethane (C_2F_6) both are produced during electrolysis of alumina (Al_2O_3) in cryolite (Na_3AlF_6) at carbon electrodes. The release of the gases has been estimated to be at about 0.77 and 0.1 kg respectively per ton of aluminum produced. Yet another gas, sulfur hexafluoride (SF_6) is formed during magnesium production.

Clouds are the most important atmospheric aerosol in terms of reflecting and absorbing incoming radiation and emitted radiation from the earth. We know the cooling effects of clouds on warm days, and warming effect on cooling nights.

Q.61. What is global warming? Discuss its causes and effects.

Ans. The four major greenhouse gases, which cause adverse effects, are carbon dioxide (CO_2), methane (CH_4), nitrous oxide (N_2O) and chlorofluorocarbons (CFCs). Among these CO_2 is the most common and important greenhouse gas. Carbon dioxide contributes about 58% to global warming. The other greenhouse gases such as methane and chlorofluorocarbons contribute about 18% and 14% respectively to the global warming.

Since the beginning of the industrial revolution, three human activities have contributed to significant rise of concentration of greenhouse gases. The activities are:

Use of fossil fuels: It releases huge amounts of carbon dioxide into the atmosphere.

Deforestation and burning of forests and grasslands to convert into cropland: Forests and grasslands are cleared and/or burned for converting them into cropland. Burning of biomass produces large quantities of CO_2. Clearance of forests also deprives available vegetation for absorption of CO_2 through photosynthesis.

Cultivation of rice in paddies and use of inorganic fertilizers: Cultivation of rice in paddies generates methane and use of fertilizers release N_2O into the atmosphere. Since, 1860, the concentration of greenhouse gases, CO_2, CH_4 and N_2O have risen sharply especially since, 1950. Burning of coal for power generation and for industrial purposes and burning of petroleum products by millions of vehicles are the two major contributors of CO_2 emissions.

As the greenhouse gases increase dramatically, by human activities, the greenhouse effect is enhanced. That is more amount of outgoing long wave radiation is retained than required amount. It results in greater warming of the atmosphere than normal. It is called "enhanced greenhouse effect". Due to this enhancement, the earth's atmosphere is warming up gradually more and more. This phenomenon is called ***global warming***. Since 1960 total atmospheric carbon dioxide has increased from about 320 to over 390 ppm. Over the same period, the average global temperature has increased very slightly, 0.6 ± 0.2 °C. Nine warmest years have occurred since 1990. The hottest year was 1998, followed in order by 2002 and 2001. There is an apparent correlation between increases in fossil fuel use, atmospheric concentration of CO_2 and global temperature between 1970 and 2002.

At earth's poles and in Greenland, the temperature rise has been noticed, so also some melting of land-based ice caps and floating ice. Some glaciers on the top of mountains in Alps, Andes, Himalayas, northern Cascades of Washington, and Mount Kilimanjaro in Africa have begun shrinking due to melting of ice.

In addition, some warm-climate fish and other species have migrated northward. The season, spring arrives earlier and autumn arrives later than normal in many parts of the globe. All the above suggest that the global warming is real and happening. If the warming continues, the ice from polar region and from mountain glacier will melt (not all the ice, but some) which will cause coastal flooding. Coastal flooding will result in rise of the sea level and consequently inundation of coastal areas (villages, towns, cities). It is estimated that sea level will rise by 1.5 m by 2050.

Q.62. What is acid rain?

Ans. Acid rain means, the presence of excessive acids in rain waters. Acid rain is in fact, a mixture of mainly H_2SO_4 and HNO_3 where the ratio of these two acids may vary depending upon the relative quantities of oxides of sulfur and nitrogen emitted. H_2SO_4 is the major contributor (60 – 70%) to acid precipitation, HNO_3 ranks second (30-40%) and HCl third. These acids tend to lower the pH of the precipitation.

When the pH of rain water or snow is less than 5.6, it is called acid rain. Neutral pH is 7.0. Rain having pH value below 5.6 is called acid rain. It is because of lowering pH by dissolution of atmospheric carbon dioxide in rain water thereby formation of carbonic acid. Hence, pH value at 5.6 and above is considered as the acidity by CO_2. Below 5.6, it is considered to be man-made acidity.

Although the oxides of sulfur as well as that of nitrogen have been recognized as the main components responsible for acid rain, the relative contributions to acid rain is still not very clear because of highly complex nature of their transport and complexity involved in their removal from rain water. Once these oxides have fallen, along with rain water, it is difficult to remove them from the environment. These oxides may travel long distances in the atmosphere and during this journey, they may undergo several physical and chemical transformations to produce more hazardous products which may also fall with rain. Hence, acid rain is not simply a man made problem but is a global ecological problem

Q.63. Discuss the history of acid rain.

Ans. Most acids come from human activities. However, there has always been some acid in rain, coming from volcanoes, swamps and the plankton in the oceans. It has increased very sharply over the past 200 years. The acidity is mainly associated with the transport and subsequent deposition of oxides of sulfur, nitrogen and their oxidative products. These oxides are produced by combustion of fossil fuels, smelter, power plants, automobile exhausts and domestic fires etc. The accumulation of these oxides was first reported to be responsible for increasing the acidity in **Swedish lakes and rivers.** The acid rain problem has increased due to industrialization. Burning of fossil fuels for power generations contributes to 60-70% of total SO_2 emitted globally.

Emission of NO_x from anthropogenic sources ranges between 50– 90 million tons annually over the globe.

Acid rain was first found in Manchester, England. In 1852, Robert Angus Smith found the relationship between acid rain and atmospheric pollution. Though acid rain was discovered in 1852, it wasn't until the late 1960s that scientists began widely observing and studying the phenomenon. Canadian Harold Harvey was among the first to research a "dead" lake due to acid rain.

In developed countries, tall stacks are used for releasing the pollutants above the inversion layer to solve the air pollution problem locally. But, the primary air pollutants, SO_2 and nitrogen oxides emitted above the inversion layer are transported as much as 1,000 km by prevailing winds.

During their transportation, they are converted into acid vapor, droplets of acid and particles of sulfate and nitrate salts by chemical transformations.

Acid deposition is a regional problem in the eastern United States and in other parts of the world. Most of these regions are downwind from coal-burning power plants, smelters, or factories or are major urban areas with large number of vehicles. The pH of rain water in eastern United States was reported to be in the range of 4.2-4.7. In some parts of eastern United States and east of Los Angeles the pH of precipitation was reported as close to 2.3. This pH value is equivalent to that of lemon juice.

It has been estimated that coal-burning power plants release about 66% of the SO_2 and 25% of the NO_x of the total SO_2 and NO_x responsible for acid deposition. Acidic emissions from industrialized areas of Western Europe and Eastern Europe are transported into Norway, Switzerland, Austria, Sweden, the Netherlands and Finland. During transportation these gases are converted into acids resulting in acid deposition.

Q.64. What is wet and dry acid deposition?

Ans. The acidic substances formed in the atmosphere remain in the atmosphere for 2-14 days, depending mainly on prevailing winds, precipitation, and other weather patterns. During this period, they descend to earth's surface either as wet deposition or as dry deposition. The resulting mixture is called acid deposition.

Precipitation is water falling on earth in either liquid or solid form. Precipitation occurs in the form of ***rain, drizzle, snow, hail*** and/or their modifications. The term, "**Acid deposition**" is used to refer to the process of bringing down the acidic substances along with precipitation or with particles coming down. The first one is called "**wet deposition**" and the latter "**dry deposition**". In dry deposition, the acidic particles are brought either by gravitational settling or by adsorption of the particles on the surfaces they encounter with. Precipitation by wet deposition refers to the deposition of water-based particles in liquid or solid form on the earth's surface. Precipitation of dry species including gases, such as SO_2 and solid acidic particles such as sulfates and nitrates is known as **precipitation by dry deposition.** However, the wet deposition is much more common.

Q.65. Describe the impact of acid rain?

Ans. Impacts of acid rain are described below:

Impact on Animals: Most aquatic animals cannot survive a drop in pH, especially when it is lower than 4. Some species of fish, such as salmon, die even when the pH drops to 5.5. Generally, at pH < 6, several susceptible species of algae and zooplankton are eliminated but the resistant ones become more dominant. A reduction in the zooplankton and bottom fauna will ultimately affect the food availability for the fish population.

Impact on Plants: Corrosive nature of the acidic water itself is dangerous to plants. It decolorizes the leaf pigments, thus rendering them chlorophyll–less; such an yellowing would result in a decreasing agricultural productivity. Alp alpha, barley, cotton, lettuce and spinach are the non woody plants most susceptible to acid-rain damage. Similarly, apple, pear and pine trees figure among the most vulnerable trees species.

Chemical Imbalance: The normal growths and development of plant and animal organisms are favored in balanced ecosystems, where an equilibrium exists between available and non-available forms of an element; this balance occurs among the various chemical forms of an element and among the various elements present in different compartments of the ecosystem. Acidic rainfall may upset a chemical balance. Since the acid rain water flushing through the atmosphere may dissolve some metallic air pollutants, it can enhance their toxic action more directly, immediately and drastically, thus posing a serious threat to plant and/or animal life. The integrity of nutrient forms in various environmental compartments is altered.

Impact on Material: Metallic surfaces exposed to acid rain are susceptible to corrosion damage. Thus, the metals, with which H_2SO_4 reacts, form their respective sulfates. For example, blue $FeSO_4$ results from iron, green $CuSO_4$ from Copper and white $Al_2(SO_4)_3$ from aluminum. Surfaces made of steel and zinc are also susceptible. On contact with acid rain with water, textile fabrics, paper and leather products may lose their material strength, get discolored or simply disintegrate depending on the degree of acidity. Building materials (limestone, marble, dolomite, mortar and slate) are especially weakened on reaction with acidic water. Water, whose pH is acidic, can react with these water-insoluble carbonates to yield water-soluble sulfates which can easily be removed from the building structures:

Q.66. What is photochemical smog? Discuss its formation mechanism.

Ans. Nitrogen oxides (NO_x) in combination with unsaturated hydrocarbons, CO, O_3, hydrogen peroxides as well as hydroperoxides and organic peroxides in presence of sunlight lead to the formation of photochemical smog. When oxides of nitrogen, Volatile Organic Chemicals (VOCs) and sunlight come together, they can initiate a complex set of reactions that produce a number of secondary pollutants known as photochemical oxidants. Ozone (O_3) is the most abundant of the photochemical oxidants. Photochemical reactions are initiated by ultraviolet light in air. Peroxybenzoyl nitrate (PBxN) is also produced in photochemical smog in presence of NO_2 and olefins. The smog may have a 50 ppb of PAN(Peroxyacetylnitrate) also. PBxN is 100 times more toxic than PAN and ~200 times as powerful as formaldehyde.

Cities like Los Angeles, London, Tokyo, Sydney, Athens, Madrid, Delhi, Calcutta, Madras and Bombay are suffering from this smog because of abundant sunshine, atmospheric temperature inversion and growing vehicular emissions.

The formation of photochemical smog is represents as below:

$$VOCs + NO_x + Sunlight = Photochemical\ smog$$

Q.67. What is the effect of photochemical smog on human beings and plants?

Ans. Effects of smog on human beings: Photochemical smog is characterized by yellowish grey haze in the atmosphere. This smog along with ozone irritates eyes and nose, change lung function, degrade human physical performance and exacerbate asthma. Symptoms like nasal-block, discharge of sneezing, hacking cough, reduction or changes of air way by obstruction may persist in highly vehicular density areas. Prevalence of chest problems and upper air way morbidity in areas with higher levels of oxidants in Bombay and Delhi have been reported. Occurrence of cold, cough and sputum and breathlessness are closely related to peaking of vehicular pollution. Photochemical smog prompts respiratory problems, reduces visibility, alters various blood parameters and aggravates diseases like headache and bronchitis.

Effects of smog on plants: Photochemical smog at a very low concentration of 0.01 ppm causes injuries to petunia, lettuce, pinto bean, citrus, salad crops and coniferous trees. PAN causes injury in beets, celery, pepper, aster. It causes silvering of leaves and death to forest trees. PAN inhibits important Hill reaction of photosynthesis. Smog produces early maturity or senescence in plants. Exposure of 4 ppb of PAN for four hours creates visible damage in plants. Vegetation damage may occur in several forms such as chlorosis, leaf abscission and curling etc. Photochemical smog in conjunction with O_3, PAN and NO_x cause damage to forests, agriculture and other materials-rubber, paint, fibers and polymers.

Q.68. What are chlorofluorocarbons (CFCs) and their effects?

Ans. Chlorofluorocarbons are popularly known for their role in ozone depletion and in greenhouse effect. When they are present in troposphere, they absorb the outgoing radiation and cause greenhouse effect; when present in stratosphere, they deplete ozone layer.

Chlorofluorocarbons abbreviated as CFCs are members of the group of chemicals, haloalkanes (also known as halogenoalkanes or alkyl halides). A **haloalkane** is a chemical compound derived from an alkane by substituting hydrogen atoms with halogen atoms. Substitution with fluorine, chlorine, bromine andiodine results in **fluoroalkanes**, **chloroalkanes**, **bromoalkanes** and **iodoalkanes**, respectively. Mixed compounds are also possible, the best-known examples being the chlorofluorocarbons (CFCs) which are mainly responsible for ozone depletion. Haloalkanes are used in semiconductor device fabrication, as refrigerants, foam blowing agents, solvents, aerosol spray propellants, fire extinguishing agents, and chemical reagents. **Freon** is a trade for a group of chlorofluorocarbons used primarily as a refrigerant. The word *Freon* is a registered trademark belonging to DuPont.

Q.69. Describe the origin of CFCs.

Ans. Chlorofluorocarbons (CFC) are compounds containing chlorine, fluorine and carbon only, that is they contain **no** hydrogen. They were formerly used widely in industry, for example as refrigerants, propellants, and cleaning solvents. An American, Thomas Midgley developed chlorofluorocarbons (CFC) in 1928 as a replacement for ammonia (NH_3), chloromethane (CH_3Cl), and sulfur dioxide (SO_2), which are toxic but were in common use at that time as refrigerants. The new compound developed had to have a low boiling point and be non-toxic and generally non-reactive.

Midgley specifically developed CCl_2F_2. However, one of the features is that, there exists a whole family of the compounds, each having a unique boiling point which suit different applications. In addition to their original application as refrigerants, chlorofluoroalkanes have been used as propellants in aerosol cans, cleaning solvents for circuit boards, and blowing agents for making expanded plastics (such as the expanded polystyrene used in packaging materials and disposable coffee cups).

Q.70. Discuss various applications of CFCs.

Ans. The CFCs have many applications. They are used as propellants, fire extinguishers, and refrigerants.

Propellant: One major use of CFCs has been as propellants in aerosol inhalers for drugs used to treat asthma.

Fire extinguisher: At high temperatures, halons decompose to release halogen atoms that combine readily with active hydrogen atoms, quenching the flame propagation reaction even when adequate fuel, oxygen and heat remains. The chemical reaction in a flame proceeds as a free radical chain reaction; by sequestering the radicals which propagate the reaction, halons are able to "poison" the fire at much lower concentrations than are required by fire suppressants using the more traditional methods of cooling, oxygen deprivation, or fuel dilution.

For example, Halon 1301 total flooding systems are typically used at concentrations no higher than 7% v/v in air, and can suppress many fires at 2.9% v/v. By contrast, carbon dioxide fire suppression flood systems are operated from 34% concentration by volume (surface-only combustion of liquid fuels) up to 75% (dust traps). Carbon dioxide can cause severe distress at concentrations of 3 to 6%, and has caused death by respiratory paralysis in a few minutes at 10% concentration. Halon 1301 causes only slight giddiness at its effective concentration of 5%, and even at 15% persons remain conscious but impaired and suffer no long-term effects.

Halon 1211 also has low toxicity, although it is more toxic than Halon1301, and thus considered unsuitable for flooding systems. However, Halon 1301 fire suppression is not completely non-toxic; very high temperature flame, or contact with red-hot metal, can cause decomposition of Halon 1301 to toxic byproducts. The presence of such byproducts is readily detected because they include hydrobromic acid and hydrofluoric acid, which are intensely irritating. Halons are very effective on Class A (organic solids), B (flammable liquids and gases) and C (electrical) fires, but they are totally unsuitable for Class D (metal) fires, as they will not only produce toxic gas and fail to halt the fire, but in some cases pose a risk of explosion. Halons can be used on Class K (kitchen oils and greases) fires, but offer no advantages over specialized foams.

Halon 1211 is typically used in hand-held extinguishers, in which a stream of liquid halon is directed at a smaller fire by a user. The stream evaporates under reduced pressure, producing strong local cooling, as well as a high concentration of halon in the immediate vicinity of the fire. In this mode, extinguishment is achieved by cooling and oxygen deprivation at the core of the fire, as well as radical quenching over a larger area. After fire suppression, the halon moves away with the surrounding air, leaving no residue. However, use of Halons as fire extinguisher has been discontinued due to its high ozone depletion potential.

Refrigerants: CFCs have been used as refrigerants in cooling systems. The commonly used refrigerants are:

- **CFC**
- **HCFC**
- **HFC**
- **FC**
- **PFC**

Q.71. Describe the discovery and history of ozone layer depletion by CFCs.

Ans. James Lovelock was the first to detect the presence of CFCs in air, finding a concentration of 60 parts per trillion (ppt) of CFC-11 over Ireland in 1973.But he incorrectly concluded that CFC's are not hazardous to the environment.

The damage caused by CFC's was discovered by Sherry Rowland and Mario Molina One of CFCs' most attractive features -their unreactivity-has been instrumental in making them one of the most significant pollutants. CFCs' lack of reactivity gives them a lifespan which can exceed 100 years in some cases. This gives them time to diffuse into the upper stratosphere. Here, the sun's ultraviolet radiation is strong enough to break off the chlorine atom, which on its own is a highly reactive free radical. This catalyzes the breakup of ozone into oxygen by a variety of mechanisms, of which the simplest is:

$$Cl\cdot + O_3 \longrightarrow ClO\cdot + O_2$$

$$ClO\cdot + O_3 \longrightarrow Cl\cdot + 2\,O_2$$

Since the chlorine is regenerated at the end of these reactions, a single Cl atom can destroy many thousands of ozone molecules. Reaction schemes similar to this one (but more complicated) are believed to be the cause of the ozone hole observed over the poles and upper latitudes of the earth. Decreases in stratospheric ozone may lead to increases in skin cancer.

The history of discovery of ozone layer is summarized as follows:

- In 1974, Sherwood Roland and Mario Molina suggested that CFCs were depleting the ozone in stratosphere.
- In 1985, scientists discovered the annual seasonal thinning of ozone layer over Antarctica.
- In 1987, 24 countries agreed to halve their CFC emissions in Montreal Protocol.
- In 1995, Paul Cruzen, Mario Molina and Sherwood Rowland won Nobel Prize for their work on ozone layer depletion by CFCs.

Q.72. Summarise the chemistry of ozone layer depletion.

Ans. Rowland and Molina undertook the research and concluded the following facts about ozone layer depletion:

- CFCs remain in the troposphere because they are insoluble in water and are chemically unreactive.
- Within 11 to 20 years of their release they rise to stratosphere mostly through convection, random drift and the turbulent mixing of air in the troposphere.
- Once they reach the stratosphere, the CFCs break down under the influence of high energy UV radiation. This releases highly reactive chlorine atoms, which seed up the breakdown of very reactive ozone into O_2 and O in a cyclic chain of chemical reactions. This causes ozone in various parts of the atmosphere to be destroyed faster than it is formed.
- Each CFC molecule can last in the stratosphere for 65 to 385 years, depending on its type. During that time, each chlorine atom released can convert about 100,000 molecules of O_3 into O_2.

Q.73. Describe the phenomenon of ozone layer depletion over polar regions and upper latitudes.

Ans. In 1984, after analyzing the satellite data, scientists discovered that 40-50% of the ozone in upper stratosphere over Antarctica was destroyed during the Antarctic spring and early summer (September to December) especially since 1974. The observed seasonal loss has been incorrectly called an ***ozone hole***. A more accurate term is ***ozone thinning*** because the ozone depletion varies with altitude and location. The total area of the atmosphere above Antarctica that suffers from ozone thinning during the peak season varies from year to year. In 2000, seasonal ozone thinning above Antarctica was the largest ever and covered an area three times the size of the continental United States. However, in 2001 and 2002 its size decreased somewhat. CFCs are attributed for this ozone depletion.

After a dark winter, sunlight returns to Antarctica. This sets into motion reactions that release large numbers of Cl atoms that initiate the catalytic chlorine cycle. Within weeks, this typically destroys 40-50% of the ozone above Antarctica (100% in some places). Huge mass of ozone-depleted air above Antarctica then flow northward and linger for few weeks over parts of Australia, New Zealand, South America, and South Africa. This raises biologically damaging UV-B levels in these areas by 3-10% and in some years by as much as 20%.

In 1988, it was discovered that similar but usually less severe ozone thinning occurs over the Arctic during the arctic spring and early summer (February-June) with a seasonal ozone loss of 11-38%. When this mass of air above the Arctic breaks up each spring, large mass of ozone-depleted air flow south to linger over parts of Europe, North America and Asia. However, scientists believe that the Arctic is unlikely to develop the large-scale ozone thinning found over Antarctic. According to a 1988 model developed by scientists of NASA, ozone depletion over

Antarctic and Arctic will be at its worst between 2010 and 2019.

Q.74. What are the global measures initiated to protect ozone layer?

Ans. In 1975, the US state of Oregon enacted the world's first ban of CFCs. The United States and several European countries banned the use of CFCs in aerosol spray cans in 1978, but continued to use them in refrigeration, foam blowing, and as solvents for cleaning electronic equipment. Scientists concluded that even with immediate action, it will take 50 years for the recovery of ozone layer to pre-1950 levels.

By 1985, scientists observed a dramatic seasonal depletion of the ozone layer over Antarctica. International attention to CFCs resulted in a meeting of world diplomats in Montreal in 1987. Representatives of 36 nations met in Montreal, Canada and developed a treaty, called ***Montreal Protocol***. It set the goal of cutting emissions of CFCs into the atmosphere by about 35% between 1989 and 2000.

On March 2, 1989, 12 European community nations agreed to ban the production of all CFCs by the end of the century. Representatives of 93 courtiers met in London in 1990 and in Copenhagen, Denmark in 1992 and adopted the ***Copenhagen Protocol*** to accelerate the measures of phasing out of key ozone-depleting chemicals. They decided that by the year 2010, CFCs should be completely eliminated all over the globe including developing countries.

Use of certain chloroalkanes as solvents for large scale application, such as dry cleaning, have been phased out, for example, by the IPPC directive on greenhouse gases in 1994 and by the Volatile Organic Compounds (VOC) directive of the EU in 1997. Permitted chlorofluoroalkane uses are medicinal only. Finally, Bromo fluoroalkanes have been largely phased out and the possession of such equipment is prohibited in some countries like the Netherlands and Belgium, from 1 January 2004, based on the Montreal Protocol and guidelines of the European Union.

Production of new stocks ceased in most (probably all) countries as of 1994. However, many countries still require aircraft to be fitted with halon fire suppression systems because no safe and completely satisfactory alternative has been discovered for this application. There are also a few other, highly specialized, uses. These programs recycle halon through "halon banks" coordinated by the Halon Recycling Corporation to ensure that discharge to the atmosphere occurs only in a genuine emergency and to conserve remaining stocks.

On September 21, 2007, approximately 200 countries agreed to accelerate the elimination of hydrochlorofluorocarbons entirely by 2020 in a United Nations-sponsored Montreal summit. Developing nations were given until 2030. Many nations, such as the United States and China, who had previously resisted such efforts, signed the treaty.

Q.75. What initiative and commitment has been undertaken by India to protect ozone layer.

Ans. India became party to the Montreal Protocol in 1992 and also ratified the Copenhagen, Montreal and Beijing Amendments on 3rd March 2003. India's per capita consumption of ODSs (Ozone Depleting Substances) is at present less than 3 g and did not cross 20 g during 1995-97 as against 300 g permitted under the Protocol. India prepared a detailed Program to phase out ODS in accordance with its national industrial development strategy in 1993. The objective was to phase out ODS without undue economic burden to both consumers and industry manufacturing equipment using ODSs and provided India with an opportunity to access the Protocol's Financial Mechanism.

The government of India has entrusted the work relating to ozone layer protection and implementation of the Montreal Protocol to the Ministry of Environment and Forests (MOEF). The MOEF has set up an ozone cell as a national unit to look after and to render necessary services to implement the protocol and its ODS phase out program in India.

Q.76. Describe the suggested alternatives to the use of CFCs.

Ans. Work on alternatives for chlorofluorocarbons (CFCs) in refrigerants began in the late 1970s after the first warnings of damage to stratospheric ozone by Molina and Rowland (who shared the 1995 Nobel Prize for Chemistry for their work).

Adding hydrogen and thus creating hydrochlorofluorocarbons (HCFC), chemists made the compounds less stable in the lower atmosphere, enabling them to break down before reaching the ozone layer. Later alternatives dispense with the chlorine, creating hydrofluorocarbons (HFCs) with even shorter lifetimes in the lower atmosphere.

By 1980s, Bromo fluoroalkanes were in common use on aircraft, ships and large vehicles as well as in computer facilities and galleries. However, concern was beginning to be felt about the impact of chloroalkanes and bromoalkanes on the ozone layer. The Vienna Convention on Ozone Layer Protection did not cover Bromo fluoroalkanes as it was thought that emergency discharge of extinguishing systems was too small in volume to produce a significant impact,and too important to human safety for restriction. However, by the time of the Montreal Protocol it was realized that deliberate and accidental discharges during system tests and maintenance accounted for substantially larger volumes than emergency discharges, and consequently halons were brought into the treaty with many exceptions.

Q.77. Describe the green house effect caused by CFCs.

Ans. According to an estimate, the greenhouse effect caused by CFCs will have a far reaching disastrous effect, since CFCs are the major component of greenhouse gases and are responsible for global warming.

In addition to their role as agents for the catalytic decomposition of stratospheric ozone, chlorofluorocarbons (CFCs) are also important greenhouse gases. They absorb in the range 8 to 12 μm with each CFC having specific absorption bands in this region. Thus CFC-11 absorbs at 9.5 μm and 11.5 μm, and CFC-12 at 9.5 μm and 11.0 μm. The hydrochlorofluorocarbons (HCFCs) also attenuate radiation within the same range, but their residence time in the troposphere is much shorter than that of the CFCs. The rate of increase of the CFCs has declined by a factor greater than two in the past decade, but HCFCs concentrations are increasing at a much higher rate.

Three fully fluorinated gases of industrial origin have been recognized as potential contributors to global warming. They are present in trace amounts, but have lifetime of thousands or tens of thousands of years. Tetrafluoromethane (CF_4) and hexachloroethane (C_2F_6) both are produced during electrolysis of alumina (Al_2O_3) in cryolite (Na_3AlF_6) at carbon electrodes. The release of the gases has been estimated to be at about 0.77 and 0.1 kg respectively per ton of aluminum produced. Yet another greenhouse gas, sulfur hexafluoride (SF_6) is formed during magnesium production.

Q.78. What are various methods of control of gaseous pollutants?

Ans. The control methods available for removal of gaseous pollutants/contaminants are its removal from the gaseous effluents by adsorption, absorption, condensation and combustion.

Gaseous pollutants can be controlled by a wide variety of devices working based on the above principles. A suitable device is chosen considering cost effective, and collection efficiency factors. Even after selecting the best technique, monitoring should be done to ensure that emissions meet the emission standards.

Q.79. What is the principle of control of gaseous pollutants by adsorption?

Ans. Adsorption involves passing a stream of effluent gas through a porous solid material (adsorbent) contained in an adsorption bed. The surfaces of the solid material attract and hold the pollutant gas (the adsorbate) by either physical or chemical adsorption.

In adsorption, preferential affinity between the adsorbent and adsorbate is crucial for achieving maximum absorption efficiency. The surface to volume ratio of the adsorbent is important in determining its adsorption capacity. Large surface to volume ratios provides increased adsorption which can be increased by activating the adsorbent.

Activated carbon, the most common adsorbent, is prepared by carbonizing wood, fruit pits, or coconut shells at very high temperatures and treating them with steam to burn away the carbon material. As a result, a large internal pore structure is created providing exponentially large surface area.

Q.80. What is meant by physical adsorption?

Ans. ***In physical adsorption,*** gases and vapors condense on solids at temperatures above dew point, which depends upon Van der Waals force which is an attractive force existing between atoms or molecules of all substances. The amount of gas adsorbed depends on the ease of condensation of the gas – the higher the boiling point, the greater the amount adsorbed.

Physical adsorption is proportional to the amount of solid surface available for adsorption. The total surface area can be increased by building up a number of molecular layers on the surface. For example, use of the activated carbon provide increased surface area by creating large internal pore structure. Physical adsorption is accompanied by capillary condensation within the pores, which substantially increases the amount of gas that can be adsorbed. Physical adsorption liberates small amount of heat during adsorption of gases. By lowering pressure or raising temperature, the adsorbed gas can be desorbed without any chemical change.

Q.81. What is the principle of chemical adsorption which is also called chemisorption?

Ans. In this process, the pollutant gas reacts chemically with the adsorbent to form a chemical bond. Chemisorption is much slower than physical adsorption. It results in the formation of a single layer of molecules on the solid surface, and the process is usually irreversible due to occurrence of chemical change. The amount of gas adsorbed depends upon pressure and temperature.

Q.82. What are various types of adsorbents and their uses?

Ans. In general, there are two types of adsorbents: having affinity towards polar and non-polar vapors/gases. For example, water vapor, a polar vapor is attracted by alumina, bauxite and silica gel. Hence, these adsorbents are used as drying agents. On the other hand, the organic vapors (non polar) are attracted by activated charcoal.

Molecular sieves such as synthetic silicates or zeolites can be used and they can be tailored in such a way to adsorb only certain vapors/gases. Molecular sieves made from crystalline metal aluminosilicates are capable of adsorbing/recovering SO_2 even up to 99%. Certain other molecular sieves are proven to remove NO_x to an extent of 50%. However, use of molecular sieves is highly expensive.

A number of materials possesses adsorptive properties. Some of the adsorbents available for use in air pollution control and their specific uses are given below :

Activated carbon -To eliminate odors, to purify gases and to recover solvents.

Alumina - To dry air, gases and liquids.

Bauxite - In treating petroleum fractions, drying gases and liquids.

Magnesia - To treat gasoline and solvents, to remove metallic impurities from caustic solutions.

Molecular sieves - To control and recover Hg, SO_2 and NO_x.

Silica gel - To dry and purify gases.

Q.83. What are various types of adsorption equipment available for air pollution control?

Ans. There are three types of adsorbing equipment available for air pollution control, viz. fixed bed adsorbers, moving bed adsorber and fluidized bed adsorber.

Fixed bed adsorber: It is simple vertical or horizontal cylindrical shell in structure. The adsorbent is arranged on beds or trays in layers 1.3 cm thick in thin-bed adsorbers and greater in deep bed adsorbers.

Moving bed adsorber: In this unit, the adsorption bed is contained in a rotating drum. The filtered air containing gaseous pollutant is forced by the fan into the rotating drum section. The air enters into the adsorption bed and passes through it then leaves at the ends of the drum.

Fluidized bed adsorber: It contains a shallow, floating bed of adsorbent. When the air flows upward, the bed expands and get fluidized providing intimate contact between the contaminant gas and the adsorbent.

Adsorption units are highly efficient until a breakpoint occurs when the adsorbent becomes saturated. When saturation occurs, it can be detected by increased concentration of the pollutant gas. At this point, the adsorber must be regenerated or changed.

Adsorbers can be classified as *regenerative* and *non-regenerative.* The adsorbed gas can be desorbed either for recovery or for disposal in regenerative adsorber and the adsorber is regenerated. In non-regenerative adsorber, the gas cannot be desorbed and hence, adorber is not regenerated. Due to this, adsorbent of the non-regenerative adsorber must be discarded after saturation and it must be replaced with new material.

Q.84. What is the principle of control of gaseous pollutants by absorption?

Ans. In absorption, the contaminated effluent gas (solute)is brought into contact with a liquid absorbent (solvent). During this contact, the effluent gas is removed, treated or modified by the liquid absorbent. Liquid absorbent may either react chemically with the gas or simply dissolve the gas thereby remove the pollutant gas.

For example, water and limestone can be used to remove SO_2 from flue gases. Water reacts with the limestone to form calcium hydroxide [$Ca(OH)_2$] which then reacts with SO_2 to form calcium sulfate salt. It is scrubbed by the addition of more water. In this way water and limestone serve as reactive absorbent. The non-reactive absorbent does not react with the gas but dissolve the gas and thereby removes from the gas stream.

Absorption is used primarily in the control of gases such as sulfur dioxide, oxides of nitrogen, hydrogen sulfide, hydrogen chloride, chlorine, ammonia and some light hydrocarbons.

Q.85. What are the desirable properties of absorbent used in air pollution control?

Ans. **Pollutant gas** solubility differs from solvent to solvent. Solvents that are chemically similar to the solute to be absorbed generally provide good solubility. Viscosity of the solvent also determines the solubility of the gas. Lower the viscosity; the greater will be the solubility.

Other desirable factors for a good absorbing solvent are as follows:

- Ideally a solvent should have low freezing point.
- It should be relatively non-volatile.
- It should be relatively inexpensive.
- It should be non-corrosive.
- It should be nonflammable and chemically stable.
- It should be readily available.

Q.86. What are various absorption equipment/devices used in air pollution control?

Ans. A number of gas absorption devices are available to provide intimate contact between the gas and the liquid. The available absorption units are: *spray towers, plate* or *tray towers, packed towers,* and *venturi scrubbers.*

Spray towers: They are capable of handling large volumes of gas with relatively little pressure drop and reasonably with high efficiency. Spray towers can also remove particulates and serve the dual purpose.

For efficient absorption of gas, smaller droplet size of absorbing liquid and its greater turbulence is required. For production of fine droplets, high pressure nozzles are to be used which consume more energy than do the low-pressure nozzles.

In a spray tower, the absorbing liquid is sprayed from the top of the tower. As the absorbing liquid falls downward, the pollutant laden gas is passed from the bottom counter currently. During this counter-current flow, the gaseous contaminant comes into contact with the falling droplets of absorbing solution and is absorbed. Moisture eliminators placed near the outlet remove the moisture content of the cleaned gas. The absorbent with contaminant gas is collected at the bottom and disposed either with or without further treatment. Since spray towers provide much less gas-liquid interfacial area they are generally less effective when compared with that of other absorbing devices. However, they are less expensive.

Plate or tray towers: They contain horizontal trays or plates which provide large liquid-gas interfacial areas. The absorbent enters from the side of the column near the top and spills across the top sieve tray and then descends to other trays one after the other. The trays are spaced 0.3 to 0.9 m apart and arranged alternately inside the tower. This facilitates the liquid to flow in a zigzag manner to reach the bottom of the column. The contaminant laden gas is passed from the bottom upward and as it moves upward, it comes into contact repeatedly with the falling liquid and the contaminant gas is absorbed.

The plate towers can be designed with either sieve plates or bubble-cap trays. In bubble-cap trays, the contaminated gases rise upward until they strike the caps, at which point they are diverted downward and discharged as small bubbles from slots at the bottom of the caps. As the gas moves upward, the liquid-gas contact continues and the contaminant gas is absorbed.

Packed towers: In a packed tower, packing is used to increase the contact time between contaminant gas and liquid. The material chosen for packing has a large surface to volume ratio and large void ratio that offers minimum resistance to gas flow. Light weight unbreakable packing is used. The contaminant laden gas and the liquid are passed in counter-current flow as done in other towers. The falling liquid from the top flows through the packing and spreads on the surface of the packing as thin film thus providing more contact area to the rising gas.

Venturi scrubbers: The pollutant laden gas is passed through the venturi throat and the liquid is introduced in the throat region and both the gas and liquid come into close contact as they move together. During this process, the contaminant gas is absorbed and removed. The clean gas stream moves to the exit.

Q.87. How removal of sulfur dioxide (SO_2) is done by absorption?

Ans. By using absorption technique, SO_2 is removed with the efficiency up to 80 to 90%. The main absorbents used in the various SO_2 absorption processes are aqueous solutions of the alkalies (sodium and ammonia) and the alkaline earths (calcium and magnesium).m Use of sodium is advantageous as it is non-volatile and does not produce any fume problems. Use of ammonia has another advantage that the SO_2 reacts with ammonia to form ammonium sulfate, a useful by-product (a fertilizer).

Magnesium oxide (MgO), calcium oxide (CaO) and calcium carbonate ($CaCO_3$) are the alkaline earth compounds used as absorbents for removal of SO_2. In magnesia process, use of MgO forms $Mg(OH)_2$ with water and this in turn reacts with SO_2 to form $MgSO_3$.

Calcination will yield a stream of SO_2 and regenerate MgO. The SO_2 can be concentrated and used for manufacture of sulfuric acid while the MgO regenerated can be used once again, as absorbent.

Q.88. Describe the principle of condensation technique for control of gaseous pollutants?

Ans. A substance will condense at a given temperature, if its partial pressure is increased until it is equal to or greater than its vapor pressure at that temperature. If the temperature of a gas is reduced to its saturation temperature, its vapor pressure equals its partial pressure and condensation will occur. In air pollution control system, the temperature of the gaseous mixture is reduced so that it is equal to the saturation temperature of the pollutant gas. As temperature of the system becomes equal to the saturation temperature, the contaminant gas starts condensing.

Q.89. What are various gas condensation equipment?

Ans. There are two types of condensation equipment: *surface condenser* and *contact condenser.*

***Surface condenser*:** In surface condensers, contaminants are adsorbed onto a cool surface as the gaseous compound condenses. Air or water is used as a cooling medium and it is circulated through the tubes. The gaseous pollutant vapor condenses on the outer surface of the tubes. Thus, condensed vapor forms as a film of liquid and the liquid drains off through outlet for condensate.

***Contact condenser*:** Here, the vapor and liquid are brought into direct contact. The cold water is sprayed from the top and as this cold-water droplets fall downwards. The contaminant laden gas is passed from the bottom of the condenser and it comes into contact with the falling droplets of cold water. During this contact the vapor is cooled down to the saturation temperature and condenses.

Q.90. Describe the principle of combustion technique for control of gaseous pollutants?

Ans. Although combustion is a major source of air pollution it can also be used to control certain air pollutants. In this method of air pollution control, partially combusted products like CO and hydrocarbons are converted into innocuous carbon dioxide and water. Combustion equipment for control of these gases are designed to bring the oxidation process near completion to a maximum extent. Four important factors are essential for complete combustion to occur:

- sufficient amount of oxygen
- temperature of the system
- turbulence of the system
- adequate residence time for complete oxidation to occur

Soot, carbon monoxide and other hydrocarbons are produced by incomplete combustion due to insufficient amounts of oxygen available. Therefore, a sufficient amount of oxygen is essential for complete oxidation.

Temperature of the system will determine the rate of oxidation. Hence, the temperature must be kept at ignition temperature. Turbulence keeps the oxygen well mixed in the combustion zone. By providing baffles or injection nozzles, the turbulence is created inside the system. Combustion chambers are designed to provide enough time for complete burning to occur; it can be accomplished by increasing the height of the chamber.

There are three types of combustion methods are available. Based on the type of contaminant and its concentration, the combustion method can be chosen. They are:

- direct flame combustion
- thermal combustion and
- catalytic combustion

Q.91. Describe the various combustion methods used for air pollution control?

Ans. Various combustion methods for air pollution control are described below.

***Direct flame combustion*:** In this method, the flares are usually open ended combustion units located at the top of a stack and equipped with pilot flame to ensure continuous burning.

Waste gases are burned directly in a combustor with or without supplementary fuel. Though flare burning is a relatively safe means of disposing of the large quantities of highly combustible waste gases, they may cause formation of oxides of nitrogen due to high temperatures at the flare. Thus, new air pollutant is created. If the fuel-air ratios and other factors are not kept at desired levels, the flares may produce visible smoke or soot. The heat produced by the flares go waste in the atmosphere.

Thermal combustion: The waste gas is preheated by use of a heat-exchanger utilizing heat produced by the thermal incinerator itself. The preheated gas is directed into the combustion zone, where the waste gas is burned and completely oxidized. The burner may be supplied with supplementary fuel for burning. The temperature of operation depends upon the nature of contaminant gases present. Thermal afterburners must be carefully designed to provide safe, efficient operation.

Thermal afterburners are capable of burning the malodorous gases of visible plumes and converting them into odorless plumes. Hence, they are useful in controlling emissions from industries that produce malodorous gases.

Catalytic combustion: When combustible materials in the waste gas are too low, this method is preferred. A catalyst is used in this method which accelerates the rate of oxidation without itself undergoing any chemical change. Use of catalyst reduces the residence time for complete oxidation to occur. Catalytic combustion brings down the residence time to the extent of 20 to 50 times. A catalytic incinerator consists of pre-heating section and catalytic section. A fan is used to mix the gases and distribute them evenly over the catalyst. As preheated gas is forwarded towards the catalyst, the oxidation is pushed forward. During this, the catalytic surface glows. Well-designed catalytic incinerator can remove the waste gases with efficiency between 95 and 98 percent.

Catalytic combustion process is used to control SO_2, NO_x, hydrocarbons and CO.

Monsanto Corporation has designed the catalytic incinerator for removal of SO_2. The dust-free waste gas is passed through vanadium pentoxide (SO_2-oxidation catalyst) at 450 °C. This process yields sulfuric acid mist which is collected as a useful product. A palladium (II)/copper (II) catalyst has been developed to oxidize CO to CO_2 at ambient temperatures.

Q.92. Describe various methods of control of nitrogen oxides (NO_x) pollution

Ans. The following steps help to minimize NO_x emissions:

- To decrease the generation of NO_x by decreasing the flame temperature by injecting re-circulated flue gases, water or steam.
- Two-stage combustion process has been suggested to reduce NO_x formation.
- To remove NO_x formation by catalytic decomposition by:
  (*a*) **decomposing** NO into N_2 and O_2 catalytically ($2\,NO \longrightarrow N_2 + O_2$) NO is unstable thermo-dynamically but kinetically it is stable.
  (*b*) **combining** NO with CO catalytically, thus getting rid of both the noxious pollutants at the same time as: $NO + CO \longrightarrow 1/2\,N_2 + CO_2$
- NO_x can be removed from stack gas by chemical sorption employing alkaline scrubbing solutions or sulfuric acid solutions. NO is converted into N_2O_3 which is easily absorbed ($NO_2 + NO \longrightarrow N_2O_3$). Alkaline scrubbing process includes following steps:
  (*a*) **Oxidizer:** Here flue gas and NO_2 are injected ($NO_2 + SO_2 + H_2O \longrightarrow H_2SO_4 + NO$).
  (*b*) **Scrubber:** NO and NO_2 react to form N_2O_3 which is then scrubbed with H_2SO_4 in a scrubber ($NO_2 + NO \longrightarrow N_2O_3$; $N_2O_3 + 2\,H_2SO_4 \longrightarrow 2\,NOHSO_4 + H_2O$)
  (*c*) **Decomposer:** The reaction product obtained from scrubber is then decomposed ($2\,NOHSO_4 + 1/2\,O_2 + H_2O \longrightarrow 2\,H_2SO_4 + 2\,NO_2$).
  (*d*) **Nitric acid reactor:** NO_2 reacts with H_2O to produce nitric acid ($3\,NO_2 + H_2O \longrightarrow 2HNO_3 + NO$). Excess NO and NO_2 are circulated again through the oxidizer

- NO_x from the stack gas can be removed by catalytic reduction of NO with CH_4 (methane), NH_3 (ammonia), and CO (carbon monoxide):

$$4\,NO + CH_4 \longrightarrow 2\,N_2 + CO_2 + 2H_2O$$
$$6\,NO + 4\,NH_3 \longrightarrow 5\,N_2 + 6\,H_2O$$
$$2\,NO + 2\,CO \longrightarrow N_2 + 2\,CO_2$$

Disadvantage of the process is that undesirable byproducts like SO_2 so formed react with CO to form highly toxic carbonyl sulphide (COS) ($3\,CO + SO_2 \longrightarrow 2\,CO_2 + COS$).

- Spark retard is an effective measure to control NO_x and hydrocarbon emissions. NO_x level from 5000 ppm can be lowered upto 1000 ppm by using this technique.
- Catalytic exhaust reactions can be used to eliminate nitric oxide from vehicular exhaust.

Q.93. Describe various methods of control of SO_x pollution.

Ans. SO_x can be controlled by the following methods:

- Removal of sulfur from fuel before burning.
- Use of fuels with low sulfur content.
- Removal of SO_x from fuel gases.
- Using other energy sources for fuel gases.
- Use of natural gas.

Lots of SO_2 emission is caused by coal and oil combustion which contains sulfur. Substitution of energy sources is a better technique which can solve SO_x pollution problem. For example, hydroelectric plants need no fuel, so they are free of SO_x pollution. The removal of three forms of sulfur (pyrites, sulfates and organic sulfides) from fuel before burning can be achieved by adopting suitable techniques.

However, coal can be converted to gas by gasification process. In this process, powdered coal reacts with steam and oxygen in a fluidized bed at a high pressure ranging from 40-70 kg/cm^2.

The gaseous products formed comprise of H_2, CO_2, CO, CH_4 and H_2S. CO and H_2 react to form CH_4 thereby reducing sulfur contents.

Some refining processes are also operated in petroleum industries which lower the sulfur content by 2.5% to 3%. Huge investment and heavy recurring costs are involved in the technology for removal of SO_x from flue gases. However, double alkali process has been used commercially to some extent. SO_2 is scrubbed off by two alkalis *i.e.*, $Ca(OH)_2$ and NaOH, in two steps.

Removal of SO_2 from flue gases can also be done by introducing limestone into combustion zone of the furnace.

$$2\,CaCO_3 + 2\,SO_2 + O_2 \longrightarrow 2\,CaSO_4 + 2\,CO_2$$

Flue gases can be passed through slurry of milk of lime. $CaCO_3$ absorbs SO_2 very efficiently. However, $CaSO_4$ in large amounts possess a waste disposal problem. SO_2 has also been removed from the flue gases by the use of a reaction between bisulphate ions (from SO_2 and citrate ions). About 99% SO_2 from flue gas can be removed by this procedure.

The smelting industry is also adopting some catalytic oxidation methods to convert SO_2 into H_2SO_4. The acid so produced can be used internally during smelting operations. Thus, by applying any of these methods, SO_x emission can be controlled to a great extent.

Q.94. Describe various methods of control of carbon monoxide (CO) pollution.

Ans.

- The major source of CO emission is vehicular exhaust which can be controlled by fitting catalytic converter in all the vehicles which converts CO to CO_2.
- The possible sinks which oxidize CO into CO_2 in the atmosphere are atomic oxygen, hydroxyl radical, NO_2, N_2O, O_3 and excited oxygen molecules.

- The major CO sinks are some soil microorganisms, which are capable of removing CO from ambient air. About 16 fungi are playing an acting role in the conversion of CO to CO_2 out of 200 microorganisms, isolated from soil. Today the annual input of CO in the air by man's activities is expected to double its concentration in the ambient air every 5 years. But the actual rise in ambient (0.1 ppm) global CO content is much below due to the presence of soil bacteria.
- Plants are world's natural pollutant sink of CO and CO_2. They fix and metabolize CO also, with the help of chlorophyll in light and in dark photosynthetically and non-photosynthetically. Fixation of CO by plants is of immense importance as the green plants are major global sink of CO. This CO absorption by plants increases linearly with the increase of CO concentrations. Therefore, in cities where CO concentration is higher, the rate of CO absorption by plant may be greater by a factor of 10 to 100. Thus, plants play a vital role in global CO sink. Planting of trees and green vegetation will solve the problem of CO to a considerable extent. Proper combustion technology can be adopted in industries to control the CO which will oxidize the CO into CO_2. Supplementary fuels may be used in the combustion chamber.

Q.95. Describe the methods of control of hydrocarbons.

Ans. Hydrocarbons and NO_x produce PAN (Peroxyacetyl nitrate) and O_3 in atmosphere in presence of sunlight, which are chronic secondary pollutants. So their control ultimately depends on the control of the primary precursors *i.e.*, hydrocarbons and NO_x which are the main culprit of air pollution. Hydrocarbons from auto exhaust emission can be controlled by catalytic converter. By adopting this method, all the three pollutants (hydrocarbons, NO_x and CO) can be converted into less harmful end products.

❑❑❑

Section

A 3

Water Pollution

SHORT QUESTIONS AND ANSWERS ON WATER POLLUTION

Q.1. What are main sources of water?

Ans. The main sources of water include rainwater, surface water and groundwater. The distribution of water varies as per its locations. Water also exists on earth as a solid, liquid or gas, oceans, rivers, clouds, and rain, all of which contain water, are in a frequent state of change.

Water Cycle recycles the earth's water supply. The sun is the energy that powers this process. Its energy in the form of light, and heat causes water to *evapor*ate from oceans, rivers, lakes etc. Warm air currents rising from the earth's surface lift this water vapour up into the atmosphere.

When the air currents reach the cooler layers of the atmosphere, the water vapour condenses around and clings on to fine particles in the air. This step is called condensation. When enough vapours attach itself to tiny pieces of dust, pollen or pollutants, it forms a cloud. As the air gets more and more moist, the droplets that form the clouds grow larger and larger. Eventually they will get so big that the swirling atmospheric winds can no longer hold them up. The droplets then fall from the sky as precipitation. Precipitation can be in the form of rain, snow, sleet or hail depending on other atmospheric conditions such as temperature. Some of the precipitation will be absorbed into the ground. This is called infiltration. Once in the ground, the water can join the earth's groundwater supply. This is one of the world's largest storehouses of water. The water could also be absorbed from the ground by the roots of plants.

Q.2. Describe the water quality classification system followed in India.

Ans. In India, five water quality classes have been designated (A-E) on the basis of the water quality requirements for a particular use:

***Class A* waters:** for use as drinking water source without conventional treatment but after disinfection.

***Class B* waters:** for use for organized outdoor bathing.

***Class C* waters:** for use as drinking water source with conventional treatment followed by disinfection.

***Class D* waters:** to maintain aquatic life (*i.e.*, propagation of wildlife and fisheries).

***Class E* waters:** for use for irrigation, industrial cooling and controlled waste disposal.

The five classes have been used to set quality objectives for stretches of the Yamuna and Ganga rivers, and surveys have been carried out to compare the actual river-quality classification with that required to sustain the designated best use. Where a river has multiple uses, the quality objectives are set for the most stringent (best) use requirements. After comparing ambient water quality with the designated water quality objective, any deficiencies will require appropriate pollution control measures on the discharges, including discharges in upstream stretches. This system is also helpful for the planning and siting of industry. No industries are permitted to discharge any effluent in stretches of rivers classified in Class A.

Q.3. Describe the features and properties of water.

Ans. Features and properties of water are described below:

- **Water has a high specific heat:** Since water has a high specific heat, it can absorb large amounts of heat energy before it begins to get hot. It also means that water releases heat energy slowly when situations cause it to cool. Water's high specific heat allows for the moderation of the Earth's climate and helps organisms regulate their body temperature more effectively.
- **Water in a pure state has a neutral pH:** Pure water is neither acidic nor basic. Water changes its pH when substances are dissolved in it. Rain has a naturally acidic pH of about 5.6 because it contains natural derived carbon dioxide, nitrogen oxides and sulfur dioxide.
- **Water conducts heat more easily than any other liquid except mercury:** This feature causes large bodies of liquid water like lakes and oceans to have a uniform vertical temperature profile.
- **Water molecules exist in liquid form over a wide range of temperature from 0 to 100°C:** This range allows water molecules to exist as a liquid in most places on our planet.
- **Water is a universal solvent:** It is able to dissolve a large number of different chemical compounds. This feature also enables water to carry solvent nutrients in runoff, infiltration, groundwater flow, and also in living organisms.
- **Water has a high surface tension:** Water is adhesive and elastic, and tends to aggregate in drops rather than spread out over a surface as a thin film. This phenomenon also causes water to stick to the sides of vertical structures despite gravity's downward pull. Water's high surface tension allows for the formation of water droplets and waves, allows plants to move water (and dissolved nutrients) from their roots to their leaves, and the movement of blood through tiny vessels in the bodies of some animals.
- **Water molecules are the only substance on Earth that exists in all three physical states of matter – solid, liquid, and gas:** This feature plays an important role in the redistribution of heat energy in the Earth's atmosphere. In terms of heat being transferred into the atmosphere, approximately 75 % of this process is accomplished by the *evapor*ation and condensation of water.
- **The freezing of water molecules causes their mass to occupy a larger volume:** When water freezes it expands rapidly adding about 9% by volume. Fresh water has a maximum density at around 4°C. Water is the only substance on this planet where the maximum density of its mass does not occur when it becomes solidified.

Q.4. What are five main physical actions involved in hydrological/water cycle?

Ans. The water cycle: Technically known as the hydrologic cycle - is the circulation of water within the earth's hydrosphere, involving changes in the physical state of water between liquid, solid, and gas phases. The hydrologic cycle refers to the continuous exchange of water between atmosphere, land, surface and subsurface waters, and organisms. In addition to storage in various compartments (the ocean is one such compartment); the multiple cycles that make up the earth's water cycle involve five main physical actions: ***evaporation, precipitation, infiltration, runoff, and subsurface flow.***

*Evapor*ation**:** It occurs when radiant energy from the sun heats water, causing the water molecules to become so active that some of them rise into the atmosphere as vapour. It is the transfer of water from bodies of surface water into the atmosphere. This transfer entails a change in the physical nature of water from liquid to gaseous phases. Along with *evapor*ation can be counted transpiration from plants. Thus, this transfer is referred to as evapotranspiration. About 90% of atmospheric water comes from *evapor*ation, while remaining 10% is from transpiration. Transpiration occurs when plants take in water through roots and release it through the leaves, a process that can clean water by removing contaminants and pollution. Evapotranspiration is water *evaporating* from the ground and *transpiration* by plants. Evapotranspiration is also the way water vapour re-enters the atmosphere.

Precipitation: In cold air way up in the sky, rain clouds form. Rising warm air carries water *vapour* high into the sky where it cools, forming water droplets around tiny bits of dust in the air. Some *vapour* freezes into tiny ice crystals which attract cooled water drops. The drops freeze to the ice crystals, forming larger crystals called snowflakes. When the snowflakes become heavy, they fall. When the snowflakes meet warmer air on the way down, they melt into rain drops. In tropical climates, cloud droplets combine together around dust or sea salt particles. They bang together and grow in size until they're heavy enough to fall. Sometimes there is a layer of air in the clouds that is above freezing, or 0°C. Then closer to the ground the air temperature is once again below freezing. Snowflakes partially melt in the layer of warmer air, but then freeze again in the cold air near the ground. This kind of precipitation is called sleet. It bounces when it hits the ground. If snowflakes completely melt in warmer air, but temperatures are below freezing near the ground, rain may freeze on contact with the ground or the streets. This is called freezing rain, and a significant freezing rain is called an ice storm which are extremely dangerous because the layer of ice on the streets can cause traffic accidents. Ice can also build up on tree branches and power lines, causing them to break and our lights to go out. There is another kind of precipitation that comes from thunder storms called hail.

Infiltration: Under some circumstances precipitation actually *evapor*ates before it reaches the surface. More often, though, precipitation reaches the Earth's surface, adding to the surface water in streams and lakes, or infiltrating a portion of the precipitation that reaches the earth's surface seeps into the ground through the process called infiltration. Infiltration into the ground is the transition from surface water to groundwater. The infiltration rate will depend upon soil or rock permeability as well as other factors. Infiltrated water may reach another compartment known as groundwater (*i.e.*, an aquifer). Groundwater tend to move slowly, so the water may return as surface water after storage within an aquifer for a period of time that can amount to thousands of years in some cases. Water returns to the land surface at lower elevation than where it infiltrated, under the force of gravity or gravity induced pressures.

Runoff: The amount of water that infiltrates the soil varies with the degree of land slope, the amount and type of vegetation, soil and rock type, and whether the soil is already saturated by water. The more is openings in the surface (cracks, pores, joints), the more infiltration occurs. Water that doesn't infiltrate the soil flows on the surface as runoff. Precipitation that reaches the surface of the earth but does not infiltrate the soil is called **runoff**. Runoff can also come from melted snow and ice. It also includes the variety of ways by which land surface water moves down slope to the oceans. Water flowing in streams and rivers may be delayed for a time in lakes. Not all precipitated water returns to the sea as runoff; much of it *evapor*ates before reaching the ocean or reaching an aquifer.

Subsurface flow: Surface flow incorporates movement of water within the earth, either within the recharge zone or aquifers. After infiltrating, subsurface water may return to the surface or eventually seep into the ocean.

Q.5. What are ground water resources?

Ans. About 70% of the rainfall is lost to *evapor*ation. Most of the rest runs off the surface into streams and lakes, and eventually returns to the sea. Some, however, slowly percolates down through the soil until it reaches a layer of impermeable rock. The water which builds up on the top of the impermeable rock fills all the openings, crevices, and pores in the overlying sand and rock, to a level called the *water table.* It then very slowly flows to the oceans. When the ground level drops below the water table, the ground water helps supply streams and lakes giving a continuity of discharge they would not otherwise have.

Groundwater resources are extremely important because they supply to water withdrawal. Difficulties exist with this in that the rate of replenishment of these supplies is so slow that they must be considered to be essentially nonrenewable resources.

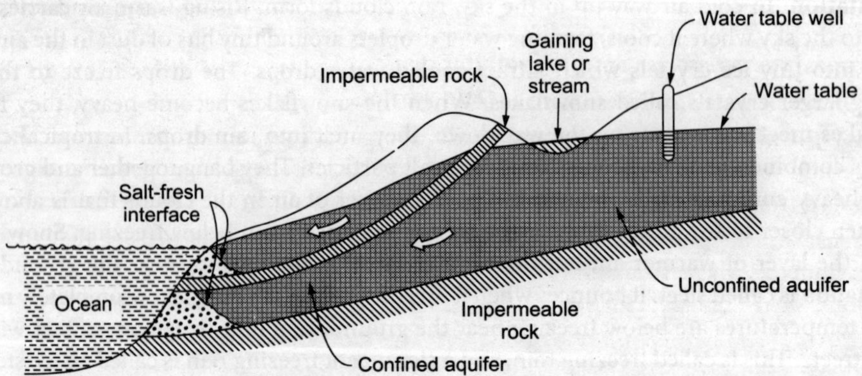

Fig. 3.1: Groundwater resources

One problem that can result from a drop in the water table is a corresponding sinking of the ground level in a process called *subsidence.* Another problem that can result from overexploitation of groundwater is that the water table can drop so low that it becomes uneconomical to pump it out. When this happens, the area involved may lose its principal water supply. Coastal areas face another sort of groundwater problem because of the close proximity of saline and fresh water. Under natural conditions there is a balance between the inflow of salt water and the outflow of fresh water, which establishes a boundary between the two regimes. However, as freshwater is pumped out of the area, the water table drops and the boundary moves inland. After a period of time, a coastal freshwater well may need to be abandoned because of this intrusion.

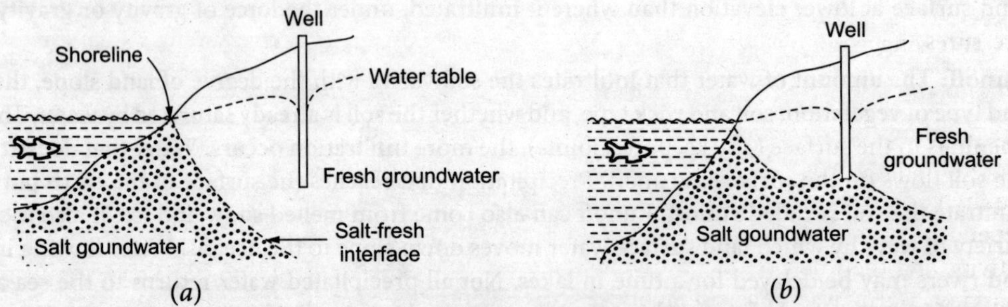

Fig. 3.2: Saltwater intrusion into a freshwater well due to over-pumping. (a) before, (b) after.

Q.6. What do you mean by point sources and non-point sources of water pollution?

Ans. Point sources of water pollution: When pollutants enter the river though a specific entry point, such as a drainpipe draining directly into a river, it's called point source pollution. Industrial water discharges and sewage treatment plants are the main sources of this type of pollution. Point source pollutants can include many different organic and inorganic substances, including human waste and toxic metals.

Point source pollution can be traced to a specific discharge point and therefore, it is the easiest source of pollution to control and regulate.

Non-point sources of water pollution: In contrast to point source pollution, non-point source (NPS) of water pollution comes from many different diffuse sources and is extremely difficult to regulate and control. NPS pollution is mainly caused by runoff, when rain water and snow melt move over the land, picking up pollutants along the way and dumping the pollutants into rivers, tanks and lakes. Some common NPS pollutants include fertilizers and pesticides from agricultural lands and homeowners; oil, grease and salt from highways; sediment from construction sites and eroding shorelines; and animal and human waste.

Q.7. What are various categories of water pollutants?

Ans. Various Types of water Pollutants are as follows:

- ***Oxygen-demanding wastes*** are biodegradable organic compounds contained in domestic sewage or certain industrial effluents. When these compounds are decomposed by bacteria, oxygen is removed from the water. If the oxygen level drops low enough, the fish will die.
- ***Disease-causing agents*** are various pathogenic microorganisms which usually enter the water with human sewage. Contact with these microbes can be made by drinking the water or through various water-contact activities.
- ***Synthetic organic compounds*** include detergents and other household aids, pesticides, and various synthetic industrial chemicals. Many of these compounds are toxic to aquatic life and may be harmful to humans.
- ***Plant nutrients*** such as the nitrogen and phosphorus which drains from fertilized lands, as well as the effluent from most sewage treatment plants, stimulate the growth of algae and water weeds.
- ***Inorganic chemicals and mineral substances*** include the acids which form when water drains from abandoned mines, as well as the heavy metals such as mercury and cadmium.
- ***Sediments*** are particles of soils, sands, and minerals washed from the land. They can smother bottom life such as shellfish and coral, as well as fill in reservoirs and harbors. Improper soil management leading to erosion is a major contributor.
- ***Radioactive substances*** can enter the water from the mining and processing of radioactive ores, from various nuclear power operations, from medical facilities, and from nuclear weapons testing.
- ***Thermal discharges of hot water from*** steam-electric power plants raise the temperature of the receiving water by 7-10°C resulting in various changes in local eco systems.

Q.8. What is the importance and implication of dissolved oxygen in water?

Ans. The amount of dissolved oxygen (DO) in water is an important parameter of water quality. Fish, for example, require certain minimum amounts of DO for their survival depending upon their species, stage of development, level of activity, and the water temperature. For example, for a well-rounded, warm water fish population, it is recommended that the DO remains above 5 ppm for at least 16 hours of the day and during the other 8 hours it should not drop below 3 ppm.

There are four processes which affect the amount of oxygen in the water: **Reaeration, photosynthesis, respiration, and the oxidation of wastes.**

Reaeration is the process wherein oxygen enters the water through the contact the water surface makes with the atmosphere. The solubility of oxygen in fresh water, at a pressure of 1 atmosphere, decreases with temperature as shown in Fig. 3.3.

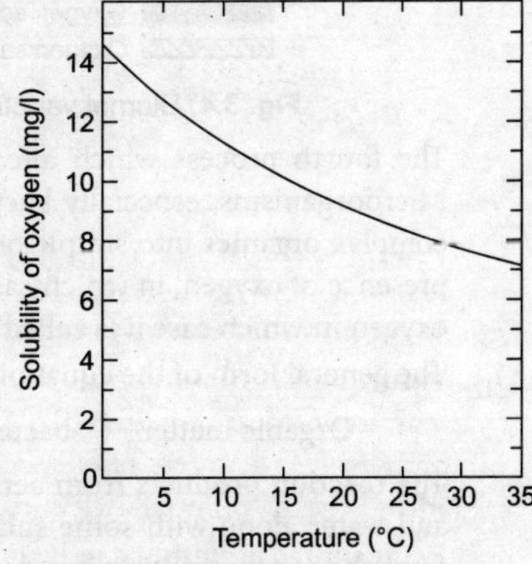

Fig. 3.3: Solubility of oxygen in fresh water exposed to water saturated air at 1 atmosphere pressure.

When the actual amount of oxygen in the water is less than the saturation value given in the figure, atmospheric oxygen passes into the water at a rate which is proportional to the deficit. By increasing the surface area in contact with the atmosphere the transfer of oxygen in contact with the atmosphere is increased, so a bubbling stream will pick up oxygen easier than a stagnant pond.

During photosynthesis, which of course occurs only during the daylight hours, oxygen is liberated, thereby increasing the DO level in the water. Respiration, however, is a process which continuously removes oxygen from the water. Combining the three effects of photosynthesis, respiration, and reaeration produces a diurnal variation of DO as shown in Figure below. In this figure, photosynthesis is assumed to occur from 6 AM to 6 PM, bringing the DO level to slightly above saturation in the afternoon. While the water is supersaturated, oxygen diffuses out of the water instead of into it.

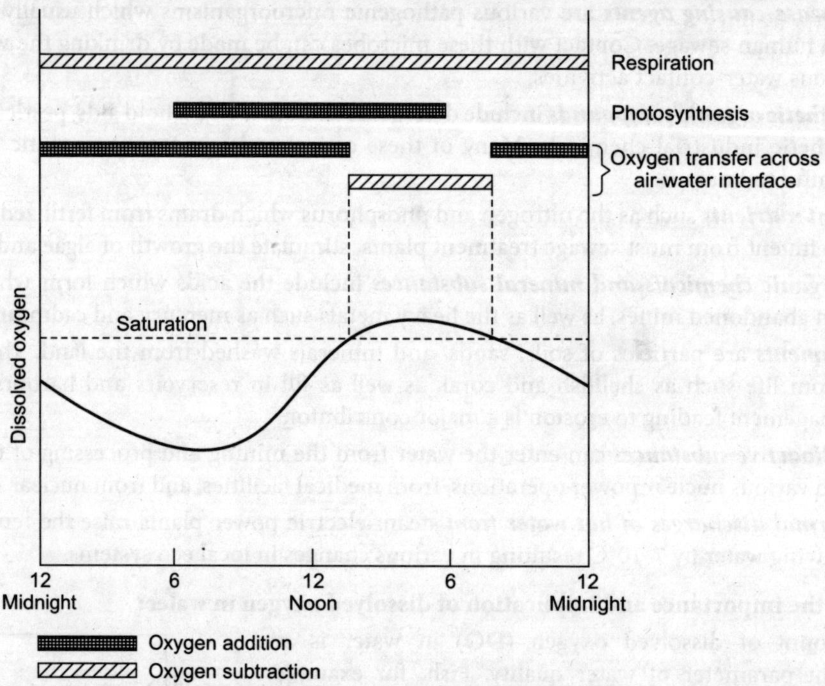

Fig. 3.4: Diurnal variation of DO assuming photosynthesis from 6 AM to 6 PM.

The fourth process which affects the amount of oxygen in the water is the oxidation of wastes. Microorganisms, especially bacteria, use organic wastes as food and in the process break down the complex organics into simple organic and inorganic materials. This decomposition may occur in the presence of oxygen, in which case it is called *aerobic* decomposition, or it may occur in the absence of oxygen in which case it is called *anaerobic* decomposition.

The general form of the equation for aerobic decomposition of organic material is

$$\text{Organic matter} + \text{bacteria} + O_2 \longrightarrow CO_2 + H_2O + \text{new bacterial cells}$$

The reaction products from aerobic decomposition are not obnoxious, being simply carbon dioxide and water along with some sulfates and nitrates. During aerobic decomposition oxygen is removed from the water, resulting in a drop in DO.

If too much organic matter must be decomposed, the DO supply in the water may fall to zero. If this happens, not only will the oxygen-dependent aquatic life be killed, but the end products from the resulting anaerobic decomposition will be toxic and malodorous. Typical anaerobic decomposition reaction products include ammonia, methane, hydrogen sulfide, carbon dioxide, and water.

Q.9. What is Biochemical Oxygen Demand (BOD) and its importance in water pollution?

Ans. One of the most widely used measurements for estimating the water pollution potential of a given amount of organic waste is called the *biochemical oxygen demand*, or BOD. The BOD is a measure of the amount of oxygen required by bacteria to oxidize waste aerobically to carbon dioxide and water.

If we were to introduce an amount of organic matter into some water and monitor its decomposition, we would observe that the amount of organic material remaining would decrease exponentially with time. The amount of BOD which remains after some period of time is proportional to the remaining organic material, which results in a curve as shown in Fig. 3.5.

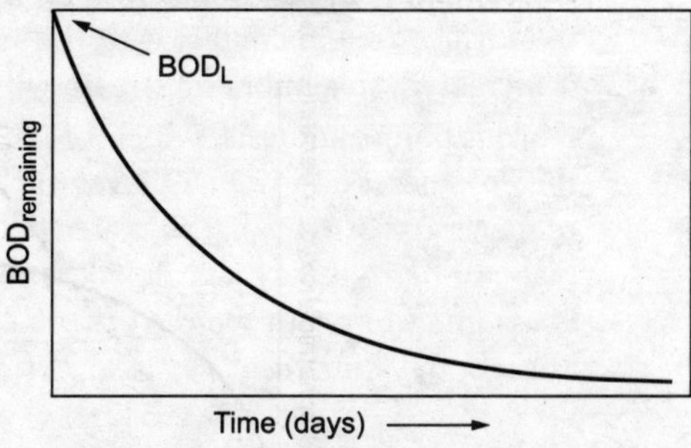

Fig. 3.5: BOD remaining as a function of time with initial value BOD_L

The initial value of BOD will of course be the total oxygen requirement to oxidize the organic material. This quantity is called the *ultimate BOD*, or BOD_L.

Usually when curves of BOD versus time are drawn, they are actually made for BOD utilized and not BOD remaining. BOD utilized will be proportional to the amount of organic matter which has been oxidized and hence will have the shape as shown in Fig. 3.6.

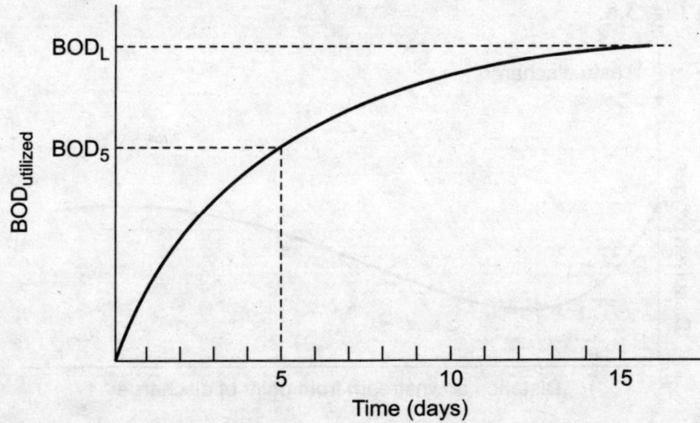

Fig. 3.6: This is the usual curve for BOD, and is actually BOD utilized. Also shown is the 5 day BOD.

A standard way to measure BOD is to determine the amount of oxygen required by the bacteria during the first 5 days of decomposition (at 20°C). This is the 5-day BOD called BOD_5, which is shown in Fig. 3.6.

The importance of the concept of BOD in water quality management must be emphasized. Consider that normal domestic sewage may have a 5-day BOD of around 200 mg/1 and that industrial sewage may have a BOD of several thousands of milligrams per liter. Compare these requirements for oxygen to the amounts actually dissolved in water.

It is thus easy to anticipate the quick depletion of oxygen in any receiving water unless the dilution factor is quite high. Depleting the water of oxygen, will kill the fish and the anaerobic decay products will be extremely objectionable. There is a secondary effect which causes the demand for oxygen to suddenly increase after about 8-10 days. The organic nitrogen in the wastes is converted to ammonia during decomposition and the subsequent oxidation of the ammonia to nitrite (NO_2) and then nitrate (NO_3) requires oxygen.

The total oxygen demand curve showing both the carbonaceous demand and the nitrification requirement is shown in Fig. 3.7.

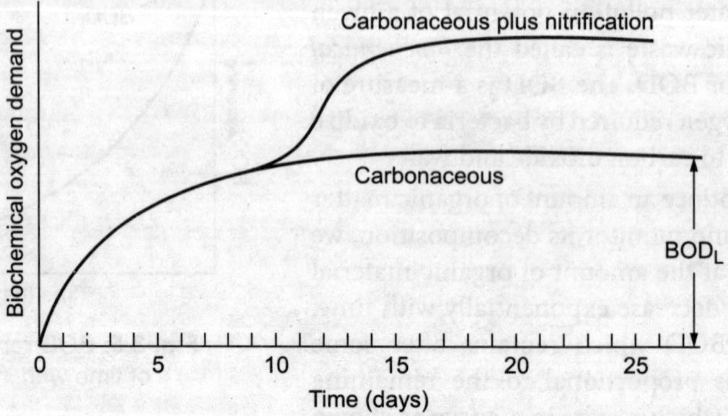

Fig. 3.7: The complete BOD curve includes the carbonaceous and nitrification demands

Q.10. What is oxygen sag curve and its importance in water pollution?

Ans. When wastes are discharged into a body of water, the amount of dissolved oxygen will decrease due to oxidation by bacteria. Opposing this drop in dissolved oxygen is reaeration which replaces oxygen through the surface, at a rate which is proportional to the depletion of oxygen below the saturation value. The simultaneous action of deoxygenation and reaeration produces what is called the oxygen sag curve, as shown in Fig. 3.8.

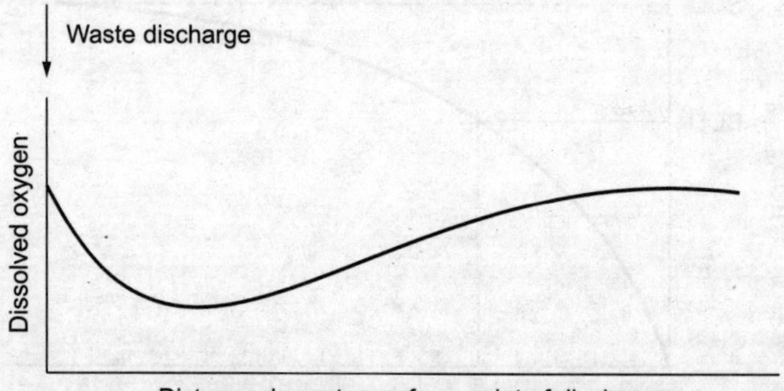

Fig. 3.8: The oxygen sag curve

The DO curve initially drops as the wastes deplete the oxygen faster that it can be replaced. At the point where the DO is a minimum, the rate of reaeration becomes equal to the rate of oxygen utilization. Beyond that point the rate of reaeration exceeds the rate of utilization and the DO level eventually returns to normal. This sequence is referred to as the natural self-purification ability of water.

The horizontal axis of the oxygen sag curve may be either time or distance. If for example, a certain amount of waste is released all at one time into some impounded water, the DO level will be a function of time. If, however, there is a continuous discharge of wastes into a stream, then the oxygen sag curve will be a function of the distance downstream from the point of discharge. As we move downstream from the point of discharge, and the dissolved oxygen begins to drop, there will be a corresponding change in the biota. Normal game fish will be replaced by fishes tolerant of the turbid, low DO waters. At the point downstream where the DO reaches its lowest value, conditions will be at their worst. If the stream goes anaerobic, there will be no fish and the only organisms present will be those able to obtain their oxygen from the surface, or those which are tolerant of low oxygen conditions.

Q.11. What are the health hazards and diseases associated with polluted/contaminated water?

Ans. There are many diseases which are associated with the contamination of water supplies by animal or human wastes. They include cholera, typhoid fever, paratyphoid fever, dysentery and infectious hepatitis.

The basic sanitation techniques which are essential to the control of water related diseases are the purification of water, and the treatment of sewage. The emphasis in water-quality control has shifted from the area of water-borne diseases, to control of oxygen consuming wastes, and now to nutrient removal and water reclamation.

The communicable water-associated diseases, are caused by the ingestion of biological agents. However, other modes of transmission including simple water contact, as in the case of schistosomiasis, or insect vectors, as in the case of malaria, are also important. The contamination of water by sewage is the principal cause of waterborne diseases. Intestinal discharges of patients or carriers of such diseases contain the biological agents which are responsible for the disease's spread. The disease-causing organisms can be transferred from the excrement of an infected person into the mouths of healthy people by means of contaminated food or water. The spread of such infectious diseases can be controlled both by proper treatment and disposal of human wastes, and by purification of drinking water supplies.

The principal biological agents of concern are pathogenic (disease-causing) bacteria, viruses, and parasites. Pathogenic bacteria are the causative agents of the great epidemic disease's cholera and typhoid - as well as bacillary dysentery, paratyphoid fever, and tularemia. Fortunately, intestinal bacteria that are discharged into natural waterways usually survive for only a few days.

These bacteria exhibit an initial rapid decrease that results in about a 90% reduction in 2 days. However, thereafter the rate of decrease is considerably slower. This reduction is faster in warm, polluted water than in cold, clean water.

The only viral disease that has been proven to be transmitted through drinking water is infectious hepatitis, although there are others for which transmission through water is theoretically possible. One of the particular difficulties with viruses is that some can survive the normal dosages of chlorine which are applied at water and sewage treatment plants to kill pathogens. Among the parasites that may be ingested is the protozoa *Entamoeba hystolytica,* which causes amebic dysentery. Another parasite is the guinea-worm which causes drantiasis, a common disease among the rural population.

Q.12. What is water quality? Discuss important water quality parameters and their regulatory limits.

Ans. "The water quality is that which a body of water should be able to support its designated uses, which include drinking, industrial uses,shell fishing, swimming, other body contact, recreational boating, or fishing".

Any physical, chemical, or biological condition that prevents the designated use(s) of a body of water represents poor water quality.

Some of the important water quality parameters and regulatory limits for them are as follows:

Dissolved oxygen: Dissolved oxygen (DO) is critical for aquatic life and also controls chemical processes. DO is important to aquatic organisms to perform their decomposition function ("aerobic" decomposers). DO enter the water from the air and from aquatic plant photosynthesis. Oxygen is consumed by respiration (Biological Oxygen Demand) and some chemical reactions. Values of DO less than a daily mean of 5.0 mg/liter (ppm) or instantaneous value of 4.0 mg/l is considered to be low and less than 2.0 mg/l to be dangerous to aquatic life.

Turbidity: Turbidity is caused by suspended material that blocks light. High levels of turbidity can slow the growth of aquatic plants and reduce visibility for aquatic animals that need to see their prey. Turbidity is a measure of how much light is scattered by materials suspended in the water. Scattered light is much less available for aquatic plants, so high levels of turbidity can limit the growth of aquatic plants as well as restrict the visibility of prey and enemies to aquatic animals. Consequently, there are standards for turbidity in fresh (50 NTU) and salt (25 NTU) waters. An NTU is a Nephelometric Turbidity Unit.

Fecal coliform bacteria: Fecal Coliform Bacteria are indicators of recent contamination by fecal matter from warm blooded creatures, including humans, other mammals and birds. The MFC membrane filtration technique/ method, enumerate fecal coliforms in water samples. Most of the bacteria detected by this method are E. coli, the most abundant bacterium in the guts of humans. While E. coli itself is usually not a pathogen, its presence is an indicator of the possible presence of real pathogens that can cause disease.

Coliform bacteria themselves are not necessarily pathogenic, but they indicate the likelihood that more dangerous organisms, such as the organisms that cause salmonellosis, cholera or hepatitis, are present. Although contamination of waters by human wastes is especially serious, animal wastes can also transmit disease. Consequently, high levels of fecal coliform bacteria, regardless of the source, imply a high risk of disease transmission. Two additional warnings apply: (*a*) some pathogens live longer in surface waters than fecal coliform bacteria, and (*b*) many bacteria are alive in the water but cannot be cultured and detected by the standard techniques. Owing to the risk of disease transmission by contact with waters contaminated by fecal wastes, standards for this water quality parameter are quite stringent.

As guidelines, waters designated for shell fishing should not have average fecal coliform levels above 14 colony forming units (CFU) per 100 ml. Waters designated for most other uses, e.g., swimming, should not average above 200 CFU/100 ml.

Total nitrogen and phosphorus: Total nitrogen is a measure of the quantity of the important plant nutrient, nitrogen, present in the water in dissolved and particulate forms. It is one measure of the degree of nitrogen loading to surface waters.

Total phosphorus is a measure of the important plant nutrient, phosphorus, which not only stimulates aquatic weed growth, but also blooms of nitrogen-fixing blue-green algae (cyanobacteria) and bacteria. These plant nutrients occur at variable but typically low levels in natural waters. Elevated levels of nitrogen and/or phosphorus can stimulate excessive plant growth ("eutrophication"), which can be very harmful to aquatic ecosystems. There are no general standards for nitrogen or phosphorus. Non-point sources of nitrogen and phosphorus are not currently regulated. Some countries have banned high-phosphate detergents, in order to control algae blooms in rivers receiving sewage effluents.

Total suspended solids: Sediment loading to surface waters raises turbidity and suspended solids levels and causes siltation, other pollutant loadings, and problems for aquatic life. Sedimentation is a serious problem in areas downstream of improperly managed land disturbing activity. Suspended solids cause siltation, clog the gills of aquatic animals, and carry toxic chemicals and nutrients with them when they enter the water. High levels of suspended solids are considered to be one of the most widespread water quality problems, and are frequently caused by human activities.

Chlorophyll: Chlorophyll a is a measure of phytoplankton biomass. All plants have the photosynthetic pigment chlorophyll, so its concentration in the water is a measure of plant abundance. It is a good measure for phytoplankton, which is otherwise hard to quantify. Standard for chlorophyll is 40 micrograms per liter (μg/l), above which water quality is considered to be impaired by excess phytoplankton. However, chlorophyll measures only the biomass of small phytoplankton. Larger aquatic plants that grow on surfaces or the big aquatic weeds that fill many ponds are not measured by this technique. There is no accepted standard method for the measurement of these larger plants, so measures of chlorophyll in a body of water may grossly underestimate the actual amount of plant growth there. High levels of plant growth indicate excessive nutrient loading and can cause problems with low dissolved oxygen at night.

Because shallow waters usually have enough light for plants to grow, unusually high levels of chlorophyll a usually indicate excessive nutrient loading.

Q.13. What are primary water quality standards in India for various water uses?

Ans. Indian water quality standards are tabulated below.

A. Drinking water source without conventional treatment but after disinfections:

- Total coliform organism MPN/100 ml shall be 50 or less
- pH between 6.5 and 8.5
- Dissolved Oxygen 6 mg/l or more
- Biochemical Oxygen Demand 5 days 20°C 2 mg/l or less

Outdoor bathing (Organized)

- Total coliforms organism MPN/100 ml shall be 500 or less
- pH between 6.5 and 8.5
- Dissolved Oxygen 5 mg/l or more

B. Drinking water source after conventional treatment and disinfection:

- Total coliforms organism MPN/100ml shall be 5000 or less
- pH between 6 and 9
- Dissolved Oxygen 4 mg/l or more
- Biochemical Oxygen Demand 5 days 20°C 3 mg/l or less

C. Propagation of wild life and fisheries:

- pH between 6.5 to 8.5
- Dissolved Oxygen 4 mg/l or more
- Free Ammonia (as N) 1.2 mg/l or less

D. Irrigation, industrial cooling, controlled waste disposal:

- Electrical Conductivity at 25°C micro mhos/cm Max. 2250
- Sodium absorption Ratio Max. 26
- Boron Max. 2 mg/l

Q.14. What are deleterious effects of industrial effluents?

Ans. Industries need a wide variety of raw materials and chemicals which are later discharged as effluents. Acids, alkalis, toxic metals, pesticides and other poisonous substances such as cyanide, dyes, oils, detergents, resins, rubbers are a few to mention. Heated effluents that impart thermal loading on receiving waters and effluents containing radioactive materials are also of prime concern. Some of the effluents such as from tanning and meat packing may also contain pathogenic bacteria. The nature and extent of pollution depends on the materials present in the effluent and on the quantity discharged.

Effects on water courses:

- **Colour:** The effluents contain dyes in higher concentrations which impart colour to the receiving streams and they persist for longer distances. Photosynthesis of phytoplankton is affected by these colours.
- **pH value:** The extreme alkalinity makes the receiving water unfit for any purpose. Further, it is deleterious to most of the aquatic life.
- **Suspended impurities:** The colloidal and suspended impurities produce turbidity in the receiving waters. The turbidity and colour along with the oil and scum create an unsightly appearance.

- **Depletion of oxygen:** Natural substances such as starch and dextrin and inorganic substances such as sulfide and nitrite present in the effluent exert an immediate oxygen demand. The stream will then be devoid of oxygen and the aquatic life are affected adversely.
- **Toxic substances:** Chromium, sulfide, chlorine and aniline dyes present in these wastes are directly toxic to fish and microbial organisms which carry out purification. Thus the self-purification of the water body is affected.
- **Oils:** Various oils (mineral) in the effluent interfere with the oxygenation of stream as they form a blanket on the surface and prevents the entry of oxygen at air/water interface.
- **Dissolved minerals:** The mineral materials, mostly sodium salts increase the salinity of the water and consequently it becomes unfit for irrigation.

Effects on land:

- The excess content of sodium (60%) and boron (2 mg/l) are deleterious to crops.
- The high sodium alkalinity combined with salinity impairs the growth of plants.
- Texture of the soil is affected by sodium and penetration of roots is prevented.
- Soil permeability is also affected by sodium and ultimately the soil will lose its productivity.
- Suspended and colloidal impurities clog the pores and form a mat on the surface of soil preventing the passage of air, water etc.

Q.15. What are the effects of various water pollutants?

Ans. Effects of various water pollutants are tabulated below.

Water Pollutants	Effects
Detergent	Complex formation between DNA and mercury or cadmium results in birth defects. Although, detergents are not highly toxic to fishes, they do cause damage to gills and remove the protective mucous from the skin and the intestine.
Domestic waste	Water borne diseases like cholera (Vibrio cholera) typhoid, dysentery and various health problems, depletion of dissolved oxygen, objectionable odour.
Industrial effluents	It causes deleterious effects on living things and may bring death or sub lethal pathology of kidneys, liver, lungs, brain and reproductive system.
Agricultural waste	Excessive fertilizer leads to accumulation of nitrates in children called methemoglobinemia. Richness of nutrients results in eutrophication.
Eutrophication	During eutrophication, algal bloom release toxic chemicals into the aquatic system. Algal Bloom leads to oxygen depletion and an increase in CO_2 level. Thus, aquatic organisms begin to die which leads to succession.
Bioaccumulation (or) Biological magnification	Aquatic plants and animals can accumulate certain pesticides in their body tissues in greater concentration than in water. This phenomenon is commonly referred to as biological magnification or biological amplification e.g., DDT. It is more threatening as its concentration continuously increases in successive trophic levels in a food chain which results in many health hazards. DDT absorbed by fish eating birds 25 ppm ↑ DDT in large fish 2 ppm ↑ DDT in small fish 0.5 ppm ↑ DDT in zooplankton 0.003 ppm ↑ DDT in water 0.000003 ppm or 0.003 ppb

Water Pollutants	Effects
Lead (pb)	Anemia, vomiting, damage of liver, brain and kidney
Arsenic(A)	Mental disturbance, lung cancer, ulcer, kidney damage
Mercury (Hg)	Abdominal pain, headache, diarrhea, chest pain
Cadmium (Cd)	Growth retardation, diarrhea, bone deformation, kidney damage, anemia, damage to lever
Barium (Ba)	Excessive salvation, diarrhea, paralysis
Chromium (Cr)	Gastro intestinal ulceration, diseases of central nervous system, cancer, nephritis
Zinc (Zn)	Vomiting, renal damage
Copper (Cu)	Hypertension, uremia, coma
Temperature	Reduction of dissolved oxygen, increase in biological oxygen demand Early hatching of fish eggs and fish mortality, mitigation of aquatic biota
Radioactivity	Serious skin cancer, carcinoma, melanoma, breast cancer, leukemia, DNA breakage and cataract
Siltation	Reduced visibility, reduction in direct light penetration, decrease in photosynthetic rate, chances of anaerobic digestion in the benthos zone.
Oil	Reduces visibility, birds ingest it.

Q.16. Describe in detail major types of water pollutants and their effects.

Ans. Human activities give rise to water pollution by introducing various categories of substances or waste heat into a water body. The more common types of polluting substances include pathogenic organisms, oxygen demanding organic substances, plant nutrients which stimulate algal blooms, inorganic and organic toxic substances and oil and grease.

Pathogenic organisms: Many serious human diseases such as cholera, typhoid, bacterial and amoebic dysentery, enteritis, polio and infectious hepatitis are caused by water-borne pathogens. In addition, malaria, yellow fever and filariasis are transmitted by insects that have aquatic larvae. Faecal pollution of water resources by untreated or improperly treated sewage is a major cause for the spread of water-borne diseases. To a lesser extent, disease causing organisms may also be derived from animal rearing operations and food processing factories with inadequate waste water treatment facilities.

The spread of water-borne infectious diseases can be controlled through the introduction of water and sewage treatment facilities and through improved hygiene.

Biodegradable organic substances: Human and animal wastes as well as effluents from industries processing plant or animal products contain a mixture of complex organic substances such as carbohydrates, proteins and fats as their major pollution load. These substances are readily biodegradable and when introduced into the environment are quickly decomposed through the action of natural microbial populations. Some of the organic matter is oxidised to carbon dioxide and water while the rest is assimilated and used for the synthesis of new microbial cells. In due course, these organisms will also die and become food for other decomposers. Eventually virtually all of the organic carbon will be oxidised.

When a biodegradable organic waste is discharged into an aquatic ecosystem such as a stream, estuary or lake, oxygen dissolved in the water is consumed due to respiration of micro-organisms that oxidise the organic matter. The more biodegradable a waste, the more rapid is the rate of its oxidation and the corresponding consumption of oxygen. In an aquatic ecosystem, a greater number of species of organisms are supported when the dissolved oxygen concentration is high. Oxygen depletion due to waste discharge has the effect of increasing the numbers of decomposer organisms at the expense of others.

When oxygen demand of a waste is so high as to eliminate all or most of the dissolved oxygen from a stretch of a water body, organic matter degradation occurs through the activities of anaerobic organisms which do not require oxygen. Not only does the water then become devoid of aerobic organisms, but anaerobic decomposition also results in the formation of a variety of foul smelling volatile organic acids and gases such as hydrogen sulfide and mercaptans. The stench from these is quite unpleasant.

Plant nutrients: The availability of plant nutrients, e.g., nitrogen and phosphorus are important determinants of the biological productivity of aquatic ecosystems. Nutrient deficient aquatic environments are called "oligotrophic" and those rich in nutrients, "eutrophic". Young lakes are generally oligotrophic, but they naturally accumulate nutrients over time, derived from drainage and sediment run off from its catchment. When human activities greatly accelerate nutrient enrichment of water bodies, the process is called "cultural eutrophication". Sewage, animal wastes and many industrial effluents contain high levels of nitrogen and phosphorus. Another major source is fertiliser run off from urban and agricultural catchments.

While in the long term, cultural eutrophication accelerates the natural successional progress of aquatic ecosystems towards a terrestrial system, in the short-term problems arise due to cyclic occurrences of algal blooms and decay. In warm weather, nutrients stimulate rapid growth of algae and floating aquatic weeds. The water often becomes opaque and has unpleasant tastes and odours. When these organisms die they become food for decomposer bacteria. Depletion of dissolved oxygen leads to anaerobic conditions and a general decline in the ecological and aesthetic qualities of the water body. Algal blooms also reduce light penetration into the water making it impossible for sea grasses and other bottom anchored plants to survive.

Toxic inorganic pollutants: Many inorganic substances are released by natural weathering of rocks and are washed into water courses. However, human activities such as mining and mineral processing as well as wastage are responsible for far greater quantities of toxic inorganic pollutants entering water supplies and aquatic ecosystems. Of particular concern among these are arsenic, an ingredient of some pesticides, and heavy metals such as mercury, lead, tin and cadmium as they tend to accumulate in tissues. Mine drainage and leaching of mine tailings as well as metal finishing and inorganic chemical industries are major sources of metal pollution in the water environment.

Mercury poisoning causes birth defects and permanent brain damage. The worst case of mercury poisoning of a community to date occurred in Minamata in Japan, due to the consumption of contaminated fish and shell fish from the bay which received discharges of chemical industry effluents over a long period of time. Insoluble metallic mercury is converted by bacteria in the marine sediments into water soluble methyl mercury. It is then concentrated through the trophic levels of the food chain due to selective retention in tissues.

Lead causes damage to the nerve system, and experts recommend a tolerance limit of less than 10 parts per billion in drinking water. Prior to the introduction of PVC water pipes, lead piping and cast-iron pipes using lead solder were widely used for public water supplies. The main source of environmental lead is automobile exhausts. Lead has long been an additive to petrol to boost its octane rating. Exhaust fumes from cars and other vehicles are eventually washed down by rain and enter watercourses. Some inorganic industrial effluents also contain lead.

Other heavy metals such as **Chromium** and **Zinc** are present in effluents from metal finishing industries. Discharge of waste waters from many small metal finishing industries causes ground water pollution. Chromium salts are added to cooling waters of air conditioning systems for corrosion prevention. Significant contribution of this metal therefore comes from this source as well.

Production of alumina from bauxite gives rise to highly caustic effluents which also contain dissolved aluminium and other metals. **Aluminium** has recently been implicated in the Alzheimer's disease, although there is still uncertainty about the link. **Caustic Effluents** can raise the pH of receiving waters to levels unsuitable for many organisms. **Acidic Effluents** are produced from mine drainage and by many industries such as ore smelting, metal finishing, leather tanning and petroleum processing. Many lakes are acidified due to acid precipitation. Increased levels of sulfur and nitrogen oxides in the atmosphere from the combustion of fossil fuels such as coal and petroleum are the principal causes of acid rain.

Acidification lowers the pH of the water, especially when there is little buffering capacity in the form of alkalinity to neutralize it. Fish, amphibians and many insects will be killed by increased acid levels and in severe cases only a few resistant species such as fungi may survive.

Toxic organic chemicals: Many thousands of natural and synthetic organic chemicals are in use today for the manufacture of a variety of products ranging pesticides, pigments, pharmaceuticals and plastics. Several of these are known to cause birth abnormalities, genetic defects and cancer. Some chemicals like DDT and PCB's (Poly Chloro Biphenyls) are concentrated in tissues to dangerous levels. Many are only very slowly biodegradable and persist in the environment for long periods of time.

The major causes of organic chemical pollution are improper disposal of domestic and industrial wastes and herbicide and pesticide run off from farming areas, where they are used in substantial quantities. In addition, large quantities of hazardous wastes have, in the past, been disposed of in landfills with inadequate containment from where they are slowly leached into surface and ground water supplies, eventually finding their way into the food chain.

Oil pollution: Petroleum is one of the major energy sources today and huge volumes of oil are transported between points of production and consumption around the globe. All along these major transportation routes oil spills happen regularly and oil slicks are ever present. With serious spills, many marine birds and other animals are choked to death by the oil slick. Even when dispersed, many hydrocarbons in the oil are toxic to aquatic organisms. Some are thought to be carcinogenic. Oil being lighter than water floats on the surface as a thin film which can interfere with the transfer of gases such as oxygen and carbon dioxide, as well as heat, between the water and the atmosphere.

Routine petroleum refining, storage and use also results in pollution by leaking oil, oily waste water and sludge. Underground fuel storage tanks in service also leak oil into the ground water.

Thermal pollution: Tremendous quantities of waste heat is produced by power plants and to a lesser extent by a broad spectrum of other industries. Cooling water drawn from the ocean, river, lake or aquifer is passed through heat exchangers where it absorbs the waste heat and is subsequently discharged back into the water environment.

Significant rises in water temperature can be caused in the receiving water in the vicinity of cooling water disposal sites. Such increases in temperature alter the species composition in ecosystems as organisms normally tolerate temperature variations only over a very small range. At higher temperatures, oxygen solubility is reduced, but bacterial respiration rate will increase, making the water more prone to deoxygenation. Temperature variations will also cause alterations in pH due to changes in the degree of ionization and increased solubility or precipitation of bottom deposits.

Q.17. Describe various methods of control of water pollution.

Ans. With increasing urbanization and expanding agricultural and industrial production, water pollution problems have become serious necessitating the adoption of suitable control measures for reducing water pollution.

Approaches to controlling sources of water pollution may be grouped into three broad categories: (1) minimization of waste or pollutant generation, (2) Treatment prior to disposal of waste water streams at source, and (3) "in-situ" reduction or elimination of pollution.

Minimisation of pollutant generation: Reduction of quantity of waste /pollutants generated is the most desirable approach to pollution control. Since it conserves resources that would otherwise be wasted, and at the same time eliminates the cost of removing pollutant after they are produced, it is the cheapest and the most effective alternative. For non-point pollution sources, this is the only practicable method of pollution control. Yet, this approach has not been exploited to its fullest extent.

As a general rule, a resource becomes a waste when it can no longer be economically utilized or recovered. It is then disposed of in the environment in the cheapest manner possible. Availability of economical technology for resource processing and usage has been a main determinant of when the resource is discarded as waste.

Minimizing soil erosion by improved agricultural practices (e.g. by minimizing surface runoff and leaving crop residues in the ground), more efficient use of nutrients (e.g., though the use of slow release fertilizers) and the development and use of biological pest control techniques in preference to the use of non-biodegradable toxic chemicals are some of the measures for minimizing water pollution from agriculture.

Considerable potential also exists in many industries to reduce waste generation. Development and use of non-polluting technology to modify or replace existing manufacturing processes, and recycling or recovering materials that would otherwise be wasted are two approaches which not only reduce pollutant generation, but can even result in a saving for the industry by minimizing or eliminating the need for waste treatment for pollutant removal.

In some cases, it may be more practical to segregate strong and weak waste streams to facilitate materials or energy recovery. Good housekeeping practices, like minimizing spillage and materials wastage, also leads to waste reduction and savings in production cost.

Waste water treatment at source. In nature, a variety of different mechanisms operate to degrade and transform waste materials into stable, harmless end products such as carbon dioxide. This cleansing ability is referred to as the "self-purification" or "assimilative" capacity. When the quantities of wastes to be disposed of are large, however, the natural purification processes become overloaded and can no longer assimilate the wastes without adversely affecting environmental quality. Man-made treatment systems are then needed to reduce pollutant loads to acceptable levels for discharge. For the most part, these purification systems make use of the same mechanisms as in the natural environment to bringing about waste stabilization.

The waste water treatment technologies are classified as physical, chemical and biological processes, depending on the nature of the purification mechanism employed. The character of the pollutants and the form (suspended or dissolved) in which they are present usually determine the most suitable process for their removal. For example, total suspended solids and floatable materials such as oil and fat are readily removed by physical processes such as sedimentation or flotation respectively.

Biological methods are effective and economical when the waste water contains mostly biodegradable pollutants such as organic matter. An advantage of biological processes is that the microorganisms involved in waste stabilization are themselves produced in the process. For dilute wastes - including general domestic waste waters, "aerobic" biological processes (activated sludge, oxidation ponds and aerobic biofilter) are usually favoured since they are capable of producing an effluent with very low residual pollutant concentrations. These processes, require oxygen, in proportion to the pollutant load present. Oxygen is supplied through aeration, which is a significant cost component. For strong wastes, "anaerobic" biological treatment in enclosed vessels is generally preferred as they proceed in the absence of oxygen, and in addition produce a useful, energy-rich by-product in the form of methane.

The effluent from anaerobic processes, however, contain higher levels of residual organic materials and may require further polishing treatment (often in aerobic processes).

Chemical treatment is used when the pollutant is non- biodegradable and is not amenable to removal by simple physical means (e.g., when it occurs in dissolved form). Heavy metals are typically removed by chemical precipitation, while toxic substances such as cyanide may be chemically oxidized. An important disadvantage of chemical treatment methods is that they generally require dosing with a chemical which can be quite expensive. In addition, disposal of the chemical sludge produced in these processes may also pose some problems.

When a community-based treatment system is impractical, it is possible to provide a degree of treatment prior to discharging sewage into the environment. A popular method used for individual homes and small groups of residences is the SEPTIC TANK. It consists of a simple baffled tank which traps most of the solids in the waste water and also affords some decomposition of soluble organic matter. The effluent is disposed of into the ground through a system of leach drains. As solids progressively accumulate in the tank, it is necessary to periodically desludge the system, typically every 3 to 7 years. As deep sewering in built-up areas is very expensive, other more efficient alternatives to the septic tank are also desirable for on- site use.

In-situ pollution control: Waste minimization and treatment help prevent pollution from occurring and should be the principal approaches to water quality maintenance. Occasionally, however, when a water body is already adversely affected, it will be necessary to consider action aimed at helping the ecosystem recover from the impact of pollution. Methods to facilitate this are collectively grouped under in-situ control techniques.

Aeration of lakes and reservoirs, especially when they are thermally stratified (in summer), has been used to prevent anaerobic conditions from occurring. Forced circulation of water in stratified lakes is an alternative method. Dredging nutrient rich superficial sediments from highly eutrophic lakes, while very expensive, helps reduce occurrence of severe algal blooms. Addition of aluminium or iron salts to assist the precipitation of phosphorus has also been practiced in some lakes to control dissolved phosphorus levels in the water.

Q.18. Discuss the composition of water in rivers and lakes.

Ans. The composition of a river, and a lake, is a reflection of four inputs: atmospheric input of gases and solutes, biologic processes, discharge of groundwater, and local interactions with mineral components in the soil or stream bed.

At base flow, the water in a gaining river is derived from groundwater, so the composition of the river is a reflection of the aquifer, while for a losing stream the reverse holds good - the composition of the shallow groundwater reflects the river, for example an alluvial aquifer.

The composition of a river is a dynamic thing, with concentrations generally (but not always) increasing downstream, but pH generally increasing.

Q.19. What is hard water?

Ans. Hard water occurs when excess minerals in the water create certain problems but water hardness is not a safety issue. Hard water is safe for drinking, cooking, and other household uses.

Hard water can cause problems for like decreased life of household plumbing and water-using appliances, increased difficulty in cleaning and laundering, decreased efficiency of water heaters, and white/chalky deposits on items such as plumbing, tubs, sinks, pots and pans.

Q.20. Describe the causes of hard water.

Ans. Approximately 22 % of the earth's fresh water is groundwater, and naturally, as it flows through soil and rock, it picks up minerals. Hard water results when an excessive amount of calcium and magnesium are present. Total hardness is measured in milligrams per litre (mg/l) or in parts per million (ppm). Parts per million (ppm) measures the unit(s) of a substance for every one million units of water. Milligrams per litre (mg/l) and parts per million (ppm) are roughly equal in water analysis. When conducting chemical analysis, laboratories usually measure hardness minerals in milligrams per liter of water (mg/l).

Identifying hard water

- Decreased cleaning capabilities of soaps and detergents, resulting in dingy laundry and reduced life of fabrics.
- Increased buildup of scale on plumbing fixtures and cooking utensils.
- Film left on the body resulting in dry skin and dull, limp hair.
- Soap scum on bathtubs, shower tiles, and basins.
- Increased water heating costs due to scale buildup and mineral deposits, and more frequent replacement of hot water heating elements.
- Clogged pipes or appliances resulting in reduced water flow and increased repairs.

Q.21. What are different types of water hardness?

Ans. There are two types of water hardness-temporary and permanent hardness.

Temporary hardness: This refers to hardness whose effects can be removed by boiling the water. Such waters have usually percolated though limestone formations and contain bicarbonate HCO_3 along with small amounts of carbonate CO_3 as the principal negative ions. Boiling the water promotes the reaction:

$$2\,HCO_3 \longrightarrow CO_3 + CO_2$$

by driving off the carbon dioxide gas. The CO_3 reacts with Ca^{2+} or Mg^{2+} ions, to form insoluble calcium and magnesium carbonates which precipitate out. By tying up the metal ions in this way, the amounts available to form soap scum are greatly reduced.

Permanent hardness: Waters that contain other anions such as chloride or sulphate cannot be remediated by boiling, and are said to be 'permanently' hard. The only practical treatment is to remove all the ions by various methods.

Q.22. What do you mean by water quality? What is its significance?

Ans. Water quality is a term used to express the suitability of water for various uses or processes. Any particular use of water will have certain requirements for the physical, chemical or biological characteristics of water; for example, limits on the concentrations of toxic substances for drinking water use, or restrictions on temperature and pH ranges for water supporting animals. Consequently, water quality can be defined by a range of variables which limit the water. Although many uses have some common requirements for certain variables, each use will have its own demands and influences on water quality. It should be free from color, turbidity, taste and odour, of moderate temperature and aerated.

The quality of water may be described in terms of the concentration and state (dissolved or particulate) of some or all of the organic and inorganic material present in the water, together with certain physical characteristics of the water.

Water temperature, pH, specific conductance, turbidity, dissolved oxygen and suspended sediments are important parameters if we discuss about water quality. Potable water is an important parameter for human beings, because it is useful for maintaining human health.

Q.23. What are the main water quality parameters?

Ans. Water quality is determined by various water parameters like temperature,pH,specific conductance,turbidity,dissolved oxygen, hardness,suspended sediments etc.

Water temperature: It is important to fisherman, industries, and growth of fish and algae. A lot of water is used for cooling purposes in power plants that generate electricity. They need cool water to start with, and they generally release warmer water back to the environment. The temperature of the released water can affect downstream habitats. Temperature also can affect the ability of water to hold oxygen as well as the ability of organisms to resist certain pollutants.

pH of water: pH is a measure of how acidic/basic a water sample is. The range goes from 0–14, with 7 being neutral. The pH of less than 7 indicates acidity, whereas a pH of greater than 7 indicates a base. Since pH can be affected by chemicals in the water, pH is an important indicator of water that is changing chemically. The pH is reported in logarithmic units. Each number represents a 10-fold change in the acidity/basicity of the water. Water with a pH of 5 is ten times more acidic than water having a pH of 6. Pollution can change a water's pH, which in turn can harm animals and plants living in the water. For instance, water coming out of an abandoned coal mine can have a pH of 2, which is very acidic and would definitely affect any fish and other micro organisms.

Specific conductance: Specific conductance is a measure of the ability of water to conduct an electrical current. It is highly dependent on the amount of dissolved solids (such as salt) in the water. Pure water, such as distilled water, will have a very low specific conductance, and sea water will have a high specific conductance. Rainwater often dissolves airborne gases and dust while it is in the air, and thus often has a higher specific conductance than distilled water. Specific conductance is an important water-quality measurement because it gives a good idea of the amount of dissolved material in the water.

Turbidity: Turbidity is the amount of particulate matter that is suspended in water. Turbidity measures the scattering effect that suspended solids have on light: the higher the intensity of scattered light, the higher the turbidity. Material that causes water to be turbid include:

- clay
- finely divided organic and inorganic matter
- plankton
- silt
- soluble colored organic compounds
- microscopic organisms

Turbidity makes the water cloudy or opaque. Turbidity is measured by shining a light through the water and is reported in Nephelometric Turbidity Units (NTU). During periods of low flow (base flow), many rivers are a clear green color, and turbidities are low, usually less than 10 NTU. During a rainstorm, particles from the surrounding land are washed into the river making the water a muddy brown colour, indicating water that has higher turbidity values. Also, during high flows, water velocities are faster and water volumes are higher, which can more easily stir up and suspend material from the stream bed, causing higher turbidities.

Dissolved oxygen: Although water molecules contain oxygen atom, this oxygen is not what is needed by aquatic organisms living in natural waters. A small amount of oxygen, up to about ten molecules of oxygen per million of water, is actually dissolved in water. This dissolved oxygen is breathed by fish and zooplankton and is needed by them to survive. Rapidly moving water, such as in a mountain stream or large river, tends to contain a lot of dissolved oxygen, while stagnant water contains little. Bacteria in water can consume oxygen as organic matter decays. Thus, excess organic material in the tanks, lakes and rivers can cause an oxygen-deficient situation to occur. Aquatic life can have a hard time in stagnant water that has a lot of rotting, organic material in it, especially in summer, when dissolved oxygen levels are at a seasonal low.

Hardness: The amount of dissolved calcium and magnesium in water determines its 'hardness. Where the water is relatively hard, it is difficult to get lather up when washing hands or clothes. And, industries in these areas have to spend money to soften their water, as hard water can damage equipment. Hard water can even shorten the life of fabrics and clothes.

Suspended sediment: It is the amount of soil moving along in a stream. It is highly dependent on the speed of the water flow, as fast-flowing water can pick up and suspend more soil than calm water. During storms, soil is washed from the stream banks into the stream. The amount that washes into a stream depends on the type of land in the river's watershed and the vegetation surrounding the river. If land is disturbed along a stream and protection measures are not taken, then excess sediment can harm the water quality of a stream. These bunds and fences are supposed to trap sediment during a rainstorm and keep it from washing into a stream, as excess sediment can harm the creeks, rivers, lakes, and reservoirs. Sediment coming into a reservoir is always a concern; once it enters it cannot get out - most of it will settle to the bottom. Reservoirs can 'silt in' if too much sediment enters them. The volume of the reservoir is reduced, resulting in less area for water storage for agriculture as well as reducing the power-generation capability of the hydel power plant in the dam.

Q.24. What are various impurities found in water?

Ans. As water goes through the hydrologic cycle, it gathers many impurities. Dust, smoke, and gases in the air contaminate rain, snow, hail, and sleet. Rain water picks up silt, chemicals, and disease organisms. As it enters the earth through seepage and infiltration, some of the suspended impurities may be filtered out. However, other minerals and chemicals are dissolved and carried along. As groundwater it may contain disease organisms and harmful chemicals. In addition to the impurities in water resulting from infiltration, many are contributed by industries. The garbage, sewage, industrial waste, pesticides etc., are all possible contaminants of raw water. Impurities in raw water are either suspended or dissolved. Suspended impurities include diseases organisms, silt, bacteria, and algae. They must be removed/destroyed before the water is consumed. Dissolved impurities include salts, (calcium, magnesium, and sodium), iron, manganese, and gases (oxygen, carbon dioxide, hydrogen sulphide, pH and nitrogen). These impurities must be reduced to levels acceptable for human consumption.

Q.25. How are raw water classified?

Ans. Raw water is classified as fresh, brackish, or salt water (seawater) based on the concentration of TDS (Total Dissolved Solids). Fresh water has a TDS concentration of less than 1,500 ppm. Brackish water is high in minerals and has a TDS concentration between 1,500 ppm and 15,000 ppm. Salt water has a TDS concentration greater than 15,000 ppm. Generally, groundwater (subsurface) has less chemical or biological contaminants than surface water, provided reasonable care is exercised in the selection of the well site. Harmful microorganisms are usually reduced to tolerable levels by passage through the soil.

Q.26. How are treated water classified?

Ans. Treated water is classified as palatable water and potable water.

Palatable water: Palatable water is water that is pleasing in appearance and taste. It is significantly free from colour, turbidity, taste, and odour. It should also be cool and aerated.

Potable water: Potable water means water that is fit for drinking. Quality standards for treated water reflect the values of substances allowed in potable water. Standards exist to measure the physical, chemical, microbiological, and radiological **quality** of water and to test for the presence of chemical agents.

Q.27. What are physical quality characteristics of water?

Ans. The main physical quality characteristics of water are colour, odour, taste, turbidity, and temperature. These physical quality characteristics and their related quality standards are described below.

Colour: Colour in water is derived from coloured substances, such as vegetable matter, dissolved from roots and leaves, from humus, or from inorganic compounds such as iron and manganese salts. The colour standard is designed to make drinking water more palatable.

Odour and taste: There is no set standards for odour and taste as there are no specific tests for these. Odour and taste found in water are most commonly caused by algae, decomposed organic matter, dissolved gases, or industrial waste.

Turbidity: Turbidity refers to a muddy or unclear condition of water caused by suspended clay, silt, organic and inorganic matter, plankton and other microorganisms. The turbidity standard was established to improve the efficiency of disinfection by reducing particles to which microorganisms could attach.

Temperature: Warm water tastes flat. Cooling water suppresses odours and tastes and makes it more palatable. Temperature also effects the chlorination and purification of water. Disinfection takes longer when water is colder, and purification capacity is reduced with reverse osmosis treatment equipment. Water having physical characteristics exceeding the limits or making it less palatable should not be used for drinking. Otherwise, reduced consumption and increased risk of dehydration may result. High water temperature in rivers and stream reduces dissolved oxygen content posing threat to survival of aquatic creatures, fishes etc.

Q.28. What are chemical quality characteristics of water?

Ans. The chemical quality of water depends on the chemical substances it contains. These substances include TDS, chlorides, sulphates and other ions. The chemical quality of water involves its hardness, alkalinity, acidity, and corrosiveness. Chemical substances having an adverse health effect have established standards that should not be exceeded.

Potential hydrogen (pH): The pH is a measure of the acidic or alkaline nature of water. It is defined as the negative logarithm of the hydrogen ion concentration. It ranges from 0 to 14 and pH of 7 is neutral. The pH influences the corrosiveness of the water, the amount of chemicals needed for proper disinfection, and the ability of an analyst to detect contaminants. Water with a pH below 7 is regarded as acidic while that with a pH above 7 is regarded as alkaline. The pH standard was established to ensure effective purification and disinfection.

Arsenic: Arsenic can be present in natural water sources in a wide range of concentrations. It can come from either natural or industrial sources. Ingestion of low concentrations of arsenic can cause nausea, vomiting, abdominal pain, or nerve damage.

Chloride: Chloride exists in most natural waters. It is the main anion found in seawater. Chloride comes from natural salt deposits, domestic and industrial waste, and agricultural runoff. Even in low concentrations, chloride can produce an objectionable taste in water.

Calcium, magnesium and hardness: Calcium and magnesium in drinking water, is an important parameter is to be taken into account Water hardness is a measure of the capacity of water to precipitate soap, which is in practice the sum of concentrations of all polyvalent cations present in water (Ca, Mg, Sr, Ba, Fe, Al, Mn, etc.); nevertheless, since the other ions (apart from Ca and Mg) play a minor role, generally hardness is defined as the sum of the Ca and Mg concentrations.

Water calcium and magnesium result from decomposition of calcium and magnesium aluminosilicates and, at higher concentrations, from dissolution of limestone, magnesium limestone, magnesite, gypsum and other minerals. Drinking water may be intentionally supplemented with these elements while treated, as happens with deacidification of underground waters by means of calcium hydroxide or filtration through different compounds counteracting acidity such as $CaCO_3$, $MgCO_3$ and MgO, and also with stabilization of low-mineralized waters by addition of CaO and CO_2. In low and medium mineralized underground and surface waters (as drinking waters are), calcium and magnesium are mainly present as simple ions Ca^{2+} and Mg^{2+}, the Ca levels varying from tens to hundreds of mg/l and the Mg concentrations varying from units to tens of mg/l.

Magnesium is usually less abundant in waters than calcium, which is easy to understand since magnesium is found in the Earth's crust in much lower amounts as compared with calcium. In common underground and surface waters the weight concentration of Ca is usually several times higher compared to that of Mg, the Ca to Mg ratio reaching up to 10. However, a common Ca to Mg ratio is about 4.

Sulphates: Sulphates occur naturally in water as the result of dissolution of sulfur bearing minerals. High concentrations also result from industry sources, such as coal mine drainage, pulp paper mills, tanneries, textile mills, and domestic waste water. When ingested, sulphates have a laxative effect. They also can produce a bad taste in water. The sulphate standard was established to prevent chemically induced diarrhea.

Total Dissolved Solids (TDS): The TDS of water is composed of mineral salts and small amounts of other inorganic and organic substances. The proportion of each constituent is the result of weathering of rocks found in the drainage basin and of any industrial contributions. Since TDS is composed of chloride, magnesium, sulfate, and other ions, its ingestion in water has the same effects. Therefore, the TDS standard was established to prevent chemically induced diarrhea. Chemical water quality standards are based on the effect the water will have on the health of living organisms. The effect of a particular chemical substance determines if a limit is established for that substance. Chemical substances having a negative physical effect will have a mandatory limit that should not be exceeded. Some substances, such as iron and manganese, have no significant negative physical effect, but may restrict the use of the water, such as for the laundering of clothes.

Q.29. What are microbiological quality characteristics of water?

Ans. The microbiological quality of potable water shows its potential for transmitting waterborne diseases which are caused by viruses, bacteria, protozoa, or higher organisms. The testing for microorganisms in water is extremely difficult. The number of these organisms is usually very low, even in a badly polluted water supply, and the test used to find them is difficult. For these reasons, indicator organisms are used to detect the presence of contamination. The bacterial organisms used as an indicator of possible contamination are total coliform. These organisms occur in large quantities in the intestines of warm-blooded animals. The presence of coliform organism in potable water is an indication of either inadequate treatment or the introduction of undesirable materials to the water after treatment. While the detection of many disease-causing microbes is difficult, the test to detect a surrogate organism, E. coli, is simple and effective for field use. Because of its relative simplicity and field adaptability, the membrane filter technique has gained wide acceptance as the preferred technique for the presumptive determination of the presence of coliform organisms in potable water. The microbiological standard was established to ensure infectious microorganisms would not cause diseases.

Q.30. Describe the acceptable water quality standards.

Ans. Water quality management should ensure that water supplied is free from pathogenic organisms, clear, potable, free from undesirable taste and odour, of reasonable temperature, neither corrosive nor scale forming and free from minerals which could produce undesirable physiological effect.

Physical and chemical standards for potable and safe water is tabulated below.

S.N.	Characteristics	Acceptable	Cause for Rejection
1.	Turbidity (Units on JTU scale)	2.5	10
2.	Colour (units on platinum-cobalt scale	5.0	25
3.	Taste and odour	unobjectionable	
4.	pH	7.0 to 8.5	6.5 to 9.2

5.	Total dissolved solids(mg/l)	500	1500
6.	Total hardness (mg/l as $CaCO_3$)	200	600
7.	Chlorides (mg/l as Cl)	200	1000
8.	Sulphates (as SO_4, mg/l)	200	1000
9.	Fluorides (as F, mg/l)	1.0	1.5
10.	Nitrates (as NO_3, mg/l)	45	100
11.	Calcium (as Ca, mg/l)	75	200
12.	Magnesium (as Mg, mg/l)	30	150
13.	Iron (as Fe, mg/l)	0.1	1.0
14.	Manganese (as Mn, mg/l)	0.05	0.5

Bacteriological standards

(*a*) Coliform count in any water sample of 100 ml should be zero. A sample of water entering the distribution system that does not conform to this standard call for an immediate investigation into both efficacy of the purification and the method of sampling.

(*b*) Water in the distribution system should satisfy all the three criteria indicated below:

— *E,Coli* count in 100 ml of any sample should be zero.

— Coliform organisms not more than 10 per 100 ml shall be present in any sample.

— Coliform organisms should not be detectable in 100 ml of any two consecutive samples or more than 50% of the samples collected for the year.

World Health Organization (WHO) Guideline Values for chemicals (in **mg/l**) that are of health significance in drinking water is as follows ;

- Arsenic - 0.01
- Fluoride - 1.5
- Manganese - 0.4
- Nitrate - 50
- Nitrite - 3
- Chlorine - 5
- Copper - 2000
- Lead - 10
- Nickel - 20

Q.31. What is water pollution? What causes it?

Ans. Water pollution occurs when a body of water is adversely affected due to the addition of large amounts of foreign materials to the water. When it is unfit for its intended use, water is considered polluted. Two types of water pollutants exist; point source and non-point source. Point sources of pollution occur when harmful substances are emitted directly into a body of water.

Many causes of pollution including sewage and fertilizers contain nutrients such as nitrates and phosphates. In excess levels, nutrients over stimulate the growth of aquatic plants and algae. Excessive growth of these types of organisms consequently clogs waterways, use up dissolved oxygen as they decompose, and block light to deeper waters.

Q.32. How are different types of water pollution sources classified?

Ans. Water pollution can be classified into three types as domestic pollution, industrial pollution and agricultural pollution.

Domestic pollution: Important source of water pollution is domestic sewage system, which pollutes well and rivers, which are important source of drinking water. The solution to excreta disposal and its proper management is required to prevent the spread of pathogens in the environment which directly influences disease transmission through person to person contact, water and the food chain. Inadequate system for disposing of faeces, faeces deposited near the homes threaten household hygiene, by direct contact with people or by being carried into homes and kitchens by children, domestic animals or insects.

Domestic neighbourhood or district water supplies such as wells, tanks and reservoirs may be contaminated by poorly designed or maintained sewage disposal systems. The potable water contaminated with faeces is the chief cause of some important diseases of human beings. Use of open field for defecation results in faeces causing pollution of potable water as they consist primarily of intestinal bacteria.

Since a primary cause of contamination of water is inadequate or improper disposal of human and animal excreta, better water quality only improves health when sanitation is improved as well and when the quantity of water is sufficient. Deteriorating water quality is a particular threat in developing countries, where people lack access to clean drinking water and the majority of sewage are discharged into surface waters without waste water treatment. When untreated or inadequately treated waste water or excreta (faecal sludge) is applied to soil and crops, disease transmission can occur. The persons at risk are the farmers, farm workers and their families as well as consumers of crops produced in such a way.

Industrial pollution: Sewage is not the only the cause of water pollution, industrial waste is also a significant polluter – giving rise to contamination with heavy metals. The chemicals discharged by the factories are more harmful than the sewerage that flows into the river. The effluent discharged by the factories contains detergents, non-biodegradable materials and toxic chemicals hazardous to health and hygiene. Other sources of water pollution are distillery, potassic fertilizers, electroplating plant, which contains harmful heavy metals, and cyanides which causes excessive acidity in water, which is harmful for aquatic life. Water pollution ranged from moderate to very high in case of textiles, jute, leather, vegetables fats and brewery. Moderate pollution results from dairy and canning establishments.

Agricultural pollution: In addition to problems of waterlogging, desertification, salinization, erosion, etc., that affect irrigated areas; the problem of downstream degradation of water quality by salts, agrochemicals and toxic leachates is a serious environmental problem. 'Intensive use of pesticides and fertilisers has increased the level of nitrates in water. This can lead to brain damage and even death among children. Salinisation of water resources is a major concern to the sustainability of irrigation than is that of the salinisation of soils. Trace toxic constituents, such as Se, Mo and As in agricultural drainage waters may cause pollution problems that threaten the continuation of irrigation in some projects.

Q.33. Describe various physical parameters of water pollution.

Ans. Various water pollution physical parameters are described below.

pH: pH is a measurement of how **acidic** or how **basic (alkaline)** a solution is. When substances dissolve in water they produce charged molecules called ions. Acidic water contains extra hydrogen ions (H^+) and basic water contains extra hydroxyl (OH^-) ions. Pollution can change pH in water, which in turn can harm animals and plants living in the water. For instance, water coming out of an abandoned coal mine can have a pH of 2, which is very acidic and would definitely affect any fish that live in it. The mine-drainage water would be 100,000 times more acidic than neutral water.

Total suspended solids (TSS): The TSS is solid materials, including organic and inorganic, that are suspended in the water. These would include silt, plankton and industrial wastes. High concentrations of suspended solids can lower water quality by absorbing light.

Waters then become warmer which lessen the ability of the water to hold oxygen necessary for aquatic life. Because aquatic plants also receive less light, photosynthesis decreases and less oxygen is produced. The combination of warmer water, less light and less oxygen makes it impossible for some forms of aquatic life to exist.

Suspended solids affect aquatic life in other ways. They can clog fish gills, reduce growth rates, decrease resistance to disease, and prevent egg and larval development. Particles that settle out can smother fish eggs and those of aquatic insects, as well as suffocate newly-hatched larvae. The material that settles also fills the spaces between rocks and makes these microhabitats unsuitable for various aquatic insects. Suspended solids can result from erosion from urban runoff and agricultural land, industrial wastes, river bank erosion, bottom feeders (such as carp), algae growth or waste water discharges.

Conductivity: Conductivity is a measurement of the ability of an aqueous solution to carry an electrical current. An ion is an atom of an element that has gained or lost an electron which will create a negative or positive state. For example, sodium chloride consists of sodium ions (Na^+) and chloride ions (Cl^-) held together in a crystal. In water it breaks apart into an aqueous solution of sodium and chloride ions. This solution will conduct an electrical current. Degree to which water will carry an electrical current depends upon:

- the concentration or number of ions;
- oxidation state (valency) and;
- mobility of the ion;
- temperature of the water.

Resistance, expressed in ohms, is the opposite of conductivity. Conductivity is then expressed in reciprocal ohms. A more convenient unit of measurement in the chemical analysis of water is micromhos. The specific conductance or conductivity measurement is related to ionic strength and does not indicate what specific ions are present. Conductivity measurement is used to determine a number of applications related to water quality. These are as follows:

- **Determining mineralization:** This is commonly called total dissolved solids (TDS). Total dissolved solids information is used to determine the overall ionic effect in a water source. Certain physiological effects on plants and animals are affected by the number of available ions in the water.
- Noting variation or changes in natural water and waste waters quickly.
- Estimating the sample size necessary for other chemical analyses.
- Determining amounts of chemical reagents or treatment chemicals to be added to water sample.

Elevated dissolved solids can cause 'mineral tastes' in drinking water. Corrosion or encrustation of metallic surfaces by waters high in dissolved solids causes problems with industrial equipment and boilers as well as domestic plumbing, hot water heaters, toilet flushing mechanisms and washing machines. Indirect effects of excess dissolved solids are primarily the elimination of desirable food plants and habitat forming plant species. Agricultural uses of water for livestock watering are limited by excessive dissolved solids and high dissolved solids can be a problem in water used for irrigation.

Hardness: Hardness is due to the presence of multivalent metal ions which come from minerals dissolved in the water. Hardness is based on the ability of these ions to react with soap to form a precipitate or soap scum. In fresh water the primary ions are calcium and magnesium; however, iron and manganese may also contribute. Carbonate hardness is equal to alkalinity but a non-carbonate fraction may include nitrates and chlorides. Hardness is expressed in mg/l of $CaCO_3$ (even though all the hardness may not be due to $CaCO_3$).

Classification of water by hardness content (Hardness unit: mg/l) is as follows:

Concentration mg/l - Description

- 0-75 Soft water
- 75-150 Moderately hard water
- 150-300 Hard water
- 300 and more Very hard water

The most important impact of hardness on fish and other aquatic life appears to be the affect the presence of these ions has on the other more toxic metals such as lead, cadmium, chromium and zinc. Generally, the harder the water, the lower the toxicity of other metals to aquatic life. In hard water some of the metal ions form insoluble precipitates and drop out of solution and are not available to be taken in by the organism. Large amounts of hardness are undesirable mostly for economic or aesthetic reasons. If a stream or river is a drinking water source, hardness can present problems in the water treatment process. Hardness must also be removed before certain industries can use the water.

Total dissolved solids (TDS): The TDS comprise inorganic salts (principally calcium, magnesium, potassium, sodium, bicarbonates, chlorides and sulphates) and small amounts of organic matter that are dissolved in water. The TDS in drinking-water originate from natural sources, sewage, urban runoff and industrial waste water. Concentrations of TDS in water vary considerably in different geological regions owing to differences in the solubilities of minerals. The presence of high levels of TDS in drinking-water may be objectionable to consumers.

Q.34. Describe various chemical parameters of water pollution.

Ans. Various chemical parameters of water pollution is described below.

Biological oxygen demand (BOD): Biochemical oxygen demand (BOD) refers to the amount of oxygen that would be consumed if all the organics in one litre of water were oxidised by bacteria and protozoa. The first step in measuring BOD is to obtain equal volumes of water from the area to be tested and dilute each specimen with a known volume of distilled water which has been thoroughly shaken to insure oxygen saturation.

After this, an oxygen meter is used to determine the concentration of oxygen within one of the vials. The remaining vial is than sealed and placed in darkness and tested five days later. BOD is then determined by subtracting the second meter reading from the first. The range of possible readings can vary considerably: water from an exceptionally clear lake might show a BOD of less than 2 ml/l of water. Raw sewage may give readings in the hundreds and food processing wastes may be in the thousands.

Generally, when BOD levels are high, there is a decline in DO (Dissolved Oxygen) levels. This is because the demand for oxygen by the bacteria is high and they are taking that oxygen from the oxygen dissolved in the water. If there is no organic waste present in the water, there would not be as many bacteria present to decompose it and thus the BOD will tend to be lower and the dissolve oxygen level will tend to be higher. BOD vs. water quality is shown below.

BOD(ppm) **Water Quality**

1-2 very good: There will not be much organic waste present in the water supply.

3-5 fair: Moderately clean water

6-9 poor: Somewhat polluted water.

Usually indicates organic matter is present and bacteria are decomposing this waste.

100 or greater very poor: Highly polluted water. Contains organic waste.

Chemical oxygen demand (COD). The COD test will give a good estimate of the first stage oxygen demand for most waste waters. An advantage of the COD test over the biochemical oxygen demand (BOD) test is 2 to 3 hours versus 5 days. The COD test also is used to measure the strength of wastes that are too toxic for the BOD test. The COD test should be considered an independent measurement and not a quick substitute for the BOD test. The COD is usually higher than the BOD, but the amount will vary from waste to waste. Chemical Oxygen Demand measures the ability of hot chromic acid solution to oxidize organic matter. This analyzes both biodegradable and non-biodegradable organic matter. expressed as O_2. The results of the COD tests are usually higher than the corresponding BOD test for several reasons. Many organic compounds which are dichromate oxidizable are not biochemically oxidizable; Certain inorganic substances, such as sulphides, sulphates, thiosulphates, nitrites and ferrous iron are oxidized by dichromate, creating an inorganic COD, which is misleading when estimating the organic content of the waste water.

The BOD to COD ratio is nothing more than the BOD concentration divided by the COD concentration for the same sample (e.g., if BOD is 60 mg/L, and COD is 100 mg/L for a sample, the ratio is 60/100, or 0.60). There are three fairly reliable 'rules of thumb' correlations between COD/BOD. Ratios for COD to BOD of 1 to 2 are found in potable water or clean surface or groundwater.. Ratios of COD to BOD of 2 to 4 are usually seen in routine domestic/municipal sewage wastes. Ratios of COD to BOD of 4 to 6 are usually indicative of industrial type wastes.

In some industrial effluents (pretreatment program), BODs *can* be higher than CODs (for example, some effluents which are high in sugars, as can be found in the bakery industry, or soda bottling. Some industrial effluents will have higher demand because of the higher quantities of chemicals that demand oxygen).

Most applications of COD determine the amount of organic pollutants found in surface water (e.g., lakes and rivers), making COD a useful measure of water quality. It is expressed in milligrams per liter (mg/ l) or ppm, which indicates the mass of oxygen consumed per liter of solution.

Chloride: Chloride in drinking water originates from natural sources, sewage and industrial effluents, urban runoff containing deicing salt and saline intrusion. The main source of human exposure to chloride is the addition of salt to food, and the intake from this source is usually greatly in excess of that from drinking water. Excessive chloride concentrations increase rates of corrosion of metals in the distribution system, depending on the alkalinity of the water. This can lead to increased concentrations of metals in the water supply. Chloride concentrations in excess of about 250 mg/litre can give rise to detectable taste in water.

Sulphate: Sulphates occur naturally in numerous minerals and are used commercially, principally in the chemical industry. They are discharged into water in industrial wastes and through atmospheric deposition; however, the highest levels usually occur in groundwater and are from natural sources. In general, the average daily intake of sulphate from drinking water, air and food is approximately 500 mg, food being the major source. However, in areas with drinking water supplies containing high levels of sulphate, drinking-water may constitute the principal source of intake.

Q.35. What are the biological parameters of water pollution.

Ans. Various biological parameters are discussed below.

Coliforms: Coliforms are bacteria that are always present in the digestive tracts of animals, including humans, and are found in their wastes. They are also found in plant and soil material.

Total coliforms, Faecal coliforms, and E. coli: The most basic test for bacterial contamination of a water supply is the test for total coliform bacteria. Total coliform counts give a general indication of the sanitary condition of a water supply. Total coliforms include bacteria that are found in the soil, in water that has been influenced by surface water, and in human or animal waste.

Fecal coliforms are the group of the total coliforms that are considered to be present specifically in the gut and feces of warm-blooded animals. Because the origins of fecal coliforms are more specific than the origins of the more general total coliform group of bacteria, fecal coliforms are considered a better indication of animal or human waste than the total coliforms.

The major species in the fecal coliform group is *Escherichia coli (E. coli)*. Of the five general groups of bacteria that comprise the total coliforms, only *E. coli* cannot be found growing and reproducing in the environment. Consequently, *E. coli* is considered to be the species of coliform bacteria that is the best indicator of fecal pollution and the possible presence of pathogens.

Coliform bacteria may harm us by causing diseases and food spoilage. Of environmental concern are the many types of coliform bacteria. Fecal coliform bacteria are a group of bacteria that are passed through the fecal excrement of humans, livestock and wildlife. They aid in the digestion of food. A specific subgroup of this collection is the fecal coliform bacteria, the most common member being Escherichia coli. These organisms may be separated from the total coliform group by their ability to grow at elevated temperatures and are associated only with the fecal material of warm-blooded animals. Bacteria reproduce rapidly if conditions are right for growth. Most bacteria grow best in dark, warm, moist environments with food. Some bacteria form colonies as they multiply which may grow large enough to be seen. By growing and counting colonies of fecal coliform bacteria from a sample of stream water, we can determine approximately how many bacteria were originally present. Most coliform bacteria do not cause disease. However, some rare strains of *E. coli*, are pathogenic and can cause serious illness. Recent outbreaks of disease caused by E. coli 0157:H7 have generated much public concern about this organism. *E. coli* 0157:H7 has been isolated from cattle, chickens, swine, and sheep. Most of the reported cases have been attributed to the consumption of partially cooked hamburger. Cases of E. coli 0157:H7 caused by contaminated drinking water supplies are extremely rare.

Standards

- The first standard is a mandatory one that must not be exceeded; there should be no more than 10,000 total coliforms per 100 ml of water or 2,000 faecal coliforms per 100 ml of water.
- The second standard is a guideline that should be achieved where possible; within 100 ml of water there should be no more than 500 total coliforms, 100 faecal coliforms and 100 faecal streptococci.

Q.36. What are the effects of water pollution?

Ans. Water pollution affects the health of the waterway, the health of the organisms living in and around the water bodies, and, eventually, the health of humans. The effects of water pollution can range from aquatic deformities to contaminated fish to 'dead' lakes. Toxic pollutants can also alter the genetic makeup of an organism, resulting in either death or extreme deformities..

Persistent Organic Pollutants (POPs), such as dioxin, PCBs and DDT, are chemical substances that persist in the environment and bioaccumulate through the food web. Therefore, POPs can also cause sickness and disease in humans, who are at the end of the food chain. People who regularly consume a lot of fish will have larger levels of toxic chemicals in their bodies than those who only eat fish occasionally. Toxic chemicals can lead to reproductive problems, cancer and neurological disorders. People who are most at risk of health problems due to contaminated fish consumption are those with weakened immune systems, including children, pregnant women and the elderly.

Water-borne diseases are caused by drinking water contaminated by human or animal faeces, which contain pathogenic microorganisms. There are many types of water-borne diseases, like Cholera, Dengue, Diarrhoea, Leptospirosis, Malaria, Hepatatis, Scabies, Typoid and Fluorosis. **Fluorosis** is a dental problem which attacks the children by making the brownish tinch on their teeth. The amount of nutrients entering the water bodies has intensified due to increased urbanization and agriculture, leading to increased biological growth, or eutrophication. Under eutrophic conditions, nutrient loading (more nutrients than the water body can handle) stimulates excessive plant growth, which in turn decreases the amount of oxygen in the water and kills off certain species of animal life.

Freshwater resources all over the world are threatened not only by over exploitation and poor management but also by ecological degradation. The main source of freshwater pollution can be attributed to discharge of untreated waste, dumping of industrial effluent, and run-off from agricultural fields. Industrial growth, urbanization and the increasing use of synthetic organic substances have serious and adverse impacts on freshwater bodies. The developed countries suffer from problems of chemical discharge into water sources mainly groundwater, while developing countries face problems of agricultural run-off in water sources.

Q.37. Describe the causes, sources, effects and control measures of water pollution.

Ans. Water pollution is defined as alteration in physical, chemical or biological characteristics of water making it unsuitable for designated use in its natural state.

Causes of water pollution: Water is an essential commodity needed for drinking, cooking, bathing, washing, irrigation, and for industrial operations. Most water for such uses comes from rivers, lakes or ground water sources. Water dissolves many substances in it, therefore, it can easily get polluted. Pollution of water can be caused by point sources or non-point sources. Point sources are specific sites near water bodies which directly discharge effluents into them. Major point sources of water pollution are industries, power plants, underground coal mines, offshore oil wells etc. The discharge from non-point sources is not at any particular site, rather, these sources are scattered, which individually or collectively pollute water. Surface run-off from agriculture fields, overflowing small drains, rain water sweeping roads and fields, atmospheric deposition etc. are the non-point sources of water pollution.

Sources of water pollution

Ground water pollution: Ground water forms about 6 % of the total water available on earth and is about 30 times more than surface water. Ground water seems to be less prone to pollution as the soil mantle through which water passes helps to retain various contaminants due to its caption exchange capacity. However, there are a number of potential sources of ground water pollution. Septic tanks, industry, deep well injection, mining etc. are mainly responsible for ground water pollution, which is irreversible. Ground water pollution with arsenic, fluoride and nitrate are posing serious health hazard.

Surface water pollution: The major sources of surface water pollution are:

Sewage: Pouring the drains and sewers in fresh water bodies causes water pollution. The problem is severe in cities.

Industrial effluents: Industrial wastes containing toxic chemicals, acids, alkalis, metallic salts, phenols, cyanides, ammonia, radioactive substances, etc. are sources of water pollution. They also cause thermal pollution of water.

Synthetic detergents: Synthetic detergents used in washing and cleaning products foam and pollute water.

Agrochemicals: Agrochemicals like fertilizers and pesticides washed by rain water and surface run off pollute water.

Oil: Oil spillage into sea-water during drilling and shipment pollute it.

Waste heat: Waste heat from industrial discharges increases the temperature of water bodies and affects distribution and survival of sensitive species.

Effects of water pollution. Organic matter which reaches water bodies is decomposed by micro-organisms present in water. For this degradation oxygen dissolved in water is consumed. Dissolved oxygen (DO) is the amount of oxygen dissolved in a given quantity of water at a particular temperature and atmospheric pressure. Amount of dissolved oxygen depends on aeration, photosynthetic activity in water, respiration of animals and plants and ambient temperature.

The saturation value of DO varies from 8-15 mg/l. For active fish species 5-8 mg/l of DO is required whereas less desirable species like carp can survive at 3.0 mg/l of DO.

Lower DO may be harmful to animals especially fish population. Oxygen depletion helps in release of phosphates from bottom sediments and causes eutrophication.

Control of water pollution. Following steps will help reduce water pollution:

(*i*) Prevent run-off of manure. Divert such run off to basin for settlement. The nutrient rich water can be used as fertilizer in the fields.

(*ii*) Separate drainage of sewage and rain water should be provided to prevent overflow of sewage with rain water.

(*iii*) Planting trees would reduce water pollution by sediments and will also prevent soil erosion.

For controlling water pollution from point sources, treatment of waste waters is essential before being discharged. Parameters which are considered for reduction in such water are- Total solids, biological oxygen demand (BOD), chemical oxygen demand (COD), nitrates and phosphates, oil and grease, toxic metals etc.

Waste waters should be properly treated by primary and secondary treatments to reduce the BOD, COD levels up to the permissible levels for discharge.

Advanced treatment for removal of nitrates and phosphates will prevent eutrophication. Before the discharge of waste water, it should be disinfected to kill the disease-causing organisms like bacteria. Proper chlorination should be done to prevent the formation of chlorinated hydrocarbons or disinfection should be done by ozone or ultra violet radiations.

Q.38. Describe the causes, effects and control of thermal pollution of water.

Ans. Thermal pollution is defined as presence of waste heat in the water which increases its temperature can cause undesirable changes in the natural water environment.

Causes of thermal pollution: Heat producing industries *i.e.*, thermal power plants, nuclear power plants, refineries, steel plants etc. are the major sources of thermal pollution. Power plants utilize only 1/3 of the energy provided by fossil fuels for their operations. Remaining 2/3 is generally lost in the form of heat to the water used for cooling. Cold water, generally, is drawn from some nearby water-body, passed through the plant and returned to the same water-body, with temperature 10-12°C higher than the initial temperature. Excess of heat reaching such water bodies causes thermal pollution of water.

Effects of thermal pollution:

(*i*) The dissolved oxygen content of water is decreased as the solubility of oxygen in water is decreased at high temperature. High temperature becomes a barrier for oxygen penetration into deep cold waters.

(*ii*) Toxicity of pesticides, detergents and chemicals in the effluents increases with increase in temperature.

(*iii*) The composition of flora and fauna changes because the species sensitive to increased temperature due to thermal shock will be replaced by temperature tolerant species.

(*iv*) Metabolic activities of aquatic organisms increase at high temperature and require more oxygen, whereas oxygen level falls under thermal pollution.

(*v*) Discharge of heated water near the shores can disturb spawning and can even kill young fishes.

(vi) Fish migration is affected due to formation of various thermal zones.

Control of thermal Pollution: Following methods can be employed for control of thermal pollution:

(*i*) Cooling ponds (*ii*) Spray Ponds

(*iii*) Cooling towers

Water from condensers is stored in ponds where natural *evapor*ation cools the water which can then be re circulated or discharged in nearby water body. Spray ponds and cooling towers can be used to cool heated water before being discharged into water bodies e.g., rivers, sea etc.

Q.39. Describe the sources, effects and control of marine pollution.

Ans. Sources: The main sources of marine pollution are (*i*) rivers, which bring pollutants from their drainage basins, (*ii*) Catchment area *i.e.*, coastline where human settlements in the form of hotels, industry, agricultural practices have been established, and (*iii*) oil drilling and shipment.

Most of the rivers ultimately join the ocean. The pollutants which these rivers carry from their drainage basins are finally poured into the sea. These include sewage sludge, industrial effluents, synthetic detergents, agro-chemicals, solid wastes, plastics, metals and waste heat released by industries. *I*n the sea the pollutants get diluted and the organic matter is further broken down as in river water. Still many pollutants specially the recalcitrant once remain unchanged or are partially degraded causing marine pollution. These pollutants get bio-magnified and affect fisheries and other marine life. Another important source of marine pollution is the leaking toxic substances, radioactive wastes etc. which are stored in large containers and dumped in deep sea considering sea to be a better disposal site than land.

Tankers and other shipping means, industries, lubricating oil using industry, metal industry, paint industry, automotive wastes, refineries, ship-accidents and off shore production add to marine pollution. Tankers transporting oil contribute to oil pollution significantly. After delivering the oil through sea-route, earlier empty tankers used to be filled with water called ballast water to maintain balance. The ballast-water containing residual oil from tankers are released into the sea on completion of return journey. Now-a-days the oil floating on the ballast water is removed in the newly designated 'load on top tankers' before ballast water is let off.

Effects: Oil in sea water can spread over a large area of the sea, remain dispersed or get adsorbed on sediments. It can cause adverse effects on marine life.

Oil in sea water affects sensitive flora and fauna. Phytoplankton, zooplankton, algal species, and various species of invertebrates, coral reefs, fish, birds and mammals are after the slimy mucus and gills is affected. Oil disrupts the installing capacity of feathers. Death occurs due to loss of buoyancy and subsequent drowning of birds. Leakage from oil tanker near Alaska in 1989 caused damage to coral reefs and resulted in death of lakhs of birds. Some important incidents of bird mortality due to oil have been reported at Brittany, France where 20 thousand birds died due to more than 220 tons of oil spillage in 1978. At Elbe, Germany 500 thousand birds in 1955. During the 1991 Gulf War 200 million gallons of oil spread in the Persian Gulf badly affected the marine ecosystem adversely.

Control:

(*i*) Toxic pollutants from industries and sewage treatment plants should not be discharged in coastal waters.
(*ii*) Run off from non-point sources should be prevented to reach coastal areas.
(*iii*) Sewer overflows should be prevented by having separate sewer and rain water pipes.
(*iv*) Dumping of toxic, hazardous wastes and sewage sludge should be banned.
(*v*) Development activities on coastal areas should be minimized.
(*vi*) Oil and grease from service stations should be processed for reuse.
(*vii*) Oil ballast should not be dumped into sea.
(*viii*) Ecologically sensitive coastal areas should be protected by not allowing drilling.

Q.40. What are major water borne diseases?

Ans. Major water borne diseases are described below:

Cholera: Cholera is an acute, diarrheal illness caused by infection of the intestine with the bacterium Vibrio cholerae. A person may get cholera by drinking water or eating food contaminated with the cholera bacterium. This disease spreads very rapidly in areas with inadequate treatment of sewage and drinking water. The cholera bacterium may also live in the environment in brackish rivers and coastal waters.

Approximately 1 in 20 infected people has severe disease characterized by profuse watery diarrhoea, vomiting and leg cramps. In these people rapid loss of body fluids leads to dehydration. Without treatment death can occur within hours.

Dengue: Dengue is an acute infectious disease caused by a virus and transmitted by the bite of the *Aedes* mosquito, also known as *break bone fever* and *bone-crusher disease*. Dengue fever is a severe, flu-like illness that affects infants, young children and adults but rarely causes death.

The symptoms, following an incubation period of five to eight days, are fever, chills, and severe headache, pain in the joints, sweating, and prostration. The clinical features vary according to the age of the patient. Infants and young children may have a feverish illness with rash. Older children and adults may have either a mild feverish illness, or the classical incapacitating disease with abrupt onset and high fever, severe headache, pain behind the eyes, muscle and joint pains and rash.

Dengue hemorrhagic is a potentially lethal complication and is a leading cause of childhood death in several Asian countries. It is characterized by high fever, hemorrhage, enlargement of the liver, and in the most severe cases circulatory failure. Today, dengue is found in tropical and subtropical regions around the world, predominately in urban and per urban areas, where Aedes mosquitoes are prevalent.

Diarrhoea: Diarrhea is frequent discharge of watery faeces from the intestines, sometimes containing blood and mucus. Diarrhoea due to infection may last a few days, or several weeks. Persistent diarrhoea results in severe dehydration. It is therefore necessary to replace the fluid lost by the body. Severe diarrhoea may be life threatening due to fluid loss in watery diarrhoea, particularly in infants and young children, the malnourished and people with impaired immunity.

Diarrhoea is a concomitant of many infectious diseases, especially typhoid fever, bacillary or amoebic dysentery, and cholera. Diarrhoea is a symptom of infection by a host of bacterial, viral and parasitic organisms most of which can be spread by contaminated water. Diarrhoea due to infection is widespread throughout the developing world.

Fluorosis: Fluorosis is an abnormal condition caused by excessive intake of fluorine from fluoridated drinking water, characterized by mottling of the teeth. Moderate level chronic exposure (above 1.5 mg/l of water) is quite common. Long-term ingestion of large amounts can lead to potentially severe skeletal problems (*skeletal fluorosis*). The early symptoms of skeletal fluorosis include stiffness and pain in the joints. In severe case, the bone structure may change and ligaments may calcify, with resulting impairment of muscles and pain. Acute high-level exposure to fluoride causes immediate effects of abdominal pain, excessive saliva, nausea and vomiting. Seizures and muscle spasm may also occur.

Fluorosis affects millions of people around the world, but as regards dental fluorosis the very mild forms are the most frequent. Waters with high levels of fluoride content are mostly found at the foot of high mountains and in areas where the sea has made geological deposits.

Hepatitis: Hepatitis is any disease featuring inflammation of the liver. Two of the viruses that cause hepatitis (hepatitis A and E) can be transmitted through water, food and from person to person. Hygiene is therefore important in their control. The illness starts with an abrupt onset of fever, body weakness, loss of appetite, nausea and abdominal discomfort, followed by jaundice within a few days. The disease may range from mild (lasting 1-2 weeks) to severe disabling disease (lasting several months).

Both hepatitis A and E are found worldwide. Hepatitis A is particularly frequent in countries with poor sanitary and hygienic conditions.

Leptospirosis: Leptospirosis is an infectious disease of domestic animals, especially cattle, swine, and dogs, caused by spirochetes of the genus Leptospira and characterized by jaundice and fever. It is transmissible to humans through direct contact with the urine of infected animals or by contact with a urine-contaminated environment, such as surface water, soil and plants.

In humans it causes a range of symptoms, while some infected people may have no symptoms at all. Symptoms of leptospirosis include high fever, severe headache, chills, muscle aches, and vomiting, and may include jaundice, red eyes, abdominal pain, diarrhoea, or a rash. Complications include renal failure, meningitis, liver failure, and respiratory distress. In rare cases death occurs. Leptospirosis occurs worldwide, in both rural and urban areas and in temperate and tropical climates.

Malaria: Malaria is the world's most important parasitic disease transmitted from one person to another through the bite of female *Anopheles* mosquitoes, which breed in fresh or occasionally brackish water.

Its symptoms are at the onset of malaria, bouts of chills and fever lasting several hours and occurring every three or four days. If the disease is not treated, the spleen and the liver become enlarged, anaemia develops, and jaundice appears. General debility, anaemia, or clogging of the vessels of cerebral tissues by affected red blood cells can be followed by death. Malaria occurs mostly in tropical and subtropical countries.

Scabies: Scabies is a highly contagious parasitic skin disease caused by the itch mite (*Sarcoptes scabiei*). It is acquired through close contact with an infested individual or contaminated clothing and is most prevalent among those living in crowded and unhygienic conditions. Personal hygiene is an important preventive measure and access to adequate water supply is important in control. The principal sign of the disease is a pimple-like rash that is most commonly found on hands, especially the webbing between the fingers, the skin folds of the wrist, elbow or knee, the penis, the breast or the shoulder. Infestation causes intense itching all over the body, especially at night.

Scabies mites are found worldwide, affecting all socioeconomic classes and in all climates. Epidemics have been linked to poverty, poor water supply, sanitation, and overcrowding.

Typhoid: Typhoid and paratyphoid enteric fever are acute, generalized infections caused by *Salmonella typhi* and *Salmonella paratyphoid* respectively. The main sources of infection are contaminated water or milk and, food handlers who are carriers. Their germs are passed in the faeces and urine of infected people. People become infected after eating food or drinking beverages that have been handled by a person who is infected or by drinking water that has been contaminated by sewage containing the bacteria.

Once the bacteria enter the person's body they multiply and spread from the intestines, into the bloodstream. The symptoms of typhoid appear 10 to 14 days after infection; and include high fever, rose-colored spots on the abdomen and chest, diarrhoea or constipation, and enlargement of the spleen and liver. In untreated patients, complications may be numerous, affecting practically every body system, and can even include perforation of the intestine with haemorrhage. Typhoid and paratyphoid fevers are common in less-industrialized countries, mainly due to the problem of unsafe drinking water, inadequate sewage disposal and flooding.

Q.41. What are the major sources of water pollution?

Ans. Principal sources of water pollution are:

- industrial discharge of chemical wastes and by-products.
- discharge of poorly-treated or untreated sewage.
- surface runoff containing pesticides.
- slash and burn farming practice, which is an element within shifting cultivation agricultural systems.
- surface runoff containing spilled petroleum products.
- surface runoff from construction sites, farms, or paved and other impervious surfaces e.g., silt.
- discharge of contaminated and/or heated water used for industrial processes.
- acid rain caused by industrial discharge of sulfur dioxide (by burning high-sulfur fossil fuels).
- excess nutrients added by runoff containing detergents or fertilizers.
- underground storage tank leakage, leading to soil contamination, thence aquifer contamination.

Q.42. What are the organic and inorganic pollutants present in polluted water?

Ans. Some organic water pollutants are:

- insecticides and herbicides, a range of organohalide and other chemicals.
- bacteria, often is from sewage or livestock operations;
- food processing waste, including pathogens.
- tree and brush debris from logging operations.
- VOCs (Volatile Organic Compounds, industrial solvents) from improper storage.

Some inorganic water pollutants are:

- heavy metals including acid mine drainage.
- acidity caused by industrial discharges (especially sulfur dioxide from power plants).
- chemical waste as industrial by products.
- fertilizers, in runoff from agriculture including nitrates and phosphates.
- silt in surface runoff from construction sites, logging, slash and burn practices or land clearing sites.

Q.43. What are main point sources and non-point sources of water pollution?

Ans. Point Sources: They include transfer of pollutants from municipal and industrial liquid waste disposal sites and from municipal and household hazardous waste and refuse disposal sites.

Non-point Sources: They include:

Diffuse agricultural sources: Wash off and soil erosion from agricultural lands carrying materials applied during agricultural use, mainly fertilizers, herbicides and pesticides.

Diffuse urban sources: Run off from city streets, from horticultural, gardening and commercial activities in the urban environment and from industrial sites and storage areas.

Q.44. Describe the transport and chemical reactions of water pollutants?

Ans. Most water pollutants are eventually carried by the rivers into the oceans. There are areas of oxygen depletion, caused by chemicals using up oxygen and by algae blooms, caused by excess nutrients from algal cell death and decomposition. Fish and shellfish kills have been reported, because toxins climb the food chain after small fish consume copepods, then large fish eat smaller fish, etc. Each step up the food chain concentrates certain toxins like heavy metals and DDT by approximately a factor of ten.

Many chemicals undergo reactive decay / change over long periods of time in groundwater reservoirs. Such chemicals include chlorinated hydrocarbons such as trichloroethylene (used in industrial metal degreasing) and tetrachloroethylene used in the dry-cleaning industry. Both of these chemicals, which are carcinogens, undergo partial decomposition reactions leading to new hazardous chemicals.

Groundwater pollution is much more difficult to abate than surface pollution because ground water can move great distances through unseen aquifers. Non-porous aquifers such as clays partially purify water of bacteria by simple filtration (adsorption and absorption), dilution, and, in some cases, chemical reactions and biological activity. However, in some cases, the pollutants merely transform to soil contaminants. Ground water that moves through cracks and caverns is not filtered and can be transported as easily as surface water.

There are a variety of secondary effects stemming not from the original pollutant, but a derivative condition. Some of these secondary impacts are:

- Silt bearing surface runoff can inhibit the penetration of sunlight through the water column, hampering photosynthesis in aquatic plants.
- Thermal pollution of water can induce fish kills and invasion by new thermophilic species.

Q.45. What are the effects of water pollution on aquatic system?

Ans. Water pollution affects the health of the waterway, the health of the organisms living in and around the waterway, and, eventually, the health of humans. The effects of water pollution on aquatic systems can range from aquatic deformities to contaminated fish to 'dead' lakes. Heavy metals like mercury and lead, and human-made organic chemicals such as pesticides, bio magnify as they move up the food chain, resulting in tumors and death for predatory animals, such as lake trout, herring gulls, and even humans.

Toxic pollutants can also alter the genetic makeup of an organism, resulting in either death or extreme deformities. Cormorants suffer from cross-billed syndrome, while terns exhibit birth defects from dioxin, PCBs (Poly Chlorinated Biphenyls) and furan exposure Other examples of deformities include large fish tumors and three-legged frogs.

Q.46. What is eutrophication and its effects?

Ans. Before the introduction of chemical fertilizers the water contained little plant nutrients and were clear due to their size and depth. These lakes support high levels of animal life and receive proper amounts of nutrients, mainly phosphorous and nitrogen, from natural sources, such as decomposing plant matter. Urbanization, agricultural revolution and industrialization has changed the situation. The amount of nutrients entering the water bodies has increased greatly, mainly due to increased urbanization and agriculture, leading to increased **biological growth, or eutrophication**.

Under eutrophic conditions, nutrient loading (more nutrients than the waterbody can handle) stimulates excessive plant growth, which in turn decreases the amount of dissolved oxygen in the water and eventually kills off certain species of animal life. Other pollution-tolerant species, such as worms and carp, grow more rapidly; thus, the ecological balance of the lake is altered.

Q.47. What are various water treatment techniques employed for potable (drinking) water?

Ans. Any physical, biological, or chemical change in water quality that adversely affects living organisms or makes water unsuitable for desired uses is considered pollution. The most serious water pollutants for human health are pathogenic organisms from human and animal wastes. We have traditionally taken advantage of the capacity of ecosystems to destroy these organisms, but as the population density has increased, these systems have become overloaded and ineffective necessitating water treatment to make it usable.

In industrialized nations, toxic chemical wastes have become an increasing problem. Agricultural and industrial chemicals have been released or spilled into surface waters and are seeping into groundwater supplies.

The water treatment processes are classified as:

- **aeration;**
- **sedimentation;**
- **coagulation and flocculation;**
- **filtration;**
- **disinfection.**

Potential industrial and agricultural contaminants (chemicals such as oil, phosphates, sulphates, heavy metals, etc.) end up in water resources in increasing amounts causing water pollution. These contaminants must be removed if the water is to be made potable. Since water treatment presents the most demanding component of a water supply system, it must be examined whether alternative methods exist that yield a measure of water quality improvement: protection of the water source from contamination, as well as rehabilitation, upgrading and systematic monitoring of already existing works, and construction of efficient sanitation facilities.

Q.48. What is the basic purpose and range of applications of aeration in water treatment?

Ans. The basic purpose of aeration is the reduction of substances which cause unpleasant tastes and odours as well as discoloration in water. Aeration is used for treatment of groundwater where it also has additional positive side effects (precipitation of iron and manganese). When treating surface water aeration is useful in adding oxygen to the raw water. Aeration always precedes some other treatment process.

Aeration equipment is used to intensively mix air and water to facilitate the transfer of gases into or out of the water. The following effects can be achieved by aeration of water:

— **Addition of oxygen:** this may be necessary for surface water where the natural oxygen content was depleted due to the presence of large amounts of organic substances. Aeration contributes positively to subsequent biological treatment (e.g., slow sand filtration).

— **Removal of dissolved iron and manganese:** These are oxidized and form nearly insoluble hydroxide sludges. They can be removed in a settling tank or by means of a coarse filter.

— **Removal of excess carbon dioxide** (CO_2) to prevent corrosion of metal and concrete surfaces.

— **Reduction of H_2S, CH_4 and other volatile compounds** which produce objectionable taste and odour.

— **Temperature reduction.**

Q.49. Describe the role and area of applications of sedimentation in water treatment.

Ans. Sedimentation occurs in nature perpetually and aids the natural purification of lakes and rivers. Use is made of this physical process in the treatment of water by passing it through settling basins or storage tanks at low and uniform velocities. This is a simple means of reducing the contents of suspended matter and partially of bacteria. It can be combined by preceding it with coagulation and flocculation, and succeeding it with slow sand filtration. Following these procedures, disinfection is required for high bacteria contents.

Areas of application of sedimentation

Turbidity control: Under the influence of gravity, suspended matter in rain water settles out if it has a density greater than that of the water itself. The efficiency of a settling basin depends on the nature (shape, size, density) of the particles that are accountable for the turbidity; gravity, sand and silt, which pollute surface waters heavily and settle easily, especially during the rainy season. Colloidal matter which contributes much to turbidity is held in suspension mainly by electrostatic forces and because of its low density. Colloidal particles, when brought in contact with coagulants, form flocculent material that can be settled or filtered out..

Pathogenic organisms removal: Simple sedimentation by passing water through a settling tank does not achieve a significant removal of pathogens. Two to four weeks storage, though, can reduce bacteria populations considerably (50- 90%) by means of biological processes. Storage in excess of one month can reduce the viral count. The degree of purification depends on the severity of pollution and on the presence of other pollutants. Storage induced contamination (mosquito breeding due to algal growth) must be avoided by covering tanks. Schistosoma larvae, infectious agents of Bilharzia, usually cannot survive more than two days of protected storage, provided suitable hosts (snails) are not present.

Color removal: Removal of color without the use of chemical procedures can only be achieved by very long storage times.

Q.50. What is the purpose of coagulation and flocculation in water treatment?

Ans. Finely dispersed, suspended and colloidal particles causing turbidity and color of the water cannot be removed sufficiently by the ordinary sedimentation process. Adding a coagulant and mixing and stirring the water causes the formation of settleable particles. These flocs are large enough to settle rapidly under the influence of gravity, and may be removed from suspension by filtration. The choice and dose rates of coagulants depend on the characteristics of the water to be treated. The coagulation chemicals like aluminium sulfate should be readily available and their application must be closely monitored. On the household level, coagulation by means of natural coagulants of plant and soil origin and simple devices has been practiced traditionally in developing countries.

Q.51. Describe the mechanisms of coagulation.

Ans. Colloidal particles (dia: 10^{-4} to 10^{-6} mm) generally carry a negative electrical charge and are surrounded by an electrical double layer (due to attachment of positively charged ions from the ambient solution) and thus inhibit the close approach of each other. They remain finely divided and don't agglomerate. Due to their low specific gravity, they don't settle out. A coagulant (generally positively charged) causes compression of the double layer and thus the neutralization of the electrostatic surface potential of the particles. The resulting destabilized particles stick sufficiently together when contact is made. Rapid mixing (a few seconds) is important at this stage to obtain uniform dispersion of the chemical and to increase the opportunity for particle-to-particle contact. Subsequent gentle and prolonged (several minutes) mixing cements the still microscopic coagulated particles into larger flocs. These flocs then are able to aggregate with suspended polluting matter. When increased sufficiently in size and weight, the particles settle to the bottom.

Q.52. What are various coagulants commonly used in water treatment?

Ans. Coagulants used include chemicals, materials of soil origin, materials of plant origins and natural coagulants.

Chemicals: It is most common practice to use aluminium and iron salts as coagulant in water treatment.. Both salts hydrolyze when added to water and form insoluble materials: aluminium and ferric hydroxides – when reacting with calcium and magnesium carbonates, which are almost always present in water. If these carbonates are not present in sufficient **quantity** (as in soft water), hydrated lime $Ca(OH)_2$ or sodium carbonate Na_2CO_3 may be added also. The formation of the insoluble hydroxides depends on the pH. Aluminium sulphate coagulates best in a pH range between 4.4 and 6. At higher pH values, higher rates of soluble aluminate ions form. Sodium aluminate is generally used at medium pH values (6.5 to 8). Irons salts are effective over a wide range of pH values (except for values between 7 and 8.5). Whereas turbidity is best removed within a pH range of 5.7 to 8.0, color removal is generally obtained at acid pH of about 4.4 to 6.0. To improve the coagulation and flocculation process and to reduce the dose of coagulants, **flocculation aids** may be used. The most commonly used material is activated silica. Yet diatoms (kieselgur), activated carbon in powder form, bentonite and certain other types of adsorbtive clays, organic substances and derived materials are also used.

Materials of soil origin: Mineral substances are used as flocculation aids in modern water treatment. A dose of 10 mg/l of bentonite, for instance, together with 10 mg/l of aluminium sulphate, yield significantly better results than a higher dose of aluminium sulphate alone. In rural households in developing countries, however, various naturally occurring materials are traditionally used as coagulants: e.g., fluvial clays from rivers, clarifying rock material from desert regions, and earth from termite hills. Their main constituents are quartz, montmorillonite, kaolinite, calcite and feldspar. Their coagulating mechanisms differ greatly from those of metal salts.

Application of-clay as a coagulant yields the following results:

— reduction of turbidity;
— no effect on pH value;
— an initial mineral taste, later on normal;
— no effect on bacteria count.
— Clays contain traces of heavy metals (mostly chromium and manganese). High intakes of these metals may have toxic effects;
— viruses survive in the settled sludge.

Coagulants of plant origin. Such substances are widely used in developing countries to purify water. Usually the plants are not cultivated. Rather, according to passed on experience, certain substances are gathered, prepared and added to the water that is to be purified. Some examples of traditionally used coagulants and coagulant aids are:

— seeds from the Indian Nirmali tree (strychnos potatorum);
— seeds of the trees of the family of the Moringaceae: Moringa Olifeira, occurring in India, Senegal, Sudan (Behenus tree) and Moringa Stenopetala, Kenya;

— sap from the stem of the tuna cactus (opuntia ficus indica) occurring in Peru and Chile.
— the bark of the south American tree Schinopsis Quebracho-Colorado which contains tannin: it is known commercially as "Floccatan ;"
— potato starch.

For most of these plant materials, it is not known which particular substance actually triggers the coagulation. Neither is it known whether there are toxic side effects from frequent use. To obtain the optimal dose for various substances and raw water qualities, coagulation experiments must be carried out; generally this dose is smaller than that of aluminium sulphate.

For coagulation with Moringa Olifeira seeds, the following effects have been achieved.

— significant reduction of turbidity; — pleasant taste;
— unchanged pH value;
— initial reduction of the bacteria count, followed by a secondary rise after only 24 hours, reaching or even surpassing the initial concentration;
— antibiotic effect on various bacteria and fungi.

Nirmali seeds and Tunaflex as natural coagulants and aid substances combined with alum salts have been successfully used in municipal water treatment. Substantial savings in primary coagulants can be achieved which, reduces the overall cost considerably.

Other natural coagulants include:

— Algae-derived substances;
— Chitosan, acting faster than any known coagulant from plant materials (produced from the shells of shrimp and lobster);
— dough from millet bread (Sudan) or curds (thin layers).

Q.53. What is filtration and types of filters used in water treatment?

Ans. Filtration is the passage of polluted water through a porous medium, thus utilizing the principle of natural cleansing of the soil. This technique used in water treatment is based on several simultaneously occurring phenomena:

— mechanical straining of undissolved suspended particles (screening effect);
— charge exchange, flocculation, adsorption of colloidal matter;
— bacteriological-biological processes within the filter.

Filters may be divided into two principally different types:

— slow sand (or biological) filtration ($v = 0.1$ to 0.3 m/h),
— rapid filtration ($v = 4$ to 15 m/h).

In-between types also exist. Depending on the filtration rate, different mechanisms are operative within the filter. Generally, a filter consists of the following components:

— filter. medium (inert medium: quartz sand; or chemically activated medium: burnt material),
— support bed (gravel) and under-drain system,
— influent and effluent pipes, -wash and drain lines, -control and monitoring appurtenances.

Q.54. Describe the principle, process and mechanism of rapid filtration?

Ans. Rapid filtration is based on the principle of mechanical straining of suspended matter due to the screening effect of the filter bed (sand, gravel, etc.). The particles in the water pass into the filter bed and lodge in the voids between grains of the filter medium. The cleaning of the rapid filter is facilitated by backwashing *i.e.*, by reversing the flow direction; a backwash may be done simply with water or by use of a water-air mix (upward air scour). The impurities are thus dislodged and removed from the filter bed.

The performance of a rapid filter regarding the removal of suspended matter is determined by the following filtration process variables and parameters:

— filtration rate (v),
— influent characteristics, *i.e.*, particle size, distribution, etc.,
— filter medium characteristics which control the removal of the particles and their release upon backwashing, respectively.

Generally, the treatment effect can be improved by:

— reduced filtration rates,
— smaller granulation size of the filter medium,
— increasing depth of the filter bed,
— decreasing concentration of particles to be retained.

Q.55. What are various types of rapid filters used in water treatment?

Ans. Various types of rapid filters used in water treatment are described below:

(*a*) Conventional (down flow) filters: Rapid filtration is complex, demanding and expensive process. This is due to the need for frequent filter washing which requires elaborate back washing systems. Additional complexities associated with the generation of pressure arise for pressure filters. Combined with coagulation, flocculation and sedimentation, rapid filtration is a very efficient treatment process for the removal of impurities. However, it should only be used in larger plants. For smaller plants in rural areas, simple rapid filters -without backwashing facilities are recommended. A number of filter types operating at filtration rates lower than those for conventional filers are described below. Generally, they serve as pretreatment units to reduce the turbidity of water. The removal of pathogens requires, in addition, either slow sand filtration and/or disinfection.

(*b*) Up flow filters: In up flow filters, the direction of flow of the raw water is upwards through the filter bed. Backwashing is done by reversal of the flow direction. The effect of the filter depends on the type of the filter medium, the filtration rate, and preceding aeration or addition of a coagulant. For coarse organic and inorganic substances, the filter may act as a simple screen. Or else it may retain precipitated iron compounds. At low filtration rates and sufficient oxygen content of the raw water, biological activity can be observed.

The advantages of up flow filters as compared with gravity down flow rapid filters are:

— can be constructed from locally available materials,
— quality requirements (uniformity and gradation) and volume of the filter medium are lower. Instead of sand, gravel, crushed bricks, coconut and other type fibers can be used,
— longer filter runs,
— better turbidity removal.

Up flow filters can be constructed at a variety of degrees of complexity. A rather simple type can be built from a drum equipped with a raw water inlet pipe, a somewhat larger size drain at the bottom, and an outlet pipe for the clarified water near the top of the drum

Filtration effect: Reduction of between 50 and 70% of organic and inorganic coarse and fine particles, slight reduction of bacteria.

Filtration rate: 0.5 to 1.5 m/in.

Filter medium: Coarse sand, grain size between 3- and 4-mm diameter.

Filter bed depth: 0.3 m.

Support layer and under drain: gravel covered by perforated metal tray.

Cleaning: Shut off of the inlet. Quick removal of drain stopper so that supernatant as well as water in the filter bed drain out together with retained particles.

Cleaning of the filter which takes ten minutes should be done every day for preventing the filter bed from clogging.. As bacteria cannot be sufficiently removed, subsequent disinfection is indispensable in case of bacterial water contamination. Higher filtration rates result in higher buoyancy forces on the filter medium. The top layer of the sand may be spewn up. This can be avoided by covering the filter bed with a metal grate or by raising the depth of the filter bed. In the latter case though, backwashing by means of simply draining the water in a reversed direction may become increasingly impossible. Conventional backwashing facility have to be added. Better results may be obtained by using smaller grains and stratified filter beds with decreasing grain size from bottom to top (e.g., 0.7 to 2 mm over a depth of 1 to 1.5 m).

(*c*) **Coarse filters:** Rapid filters preceding slow sand filters are frequently used to retain coarse particles and to sufficiently reduce turbidity. Coarse sand, gravel or plant fibres are used as a filter medium. It can be replaced upon cleaning. Such pre-filtration can be done either horizontally or vertically. The filtration rates for a coarse filter are lower than those for a conventional rapid filter.

Filtration effect: Reduction of turbidity by between 50 and 80% (max. load 250 NTU)

Filtration rate: 0.5 to 1.0 m/h

Filter box: same as slow sand filter

Operating head: 1.0 to 1.5 m.

Filter medium: coarse sand, gravel, shredded coconut fibres. Two or more layers of different material possible (coarser material up top and finer material below).

Filter bed depth: 0.8 to 1.4 m.

Drainage system: same as slow sand filter.

Cleaning: replace medium completely when head loss exceeds certain value, *i.e.*, when too big (approximately once every 3-4 months).

(*d*) **Household size rapid filter:** Household filters can be made from sand or gravel of different grain sizes, from ceramics, porcelain or other fine porosity materials. They operate on the principle of mechanical straining of the particles contained in the water. The filter performance depends on the porosity of the filter medium. Through additives in the filter material, additional effects can be obtained (eg, adsorption, disinfection).

(*e*) **Multiple layer filter:** Using metal drums, plastic containers or clay vessels and filling them with several layers of sand, gravel or charcoal, simple household filters can be put together. They do not perform well at removing pathogens, though. After filtration, the water therefore needs to be disinfected. Charcoal adsorbs organic substances which cause disagreeable color and taste. This effect can only be sustained, however, if the charcoal is frequently renewed. If this is not possible, or if the filter (empty or filled with water) is left unused for some time, the charcoal can become a breeding ground for bacteria. The result is that the filtered water exhibits a higher bacterium count than the raw water. Monitoring of the filter condition is rendered more difficult by the fact that there is no visual indication given for the point when charcoal should be replaced. Charcoal cannot be regenerated. It is for these reasons that the use of filters with charcoal media is not recommended.

(*f*) **Ceramics filter:** For the household purpose ceramics filters may be used for the purification of drinking water The purifying agent is a filter element, also called candle, through which the water is passed. Suspended particles are thus mechanically retained, and, depending on the size of the pores, also pathogens. Ceramics filters should only be used if the water is not too turbid, as the pores clog quickly. Ceramics filter elements can be made from various different material compositions (e.g., diatomaceous earth, porcelain); they have pore sizes of between 0.3 and 50 μ. If the pore size is smaller than or equal to 1.5 μ all pathogens get removed with certainty. Post treatment of the water prior to consumption is rendered unnecessary. Filters with larger pores only retain macro organisms such as cysts and worm eggs. The filtered water must be boiled subsequently or otherwise disinfected.

The impurities held back by the candle deposit on the candle's surface. At regular intervals, this coating can be brushed off under running water. After the cleaning, candle should be boiled. Candles made from diatomaceous earth which contain silver, avoid recontamination of purified water due to infestation of the filter material with bacteria laden washing water can be avoided.

Filters operating at atmospheric pressure exhibit a very slow rate of percolation. This can be increased considerably by forcing the water through the medium. Ceramics filters must be checked from time to time for fissures so as to prevent the water from passing through the medium without being filtered.

(*g*) **Cartridge microfilter:** Besides ceramics filters, other microfilters made from fine porosity materials are also available, synthetics, paper, felt like material (pore size between 25 and 50 μ). They are inserted into a bell-like filter device, which is mounted on the top of a water pipe. When the filter material becomes clogged, *i.e.*, used up, it must be discarded and replaced new. Even though these filters are cheaper than ceramics filters, their use is more expensive, since the filter material cannot be regenerated.

Q.56. What is slow sand filtration and its mechanism?

Ans. Slow sand filtration is accomplished by passing raw water slowly - driven by gravity through a medium of fine sand. On the surface of the sand bed, a thin biological film develops after some time of ripening (different from the rapid filter). This film consists of active microorganisms and is called filter skin. It is responsible for the bacteriological purification effect. The slow sand filter is therefore also called "surface filter" or biological filter. The principle purification processes taking place during slow sand filtration are:

Sedimentation: The water body sitting on top of the filter bed acts as a settling reservoir. Settleable particles sink to the sand surface.

Mechanical straining: The sand acts as a strainer. Particles too big to pass through the interstices between the sand grains are retained.

Adsorption: The suspended particles and colloids that come in contact with the surface of the sand grains by following the passage of the water are retained by:

— adhesion to the biological layer,
— physical mass attraction (Van der Waals force), and
— electrostatic and electro kinetic attractive forces (Coulomb forces).

On account of these forces, an agglomerate of opposite charged particles forms within the top layer of sand. This process needs some time of ripening to fully develop.

Biochemical processes in the biological layer are:

— partial oxidation and breakdown of organic substances forming water, CO_2 and inorganic salts,
— conversion of soluble iron and manganese compounds into insoluble hydroxides which attach themselves to the grain surfaces,
— killing of E. Coli and of pathogens.

Organic substances are deposited on the upper layer of sand, where they serve as a breeding ground and food for bacteria and other types of microorganisms. These produce a slimy, sticky, gelatinous film which consists of active bacteria, their wastes and dead cells and partly assimilated organic materials. The dissimilation products are carried away by the water to greater depth. Similar processes occur there. The bacterial activity gradually decreases with depth. Different types of bacteria are normally found at various depths.

Algae can contribute to the breakdown of organic material and bacteria. They can improve the formation of the biological layer (filter skin). In uncovered filters, growth of algae is driven by photosynthesis. The presence of large amounts of algae in the supernatant reservoir of a filter generally impedes the functioning of the filter. Dead cell material may clog the filter. Increased consumption of oxygen due to the presence of dead cell material increases the possibility that anaerobic conditions will occur. There is always a diurnal variation in the oxygen content due to growth and decay of the algae mass. When algae growth is strong, the algae must be either removed regularly or the filter must be covered.

The conditions necessary for those biochemical processes are:

— sufficient ripening of the biological layers,

— uniform and slow flow of water through the filter,

— a depth of the filter bed of 1 m (0.5 m is needed solely for the biochemical process) of specific grain sizes,

— sufficient oxygen in the raw water (at least 3 mg/l) to induce biological activity.

At MPN (Most Probable Number) Contents Greater than 1000 *E. Coli*/100 ml, Raw Water Should Subsequently Be Disinfected. Microbiological processes and chemical activity are very sensitive to changes in temperature. Both slow down under conditions of low temperature. A reduction in filtration rate can compensate for this effect. Under prolonged cold conditions, the filter should be covered to prevent heat loss, and subsequent disinfection should be provided.

Q.57. What is the purpose of disinfection of water in water treatment?

Ans. Water disinfection processes are designed to destroy disease producing organisms by means of disinfectants. Drinking water should be free of pathogenic organisms. Storage, sedimentation, coagulation, flocculation and filtration of water both individually and jointly reduce the contents of bacteria in water to a certain extent. None of these methods can guarantee the complete removal of germs. Disinfection is needed at the end. Water with low turbidity may even be disinfected without any additional treatment for bacteria removal. Groundwater abstracted from deep wells is usually free of bacteria. Surface water and water obtained from shallow wells and open dug wells generally need to be disinfected.

The degree or efficiency of disinfection depends on the method employed and on the following factors influencing the process:

— kind and concentration of microorganisms in the water,

— other constituents of the water which may impede disinfection or render it impossible,

— contact time provided (important for chemical disinfectants, since their effect is not instantaneous, a minimum time of contact is necessary),

— temperature of the water (higher temperatures speed up chemical reactions).

Q.58. What are various means/agents of water disinfection?

Ans. Water disinfection can be accomplished by several means:

— physical treatment: removal of bacteria through slow sand filtration, straining of macro organisms by means of micro screening, application of heat (boiling), storage, etc.

— irradiation, such as UV (Ultra Violet) light,

— metal ions, such as silver (and copper),

— chemical treatment, use of oxidants (halogens and halogen compounds: chlorine, iodine, bromine, ozone, potassium permanganate, hydrogen peroxide, etc.).

Q.59. What are the requisites of a good chemical disinfectant?

Ans. A good chemical disinfectant should have the following capabilities:

— destroy all organisms present in the water within reasonable contact time, the range of water temperature encountered, and the fluctuation in composition, concentration and condition of the water to be treated;

— accomplish disinfection without rendering the water toxic or carcinogenic;

— permit simple and quick measurement of strength and concentration in the water,

— persist in residual concentration as a safeguard against recontamination;

— allow safe and simple handling, application and monitoring;

— ready and dependable availability at reasonable cost.

Though there are potential hazards for the human organism associated with prolonged ingestion of chemicals, the application of chlorine and its compounds for the purpose of water disinfection is the best and most tested compromise when evaluated according to the aforementioned criteria. The other methods differ significantly from each other in terms of their effect, the technological level and particularly in their applicability. They are introduced only briefly.

Q.60. Describe pros and cons of chlorination as a means of water disinfection.

Ans. Chlorination is the addition of chlorine gas or some other oxidizing chlorine compound (sodium or calcium hypochlorite, chlorinated lime, chlorine dioxide etc.) to the water to be treated. The actual disinfection agent is hypochlorous acid (HOCl) which forms when chlorine is added to water: Hypochlorous acid also forms subsequent to dissociation, when chlorinated lime or hypochlorite are added. At pH levels between 3 and 6, hypochlorous acid dissociates poorly. Chlorination is most effective in that range of pH. At pH levels greater than 8, hypochlorite ions predominate or exist almost exclusively. Hence the disinfecting effect drops off rapidly as the pH level increases. Simultaneously with the dissociation, hypochlorous acid partly breaks up, forming monatomic oxygen, which contributes to the oxidizing effect:

Chlorination is the most widely used method for drinking water disinfection. It is effective and economical. The fraction that becomes effective as an oxidizing agent when chlorine or some of its compounds is added to raw water is called "free available" or "active" chlorine. Small amounts of chlorine, due to its ability to penetrate cells of microorganisms, are sufficient to destroy many different strains of bacteria. Many types of viruses and macro-organisms like schistosoma larvae can be killed. A contact time of at least 30 minutes is required, at the end of which the residual chlorine concentration in the water must still be between 0.1 and 0.5 mg/l (= ppm). Amoebic cysts and spores with resistant cell membranes require higher doses and longer contact times.

Chlorine also reacts with many other oxidizable water constituents such as iron and manganese compounds, ammonia, and compounds thereof (forming chloramines), as well as numerous types of organic particles. The presence of these substances reduces the germicidal effect considerably. Sufficient chlorine must be added to the water to make sure that there is a residual concentration to prevent recontamination.

It is advisable to remove or reduce prior to chlorination, those substances by sedimentation and/or filtration which would impede disinfection. Through such pretreatment, helminth eggs (parasitic worms) can be removed which are insensitive to chlorination. Through chlorination, certain undesirable side effects may occur. Particularly in industrialized areas, synthetic organic compounds may enter the hydrologic cycle in high concentrations. The presence of chlorine enhances the danger of the formation of carcinogenic compounds (e.g., chloroform and other trihalomethanes).

Q.61. Describe the merits and demerits of various chemicals used for chlorination of water for its disinfection.

Ans. Chlorine gas and chlorine dioxide are the most widely used in water treatment on account of their high efficiency and ease of application. Handling and transport, however, are considered too demanding and hazardous (explosive, toxic). Several other chlorine compounds which have various active chlorine contents are more easily applicable. In some form or another they are available virtually anywhere.

Sodium hypochlorite (NaOCl), commonly known as bleach is generally available in dissolved form. Its commercial strength in terms of active chlorine is between 1 and 15%. It is stored in dark glass or plastic bottles. The solution loses some of its strength during storage. Prior to use, active chlorine content should be tested. Sunlight and high temperatures accelerate deterioration of the solution. The containers therefore should be stored in cool darkened areas. The stability of the solution decreases with increasing contents of available chlorine. A 1% solution is relatively stable. But it is not economical to store. Even though hypochlorite solutions are less hazardous than chlorine gas, precaution should be taken to avoid skin contact and to protect containers against physical damage.

Chlorinated lime or bleaching powder ($CaO \cdot 3\ CaOCl_2 \cdot 3\ H_2O$) in the powder form is readily available and inexpensive. It is stored in corrosion resistant cans. When fresh, it contains 35% active chlorine. When exposed to air, it quickly loses its effectiveness. It is usually applied in solution form which is prepared by adding the powder to a small amount of water to form a soft cream. Stirring prevents lumping when more water is added. When the desired volume of the solution has been prepared, it is allowed to settle before decanting. Solutions should have concentrations between 5 and 1% of free chlorine, the latter being the most stable solution. Some 10% of the chlorine remains in the settled **sludge.**

High test hypochlorite (HTH) is a stabilized version of calcium hypochlorite ($Ca(OCl)_2$) containing between 60 and 70% available chlorine. Under normal storage conditions, commercial preparations will maintain their initial strength with little loss. Even though HTH is expensive, it may be economical, due to its properties. It is available in tablet or granular form.

These chemicals must be handled with great caution. They are caustic, corrosive and sensitive to light. They should be stored in tightly closed containers and in darkened spaces. When handling the material, contacts with skin, eyes and other body tissues must be avoided. Chlorine corrodes metal and to a less extent, wood and some synthetic materials. Metal parts which come in contact with the chemicals should be resistant.

Q.62. What are the factors for determination of chlorine dose during disinfection by chlorination?

Ans. Chlorine of any type must be added to water in closely controlled concentrations which depend on the characteristics of the water. As the use of dry chemicals doesn't always permit sufficient accuracy of dosing, solutions are preferred. Chlorine is usually added to the water for disinfection at the end of the treatment process. This allows the most effective treatment at the lowest level of chlorine application. Measurements of the chlorine demand and residual chlorine must be taken to assure that sufficient free chlorine is available to accomplish disinfection. Water characteristics and, hence, the chlorine demand may vary due to external influences (e.g., rainy season, etc.). It is therefore necessary to monitor the water quality from time to time, at the points of consumption in cases where the chlorine dosage is fixed. The objective of disinfection via chlorination can only be obtained if the chlorine dosage is adjusted to the changed water characteristics.

The chlorine demand of water of a given quality can be determined as follows: One litre samples of the water are taken. Chlorine solution of a known concentration is added and mixed with the water. After 30 minutes of contact time, the residual chlorine content is measured. The difference to the amount added then yields the chlorine consumption.

$$\text{Chlorine demand} = \text{chlorine consumption} + \text{desired residual chlorine}$$

Usually 1% chlorine solutions are applied. The chlorine flow is set such that a chlorine residual level of between 0.1 and 0.3 mg/l is obtained. Higher levels are used if rapid recontamination is likely.

Colorimetric tests are employed to determine total chlorine residuals. Chemical agents are used which are oxidized by chlorine to produce a coloured complex, intensity of which is proportional to the amount of chlorine present. Reading the colours and matching colour standards by means of a comparator and disks, gives the amount of free, available and residual chlorine. Various simple test kits are available, using permanent glass and containing DPD reagents in liquid /compressed tablet form.

Q.63. Discuss various factors to be considered for practical application of chlorination for disinfection of water.

Ans. Chlorination should never be performed prior to slow sand filtration (residual chlorine destroys biological agents). Sedimentation and filtration preceding chlorination enhance the disinfection effect. The lower the turbidity, the smaller the amount of chlorine necessary for effective disinfection. The chlorine solution can either be added to a batch of water (non-continuous or diffusion chlorination) or alternatively, it can be fed continuously to a constant flow of water.

Batch chlorination: Where tanks are used for storage of drinking water, the required amount of chlorine can be added to the tank periodically. It is advantageous to alternate between two tanks. While one tank is in use, the other one is refilled and treated with chlorine. The water can be used after a minimum of 30 minutes contact time. This procedure allows uninterrupted supply.

If the water quality of a given source varies, the chlorine demand must be reevaluated from time to time. Before a tank is used the first time for storing water, it must be cleaned carefully and disinfected (application of between 50 and 100 ppm active chlorine). Once the water has been disinfected, recontamination must be carefully prevented. Tanks should be covered. A tap should be used to release the water so as to avoid scooping out the water with unclean jars and the like. If the water is not used immediately but left in the house for a while, only well cleaned and covered jars should be used.

Chlorine tablets: In certain situations, e.g., while traveling, chlorine tablets can be used.

They are used for periodic chlorination of small batches of water.

Diffusion chlorination: Open wells are often bacteriologically contaminated because of non hygienic methods for lifting the water, or due to careless use of the surroundings of the well.

Various devices for diffusion chlorination include:

Type I which is a clay jar (12 to 15 l volume), filled nearly half-way with a mix of 1.5 kg bleach powder and 3 kg coarse sand (grain size 1.4 to 1.6 mm). It has two holes above the sand surface. The jar is covered with a plastic foil. The jar is suspended approximately 1 m below the water surface in the well. The chlorine can thus diffuse through the two holes into the well water.

Type II also consists of a clay jar (volume 7 to 10 l). It has 6 to 8 holes in the bottom. These are covered with stones on top of which a layer of gravel is placed. On top of that is put a mix of 1.5 kg bleaching powder and 3 kg of coarse sand. Stones are filled to the rim of the jar, which is then lowered into the water. For larger wells and higher rates of water use, two jars should be used which are interchangeably refilled.

In **Type III** used for small household wells, a double jar is recommended which releases less chlorine per time unit. The inner jar contains a mix of 1 kg bleaching powder and 2 kg coarse sand. The diffusion openings are provided. As these devices are not fit for large variations in water use, insufficient chlorination may occur at higher rates of water use.

Continuous chlorination: Simple chlorine dosing instruments can be installed in piped water supply systems. Chlorine is fed to the water in proportion to the flow rate. A pipeline transmits the water from the source to the reservoir, passing through some sort of pretreatment (e.g., coagulation/flocculation and settling). Before entering the reservoir, the water is passed through a mixing chamber where a dosing apparatus introduces droplets of chlorine into the water.

A variety of types of chlorine dosing instruments are commercially available. They range from manually controlled types to fully automated ones. Usually a unit consists of a storage tank and a diaphragm pump for feeding the hypochlorite solution. The feed rate is proportional to the water flow rate and, thus depends on the consumption. The use of these devices is limited to piped water supply systems.

Q.64. Describe the use of iodine as a water disinfectant.

Ans. Iodine is an excellent water disinfectant, effective against bacteria, amoeba cysts, cercerea and some viruses. It is added to the water mostly in the form of an aqueous solution @ 2 droplets per liter of water of a 2% iodine tincture. Iodine preparations are also available in tablet form.

In comparison to chlorination, the use of iodine has the following **advantages:**

— effectiveness over a wider range of pH values (up to pH 10), except at very low temperatures;
— ammonia and organic nitrogenous compounds have little effect on germicidal efficiency because they do not form substitution compounds with iodine;
— action depends less on contact time and temperature;
— effectiveness against more pathogenic organisms within short times;
— use and handling is simpler.

Since operating costs are too high, the use of iodine is not expected to ever become an important widely applied disinfectant. The applicability is limited due to the following **disadvantages:**

— higher concentrations than chlorine (on a ppm basis) are necessary for effective action;
— muddy or turbid water substantially affect germicidal action;
— iodine is about 20 times as expensive as chlorine per unit of germicidal effectiveness;
— taste and slight color produced by the iodine affect palatability and aesthetic quality;
— physiological effect of prolonged use of iodine (especially in children) is suspected. Allergies were ascertained.

In view of these economic and health implications the use of iodine for disinfection is recommended only for occasional application (e.g., in case of catastrophe or while traveling). Aside from that, iodine is a highly effective and technically widely applicable disinfectant.

Q.65. Describe ozonation (application of ozone) as a disinfectant in water treatment.

Ans. Ozone (O_3) is one of the most effective water disinfectants. As a powerful oxidant, it reduces the contents of iron, manganese, and lead, and eliminates most of the objectionable taste and odour present in water. Its effectiveness does not depend on pH value, temperature or ammonia content of the water. Since ozone is relatively unstable, it is generated almost invariably at the point of use. Ozone is obtained by passing a current of dried and filtered air (or oxygen) through between two electrodes (plates or tubes) subjected to an alternating current potential difference. A portion of the oxygen is then converted into ozone.

This principle of ozone production has been used in Europe for a long time, since it has the advantage of being applicable under a wide range of conditions. It leaves no chemical residuals behind in the treated water. On the other hand, no lasting protection against recontamination is provided either. Capital costs for the instrumentation of ozone production and feeding, as well as operating costs due to the electrical energy requirements, are very high. Moreover, operation of ozonizers requires continuous and skilled monitoring.

Q.66. Describe the use of potassium permanganate as a disinfectant for water.

Ans. Potassium permanganate ($KMnO_4$), a powerful oxidant, is rarely applied in water treatment for disinfection. It is effective against cholera bacteria, but not against other pathogenic germs. A dose of 1 to 5 ppm $KMnO_4$ is recommended for application. It creates a purple brown precipitate which coats the walls of the storage tank. It cannot be removed easily. In recent years, potassium permanganate has gained steadily in the application in pretreatment since it has proved effective at:

— removing objectionable odour and taste by means of oxidation of organic material, and hydrogen sulfide;

— preventing algal growth;

— removing iron and manganese compounds by means of oxidation and subsequent separation by filtration.

Q.67. Discuss the role of silver in disinfection of water?

Ans. Preparations containing silver (available as liquid/powder) may be used to reduce the germ count of water.. They are readily soluble in water and can be dosed easily. Effectiveness of silver can be explained by the oligo-dynamic properties of silver ions (silver nitrate or salt compounds). Even minute concentrations (0.03 to 0.04 ppm) are notably effective. The silver ions curb the growth of germs. After contact of between 30 minutes and 6 hours, depending on the level of bacteriological contamination, water of a very low germ content may be obtained. Odour and taste of the water are not affected by the application of silver. Disinfection by silver is a simple and very effective method. Its major advantage is that it provides already treated water with long lasting protection against recontamination by germs.

The effect of silver and other metals has been known to many peoples for a long time. The tradition of storing drinking water in silver vessels is still maintained in wealthier families. Although there is a tendency at present to exchange the metallic containers for plastic ones, even Indian villagers can still be seen fetching water from the well in brass or copper vessels. The metallic vessels are believed to have antiseptic qualities. Silver preparations are also used in ceramic filters. Major disadvantages of silver for the purpose of disinfection are the costs of treatment (about 200 times higher than gaseous chlorine), relatively long contact periods are required, organic substances and iron, Sulphur, etc., inhibit action, thus limiting the applicability of the technique. If combined with chlorine, silver preparations are more widely applicable (direct disinfection and protection against recontamination).

Q.68. Describe the role of water boiling as a means of water disinfection.

Ans. Boiling water is a very effective though energy consuming method to destroy pathogenic germs: bacteria, viruses, spores, cercerea and amoeba cysts, worm eggs, etc. The presence of turbidity or other impurities has little effect on germicidal effectiveness. If boiling is the only type of treatment available, it is recommended to let the water settle before, and decant it or filter it through a fine-meshed cloth so as to remove coarse impurities and suspended particles. The water is then brought to a strong boil which is maintained for at least five, preferably twenty minutes. For storing, it must not be transferred to a different vessel, but left in the former one and covered, so as to protect it from recontamination.

Boiling, together with the associated release of gases, especially CO_2, alters the taste of water. But through stirring while boiling and by letting the water sit in the partially filled vessel for a few hours afterward, the water picks up air and loses its bland taste. To improve the taste of the water, flavoring plant materials may be added during boiling. If done properly, boiling is a very effective and simple disinfection method. Since it requires a significant amount of energy, this method is only recommended in exceptional cases

Q.69. Describe the use of ultra violet (UV) radiation for disinfection of water.

Ans. The germicidal effect of UV rays is well known to harness it for water disinfection. The effect of sunlight on surface water is initiated in a more intense and controllable way. The most commonly used source of

UV radiation is a low pressure quartz mercury *vapour* lamp which emits invisible light at a wavelength in the range of 200 to 300 nm (nanometer).The germicidal effect depends on the electric power of the lamp and on the time of exposure of the water to the UV radiation. It decreases with increasing distance between water and lamp. Also, many substances present even in pretreated water (e.g., small amounts of dissolved iron) absorb UV light. Other constituents (turbidity, suspended matter) inhibit or prevent the transmission of radiation. A disinfection unit is built such that the water is made to flow through a pipe in a thin film around the lamp, which is located at the pipe's center, emitting radiation. The flow rate is adjusted as required. The water must be pre-filtered. Disinfection by UV radiation is a "clean" process, since no chemical additives are used. Residual matter doesn't occur, and tastes and odours are neither produced in the water nor altered. Automatic devices are available which indicate when the lamp's output is not sufficient.

Some of disadvantages as given below, forbid its widespread use for water disinfection:

— commercially available devices are relatively expensive,

— there is a dependence on steady power supply,

— the lamp's powers of penetration are limited; thin water films are necessary,

— turbidity, and impurities reduce the effectiveness notably,

— the lamps gradually lose their radiation power, accelerated by a coating of dirt. The lamp's average life is 1000 to 5000 hours,

— disinfection occurs rather quickly and effectively (up to 99.9%), though no protection for recontamination is provided.

Q.70. Why waste water treatment is necessary. Discuss its importance.

Ans. Waste water treatment aims at removal of unwanted components in waste stream. Its objective is to produce a waste stream (or treated effluent) and a solid waste or sludge also suitable for discharge or reuse back into the environment. This material is contaminated with toxic organic and inorganic compounds. Waste water treatment can be done by physical, chemical and biological methods either alone or in combination. However, most of the treatment processes are not completely effective and also do not offer complete solutions for the removal of water contaminants as it transfers the contaminant to a different component. The waste water generated by residential, commercial, and industrial sources is collected by sewers and transported to the Waste water Treatment Plant. A lot of sewage also includes some surface water from roofs or hard-standing areas. Municipal waste water includes residential, commercial, and industrial liquid waste discharges, and may include storm water runoff.

The waste water contains many of the physical, chemical and biological impurities which may impair quality of the receiving water body as well as interfere with the aquatic ecosystem in many ways. In order to render it safe, it would be appropriate to treat the waste water prior to disposal by adopting a systematic logical sequence of treatment.

Sewage systems capable of handling storm water are known as combined systems. Such systems are usually avoided since they complicate and thereby reduce the efficiency of sewage treatment plants owing to their seasonality. Besides, heavy storms may overwhelm the sewage treatment system, causing a spill or overflow. It is preferable to have a separate storm drain system for storm water. Sewerage systems that transport liquid waste discharges and storm water together to a common treatment facility are called combined sewer systems.

As rainfall runs over the surface of roofs and the ground, it may pick up various contaminants including soil particles, (sediment), heavy metals, organic compounds, animal waste, and oil and grease. Some jurisdictions require storm water to receive some level of treatment before being discharged directly into waterways.

Q.71. What are the major constituents of preliminary, primary and secondary treatment of waste water?

Ans. The waste water contains many floatable materials that may interfere with the downstream processes or be detrimental to the plant equipment. To increase the waste water treatment efficiency of the secondary treatment units it is essential to have **preliminary and primary treatment** of waste water.

Materials removed may include rags, plastic, lumber, and grit. This is done by bar screens which remove large objects, rags, and plastic that would cause plugging problems in the downstream lines. The grit basins remove the grit by sedimentation. Grit is abrasive materials such as sand and gravel that must be removed to minimize wear on the downstream equipment. Materials removed by the bar screens and grit removal system are de-watered, discharged to a dump trailer, and transported to the landfill for disposal.

Secondary treatment helps in reducing the organic load present in waste water. The commonly used units in this treatment are Trickling filter, activated sludge, Water treatment ponds etc.

The secondary treatment of water is based on the natural self-purification capacity of natural waters. This treatment is designed to substantially degrade the biological content of the sewage such as are derived from human waste, food waste, soaps and detergent. The majority of the municipal and industrial water treatment plants treat the settled sewage liquor using aerobic biological processes. For this to be effective, the biota requires nutrients and a substrate on which to live. There are number of ways in which this is done. In all these methods, the bacteria and protozoa consume biodegradable soluble organic contaminants (e.g., sugars, fats, organic short-chain carbon molecules, etc.) and bind much of the less soluble fractions into flocs.

Secondary treatment systems are classified as **fixed film** or suspended growth, such as rock filters where the biomass grows on media and the sewage passes over its surface. In **suspended growth systems** - such as activated sludge - the biomass is well mixed with the sewage and be operated in a smaller space than fixed film systems that treat same amount of water. However, fixed film systems are more able to cope better with drastic changes in the amount of biological material and can provide higher removal rates for organic material and suspended solids than suspended growth systems. This is reflected in two different possibilities for water treatment. The first type is a so-called fixed bed bio-reactor, usually referred to as a trickling filter.

Q.72. What are different stages in preliminary treatment of waste water? What is its use and necessity?

Ans. It is used to reduce Solids (50-60%), and BOD (30- 50%). The commonly used units in the preliminary treatment of water are bar screens, Comminutor, and Grit chamber (Settling tank).

Preliminary treatment, the first treatment process, consists of the removal of substances that may interfere with the downstream processes or be detrimental to the plant equipment. Primary treatment is to reduce oils, grease, fats, sand, grit, and coarse (settleable) solids. This step is done entirely with machinery, hence the name mechanical treatment.

Influx (influent) and removal of large objects: In the mechanical treatment, the influx (influent) of sewage water is strained to remove all large objects that are deposited in the sewer system, such as rags, sticks, condoms, sanitary towels (sanitary napkins) or tampons, cans, fruit, plastic, lumber, grit, etc. This is most commonly done with a manual or automated mechanically raked screen. This type of waste is removed because it can damage the sensitive equipment in the sewage treatment plant.

Sand and grit removal: This stage includes a sand or grit channel where the velocity of the incoming waste water is carefully controlled to allow sand grit and stones to settle but still maintain the majority of the organic material within the flow. This equipment is called a detritor or sand catcher. Sand grit and stones need to be removed early in the process to avoid damage to pumps and other equipment in the remaining treatment stages. Sometimes there is a sand washer (grit classifier) followed by a conveyor that transports the sand to a container for disposal. The contents from the sand catcher may be fed into the incinerator in a sludge processing plant, but in many cases, the sand and grit is sent to a land-fill. Materials removed may include rags, plastic, lumber, and grit. This is done by bar screens which remove large objects, rags, and plastic that may cause plugging problems in the downstream lines. The grit basins remove the grit by sedimentation. Grit is abrasive materials such as sand and gravel that must be removed to minimize wear on the downstream equipment. Materials removed by the bar screens and grit removal system are de-watered, discharged to a dump trailer, and transported to the landfill for disposal.

Grit and screen facility: The Grit and Screen Facility is where sewage first enters the plant. The speed of the waste water is reduced to 0.3 metre per second and air is pumped in to set up a rolling motion. This movement separates the heavier grit and sand materials from the suspended organic matter. The larger sand particles settle to the bottom of the tanks, thus preventing harm to equipment within the Plant or clogging of pipes during the treatment process. The settled sand is removed once a month and buried. Waste water continues through the grit tanks and passes through bar screens to remove any material larger than 12.5mm in size. The waste water then flows by gravity to the Primary Settling Basins.

Q.73. What are different stages in primary treatment of waste water? What is its purpose?

Ans. The second step in the waste water treatment process is primary treatment. The waste water enters two primary clarifiers (sedimentation basins) which remove suspended and floating materials. The primary clarifiers remove about 60% of the Total Suspended Solids and about 30% of the Biochemical Oxygen Demand (BOD) in the incoming waste water. BOD is a measure of the amount of oxygen needed to biochemically degrade the organic matter in the waste water. Primary treatment does not remove colloidal or dissolved solids.

Sedimentation: Many plants have a sedimentation stage where the sewage is allowed to pass slowly through large tanks, commonly called "primary clarifiers" or "primary sedimentation tanks". The tanks are large enough that faecal solids can settle and floating material such as grease and plastics can rise to the surface and be skimmed off. The main purpose of the primary stage is to produce a generally homogeneous liquid capable of being treated biologically and a sludge that can be separately treated or processed. Primary settlement tanks are usually equipped with mechanically driven scrapers that continually drive the collected sludge towards a hopper in the base of the tank from where it can be pumped to further sludge treatment stages.

Primary settling basins: This is where suspended solids settle out and floating scum is removed for further treatment. Upon entering the Primary Settling Basins, the velocity of the waste water is reduced to 3 meters per minute, allowing fine particles to settle to the bottom. Settled sludge on the bottom of the basins is continuously scraped into hoppers at the end of tanks. It takes about 4 hours for waste water to flow through the Primary Settling Basins. Upon completion, the primary effluent is pumped to the Bioreactors and the settled sludge is pumped to the Fermenters. The scum from the top of the basins is collected in hoppers and pumped to the Digesters.

Q.74. What are different stages and purpose of secondary treatment of waste water?

Ans. Secondary treatment usually consists of two steps which remove the dissolved and colloidal organic material not removed by the primary treatment. There are various secondary treatment processes in use. This treatment use microorganisms to convert organic wastes into stabilized compounds - similar to self-purification process in streams.

It is used to reduce Solids (85-95%), BOD (80-95%), and Coliform (90-95%). The commonly used units in this treatment are Trickling filter, Activated sludge, and Water treatment ponds.

The biological treatment of water is based on the natural self-purification capacity of natural waters. This treatment is designed to substantially degrade the biological content of the sewage which are derived from human waste, food waste, soaps and detergent. The majority of the municipal and industrial water treatment plants treat the settled sewage liquor using aerobic biological processes. For this to be effective, the biota requires nutrients and a substrate on which to live. There are number of ways in which this is done. In all these methods, the bacteria and protozoa consume biodegradable soluble organic contaminants (e.g., sugars, fats, organic short-chain carbon molecules, etc.) and bind much of the less soluble fractions into floc.

Secondary treatment systems are classified as **fixed film** or **suspended growth**, such as rock filters where the biomass grows on media and the sewage passes over its surface. In **suspended growth systems** - such as activated sludge - the biomass is well mixed with the sewage and can be operated in a smaller space than fixed film systems that treat same amount of water. However, fixed film systems cope better with drastic changes in the amount of biological material and can provide higher removal rates for organic material and suspended solids than suspended growth systems. This is reflected in two different possibilities for water treatment. The first type is a so-called fixed bed bio-reactor, usually referred to as a trickling filter.

Roughing filters are intended to treat particularly strong or variable organic loads, typically industrial, to allow them to then be treated by conventional secondary treatment processes. Characteristics include typically tall, circular filters filled with open synthetic filter media to which sewage is applied at a relatively high rate. Designed to allow high hydraulic loading and a high flow-through of air. On larger installations, air is forced through the media using blowers. The resultant liquor is usually within the normal range for conventional treatment processes.

Q.75. What do you mean by biological treatment of waste water?

Ans. Treatment of industrial effluents using microorganisms is is called biological treatment of waste water. Secondary treatment typically utilizes biological treatment processes, in which microorganisms convert non settleable solids to settleable solids. Biological treatment processes followed by settling tanks removes approximately 85% of the BOD and TSS in waste water. Secondary treatment for municipal waste water is the minimum level of treatment required. Its objective is to produce a stream (or treated effluent) and a solid waste or sludge also suitable for discharge or reuse back into the environment.

Q.76. Describe flocculation technique of waste water treatment.

Ans. Generally aluminium compounds are used as flocculants. The aluminium ion, when released into the water, will form aluminium hydroxide, $Al(OH)_3$, flocs, which will partly block the pores of a sand filter and thus enhance filtering action. A flocculant should in principle also be able to destabilize colloid solutions. A colloid in water containing a low concentration on some ions, will be surrounded by charged particles, that create an electrical double layer around it. The electrical potential of that layer is called the zeta-potential. The higher this zeta potential is, the stronger the repulsion between the particles and thus the smaller the chance that two particles will collide and then coagulate. If to such a solution aluminium ions are added, the zeta-potential of the colloid will be strongly reduced, the colloid solution becomes unstable and flocculation will occur. This process is called salt flocculation.

Q.77. Describe the activated sludge process of waste water treatment?

Ans. The activated sludge process is a biological process in which air or oxygen is forced into sewage liquor to develop a biological floc utilizing an active biomass. The activated sludge water treatment plants use a variety of mechanisms and processes to use dissolved oxygen to promote the growth of biological floc that substantially removes organic material. It also traps particulate material and can, under ideal conditions, convert ammonia to nitrite and nitrate and ultimately to nitrogen gas. The biomass is largely composed off saprophytic bacteria but also has an important protozoan flora mainly composed of amoebae, Spirotrichs, Peritrichs including Vorticellids and a range of other filter feeding species. In poorly managed activated sludge, a range of mucilaginous filamentous bacteria can develop including *Sphaerotilus natans* which produces a sludge that is difficult to settle and can result in the sludge blanket decanting over the weirs in the settlement tank to severely contaminate the final effluent quality. This material is often described as sewage fungus but true fungal communities are relatively uncommon to neutralize the organic matter in the waste water.

The active biomass (called return activated sludge) is mixed with the waste water leaving the primary clarifiers and with air in the aeration basins. Part of the settled material, the sludge, is returned to the head of the aeration system to re-seed the new sewage entering the tank. This fraction of the floc is called R.A.S - Return Activated Sludge. The remaining sludge, also called W.A.S - Waste Activated Sludge, is further treated prior to disposal. (W.A.S is also sometimes called S.A.S - Surplus Activated Sludge) The microorganisms, in the presence of oxygen, use organic matter and nutrients in waste water to sustain their life processes, including reproduction. Effluent from the aeration basins contains large quantities of suspended biological solids (called activated sludge) that must be removed prior to discharge. Therefore, the second step of the secondary treatment process is sedimentation in the secondary clarifiers. Most of the solids removed in the secondary clarifiers are returned to the aeration basins as returned activated sludge to maintain the appropriate population of microorganisms needed to assimilate the organic matter entering the aeration basins.

Q.78. Explain the trickling filter method of waste water treatment.

Ans. In the trickling filter method of waste disposal, a bed of crushed rock or synthetic media supports a film of aerobic microorganisms. The liquid waste water is sprayed over the bed and the microorganisms act on the contaminants as they pass through the filter. This is a common method used in sewage treatment plants to treat acetaldehyde, acetic acid, benzene, chlorinated hydrocarbons, cyanides, formaldehyde, ketones, and resins. Trickling filterization is generally effective for aqueous wastes where concentrations are < 1%. Like all biological treatment, it requires stable, consistent operating conditions. It is not suitable for removal of highly chlorinated organics, aliphatics, amines, and aromatic compounds. Heavy metals and organic chemicals may kill the microorganisms. Heavy metals and non-biodegradable organics may also concentrate in the sludge. Hydrogen sulfide gas may also be released.

In a trickling filter, water is divided over the top and is allowed to flow slowly over the filter bed. In the course of that flow organic matter, both dissolved and suspended, is attacked by micro-organisms. In the first part of the filter (the top) mainly carbon decomposition takes place. The depth of that carbon decomposition zone depends on the load: if the water contains much organic carbon the zone will be very deep. In heavily loaded filters carbon decomposition may be the only process taking place. In the lower part of the trickling filter nitrification can occur. It will take rather long before a population of nitrifying bacteria has been established, since they grow slowly. It will take several weeks before a filter is properly conditioned.

If a trickling filter is just put into operation the first organisms that will grow in it will be bacteria and fungi. These will coat the filling material: a bio-film is formed. On this film the secondary colonizers will start growing. To this group belong among others protozoa and nematodes. The growth of organisms in and on the bio-film will increase its thickness. If a film becomes too thick it will start sloughing. This sloughed film will appear in the water as flocs. These flocs settle easily, so if water from the trickling filter is led through a settling area first, nearly all organic matter can be removed.

Processes happening in a trickling filter are the following:

- dissolved organic matter is mineralized.
- particulate organic matter and colloids are either converted into dissolved organic matter and then mineralized, or are adsorbed onto the bio-film and removed from the water as settleable flocs.

- inorganic matter, particulate, colloidal or dissolved can react with or be adsorbed onto the bio-film and can thus be removed with the flocs.

The organisms that can be found in a trickling filter do not differ much from the ones found in activated sludge. No differences have been found between the bacterial flora of activated sludge and that of trickling filters. In activated sludge most are attached to suspended flocs, whereas in a trickling filter they are attached to the filter bed.

Q.79. Describe the process of anaerobic digestion of sewage sludge.

Ans. Waste treatment in this fashion uses the same process as that which naturally occurs in landfills. In an oxygen free environment, it reduces organic waste to a relatively stable solid residue (digestate) similar to compost. Anaerobic digestion is particularly suited to wet, organic material and as such has been used for the treatment of sewage sludge.

The organic portion is separated from the waste to remove plastic, glass and metals and then placed in a sealed reactor. The conditions necessary for biological degradation are created which produces biogas, a mixture of methane and carbon dioxide. This can then be used as a fuel. The methane is generally burnt on site for heating or to produce electricity on a small scale. The digester will require an energy input to retain the material at elevated temperatures (between 20 and 40°C - the mesophilic range, or up to 60°C - the thermophilic range) in order to speed digestion. Some of the energy available from the methane generated can be recycled for this purpose, typically 20-40% of the total energy produced. Alternatively, the process can produce solid fuel, a fiber product for board manufacture, or more commonly the residue is used as a soil conditioner.

The waste water exiting the anaerobic digestion facility have elevated levels of BOD and COD. Some of this material is termed as 'hard COD' meaning it cannot be accessed by the anaerobic bacteria for conversion into biogas. If this effluent was put directly into water courses it would negatively affect them by causing eutrophication. As such further treatment of the waste water is often required. This treatment will typically be an oxidation stage where air is passed through the water in a sequencing batch reactors or reverse osmosis unit.

Q.80. Describe various pond treatment processes for waste water treatment.

Ans. The treatment of waste water could also be accomplished by various biological pond treatment processes which are listed below :

Aerobic stabilization ponds: Treatment is done with aerobic bacteria; oxygen is supplied by algal photosynthesis and natural surface reaeration; depth of 0.15 to 1.5 m is maintained, Carbonaceous BOD removal is achieved.

Maturation (tertiary) ponds: Use aerobic treatment; applied loadings are low to preserve aerobic conditions, Secondary effluent polishing and seasonal nitrification.

Facultative ponds: Treatment with aerobic, anaerobic and facultative bacteria; the pond has 3 zones: a surface aerobic zone, a bottom anaerobic zone, and an intermediate zone partly aerobic-anaerobic, Carbonaceous BOD removal is achieved.

Anaerobic ponds: Treatment done with anaerobic bacteria; depths of up to 9.1 m to conserve anaerobic conditions, Carbonaceous BOD **removal** (waste stabilization) is achieved.

Q.81. Discuss the role of fermenters, bioreactors and secondary clarifiers in biological treatment of waste water.

Ans. Fermenters are where the primary sludge settling out from the primary settling basins is broken down. The fermentation process converts the organic material into volatile acids called volatile fatty acids (VFA's). An example of a VFA is acetic acid or vinegar. The VFA's are sent to the bioreactors and utilized by the nutrient consuming microorganisms.

Bioreactors are where the biological nutrient removal (BNR) process takes place. This natural process greatly reduces the concentration of dissolved organic compounds in the effluent and removes unwanted carbon, phosphorus, and nitrogen from the waste water without the addition of chemicals. In the bioreactor, the effluent from the Primary Settling Basins is mixed with micro-organisms and Volatile Fatty Acids (VFA's). The microorganisms come from the secondary clarifiers, where they have settled out and are then returned to the bioreactors. The VFA's are created in the Fermenter. The micro-organisms naturally break down excess carbon and nutrients present in the waste water. The effluent is moved through carefully controlled anaerobic (absence of available oxygen), anoxic (chemically available oxygen only) and aerobic (abundance of free oxygen) zones of the bioreactors. Each of these areas remove specific organic compounds as the micro-organisms continue to grow and flourish, consuming impurities in the waste water. After approximately 9 hours in the bioreactors, the effluent flows to the secondary clarifiers.

Secondary clarifiers are where any remaining solids along with the microorganisms from the bioreactors settle to the bottom and the clear effluent flows out the top of the basins. Of the settled micro-organisms, about 97% are returned to the Bioreactors to be used again in the BNR process and 3% are sent to the DAF Thickener and then to the digesters for further treatment. The clarified final effluent flows from the secondary clarifiers to the chlorine contact chamber before being released to the river.

Q.82. Describe the working of rotating biological contactors (RBCs) used in secondary waste water treatment systems.

Ans. Rotating biological contactors (RBC's), used in waste water treatment plants are mechanical secondary treatment systems, which are robust and capable of withstanding surges in organic load. The rotating disks support the growth of bacteria and micro-organisms present in the sewage, which breakdown and stabilize organic pollutants. To be successful, micro-organisms need both oxygen to live and food to grow. Oxygen is obtained from the atmosphere as the disks rotate. As the micro-organisms grow, they build up on the media until they are sloughed off due to shear forces provided by the rotating discs in the sewage.

The primary function of these bio-reactors is the reduction of organic matter. It consists of series of closely spaced discs mounted on a horizontal shaft and rotated while 40% of each disc is submerged in waste water. The discs are made up of light-weight plastic and are covered with a biofilm. The slime is 1-3 mm in thickness on disc. As the disk rotates the biofilm is exposed to the waste water only part of the time then with atmosphere for adsorption of oxygen. The rotation in and out of the waste water serves to vary the feeding cycle of the microorganisms that make up the biofilm. The shaft rotates at 1-2 rpm (slowly). Excess solids are removed by shearing forces created by the rotation mechanism. Effluent from the RBC is then passed through final clarifiers where the micro-organisms in suspension settle as a sludge. The sludge is withdrawn from the clarifier for further treatment.

The primary and secondary treatment processes generally remove at least 85% of the total suspended solids and biochemical oxygen demand. The secondary process is a very sensitive biological process and can be adversely impacted by the discharge of incompatible or toxic wastes into the sewer system. Precise control of this process is necessary to effectively treat the waste water.

Q.83. Define advanced waste water treatment. Why is it needed?

Ans. Advanced waste water treatment is defined as any process designed to produce an effluent of higher quality than normally achieved by secondary treatment processes or containing unit operations not normally found in Secondary Treatment units. Primary and secondary waste water treatment processes remove the majority of BOD and Suspended Solids found in waste waters. However, in many cases this level of treatment is insufficient to protect the receiving waters or to provide reusable water for industrial and/or domestic recycle. Thus, additional treatment steps have been added to waste water treatment plants to provide for further organic and solids removals or to provide for removal of nutrients and toxic materials.

Once freshwater has been used for an economic/beneficial purpose, it is generally discarded as waste. These waste waters are discharged, either as untreated waste or as treated effluent, into natural watercourses, from which they are abstracted for further use after undergoing "self-purification" within the stream. Through this system of indirect reuse, waste water may be reused up to a dozen times or more before being discharged to the sea. Such indirect reuse is common in the larger river systems. However, more direct reuse is also possible: the technology to reclaim waste waters as potable or process waters is a technically feasible option for agricultural and some industrial purposes (such as for cooling water or sanitary flushing), and a largely experimental option for the supply of domestic water. Waste water reuse for drinking raises public health, and religious, concerns among consumers. Due to stringent requirement of the quality of potable and process water, tertiary treatment of water is not a luxury any more rather it has emerged out as the only way to provide adequate safe guard for community health.

Q.84. What are different types, purpose and stages of advanced waste water treatment?

Ans. Advanced Waste water Treatment is categorised by the type of process flow scheme as:

- Tertiary treatment
- Physical-chemical treatment
- Combined biological-physical treatment

Tertiary treatment is defined as any treatment process in which unit operations are added to the flow scheme following conventional secondary treatment. Additions to conventional secondary treatment could be the addition of a filter for suspended solids removal or the addition of many other unit processes for organic, suspended solids, nitrogen and phosphorous removal.

Physical-chemical treatment is defined as a treatment process in which biological and physical-chemical processes are intermixed to achieve the desired effluent characteristics.

Combined biological-physical-chemical treatment is differentiated from tertiary treatment in that in tertiary treatment any unit processes are added after conventional biological secondary treatment, while in combined treatment, biological and physical –chemical treatment are mixed.

Another way to classify advanced waste water treatment is to differentiate on the basis of desired treatment goals. Advanced waste water treatment is used for:

- Additional organic and suspended solids removal
- Removal of nitrogenous oxygen demand (NOD)
- Nutrient removal
- Removal of toxic materials

The first three reasons for additional organic removal through advanced waste water treatment are simple. The fourth requires explanation. The performance of secondary treatment plants is measured in terms of BOD and SS removals. A well designed and operated secondary water treatment plant will remove from 85 to 95% of the influent BOD and SS. However, the BOD test does not measure all of the organic material present in the waste water. An average secondary effluent may have a BOD of 20 mg/L and a COD of 60 to 100 mg/L. The average secondary plant removes approximately 65% of the influent COD. Thus, when high quality effluents are required, additional organic removals must be accomplished. In addition to the organic materials remaining in most secondary effluents, there is an additional oxygen demand resulting from the nitrogen present in the waste water.

In waste waters, much of the nitrogen is found in the form of ammonia. When secondary treatment is used, a great deal of this ammonia is discharged in the effluent. Bacteria can utilize this ammonia as an energy source and convert ammonia to nitrite and nitrate. Another reason for advanced waste water treatment may be to remove nutrients contained in discharges from secondary treatment plants. The effluents from secondary treatment plants contain both nitrogen (N) and phosphorous (P). which are ingredients of fertilizers. When excess amounts of N and P are discharged, plant growth in the receiving waters may be accelerated. Algae growth may be stimulated causing blooms which are toxic to fish life as well as aesthetically unpleasing. Fixed plant growth may also be accelerated causing the eventual process of a lake becoming a swamp to be speeded up. Therefore, is necessary to remove nitrogen and phosphorous prior to discharge in some cases.

Toxic materials, both organic and inorganic are discharged into many sewage collections systems. When these materials are present in sufficient quantities to be toxic to bacteria, it is necessary to remove them prior to biological treatment. In other cases, it is necessary to remove even small amounts of these materials prior to discharge to protect receiving waters or drinking water supplies. Thus, advanced waste water treatment processes have been used in cases where secondary treatment was not possible due to materials toxic to bacteria entering the plant as well as in cases where even trace amounts of toxic materials are unacceptable in plant effluents.

Q.85. Describe ion exchange method of water purification.

Ans. **Ion exchange** is a process for water purification in which ions are exchanged between a solution and an **ion exchanger**, a non-aqueous solid or gel. Typical ion exchangers are ion exchange resins, zeolite, montmorillonite, clay, and humus. Ion exchangers are either **cation exchangers** for positively charged cations or **anion exchangers** for negatively charged anions. There are also ion exchangers called mixed beds which have both anion and cation exchanging resins within them. Ion exchange is a reversible process and the ion exchanger can be *regenerated* or *loaded* by washing with an excess of the ions to be exchanged.

Ion exchange method is widely used in household and industrial water purifications to produce soft water. This is accomplished by exchanging calcium Ca^{2+} and magnesium Mg^{2+} cations against sodium Na+ or hydrogen H+ cations. Ion exchange is also applied in water softeners. Ion Exchange can be used in the food and beverage, hydrometallurgical, metals finishing, chemical and petrochemical, pharmaceutical, sugar and sweeteners, ground and potable water, nuclear, softening and industrial water, semiconductor, power, and a host of other industries.

Q.86. Describe the features of ion exchange resins used in water purification.

Ans. An **ion exchange resin** is an insoluble matrix (on support structure) normally in the form of small (1-2 mm diameter) beads, usually white or yellowish, fabricated from an organic polymer substrate on the surface of which are sites with easily trapped and released ions in a process called ion exchange. There are different types of ion exchange resin which are fabricated to selectively prefer one or several different types of ions. This resin is used extensively for water softening during water purification. They were developed as a more flexible alternative to the use of natural or artificial zeolites.

Most ion exchange resins are based on crosslinked polystyrene. The required active groups can be introduced after polymeration, or substituted monomers can be used. The crosslinking is usually achieved by adding a small proportion of divinyl benzene to styrene. Non-crosslinked polymers are used only rarely because of their tendency to change dimensions in dependence on the ions bonded. However, crosslinking somewhat decreases the capacity of the resin and prolong the time for reaching the equilibrium for the ions in solution and in the resin. Particle size also influences the resin parameters; smaller particles have larger surface, but cause larger head loss in the column. Membranes made of highly cross-linked ion exchange resins that allow passage of ions but not of water are used for electrodialysis.

There are four main types of ion exchange resins differing in their functional groups:

- strongly acidic (sulfonic acid groups, e.g., sodium polystyrene sulfonate or polyAMPS)
- strongly basic, (trimethylammonium groups, e.g., polyAPTAC)
- weakly acidic (carboxylic acid groups)
- weakly basic (amino groups, e.g., polyethylene amine)

There are also specialized types:

- chelating resin

Q.87. What is the difference between normal osmosis process and reverse osmosis process?

Ans. In the normal osmosis process, more dilution of concentrated solution occurs by the movement of molecules from the less concentrated side. But in case of reverse osmosis process, due to the application of high external pressure, the molecules of solution move from the more concentrated side to the less concentrated side through the semi-permeable membrane.

Q.88. Describe the reverse osmosis process used for desalination of brackish water and sea water.

Ans. To desalinate water, we must create a flow through a membrane, causing the water to leave the salty side of the membrane, flowing into the unsalted side. To achieve this, pressure must be created upon the water column on the salt side of the membrane; firstly, to remove the natural osmotic pressure and secondly, to create extra pressure on the water column, in order to push the water through the membrane. For the desalination of seawater, the pressure must be about 50-60 bars.

There are several different techniques that can be applied for water desalination. Examples are reverse osmosis, electro dialysis, distillation and ion exchange: **Reverse osmosis** is the most economic process for the **desalination** of brackish water and seawater. When we compare this process to the traditional thermic process of distillation, the capital investments and the energy use are much lower. Membranes occupy through a selective separation wall. Certain substances can pass through the membrane, while other substances are caught.

Membrane filtration: This can be used as an alternative for flocculation, sediment purification techniques, adsorption (sand filters and active carbon filters, ion exchangers), extraction and **distillation.** Membrane filtration can be divided up between micro and ultra-filtration on the one hand and Nano filtration and Reverse Osmosis (RO or hyper filtration) on the other hand. When membrane filtration is used for the removal of larger particles, micro filtration and ultra-filtration are applied. Because of the open character of the membranes the productivity is high while the pressure differences are low. When salts need to be removed from water, nano filtration and Reverse Osmosis are applied. Nano filtration and RO membranes do not work according to the principle of pores; separation takes place by diffusion through the membrane. The pressure that is required to perform nano filtration and Reverse Osmosis is much higher than the pressure required for micro and ultra-filtration, while productivity is much lower.

Q.89. Describe the role of electrodialysis in treating waste water.

Ans. **Electrodialysis** is used to transport salt ions from one solution through ion exchange membranes to another solution under the influence of an applied electric potential difference. This is done in an electrodialysis cell. The cell consists of a feed (dilute) compartment and a concentrate (brine) compartment formed by an anion exchange membrane and a cation exchange membrane placed between two electrodes. In practical electrodialysis processes, multiple electrodialysis cells are arranged into a configuration called an electrodialysis stack, with alternating anion and cation exchange membranes forming the multiple electrodialysis cells.

Electrodialysis processes are unique compared to distillation techniques and other membrane-based processes (such as reverse osmosis) in that dissolved species are moved away from the feed stream rather than the reverse. Because the quantity of dissolved species in the feed stream is far less than that of the fluid, electrodialysis offers the practical advantage of much higher feed recovery in many applications.

Q.90. Describe the working principle of electrodialysis cell used for water *purification*.

Ans. In an electrodialysis stack, the dilute (D) feed stream, brine or concentrate (C) stream, and electrode (E) stream are allowed to flow through the appropriate cell compartments formed by the ion exchange membranes. Under the influence of an electrical potential difference, the negatively charged ions (e.g., chloride) in the dilute stream migrate toward the positively charged anode. These ions pass through the positively charged anion exchange membrane, but are prevented from further migration toward the anode by the negatively charged cation exchange membrane and therefore stay in the C stream, which becomes concentrated with the anions. The positively charged species (e.g., sodium) in the D stream migrate toward the negatively charged cathode and pass through the negatively charged cation exchange membrane. These cations also stay in the C stream, prevented from further migration toward the cathode by the positively charged anion exchange membrane. As a result of the anion and cation migration, electric current flows between the cathode and anode. Only an equal number of anion and cation charge equivalents are transferred from the D stream into the C stream and so the charge balance is maintained in each stream.

The result of the electrodialysis process is an ion concentration increase in the concentrate stream with a depletion of ions in the dilute solution feed stream.

The E stream is the electrode stream that flows past each electrode in the stack. This stream may consist of the same composition as the feed stream (e.g., sodium chloride) or may be a separate solution containing a different species (e.g., sodium sulfate). Depending on stack configuration, anions and cations from the electrode stream may be transported into the C stream, or anions and cations from the D stream may be transported into the E stream. In each case, this transport is necessary to carry current across the stack and maintain electrically neutral stack solutions.

Anode and cathode reactions

Reactions take place at each electrode.

At the cathode,

$$2e^- + 2\,H_2O \longrightarrow H_2\,(g) + 2\,OH$$

While at the anode

$$H_2O \longrightarrow 2\,H^+ + 1/2\,O_2\,(g) + 2e^- \quad \text{or} \quad 2\,Cl^- \longrightarrow Cl_2\,(g) + 2e^-$$

Small amounts of hydrogen gas are generated at the cathode and small amounts of either oxygen or chlorine gas (depending on composition of the E stream and end ion exchange membrane arrangement) at the anode. These gases subsequently dissipated as the E stream effluent from each electrode compartment is combined to maintain a neutral pH and discharged or re-circulated to a separate E tank.

Q.91. Discuss various practical applications of electrodialysis.

Ans. The major application of electrodialysis is for desalination of brackish water/seawater as an alternative to RO for potable water production and seawater concentration for salt production. In normal potable water production without the requirement of high recoveries, RO is generally believed to be more cost effective when total dissolved solids (TDS) are 3,000 parts per million (ppm) or greater, while electrodialysis is more cost-effective for TDS feed concentrations less than 3,000 ppm or when high recoveries of the feed are required.

Another important application for electrodialysis is the production of pure water and ultrapure water by electro deionization (EDI). In EDI, the compartments (dilute, concentrate, or both) of the electrodialysis stack are filled with ion exchange resin. When fed with low TDS feed (e.g., feed purified by RO), the dilute can reach very low levels (e.g., 18 Megaohms). The ion exchange resins act to retain the ions, allowing these to be transported across the ion exchange membranes. The main usage of EDI systems are in electronics, pharmaceutical, and cooling tower applications.

Some practical applications of electrodialysis include:

- Large scale brackish and seawater desalination and salt production.
- Small and medium scale drinking water production (e.g., towns and villages, construction and military camps, nitrate reduction, hotels and hospitals).
- Water reuse (e.g., industrial laundry waste water, produced water from oil/gas production, cooling tower makeup and blowdown, metals industry fluids, wash-rack water).
- Pre-demineralization (e.g., boiler makeup and pretreatment, ultrapure water pre treatment, process water desalination, power generation, semiconductor, chemical manufacturing, food and beverage).
- Food processing.
- Agricultural water (e.g., water for greenhouses, hydroponics, irrigation, livestock).
- Glycol desalting (e.g., antifreeze or engine-coolants, capacitor electrolyte fluids, oil and gas dehydration, conditioning and processing solutions, industrial heat transfer fluids, secondary coolants from heating, venting, and air conditioning (HVAC).
- Glycerin Purification.

In application, electrodialysis systems can be operated as continuous production or batch production processes. In a continuous process, feed is passed through a sufficient number of stacks placed in series to produce the final desired product quality. In batch processes, the dilute or concentrate streams are re-circulated through the electrodialysis systems until the final product or concentrate quality is achieved. Electrodialysis is usually applied to deionization of aqueous solutions. However, desalting of sparingly conductive aqueous organic and organic solutions is also possible.

Q.92. What are the limitations of electrodialysis?

Ans. Electrodialysis has inherent limitations, working best at removing low molecular weight ionic components from a feed stream. Non-charged, higher molecular weight, and less mobile ionic species will not typically be significantly removed. Also, in contrast to RO, electrodialysis becomes less economical when extremely low salt concentrations in the product are required and with sparingly conductive feeds: current density becomes limited and current utilization efficiency typically decreases as the feed salt concentration becomes lower, and with fewer ions in solution to carry current, both ion transport and energy efficiency greatly declines. Hence, comparatively large membrane areas are required to satisfy capacity requirements for low concentration (and sparingly conductive) feed solutions.

As with RO, electrodialysis systems require feed pretreatment to remove species that coat, precipitate onto, or otherwise "foul" the surface of the ion exchange membranes. This fouling decreases the efficiency of the electrodialysis system. Species of concern include calcium and magnesium hardness, suspended solids, silica, and organic compounds. Water softening can be used to remove hardness, and micron or multimedia filtration can be used to remove suspended solids. Hardness in particular is a concern since scaling can build up on the membranes. Various chemicals are also available to help prevent scaling. Also, electrodialysis reversal systems seek to minimize scaling by periodically reversing the flows of dilute and concentrate and polarity of the electrodes.

Q.93. What is sewage? Describe various sewage treatment methods.

Ans. Sewage is created by residences, institutions, and commercial and industrial establishments. It can be treated close to where it is created (in septic tanks, biofilters or aerobic treatment systems), or collected and transported via a network of pipes and pump stations to a municipal treatment plant. Sewage collection and treatment is typically subject to local, state and federal regulations and standards.

Industrial sources of waste water often require specialized treatment processes. Raw influent (sewage) is the liquid waste from toilets, baths, showers, kitchens, sinks etc. Household waste is that which is disposed of via sewers. Municipal waste water therefore includes residential, commercial, and industrial liquid waste discharges, and may include stormwater runoff. Sewage systems capable of handling storm water are known as combined systems.

Typically, sewage treatment involves three stages, called *primary*, *secondary* and *tertiary treatment*. Sewage Treatment Methods include:

- **Mechanical treatment**;
  - — Removal of large objects — Removal of sand and grit
  - — Pre-precipitation
- **Biological treatment**;
  - — Oxidation bed (oxidizing bed) or aeration system
  - — Post precipitation — Effluent
- **Chemical treatment** (this step is usually combined with settling and other processes to remove solids, such as filtration.)

Primary treatment: It involves removal of large objects, sand and grit. Primary treatment removes the materials that can be easily collected from the raw waste water and disposed of. The typical materials that are removed during primary treatment include fats, oils, and greases, sand, gravels and rocks (also referred to as grit), larger settleable solids including human waste and floating materials. This step is done entirely with machinery, hence the name mechanical treatment.

Secondary treatment: It is designed to substantially degrade the biological content of the sewage such as are derived from human waste, food waste, soaps and detergent. The majority of municipal and industrial plants treat the settled sewage liquor using aerobic biological processes. For this to be effective, the biota require both oxygen and a substrate on which to live. There are number of ways in which this is done. In all these methods, the bacteria and protozoa consume bio degradable soluble organic contaminants (e.g. sugars, fats, organic short chain carbon molecules, etc.) and bind much of the less soluble fractions into floc.

Secondary treatment systems are classified as **fixed film** or suspended growth. Fixed-film treatment process including trickling filter and rotating biological contactors where the biomass grows on media and the sewage passes over its surface. In **suspended growth systems**—such as activated sludge—the biomass is well mixed with the sewage and can be operated in a smaller space than fixed-film systems that treat the same amount of water.

However, fixed-film systems are more able to cope with drastic changes in the amount of biological material and can provide higher removal rates for organic material and suspended solids than suspended growth systems.

Tertiary treatment: This treatment provides a final stage to raise the effluent quality before it is discharged to the receiving environment (sea, river, lake, ground, etc.). More than one tertiary treatment process may be used at any treatment plant. If disinfection is practiced, it is always the final process. It is also called "effluent polishing".

Q.94. Describe various primary sewage treatment units.

Ans. Primary treatment removes the materials that can be easily collected from the raw waste water and disposed off eg; fats, oils, and greases, sand, gravels and rocks, larger settleable solids including human waste and floating materials.

Influx (influent) and removal of large objects: In the mechanical treatment, the influx (influent) of sewage water is strained to remove all large objects that are deposited in the sewer system, such as rags, sticks, condoms, sanitary towels/napkins/tampons, cans, fruit, etc. This is most commonly done with a manual or automated mechanically raked screen. This type of waste is removed because it can damage or clog the equipment in the sewage treatment plant.

Sand and grit removal: Primary treatment typically includes a sand or grit channel or chamber where the velocity of the incoming waste water is carefully controlled to allow sand, grit and stones to settle, while keeping the majority of the suspended organic material in the water column. This equipment is called a detritor or sand catcher. Sand grit and stones need to be removed early in the process to avoid damage to pumps and other equipment in the remaining sewage treatment stages. Sometimes there is a sand washer (grit classifier) followed by a conveyor that transports the sand to a container for disposal. The contents from the sand catcher may be fed into the incinerator in a sludge processing plant, but in many cases, the sand and grit is sent to a landfill.

Sedimentation: Many plants have a sedimentation stage where the sewage is allowed to pass slowly through large tanks, commonly called "primary clarifiers" or "primary sedimentation tanks". The tanks are large enough that fecal solids can settle and floating material such as grease and oils can rise to the surface and be skimmed off. The main purpose of the primary stage is to produce a generally homogeneous liquid capable of being treated biologically and a sludge that can be separately treated or processed. Primary settlement tanks are usually equipped with mechanically driven scrapers that continually drive the collected sludge towards a hopper in the base of the tank from where it can be pumped to further sludge treatment stages.

Q.95. How does secondary treatment help in treating sewage water? What are various treatment units involved in secondary treatment of sewage?

Ans. Secondary treatment degrades the biological content of the sewage such as are derived from human waste, food waste, soaps and detergent. The majority of municipal and industrial plants treat the settled sewage liquor using aerobic biological process. For this to be effective, the biota requires both oxygen and a substrate on which to live. There are number of ways in which this is done. In all these methods, the bacteria and protozoa consume biodegradable soluble organic contaminants (e.g., sugars, fats, organic short-chain carbon molecules, etc.) and bind much of the less soluble fractions into floc.

Secondary treatment systems are classified as **fixed film** or **suspended growth**. Fixed-film treatment process including trickling filter and rotating biological contactors where the biomass grows on media and the sewage passes over its surface.

In **suspended growth** such as activated sludge, the biomass is well mixed with the sewage and can be operated in a smaller space than fixed-film systems that treat the same amount of water.

However, fixed-film systems are more able to cope with drastic changes in the amount of biological material and can provide higher removal rates for organic material and suspended solids than suspended growth systems.

Roughing filters: These are intended to treat particularly strong or variable organic loads, typically industrial, to allow them to then be treated by conventional secondary treatment processes. Characteristics include typically tall, circular filters filled with open synthetic filter media to which waste water is applied at a relatively high rate. They are designed to allow high hydraulic loading and a high flow-through of air. On larger installations, air is forced through the media using blowers. The resultant waste water is usually within the normal range for conventional treatment processes.

Activated sludge: Activated sludge plants encompasses a variety of mechanisms and processes that use dissolved oxygen to promote the growth of biological floc that substantially removes organic material. It also traps particulate material and can, under ideal conditions, convert ammonia to nitrite and nitrate ultimately to nitrogen gas.

Fluidized bed reactors: The carbon adsorption following biological treatment are particularly effective in reducing both the BOD and COD to low levels. A fluidized bed reactor is a combination of the most common stirred tank packed bed, continuous flow reactors and has excellent heat and mass transfer characteristics.

In a fluidized bed reactor, the substrate is passed upward through the immobilized enzyme bed at a high velocity to lift the particles. However, the velocity must not be so high that the enzymes are swept away from the reactor entirely. This causes low mixing; these types of reactors are suitable for the exothermic reactions. It is most often applied in immobilized enzyme catalysis.

Filter beds (oxidizing beds): In older plants and plants receiving more variable loads, trickling filter beds are used where the settled sewage liquor is spread onto the surface of a deep bed made up of coke (carbonized coal), limestone chips or specially fabricated plastic media. Such media must have high surface areas to support the biofilms to that form.

The liquor is distributed through perforated rotating arms radiating from a central pivot. The distributed liquor trickles through this bed and is collected in drains at the base. These drains also provide a source of air which percolates up through the bed, keeping it aerobic. Biological films of bacteria, protozoa and fungi form on the media's surfaces and eat or reduce the organic content. This biofilm is grazed by insect larvae and worms which help maintain an optimal thickness.

Overloading of beds increases the thickness of the film leading to clogging of the filter media and ponding on the surface.

Biological aerated filters: Biological Aerated (or Anoxic) Filter (BAF) or Biofilters combine filtration with biological carbon reduction, nitrification or denitrification. BAF usually includes a reactor filled with a filter media. The media is either in suspension or supported by a gravel layer at the foot of the filter. The dual purpose of this media is to support highly active biomass that is attached to it and to filter suspended solids. Carbon reduction and ammonia conversion occurs in aerobic mode and sometime achieved in a single reactor while nitrate conversion occurs in anoxic mode. BAF is operated either in up flow or downflow configuration depending on design.

Membrane biological reactors: Membrane biological reactors (MBR) combines activated sludge treatment with a membrane liquid-solid separation process. The membrane component utilizes low pressure microfiltration or ultra-filtration membranes and eliminates the need for clarification and tertiary filtration. The membranes are typically immersed in the aeration tank (however, some applications utilize a separate membrane tank). One of the key benefits of a membrane bioreactor system is that it effectively overcomes the limitations associated with poor settling of sludge in conventional activated sludge (CAS) processes.

This technology permits bioreactor operation with considerably higher mixed liquor suspended solids (MLSS) concentration than CAS systems, which are limited by sludge settling. The process is typically operated at MLSS in the range of 8,000–12,000 mg/l, while CAS are operated in the range of 2,000–3,000 mg/l. The elevated biomass concentration in the membrane bioreactor process allows for very effective removal of both soluble and particulate biodegradable materials at higher loading rates. Thus, increased Sludge Retention Times, usually exceeding 15 days ensure complete nitrification even under extreme cold weather operating conditions.

The cost of building and operating a MBR is usually higher than conventional waste water treatment, however, as the technology has become increasingly popular and has gained wider acceptance throughout the industry, the life-cycle costs have been steadily decreasing.

Secondary sedimentation: The final step in the secondary treatment stage is to settle out the biological floc or filter material and produce sewage water containing very low levels of organic material and suspended matter.

Rotating biological contactors: Rotating biological contactors (RBCs) are mechanical secondary treatment systems, which are robust and capable of withstanding surges in organic load. The rotating disks support the growth of bacteria and micro-organisms present in the sewage, which breakdown and stabilize organic pollutants. Microorganisms need both oxygen to live and food to grow. Oxygen is obtained from the atmosphere as the disks rotate. As the micro-organisms grow, they build up on the media until they are sloughed off due to shear forces provided by the rotating discs in the sewage. Effluent from the RBC is then passed through final clarifiers where the micro-organisms in suspension settle as a sludge. The sludge is withdrawn from the clarifier for further treatment.

Q.96. Why is it essential to go for tertiary treatment of sewage?

Ans. Tertiary treatment provides a final stage to raise the effluent quality before it is discharged to the receiving environment (sea, river, lake, ground, etc.). More than one tertiary treatment process may be used at any treatment plant. If disinfection is practiced, it is always the final process. It is also called "effluent polishing".

Filtration: Sand filtration removes much of the residual suspended matter. Filtration over activated carbon removes residual toxins.

Lagooning: Lagooning provides settlement and further biological improvement through storage in large man-made ponds or lagoons. These lagoons are highly aerobic and colonization by native macrophytes, especially reeds, is often encouraged. Small filter feeding invertebrates such as Daphnia and species of Rotifera greatly assist in treatment by removing fine particulates.

Constructed wetlands: Constructed wetlands include engineered reed beds and a range of similar methodologies, all of which provide a high degree of aerobic biological improvement and can often be used instead of secondary treatment for small communities.

Waste removal: Waste water may contain high levels of the nutrients; nitrogen and phosphorus whose excessive release to the environment can lead to a buildup of nutrients, called eutrophication, which can in turn encourage the overgrowth of weeds, algae, and cyanobacteria (blue-green algae). This may cause an algal bloom, a rapid growth in the population of algae. The algae numbers are unsustainable and eventually most of them die. The decomposition of the algae by bacteria uses up so much of oxygen in the water that most or all of the animals die, which creates more organic matter for the bacteria to decompose. In addition to causing deoxygenation, some algal species produce toxins that contaminate drinking water supplies.

Different treatment processes are required to remove nitrogen and phosphorus.

Nitrogen removal: The removal of nitrogen is effected through the biological oxidation of nitrogen from ammonia (nitrification) to nitrate, followed by denitrification, the reduction of nitrate to nitrogen gas. Nitrogen gas is released to the atmosphere and thus removed from the water.

Phosphorus removal: Phosphorus can be removed biologically in a process called enhanced biological phosphorus removal. In this process, specific bacteria, called polyphosphate accumulating organisms, are selectively enriched and accumulate large quantities of phosphorus within their cells (up to 20% of their mass). When the biomass enriched in these bacteria is separated from the treated water, these biosolids have a high fertilizer value.

Disinfection: The purpose of disinfection in the treatment of waste water is to substantially reduce the number of microorganisms in the water to be discharged back into the environment. The effectiveness of disinfection depends on the quality of the water being treated (e.g., cloudiness, pH, etc.), type of disinfection being used, the disinfectant dosage (concentration and time), and other environmental variables. Cloudy water will be treated less successfully since solid matter can shield organisms, especially from ultraviolet light or if contact times are low. Generally, short contact times, low doses and high flows all militate against effective disinfection. Common methods of disinfection include ozone, chlorine, or ultraviolet light. Chloramine, which is used for drinking water, is not used in waste water treatment because of its persistence.

Q.97. Briefly explain the overall method of sewage water treatment.

Ans. Sewage can be treated close to where the sewage is created, in a decentralized system (in septic tanks, biofilters or aerobic treatment systems), or be collected and transported by a network of pipes and pump stations to a municipal water treatment plant, a centralized system. Sewage treatment generally involves three stages, called primary, secondary and tertiary treatment.

Primary treatment *consists* of temporarily holding the sewage in a quiescent basin where heavy solids can settle to the bottom while oil, grease and lighter solids float to the surface. The settled and floating materials are removed.

Secondary treatment removes dissolved and suspended biological matter. Secondary treatment is typically performed by indigenous, water-borne micro-organisms in a managed habitat.

Tertiary treatment, *Treated* water is sometimes disinfected chemically or physically (for example, by lagoons and microfiltration) prior to discharge into a stream, river, bay, lagoon or wetland, or it can be used for the irrigation of a golf course, green way or park.

Primary treatment includes Screening, Flow Equalization, Neutralization, Coagulation, Primary sedimentation.

Secondary treatment includes Biological treatment like Activated sludge process, Trickling filter, Membrane Bio Reactor, Secondary clarifier

Tertiary treatment includes Heavy metal removal, Disinfection (UV treatment).

Pretreatment: Pretreatment of sewage water removes materials that can be easily collected from the raw sewage before they damage or clog the pumps and sewage lines of primary treatment clarifiers.

Screening: The influent sewage water passes through a bar screen to remove all large objects like cans, rags, sticks, plastic packets etc. carried in the sewage stream. This is done with an automated mechanically raked bar screen in modern plants serving large populations, whilst in smaller or less modern plants, a manually cleaned screen may be used.

Grit removal: Pretreatment may include a sand or grit channel or chamber, where the velocity of the incoming sewage is adjusted to allow the settlement of sand, grit, stones, and broken glass.

Flow equalization: Clarifiers and mechanized secondary treatment are more efficient under uniform flow conditions. Equalization basins may be used for temporary storage of diurnal or wet weather flow peaks. Basins provide a place to temporarily hold incoming sewage during plant maintenance and a means of diluting.

Fat and grease removal: In some larger plants, fat and grease are removed by passing the sewage through a small tank where skimmers collect the fat floating on the surface. Air blowers in the base of the tank may also be used to help recover the fat as a froth.

Primary clarifier: In the primary sedimentation stage, sewage flows through large tanks, the tanks are used to settle sludge while grease and oils rise to the surface and are skimmed off.

Secondary treatment: Secondary treatment systems are classified as *fixed-film* or *suspended-growth* systems.

Surface-aerated basins (lagoons): Most biological oxidation processes for treating industrial waste waters have in common the use of oxygen (or air) and microbial action. Surface-aerated basins achieve 80 to 90 percent removal of BOD with retention times of 1 to 10 days.

Filter beds (oxidizing beds): In older plants and those receiving variable sewage loadings, trickling filter beds are used where settled sewage liquor is spread onto the surface of a bed made up of coke limestone chips or specially fabricated plastic media. The liquor is typically distributed through perforated spray arms. The distributed liquor trickles through the bed and is collected in drains at the base. Biological films of bacteria, protozoa and fungi form on the media's surfaces and eat or otherwise reduce the organic content.

Membrane bioreactors combine activated sludge treatment with a membrane liquid-solid separation process. The membrane component uses low pressure microfiltration or ultrafiltration membranes and eliminates the need for clarification and tertiary filtration. The membranes are immersed in the aeration tank.

Secondary sedimentation: The final step in the secondary treatment stage is to settle out the biological floc or filter material through a secondary clarifier and to produce sewage water containing low levels of organic material and suspended matter.

Tertiary treatment: The purpose of tertiary treatment is to provide a final treatment stage to further improve the effluent quality before it is discharged to the receiving environment (sea, river, lake, ground, etc.).

Filtration: Sand filtration removes much of the residual suspended matter. Filtration over activated carbon, also called *carbon adsorption,* removes residual toxins.

Disinfection: The purpose of disinfection in the treatment of waste water is to substantially reduce the number of microorganisms in the water to be discharged back into the environment for the later use of drinking, bathing, irrigation, etc. The type of disinfectant used, the disinfectant dosage (concentration and time), and other environmental variables affect the disinfection. Cloudy water will be treated less successfully, since solid matter can shield organisms, especially from ultraviolet light or if contact times are low. Generally, short contact times, low doses and high flows all act against effective disinfection. Common disinfectant include ozone, chlorine, ultraviolet light, or sodium hypochlorite. Chloramine, which is used for disinfecting drinking water, is not used in the treatment of waste water because of its persistence.

Overall process of sewage water treatment is shown in the Fig. 3.9.

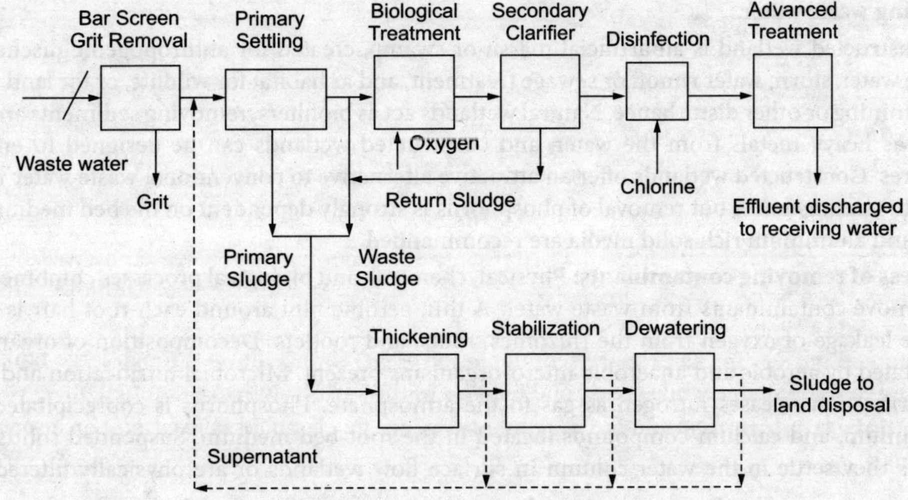

Fig. 3.9: Overall process of sewage water treatment

Q.98. Briefly describe sewage sludge treatment and disposal method.

Ans. The sludges accumulated in a waste water treatment process must be treated and disposed of in a safe and effective manner. The purpose of digestion is to reduce the amount of organic matter and the number of disease-causing microorganisms present in the solids. The most common treatment options include anaerobic digestion, aerobic digestion, and composting. The choice of a waste water solid treatment method depends on the amount of solids generated and other site-specific conditions. However, in general, composting is most often applied to smaller-scale applications followed by aerobic digestion and then lastly anaerobic digestion for the larger-scale municipal applications.

Anaerobic digestion: Anaerobic digestion is a bacterial process carried out in the absence of oxygen. The process can either be *thermophilic* digestion, in which sludge is fermented in tanks at a temperature of 55°C, or *mesophilic*, at a temperature of around 36°C. Though allowing shorter retention time (and thus smaller tanks), thermophilic digestion is more expensive in terms of energy consumption for heating the sludge.

One major feature of anaerobic digestion is the production of biogas, which can be used in generators for electricity production and/or in boilers for heating purposes.

Aerobic digestion: Aerobic digestion is a bacterial process occurring in the presence of oxygen. Under aerobic conditions, bacteria rapidly consume organic matter and convert it into carbon dioxide. The operating costs are much greater than for anaerobic digestion because of the energy costs needed to add oxygen to the process.

Composting: Composting is also an aerobic process that involves mixing the waste water solids with sources of carbon such as sawdust, straw or wood chips. In the presence of oxygen, bacteria digest both the waste water solids and the added carbon source and, in doing so, produce a large amount of heat.

Sludge disposal: When a liquid sludge is produced, further treatment may be required to make it suitable for final disposal. Typically, sludges are thickened (dewatered) to reduce the volumes and transported off site for disposal. No process completely eliminates the need to dispose of biosolids. However, an additional step some cities are taking to superheat the waste water sludge and convert it into small pelletized granules that are high in nitrogen and other organic materials. This product is then sold to local farmers and turf farms as a soil amendment or fertilizer, reducing the amount of space required to dispose of sludge in landfills.

Q.99. Describe the role and removal mechanism of contaminants in constructed wetlands used for treating waste water.

Ans. A constructed wetland is an artificial marsh or swamp, created for anthropogenic discharge such as waste water, storm water runoff or sewage treatment, and as habitat for wildlife, or for land reclamation after mining or other disturbance. Natural wetlands act as biofilters, removing sediments and pollutants such as heavy metals from the water, and constructed wetlands can be designed to emulate these features. Constructed wetlands offer an attractive alternative to conventional waste water treatment in certain circumstances but removal of phosphorus is strongly dependent on the bed medium. Calcium, iron and aluminium rich solid media are recommended.

Process of removing contaminants: Physical, chemical, and biological processes combine in wetlands to remove contaminants from waste water. A thin aerobic film around each root hair is aerobic due to the leakage of oxygen from the rhizomes, roots, and rootlets. Decomposition of organic matter is facilitated by aerobic and anaerobic micro-organisms present. Microbial nitrification and subsequent denitrification releases nitrogen as gas to the atmosphere. Phosphorus is coprecipitated with iron, aluminium, and calcium compounds located in the root-bed medium. Suspended solids are filtered out as they settle in the water column in surface flow wetlands or are physically filtered out by the medium within subsurface flow wetland cells. Harmful bacteria and viruses are reduced by filtration and adsorption by biological films on the rock media in subsurface flow and vertical flow systems.

Q.100. Define watershed management. What are its types?

Ans. Watershed is defined as land area bounded by divide line from which water drains under influence of gravity in to stream, lakes, reservoir. e.g., Pits, dams, Farm, ponds etc. Watershed Management is the management of rainfall and resultant run off.

Its various types are micro, mini and macro.

Q.101. What is the aim, objectives and approaches to watershed management?

Ans. Watershed management focuses on how to manage the river basin for optimum utilization of water resource and thereby sustainable development of the society. As the development goals have widened from increasing food production to include poverty reduction and environmental sustainability, protecting the environment is a big challenge and greater emphasis should be given to check the exploitation of the natural resources base. Soil erosion and land degradation coupled with declining per capita availability of land and freshwater are posing serious threat to environment. This is becoming more intense with the increasing population causing food security problems. Hence efforts are needed for management of natural resources for increased productivity of the soils. Several government and non-government agencies have launched watershed development and management projects to tackle some of these problems with the objectives of soil conservation, improving the land productivity and promoting appropriate technologies for efficient and sustainable use of natural resources. However, many watershed projects around the world have not performed well because of the poor community participation.

To ensure their sustainability, water resources must be viewed holistically, both in their natural state and in balancing competing demands on them—domestic, agricultural, industrial (including energy), and environmental. Sustainable management of water resources requires systemic, integrated decision-making that recognizes the interdependence in three areas. First, decisions on land use also affect water, and decisions on water also affect the environment and land use. Second, decisions on our economic and social future currently organized by socioeconomic sectors and fragmented, affect the hydrology and ecosystems in which humans live. Third, decisions at the international, national, and local levels are interrelated.

Q.102. Discuss in detail the water shed management practices.

Ans. The watershed is defined as the land area from which water drains under gravity to a common drainage channel. Thus watershed is a delineated area with a well-defined topographic boundary and one water outlet. The watershed can range from a few square kilometers to few thousand square kilometers in size.

Watershed management practices:

In watershed management the aspects of development are considered with regard to availability of the resources. The practices of conservation and development of land and water are taken up with respect to their suitability for people's benefit as well as sustainability. Various measures taken up for watershed management include the following:

Water harvesting: Proper storage of water is done with provision for use in dry seasons in low rainfall areas. It also helps in moderation of flood.

Afforestation and agro-forestry: In watershed development, afforestation and crop plantation play a very important role. They help to prevent soil erosion and retention of moisture. In high rainfall areas, woody trees are grown in between crops to substantially reduce the runoff and loss of fertile soil. In

Dehradun tree like Eucalyptus, Leucaena and grasses like chrysopogon are grown along with maize or wheat to achieve the objectives. Woody trees grown successfully in such agro-forestry programs include Sheesham, Teak and Keekar which have been used in watershed areas of river Yamuna.

Mechanical measures for reducing soil erosion and runoff losses: Several mechanical measures like terracing, bunding, bench terracing, no till farming, contour cropping, strip cropping etc. are used to minimize runoff and soil erosion particularly on the slopes of watersheds. Bundling has proved to be a very useful method in reducing runoff, peak discharge and soil loss in Dehradun.

Scientific mining and quarrying: Due to improper mining, the hills lose stability and get disturbed resulting in landslides, rapid erosion etc. Contour trenching at an interval of one meter on overburdened dump, planting some soil binding plants land draining of water courses in the mined area are recommended for minimizing the destructive effects of mining in watershed areas.

Public participation: People's involvement including the farmers and tribal's is the key to the success of any watershed management program, particularly the soil and water conservation. The communities are to be motivated for protecting a freshly planted areas and maintaining a water harvesting structure implemented by the government or some external agency (NGO) independently or by involving the locale people. Properly educating the people about the campaign and its benefits or sometimes paying certain incentives to them can help effective people's participation. Watershed management in Himalayan region is of vital importance since most of the watersheds of our country lie there. Several anthropogenic activities accelerate its slope instability which should be prevented and efforts should be made to project the watershed by preventing overgrazing, terracing and contour farming to check runoff and erosion etc. On steeper slopes with sliding faces, straw mulching tied with thin wires and ropes helps in establishing the vegetation and stabilizing the slopes.

Q.103. How do the nutrients from agricultural fields affect the watershed?

Ans. Enrichment of nutrients (Eutrophication) from surrounding watershed affects the penetration of light through the water, causing damage to the characteristic of water and aquatic life.

Q.104. Describe the history of watershed development in india?

Ans. A watershed is a logical, natural planning unit for sustainable agricultural research and development particularly when environmental considerations are emphasized. Hydrologically, watershed could be defined as an area from which the runoff drains through a particular point in the drainage system.

India began to look at the watershed development programs in the 1970s for increasing land controlling land degradation and increasing the productivity of soils. In the 1970s, watershed development held no special significance for the development community in India, however by the end of the 1980s the situation changed radically. Initially watershed projects concentrated on soil and water conservation issues. A decade later, it was felt that technical and physical works alone would not lead to the desired objectives of watershed development and it must also take into account the social, financial and institutional aspects of rural development. Watershed programs have been established over a diverse range of rain-fed agro eco regions in India.

Watersheds in India are broadly grouped into five agroclimatic zones:

(*i*) Trans-Gangetic Plain zone,
(*ii*) Western Himalayan zone,
(*iii*) Western Plateau and Hill zone,
(*iv*) Gujarat Plains and Hill zone, and
(*v*) Southern zone.

Due to inherent heterogeneity of agro-climatic characteristics over different regions, they have divergent potentials and opportunities. Depending upon the size of the watersheds, these are broadly divided into micro and macro watersheds. Watersheds with areas up to 1250 hectares were classified as micro-watersheds, whereas the macro-watersheds were those, of greater area.

In 1994, the Ministry of Rural Development (MORD) produced a set of guidelines for implementing its watershed programmes, which aimed to tackle the concerns related to the realization of the full benefits of watershed work. This progressive policy was essentially people-centered and it incorporated good practice from NGO and government policy, such as awareness raising, bottom-up planning, partnerships with NGOs, and community participation. Since 1994-95 Ministry of rural areas and employment, government of India has implemented nearly 10000 watersheds.

Q.105. What are the advantages of watershed projects?

Ans. Advantages include:

- Improved access to drinking water in project areas during drought.
- Increase in cultivation area leading to increase in employment.
- Increase in crop yield, resulting better income to rural population.
- Improved availability of fodder for animals and increase in milk yield.
- Increase in employment and involvement of women.
- Increase in net returns from all crops.
- Decrease in soil erosion.
- Restoration of ecological balance.

Q.106. Describe current status of watershed management in India.

Ans. Food production increased markedly in many parts of the world during the twentieth century. In India, the rate of growth of food production surpassed the rate of population growth, resulting in increased food per capita. Increased food production has required rapid expansion of irrigated area and water diversion to agriculture, which has resulted in groundwater overdraft and stream flow depletion in some basins. Water scarcity resulting from basin closure has numerous implications for water management, including increased competition and an imperative to improve water productivity in order to maintain growth in the agricultural economy. Basin closure also occurs in the context of global climate change, which may impact water availability and introduce additional uncertainties. Water resources analysis in regions experiencing closure should be carried out in a basin context, because uses in one area affect water availability in downstream areas.

In India, water management has been a government priority for irrigation and drinking water supplies. Today the scope for expanding surface and ground water sources is limited. Thus better management of existing water resources, through a judicious mix of delegation of responsibility and authority to local institutions and large-scale investment in re-directing surplus water to deficit areas is needed.

Q.107. What is water harvesting and its necessity?

Ans. Water harvesting is the capture, diversion, and storage of rainwater for plant irrigation and other uses. It is appropriate for large scale landscapes such as parks, schools, commercial sites, parking lots, and apartment complexes, as well as small scale residential landscapes. The problem of water shortage in arid and semi-arid regions is one of low rainfall and uneven distribution throughout the season, which makes rainfed agriculture a risky enterprise. The irrigation water are often at the break of overuse and therefore untapped sources of (irrigation) water have to be sought for increasing agricultural productivity. Water harvesting for dry-land agriculture is a traditional water management technology to ease future water scarcity in many arid and semi-arid regions of world.

Historically, harvested rain water provided water for drinking, landscape watering, and for agricultural uses. Once urban areas started to develop, centralized water supply systems replaced the need to harvest water. Harvesting rainwater can reduce the use of drinking water for land scape irrigation. It is also an effective water conservation tool and proves more beneficial when coupled with the use of native, low-water-use and desert-adapted plants. Additionally, rainwater is available free of charge and puts no added strain on the municipal supply or private wells.

As the appropriate choice of technique depends on the amount of rainfall and its distribution, land topography, soil type and soil depth and local socio-economic factors, these systems are very site specific. The water harvesting methods applied strongly depend on local conditions and include such widely differing practices as bunding, pitting, micro-catchments water harvesting, flood water and ground water harvesting.

Q.108. What are various types of water harvesting?

Ans. Three groups of water harvesting techniques is described below:

Rainwater harvesting: Rainwater harvesting is defined as a method for inducing, collecting, storing and conserving local surface runoff for agriculture in arid and semi-arid regions. Three types of water harvesting are covered by rainwater harvesting.

(*a*) Water collected from roof tops, courtyards and similar compacted or treated surfaces is used for domestic purpose or garden crops.

(*b*) Micro catchment water harvesting is a method of collecting surface runoff from a small catchment area and storing it in the root zone of an adjacent infiltration basin. The basin is planted with a tree, a bush or with annual crops.

(*c*) Macro-catchment water harvesting, also called harvesting from external catchments, is the case where runoff from hill-slope catchments is conveyed to the cropping area located at hill foot on flat terrain.

Flood water harvesting can be defined as the collection and storage of creek flow for irrigation use. Flood water harvesting, also known as 'large catchment water harvesting' or 'Spate Irrigation', may be classified into following two forms:

(*a*) In case of '**floodwater harvesting within stream bed**', the water flow is dammed and as a result, inundates the valley bottom of the flood plain. The water is forced to infiltrate and the wetted area can be used for agriculture or pasture improvement.

(*b*) In case of '**floodwater diversion**', the wadi water is forced to leave its natural course and conveyed to nearby cropping fields. It is difficult to give exact figure on the present area under various forms of floodwater harvesting systems.

Groundwater harvesting is employed to cover traditional as well as unconventional ways of ground water extraction. Qanat systems, underground dams and special types of wells are few examples of the groundwater harvesting techniques. For example, **Qanats**, widely used in Iran, Pakistan, North Africa and in Spain, consists of a horizontal tunnel that taps underground water in an alluvial fan, brings it to the surface due to gravitational effect. Qanat tunnels have an inclination of 1-2% and a length of up to 30 km. Many are still maintained and deliver steadily water to fields for agriculture production and villages for drinking water supply.

Groundwater dams like 'subsurface dams' and '**sand storage dams**' are other examples of groundwater harvesting. They obstruct the flow of ephemeral streams in a river bed; the water is stored in the sediment below ground surface and can be used for aquifer recharge. Sand filled reservoirs have the following advantages:

- *Evapor*ation losses are reduced,
- No reduction in storage volume due to siltation,
- Stored water is less susceptible to pollution, and
- Health hazards due to mosquito breeding are avoided.

Groundwater harvesting does not play the same role globally as rain and floodwater harvesting.

Q.109. Describe the objectives and methods of rainwater harvesting.

Ans. Rainwater harvesting is a technique of increasing the recharge of groundwater by capturing and storing rainwater. This is done by constructing special water-harvesting structures like dug wells, percolation pits, lagoons, check dams etc. Rainwater, wherever it falls, is captured and pollution of this water is prevented.

The annual average rainfall in India is 1200 mm; however, in most places it is concentrated over the rainy season, from June to September. In fact Cherapunji in Assam (the place receiving the second highest annual rainfall as11000 mm) still suffers from water scarcity. The water flows with runoff and there is little vegetation to check the runoff and allow infiltration. Till now there is hardly any rain-water harvesting being done in this region, thereby losing all the water that comes through rainfall.

Rainwater harvesting has the following objectives:

- To reduce run off loss.
- To avoid flooding of roads.
- To meet the increasing demands of water.
- To raise the water table by recharging ground water.
- To reduce ground water contamination.
- To supplement ground water supplies during lean seasons.

Rainwater can be mainly harvested by anyone of the following methods:

- By storing in tanks or reservoirs above or below ground.
- By constructing pits, dug wells, lagoons, trench or check dams on small rivulets.
- By recharging the ground water.

Q.110. Describe the modern techniques of rain water harvesting.

Ans. Traditional method of rainwater harvesting: In India, it is an old practice in high rainfall areas to collect rainwater from roof tops into storage tanks. In foot-hills water flowing from springs are collected by embankment type water storage. In Himalayan foot-hills people use the hollow bamboos as pipelines to transport the water of natural springs. Rajasthan is known for its "Tankas" (underground tanks) and "Khadins" (Embankments) for harvesting rainwater. In our ancient times, we had adequate Tellabs, Baawaris, Johars, Hauz etc. in every cities, village and capital cities of our Kings which were used to collect rainwater and ensure adequate water supply in dry periods.

Modern techniques of rainwater harvesting: In arid and semi-arid regions artificial ground water recharging is done by constructing shallow percolation tanks. Check-dams made of any suitable native material (brush, polls, rocks, plants, loose rocks, wire nets, stones, slabs, sacks etc.) are constructed for harvesting runoff from large catchment areas.Rajendra singh of Rajasthan popularly known as "Waterman" has been doing a commendable job for harvesting rainwater by building check-dams in Rajasthan and he has been honored with the prestigious Magsaysay Award for his work.

Ground water flow can be intercepted by building ground water dams or storing water underground. As compared to surface dams, ground water dams have several advantages like minimum *evapor*ation loss, reduced chances of contamination etc.

In roof top rainwater harvesting which is a low cost and effective technique for urban houses and buildings, the rainwater from the top of the roofs is diverted to some surface tank or pit through a delivery system which can be later used for several purposes. Also, it can be used to recharge underground aquifers by diverting the stored water to some abandoned dug well or by using a hand pump.

All the above techniques of rainwater harvesting are low cost methods with little maintenance expenses. Rainwater harvesting helps in recharging the aquifers, improves ground water quality by dilution, improves soil moisture and reduces soil erosion by minimizing run-off water.

Q.111. What are the benefits of harvesting rainwater?

Ans.
- Water harvesting not only reduces potable water use and related costs, but also reduces off-site flooding and erosion by holding rainwater on the site.
- If large amounts of water are held in pervious areas where water penetrates easily, some of the water may percolate to the water table.
- Rainwater is a clean, salt-free source of water for plants.
- Rainwater harvesting can reduce salt accumulation in the soil which can be harmful to root growth. When collected, rainwater percolates into the soil, forcing salts down and away from the root zone area. This allows for greater root growth and water uptake, which increases the drought tolerance of plants.
- Limitations of water harvesting are few and are easily met by good planning and design.

Q.112. What is water recycling?

Ans. Though recycling is a term generally applied to aluminium cans, glass bottles, and newspapers, water can be recycled as well. *Water recycling is reusing treated waste water for purposes such as agricultural and landscape irrigation, industrial processes, toilet flushing, and replenishing a ground water basin (referred to as ground water recharge). Water* is sometimes recycled and reused onsite; for example, when an industrial facility recycles water used for cooling processes. A common type of recycled water is water that has been reclaimed from municipal waste water, or sewage. The term water recycling is generally used synonymously with water reclamation and water reuse.

A common example of unplanned water recycling occurs when cities draw their water supplies from rivers, which receive waste water discharges upstream from those cities. Water from these rivers has been reused, treated, and piped into the water supply a number of times before the last downstream user withdraws the water. Planned projects are those that are developed with the goal of beneficially reusing a recycled water supply.

Q.113. What are the benefits of recycled water?

Ans. Recycled water can satisfy most water demands, as long as it is adequately treated to ensure water quality appropriate for the use. As for any water source that is not properly treated, health problems arise from drinking or being exposed to recycled water if it contains disease-causing organisms or other contaminants.

Recycled water is most commonly used for non-potable (not for drinking) purposes, such as agriculture, landscape, public parks, and golf course irrigation. Other non-potable applications include cooling water for power plants and oil refineries, industrial process water for such facilities as paper mills and carpet dyers, toilet flushing, dust control, construction activities, concrete mixing, and artificial lakes.

Although most water recycling projects have been developed to meet non-potable water demands, a number of projects use recycled water indirectly1 for potable purposes. These projects include recharging ground water aquifers and augmenting surface water reservoirs with recycled water. In ground water recharge projects, recycled water can be spread or injected into ground water aquifers to augment ground water supplies, and to prevent salt water intrusion in coastal areas.

Indirect potable reuse refers to projects that discharge recycled water to a water body before reuse. Direct potable reuse is the use of recycled water for drinking purposes directly after treatment. While direct potable reuse has been safely used in Namibia (Africa), it is not a generally accepted practice elsewhere.

Q.114. Describe the environmental benefits of water recycling.

Ans. In addition to providing a dependable, locally-controlled water supply, water recycling provides tremendous environmental benefits. By providing an additional source of water, water recycling can help find ways to decrease the diversion of water from sensitive ecosystems.

Other benefits include decreasing waste water discharges and reducing and preventing pollution. Recycled water can also be used to create or enhance wetlands and riparian habitats. Environmental benefits of water recycling is listed below:

- ***Water recycling can decrease diversion of freshwater from sensitive ecosystems:*** Plants, wildlife, and fish depend on sufficient water flows to their habitats to live and reproduce. The lack of adequate flow, as a result of diversion for agricultural, urban, and industrial purposes, can cause deterioration of water quality and ecosystem health.

  Water users can supplement their demands by using recycled water, which can free considerable amounts of water for the environment and increase flows to vital ecosystems.
- ***Water recycling decreases discharge to sensitive water bodies:*** In some cases, the impetus for water recycling comes not from a water supply need, but from a need to eliminate or decrease waste water discharge to the ocean, an estuary, or a stream. By avoiding the conversion of salt water marsh to brackish marsh, the habitat for endangered species can be protected.
- ***Recycled water may be used to create or enhance wetlands and riparian (stream) habitats:*** Wetlands provide many benefits, which include wildlife and wildfowl habitat, water quality improvement, flood diminishment, and fisheries breeding grounds. For streams that have been impaired or dried from water diversion, water flow can be augmented with recycled water to sustain and improve the aquatic and wildlife habitat.
- ***Water recycling can reduce and prevent pollution:*** When pollutant discharges to oceans, rivers, and other water bodies are curtailed, the pollutant loadings to these bodies are decreased. Moreover, in some cases, substances that can be pollutants when discharged to a body of water can be beneficially reused for irrigation. For example, recycled water may contain higher levels of nutrients, such as nitrogen, than potable water. Application of recycled water for agricultural and landscape irrigation can provide an additional source of nutrients and lessen the need to apply synthetic fertilizers.

Q.115. Discuss various water sources and need and ways/methods of water conservation.

Ans. Water sources include:

- Fresh water
- River
- Stream
- Pond
- Ocean

Need for water conservation:

- Population increases water requirement also increases.
- Due to deforestation annual rainfall decreases.
- Over exploitation of ground water.

Ways of water conservation:

- Reducing *evapor*ation loss.
- Reducing irrigation loss.
- Reuse water.
- Avoid sewage discharge.

Water conservation method:

- Rain water harvesting
- Watershed management

Q.116. Describe in detail about the water conservation methods.

Ans. Water being one of the most precious and indispensable resources needs to be conserved. The following strategies can be adopted for conservation of water.

- **Decreasing run-off losses:** Huge water-loss occurs due to run-off on most of the soils, which can be reduced by allowing most of the water to infiltrate into the soil. This can be achieved by using contour cultivation, terrace farming, water spreading, chemical treatment or improved water-storage system.

(*a*) **Contour cultivation:** On small furrows and ridges across the slopes trap rainwater and allow more time for infiltration. Terracing constructed on deep soils have large water-storage capacity. On gentle slopes trapped run off is spread over a large area for better infiltration.

(*b*) **Conservation-bench terracing:** It involves construction of a series of benches for catching the runoff water.

(*c*) **Water spreading is done by channeling or lagoon-leveling:** In channeling, the water flow is controlled by a series of diversions with vertical intervals. In lagoon leveling, small depressions are dug in the area so that there is temporary storage of water.

(*d*) **Chemical wetting agents (Surfactants):** These seem to increase the water intake rates when added to normal irrigated soil.

(*e*) **Surface crop residues, tillage, mulch, animal residues etc**. help in reducing run-off by allowing more time for water to penetrate into the land.

(*f*) **Chemical conditioners like gypsum ($CaSO_4.2H_2O$)** when applied to sodic soils improve soil permeability and reduce run off. Another useful conditioner is HPAN (hydrolyzed polyacrylonitrile)

(*g*) **Water-storage structures like farm ponds**, dug-outs etc. build by individual farmers can be useful measures for conserving water through reduction of runoff.

- **Reducing evaporation losses:** This is more relevant in humid regions. Horizontal barriers of asphalt placed below the soil surface increase water availability and increase crop yield This is more effective on sandy soil but less effective on loamy sand soils. A co-polymer of starch and acrylonitrile called 'super slumper' has been reported to absorb water up to 1400 times its weight. This chemical has been found to be useful for sandy soils.
- **Storing water in soil:** Storage of water takes place in the soil root zone in humid regions when the soil is wetted to field capacity. By leaving the soil fallow for one season water can be made available for the crop grown in next season.
- **Reducing irrigation losses:**

(*a*) Use of lined or covered canals to reduce seepage.

(*b*) Irrigation in early morning or late evening to reduce *evapor*ation losses.

(*c*) Sprinkling irrigation and drip irrigation to conserve water.

(d) Growing hybrid crop varieties with less water requirements and tolerance to saline water help conserve water.

- **Reuse of water:**

(*a*) treated waste water can be used for ferti-irrigation.

(*b*) using grey water from washings, bath-tubs etc. for watering gardens, washing cars or paths help in saving fresh water.

- **Preventing wastage of water:** This can be done in households, commercial buildings and public places.

(*a*) Closing taps when not in use. (*b*) Repairing any leakage from pipes.

(*c*) Using small capacity flush in toilets.

- **Increasing block pricing:** The consumer has to pay a proportionately higher bill with higher use of water. This helps in economic use of water by the consumers.

❑❑❑

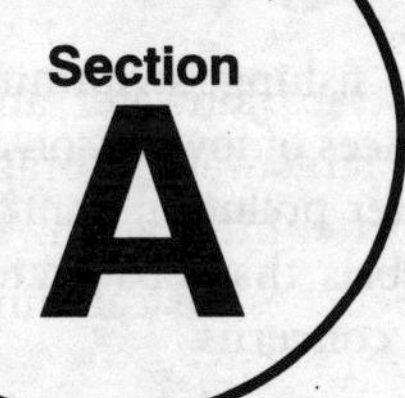

Section A 4 Waste Water Treatment

SHORT QUESTIONS AND ANSWERS ON WASTE WATER TREATMENT

Q.1. What is waste water?

Ans. Waste water is "used" water that is deposited into drains by residents and businesses. Flushing the toilet, taking a shower, and running the washing machine are all examples of everyday activities that produce waste water. This waste water drains down the sink or toilet and travels through a network of underground pipes known as the sanitary sewer system. It includes substances such as human waste, food scraps, oils, soaps and chemicals. In homes, this includes water from sinks, showers, bathtubs, toilets, washing machines and dishwashers. Businesses and industries also contribute their share of used water that must be cleaned.

Q.2. What are the effects of waste water pollutants?

Ans. If waste water is not properly treated, then the environment and human health can be negatively impacted. These impacts can include harm to fish and wildlife populations, oxygen depletion, beach closures and other restrictions on recreational water use, restrictions on fish and shellfish harvesting and contamination of drinking water.

Q.3. Give some examples of waste water pollutants and their harmful effects.

Ans. Some examples of pollutants found in waste water and the potentially harmful effects these substances can have on ecosystems and human health are listed below:

- decaying organic matter and debris can use up the dissolved oxygen in a lake so fish and other aquatic biota cannot survive;
- excessive nutrients, such as phosphorus and nitrogen (including ammonia), can cause eutrophication, or over-fertilization of receiving waters, which can be toxic to aquatic organisms, promote excessive plant growth, reduce available oxygen, harm spawning grounds, alter habitat and lead to a decline in certain species;
- chlorine compounds and inorganic chloramines can be toxic to aquatic invertebrates, algae and fish;
- bacteria, viruses and disease-causing pathogens can pollute beaches and contaminate shellfish populations, leading to restrictions on human recreation, drinking water consumption and shellfish consumption;
- metals, such as mercury, lead, cadmium, chromium and arsenic can have acute and chronic toxic effects on species.
- other substances such as some pharmaceutical and personal care products, primarily entering the environment in waste water effluents, may also pose threats to human health, aquatic life and wildlife.

Q.4. What is the need for waste water treatment?

Ans. Waste water treatment is needed so that we can use our rivers and streams for fishing, swimming and drinking water. Pollution in the urban waterways results in frequent occurrences of low dissolved oxygen, fish kills, algal blooms and bacterial contamination. Early efforts in water pollution control prevented human waste from reaching water supplies or reduced floating debris that obstructed shipping. Pollution problems and their control were primarily local, not national, concerns.

Since then, population and industrial growth have increased demands on our natural resources, altering the situation dramatically. Progress in abating pollution has barely kept ahead of population growth, changes in industrial processes, technological developments, changes in land use, business innovations, and many other factors.

Increases in both the quantity and variety of goods produced can greatly alter the amount and complexity of industrial wastes and challenge traditional waste water treatment technology. The application of commercial fertilizers and pesticides, combined with sediment from growing development activities, continues to be a source of significant pollution as runoff washes off the land.

The basic function of the waste water treatment plant is to speed up the natural processes by which water purifies itself. In earlier years, the natural treatment process in streams and lakes was adequate to perform basic waste water treatment. As our population and industry grew to their present size, increased levels of treatment prior to discharging domestic waste water became necessary.

Q.5. How waste water is collected for treating?

Ans. The most common form of pollution control consists of a system of sewers and waste water treatment plants. The sewers collect municipal waste water from homes, businesses, and industries and deliver it to facilities for treatment before it is discharged to water bodies or land, or reused.

During the early days people living in both the cities and the countryside used cesspools to dispose of domestic waste water. Cities began to install waste water collection systems later because of an increasing awareness of water borne disease.

The use of sewage collection systems has brought improvements to public health.

Q.6. Briefly describe the conventional waste water treatment processes.

Ans. Conventional waste water treatment process is described below.

1. **Preliminary treatment: Screening:** Upon arrival via the sewer system, the waste water is sent through a bar screen, which removes large solid objects such as sticks and rags. Leaving the bar screen, the waste water flow is slowed down entering the grit tank. This allows sand, gravel, and other heavy material that was small enough not to be caught by the bar screen to settle to the bottom. All the collected debris from the grit tank and bar screen is disposed of at a sanitary landfill or recycled.

   Waste water entering the treatment plant includes items like wood, rocks, and even dead animals. Unless they are removed, they could cause problems later in the treatment process. Most of these materials are sent to a landfill.

2. **Pumping:** The waste water system relies on the force of gravity to move sewage from home to the treatment plant. So waste water-treatment plants are located on low ground, often near a river into which treated water can be released. If the plant is built above the ground level, the waste water has to be pumped up to the aeration tanks. From here on, gravity takes over to move the waste water through the treatment process.

3. **Aerating:** One of the first steps that a water treatment facility can do is to just shake up the sewage and expose it to air. This causes some of the dissolved gases (such as hydrogen sulfide, which smells

like rotten eggs) that taste and smell bad to be released from the water. Waste water enters a series of long, parallel concrete tanks. Each tank is divided into two sections. In the first section, air is pumped through the water.

4. **Primary treatment:** Primary treatment is the second step in waste water treatment. It allows for the physical separation of solids and greases from the waste water. The screened waste water flows into a primary settling tank where it is held for several hours. This allows solid particles to settle to the bottom of the tank and oils and greases to float to the top. As organic matter decays, it uses up oxygen. Aeration replenishes the oxygen. Bubbling air through the water also keeps the organic material suspended while it forces 'grit' (sand and other small, dense particles) to settle out. Grit is pumped out of the tanks and taken to landfills.

   Waste water then enters the second section or sedimentation tanks. Here, the sludge (the organic portion of the sewage) settles out of the waste water and is pumped out of the tanks. Some of the water is removed in a step called thickening and then the sludge is processed in large tanks called digesters.

5. **Secondary treatment:** Secondary treatment is a biological treatment process that removes dissolved organic material from waste water. The partially treated waste water from the settling tank flows by gravity into an aeration tank. Here it is mixed with solids containing micro-organisms that use oxygen to consume the remaining organic matter in the waste water as their food supply. The aeration tank uses air bubbles to provide the mixing and the oxygen, both of which are needed for the micro-organisms to multiply.

   From here the liquid mixture, composed of solids with micro-organisms and water, is sent to the final clarifier. Here the solids settle to the bottom where some of the material is sent to the solids handling process, and some is recirculated to replenish the population of micro-organisms in the aeration tank to treat incoming waste water.

   As sludge is settling to the bottom of the sedimentation tanks, lighter materials are floating to the surface. This 'scum' includes grease, oils, plastics, and soap. Slow-moving rakes skim the scum off the surface of the waste water. Scum is thickened and pumped to the digesters along with the sludge.

   Many plants also use filtration in sewage treatment. After the solids are removed, the liquid sewage is filtered through a substance, usually sand, by the action of gravity. This method gets rid of almost all bacteria, reduces turbidity and colour, removes odours, reduces the amount of iron, and removes most other solid particles that remained in the water. Water is sometimes filtered through carbon particles, which removes organic particles.

6. **Final treatment: Killing pathogens:** Waste water that remains is disinfected to kill harmful micro-organisms before being released into receiving waters. Although there are many methods available to kill these micro-organisms, chlorine and ultraviolet disinfection are the most widely used.

   Dechlorination occurs in the final waste water treatment step. A solution of sodium bisulfite is added to the chlorinated effluent to remove residual chlorine.

   Following disinfection and dechlorination, the treated waste water (now called final effluent) can be returned to the receiving waters from which it came. The flow is conveyed to an outfall and discharged through a series of diffusers into a surface water body or stream.

   Finally, the waste water flows into a 'chlorine contact' tank, where the chemical chlorine is added to kill bacteria, which could pose a health risk, just as is done in swimming pools. The chlorine is mostly eliminated as the bacteria are destroyed, but sometimes it must be neutralized by adding other chemicals. This protects fish and other marine organisms, which can be harmed by the smallest amounts of chlorine.

The production of water of satisfactory quality is dependent upon the elimination or destruction of two groups of living organisms (1) the pathogenic microorganisms, which may infect humans through the use of contaminated water and (2) the algae and other aquatic growths, which may render water aesthetically unfit for human consumption. The treated water (called effluent) would then be discharged to a local river or the ocean.

Waste water residuals: Solids processing.

Primary solids from the primary settling tank and secondary solids from the clarifier are sent to the digester. During this process, micro-organisms use the organic material present in the solids as a food source and convert it to by-products such as methane gas and water. Digestion results in a 90% reduction in pathogens and the production of a wet soil-like material called "biosolids" that contain 95-97% water. To remove some of this water and reduce the volume, mechanical equipment such as filter presses or centrifuges are used to squeeze water from the biosolids. The biosolids are then sent to landfills, incinerated, or beneficially used as a fertilizer or for soil amendment.

Another part of treating waste water is dealing with the solid-waste material. These solids are kept for 20 to 30 days in large, heated and enclosed tanks called 'digesters.' Here, bacteria break down (digest) the material, reducing its volume, odours, and getting rid of organisms that can cause disease. The finished product is mainly sent to landfills, but sometimes can be used as fertilizer.

Our home is serviced by either a sanitary and a storm sewer or a combined sewer, a single pipe which conveys both sanitary sewage and storm water. It is 99.90 percent water, with only 0.01 percent of the waste water being dissolved and suspended solid material. The cloudiness of sewage is caused by suspended particles which, in untreated sewage, range from 100 to 350 mg/l. A measure of the strength of the waste water is biochemical oxygen demand, or BOD_5. The BOD_5 measures the amount of oxygen microorganisms require in five days to break down sewage. Untreated sewage has a BOD_5 ranging from 100 mg/l to 300 mg/l. Pathogens or disease-causing organisms are present in sewage. Coliform bacteria are used as an indicator of disease-causing organisms. Sewage also contains nutrients (such as ammonia and phosphorus), minerals, and metals. Ammonia can range from 12 to 50 mg/l and phosphorus can range from 6 to 20 mg/l in untreated sewage.

The sewer systems are built to follow the natural slope of the land. This design allows gravity to do most of the work transporting the waste water to treatment facilities. In some situations, pump stations are constructed in low-lying areas to force the waste water uphill to the treatment facility.

The organic and inorganic material found in this waste water is too abundant to be handled by the environment. Therefore, treatment of waste water is an essential process that prevents contamination and destruction of waterways, drinking water sources, and natural water resources.

Sewage treatment is a multi-stage process to renovate waste water before it reenters a body of water, is applied to the land, or is reused. The goal is to reduce or remove organic matter, solids, nutrients, disease causing organisms, and other pollutants from waste water. Each receiving body of water has limits to the amount of pollutants it can receive without degradation. Therefore, each sewage treatment plant must treat it to allowable levels of BOD_5, suspended solids, coliform bacteria, and other pollutants.

Q.7. Differentiate between domestic sewage and industrial waste water.

Ans. The term waste water is commonly used to describe liquid wastes that are collected and transported to a treatment facility through a system of sewers. **Waste water is generally divided into two broad classifications: domestic waste water and industrial waste water.**

Domestic waste water: Domestic sewage is waste water discharged from sanitary conveniences in residential, office, commercial, factories and various institutional properties. It is a complex mixture containing primarily water (approximately 99%) together with organic and inorganic constituents.

These constituents or contaminants comprise of suspended, colloidal and dissolved materials. Domestic sewage, since it contains human wastes, also contains large numbers of micro-organisms and some of these can be pathogenic (disease causing). Waterborne bacterial diseases that can be present in sewage include cholera, typhoid and tuberculosis. Viral diseases can include infectious hepatitis. Inorganic constituents include chlorides and sulphates, various forms of nitrogen and phosphorous, as well as carbonates and bicarbonates. Proteins and carbohydrates constitute about 90% of the organic matter in domestic sewage. These arise from the excreta, urine, food wastes, and waste water from bathing, washing, and laundering, and because of the latter, soaps, detergents, and other cleaning products can be found as well.

Domestic sewage has a flow pattern which typically shows two peaks — in the morning before the start of working hours and in the evening after the population has returned from work. Typically, these hydraulic peaks would also become more distinct as the sewage flows considered come from smaller populations and consequently smaller sewer networks. Variations in sewage characteristics across a given community tend to be relatively small although variation across communities can be more readily detected. Notwithstanding these variations, the composition of domestic sewage is such that it lends itself well to biological treatment in terms of the availability of and balance between carbonaceous components and nutrients.

The biodegradability of sewage can be estimated by considering its Chemical Oxygen Demand (COD) and the corresponding BOD_5 (5 day BOD), and is indicated by its COD:BOD_5 and BOD_5:N:P ratios. This would typically be about 1.5:1 and 25:4:1 respectively. The nitrogen, N, would typically be in the form of organic nitrogen and ammonia nitrogen (Amm-N). Nitrates (NO_3-N) would not be expected to be present as conditions in the sewers would be such that nitrate formation is unlikely while nitrate degradation because of anoxic reactions is likely. The phosphorous (P) would be a combination of organic and phosphate (PO_4) forms. The pH of sewage would be within the range of 6–9 and this is generally considered suitable for biological processes. Examples of values of BOD_5, TSS (Total Suspended Solids), and TKN (Total Kjeldhal Nitrogen) which have been used for purposes of plant design are 250, 300 and 40 mg/ l respectively. As indicated earlier in this paragraph, sewage characteristics can vary across communities and a raw sewage BOD_5 of 500mg/l has been encountered.

Industrial waste water: Industrial waste water characteristics vary with each type of discharge. Food processing, dairy operations and meat packing all have wastes that are very high in BOD, sometimes over 1000 mg/l. This BOD is high in dissolved sugars, fats, and proteins and low in suspended material. It will dramatically increase the loading on the secondary treatment processes. Some food process wastes can also have a high pH, like potato processing and some vegetable canning, or a low pH, like green chili processing or fruit canning.

Toxins, like heavy metals coming from metal plating, battery shops and heavy equipment manufacturing, and solvents from industries like body shops, dry cleaners, and furniture manufacturing can kill the microorganisms that are needed to treat the waste. Copper, chrome, lead, and cyanide are all chemicals that are commonly found in heavy industrial manufacturing discharges.

Hospitals and other medical facilities have the potential to discharge wastes that are radioactive or represent a biohazard.

In some cases, pretreatment of these industrial discharges may be required. Systems that deal with industrial discharges should adopt a pretreatment ordinance to cover conditions that the system may need to impose on the industrial dischargers in order to maintain their NPDES requirements on the treatment plant effluent. If local ordinances are different from specific categorical limits imposed in the Federal Regulations, the most stringent limits of the two must be followed. Pretreatment ordinances can require actual treatment of the industrial waste prior to discharge to the system, or it may levy a monetary surcharge based on the addition loading that results from the discharge.

Any industrial discharger is considered to be a "Significant Industrial User" or SIU if:

(*a*) The discharge requires pretreatment. (*b*) The flow is over 25,000 gpd (gallon per day), or

(*c*) The flow is more than 5% of total average plant BOD or hydraulic load

Q.8. What is the impact of industrial waste water on water bodies?

Ans. The impacts of industrial waste water on water bodies is as follows:

(*a*) **Physical effects:** These include impact on clarity of the water and interference to oxygen dissolution in it. Water clarity is affected by turbidity which may be caused by inorganic suspended Solids and/or organic particulates suspended in the water (Volatile Suspended Solids). The latter may undergo biodegradation and thereby also have oxidation effects. Turbidity reduces light penetration and this reduces photosynthesis while the resultant loss in clarity would adversely affect the food gathering capacity of aquatic animals because these may not be able to see their prey. Very fine particulates may also clog the gill surfaces of fishes and thereby affecting respiration and eventually killing them. Settleable particulates may accumulate on plant foliage and bed of the water body forming sludge layers which would eventually smother benthic organisms. As the sludge layers accumulate, they may eventually become sludge banks and if the material in these is organic then its decomposition would give rise to malodors.

In contrast to the settleable material, particulates lighter than water eventually float to the surface and form a scum layer. The latter also interferes with the passage of light and oxygen dissolution. Because of the former, these scum layers affect photosynthesis. Discharge limits on waste water or treated waste water discharges typically have a value for TSS such as 30-50 mg/l. Many industrial waste waters contain oil and grease (O and G). While some of the latter may be organic in nature, there are many which are mineral oils. Notwithstanding their organic or mineral nature, both types cause interference at the air-water interface and inhibit the transfer of oxygen. Apart from their interference to the transfer of oxygen from atmosphere to water, the O and G (particularly the mineral oils) may also be inhibitory. Unlike domestic sewage, industrial discharges can have temperatures substantially above ambient temperatures. These raise the temperatures of the receiving water and reduce the solubility of oxygen. Apart from this, rapid changes in temperature may result in thermal shock and this may be lethal to the more sensitive species. Heat, however, does not always have a negative impact on organisms as it may positively affect growth rates although there are limits here too, since the condition may favour certain species within the population more than others and over time biodiversity may be negatively affected.

(*b*) **Oxidation and residual dissolved oxygen:** Waterbodies have the capacity to oxygenate themselves through dissolution of oxygen from the atmosphere and photosynthetic activity by aquatic plants. Of the latter, algae often play an important role. However, there is a finite capacity to this re-oxygenation and if oxygen depletion, as a result of biological or chemical processes induced by the presence of organic or inorganic substances which exert an oxygen demand (*i.e.*, as indicated by the BOD or COD), exceeded this capacity then the dissolved oxygen (DO) levels would decline. The latter may eventually decline to such an extent that septic conditions occur. A manifestation of such conditions would be the presence of malodours released by facultative and anaerobic organisms. An example of this is the reduction of substances with combined oxygen such as sulphates by facultative bacteria and resulting in the release of hydrogen sulphide. The depletion of free oxygen would affect the survival of aerobic organisms. DO levels do not, however, need to drop to zero before adverse impacts are felt. A decline to 3–4 mg/l, which still means the water contains substantial quantities of oxygen, may already adversely affect higher organisms like some species of fish. If inhibitory substances are also present, then the DO level at which adverse effects may be felt can be even higher than before. The case of elevated water temperatures due to warm discharges is somewhat different. The elevated temperatures can affect metabolic rates positively (possibly two-fold for each 10°C rise in temperature) but elevated temperatures also reduce the solubility of oxygen in water. This would mean increasing demand for oxygen while its availability declines. Because of the impact of DO levels on aquatic life, much importance has been placed on determining the BOD value of a discharge. Typical BOD_5 limits set values are 20 and 50mg/l.

(*c*) **Inhibition or toxicity and persistence:** These effects may be caused by organic or inorganic substances and can be acute or chronic. Examples of these include the pesticides and heavy metals. Many industrial waste waters do contain such potentially inhibitory or toxic substances. The presence of such substances in an ecosystem may bias a population towards members of the community which are more tolerant to the substances while eliminating those which are less tolerant and resulting in a loss of biodiversity. For similar reasons, an awareness of the impact such substances have on biological systems is not only relevant in terms of protection of the environment but is of no less importance in terms of their impact on the biological systems used to treat industrial waste waters. Even successful treatment of such a waste water may not necessarily mean that the potability of water in a receiving waterbody would not be affected. For example, small quantities of residual phenol in water can react with chlorine during the potable water treatment process giving rise to chlorophenols which can cause objectionable tastes and odours in the treated water. Apart from the organic pollutants which are potentially inhibitory or toxic, there are those which are resistant to biological degradation. Such persistent compounds can be bioaccumulated in organisms resulting in concentrations in tissues being significantly higher than concentrations in the environment and thereby making these organisms unsuitable as prey/food for organisms (including Man) higher up the food chain. While some organic compounds may be persistent, metals are practically non-degradable in the environment.

(*d*) **Eutrophication:** The discharge of nitrogenous and phosphorous compounds into receiving waterbodies may alter their fertility. Enhanced fertility can lead to excessive plant growth. The latter may include algal growth. The subsequent impact of such growth on a waterbody can include increased turbidity, oxygen depletion, and toxicity issues. Algal growth in unpolluted water bodies is usually limited because the water is nutrient limiting. While nutrients would include marco-nutrients like nitrogen, phosphorous, and carbon, and micro-nutrients like cobalt, manganese, calcium, potassium, magnesium, copper, and iron which are required only in very small quantities, the focus in concerns over eutrophication would be on phosphorous and nitrogen as quantities of the other nutrients in the natural environment are often inherently adequate. In freshwaters, the limiting nutrient is usually phosphorous while in estuarine and marine waters it would be nitrogen. Treatment of industrial waste water (or domestic sewage for that matter) can then target the removal of either phosphorous or nitrogen, depending on the receiving waterbody, to ensure that the nutrient limiting condition is maintained. When the nutrient limiting condition is no longer present in the water body, and when other conditions such as ambient temperature are appropriate, excessive algal growth or algal blooms (e.g., the red tide) may occur. Apart from aesthetic issues, such algal blooms may affect the productivity of the fisheries in the locale. It should be noted that not all industrial waste waters contain excessive quantities of nutrients, macro and micro. This deficiency, if there is, results in process instability and/or the proliferation of inappropriate microbial species during biological treatment of the waste waters.

Bulking sludge is a manifestation of such an occurrence. To address this deficiency, nutrients supplementation is required. The quantities used should be carefully regulated so that an excessive nutrients condition is not inadvertently created and these excess nutrients subsequently discharged with the treated effluent. In terms of BOD:N:P, the optimal ratio for biotreatment is often taken as 100:5:1 while the minimum acceptable condition can be 150:5:1.

(*e*) **Pathogenic effects:** Pathogens are disease-causing organisms and an infection occurs when these organisms gain entry into a host (e.g., man or an animal) and multiply therein. These pathogens include bacteria, viruses, protozoa, and helminths. While domestic and medical related waste waters may typically be linked to such micro-organisms (and especially the bacteria and viruses), industrial waste waters are not typically associated with this category of effects. The exception to this is waste waters associated with the sectors in the agro-industry dealing with animals. The concern here would be the presence of such organisms in the waste water which is discharged into a receiving water body and diseases, if any, are then transmitted through the water. While many of these organisms can be satisfactorily addressed with adequate disinfection of the treated effluent and raw potable water supplies during the water treatment process, there are those which cannot be dealt with so easily. Two examples of such organisms, *Cryptosporidum* and *Giardia*, belong to the protozoa family. The difficulty is that the infected host does not necessarily shed the organism but is likely also to shed its eggs oroocysts. The latter can unfortunately be resistant to the usual disinfection processes.

Q.9. What is the impact of industrial waste water on municipal waste water treatment plants?

Ans. On some occasions industrial waste waters are discharged into a sewerage system serving commercial and residential premises. Such a combination of waste water streams is known as municipal waste water and the quality of such a mixture of waste waters can vary depending on the proportion of industrial waste waters in it and the type of industries contributing the industrial waste water streams. Usually the domestic and commercial components in municipal waste water can be expected to provide some buffering in terms of the characteristics of the combined flow. This is then expected to enable the combined waste water to be treated easily compared to the treatment of the industrial waste water on its own. However, even where the option of discharging into a sewerage system is available, some degree of pretreatment is frequently required at the factory before such discharge is permitted. Such pretreatment may include pH adjustment to 6–9 and BOD_5 reduction to 400 mg/l. This is to protect the receiving sewers from corrosion and also protect the performance of the receiving treatment plant from an organic substrate overload.

All treatment plants should be planned giving serious consideration for the reduction of waste volume and strength, through process of chemical substitution, chemical recovery and recycling of water. The pollutional load is dealt with operations like segregation, neutralization, equalization, chemical precipitation, chemical oxidation and biological oxidation. Several chemicals are used to reduce the BOD by chemical coagulation such as alum, ferric sulphate, ferrous sulphate and ferric chloride. Lime or H_2SO_4 is used to adjust pH in this process. The dye waste may be economically treated by biological methods prior to equalization, neutralization and chemical oxidation.

The industrial waste when discharged into sewer not only chokes the sewer due to the deposition of solids but also reduces the cross section of the sewer arising out of the lime encrustature.

Chromium compounds in excess of 10-20mg disrupt the operation of the trickling filter. Sulphides are toxic to the micro organisms and are removed along with the sludge. The sludge is dried over sand drying beds and can be used as good manure. Chemical coagulation (Alum, ferric chloride, and ferrous sulphate) with or without prior neutralization followed by biological treatment is necessary for better quality of effluent.

Q.10. Give a brief account of the following with respect to industrial waste water.

(*a*) **Volume reduction** (*b*) **Strength reduction**

(*c*) **Neutralization** (*d*) **Equalization**

Ans. (*a*) **Volume reduction:** In general the first step in minimizing the effects of industrial wastes on relieving streams and treatment plants is to reduce the volume of such wastes. This may be accomplished by:

(*i*) Classification of wastes (*ii*) Conservation of waste water

(*iii*) Change in production to decrease wastes.

(*iv*) Re-using both industrial and municipal effluent as raw water supplies.

Classification of wastes: If wastes are classified so that manufacturing process waters are separated from cooling waters, the volume of water requiring intensive treatment may be reduced considerably. Sometimes it is possible to classify and separate the process waters themselves, so that only the most polluted once are treated and the relatively uncontaminated are discharged without treatment.

The 3 main classes of wastes are:

(A) Wastes from manufacturing process: These include waters used in forming paper, discharge from plating solutions in metal fabrication, those discharge from washing of milk cans in dairy plants etc.

(B) Waters used as cooling agents in industrial process: The volume of these wastes varies from one industry to another. Although cooling water can be become contaminated by small leaks, corrosion products or the effect of heat, these wastes contain little, if any organic matter and are classed as non-pollution.

(C) Wastes from sanitary uses: The volume depends on many factors including size of the plant, amount of waste producing materials washed from the floors and the degree of cleanliness required and number of workers in the unit.

Unfortunately, in most old plants process, cooling and sanitary waste water are mixed in one pipeline and many industries are paying little attention to segregating wastes to avoid stream pollution.

(*b*) **Strength reduction:** Waste strength reduction is the second major objective for an industrial plant concerned with waste treatment. The strength of wastes may be reduced by:

1. Process changes
2. Equipment modifications
3. Segregation of wastes
4. Equalization of wastes
5. By-product recovery
6. Proportioning of wastes
7. Monitoring waste streams

1. Process changes: In reducing the strength of wastes through process changes, the sanitary engineer is concerned with wastes that are most troublesome from a pollution standpoint.

2. Equipment modification: Changes in equipment can affect a reduction in the strength of the waste, usually by reducing the amounts of contaminants entering the waste stream. An outstanding example of waste strength reduction occurred in the dairy industry. The new cans were constructed with smooth necks so that they could be drained faster and more completely. This prevented a large amount of milk waste from entering streams and sewage plants.

3. Segregation of wastes: Segregation of wastes reduces the strength and/or the difficulty of treating the final waste from industrial plant. It usually results in two wastes: one strong and small in volume and the other weaker with almost the same volume as the original unsegregated waste. The small-volume strong waste can then be handled with methods specific to the problem it presents. In terms of volume reduction alone, segregation of cooling waters and storm waters from process waste will mean a saving in the size of the final treatment plant.

4. **Equalization of wastes:** Plants, which have many products, from a diversity of processes, prefer to equalize their wastes. This requires holding wastes for a certain period of time, depending on the time taken for the repetitive process in the plant. For example, if a manufactured item requires a series of operations that take eight hours, the plant needs an equalization basin designed to hold the wastes for that eight hours period. The effluent from an equalization basin is much more consistent in its characteristics than each separate influent to that same basin.
   Stabilization of pH and B.O.D and settling of Solids and Heavy Metals are among the objectives of equalization. Stable effluents are treated more easily and efficiently, than unstable ones by industrial and municipal treatment plants.
5. **By-product recovery:** All wastes contain by products, the exhausted materials used in the process. Since some wastes are very difficult to treat at low cost, it is advisable to consider the possibility of building a recovery plant which will produce a marketable By-Product and at the same time solve a troublesome wastes problem.
6. **Proportioning wastes:** By proportioning its discharge of concentrated wastes into the main sewer, a plant can often reduce the strength of its total waste to the point where it will need a minimum of final treatment or will cause the least damage to the stream or treatment plant. It may prove less costly to proportion one small but concentrated waste into the main flow. According to the rate of the main flow, then to equalize the entire waste of the plant in order to reduce the strength.
7. **Monitoring waste streams:** Accidental spills are often the sole cause of stream pollution or malfunctioning of treatment plants and these can be controlled and often eliminated completely, if all significant sources of wastes are monitored.

(*c*) Neutralization: Excessively acidic or alkaline wastes should not be discharged without treatment into a receiving stream. A stream is adversely affected by low or high pH values. This adverse condition is even more critical when sudden sludge of acids or alkalis are imposed upon the stream.

Acceptable methods of neutralization are:

1. Mixing wastes so that the net effect is a neutral pH.
2. Passing acid wastes through beds of limestone.
3. Mixing acid wastes with lime slurries.
4. Adding the proper proportions of concentrated solutions of caustic soda (NaOH) or soda ash (Na_2CO_3) to acid wastes.
5. Adding compressed CO_2 to alkaline wastes.
6. Adding sulfuric acid to alkaline wastes.

The material and method used should be selected on the basis of the overall cost, since material costs vary widely and equipment for utilizing various agents will differ with the method selected. The volume, kind and quality of acid or alkali to be neutralized are also factors in deciding which neutralizing agent to use.

(*d*) Equalization: Equalization is a method of retaining wastes in a basin so that the effluent discharged is fairly uniform in its characteristics (pH, colour, turbidity, alkalinity, B.O.D etc). A secondary but significant effect is that of lowering the concentration of effluent contaminants. A retention pond serves to level out the effects of peak loadings on the plant while substantially lowering the B.O.D and suspended solids load to the aeration unit.

Air is sometimes injected into these basins to provide:

1. Better mixing
2. Chemical oxidation of reduced compounds
3. Some degree of biological oxidation
4. Agitation to prevent suspended solids from settling.

The size and shape of the basins vary with the quantity of waste and the pattern of its discharge from the industry. The capacity should be adequate to hold and render homogeneous, all the wastes from the plant. Almost all industrial plants operate on a cycle basis; thus if the cycle of operations is repeated for every two hours, an equalization tank which can hold a two -hour flow will usually be sufficient.

The mere holding of waste, however is not sufficient to equalizing it. Each unit volume of waste discharged must be adequately mixed with other unit volumes of waste discharged many hours previously.

This mixing may be brought about in the following ways:

1. Proper distribution and baffling
2. Mechanical agitation
3. Aeration
4. Combination of all three.

Q.11. In what different ways the neutralization of industrial waste water is achieved?

Ans. Excessively acidic or alkaline wastes should not be discharged without treatment into a receiving stream. A stream is adversely affected by low or high pH values. This adverse condition is even more critical when sudden sludge of acids or alkalis are imposed upon the stream.

Acceptable Methods of Neutralization are:

- Mixing wastes so that the net effect is a neutral pH.
- Passing acid wastes through beds of limestone.
- Mixing acid wastes with lime slurries.
- Adding the proper proportions of concentrated solutions of caustic soda (NaOH) or soda ash (Na_2CO_3) to acid wastes.
- Adding compressed CO_2 to alkaline wastes.
- Adding sulfuric acid to alkaline wastes.

The material and method used should be selected on the basis of the overall cost, since material costs vary widely and equipment for utilizing various agents will differ with the method selected. The volume, kind and quality of acid or alkali to be neutralized are also factors in deciding which neutralizing agent to use.

Q.12. Why equalization and proportioning is of much importance in industrial waste water treatment? How is it carried out?

Ans. Equalization: Equalization is a method of retaining wastes in a basin so that the effluent discharged is fairly uniform in its characteristics (pH, colour, turbidity, alkalinity, B.O.D etc). A secondary but significant effect is that of lowering the concentration of effluent contaminants. A retention pond serves to level out the effects of peak loadings on the plant while substantially lowering the B.O.D and suspended solids load to the aeration unit.

Air is sometimes injected into these basins to provide:

- Better mixing
- Chemical oxidation of reduced compounds
- Some degree of biological oxidation
- Agitation to prevent suspended solids from settling.

The size and shape of the basins vary with the quantity of waste and the pattern of its discharge from the industry. The capacity should be adequate to hold and render homogeneous, all the wastes from the plant. Almost all industrial plants operate on a cycle basis; thus if the cycle of operations is repeated for every two hours, an equalization tank which can hold a two -hour flow will usually be sufficient.

The mere holding of waste, however is not sufficient to equalizing it. Each unit volume of waste discharged must be adequately mixed with other unit volumes of waste discharged many hours previously.

This mixing may be brought about in the following ways:

- Proper distribution and baffling
- Mechanical agitation
- Aeration
- Combination of all three.

Proportioning: Proportioning means the discharge of industrial wastes in proportion to the flow of municipal sewage in the sewers or to the stream flow in the receiving river. In most cases it is possible to combine equalization and proportioning in the same basin. The effluent from the equalization basin is metered into the sewer or stream according to a predetermined schedule. The objective of proportioning in sewers is to keep constant the percentage of industrial wastes to domestic sewage flow entering the municipal sewage plant. This procedure has several purposes:

- To protect municipal sewage treatment using chemicals from being impaired by a sudden overdose of chemicals contained in the industrial waste.
- To protect biological treatment devices from strong loads of industrial wastes which may inactivate the bacteria.
- To minimize fluctuations of sanitary standards in the treated effluent
- The rate of flow of industrial waste varies from instant to instant, as does the flow of domestic sewage system. Therefore, the industrial waste must be equalized and retained, then proportioned to the sewer or stream according to the volume of domestic sewage or stream.

Air is sometimes injected into these basins to provide:

- Better mixing
- Chemical oxidation of reduced compounds
- Some degree of biological oxidation
- Agitation to prevent suspended solids from settling.

The size and shape of the basins vary with the quantity of waste and the pattern of its discharge from the industry. The capacity should be adequate to hold and render homogeneous, all the wastes from the plant. Almost all industrial plants operate on a cycle basis; thus if the cycle of operations is repeated for every two hours, an equalization tank which can hold a two -hour flow will usually be sufficient.

The mere holding of waste, however is not sufficient to equalizing it. Each unit volume of waste discharged must be adequately mixed with other unit volumes of waste discharged many hours previously.

This mixing may be brought about in the following ways:

- Proper distribution and baffling
- Mechanical agitation
- Aeration
- Combination of all three.

Q.13. Discuss the following with reference to removal of dissolved inorganic solids from industrial waste water: (*a*) Dialysis (*b*) Reverse osmosis (*c*) Ion exchange.

Ans. (*a*) **Dialysis:** It is the separation of solute by means of their unequal diffusion through membranes. It is most useful in recovering pure solutions for reuse in manufacturing processes for e.g., caustic soda in textile industry. Dialysis work on the simple principle of passing a concentrated, impure caustic solution upward, counter current to a downstream water supply, from which it is separated by a semi-permeable membrane. The caustic soda permeates the membranes and goes into water more rapidly than the other impurities contained in the water The quantity of NaOH diffusing through the membrane diaphragm depends upon the time, the area of the dialyzing surface, the mean concentration difference and the temperature. Dialysis is an operation requiring very little operator attention and although its main role is to conserve raw materials and to reduce plant waste, at the same time it aids in waste treatment. With the introduction of acid resistant membranes, dialysis has been used successfully in the recovery of sulphuric acid in the copper, stainless steel and other industries. Some operation can recover as much as the 70 to 75% of the acid, but recovery as little as 20% may be justifying the process. In dialysis, the driving force of separation is natural diffusion because of concentration gradient. Electro dialysis is another form of dialysis in which the natural driving force is enhanced by the application of electrical energy. Electro-dialysis can achieve as much as 44 percent reduction in concentration of dissolved solids in industrial effluents.

(*b*) **Reverse osmosis:** It is process for separating relatively pure water or some other solvent from a less pure solution. The solution is passed over the surface of a specific semi-permeable membrane at a pressure in excess of the effective osmotic pressure of the feed solution. The permeating liquid is collected as the product and the concentrated feed solution is generally discarded. The membrane must be (1) highly permeable to water (2) highly impermeable to solutes (3) capable of withstanding the applied pressure without failure, (4) chemically inert, mechanically strong. Although several types of membranes have been developed, two types of membranes are generally used in commercial equipment. (1). The first is a symmetric of skinned cellulose acetate membranes made in flat or Tubular forms. (2). The second type of membrane is an aromatic polyamide or polyamide hydrazide. Reverse osmosis is an advanced unit operation in water treatment. Reverse osmosis membranes are capable of removing at least 90 percent of the dissolved solids in water as well as organics, bacteria and others impurities. Osmosis can be defined as the spontaneous passage of a liquid from dilute to a more concentrated solution across an ideal semipermeable membrane which allows the passage of the solvent (water) but not the dissolved solids (solutes). The transport of the water from one side of the membrane to the other continues until the pressure (p) is large enough to prevent any net transfer of the solvent (water) to the more concentrated solution. At equilibrium, the quantity of water passing in the either direction is equal, and pressure (p) is then defined as the osmotic pressure of the solution having that particular concentration and make up of dissolved solids. If a piston is placed on the more concentrated solution side of a semipermeable membrane and a pressure (P) is applied to the solution, the following conditions can be realized:

(*i*) P is less them the osmotic pressure of the solution and solvent flows (in the same direction) spontaneously toward the more concentrated solution. (*ii*) P equals the osmotic pressure of the solution and solvent flows at the same rate in both directions that is no net change in water levels. (*iii*) P is greater than the osmotic pressure of the solution and solvent flows from the more concentrated solution to the pure solvent side of the membrane leaving a higher concentration behind. This condition represents the phenomenon of reverse osmosis. Substantial energy is needed to constantly pressurized the incoming salty water is one of the drawback of the reverse osmosis.

(*c*) **Ion exchange:** Ion exchange is basically a process of exchanging certain undesirable cations and anions of the waste water for sodium, hydrogen or other ions in resinous materials. The resins, both natural and artificial are commonly referred to as Zeolites. When the ability of the zeolite bed to produce soft water is exhausted, the softner is temporarily cut out of service. It is then backwash to clean and hydraulically regrade the bed, regenerated with a solution of common salt; which remove the calcium and magnesium in the form of their soluble chlorides and restores the zeolite to its original conditions, rinsed free of these and the excess salt, and finally returned to service.

Q.14. List the techniques applied in removal of suspended solids.

Ans. There are three methods for the removal of suspended solids, which are as follows:

1. Screening. 2. Flotation 3. Sedimentation

1. **Screening:** Screening of industrial wastes is generally practiced on waste containing larger suspended solids of variable sizes. For e.g., from canneries, pulp and paper mills, or poultry processing plants. It is an economical and effective means of rapid separation of these larger suspended solids from the remaining waste material. In many cases screening alone will reduce the suspended solids to a low enough concentration to be acceptable for discharge into a municipal sewer or a nearby stream.

   Screens are available in sizes ranging from coarse (10 or 20 mesh) to fine (120 to 320 mesh). In the industry, generally two types of screens are used:

   (*i*) North screen (*ii*) Sweco screen

   (*i*) **North screen:** They are generally rotary, self-cleaning gravity-type, units. It has been used successfully in treating wastes from meat-packing, grain-washing, tanning, making, woollen and sea food plants.

   (*ii*) **Sweco screen:** They are circular, vibratory screens. They have been quite effectively used for screening wastes from food-packing processes such as meat and poultry packing or fruit and vegetable canning.

2. **Floatation:** The floatation process consists of creation of air bubbles in the waste body or tank by introducing air to the system. These air bubbles increase the buoyancy of the finer particles which could be easily lifted by skimming (*i.e.*, the process of removing floating grease or scum from the surface of the effluent in a tank).

   Advantages of floatation as waste water process are:

   (*i*) Grease, and light solids rising to the tops and grit and heavy solids settling to the bottom are all removed in one unit.

   (*ii*) Odour nuisance are minimized, because of short detention period.

   (*iii*) High overflow rates and short detention periods means smaller tank sizes, resulting in decreased space requirements and possible saving in construction costs

   Disadvantages of floatation are:

   (*i*) The additional equipment required results in higher operating costs.

   (*ii*) Floatation units generally do not give as effective treatment as gravity settling units, although the efficiency varies with the waste.

   (*iii*) More skilled maintenance is required for a flotation unit than for a gravity settling unit.

3. **Sedimentation:** Sedimentation processes are very effective in removing suspended solids in industrial waste water. The efficiency of sedimentation tank depends, in general on the following factors:

   - Detention periods.
   - Waste water characteristics
   - Tank design.
   - Particle size.
   - Velocity and density of particles.

   The clarifiers/tanks either rectangular or circular in shape are most commonly used in the application of sedimentation in waste water treatment. The design of the clarifiers are based on several factors, which are as follows:

   - Influent TSS concentration.
   - Effluent TSS concentration.
   - Surface loading
   - Detention time.
   - Sludge generation

Q.15. Explain any three methods of sludge disposal.

Ans. Both primary and secondary sewage treatments involve settling of particulate matter, and thus produce sludge. The concentration of solids in the primary sewage sludge is about 5%; the activated sludge contains about 1%; and the sludge from trickling filters has about 2% solids.

Thus the sludge is composed of almost entirely of water and volume reduction is the key to economic disposal. In addition to reducing its high water content, the sludge must be stabilized so that its biological activity and tendency towards putrefaction are reduced drastically.

The sludge is concentrated by gravity settling and floatation. After concentration, the sludge is subjected to anaerobic digestion in a digester in which the organic content of the sludge decomposes to give mainly methane and carbon dioxide and at the same time the bound water is released from the sludge. The sludge is then conditioned to improve its dewatering characteristics by adding chemicals like iron salts and polyelectrolytes. These chemicals bind the sludge particles together and encourage the release of water. The sludge is then heated under pressure and after a period of time the gel structure of the sludge breaks down so that the water is released.

The thickened sludge is then dewatered for efficient handling and disposal. The dewatering is accomplished by mechanical methods, the most common being centrifugation and filtration.

The dewatered sludge is then subjected to oxidation to reduce the organic content, with the consequent destruction of bacteria and significant reduction in their volume.

Incineration and oxidation are the two common methods employed for oxidation. Several methods are employed for the ultimate disposal of sludge. The wet digested sludge may be sprayed on to a cropland where it functions as fertilized Dried sludge may be used a land fill or soil conditioner. Wet or partially dewatered sludge or ash from incineration may be transported from the shore to dumping grounds at sea. The potential drawback to the use of sewage sludge as fertilizer in agricultural fields is the presence of both organic and inorganic toxic compounds. The former compounds are oxidation-resistant organic substances, such as organochlorine species which become bound in the organic matrix of the sludge. The inorganic toxicants are represented by heavy metals, mainly arsenic, cadmium, lead, mercury and zinc. These metals can be taken up by crops and introduced into the food chains or leached to the ground water.

Drying beds/sand bed drying: For small industrial waste treatment plants, sludge can be dewatered on open or covered sand beds. Drying of the sludge occurs by percolation and evaporation. The proportion of the water removed by percolation may vary from 20- 55%. Depending on the initial solids contents of the sludge and on the characteristics of the solids. The design and use of drying beds are affected by climatic conditions (rainfall and evaporation)

Sludge drying beds usually consists of 10-23 cms of sand, 20-46 cms of gravel or stone. The sand has an effective size of 0.3 to 1.2 mm and a uniformity coefficient < 5.0. Gravel is graded from 0.32 to 2.54 cm. The beds are provided with under drains spaced from 2.7 to 6.1 m apart. The under drain piping having a mini dia of 10 cm and minimum slope of 1% The filtration is returned to the treatment plants. Wet sludge is usually applied to the drying beds at depths of 20-30 cms. Removal; of the dried sludge varies with both individual judgment and final disposal means many cases, the bed turn over can be substantially increased by the use of chemicals. Slum treatment can be reduce the sludge drying time by 50%.The use of polymers can increase the rate of bed dewatering and also increase the depth of application.

Drying and incineration: A large volume of sludge can be reduced to a small volume of ash, which is free from organic matter and therefore easily disposable, by combination of heat drying and incineration. Incineration involves drying and combustion. Various types of incineration units are available to accomplish these reactions in single or combined units. In the incineration process, a sludge temperature is raised to 100°C, at which point moisture is evaporated from the sludge. The water vapour and air temperature are increased to ignition point. Some excess air is required for complete combustion and of the sludge. Self-sustaining combustion is often possible with dewatered waste sludges ones the burning auxiliary fuel raises the incineration temperature to ignition point. The primary end products of combustion are CO_2, SO_2 and ash.

Incineration can be accomplished in multiple hearth furnaces in which the sludge passes vertically through a series of hearths. In fluidized bed sludge particle are fed into a bed of sand fluidized by upward moving air. A temperature of 760-815°C is maintained in the beds, resulting in rapid drying and burning of sludge. Ash is removed from the bed by the upward flowing combustion gases.

Sanitary landfill: Sanitary landfill is used to burry garbage, refuse and sludge in a planned and methodical manner. It is a simple, effective and inexpensive method for disposing of dry matter such as refuse, but sludge is usually too liquid for this procedure. However, vacuum filter or sand bed dried sludge can be disposed off in this manner. The area proposed for the sanitary land fill should be easily accessible, yet removed from sources of water supply and recreational areas at the same time on the land which is not too costly. The suitability of soil and possible future use of the property are also imp. Consideration sanitary landfills should be located above the ground water level and 152 m away from water sources, particularly when the soil is sandy, gravely or of limestone. At the end of each days dumping, the sludge should be covered and compacted by a bulldozer or tractor.

Sludge barging: Sludge barging is a means of final disposal of sludge to a deep body of water usually an ocean. There is little theory involved in this method of treatment. Raw precipitated, digested or filtered sludge solids are pumped into a waiting ship barge. When the barge is fully loaded it transports the sludge to a suitability site far from the shore, where it is discharged usually by being pumped out deep under the surface of the water. The sludge should be somewhat concentrated before being loaded on the barge (to conserve space) but not too much that pumping difficulties arise.

Pumping raw or digested sludge directed into the ocean and disposing it by dilution technique is usually practiced in coastal cities. These sludges however, should not be contaminated by oil and grease, odour high coliform counts, toxic materials discolouration or be likely to create sludge banks or deposits.

Q.16. What is sludge thickening? Enumerate various methods and describe the gravity sludge thickener.

Ans. Thickening is a procedure used to increase the solids content of sludge by removing a portion of the liquid fraction. To illustrate, if waste activated sludge, which is typically pumped from secondary tanks with a content of 0.8% solids, can be thickened to a content of 4% solids, then a fivefold decrease in sludge volume is achieved. Thickening is generally accomplished by physical means including co-settling, gravity settling, flotation, centrifugation, gravity belt, and rotary drum. The volume reduction obtained by sludge concentration is beneficial to subsequent treatment processes, such as digestion, dewatering, drying and combustion from the following standpoints:

- Capacity of tanks and equipment required
- Quantity of chemicals required for sludge conditioning
- Amount of heat required by digesters and amount of auxiliary fuel required for heat drying or incineration, or both.

Most common techniques of thickening in sludge processing are:

- Gravitational settling
- Dissolved air flotation (DAF)
- Basket centrifuge
- Solid-bowl centrifuge
- Gravity belt filter
- Rotary drum

Thickening is the process by which biosolids are condensed to produce a concentrated solids product and a relatively solids-free supernatant. Thickening waste water solids reduces the volume of residuals, improves operation, and reduces costs for subsequent storage, processing, transfer, end use, or disposal. For example, thickening liquid-solids (slurry) from 3 to 6 percent will reduce the volume by 50 percent.

There are several different methods for thickening biosolids, including dissolved air floatation (DAF), centrifugal thickening, gravity belt thickening, and gravity thickening. Gravity thickening uses the natural tendency of higher-density solids to settle out of liquid to concentrate the solids.

Gravity thickeners consist of a circular tank (usually with a conical bottom) that is fitted with collectors or scrapers at the bottom. Primary and/or secondary solids are fed into the tank through a center well, which releases the solids at a low velocity near the surface of the tank. The solids settle to the bottom of the tank by gravity, and the scrapers slowly move the settled, thickened solids to a discharge pipe at the bottom of the tank. A *v* notch weir located at the top of the tank allows the supernatant to return to a clarifier. In addition, many systems also use a skimmer to collect and remove any floatables and grease that have accumulated at the top of the tank.

The biosolids concentration and thickening that occurs in the tank is achieved through three different settling processes, which include gravity settling, hindered settling, and compaction settling. Gravity settling occurs when solid particles travel downward due to their weight. Settlement continues as solids begin to concentrate near the bottom of the tank, but the settlement rate decreases as the solids concentrations increase. This is known as hindered settling.

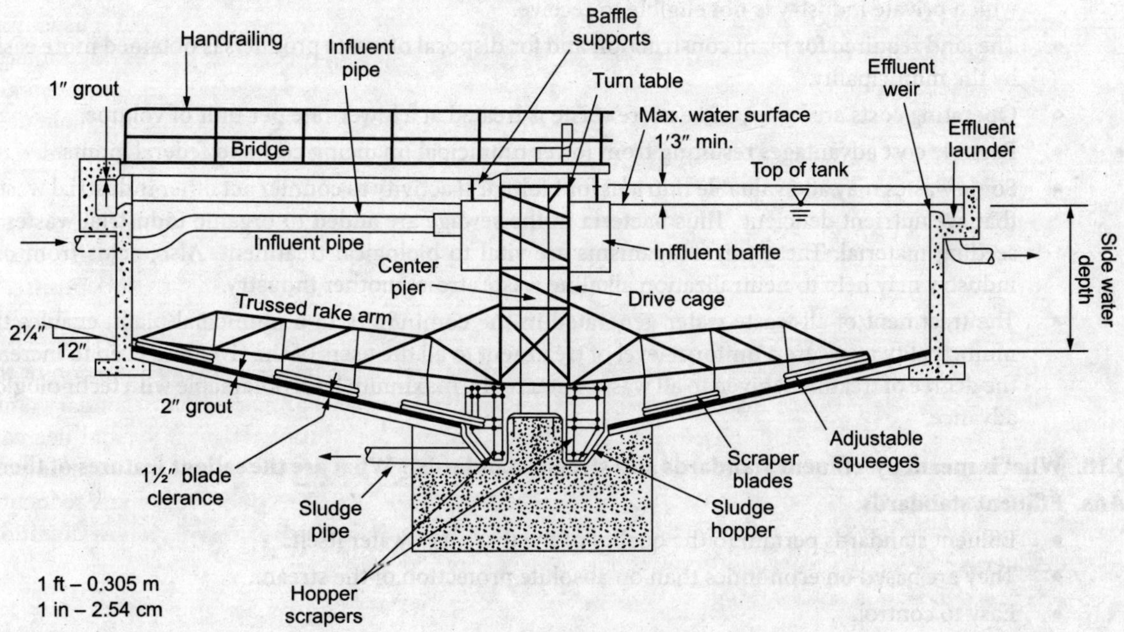

Fig. 4.1: Gravity thickener

Compaction settling occurs when bottom solids are further concentrated due to the pressure of solids on top of them.

Solids at the bottom of the tank can reach as high as 15 percent total solids (TS). A more typical result is 4 to 6 percent TS. Liquid at the surface of the tank is nearly clear, with suspended solids concentrations as low as 200 mg/l.

Q.17. What are the advantages and disadvantages of combined treatment of industrial raw waste with the domestic waste?

Ans. It is often possible and advisable for an industry to discharge its waste water directly into a municipal sewage treatment plant, where a certain portion of the pollution can be removed. A municipal sewage plant, if designed and operated properly can be handle almost any type and quantity of industrial waste. Hence one possibility that should be seriously considered is the cooperation of industry and municipality in the joint construction and operation of a municipal waste water treatment plant.

There are many **advantages** to be gained from such a joint venture:

- Here the responsibility is placed with one owner, while at the same time, the cooperative spirit between industry and municipality increases, particularly if the division of costs is mutually satisfactory.
- Only one chief operator is required, whose sole obligation is the management of the treatment plant *i.e.*, he is not burden by the miscellaneous duties often given to the industrial employee in charge of waste disposal and the chances of mismanagement and neglect which may result if industrial production men operate waste treatment plants, are eliminated.
- Since the operator of such a large treatment plant usually receives higher pay than separate domestic plant operators, better trained people are available.
- Even if identical equipment is required construction costs are less for a single plant than for two or more. Furthermore, municipalities can apply for state and or federal aid for plant construction, which private industry is not eligible to receive.
- The land required for plant construction and for disposal of waste products is obtained more easily by the municipality.
- Operating costs are lower, since more waste is treated at a lower rate per unit of volume.
- Possible cost advantages resulting from lower municipal financing cost and federal grants.
- Some wastes may add valuable nutrient for biological activity to counter act other industrial wastes that are nutrient deficient. Thus bacteria in the sewage are added to organic industrial wastes as seeding material. These microorganisms are vital to biological treatment. Also, acids from one industry may help to neutralization alkaline wastes from another industry.
- The treatment of all waste water generated in the community in a municipal plant, enables the municipality to assure a uniform level of treatment to all the users of the river and even to increase the degree of treatment given to all waste water to the maximum level obtainable with technological advance.

Q.18. What is meant by 'effluent standards and stream standards'? What are the salient features of them?

Ans. Effluent standards:

- Effluent standards pertain to the quality of the discharge water itself.
- They are based on economics than on absolute protection of the stream.
- Easy to control.
- Detailed stream analysis are not required.
- They do not establish an overall level of pollutant loading for a given water body.
- Ratio of waste water to stream flow are not considered.
- Treatment is obligatory irrespective of the size of industry.
- For effective protection of an overloaded stream, the effluent standards are required to be upgraded.
- Large industries have an edge over small industry.

Stream standards: Stream standards refer to the quality of the receiving water downstream from the origin of the waste water discharge.

- They are based on establishing classification of quality for a stream.
- The quality of the receiving water is regulated to maintain established stream classification.
- Prevention of excessive pollution/ Loading is limited to what the stream can assimilate.
- No consideration of type and location of industry.
- Allows public to establish goals for present and future water quality.
- Confusion of zone of different classification.
- Controversy over proportion of stream to be reserved for future usage (municipal, industrial, agriculture etc.).
- Opposition from industry/ public to change the established classification.
- A detailed stream analysis is required to determine the level of waste water treatment required to maintain the health of the ecosystem.
- Cost of treatment may affect the survival of industry.

Q.19. Enumerate and explain in detail the factors to be considered for stream sampling.

Ans. Before embarking on designing of any industrial waste water treatment facility a comprehensive stream-sampling program should be conducted. This will in turn help considerably for efficient operation of facility.

The factors to be considered before starting a sampling program are:

- Overall objectives of the program
- Total number of samples
- Points of collection
- Method of collection
- Data to be obtained
- Frequency of sample collection
- Time of the year for sampling
- Statistical handling of data
- Care of samples, prior to analysis

Overall objectives of the program: There are various objectives of stream sampling. In one instance the engineer may be concerned with the effect of an upstream industry on the water quality downstream, of special interest in this case might be the colour of the water in the receiving stream. In another instance, he may be attempting to ascertain the dissolved oxygen sag characteristics of the stream during the summer season only. In another case, he could concern that the stream characteristics comply with the standards stipulated by the state pollution control board. The importance of the 8 factors will depend to some degree on the overall objectives of the sampling program.

Total number of samples: The no. of samples required depends on the objectives of the programme and the amount of time allotted for completion of the same. The use of the few locations and more samples to define the results is usually much more reliable than using many stations with only a few samples from each. Also, samples are frequently taken over a long-time interval, during which the condition of the water course is subject to variation. In many instances, an attempt to test all condition by infrequent, random sampling produces no definite pattern and in fact may be misleading. It may be better to concentrate on well-defined frequent and intensive sampling

A well-planned survey with a specific objective will require a minimum sample. If for ex, someone wishes to determine the river characteristics, especially the DO profile during low flow periods, 2 or 3 samples collected at the proper time will suffice. These samples must be collected, however during extreme drought periods, if factors influencing the character of the stream to be clearly established

However, since the industrial processing is so varied and usually unpredictable, that many samples of the receiving water under all conditions must be collected in order to evaluate the effect of waste upon the stream.

Points of collection: Sampling points should be selected with great care and due consideration should be given to the sources of pollution, dilution by branch streams, changes in surrounding topography and slope of the river significant river side features also influent the choice of sampling points such as a municipal water intake, a park, an industrial area, a good fishing spot, a hotel or a camping site would each have a definite bearing on the usage of the stream. Since the acceptable pollution limits for water vary according to usage, samples should to collected and the record made of the condition of the stream just above and just below all such points of stream use are changed.

A minimum of 4 stream stations is recommended:

(*a*) An upstream site where the water is uncontaminated.

(*b*) Just below the source of pollution or dilution.

(*c*) Where the stream is in worst condition due to the source of pollution.

(*d*) A point midway between bottom of oxygen sag and the recovery of oxygen level.

Method of collection: Samples should be taken from 0.6 ft depth in streams less than 2 ft. The volume of samples depends on the number and type of analysis to be carried out. A standard type dissolved oxygen sampler is recommended for collecting most samples. Glass bottles with glass caps or polyethylene containers are most widely used. Any doubt about the cleanliness of the sample bottle can usually be dispelled by rinsing it first with some of the actual stream water, but special bottles are required for bacteriological samples.

Data to be collected: The scientific collection of data for stream analysis may be divided into three major categories *i.e.*, hydrologic factors, sources of pollution and water course sampling. The type of data to be obtained depends on the objectives of sampling programme and the amount of time and money available for the investigation. For example, if the oxygen resources of a section of a stream is required then, DO, water temperature and stream flow should be measured. If the sampling programme is of general nature, the stream analyst should undertake as many chemical, physical and biological tests as possible. For design of waste water treatment plants, the rate of flow, temperature, BOD and DO can supply sufficient information. In addition, data on pH, colour and turbidity may indicate the general physical condition of the stream. Biological analysis is required when the stream water is used for drinking, bathing or flushing. In this case, the coliform count is usually determined.

Frequency of sample collection: Samples should be collected as frequently as possible to provide a representative of total sample. The master sample should contain individual constituents of variation expected. For example, if the pH is known to vary from 4 to 10, individual samples with pH values of 4, 5, 6, 7, 8, 9 and 10 should appear in the composite sample at least once during each sampling period. If the situation requires instantaneous analysis, more individual samples with little or no composting are collected.

Time of the year for sampling: Time of the year is of at most importance when there is a deadline for producing results. In stream studies, dealing with industrial waste treatment is concerned primarily with the critical condition of pollution which generally exists when the environment is at its warmest and its stream flow slowest, and the man made pollution greatest. But many studies must be undertaken in the spring and autumn seasons due to stream conditions and man power problems. However, many times due to the urgency of the problem or unusual conditions of pollution, the investigations may have to be carried in any of the seasons. The objective of every stream analysis should however to be collect the data during critical stream conditions of temperature, flow and pollution load. It may sometime be necessary to project stream analysis to conditions which might obtain during future critical points. This is sometimes requiring even when a survey is made during the summer. Because the intensity of the problem varies to some degree from year to year and even from one period of the year to another.

Statistical handling of data: It is a well known fact that data can be manipulated to emphasize that aspect of the survey which the analyst fill the most important. For this the engineer must have a working knowledge of statistic and mathematics in order to convey this information in the best form to the layman.

The rate of flow during critical conditions is of importance to the industrial waste engineering. If he incorporates the minimum flow ever recorded in the stream, he may end up with unrealistic evaluation. On the other hand the use of the mean summer or low flow value can be dangerous, since lower than this often occurs. Therefore, some regulatory agencies make use of minimum of 7 days to review once in 10 years as the criteria for designing waste water treatment facilities.

Q.20. Explain the different zones of pollution in stream.

Ans. Zones of purification: When sewage is discharged into water, a succession of changes in water quality takes place. If the sewage is emptied into a lake in which currents about the outfall are sluggish and shift their direction with the wind, the changes occur in close proximity to each other and, as a result, the pattern of changes is not crisply distinguished. If, on the other hand, the water moves steadily away from the outfall, as in a stream, the successive changes occur in different river reaches and establish a profile of pollution which is well defined. However, in most streams, this pattern is by no means static. It shifts longitudinally along the stream and is modified in intensity with changes in season and hydrography.

When a single large charge of sewage is poured into a clean stream, the water becomes turbid, sunlight is shut out of the depths, and green plants, which by photosynthesis remove carbon dioxide from the water and release oxygen to it, die off. Depending on the stream velocity, the water soon turns nearly black. Odorous sulfur compounds are formed and solids settle to the bottom, forming a sludge. The settled solids soon decompose, forming gases such as ammonia, carbon dioxide, and methane or marsh gas. Scavenging organisms increase in number until they match the food supply.

The oxygen resources are drawn upon heavily and, when overloaded, become exhausted. Life in such waters is confined to anaerobic bacteria (which exist when no oxygen is available), larvae of certain insects such as mosquitoes, and a few worms. There are no fish; turtles are generally the only forms of higher life present. This condition is known as the **zone of degradation**.

In a second zone, or **zone of decomposition**, more solids settle out, the water becomes somewhat clearer, and sunlight penetrates the surface. Oxygen is absorbed from the atmosphere at the air-water interface permitting the establishment of aerobic (oxygen available) conditions. The aerobic bacteria continue the conversion of organic matter into nitrates, sulfates. and carbonates. These, together with the carbon dioxide produced by decomposition as well as by bacteria and plant life, are food sources. With sunlight now penetrating the water, and with abundant food, algae begin to flourish and form a green scum over the surface.

In the third zone, or **zone of recovery**, algae become more numerous and self-purification proceeds more rapidly. Green plants utilizing carbon dioxide and oxygen will liberate in the say time more oxygen than is consumed, thus hastening the recovery of the stream. Simultaneously, the fish that require little oxygen such as catfish and carp, are also found. As the dissolved oxygen increases, more types of fish appear.

After recovery, in the **zone of cleaner water**, fish find the stream highly favorable, as the algae support various aquatic insects and other organisms on which fish feed. The water is clear or turbid according to concentration of algae, and may have odour for the same reason.

Throughout the stages of recovery of self-purification, disease organisms are greatly reduced in number because they lack proper food, and experience unfavorable temperatures and pH values of water. However, the water is still dangerous since all disease organisms have not perished.

Q.21. What does the oxygen sag curve represent? Explain self-purification of stream with sag curve.

Ans. The oxygen deficit (D) at the time in a polluted river stream is the difference between the actual DO content of water at that time and the saturation DO content. (Saturation DO means the amount of DO which will be present in water if it is fully saturated with oxygen. The normal saturation DO value for fresh water varies between 7.6mg/l to 14.6mg/l for temperature varying between 0° to 30°C) at the water temperature.

Oxygen deficit (D) = Saturation DO – Actual DO

In order to maintain clean condition in a river stream, the oxygen deficit must be will and this can be found out by knowing the rates of deoxygenation and reoxygenation.

Curve I: Deoxygenation curve or curve showing depletion of DO in the absence of aeration.

Curve II: Reoxygenation curve (DO content supplied)

Curve III: Oxygen sag curve or oxygen deficit curve (Actual DO Content or oxygen is balance)

Deoxygenation curve: In a polluted stream DO content goes on reducing due to decomposition of organic matter. The rate of deoxygenation depends upon the amount of the organic matter remaining to be oxidized at a given time and as well as on the temperature of the reaction. Hence at a given temperature, the curve showing depletion of DO with time i.e., deoxygenation curve is similar to the I[st] stage BOD curve.

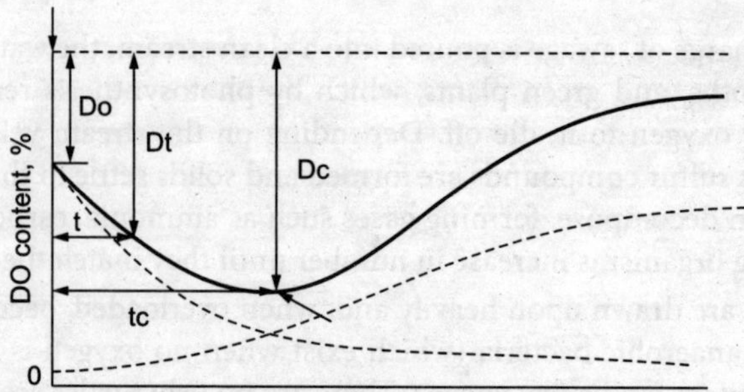

Fig. 4.2: Oxygen sag curve

Reoxygenation curve: In order to counter balance, the consumption of DO due to deoxygenation atmosphere supplies oxygen to water and the process is called reoxygenation. The rate at which the oxygen is supplied by the atmosphere to the polluted water depends upon:

(*a*) The depth of the receiving water (rate is more in a shallow depth)

(*b*) The condition of the body of the water (rate is more in a running stream than in a quiescent pond)

(*c*) The saturation deficit or the oxygen deficit (ie. the deficit of DO below the saturation value) and

(*d*) The temperature of water.

Depending upon these factors, the rate of reoxygenation can also be expressed mathematically and plotted in the form of the curve called reoxygenation curve.

Oxygen deficit curve (oxygen sag curve): In a running polluted stream exposed to the atmosphere, the deoxygenation as well as reoxygenation go hand in hand. If the deoxygenation is more rapid than the reoxygenation, as oxygen deficit results.

The amount of resultant oxygen deficit can be obtained by adding the deoxygenation and reoxygenation curves. The resultant curve so obtained is called the oxygen sag curve or the oxygen deficit curve. From this curve, the oxygen deficit and oxygen balance isn't the stream after a certain elapse of time can be find out.

It can also be seen that when the deoxygenation rate exceeds the reoxygenation rate, the oxygen sag curve shows increase in deficit of oxygen, but when both the rates become equal, the critical point is reached and then finally when the rate of deoxygenation fall below of reoxygenation, the oxygen deficit goes on decreasing till it becomes zero.

Q.22. Explain the sources, characteristics and treatment of sugar mill effluent.

Ans. Sources and characteristics of wastes: Waste from the sugar mill house includes the water used as splashes to extract maximum amount of juice and those used to cool the roller bearings. As such the mill house waste contains high BOD due to the presence of sugar and oil from the machineries. The filter cloths used for filtering the juice needs occasional cleaning. The wash waster thus produced though small in volume, contains high BOD and suspended solids.

The characteristic of waste water is tabulated below.

Table 4.1

Characteristics	Value
pH	4.6-7.1
Total solids	870-3500 mg/lt
BOD	300-200 mg/lt
COD	600-4380 mg/lt
Total suspended solids	220-800 mg/lt
Total Nitrogen	10-40 mg/lt

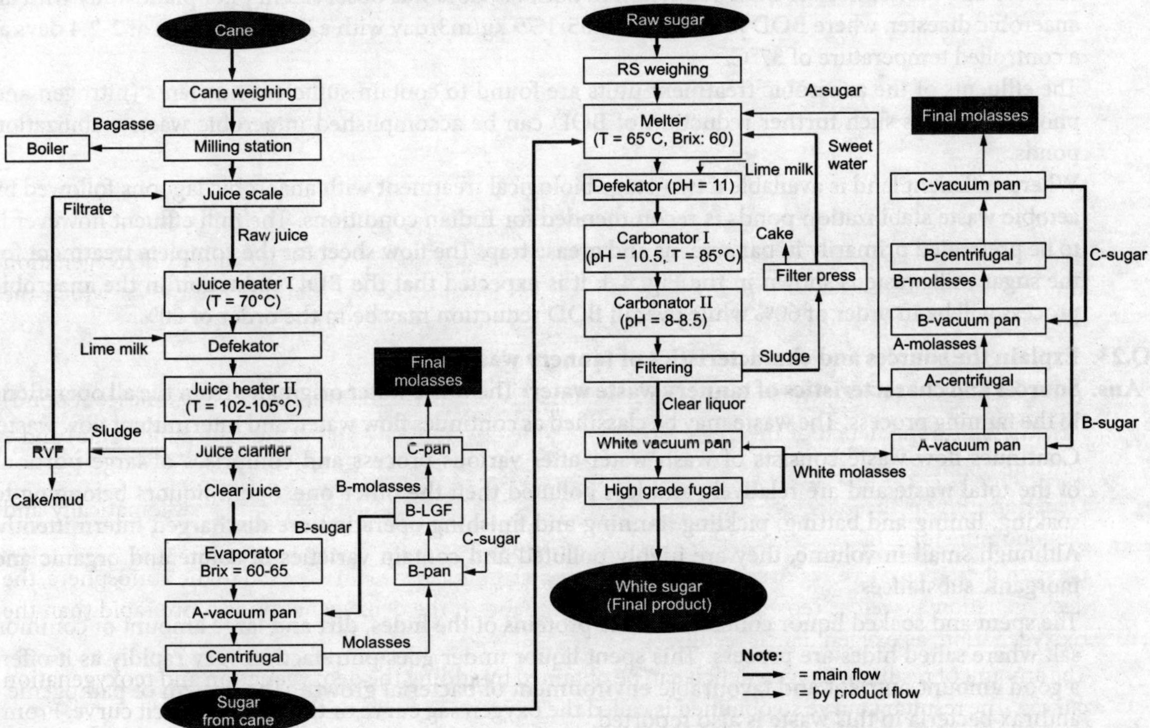

Fig. 4.3: Schematic block diagram of sugar processing

A large volume of water is required in the barometric condensers of the evaporators and vacuum pan. The water is usually partially or fully circulated after cooling through a spray pond. This cooling water gets polluted as it picks up some organic substances from the vapour of boiling syrup in evaporation and vacuum pans. This polluted water, instead of re-circulated is discarded as excess condenser water. These discharges contribute substantially to the waste volume and modulated to BOD in many sugar mills.

Additional waste originates due to the leakages and spillages of juice, syrup and molasses in different sections. The periodical washings of the floor through small in volume have got very high BOD. The periodic blow off of the boilers produces another intermittence waste discharge. This waste is high in suspended solids,low in BOD and usually alkaline.

Treatment of the waste water: Like any other industry the pollution low in sugar mills can also be reduced with a better water and material economy practiced in the plant. Judicious use of water in various plant practices and it recycle, wherever practicable, will reduced by recycling cleaning of floors or floor washings using controlled quantity of water will also reduce the volume of waste to certain extent.

The organic load of the waste cam only be reduced by a proper control of the operations. Over loading of the evaporates and the vacuum pans and the extensive boiling of the syrup leads to a loss of sugar through condenser water and this in turn increases both volume and strength of the waste effluent. Disposal of the effluent on land as irrigation water is practiced in many sugar mills, but it is associated with odour problems.

Anaerobic treatment of the effluent using both digesters and lagoons has been found to be very effective and economical. A BOD reduction of both 70- 88% was observed in pilot plant study with an anaerobic digester, where BOD loading was 0.65-1.79 kg/m3/day with a detention time of 2-2.4 days at a controlled temperature of 37°C.

The effluents of the anaerobic treatment units are found to contain sufficient nutrients (nitrogen and phosphorous) as such further reduction of BOD can be accomplished in aerobic waste stabilization ponds.

Where sufficient land is available a two-stage biological treatment with anaerobic lagoons followed by aerobic waste stabilization ponds is recommended for Indian conditions. The mill effluent however is to be pretreated primarily in bar screens and grease trap. The flow sheet for the complete treatment for the sugar mill waste is shown in the Fig. 4.3, it is expected that the BOD reduction in the anaerobic process will be in order of 60% while overall BOD reduction may be in the order of 90%.

Q.23. Explain the sources and characteristics of tannery waste water.

Ans. Sources and characteristics of tannery waste water: The waste water originates from the all operations in the tanning process. The waste may be classified as continues flow water, and intermittent flow waste. Continues flow waste consists of wash water after various process and comprises of large portions of the total waste and are relatively and less polluted then the other one. Spent liquors belonging to soaking, liming and batting, pickling, tanning and finishing operation are discharged intermittently. Although small in volume, they are highly polluted and contain varieties of solute and organic and inorganic substances.

The spent and soaked liquor contains soluble proteins of the hides, dirt and large amount of common salt where salted hides are process. This spent liquor under goes putrefaction very rapidly as it offers a good amount nutrient and favourable environment of bacterial growth. The growth of pathogenic - anthrax bacteria in this waste is also reported.

The spent lime liquor contains dissolved and suspended lime, colloidal proteins and their degradation products, sulphides emulsified fatty matters and also carrying a sludge composed of unreacted lime,

Calcium sulphide and calcium carbonate. As such, the spent lime liquor as a high alkalinity and moderate BOD and high ammonia nitrogen content.

Flowsheet of tanning industry is shown below.

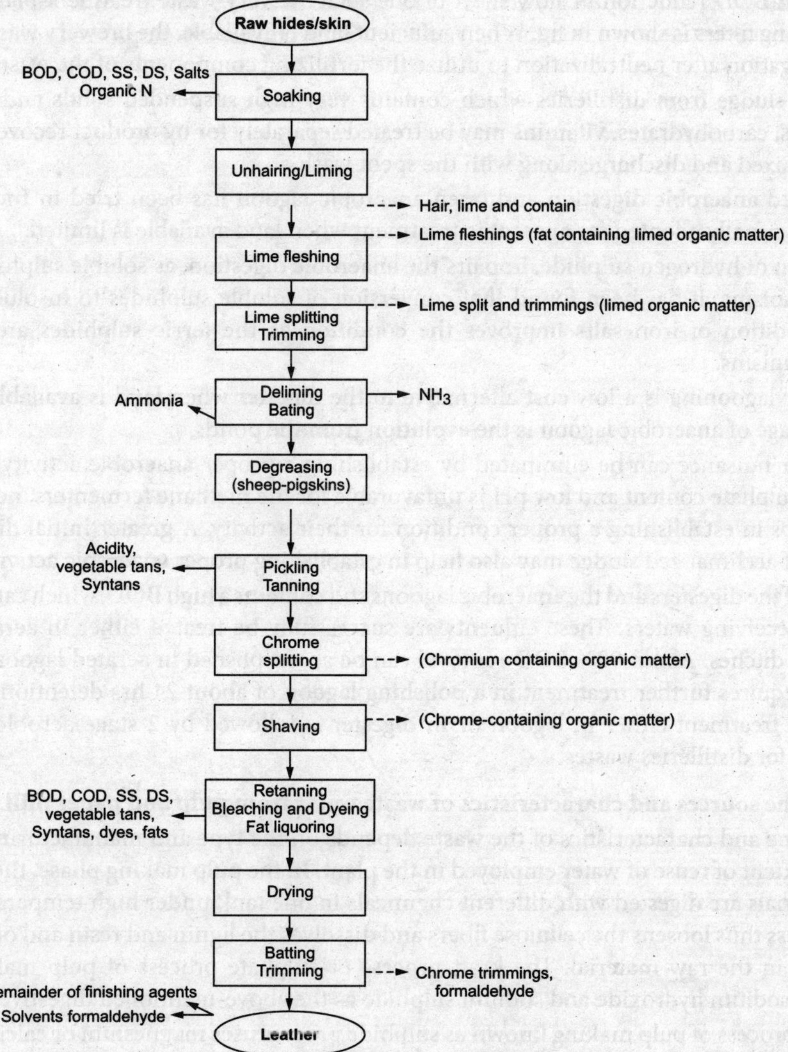

Fig. 4.4: Flowsheet of tanning industry

The spent bath liquor contains high amount of organic and ammonium nitrogen due to the presence of soluble skin proteins and ammonia salts used in bating.

The vegetable tan exact contains tannins and also non-tannins. Tannins are of high COD but relatively low BOD value. While non-tannins including inorganic salts, organic acids and salts and sugar are of high BOD and COD. The spent vegetable tanning liquor is the strongest individual waste in the vegetable tannin having the highest BOD and very strong dirty brown colour.

The spent pickling and chrome tanning waste comprises of as small volume, having a low BOD and contains traces of proteins impurities, sodium chloride and mineral acids and chromium salts. Chromium is highly toxic to the living aquatic creatures.

Q.24. Discuss the origin and treatment of distillery waste water.

Ans. Brewery wastes being comparatively less strong can be treated by aerobic biological treatment, after screening and neutralization. Usually, the biological treatment is accomplished by two stage process for 90-94% BOD reduction. A flow sheet of one such brewery waste treatment plant employing high rate trickling filters is shown in fig. When sufficient land is available, the brewery waste may be used for broad irrigation after neutralization to utilize the fertilizing components of the waste.

The yeast sludge from distilleries which contains very high suspended solids and BOD and in rich in proteins, carbohydrates, vitamins may be treated separately for by product recovery. But in practice they are mixed and discharge along with the spent wash.

Both closed anaerobic digestion and open anaerobic lagoon has been tried in India. A single stage digester is usually adopted for anaerobic treatment when land available is limited.

Production of hydrogen sulphide, impairs the anaerobic digestion, as soluble sulphide are toxic to the microorganisms. It has been found that conversion of soluble sulphides to insoluble ferric sulphide by the addition of iron salts improves the condition as the ferric sulphides are not toxic to the microorganisms.

Anaerobic lagooning is a low cost alternative to the digester when land is available plenty. The only disadvantage of anaerobic lagoon is the evolution from the ponds.

The odour nuisance can be eliminated by establishing a proper anaerobic activity in the lagoon. As the high sulphate content and low pH is unfavorable for the methane fermenters, neutralization of the waste helps in establishing a proper condition for their activity. A greater initial dilution and greater amount of acclimatized sludge may also help in establishing proper anaerobic activities.

Effluent of the digesters and the anaerobic lagoons still contains a high BOD, which cannot be discharged into the receiving waters. These effluents are successfully be treated either in aerated lagoons or in oxidation ditches. About 90% BOD removal can be accomplished in aerated lagoons. Aerated lagoon effluent requires further treatment in a polishing lagoon of about 24 hrs detention time. Single stage anaerobic treatment either in lagoon or in digester is followed by 2 stage aerobic treatment is also proposed for distilleries wastes.

Q.25. Explain the sources and characteristics of waste water from pulp and paper mill.

Ans. The volume and characteristics of the waste depends on the type and manufacturing process adopted and the extent of reuse of water employed in the plant. In the pulp making phase, the chipped cellulose raw materials are digested with different chemicals in one tank under high temperature and pressure. The process thus loosens the cellulose fibers and dissolves the lignin and resin and other non-cellulosic materials in the raw material. The kraft process or sulphate process of pulp making uses sodium, sulphate, sodium hydroxide and sodium sulphide as the above-mentioned digestive chemicals.

Another process of pulp making known as sulphide process uses magnesium or calcium bisulphate and sulphurous acid as the digestive chemicals. The alkali process uses sodium hydroxide or lime for pulp making.

The spent liquor produced by the above process of digestion is known as Black Liquor. This is not only very rich in lignin content, but also contains a large amount of utilized chemicals. Therefore, this black liquor is treated separately for the recovery. So while the entire quantity of the liquor is up hide process makes the coloured waste from the section, in the kraft process same is produced due to the leakages, spillages or overflow only from the digester. The cellulosic fiber after being separated from the black liquor is washed and then partially dewatered.

A concentrated wash water may be sent for chemical recovery) in the kraft process. The dilute wash water form the waste water. This volume of waste water is known as Brown stock wash or Unbleached Decter waste.

A schematic diagram of pulp and paper industry is shown below.

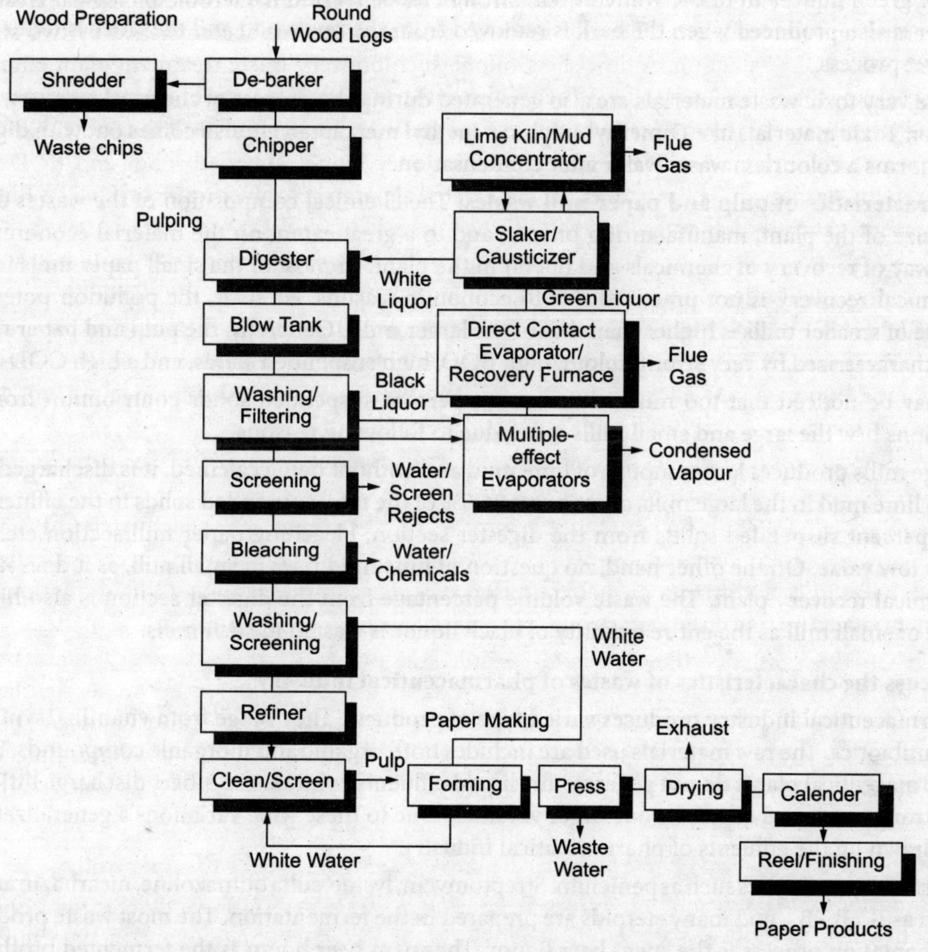

Fig. 4.5: Schematic diagram of pulp and paper industry

The washed cellulosic fibers are then sent for the bleaching in 3 stages, where chlorine, caustic and hypochlorite are used in successive stages. Waste waters from the first and the last stages are light yellow in colour while that from the caustic extraction stage is highly coloured. The bleached pulp is then sent for the paper mill. In the paper mill, the pulp is disintegrated and mixed with various filter materials like alum talc, etc and dyes on an along shaped especially pulp is refined in a machine. The refined pulp is then diluted to proper consistency for paper making and passed through a screen to remove lumps or knots. Now this pulp is carried by a travelling belt of fine screen to a series of rolls, where the final product, the paper is produced. The drained water often called as White water forms the waste water from the paper mill section. This waste water contains fine fibers, alum, talc, etc. Usually the fibers are recovered from this waste; the treated liquid is reused for the wet chipping process.

The black liquor of the Kraft process concentrated by evaporation and then incineration with the addition of sodium sulphate. The organic like lignin, resin., etc are burnt off and the smell is dissolved in water. The resulting liquid is known as green liquor; lime is now added to this liquor resulting in the formation of white liquor and lime mud containing chiefly digesting chemicals and is sent for use in digester. The lime mud is calcined (by burning) to form calcium oxide, which is reused to recausticise other green liquors in to the white liquors. Beside those mentioned about, the small volume of waste water is also produced when the bark is removed from the raw wood and the latter is reduced to chips by wet process.

Some very toxic waste materials are also generated during the process of chemical recovery from black liquor. Toxic materials like Dimethyl sulphide, methyl mercaptan etc also comes out with digester gases and forms a colourless waste water after condensation.

Characteristics of pulp and paper mill wastes: The chemical composition of the wastes depends on the size of the plant, manufacturing process and to a great extent on the material economy practiced (by way of recovery of chemicals and fibers) in the plant. In most of the small paper mills in India, the chemical recovery is not practiced due to economic reasons. As such, the pollution potential of the waste of smaller mills is higher than that of the larger mills. Generally, the pulp and paper mills wastes are characterized by very strong colour, high BOD, high suspended solids, and a high COD/BOD ratio. It may be noticed that too much difference is percent suspended solids contribution from different sections b/w the large and small mills arises due to following reasons.

Large mills produces large amount of lime mud and without being calcined, it is discharged as a waste. This lime mud in the large mills contributes 86.5% of the total suspended solids in the effluent. As such, the percent suspended solids from the digester section; bleaching paper mill section etc. assumes a very low value. On the other hand, no question of lime mud rises in small mill, as it does not have the chemical recovery plant. The waste volume percentage from the digester section is also higher in the case of small mill as the entire quantity of black liquor is wasted in such mills.

Q.26. Discuss the characteristics of wastes of pharmaceutical industry.

Ans. Pharmaceutical industry produces varied type of products. They range from vitamins, synthetic drugs to antibiotics. The raw materials used are includes both organic and inorganic compounds. Some of the pharmaceutical plants do not generate any liquid effluents, while some others discharge little quantities of strong waste and others let out larger volumes. Due to these wide variations a generalization cannot be drawn on the effluents of pharmaceutical industry.

Most of the antibiotics such as penicillin. Streptomycin, lysine, sulfa quinazoline, nicarbazin and vitamins such as B_1, B_2, B_{12} and many steroids are prepared in the fermentation. The most waste produced in the fermentation process is the spent beer liquor. The spent beer liquor is the fermented broth remaining after the recovery of antibiotics and other valuables. It contains large amounts of organic materials, proteins and other nutrients and consequently the BOD of these effluents is abnormally high.

Five main pharmaceutical wastes and their characteristics are as follows:

1. Strong fermentation beers (small in volume but having 4000 to 8000 mg/L BOD)
2. Inorganic solids (waste slurry with little BOD)
3. Washings of floor and equipment (large percentage of total volume and BOD from 600 to 2500 ppm)
4. Chemical waste – solution or solvents which exert a substantial BOD when diluted with other wastes
5. Barometric condenser waste water – resulting from solids and volatile gases being mixed with condenser waste water causing 60 to 120 ppm BOD.

The antibiotic wastes impart objectionable odours to stream and inhibit biological population and action. If they are discharged into sewer, they must be properly diluted, otherwise they affect the sewage treatment.

The volume and composition vary from unit to unit. Approximately 1000 to 3000 litres of waste will be discharged per 100 kg of products manufactured. No specific conclusion on the characteristics of the effluents can be drawn. In general, they are either highly acidic or highly alkaline and possess a high BOD and COD. Some of the effluents contain toxic substances like cyanides.

If the wastes are discharged into stream, they deplete the dissolved oxygen immediately. These are corrosive due to their high acidity/alkalinity. Further, some of the substances present in them are toxic to aquatic life.

Q.27. Discuss the wastes produced by food processing industry (breweries, wineries, etc.).

Ans. While breweries and wineries produce beer and wine respectively, a large no of products are obtained in distilleries. The range of products from distilleries includes industrial alcohols, rectified spirit absolute alcohol, spirit, beverage alcohol, etc. But two things are common in all the products mentioned above.

1. All the above products are obtained through the biochemical process of fermentation by yeast using carbohydrates as raw materials and
2. All the products contain ethyl alcohol in different proportions.

In all the industries mentioned above are all characteristically of high BOD and they present a threat to the environment when discharged into the water sources or to the land without treatment.

Making of beer essentially consists of two stages:

1. Preparation of molt from grains like barley. 2. Brewing the barley.

In the malt making the barley grain are steeped (soaked) to bleach out colour and then made to sprout under aerobic conditions. The grain malt is then dried and stored after screening the sprouts out.

The malt from the malt house is then transported to the brewing section, where the wart the medium for fermentation is prepared by mixing the course grain malt with hot water and by transforming the starch to sugars boiling in hops. The wort is then inoculated with a prepared suspension of yeast which common the sugar to alcohol.

When the fermentation is complete, the yeast and mall residue is filtered out and finally the beer is carbonated before packing for sale.

As the flavor of the product is of prime importance selection of raw materials and control of process is done accordingly.

Q.28. Discuss the origin and treatment of dairy waste.

Ans. The liquid waste from a large dairy originates from the following sections or plants. Receiving station bottling plant, cheese plant, butter plant, casein plant, condensing plant, dried milk plant and ice cream plants. Wastes also come from water softening plant and from bottle and washing plants. At the receiving station, the milk is received from the farmers and is emptied into large containers for transport to bottling of other processing plants. The empty is rinsed, washed and sterilized before returning to the farmers. At the bottling plants the raw milk delivered by receiving station is stored. The processing includes cooling, filtration, clarification, pasteurization and bottling.

In the above sections, the liquid wastes originate out of rinse and washings of bottles, cans and equipment and thus contain milk drippings and chemicals used for cleaning containers and equipment.

In a cheese plant, the milk (whole milk or skimmed milk) is pasteurized and cooled and placed in a vat, where a starter (lactic acid producing bacterial culture) and rennet are added. This separates the casein of the milk in the form of curd. The whey is then withdrawn and the curd compressed to allow excess to drain out. Other ingredients are now added and the cheese blocks are cut and packaged for sale. Waste water from this plant includes mainly the discarded whey and wash water used for cleaning vats, equipment floor.

In the creamery process the whole milk is pre-heated to about 30°C to separate the cream from the milk. In a butter plant, the cream is pasteurized and may be ripened with a selected acid and a bacterial culture. This is then churned at a temperature of about 7-10°C to produce butter granules. At the proper time the butter ilk is drained out of the churned and the butter is washed and after standardization packaged for sale. Butter milk and wash waters used to clean the churns and small quantity of butter comes out as waste from the butter plants.

The skimmed milk may now be sent for bottling for human consumption or for further processing in the dairy for other products like non-fat milk powders. Milk powders are produced by evaporation followed by drying by either roller process or spray process. The dried milk plant wastes consist of chiefly of wash waters used to clean containers and equipment.

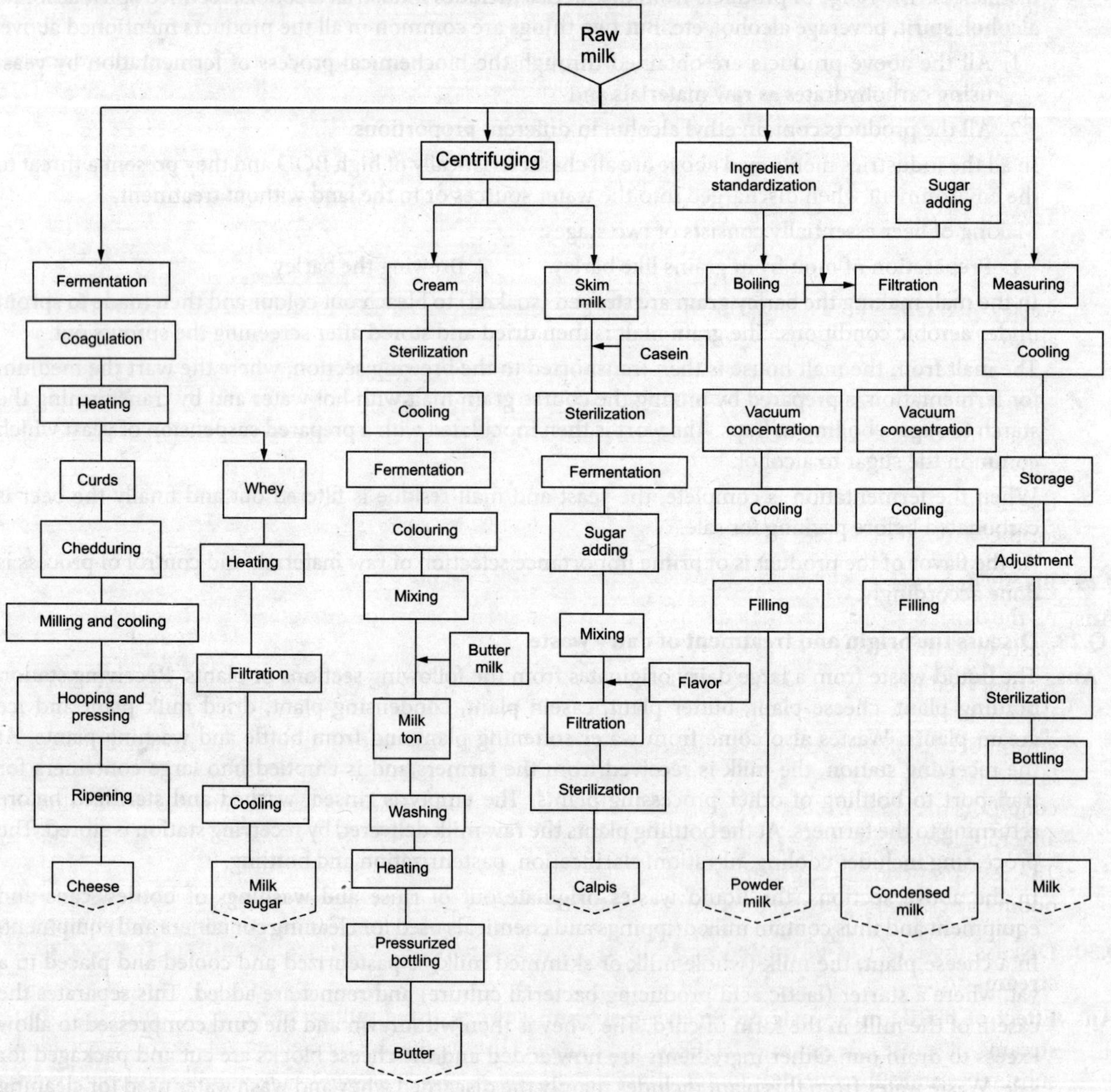

Fig. 4.6: Origin and treatment of dairy waste

The scoured or spoiled milk and some time the skimmed milk are processed to produce caseins used for preparation of some plastics. The process involves the coagulation and precipitation of the casein by the addition of some mineral acids. The wastes from this section include whey, washings and the chemicals used for precipitation. Very large dairies also produce condensed milk and ice creams.

In addition to the wastes from all the above milk processing units, some amount of uncontaminated cooling water comes as waste. These are usually re-circulated. The dairy wastes are very often discharged intermittently. The nature and composition of the waste also depends on the types of products produce and the size of the plants.

Due to low COD to BOD ratio the dairy wastes can be treated efficiently by biological processes. Moreover, these wastes contain sufficient nutrients for bacterial growth. But for economic reasons, attempt should be made to reduce the volume and strength of the waste. This can be accomplished by:

- Prevention of spills, leakages and dropping of milk from the cans.
- By reducing the amount of water for washes.
- By segregating the uncontaminated cooling water and recycling the same.
- By utilizing the buttermilk and whey for the production of dairy by-products of good market value.

Treatment of dairy waste: Due to the intermittent nature of the waste, it is desirable to provide equalization tank, with or without aeration, before the same is sent for biological treatment. A provision of grease trap is also necessary as a pretreatment to remove fat and other greasy substances from the wastes. A aeration for a day not only prevents the formation of lactic acid, but also reduces the BOD by about 50%.

Both high rate trickling filters and activated sludge plants can be employed very effectively for a complete treatment of the dairy waste. But these convention methods involve much maintenance, skilled methods like oxidation ditch, aerated lagoon, waste stabilization pond etc., can be employed with simpler type of equipment and less maintenance.

Use of the dairy waste for irrigation after primary treatment in an aerated lagoon may be a good answer for the disposal of dairy wastes.

Q.29. Briefly describe the effect of dairy waste in receiving streams.

Ans. As the dairies are usually situated in rural areas or in small towns, the question of discharging the dairy waste in to the sewers does not arise.

As observed from waste is basically organic in nature. This is also slightly alkaline when fresh. When these wastes are allowed to go into the stream without any treatment, a rapid depletion of DO of the stream occurs along with growth of sewage fungi covering the entire bottom of the stream. The waste is said to carry occasionally, the bacteria responsible for tuberculosis (TB). Though alkaline in fresh condition the milk waste becomes acidic due to the decomposition of lactose into lactic acid under anaerobic condition.

The resulting condition precipitates casein from the waste, which decomposes further into a highly odours black sludge. At certain dilution the dairy waste is found to be toxic to fishes also.

Q.30. Discuss the effect of untreated/partially treated wastes from cotton textile mill on the receiving stream.

Ans. Effect of textile mill waste on receiving streams/sewers: If the mill waste water is discharged into streams, it causes depletion of DO of the stream. This is due to the settlement of the suspended substances and subsequent decomposition of the same in anaerobic condition. The alkalinity and toxic substances like sulphides and chromium affects the aquatic life and also interferes with the biological treatment processes. Some of the dyes are also found to be toxic. The colour often renders the water unfit for use for side. The presence of sulphides makes the waste corrosive particularly to concrete structures. All treatment plants should be planned giving serious consideration for the reduction of waste volume and strength, through process of chemical substitution, chemical recovery and recycling of water.

The pollution load from a textile, mill is dealt with operations like segregation, neutralization, equalization, chemical ppt, chemical oxidation and biological oxidation. Several chemicals are used to reduce the BOD by chemical coagulation such as alum, ferric sulphate, ferrous sulphate and ferric chloride, lime or H_2SO_4 is used to adjust pH in this process. The dye waste may be economically treated by biological methods prior equalization, neutralization and chemical oxidation.

Q.31. Discuss the effect of untreated/partially treated wastes from cotton textile mill on the receiving stream.

Ans. If the mill waste water is discharged into streams, it causes depletion of DO of the stream. This is due to the settlement of the suspended substances and subsequent decomposition of the same in anaerobic condition. The alkalinity and toxic substances like sulphides and chromium affects the aquatic life and also interferes with the biological treatment processes. Some of the dyes are also found to be toxic. The colour often renders the water unfit for use for side. The presence of sulphides makes the waste corrosive particularly to concrete structures.

All treatment plants should be planned giving serious consideration for the reduction of waste volume and strength, through process of chemical substitution, chemical recovery and recycling of water. The pollution load from a textile, mill is dealt with operations like segregation, neutralization, equalization, chemical ppt, chemical oxidation and biological oxidation. Several chemicals are used to reduce the BOD by chemical coagulation such as alum, ferric sulphate, ferrous sulphate and ferric chloride, lime or H_2SO_4 is used to adjust pH in this process. The dye waste may be economically treated by biological methods prior equalization, neutralization and chemical oxidation.

A Composite waste, when free from toxic substances may be treated as efficiently as domestic sewage, as most of the textile mill wastes contain sufficient nutrients like nitrogen and phosphorous. Trickling filters, activated sludge process and stabilization ponds have been effective in treating textile mill wastes. Extended aeration is found to be very effective in treating strong wastes even without equalization and pretreatment.

Q.32. Describe the manufacturing process of cotton textile mill and waste generation.

Ans. Manufacturing process: An integrated cotton textile mill produces its own yarn from the raw cotton. Production of yarn from raw cotton includes steps like opening and cleaning picking, carding, and drawing spinning, winding and warping. All these sequences are dry operations and as such do not contribute to the liquid waste of the mill. The entire liquid waste from the textile mills comes from the following operation of slashing (sizing), scouring, desizing, bleaching, mercerizing, dyeing and finishing.

In slashing, the yarn is strengthened by loading it with starch or other substances wastes originating from the sections due to spills and floor washings. The substitution of low BOD sizes (such as carboxy methyl cellulose) for the high BOD of the mill effluent by 40 to 90%.

After slashing, the yarn goes for weaving. The prepared cloth now requires scouring and desizing to remove natural impurities and the slashing compounds. Enzymes are usually used in India to hydrolyze the starch, acids may also be used for this purpose. Caustic soda, soda ash, detergents etc. are also used in this section. Bleaching operations use oxidizing chemicals like peroxides and hyper chloride to remove natural colouring material. The section contributes about 10% of the total pollution load.

Flow diagram of a cotton textile mill is shown below.

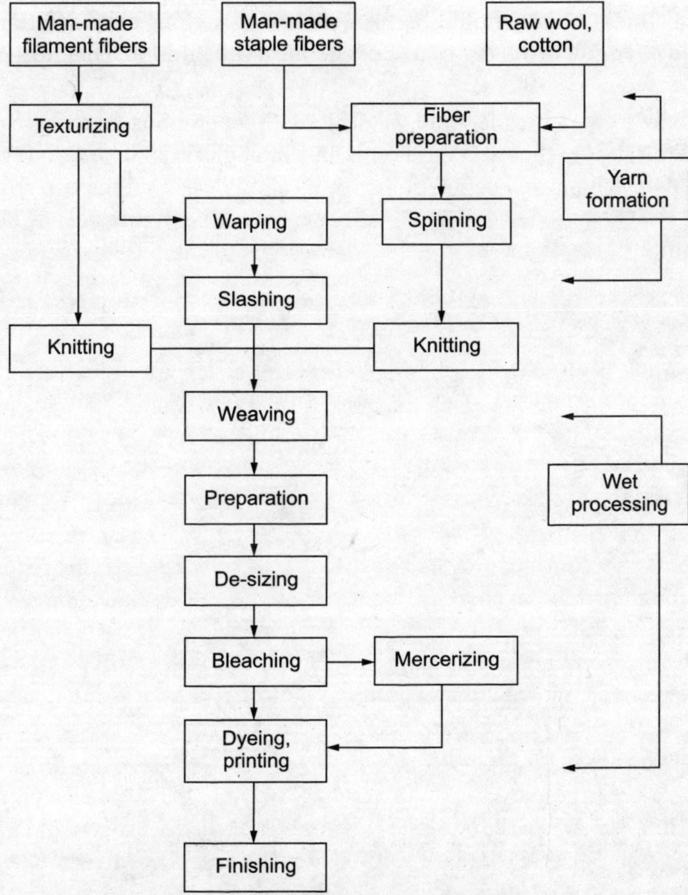

Fig. 4.7: Flow diagram of a cotton textile mill

Mercerizing consists of passing the sloth through 20% caustic soda solution. This process includes the strength elasticity luster and dye affinity. Waste from this section is recycled after sodium hydroxide recovery. Negligible waste which may come out of this section contributes little BOD but a high degree of alkalinity.

Dyeing may be done in various ways, using different types of dyes and chemical classes of dyes include Vat dyes, developing dyes etc. colour from the dyes vary widely and although these are not usually toxic, they are treated separately. Thickened dyes are used for probing and subsequent fixation. After fixation of the prints, the fabric is given a thorough wash to remove the unfixed dyes.

The finishing section of the mill imparts various types of chemicals are used for various objectives. These include starches, dextrines, natural and synthetic waxes, synthetics etc. Therefore a composite waste from an integrates cotton textile mill may include the following organic and inorganic substances starch, carboxyl methyl cellulose, sodium hydroxide, detergents, peroxides, hyperchloride dyes and pigments, sodium gums, dextrines, waxes, sulphides, soap etc. Depending on the process and predominant dye used, the characteristics of the mill waste varies widely.

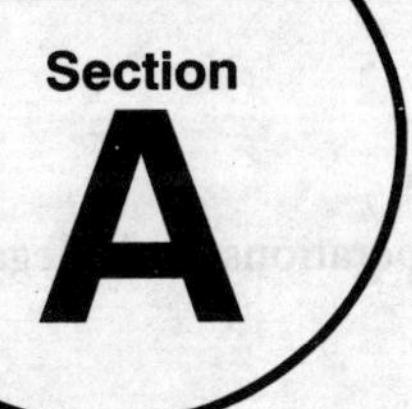

Section A

5 Marine Pollution

SHORT QUESTIONS AND ANSWERS ON MARINE POLLUTION

Q.1. What is marine pollution?

Ans. The introduction by man, directly, or indirectly, of substances or energy to the marine environment resulting in deleterious effects such as: hazards to human health, hindrance to marine activities, impairment of the quality of seawater for various uses and reduction of amenities.

Does not include natural processes like volcanic eruptions or earthquakes.

Q.2. Why marine pollution is a concern?

Ans. Marine pollution means introduction by man of substances or energy into the marine environment (including estuaries) resulting in such deleterious effects as harm to living resources, hazards to human health, hindrance to marine activities including fishing, impairment of quality for use of seawater and reduction of amenities.

We should care about the continuing environmental degradation of our oceans and costal areas because it is detrimental to human health, economic development, climate and our planet's store of biodiversity. It is interfering with the sustainability of environment and its resources.

Biodiversity or Biological Diversity" means the variability among living organisms from all sources including terrestrial, marine and other aquatic ecosystems and the ecological complexes of which they are part; this includes diversity within species, between species and of ecosystems. Sustainability is defined as the development that meet the need of the present, without compromising the ability of future generation to meet their own needs.

Q.3. What are common marine pollutants?

Ans.
- Agricultural run offs ((herbicides, pesticides and nutrients)
- Sediments
- Sewage (Faecal Coliform and Pathogens)
- Chemicals, Metals and Radioactive Substances
- Persistent toxins (PCBs, DDT, heavy metals)
- Oil
- Plastics
- Energy (Thermal and light)

Q.4. What are the sources of marine pollution?

Ans. Land sources

- 80% of non-biological marine pollution comes from land based activities
- pipes discharging directly into marine waters(sewage, industrial, chemical and food processing wastes)
- River flows into the sea carry pollutants from the entire catchment area.

From Air

- Global atmospheric inputs to the sea from air discharges.

Oil spills and offshore sources

- Oily discharges from ballast water and bilge water during routine ship operations and illegal dumping of solid waste.
- Designated dumping grounds at sea.
- Accidental spills from Ships carrying hazardous substances, oil, gas etc.

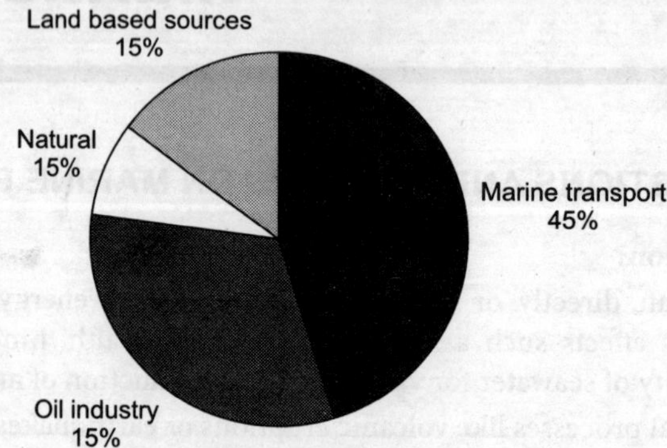

Fig. 5.1: Major marine pollutants

Q.5. What are the dangers of major marine pollutants?

Ans. Major marine pollutants – Solid waste

- A large portion and great danger is non-biodegradable plastic
- 46,000 pieces of floating plastic/sq mile of ocean surface off the N.E.U.S. coast.
- Sea turtles mistake plastic bags for jellyfish and die from internal blockages.
- Seals and sea lions starve after being entangled by nets or muzzled by six-pack rings (decomposition time 400 years).
- Plastic debris kills 100,000 marine mammals and 2 million sea birds die annually

Q.6. What are the effects of marine pollution?

Ans. Effects are as follows:

- Damages marine life to a large extent, for salt-marsh plants, oil slicks can affect flowering, fruiting and germination.
- Coral reefs
- If liquid oil contaminates a bird's plumage, its water-repellent properties are lost, drown, die
- Drill cuttings dumped on seabed create anoxic conditions and result in the production of toxic sulphides in the bottom sediment thus eliminating the benthic fauna.
- Fish and shellfish production facilities can also be affected by oil slicks. Commercial damage is tainting: imparts an unpleasant flavor to fish and seafood and is detectable even at extremely low levels of contamination.

Q.7. What are the categories of manmade effect on marine environment?

Ans. Man's effect on the marine environment may be divided into the following general categories:

- Alteration of bottom substrate through degrading, changing of shoreline structures, and filling.
- Introduction of toxic substances dangerous either to marine life or to human beings.

- Release of sewage rich in nutrients with marine microorganisms.
- Heating and release of heated water by power plants.

Q.8. How do we define coastal zone?

Ans. The coastal zone is defined as "The area extending from the coastal plains to the edge of continental shores, approximately matching the region that has been alternatively flooded and exposed during the sea level fluctuation of the late quaternary period".

In the earth's surface, 18% is represented by coastal zone, which provide space for 60% of the world's human population. 70% of the world cities with population exceeding 1.6 million are located in this zone. 90% of the world fish catch is obtained from coastal zone. The coastal zone is only 8% in the hydrosphere, but responsible for 18-33% of the total primary production. The biological wealth of this zone is very high as it serves as feeding, nursery and spawning grounds with rich biodiversity.

Q.9. What are the sources of marine pollution particularly in coastal zone?

Ans. The main sources of marine pollution are atmosphere, river runoff, agriculture, livestocks, urban runoff, automobiles, land clearing, sewage outfall, industrial waste etc.

In general the coastal zone receives pollutants from major ways viz., atmosphere, riverine and glacier. Also anthropogenic activities serve as the geological agents by ways of discharging the effluents through piped outfalls, direct dumping, operation of ships etc. Frequently preservatives have been intentionally used assuming that they are relatively harmless to the cultured species. These include antifoulants, of which the broad ecological effect of tributylin is a good example.

Pollution may arise from point source like a single sewer pipe or factory wastewater outfall, or it can arise from a variety of geographic points otherwise called as non point source. In the point source pollution the concentration of the substance or the intensity of the effect (e.g., temperature near cooling water outfalls of power plants) should decline with increasing distance from the point source. The dissipation or decline depends on the nature of substance or factor, water currents and the sedimentary environment and rate of introduction of the substance or factor. In contrast to this, the non-point source effects cannot be attributed to any single spot. Runoff arises due to rain is an example of a non-point source. The toxic substance and other fertilizer based nutrients may spread over the coast. However, these sources are more difficult to manage, because the source is widespread over a spectrum of earth surface.

Q.10. What the indicators of marine pollution?

Ans. Many organisms serve as indicators of marine pollution. For example, the ecological shift in the succession of phytoplankton communities from diatoms to dinoflagellates due to metal pollution and thermal pollution. The poor hatching of eggs and soft calcium shell of birds and mass mortalities of birds due to failure in the earlier metabolites by the interference of pesticides, large scale mass killing of birds due to casting, asphyxiation and smothering by oil pollution, reproductive failure in mammal are also some of the indicators of severe pollution.

Usually the pollutants exceeding threshold limits combined with environmental variables such as temperature, salinity, pH, dissolved oxygen, hydrogen sulphide etc., give stress to the organisms. In most cases the migratory organisms such as fishes could avoid the polluted area. Whereas the residential or sedentary organisms are exposed to the pollutants experience the stress conditions.

Q.11. Discuss the effect of pollutants on marine organisms?

Ans. Pollutants that enter into to the oceans are mostly diluted, however the organisms living in the oceans tend to concentrate the pollutants into their body by various mechanisms, like adsorption, absorption, ingestion etc. The concentration of pollutant increases with higher trophic levels, *i.e.* from primary producers to the tertiary consumer. The threshold limit of organisms to the pollutant concentration increase with increasing trophic levels. Various mode of pollution transport in organisms includes bioconcentration, bioaccumulation, biomagnifications and biodegradation.

Bioconcentration: The bioconcentration is the process by which a contaminant such as oil is directly taken up (*i.e.*, by absorption only) from water and is accumulated to levels greater than those found in the surrounding water.

Bioaccumulation: Marine organisms can accumulate chemical species in amounts far exceeding their surrounding (sea water or sediment) concentrations. The bioaccumulation is the process by which a contaminant is taken up by organisms directly through the physical exposure pathway or through consumption of food containing the contaminated substance.

Bioaccumulation incorporates the concepts of bioconcentration and biomagnification. The term 'concentrator factor' is used to quantify the bioaccumulation and may be defined as the concentration of the chemical species in the organisms or one of its components divided by the seawater or surrounding concentration in the environment from which the organism was living. Organisms may take up the chemical directly from seawater or from their food. Thus, depending upon the composition of the food, concentration factors can vary from one environment to another for a specific organism with a given chemical.

Biomagnification: Biomagnification is the increase in tissue concentrations of a bioaccumulated chemical substance such as oil as it passes up though the trophic levels.

Biodegradation: Biodegradation is the process of degradation of a contaminant such as oil because of its use as a food a source for certain microorganisms. This process is limited to a great extent by temperature, nutrients and oxygen availability.

Q.12. What are the types of pollutants discharged into marine environment?

Ans. There are two types of pollutants. They are biodegradable and non-biodegradable pollutants. Most of the materials, which reach the sea, disintegrate either through simple chemical reactions or because of the activities of bacteria and some larger organisms. There are some substances nevertheless, which are either extremely stable or else they have a very slow rate of degradation. The biodegradable pollutants are not persistent in the environment (e.g., oil, organic compound in sewage). But the non-biodegradable pollutants are persisting in environment for longer time (e.g., Heavy metals, plastics, nuclear wastes etc.).

In general most of the toxic compounds are considered as Xenobiotics. The term Xenobiotics refers to "bio accumulative chemical of concern" or "persistent organic pollutants" .Some of the compounds are organochlorine compounds, chlorodanes, hexachlorobenzene, organophosphorus compounds, polychlorinated biphenyls (PCBs), butylin compounds etc.

Q.13. Discuss the characteristics, sources, fate and effect of sewage discharged into ocean/marine environment.

Ans. Sewage is discharged into the oceans all over the world mostly from urban settlement. Sewage adds to the number of small particles suspended in the water and contributes large amounts of nutrients.

Characteristics: The raw sewage contains higher concentrations of organic particulate matter, home washings, detergents, small amount of oil, higher concentration of nutrients, toxic heavy metals, etc. Also, the sewage contains higher concentration of bacteria and viruses, which includes harmful pathogenic forms. Sewage always has the characteristic of bad odour, irritant and harmful to human and other organisms.

Sources: The main sources of sewage are the coastal outfall located near the cities. The increasing shipping activities also add higher concentration of sewage to the harbour and shipping routes. Moreover many rivers transport sewage from the inner regions of the land.

Fate and effect of sewage: In open coast, the effect of sewage is difficult to detect due to higher mixing up of tidal, wave and current actions. However in semi-enclosed areas like small bays, harbours etc., the effects are devastating such as higher Biochemical Oxygen Demand. The organic load concentration is expressed by means of a quantity called Biochemical Oxygen Demand (BOD) which is the amount of oxygen required by the microorganisms for the decomposition of the organic compounds dissolved in a given volume of sewage. High BOD values indicate high concentrations of organic material and therefore low quantities of dissolved oxygen (DO). Near sewage outflow areas in temperate waters, the benthic invertebrate communities have degraded, Kelp beds have disappeared and diseased fish have become more prevalent. Sewage input in coastal environment normally increases the nutrient concentration of the outfall area. In tropical waters, outflows near coral reefs have caused a bloom of algal species that grows over the coral, and eventually it leads to the death of coral and its associated forms. This sometimes leads to the change in the ecological succession, that is the coral reef may be replaced by an algal reef.

Q.14. What are the characteristics, sources, fate and effects of petroleum oil discharged in ocean/marine environment?

Ans. Oil is discharged in to the ocean in various forms as crude oil and as separate fractions. Most of the oil fractions are biodegradable. The spilled oil causes more devastation than sewage in the marine environment.

Characteristics of oil: Petroleum oils are very complex mixtures of large numbers of hydrocarbons, containing also small amounts of sulphur, nitrogen, oxygen and different metals combined with the hydrocarbons.

Sources of oil pollution in marine environment

The main ways in which oil enters in to the sea are:

— Natural release

— Tanker operations

— Oil tanker and other ship accidents

— Operation of ships other than tankers

— Offshore oil drilling and production platforms

— Ship-shore oil terminal operation

— Refinery operations

— Discharge of oil products on land and subsequent discharge by runoff

Source of marine oil release:

- 45% is due to marine transportation
- 32% is due to routine loading, discharging and flushing of tanker ships
- 8% is due to natural seeps
- 15% by other means

Fate of oil in the sea: As the oil enters into sea it is exposed to processes, which modify it both physically and chemically. This process is called weathering. As the oil tends to float on the surface of sea, the prevailing wind, wave, tide and current conditions make the oil to spread to a wider spectrum of area from the point of release. The majority of crude oil forms sticky layers on the surface, which prevents free diffusion of gasses, clogs adult organisms feeding structures and decreases the sunlight available for the photosynthesis. Volatile components of an oil spill eventually evaporate into the air, leaving heavier tars behind which forms into tar balls, which fall to the bottom and may be assimilated by bottom organisms or incorporated into sediments.

Weathering: The alteration of physical and chemical properties of discharged oil through a series of natural processes, which begin when the discharge occurs and continue until the oil is removed. Major processes, which contribute to weathering, include evaporation, dissolution, photooxidation and emulsification.

Very little fraction of miscible fraction of oil is dissolved in the seawater and other fractions are immiscible.

Evaporation is the single most important weathering process for the first several days of an oils discharge. Evaporation results in the loss of lighter fractions through volatization from petroleum products (e.g., benzene, naphthalene).

The process whereby oil is incorporated into the medium of concern, usually water in the form of small droplets. Emulsions in water can either be oil-in-water or water-in-oil, formed as a result of wave actions. There is some solution of hydrocarbons in the water and this also is more marked in the more volatile constituents. The oil-in-water leads to the wide dispersion of fine oil droplets in the sea. The water-in-oil composition leads to the formation of semisolid masses known as 'chocolate mousse'.

Tar ball formation: If silt or sediment is present, oil may be absorbed on to it and may fall with it to the seabed. This effect can be seen as tar ball formation, which is mostly observed over the seabed of oil spilled area.

Biodegradation of oil: Some of the oil constituents are naturally attached by bacteria present in the sea. The simpler compounds, the normal paraffin, which have straight chains of carbon atoms, are most easily broken down and consumed. The other higher and complex molecular weight fractions are broken down very slowly.

Effect of oil spill on marine organisms: When there is a major oil spill the degree of effect depends on many factors such as:

— Type oil

— Physiography of the area

— Biodiversity of the area

— Previous exposure of the area to oil

— Treatment of oil spill.

— Quantity of oil

— Weather conditions at the time of spill

— Season of the spill

— Exposure to other pollutants and

As the oil is of floating type it tends to spread and eventually reach the coastline. In many coastal areas, animals and plant situated between high and low water marks were heavily coated with oil and smothered. The exposure of organism to oil adds foreign smell in its body and tissues and the process is called tainting. The shellfish can be tainted either by deposition of oil on their shell or by ingestion of naturally emulsified oil during normal feeding. The local fish population in the immediate vicinity of oil discharge can acquire a marked taint from oil components dissolved in the water.

Almost all groups of organisms like invertebrates, vertebrates like fishes, reptiles, mammals etc., are affected by oil. When the plumage of a diving birds get oiled, the water repellent properties of the feathers are destroyed, they lose the insulation provided by the feather and its flying capacity. This leads to a condition called hypothermia, subsequently the bird may die. Birds may also ingest the oil. All these cause large scale death of marine birds. It is estimated that between 200, 000 and 500, 000 marine birds are killed annually by routine releases of oil from tankers.

The tainting can be cleared by rearing the affected organisms in clean water for a considerable period. In shellfishes, during the process of molting, the tainted shells are removed. Also, the bird's feathers and some animal's body can be cleaned with detergent. In this type of clean up, the chances of recovery is observed in many cases. Most forms of marine life recover within about five years, as crude oil is not as highly toxic as it is biodegradable.

The polynuclear aromatic hydrocarbons (PAHs) in particular 3, 4 benzopyrene are well known carcinogens (cancer producing). These compounds are present in crude oil in small amounts. The PAH assimilated by marine organisms can be passed upwards through food chain towards man. This will lead to serious concern of human health.

Environmental degradation due to oil: The environmental degradation of oil includes the destruction of algal beds, spreading of oil over rocky shores, incorporation in beach sediment and subsequent loss in scenic beauty.

Q.15. Discuss the sources and effects of various heavy metals on human beings and marine organisms.

Ans. Normally the metals having the atomic weight more than twenty is considered as heavy metals. The term "heavy metals" is used to denote elements with specific weight higher than those of Iron (Fe) and includes mainly Lead (Pb), Cadmium (Cd) and Chromium (Cr), Zinc (Zn), Arsenic (As) Mercury (Hg) and Copper(Cu).

Sources of heavy metals: Coal combustion, electric utilities, steel and iron manufacturing, fuel oils, fuel additives, and incineration of urban refuse are the major sources of oceanic and atmospheric contamination by heavy metals. Heavy metal contaminated runoff from the land, rain of pollutants from the air, and fallout from shipwrecks pollute the ocean with dangerous metals.

Effects of heavy metals on organisms: In general, the effects of heavy metals toxicity can be classified into three main categories:

— The blocking of essential functional groups of the biomolecules (e.g., proteins, enzymes).

— The displacement of a metal ion from a biomolecule.

— The modification of the structure of biomolecules in space which is very important for their function.

Some heavy metal pollutants and their effect on human beings and marine organisms is discussed below:

Lead: It affects human central nervous system. A poison by ingestion and moderately irritating. A common air contaminant due to use of bad fuel in the auto industry, now being phased out through the introduction of lead-free petrol; also in air in the vicinity of industrial plants using lead where precautions are not taken. It is a carcinogen of the lungs and kidneys. Lead is flammable in the form of dust when exposed to heat or flame. From the point of view of industrial poisoning, inhalation of lead is much more important than is ingestion. Lead is a cumulative poison. Increasing amounts build up in the body and eventually a point is reached where symptoms and disability occur. Lead can cause irreversible behavioral disturbances, neurological damage and other developmental problems in young children and babies.

Cadmium: This metal is toxic to humans by inhalation and other routes. It can enter through ingestion, intraperitoneal, subcutaneous, intramuscular and intravenous routes. It has no known biochemical or nutritional function. Also cadmium is highly toxic to freshwater and marine organisms. It is bio accumulative through the food chain. Increased exposure can increase risk of lung cancer. Cadmium was responsible for the "Itai-itai" disease in Japan.

Chromium: This metal exists in two forms, *i.e.*, trivalent and hexavalent. Hexavalent chromium in high dosages has been implicated as the cause of digestive tract cancers, cutaneous and nasal mucous membrane ulcers and dermatitis. Certain chromate salts, *i.e.*, calcium chromate are carcinogenic, at least when inhaled.

Zinc: The zinc is human skin irritant and produces pulmonary system effects. The difficulty arises from oxidation of zinc fumes prior to inhalation or presence of impurities such as cadmium, antimony, arsenic and lead. Zinc however is a needed micronutrient for humans. It is needed in agriculture. It is used in galvanizing industry. The real problem is of its association with heavy metals like lead.

Arsenic: Toxic by subcutaneous, intramuscular and intraperitoneal routes and reported to produce systemic, skin and gastrointestinal effects. It is also a human carcinogen and experimental teratogen.

Mercury: The silvery white liquid metal used in common thermometers, is a potent neurotoxin, capable of causing severe brain damage in developing foetuses and mild tremors and emotional disturbances in exposed adults. Implicated in the occurrence of Minamata disease in Japan, largest source of mercury contamination is through aquatic animals in whom it accumulates and gets bio-transferred to organo compounds, e.g., methyl mercury. Mercury and lead poisoning cause brain damage and behavioral disturbances in children.

Copper: It is also dangerous to marine organisms and it has been used in marine anti-fouling paints.

Q.16. What are the sources of pesticides in the marine environment and their effects on organisms?

Ans. Pesticides are toxins (chemical compounds) used in agricultural farm to kill the unwanted pest organisms attaching the cultivated plants. A pesticide is any substance or mixture of substances intended for preventing, destroying, repelling, or mitigating any pest. Pests can be insects, mice and other animals, unwanted plants (weeds), fungi, or microorganisms like bacteria and viruses. Though often misunderstood to refer only to insecticides, the term pesticide also applies to herbicides, fungicides, and various other substances used to control pests. A pesticide is also any substance or mixture of substances intended for use as a plant regulator, defoliant, or desiccant.

There are natural and artificial pesticides. The natural pesticides are biodegradable and are mostly derived from the extracts of plants and animals. However, artificial pesticides are manmade and produced extensively only after the industrial revolution. These pesticides belong to four main categories.

Some common kinds of pesticides and their features

1. **Organochlorine compounds** e.g., DDD – Dichlorodiphenyldichloroeathane
   - Persistent in the environment or decompose very slowly.
   - Half life is 9-116 years.
   - Accumulate many organisms.
   - Higher concentrations in higher organisms – mammals like dolphins, whales, tortoise etc.
   - Detected in Antarctic penguins.
2. **Organophosphate esters e.g.,** Parathion
   - Very toxic
   - Much toxic than previous one.
3. **Carbamates Baygon**
   - Decompose rapidly through hydrolysis.
   - Half-life period approximately one week.
4. **Chlorophenyl-acids 2, 4, 5-T**
   - Hazardous compounds.
   - Used in Vietnam War for defoliation.

Sources of pesticides in marine environment

Pesticides are the good examples of non-point source pollutants. These toxic materials used in the highland for agricultural and other purposes end up into the sea either through runoff into streams and rivers or having been transported by the wind (dust and aerial spray).

Depending on the type of the chemical, its partial and vapour pressure, evaporation, condensation and deposition occur at various places of the globe leading to their global partitioning. Because of such conditions these chemicals, which originate from lower latitudes, move to higher latitudes. Depending on various conditions, these chemicals migrate in a series of relatively short jumps termed as "grasshoffer effect". Because of this grasshopper effect, concentrations of such chemicals even in the remote parts of the world (Arctic and Antarctic) may rise even if the use of these chemicals are curtailed in source countries.

Effect of pesticides on organisms

Existence of contamination by these chemicals in the environmental media leads to potential impact on almost all species including human. All the organisms, obtain their first dosage of these chemicals from their food and water, while the mammalian species derive these toxins from theirs mothers through milk. While the plants and lower trophic level organisms also accumulate these chemical in their bodies, the effect of these chemicals on these organisms in not openly seen.

Organochlorines have been found to cause microsomal enzyme induction, accelerating body steroids such as estrogens and also to have an effect on the normal reproductive activities of many experimental animals. It is generally known that 5% egg shell thinning occurs at a concentration of 5 mg/g of DDE in the bird egg. It is also found that concentrations of 20-1000 mg/g of DDE in the liver of birds are considered to pose a threat to individual bird reproduction and therefore on the population as a whole.

Q.17. What are the sources and effects of polychlorinated biphenyls (PCBs)?

Ans. PCBs are a series of technical mixtures consisting of many isomers and compounds that vary from mobile oily liquids to white crystalline solids and hard noncrystalline resins.

Sources of PCBs: PCB's usually come from older electrical equipment like transformers, insulations etc.

Effects of PCBs on organisms: It is toxic by ingestion, inhalation and skin contact. A suspected human carcinogen, affects skin and liver. Usual signs of systemic intoxication include nausea, vomiting, weight loss, jaundice, edema and abdominal pain. Severe liver damage may cause coma and death. PCB's (polychlorinated biphenyls) typically cause reproduction problems in most marine organisms. The reduction in sperm motility in human with increasing PCB concentration was observed indicating the possible effect of these chemicals on the reproductive capability of mammalian species.

Q.18. What are the sources, fate and effects of radioactive materials in the marine environment?

Ans. The world's oceans have been a sink for radioactive waste from the production of nuclear weapons and electric power since long.

Sources: The sources of radioactivity may be natural and anthropogenic. Radioactive waste enters the ocean from nuclear weapon testing (this is banned by international conventions), the releasing or dumping of wastes from nuclear fuel cycle systems, nuclear accidents (e.g., Chernobyl accident in USSR in May 1985), ongoing pollution from smaller but more or less continuous leaks from nuclear power stations or nuclear submarines and carriers etc.

Fate of radioactivity in the marine environment: Dumping of high-level radioactive waste is no longer permitted in the ocean, but dumping of low-level wastes is still permitted. Low-level waste contains less radioactivity per gram than high-level waste. High-level wastes usually have longer half-lives. For example, one common high-level waste that is produced by spent nuclear fuel has a half-life of 24, 100 years. It has been suggested that contained nuclear waste should be disposed in the deep sea.

Half life: The time required for half the atoms in a given quantity of a radioactive element to disintegrate is called its half life.

Effects of radioactive materials on organisms: Though little is known about the deep-sea environment or the consequences of radioactive materials containment leakage and failure, the effects could be devastating. The bioaccumulation process is not associated with the radioactivity properties of a molecule but with its chemical properties. If, however, a bioaccumulated molecule happens to be radioactive, its effects on the organism are much more severe.

Radioactive materials bring about two principal sorts of consequences at the organism level:

— Direct toxicity caused by the ionization atoms and molecules of living materials (and especially water) resulting in the production of strong oxidizing agents, which are harmful to living tissues.

— The mutagen activity: Here radiation is a factor, which causes errors during the DNA transcription leading to genome alteration. These errors, called mutations, are in most cases lethal or else lead to teratogenesis when they occur during reproduction. They can also result in cancer when they take place during the division of somatic cells.

Q.19. What are the sources, fate and effects of nutrients discharged into marine environment?

Ans. The nutrients are mainly of nitrogen, phosphorus and silicates. Nutrients are used by plants and animals in a complicated cycle, returned to the environment and then re-used. They are essential for the growth of marine primary producers like phytoplankton, seaweed and seagrass. Too much nutrient in the sea and abnormal algal blooms can upset the balance of plants and animals living near the seabed.

Sources of nutrients: Agriculture in the land, aquaculture practices in the coastal areas and release of sewage as outfall are some of the major sources. Fertilizers used on land are washed into the ocean via rivers and streams. Also the aquaculture farms located near the coastal zone also release the wastewater in to the streams and estuaries or directly to the marine environment. Apart from this the sewage released into the coastal areas also contains higher concentration of nutrients.

Fate of nutrients in the sea: The higher nutrient concentrations are always utilized by phytoplankton. The release of excess nutrients into coastal waters leads to the condition called Eutrophication. Eutrophication leads to higher production of phytoplankton called bloom. The death and decay of phytoplankton leads to higher oxygen consumption, which is detrimental to the organisms.

Effects of nutrients: High nutrient concentrations cause phytoplankton blooms such as, red tides, various yellow and green foams, slimes, and slicks. Although algal blooms are natural, a higher frequency of their occurrence in the past years indicates an unhealthy ecosystem. The toxicity of recent blooms is increasing, which can have a direct effect on the organisms that feed on them. Also, phytoplankton naturally contains DMS (dimethyl sulfoxide), which is released from dead phytoplankton into the atmosphere and can be changed to sulfuric acid, which eventually contribute to acid rain.

Q.20. What is the sources, fate and effects of thermal pollution of the marine environment?

Ans. In this type of pollution, the normal temperature of marine source water column is increased by about 7-8°C from the ambient temperature.

Source of thermal pollution: Mostly the power generating plants along the ocean coastlines use the marine waters for cooling purposes, which leads to heated water expelled into the marine environment.

Fate of increased temperature in the marine environment: The excess temperature usually gets reduced by various ways. As the water has higher specific heat, it takes prolonged time for cooling unless there is an external mechanism to accelerate this. The dilution by surrounding water, evaporation and precipitation may accelerate cooling within a reasonable time.

Effects of thermal pollution: Thermal pollution seems to only affect the communities immediately adjacent to the discharge. In continuous discharge the exposure time is also continuous, which generally result in community shift that is more heat tolerant species may replace the normal biota. The longtime exposure to the heat conditions may be detrimental to the normal biota. Many studies have been done on the effects of thermal pollution on the marine environment. Thermal discharge is most noted in the tropical areas, where organisms are near their thermal maximum. For example, mangrove trees in a thermal heated bay no longer reproduce and no new seedlings can be found in the lagoon.

Q.21. What are the sources and effects of solid wastes discharged into the marine environment?

Ans. Plastic produced in large amount has danger of being discharged as solid waste into the marine environment. Plastic is not biodegradable and therefore effecting the oceans for long periods of time. When compared to other types of pollutants the biological consequences are low.

Source of solid wastes: Solid wastes are dumped into the sea mainly by the anthropogenic activities. Mostly non-biodegradable plastics made materials such as used bottles, cups, plastic bags, plastic fishing nets, metal pieces, abandoned wreckage ships, etc. are dumped into the sea. This is also called Ocean dumping. Some features of plastic dumping in ocean is as follows:

— Six-pack holders will not decompose for 400 years.

— Each square mile of ocean surface off the North East coast of the U.S. has more than 50, 000 pieces of plastic floating.

— 100, 000 marine mammals and 2 million sea birds die every year after ingesting or being caught in plastic debris.

— Adding ingredients to plastics that would hasten decomposition would add 5-7% to costs, but this price increase is unacceptable to industries.

Usefulness of solid waste: The solid waste on the other hand acts as a substratum for many fouling marine organisms. The initial biofilm formation on a new substratum and subsequent macro fouling is useful in some ways. If the solid waste is not releasing any toxic substances then it can be used build artificial reefs and has been considered as a very good way to increase the environmental heterogeneity (biodiversity) in areas which function as nursery or feeding grounds for important fishing stocks.

Effect on organisms: There have been cases that fish having been trapped inside plastic bags other plastic objects and dying of suffocation. Also some cases have been reported of fish getting entangled in plastic fishing nets which have been torn off from fishing boats and drift in the water column, causing a phenomenon known as "ghost fishing". Sea turtles mistake plastic bags for jellyfish eat and die from internal blockages. Seals and sea lions starve after being muzzled by six-pack rings or entangled by nets. Birds have become entangled in discarded fishing line and drowned, while lost drift nets are particularly dangerous to marine forms.

Q.22. What is the effect of sea level rise through global warming?

Ans. The gaseous pollutants (greenhouse gases like carbon dioxide, dimethyl sulphoxide, chlorofluorocarbon-CFC etc.) released into the air, have direct implication on the atmosphere. These gaseous pollutants are found to increase the temperature of atmosphere through depletion of ozone layer, which in turn increase the sea surface temperature resulting in melting of ice in the polar regions. Scientists are working to understand how increases in carbon dioxide and other "greenhouse" gases in the atmosphere might change the earth's climate. But so far, complex interactions between oceans, atmosphere, land, and the sun have made precise climatic predictions impossible. Oceans are crucial in shaping climate because they store and move heat around the planet, and they're a major source and storehouse for green house gases (e.g., carbon dioxide) that affect Climate.

Effects of climate change: The climate change may lead to various impacts such as:

— melting of polar ice caps.

— accelerated erosion of beaches and coastal shore areas through raging storms.

— salt water contamination of coastal aquifiers.

— salt water intrusion in to a great distance landward into estuaries and bays.

— destruction of coastal terrestrial vegetation and replacement of salt tolerant species.

— increased frequency of flooding and subsequent submergence of coastal areas.

— Organisms such as coral may bleach and submerge.

If the sea level increased few centimeter, then the existence of many coastal cities and islands will be under severe threat due to flooding and subsequent reduction in living habitat. In most islands, the area is small and the elevation from mean sea level is in few meters. These islands are under severe threat of submergence. Thus concerted efforts must be taken to save the earth from global warming.

Global warming is becoming potential threat to fisheries and biodiversity. The extinction of species has increased from 30,000 in 1970s to 50,000 in 1980s and at this rate 10-20% of the earth's 10 million species of plant and animals would disappear forever. Coastal zones, which provide habitat for 80% of the 13,200 species of marine fish in the world oceans are being affected at alarming rate. Extensive bleaching of corals is spreading fast and there is destruction of valuable fish habitats especially breeding and spawning grounds.

Q.23. What initiatives and actions are taken for marine pollution monitoring and control in India?

Ans. The Ministry of Environment and Forests under the Environment (Protection) Act, 1986 has declared the coastal stretches as Coastal Regulation Zone (CRZ) for regulating the activities in the CRZ. It declares coastal stretches of seas, bays, estuaries, creeks, rivers and backwater, which are influenced by tidal action (in the landward side) up to 500 meters from the High Tide Line (HTL) and land between Low Tide Line (LTL) and HTL as Coastal Regulating Zone and imposes restrictions on the setting up and expansion of industries operations and processes.

Categories of coastal regulation zone in India

Coastal regulation zone In india has been divided into four categories namely: CRZ-I, CRZ-II, CRZ-III and CRZ-IV

Description of various category is given below:

CRZ-I: Areas that are ecologically sensitive and important, such as national parks/marine parks, sanctuaries, reserve forests, wildlife habitats, mangroves, corals, coral reefs, areas close to breeding and spawning grounds of fish and other marine life, areas of outstanding natural beauty| historical| heritage areas, areas rich in genetic diversity; areas likely to be inundated due to rise in sea level consequent upon global warming and such other areas as may be declared by the Central Government or the concerned authorities at the State/Union Territory level from time to time

Area between the low tide line and the high tide line.

CRZ-II: The areas that have already been developed up to or close to the shoreline. For this purpose "developed area" is referred to as that area within the municipal limits or in other legally designated urban areas, which is already substantially build up and which has been provided with drainage and approach roads and other infrastructure facilities, such as water supply and sewerage mains.

CRZ-III: Areas that are relatively undisturbed and those, which do not belong to either category-I or II. These will include coastal zone in the rural areas (developed and undeveloped) and also areas within Municipal limits or in other legally designated urban areas, which are not substantially building up.

CRZ-IV: Coastal stretches in the Andaman and Nicobar, Lakshadweep and small island except those designated as CRZ-I, CRZ-II or CRZ-III.

Types of Indian marine environment based on uses

In the coastal zone marine water is subjected to several types of uses. In 1974 the water act (Prevention and control of pollution) was created to keep the coastal water clean and free from pollution, the water bodies area classified according the use. The pollution Control boards at Centre and States have adopted 5 km in to the sea from the highest high tide, as the extent, over which their authority to control waste discharges will extend for the purpose of the water (Prevention and Control of Pollution) Act, 1974.

Coastal waters of India have been used for various activities like fishing, shipping harbor activities, industrial-cooling purposes etc. Depending of the types of uses and activities, water quality criteria have been specified to determine its suitability for a particular purpose. Among the various types of uses there is one use that demands highest level of water quality/purity and that is termed as "designated best use" in that stretch of the coastal segment. Through a survey conducted during 1980-86, all the activities in the Indian coastal areas were identified and based on the "designated best use' concept (*i.e.*, exact or near quality requirement for different uses) the following classifications of seawater has been made.

Based on this, primary water quality criteria have been specified for the following five designated best uses.

Classification of sea water(SW) uses in India

- Salt pans, shell fishing, mariculture and ecologically sensitive zone – **SW-I**
- Bathing, contact water sports and commercial fishing – **SW-II**
- Industrial cooling, recreation (non-contact) and aesthetics – **SW-III**
- Harbour – **SW-IV**
- Navigation and controlled waste disposal – **SW-V**

❑❑❑

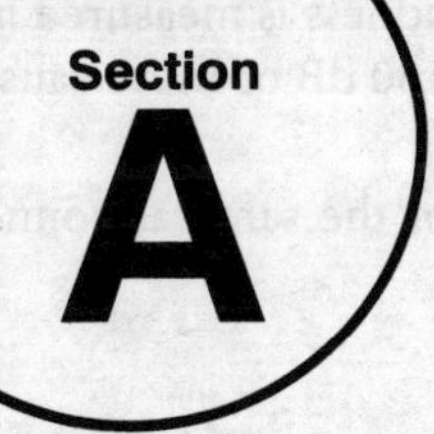

Section A

6 Noise Pollution

SHORT QUESTIONS AND ANSWERS ON NOISE POLLUTION

Q.1. What is Sound? What is the audible range of sound for human beings?

Ans. Sound is a form of energy. It is that external physical cause which produces the sensation of hearing in our ears. Sound is produced only when some material body is made to vibrate. It requires material medium for its propagation and cannot travel in vacuum because there are no material particles which can vibrate and propagate energy in vacuum. The vibrations travel through the medium in form of waves with a certain speed depending on the density and elasticity of the medium. When vibrations reach our ear, sound is heard. Human ears are sensitive only to a limited range of frequencies *i.e.*, 20 to 20,000 Hz. This range of frequency is, therefore, called the audible range of sound. The sound of frequency above 20,000 Hz is called the ultrasonic, while the sound of frequency below 20 Hz is called the infrasonic. Ultrasonic and infrasonic both are inaudible to human beings. Ultra sound is used in medical diagnosis. Infra sound is produced in earthquakes.

Q.2. What are the different ranges of audible frequencies for various animals?

Ans. Different animals have different ranges of audible frequencies. A dog can hear sounds of frequencies up to nearly 50 kHz, a bat up to about 100 kHz, while Dolphins can hear sounds of even higher frequencies up to 150 kHz. The animals can also produce ultrasonic sounds and can communicate in them. Elephants and Whales produce sounds of frequencies less than 20 Hz. Some fishes can hear sounds of frequencies in the range of 1 Hz to 25 Hz.

Table 6.1: Frequency ranges for hearing and speaking by human and animals

Object	Frequency Range of Hearing and Speaking
Bat	10 Hz – 100 kHz
Cat	80 Hz – 60 kHz
Dog	20 Hz – 50 kHz
Dolphins	200 Hz – 150 kHz
Grasshopper	90 Hz – 1.0 kHz
Human	20 Hz – 20kHz

Q.3. What is loudness of sound?

Ans. Loudness is the property by virtue of which a loud sound can be distinguished from a faint one, both having the same quality. Loudness is an aural sensation, *i.e.*, a physiological rather than a physical phenomenon. The louder a sound is, the more energy its sound waves carry. Loudness is measured in decibels (dB). Sound of 0 dB are just audible to the human ear, while sound of 130 dB or more cause pain.

Loudness of sound depends on intensity of the wave; however, loudness is not the same as sound intensity. Intensity is a measurable quantity, while loudness is a sensation.

Q.4. How are loudness and sound intensity related?

Ans. Experimentally, Weber and Fechner established a relationship between the loudness L and intensity I which is given as: $L = \text{K} \log_{10} I$, where K is a constant of proportionality. Thus, loudness is proportional to the logarithmic value of the intensity, but not in the same proportion. Relation between the two is not linear.

The intensity at any point of the medium is measured as the amount of sound energy passing per second normally through unit area at that point. Its unit is microwatt per metre^2 (which is equal to 10^{-6} joule per second per metre^2).

The intensity of normal sound is nearly 10 μW m^{-2} while minimum intensity of sound audible to ears is 10^{-6} μW m^{-2}. Greater the energy carried by a sound wave, greater is the intensity of sound, hence louder it seems to us. The intensity of a sound wave in air is proportional to (*i*) the square of the amplitude of vibration, (*ii*) the square of the frequency of vibration and (*iii*) the density of air.

Q.5. What is sound level? How it is expressed?

Ans. The sound level is usually expressed in decibel(dB).If at a given frequency, I_1 and I_0 are intensities of two sounds, which loudness are L_1 and L_0 then difference in loudness of the two sounds L is:

$$L = L_1 - L_0 = K(\log_{10} I_1 - \log_{10} I_0) \quad \text{or} \quad L = K \log_{10} \frac{I_1}{I_0}$$

If I_0 is taken as reference intensity (10^{-12} μW m^{-2}), the minimum intensity of audible sound intensity at frequency 1 kHz, then L is called the sound level.

Taking $K = 1$, the sound level is $L = \log_{10} \frac{I_1}{I_0}$

For $I_1 = I_0$, sound level $L = 0$ dB, sound of intensity 10^{-12} μW m^{-2} at 1 kHz has loudness zero dB.

For an increase in intensity at a point by 100 times the initial value, the increase in sound level will be $10 \log_{10} 100 = 20$ dB. For an increase in intensity at a point by 100,00 times the initial value, the increase in sound level will be $10 \log_{10}$ 100,00 = 40 dB. It means that for 100 times increase in intensity, the loudness get doubled.

1 dB is (1/10) th of bel where bel in the unit of sound loudness named after the inventor of telephone-the scientist Alexander Graham Bel.

1 dB is defined as the change in level of loudness when the intensity of sound changes by 26 %. Sound strength is measured as sound pressure, sound pressure level or sound power, sound power level (in decibels).

Q.6. What is sound pressure and sound pressure level? How do we express it?

Ans. The sound pressure is the force (N) of sound on a surface area (m^2) perpendicular to the direction of the sound. The SI-unit for the sound pressure is N/m^2 or Pascal.

Sound is usually measured with microphones responding proportionally to the sound pressure, *p*. The power in a sound wave goes as the square of the pressure.

The lowest sound pressure possible to hear is approximately 2×10^{-5} Pa (20 micro Pascal, 0.02 mPa), 2 ten billionths of an atmosphere.

It is therefore, convenient to express the sound pressure as a logarithmic decibel scale related to this lowest human hearable sound *i.e.*, 2×10^{-5} Pa or 0 dB.

Sound pressure level

$$L_p = 10\log\left(\frac{p^2}{p_{\text{ref}}^2}\right) = 10\log\left(\frac{p}{p_{\text{ref}}^2}\right) = 20\log\left(\frac{p}{p_{\text{ref}}}\right)$$

Where, L_p = sound pressure level (dB), p = sound pressure (Pa)

$p_{\text{ref}} = 2 \times 10^{-5}$ - reference sound pressure (Pa)

If the pressure is doubled, the sound pressure level is increased with 6 dB (20 log (2)).

Q.7. What do you mean by sound power and sound power level?

Ans. Sound power is the energy rate - the energy of sound per unit of time (J/s, W in SI-unit) from a sound source. Sound power can more practically be expressed as a relation to the threshold of hearing = 10^{-12} W in a logarithmic scale named sound power level = L_w, expressed as:

$$L_w = 10\log\left(\frac{N}{N_0}\right)$$

Where, L_w = Sound power level in decibel (dB), N = sound power (W)

$N_0 = 10^{-12}$ - reference sound power (W).

Human hearable sound power spans from 10^{-12} W to 10^{-100} W, a range of $\frac{10}{10^{-12}} = 10^{13}$.

Q.8. Describe the mechanism of hearing sound.

Ans. The Human ear consist of three main parts; the outer ear, middle ear and inner ear.

The outer ear consists of:

- The ear lobe (pinna or auricle)
- The ear canal, through which sound waves pass to the ear drum
- The ear drum (Tympanic membrane that separates the outer ear from the middle ear)

The ear lobe and the outer ear canal, which delivers sound to the middle ear, make up the outer ear. Within the outer ear canal are wax-producing glands and hairs that protect the middle ear.

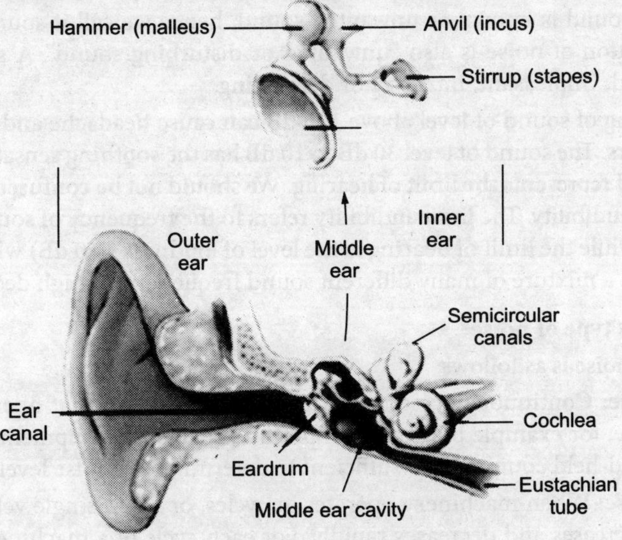

Fig. 6.1: Construction features of ear

The ear drum is a thin, semitransparent, oval-shaped membrane that separates the middle ear from the outer ear. Its purpose is to vibrate according to the frequency and amplitude of sounds that strike it.

There are about 25000 hair cells in our ear which create wave in our ear, corresponding to sound in the environment, as a response to different levels of frequencies. With increasing intensity/pitch/loudness of sound the cells get destroyed decreasing our ability to hear the high frequency sound.

The purpose of the middle ear is to:

- Transmit and amplify sounds from the eardrum to the oval window.
- Act as a dampener on loud sounds that may damage that inner ear (cochlea).

  The middle ear acts as an impedance matching device and has three small bones operating as a set of levers. These bones transfer the vibration to the inner ear which consists of two separate systems, the semi-circular canals for controlling balance and cochlea.

The inner ear consists of:

Two chambers called the vestibular labyrinth and the cochlea.

- The vestibular labyrinth consists of elaborately formed canals (3 semicircular tubes that connect to one another), which are largely responsible for the sense of balance.
- The cochlea, which begins at the oval window, curves into a shape that resembles a snail shell. Tiny hairs line the curves of the cochlea. Both the labyrinth and cochlea are filled with various fluids.

The cochlea is a fluid-filled, snailed-shaped tube which is divided longitudinally into two parts by the basilar membrane. In response to an acoustic stimulus the fluid in the cochlea is disturbed and this distorts the basilar membrane on whose upper surfaces are thousands of bery sensitive hair cells. The hair cells register this distortion and transform it into nerve impulses which are transmitted to the brain. Brain is sent these messages (translated from vibrations by the cilia) through the auditory nerve. Our brain then translates all that and tells what we are hearing. Neurologists don't yet fully understand how we process raw sound data once it enters the cerebral cortex in the brain.

Q.9. What is the difference between sound and noise?

Ans. Sound is that part of energy which excites in us, the sensation of hearing. This is because some type of energy originating from the source and is entering our ear. Sound becomes unwanted when it either interferes with normal activities such as sleeping, conversation, or disrupts or diminishes one's quality of life and such sound is known as unwanted sound; hence we call it sound pollutant or Noise. The traditional definition of noise is also "unwanted or disturbing sound". A sound might be unwanted because it is: Loud, Unpleasant, Intrusive or distracting.

A constant hearing of sound of level above 120 dB can cause headache and permanent damage to the ears of the listeners. The sound of level 30 dB to 10 dB has the soothing sensation, while the level 0 dB of loudness of sound represents the limit of hearing. We should not be confused about the limit of hearing with the limit of audibility. The limit audibility refers to the frequency of sound 20 Hz to 20 kHz which a man can hear, while the limit of hearing is the level of loudness (= 0 dB) which a man can hear. Noise is also considered a mixture of many different sound frequencies at high decibels levels.

Q.10. What are various type of noise?

Ans. Various types of noise is as follows:

Continuous noise: Continuous noise is produced by machinery that operates without interruption in the same mode, for example blowers, pumps and processing equipment. Measuring for just a few minutes with hand-held equipment is sufficient to determine the noise level.

Intermittent noise: When machinery operates in cycles, or when single vehicles or airplanes pass by, the noise level increases and decreases rapidly. For each cycle of a machinery noise source, the noise level can be measured just as for continuous noise. A single passing vehicle or aircraft is called an event. To measure the noise of an event, the sound exposure level is measured, combining level and duration into a single description.

Impulsive noise: The noise from impacts or explosions, e.g., from a pile driver, punch press or gun short is called impulsive noise. It is brief and abrupt, and its startling effects cause greater annoyance than would be expected from a simple measurement of sound pressure level.

Q.11. How do we classify noise?

Ans. Various classes of noise is as follows:

Ambient noise: It is the noise from all sources combined-factory noise, traffic noise, bird song etc.

Specific noise: Specific noise is the noise from the source under investigation. The specific noise is a component of the ambient noise and can be identified and associated with specific source. Are also major sources. Additional noise pollution is contributed by office machines, sirens, power tools, and other equipment.

Residual noise: Residual Noise is ambient noise without specific noise. The residual noise is the noise remaining at a point under certain conditions when the noise from the specific source is suppressed.

Background noise: Background noise is also a common one but should not be confused with residual noise.

Q.12. Give a broad classification of environmental noise.

Ans. Environmental noise is classified as:

- Industrial noise • Transport noise • Neighbourhood noise
- **Industrial noise:** It is sound with a high intensity sound caused by industry machines. Sources of such noise pollution is caused by machines in various factories, industries and mills. Noise from mechanical saws and pneumatic drills is unbearable and a nuisance to the public.

  The different machines of numerous industries, factories and mills produce a lot of high intensity sounds causing noise pollution. Some industrial processes like weaving, ship building, boiler making, pressing and blasting operations are noisier than others.

  The operations in pneumatic drills, milling machines, cutters, printing press with an upward, downward and sideways movement and vibrations, cause lowering of hearing capacity to a large extent.

  The noise pollution is further magnified due to the installation of industries in compact places. For example, the workers near the heavy industrial blowers in steel industry are exposed to sound of 90 dB for eight hours and hence suffer from occupational pollution. Some typical industrial noise levels are tabulated below.

Table 6.2: Some typical industrial noise levels

S.No.	Industrial Source	Noise level (dB)
1.	Steel plant riveting	130
2.	Oxygen torch	126
3.	Boiler maker's shop	120
4.	Textile loom	112
5.	Circular saw	110
6.	Farm tractor	103
7.	Newspaper press	101

S.No.	Industrial Source	Noise level (dB)
8.	Bench lathe	95
9.	Milling machine	90
10.	High speed drill	85
11.	Key press machine	82
12.	Super market	60

- **Transport noise:** Transport noise mainly consists of traffic noise from road, rail and aircraft. The number of automobiles on roads like motors, scooters, cars, motor cycles, buses, trucks and diesel engine vehicles have increased enormously in the recent past further aggravating the problem of transport noise.

  The main threat of noise comes from transport sector. The transport noise includes road traffic noise, rail traffic noise and aircraft noise.

  (*a*) **Road traffic noise:** The chief causes of road traffic noise are the number of road vehicles and their high traffic speed. Faster moving vehicles produce high noise from their gear box, exhaust system, vibrations from their body, etc.

  (*b*) **Rail traffic noise:** The noise from rail traffic is comparatively lower than that from road traffic. Introduction of diesel engines or electrical engines has reduced the intensity of rail traffic noise which was previously shown by steam engine. Besides use of welded tracks and improved coach suspension have contributed to the reduction in railways noise,

  (*c*) **Aircraft noise:** Larger and faster aircrafts produce high noise intermittently during takeoff, landing and during flight. Noise generates from compressor and turbine and near jet exhaust. Jet engines create most noise around a radius of 16 km. Sonic boom is an important aspect of aircraft noise. Sonic boom occurs when an aircraft flies supersonically overhead. Noise levels in most residential areas in metropolitan cities is hovering around the border line due to increased vehicular noise pollution. This high level of noise pollution leads to deafening in the elderly.

- **Neighborhood noise:** This type of noise includes anti-social activities of neighbors like using of loud TV, stereo, radio sets, jazz music, fireworks parties, playing of children, barking of dogs, neighborhood brawls out of intoxication and industrial neighborhood noise etc. The machines used in house construction like concrete mixes, vibrators, bulldozers, heavy diesel lorry, building demolition activities also add to the noise pollution.

  Statistically, ever since the industrial revolution, noise in the environment has been doubling every ten years.

Q.13. What is noise pollution?

Ans. Noise Pollution refers to sound in the environment that are caused by humans and that threaten the health or welfare of human or animal inhabitants. The most common source of noise pollution by far, the one that affects the most people on the planet, is loudspeaker, siren, motor vehicles. Aircraft and industrial machinery are also sources. Additional noise pollution is contributed by office machines, sirens, power tools, and other equipments. The response of ear to sound is very dependent on the frequency content of the sound. The ear has a peak response around 2.5-3 kHz and has a relatively low response at low frequencies. Noise may originate either from natural sources or from anthropogenic activities. The natural sources of noise include thunder, cyclone, roaring of sea, etc. The manmade noise are mechanized automobiles, industries, trains, aero planes, social functions etc. It may be noted that the manmade noise is now doubling after every decade.

Noise is air-borne mechanical energy striking the human eardrum. A sound of 65 dB(A) is the noise level for conversation heard at a distance of one meter. A sound of 125 dB (A) gives the sensation of pain in the ear and 150 dB (A) might kill a human being. If a noise of 90 dB in the mid-frequency range reaches the ear for more than few minutes, then the sensitivity of the ear is reduced.

Q.14. Give examples of various sources of noise.

Ans. Source is the equipment or process directly responsible for sound generation. The major sources of noise and their examples are given below:

- **Transportation sources:** Examples: Railways, Road traffic and Air traffic
- **Industrial sources:** Examples: Noise is generated in mostly all of the industrial activities such as power generation, processing, product fabrication and product assembly.
- **Public address system sources:** Examples: Use of loud speaker at any occasion like marriage function, festival, etc
- **Agricultural machine sources:** Examples: Use of tractors, tubewells, farm machines for agriculture.
- **Defence equipment sources:** Examples: Shooting practices, wars, bomb explosion, etc.
- **Household sources:** Examples: Mixer-grinder, lawn mowers, food blenders, vacuum cleaners, etc.
- **Other sources:** Examples: Rock concerts, barking dogs, construction equipments, etc

Q.15. List out various physical, physiological and psychological effects of noise pollution.

Ans. Noise affects human health in following ways:

Physical effects: Examples: Damage to ear drum, temporary impairment of hearing, permanent deafness.

Physiological effects: Examples: Muscular strain, headache, eye strain, decreased colour perception, nervous break down, pain in heart, etc.

Psychological effects: Examples: Emotional disturbance depression, fatigue, frustration, irritation, reduced efficiency, etc.

Q.16. What are auditory and non-auditory effects of noise pollution?

Ans. The various effects of noise pollution on human beings may be classified into two categories:

- Auditory effects
- Non-auditory effects.

Auditory effects: The impairing of hearing which may cause immediate auditory fatigue finally leading to deafness is known as auditory effects.

Non-auditory effects: These effects include interference with speech communication, annoyance leading to ill-temper, bickering, mental disorientation, violent behaviour and a series of health hazards.

Q.17. What are pathological, non-pathological and vibrational disorders caused by noise pollution?

Ans. In addition to serious loss of hearing, noise also causes pathological (or psychological), non-pathological (or physiological) and vibrational disorders.

(*a*) **Pathological disorder:**

(*i*) Exposure to high frequencies or ultrasonic sound above the audible range damages inner ear and induces nausea and dizziness in man.

(*ii*) Exposure to mid-audible frequency affects brain and nervous system having significant impact on thinking and coordination of limbs.

(*iii*) Moderate vibration can lead to pain, numbness and cyanosis (blue coloration) of Fingers.

(*iv*) Severe vibration results in damage to bones and joints in the hands with swelling and stiffness.

(*v*) Exposure to low frequency noise can reduce heartbeat, variation in blood pressure and breathing difficulties.

(*b*) **Non-pathological disorders:**

These disorders are mainly seen in industries and other establishments which result in lower efficiency, reduced work rate, increased absenteeism and a higher potential for accidents and injuries. Noise also disturbs sound sleep of old people. Children exposed to excessive noise show signs of behavioural disorder which in later age manifest themselves in destructive nature.

(*c*) **Vibration (physical vibration) disorder and Reynaud's phenomenon:**

(*i*) Noise from constant vibrations of hand tools like the hammer or drilling instrument creates 'dead hands' or 'white fingers' which is known as Reynaud's phenomenon.

(*ii*) Noise from moderate vibrations causes pain, numbness, and cyanosis.

(*iii*) Noise from severe vibrations causes damage to bones, joints with swelling and stiffness.

Q.18. Discuss the overall effects of noise pollution.

Ans. Noise pollution affects both human and animal health. Its effects are listed below.

- contraction of blood vessels.
- making skin pale.
- excessive adrenalin in the blood stream which is responsible for high blood pressure.
- Blaring sounds are known to cause mental distress.
- Heart attacks, neurological problems, birth defects and abortion.
- Muscle contraction leading to nervous breakdown, tension, etc.
- The adverse reactions are coupled with a change in hormone content of blood, which in-turn increases heartbeat, constriction of blood vessels, digestive spams and dilation of the pupil of the eye.
- Adversely affects health, work efficiency and behaviour. Noise pollution may cause damage to the heart, brain, kidneys, liver and may produce emotional disturbance.
- The most immediate and acute effect of noise is impairment of hearing that diminishes some part of the auditory system. Prolonged exposure to noise of certain frequency pattern leads to chronic damage to the inner ear.
- Impulsive noise may cause psychological and pathological disorders.
- Ultrasonic sound can affect the digestive, respiratory, cardiovascular system and semicircular canals of the internal ear.
- The brain is adversely affected by loud and sudden noise by jets and airplanes. People are subjected to psychiatric illness.
- Recent reports suggest that blood is thickened by excessive noise.
- The optical system of human beings is also affected by noise pollution. Severe noise pollution causes:

  (*a*) Pupullary dilation (*b*) Impairment of night vision and

  (*c*) Decrease in rate of colour perception

Q.19. Discuss the impact of noise pollution on human health.

Ans. Noise health effects are both ill health and behavioural in nature. The unwanted sound /noise can damage physiological and psychological health. Noise pollution can cause annoyance and aggression, hypertension, high stress levels, tinnitus (ringing sound in ears) , hearing loss, sleep disturbances, and other harmful effects. Furthermore, stress and hypertension are the leading causes to health problems, whereas tinnitus can lead to forgetfulness, severe depression and at times panic attacks.

High noise levels can contribute to cardiovascular effects and exposure to moderately high levels during a single eight-hour period causes a statistical rise in blood pressure of five to ten points and an increase in stress. Noise health effects are the health consequences of elevated sound levels. Elevated workplace noises or other noise can cause hearing impairment, hypertension, ischemic heart disease, annoyance, premature ejaculation, disturbed bowel movements, sleep disturbance, death, and decreased sexual performance. Changes in the immune system and birth defects have been attributed to noise exposure, but evidence is limited. Beyond these effects, elevated noise levels can create stress, increase workplace accident rates, and stimulate aggression and other anti-social behaviour. The most significant causes of noise pollution are vehicle and aircraft noise, prolonged exposure to loud music, and industrial noise. Hearing Damage is a well-established effect, caused by prolonged exposure to intense sound. It Impairs ability to understand speech. It is Permanent and irreversible, even with hearing aids.

Q.20. Discuss the effects of noise pollution on animal health.

Ans. In general, a noise impact to wildlife can be determined by the degree to which the noise disrupts a functioning ecosystem. Noise has the potential to affect wildlife in a variety of ways, varying between different types of animals. Research shows that the degree of reaction to noise often varies with age, sex, season, situation, previous exposure to noise (habituation), noise level, and frequency spectrum. Potential noise effects on wildlife include; auditory damage, physiological changes and behavioural alterations.

These effects are further characterized into primary and secondary effects. Primary effects are direct physical effects to the animal. Secondary effects are indirect changes which occur between the animal and its environment. These effects on animal health are tabulated below.

Table 6.3: Effects of noise pollution on animal health

Type of Effect	Primary	Secondary
Auditory	Hearing Loss	Change in Predator-Prey relationships
	Threshold Shift	Mating Interference
Physiological	Stress	Reduced Reproductive Capacity
	Metabolic Change	Weakened Immune System
	Hormonal Change	Reduction in Functioning
Behavioural	Signal Masking	Change in Predator-Prey relationships
	Avoidance Behaviour	Population Reduction
		Migration and Loss of Habitat
		Mating Interference

Physiological effects on animals: Physiological effects, such as metabolic and hormonal changes, are often associated with stress. Stress in wildlife in their natural setting is typically a difficult response to quantify. For wildlife, stress reactions are part of survival and a routine occurrence. Stress reactions involve what is commonly referred to as the "fight or flight" response. When this reaction is inappropriate, such as fleeing from a non-threaten noise, impacts begin to occur. Inappropriate reactions unnecessarily lead to deplete an animal's energy resources which can increase susceptibility to predators, disease, and starvation.

Behavioural effects on animals: Changes in normal behavioural patterns are the most apparent effects of noise on wildlife. When noise becomes an objectionable intrusion on wildlife habitat, these changes include alterations in habitat locations and migration patterns, and abnormal behaviour that can cause difficulty in mating and survival.

Q.21. Describe various control and prevention measures for noise pollution.

Ans. Control measures:

Source control: This includes source modification such as acoustic treatment to machine surface, design changes, limiting operational timings, etc.

Transmission path intervention: This includes containing the source inside a sound insulating enclosure, constructing a noise barrier or provision of sound absorbing materials along the path.

Receptor control: This includes protection of the receiver by altering the work schedule or provision of personal protection devices such as ear plugs for operating noisy machinery. The measure may include dissipation and deflection methods.

Machines lubrication: Proper oiling will reduce noise from the machine.

Preventive measures:

- Prescribing noise limits for vehicular traffic.
- Ban on honking (usage of horns) in certain areas.
- Creation of silence zones near schools and hospitals.
- Redesigning buildings to make them noise proof.
- Reduction of traffic density in residential areas.
- Giving preference to mass public transport system.

Q.22. Describe various noise control techniques.

Ans. There are 4 fundamental ways in which noise can be controlled:

- reduce noise at the source
- block the path of noise
- increase the path length and
- protect the recipient.

Reduce noise at the source:

(*a*) Make sure that all openings are acoustically sealed. Noise, lake water rushes out through any cracks or openings. Muffling of vehicles and machinery reduce the noise.

(*b*) In industries, different types of absorptive material can be used to control interior noise. Noise reduction can be done by using rigid sealed enclosures around machinery lined with acoustic absorbing material. Isolating machines and their enclosures from the floor using special spring mounts or absorbent mounts and pads and using flexible couplings for interior pipelines also contribute to reducing noise pollution at the source.

(*c*) Regular and thorough maintenance of operating machinery. We should reduce mechanical run out of shafts. By reducing this source of vibration excitation many components like bearing gears and cans may generate less noise and have increased life. We must improve lubrications. We should install bearings correctly. Improper installation sometimes is the reason for bearing noise problems.

(*d*) Traffic volume and speed also have significant effects on the overall sound. Example: doubling the speed increases the sound levels by about 9 dB and doubling the traffic volume (number of vehicles per hour) increases sound levels by about 3 dB. A smooth flow of traffic causes less noise than a stop-and go traffic pattern. Thus proper highway planning and design are essential for controlling traffic noise.

(*e*) Establishing lower speed limits for highways that pass through residential areas, limiting traffic volume and providing alternative routes for truck traffic are effective noise control measures.

(*f*) Using efficient flow techniques: for reducing noise associated with high fluid velocities and turbulence.

(*g*) Reducing fluid jet velocities: As jet noise is proportional to the eight power of jet velocity.

Noise control at the source is the most effective to eliminate noise at the source.

Examples:

- Reduction of noise generated by mechanical vibration of a machine by damping or isolation of the vibration by applying damping material (rubber) to the vibrating components.
- Reduction of impact force by optimizing the impact-distance and covering either or both impact surfaces by rubber.
- Modification of manufacturing design like enclosing the engine parts within proper noise insulating material.

Block the path of noise: Through construction of temporary/permanent barriers:

(*a*) Planting of trees around houses can also act as effective noise barriers.

(*b*) Highly absorptive interior finish material for walls, ceilings and floors can decrease indoor noise levels significantly.

Increasing the path length: Increasing distance from the noise source and the recipient offers a passive means of control:

- Municipal land- use ordinances pertaining to the location of airports make use of the attenuating effect of distance on sound levels.

Noise control at path: When source cannot be made quiet, noise can be controlled by modifying the path.

Examples:

- Attenuation of noise by moving source of noise away from sensitive area.
- Suppression of noise from automobiles using silencers.
- Reduction of noise around residential areas by planting trees in the form of green belt.
- Reduction of transmission of noise using acoustic screens and barriers.
- Enclosing noisy machines in isolated buildings.

Protect the recipient: If source and/or path control do not work, control at the receiver should be explored.

- Use of hearing protection devices (HPD) like ear plugs, ear muffs etc. They reduce the level of noise (by 10 dB to 55 dB) entering the outer and middle ears before it reaches the inner ear.
- Enclose receiver.
- Relocate receiver.

Besides these in general making Legislation, Educating and bringing awareness in the people is common for control of any pollution Example: Ban on loud speakers from 10 pm to 6 pm.

Q.23. Describe various methods of minimizing industrial noise and community noise.

Ans. Noise produced from different sources has created a catastrophe to the entire living world specially to man. Hence, necessary steps should be taken to minimize the level of noise thereby protecting the living world from its detrimental effects.

Industrial noise pollution control: The level of noise pollution due to industries can be lowered down by adopting the following techniques:

- **Use of improved technique:** The noise at source can be reduced by replacing noise producing machines with suitable improved technique. For example, the noise from exhaust fans can be decreased either by increasing the number of blades or by decreasing the rotational speed.
- **Sound proofing:** An insulating material may be applied on the noise producing machinery and equipment's which causes sound proofing.
- **Transmission control:** The level of noise can be minimized by covering the walls of the room by sound absorber (e.g., acoustic tiles), introducing gaskets around the doors and windows, sealing all the outlets and putting carpet, drapery and acoustic materials inside the room.
- **Creation of green vegetation cover:** Plants and trees should be planted along highways, streets and industrial areas because the vegetation covers absorb and dissipate sound energy and acts as a buffer zone.
- **Using protective devices:** The use of ear plugs or ear muffs or even cotton balls by the worker in the industry can protect them from hazardous effects of noise pollution.
- **Operations in open space:** Noisy industrial operations should be conducted at open spaces far off from residential areas.
- **Use of Building codes:** Certain codes should be enforced which require sound proofing in the construction of industries, buildings and apartments.

Community noise control: The community noise control includes minimization of noise from air traffic, transport system, building constructions etc. The noises from the above sources can be lowered by adopting the following measures:

- The air traffic noise can be reduced either by suitable technique or by zoning the area around the airport and not allowing for the construction of houses or industries within 10 meters of the airport.
- The road traffic noise includes tyre noise, engine intake noise, exhaust noise, aerodynamic noise, noise from blowing of horns etc. Such type of noise pollution can be controlled by incorporating silencers in vehicles, maintaining speed limit prescribed for the vehicle and banning air horns.
- The loudspeakers, radio and music system should be operated at threshold intensity so that the noise level should not be hazardous to living organisms.
- Public awareness should be created to educate the common man about the harmful effects of noise pollution, through radio, television, newspapers etc.
- The noise can be reduced by introduction of new regulations which mainly include lowering speed limits and designing for non-stop operation.
- Stringent laws should be implemented to control noise pollution. For example, Motor Vehicle Act of India provides restrictions on heavy vehicles using double sirens while passing through populated areas.
- The laying of ballast less tracks reduces the noise level due to railways.

Table 6.4: Materials recommended for noise reduction

S.No.	Materials	Noise reduction
1.	Cinder block, 10 cm thick	25 dB
2.	Glass, 5 to 8 cm thick	50 dB
3.	Cider block 10 cm thick, plastered on one side	42 dB

4.	Cinder block, 10 cm thick plastered on both sides, or brick 10 cm thick	45 dB
5.	Two cinder blocks, 7.5 cm thick separated by equal amount of air, with plaster on both outside surfaces	55 dB

Q.24. List out the permissible noise levels in various areas.

Ans. We have already discussed the sources of noise of different intensity and their hazardous impact on living organisms particularly on man. Still we work and live in the environment where the noise level is not optimum. Thus, it is rightly told that noise pollution is a slow agent of death. The peoples working in areas where the intensity of noise is comparatively higher suffer from progressive hearing loss, physiological (health) and psychological hazards including tension. The maximum permissible noise levels at different situations are summarized below in Table.

Table 6.5: Maximum permissible noise levels at different areas

S.No.	Situation	Permissible noise in dB(A)
1.	Road traffic near residential areas	70
2.	Ear protection to be worn	85
3.	Factory work for 8 hr in a day, 6 days in a week	90
4.	Prolonged noise causing permanent damage	100
5.	Threshold of pain duration of 30 sec (maximum)	120
6.	Maximum for impulse noise (sonic boom)	150
7.	Absolute limit with ears protected	150
8.	Ear drum rupture	180
9.	Lung damage	194

Table 6.6: Impact of noise intensity on hearing

S.No.	Noise intensity in dB	Effects
1.	Less than 25	No real difficulty in hearing.
2.	25 to 40	Difficulty with hearing soft speech.
3.	40 to 55	Difficulty with hearing normal speech.
4.	90 or more	Unable to hear even amplified speech.
5.	50 or more	Hearing aid required.

Q.25. Describe various regulating provisions under Indian air act and noise rules.

Ans. Under section 2(*a*) of Air (Prevention and Control of Pollution) Act, 1981 noise is defined as air pollutant . "Air pollutant" means any solid, liquid or gaseous substance [(including noise)] present in the atmosphere in such concentration as may be or tend to be injurious to human beings or other living creatures or plants or property or environment." Increasing ambient noise level in public places from various sources, inter-alia, industry activity, construction activity, generators sets, loud speakers, public address systems, music systems, vehicular gowns and other mechanical devices have deleterious effects on human health and the psychological wellbeing of the people. Therefore, it is considered necessary to regulate and control of noise producing and generating sources with the objective of maintaining the ambient air quality standards in respect of noise. Central Government notified the

Noise pollution (regulation and control) rules, 2000 in which following responsibilities are vested with State Governments. District Magistrate, Police Commissioner, or any other officer not below the rank or Deputy Superintendent of Police:

- Enforcement of noise pollution control measures and the due compliance of ambient air quality standards in respect of noise.
- Restriction on the use of loud speakers/public address system.
- Restriction on the use of horns, sound emitting construction equipment and bursting of fire crackers.
- Prohibition of continuance music sound or noise.
- Authority shall act on the complaint and take action against the violator in accordance with the provisions of rules.
- Disallowing sound producing instrument after 10 p.m to 6 a.m except in closed premises.
- State Government may permit loud speakers or public address system in night hours (between 10.00 p.m. to 12.00 midnight not exceeding 15 days in year.

Ambient air quality standards in respect of noise is notified under Noise pollution (regulation and control) rules, 2000 which is tabulated below.

Table 6.7: Ambient air quality standards in respect of noise

Area Code	Category of Area/Zone	Limit in dB (A) Leq*	
		Day Time	Night Time
A	Industrial Area	75	70
B	Commercial Area	65	55
C	Residential Area	55	45
D	Silence Zone	50	40

Note:

- Day time shall mean from 6 AM to 10 PM.
- Night time shall mean from 10 PM to 6 AM.
- Silence zone is defined as areas up to 100 meters around such premises as hospitals, educational institutes and courts. The silence zones are to be declared by competent authority.

  Mixed categories of areas may be declared as one of the four above mentioned categories by the competent authority.
- *dB(A) Leq denotes the time weighted average of the level of sound in decibels on scale A which is relatable to human hearing. A "decibel" is a unit in which noise is measured. "A", in dB (A) Leq, denotes the frequency weighting in the measurement of noise and corresponds to frequency response characteristics of the human ear. Leq: It is energy mean of the noise level over a specific period.

Q.26. What are various provisions for legal control of noise pollution in India?

Ans. (*a*) **Constitution of India right to life:** Article 21 of the constitution guarantees life and personal liberty to all persons. It is well settled by repeated pronouncements of the Supreme Court that right to life enshrined in Article 21 is not of mere survival or existence. It guarantees a right of persons to life with human dignity. Any one who wishes to live in peace, comfort and quiet within his house has a right to prevent the noise as pollutant reaching him.

Right to information: Every one has the right to information know about the norms and conditions on which Govt. permit the industry which effect the environment.

Right to Religion and Noise: Right to religion does not include right to perform religious activities on loud speaker and electronic goods which produce high velocity of noise.

Directive Principal of State Policy: The state has the object to make the environment pollution free.

Fundamental Duties: Every citizen of the country has the fundamental duty to clean the environment.

(*b*) **Cr.P.C. Section 133:** Here Section 133 is of great importance. Under Crpc. Section 133 the magisterial court have been empowered to issue order to remove or abate nuisance caused by noise pollution Sec 133 empower an executive magistrate to interfere and remove a public nuisance in the first instance with a conditional order and then with a permanent one. The provision can be utilized in case of nuisance of environment nature. He can adopt immediate measure to prevent danger or injury of a serious land to the public. For prevention of danger to human life, health or safety the magistrate can direct a person to abstain from certain acts.

(*c*) **I.P.C. Public Nuisance 268-295:** Chapter IV of Indian Penal code deals with offences relating to public health, safety, ...decency , morals under Sections 268, 269, 270, 279, 280, 287, 288, 290 291 294. Noise pollution can be penalized with the help of above section. Private remedies suits in the area may related to public nuisance under A299. This article punishment in case of Public nuisance law of torts covers. A person is guilty of public nuisance who does any act or is guilty of an illegal omission which causes any common injury, danger, or annoyance to the pubic or to the people in general who dwell or occupy property in the vicinity or which must necessarily cause injury, obstruction danger or annoyance to persons who may have occasion to use any public right. A common nuisance is not excused on the ground that it causes some convenience or advantage. Who ever commits a public nuisance in any case not otherwise punishable by this code, shall be punished with fine, which may extend to Rs. 200.

(*d*) **Law of Torts Noise pollution is considered as civil wrong:** Under law of torts , a civil suit can be filed claiming damages for the nuisance. For filing a suit under law of torts a plaintiff is required to comply with some of the requirement of tort of nuisance which are as follows: 1. There should be reasonable interference. 2. Interference should be with the use and enjoyment of land. 3. In an action for nuisance actual damage is required to be proved. As a general rule either the presence or absence of malice does not matter. But in some cases deviation from the rule has been made.

(*e*) **Factories Act Reduction of Noise and Oil of Machinery:** The Factories Act does not contain any specific provision for noise control. However, under the Third Schedule Sections 89 and 90 of the Act, noise induced hearing loss, is mentioned as notifiable disease. Similarly, under the Modal Rules, limits for noise exposure for work zone area have been prescribed.

(*f*) **Motor Vehicle Act:** Provision Relation to use of horn and change of Engine: In Motor vehicle Act rules regarding use horns and any modification in engine are made.

(*g*) **Noise Pollution Control Rule 2000 under Environment Protection Act 1996:** Further for better regulation for noise pollution There are The Noise Pollution (Regulation and Control) Rules, 2000 – in order to curb the growing problem of noise pollution the government of India has enacted the noise pollution rules 2000 that includes the following main provisions:

- The state government may categories the areas in the industrial or commercial or residential
- The ambient air quality standards in respect of noise for different areas have been specified.
- State government shall take measure for abatement of noise including noise emanating from vehicular movement and ensure that the existing noise levels do not exceed the ambient air quality standards specified under these rules.

- An area not less than 100 m around hospitals educations institutions and court may be declare as silence are for the purpose of these rules.
- A loud speaker or a public address system shall not be used except after obtaining written permission from the authority and the same shall not be used at night. Between 10 pm to 6 am
- A person found violating the provisions as to the maximum noise permissible in any particular area shall be liable to be punished for it as per the provision of these rules and any other law in force.

*dB(A) Leq denotes the time weighted average of the level of sound in decibels on scale A which is relatable to human hearing. A "decibel" is a unit in which noise is measured. "A", in dB(A) Leq, denotes the frequency weighting in the measurement of noise and corresponds to frequency response characteristics of the human ear. Leq: It is an energy mean of the noise level, over a specified period.|

Q.27. How can we control/reduce noise level on Diwali?

Ans. Diwali is one of the important festivals of India and bursting of crackers and noise generating fireworks are integral part of its celebration. So generation of noise pollution and air pollution during the festival cannot be denied. Though the Hon'ble Supreme Court has banned the bursting of crackers or any noise generating fireworks of high decibels to control the noise pollution but unfortunately, the same is not being followed in letter and spirit. Therefore, an unbearable noise pollution creates severe problem to the common people during Diwali Festival.

Firecrackers generate instantaneous impulsive noise, which when measured in free field condition in impulsive mode, gives peak sound pressure level. It is quite obvious that a number of crackers when bursting serially can easily form a band of continuous noise in the presence of reflecting surfaces. The reverberation of sound wave due to repeated reflection in the surfaces prolongs the time interval of prevailing sound intensity. The continuous band of noise, thus formed, definitely affects the ambient level in close proximity.

Following actions are recommended to reduce/control noise pollution during Diwali celebration:

- The use of fireworks or firecrackers should not be permitted except between 6.00 p.m. and 10.00 p.m. No fireworks of firecrackers should be used between 10.00 p.m. and 6.00 a.m.
- Firecrackers should not be used at any time in silence zone.
- Appropriate steps should be taken to educate students about the ill effects of air and noise pollution.
- The campaign against firecrackers should be initiated through Eco-Clubs of schools, which have taken up the issue in a big way by organising rallies and other programmes to create awareness among children in particular and the public in general on the ill-effects of crackers.
- The ill effects of noise pollution should also be displayed in slide show through Doordarshan and such thing should also be propagated among the people through Local Akashvani Kendra.

Q.28. How can we protect the common man from noise pollution and contribute to creating quiet environment?

Ans. A number of public complains are being made to the pollution control boards and local authority about noise pollution especially due to D.G. Sets, pressure horns and community hall for party and function. The following noise sources in the community can disturb quality of life:

- Music concerts and clubs.
- Car and home stereos.
- Personal music players.
- Bus, truck, car, motor cycles and their pressure horn etc.
- **Transportation:** subways, traffic, garbage trucks.

There are several steps one can take to reduce noise in our community and protect us:

- **Turn down the volume:** Today, many people play their television sets and personal stereos unnecessarily loud, keep the sound at safe level.
- **Reduce your exposure:** Some situations are inherently noisy, but you can decrease the risk of damage in several ways. Try to avoid going to loud bars, restaurants, or clubs for hours at a time. Limit the time you spend in noisy area.
- **Protect yourself:** Earplugs can be used to protect from noise and don't be afraid to speak up for better hearing. Politely ask to turn it down.
- **Remember your neighbours:** Noise from your personal recreation affects not only you but everyone around you. Be a good neighbour and lower the volume. This is true whether at home, the beach, or a shopping mall—anywhere your loud noises could interfere with others' peace.
- **Use silent generator only**, and
- **Educate your children:** Educate and talk to your children about what is noise and their health hazards.

❑❑❑

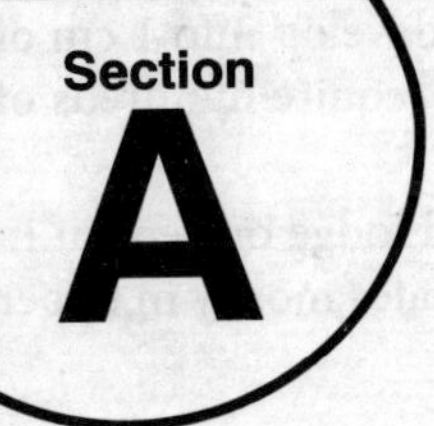

7

Soil/Land Pollution

SHORT QUESTIONS AND ANSWERS ON SOIL/LAND POLLUTION

Q.1. What is soil and its significance?

Ans. The top layer of earth is called soil. More than 80% of the land is covered by soil and it is 3 to 4 meter in depth on the earth surface. It is one of the most important natural resources. It consists of a mixture of minerals, organic material, living organism, air and water that together support the growth of plant life. The traditional meaning of soil is the natural medium for the growth of land plants. Many terms are used to refer to the non-water surface of the earth. The soil scientist calls it soil, a geologist may call fragmental rock or overburden, an engineer may call land. Land is a general term to describe the non-water portion of the earth, usually in reference to a large surface area.

Significance of soil

- The soil is the natural medium in which plants grow. It provides oxygen, water and minerals to plant-roots.
- It supports life and forms the basis of plant's and animal's life.
- We get food, clothing and building material from the plants and crops that grow in soil.
- Soil is a habitat of numerous types of organisms – plants and animals and home for a number of friendly micro-organisms that maintain and enhance the fertility of the soil.

Q.2. Describe the mechanism of the formation of soil.

Ans. Rocks of crust are continuously subjected to mechanical, chemical and biological actions. As a result, they are broken down into finer particles during long period of time. The process of breaking up of rocks into finer particles is called as weathering.

Changes in temperature, beating effects of hails, rain water, pressure of winds and streams cause mechanical weathering. Addition or removal of oxygen within rock materials and action of carbonic acids cause chemical weathering. Sometimes, plants like lichens growing in rock crevices produce some chemicals. Those chemicals help in the breaking up of rocks into finer particles. Such type of weathering is called as biological weathering. The process of soil formation is called as Pedogenesis and Study of the soil is called Pedology.

The weathered out fine particles get mixed with dead remains of plant and animals and with their excretory materials to form soil.

Several factors contribute to the formation of soil from the parent material. This includes the mechanical weathering of rocks due to temperature changes and abrasion, wind, moving water, glaciers, chemical weathering activities, and lichens. Climate and time are also important in the development of soils. In extremely dry or cold climates soils develop very slowly, while in humid and warm climates soils develop more rapidly. Under ideal climatic conditions, soft parent material may develop into 1 cm of soil within 15 days. Under poor climatic conditions, a hard parent material may require hundreds of years to develop into soil.

Fine particles(clays) in the surface soil layers are carried downward by water and lodge or deposit in layers at depth of from 15 to over 60 cm(6-24 in.) deep. Organic material accumulates mostly in upper surface layers of soils.

Q.3. What do you mean by soil horizon?

Ans. The various, distinguishable layers of soil that occur are called soil horizons.

Soil Horizon: A layer of soil or soil material approximately parallel to the land surface and differing from adjacent genetically related layers in physical, chemical, and biological properties or characteristics, such as colour, structure, texture, consistency, amount of organic matter, and degree of acidity or alkalinity. They also tell much about the characteristics of a soil which include information about hardpans, depth of organic matter accumulation, soil denseness from clay deposition, and the extent of leaching.

A cross-sectional view of the horizons in a soil is called a '**soil profile**'.

The top layer or the surface litter layer, called the '**O-horizon**', consists mostly of freshly-fallen and partially-decomposed leaves, twigs, animal waste, fungi and other organic materials. Normally, it is brown or black.

The uppermost layer of the soil, called the '**A-horizon**', consists of partially-decomposed organic matter (humus) and some inorganic mineral particles. It is usually darker and looser than the deeper layers. The roots of most plants are found in these two upper layers. As long as these layers are anchored by vegetation, the soil stores water and releases it in a trickle throughout the year instead of in a force like a flood. These two top layers also contain a large amount of bacteria, fungi, earthworms, and other small insects, which form complex food webs in the soil, help recycle soil nutrients, and contribute to soil fertility.

The '**B-horizon**', often called the subsoil, contains less organic material and fewer organisms than the A horizon.

The area below the subsoil is called the '**C-horizon**' and consists of weathered parent material. This parent material does not contain any organic materials. The chemical composition of the C-horizon helps to determine the pH of the soil and also influences the soil's rate of water absorption and retention.

Q.4. Describe the classification of various types of soil.

Ans. (A) Types of soil on the basis of particle size: On this basis of size of particles or according to their texture, soil is classified into three main types:

(*a*) Sandy soil (*b*) Clayey soil (*c*) Loamy soil.

Sandy soils: A soil having loosely packed coarse or large sized particles is called as sand. Sandy soils are well aerated, warm and are easy to work or dig. This type of soil remains dry most of the time and their nutrients are quickly washed out. This is because, it cannot hold water.

Clayey soil: Clayey soil is made up of particles of very small sizes. Those particles remain compactly arranged. It is sticky and cohesive or waterlogged when wet. It can hold greater quantity of water.

Loamy soil: A soil which is made due to mixing of clay (20-30%), silt (30-45%) and sand, which make them fertile and easy to work, is called as loamy soil. A compact soil made up of rather small sized particles as compared to sand is called as silt. Loams are the best soils for plant growth

Sometimes, soils contain dead and decomposed parts of plants and animals. Such a decomposed organic matter found in the soil is called as HUMUS. It is crucial to maintaining the balance of minerals and nutrients needed for plant growth. Presence of humus in the soil is essential for high production of crops. Farmers add humus in the soil by mixing manure (compost, wormy compost or natural manure) to the soil.

(B) Types of soil on the basis of origin: On the basis of origin, soils are classified into following major groups:

(*a*) Residual soil (*b*) Transported soil (*c*) Mountainous soil

Residual soil: The soil which is formed due to weathering of rocks and is found adjacent to those rocks is called as residual soil. This soil is basically composed of mineral particles of different sizes.

On the basis of presence of mineral components, the residual soil is further divided into: **red** soil, **black** soil and **laterite** soil.

Red soil is red in colour due to the presence of oxides of iron in it. Besides iron, it contains potassium in good quantities. The **black** soil is made up of ash and lava that come out during volcanic eruptions. This soil contains aluminium, iron, calcium and magnesium in good amounts but it lacks phosphorus and nitrogen. This soil is suitable for the cultivation of cotton crops. The **laterite** soil is formed due to weathering of open parts of rocks. It is red-yellow in colour. It contains hydrogenous oxides of iron and aluminium. This is usually acidic hence unfit for the growth of plants.

Transported soil: The soil which is transported through gravity, flowing water, wind or glacier from one place and deposited on some other place is called as transported soil. This soil has further been put into four different categories on the basis of mode of its transportation. Those four different categories are- **colluvial**, **alluvial**, **sandy** and **glacial** or drift soils.

Colluvial soil is formed due to deposit of soil in low areas due to land- slides. Alluvial soil is formed due to deposits of particles contained in the water of fast water currents. This soil often contains different layers and this is principally composed of fine-rounded particles.

The soil which is made of sand particles deposited by wind at a certain place is called as sandy or desert soil. Only **desert plants**(xerophytes) grow in this type of soil. Some soils are formed due to deposits of particles through due to deposits of particles through glaciers. Those soils are called as glacial soils.

Mountain soil: The soil which is found on mountains, hill slopes, cliffs and valleys, is called as mountain soil. This type of soil is basically composed of pebbles, clay, shell and limestone. The lower part of this soil is usually calcareous in nature. This soil is found in north-east Himalayan regions of India.

Q.5. Define soil pollution. Why it is an issue of concern?

Ans. Soil pollution is defined as the build-up in soils of persistent toxic compounds, chemicals, salts, radioactive materials, or disease-causing agents, which have adverse effects on plant growth and animal health.

Soil is the thin layer of organic and inorganic materials that covers the Earth's rocky surface.

The organic portion, which is derived from the decayed remains of plants and animals, is concentrated in the dark uppermost topsoil. The inorganic portion made up of rock fragments, was formed over thousands of years by physical and chemical weathering of bedrock. Productive soils are necessary for agriculture to supply the world with sufficient food.

Soil pollution is an issue of serious concern because it is the soil on which we live and get our food supply. Rapid industrialization, urbanization and consequent increase in population have resulted in soil pollution.

Soil pollution is quite different from air pollution and water pollution as in soil pollution, pollutants remain in direct contact with soil for a relatively longer period.

Soil pollution can also be defined as, "Undesirable changes in physical chemical and biological characteristics of soil, which are harmful for all living organisms".

Q.6. How do we categorize the sources of soil pollution?

Ans. The sources of soil/land pollution can be categorized as under:

- Agriculture
- Sewage sludge
- Household-domestic solid waste
- Industrial activities.
- Quarrying and mining
- Dredged spoils
- Demolitions and constructions

Q.7. List out the reasons of soil pollution.

Ans. Reasons are listed below:

- Increasing urbanization
- disposal of untreated wastes
- Indiscriminate use of Agrochemicals and Agricultural
- Unscientific mining
- Dumping Industrial Wastes
- Accidental Pollution/Leakages
- Outdated technology, inadequate treatment and safety management of chemicals and waste
- Lack of Designed and Engineered Landfills

Q.8. What are the different ways by which the soil can become polluted?

Ans. Soil can get polluted by:

- Seepage from a landfill
- Discharge of industrial waste into the soil
- Percolation of contaminated water into the soil
- Rupture of underground storage tanks
- Excess application of pesticides, herbicides or fertilizer
- Solid waste seepage

Q.9. What are the causes of soil pollution?

Ans. Soil pollution is caused by the presence of man-made chemicals or other alteration in the natural soil environment. This type of contamination typically arises from the rupture of underground storage links, application of pesticides, percolation of contaminated surface water to subsurface strata, oil and fuel dumping, leaching of wastes from landfills or direct discharge of industrial wastes to the soil. The most common chemicals involved are petroleum hydrocarbons, solvents, pesticides, lead and other heavy metals. This occurrence of this phenomenon is correlated with the degree of industrialization and intensities of chemical usage.

A soil pollutant is any factor which deteriorates the quality, texture and mineral content of the soil or which disturbs the biological balance of the organisms in the soil. Pollution in soil has adverse effect on plant growth.

Pollution in soil is associated with:

- Indiscriminate use of fertilizers
- Indiscriminate use of pesticides, insecticides and herbicides
- Dumping of large quantities of solid waste
- Deforestation and soil erosion

Q.10. What are the natural causes of land pollution?

Ans. Land pollution occurs massively during earth quakes, landslides, hurricanes and floods. All cause hard to clean mess, which is expensive to clean, and may sometimes take years to restore the affected area. These kinds of natural disasters are not only a problem in that they cause pollution but also because they leave many victims homeless.

Table 7.1 Sources and mechanisms of land pollution

S. No.	Sources		Reason/Cause/Mechanism
1.	Agriculture	(i)	Accumulation of animal manure.
		(ii)	Excessive input of chemical fertilizers and pesticides.
		(iii)	Farm residues.
		(iv)	Illicit dumping of tainted or harmful crops on land.
2.	Quarrying and Mining	(i)	Use of explosive to blow up mines.
		(ii)	Mining tailings and refuse.
		(iii)	Discharge of acidic effluents.
		(iv)	Use of heavy machineries and large number of vehicles.
3.	Sewage sludge	(i)	Untreated municipal sludges.
4.	Dredged spoils	(i)	Improper method of dredging disposals.
5.	Households or Domestic solid waste.	(i)	Improper and inefficient method of collection transportation and disposal of solid waste causes accumulation.
		(ii)	Improper sanitation system.
6.	Demolition and Construction	(i)	Non-biodegradable rubbles or debris create unsightly condition.
7.	Industrial	(i)	Discharge of untreated effluents into land for disposal
		(ii)	Deposition of process and effluent treatment sludges.
		(iii)	Release of gaseous pollutants such as SO_2 causing acid rains.

Q.11. What are common chemicals causing soil pollution?

Ans. The most common chemicals involved in causing soil pollution are:

- Petroleum hydrocarbons
- Heavy metals
- Pesticides
- Solvents

Q.12. What are different types of soil pollution?

Ans. Different types of soil pollution are:

- Agricultural soil pollution
  - (*i*) pollution of surface soil
  - (*ii*) pollution of underground soil
- Soil pollution by industrial effluents and solid wastes
  - (*i*) pollution of surface soil
  - (*ii*) disturbances in soil profile
- Pollution due to urban activities
  - (*i*) pollution of surface soil
  - (*ii*) pollution of underground soil

Q.13. How do biological agents, mining activities and radioactive pollutants cause soil pollution?

Ans. Biological agents: The major sources of biological agents causing soil pollution are:

- Human excreta
- Animal and bird excreta

- Faulty sanitation
- Municipal wastes
- Biomedical wastes

Industrial parasites are among the most threatening biological agents.

Mining activities: The top layer of soil is generally damaged or destroyed during mining which results in:

- Heavy metals and other toxic substances (may be found as impurities) can destroy beneficial microorganisms of the soil. There may be health hazard, since rate of retention and absorption of heavy metals is higher in infants and children.
- **Loss** of fertile land;
- Formation of huge mountains of wastes;
- Addition of harmful substances to the soil.

The uncontrolled mine fires may also destroy the productivity of the areas near mines.

Radioactive pollutants: When rain containing radionuclide's falls on the soil, the radioactivity gets transferred to the soil due to its absorption by the soil particles. Rain also adds radioactive contaminants into water bodies thus affecting the aquatic flora and fauna.

Radioactive pollutants can cause a number of undesirable diseases of digestive system if they enter our body through food chain.

Huge amounts of radioactive substances result from following sources:

- Nuclear device explosions
- Nuclear testing laboratories
- Nuclear power plants and weapons testing

All these are responsible for radioactive fallout, further enhancing soil pollution. Nuclear pollution is a kind of physical pollution of the environment. Nuclear pollution differs from air, water and soil pollution in the respect that it can bring about physiological changes in the present and future generations of all living beings.

There in no safe dose of nuclear radiation because even the smallest dose of radiation has many deleterious effects. All the living organisms are directly or indirectly affected by a large variety of radiations.

Q.14. How does indiscriminate and over use of fertilizers cause soil pollution?

Ans. Soil nutrients are important for plant growth and development. Plants obtain carbon, hydrogen and oxygen from air and water. But other necessary nutrients like nitrogen, phosphorus, potassium, calcium, magnesium, sulfur and more must be obtained from the soil. Farmers generally use fertilizers to correct soil deficiencies. Fertilizers contaminate the soil with impurities, which come from the raw materials used for their manufacture. Mixed fertilizers often contain ammonium nitrate (NH_4NO_3), phosphorus as P_2O_5, and potassium as K_2O. For instance, As, Pb and Cd present in traces in rock phosphate mineral get transferred to super phosphate fertilizer. Since the metals are not degradable, their accumulation in the soil above their toxic levels due to excessive use of phosphate fertilizers, becomes an indestructible poison for crops.

The over use of NPK fertilizers reduce quantity of vegetables and crops grown on soil over the years. It also reduces the protein content of wheat, maize, grams, etc., grown on that soil. The carbohydrate quality of such crops also gets degraded. Excess potassium content in soil decreases .

Q.15. How does indiscriminate use of pesticides, insecticides and herbicides cause soil pollution?

Ans. Plants on which we depend for food are under attack from insects, fungi, bacteria, viruses, rodents and other animals, and must compete with weeds for nutrients. To kill unwanted populations living in or on their crops, farmers use pesticides. The first wide spread insecticide use began at the end of World

War II and included DDT (dichlorodiphenyltrichloroethane) and gammaxene. Insects soon became resistant to DDT and as the chemical did not decompose readily, it persisted in the environment.

Since it was soluble in fat rather than water, it biomagnified up the food chain and disrupted calcium metabolism in birds, causing eggshells to be thin and fragile. As a result, large birds of prey such as the brown pelican, ospreys, falcons and eagles became endangered. DDT has been now been banned in most western countries. Ironically many of them including USA, still produce DDT for export to other developing nations whose needs outweigh the problems caused by it. The most important pesticides are DDT, BHC, chlorinated hydrocarbons, organophosphates, aldrin, malathion, dieldrin, furoxans, etc. The remnants of such pesticides used on pests may get adsorbed by the soil particles, which then contaminate root crops grown in that soil. The consumption of such crops causes the pesticides remnants to enter human biological systems, affecting them adversely.

An infamous herbicide used as a defoliant in the Vietnam War called Agent Orange (dioxin), was eventually banned. Soldiers' cancer cases, skin conditions and infertility have been linked to exposure to Agent Orange. Pesticides not only bring toxic effect on human and animals but also decrease the fertility of the soil. Some of the pesticides are quite stable and their bio-degradation may take weeks and even months.

Pesticide problems such as resistance, resurgence, and health effects have caused scientists to seek alternatives. Pheromones and hormones to attract or repel insects and using natural enemies or sterilization by radiation have been suggested.

Q.16. How does dumping of solid wastes cause soil pollution?

Ans. In general, solid waste includes garbage, domestic refuse and discarded solid materials such as those from commercial, industrial and agricultural operations. They contain increasing amounts of paper, cardboards, plastics, glass, old construction material, packaging material and toxic or otherwise hazardous substances. Since a significant amount of urban solid waste tends to be paper and food waste, the majority is recyclable or biodegradable in landfills. Similarly, most agricultural waste is recycled and mining waste is left on site.

The portion of solid waste that is hazardous such as oils, battery metals, heavy metals from smelting industries and organic solvents are the ones we have to pay particular attention to. These can in the long run, get deposited to the soils of the surrounding area and pollute them by altering their chemical and biological properties. They also contaminate drinking water aquifer sources. More than 90% of hazardous waste is produced by chemical, petroleum and metal-related industries and small businesses such as dry cleaners and gas stations contribute as well.

Solid Waste disposal was brought to the forefront of public attention by the notorious Love Canal case in USA in 1978. Toxic chemicals leached from oozing storage drums into the soil underneath homes, causing an unusually large number of birth defects, cancers and respiratory, nervous and kidney diseases.

Q.17. How deforestation causes soil pollution?

Ans. Soil Erosion occurs when the weathered soil particles are dislodged and carried away by wind or water. Deforestation, agricultural development, temperature extremes, precipitation including acid rain, and human activities contribute to this erosion. Humans speed up this process by construction, mining, cutting of timber, over cropping and overgrazing. It results in floods and cause soil erosion.

Forests and grasslands are an excellent binding material that keeps the soil intact and healthy. They support many habitats and ecosystems, which provide innumerable feeding pathways or food chains to all species. Their loss would threaten food chains and the survival of many species. During the past few years quite, a lot of vast green land has been converted into deserts. The precious rain forest habitats of South America, tropical Asia and Africa are coming under pressure of population growth and development (especially timber, construction and agriculture). Deforestation is slowly destroying the most productive flora and fauna areas in the world, which also form vast tracts of a very valuable sink for CO_2.

Q.18. How soil pollution is caused due to urbanization?

Ans. Pollution of surface soils: Urban activities generate large quantities of city wastes including several Biodegradable materials (like vegetables, animal wastes, papers, wooden pieces, carcasses, plant twigs, leaves, cloth wastes as well as sweepings) and many non-biodegradable materials (such as plastic bags, plastic bottles, plastic wastes, glass bottles, glass pieces, stone/cement pieces). On a rough estimate Indian city are producing solid city wastes to the tune of 60,000-80,000 metric tons every day. If left uncollected and decomposed, they are a cause of several problems such as:

- **Clogging of drains:** Causing serious drainage problems including the burst/leakage of drainage lines leading to health problems.
- **Barrier to movement of water:** Solid wastes have seriously damaged the normal movement of water thus creating problem of inundation, damage to foundation of buildings as well as public health hazards.
- **Foul smell:** Generated by dumping the wastes at a place.
- **Increased microbial activities:** Microbial decomposition of organic wastes generate large quantities of methane besides many chemicals to pollute the soil and water flowing on its surface.
- **When such solid wastes are hospital wastes they create many health problems:** As they may have dangerous pathogen within them besides dangerous medicines, and injections.

Pollution of underground soil: Underground soil in cities is likely to be polluted by:

- Chemicals released by industrial wastes and industrial wastes.
- Decomposed and partially decomposed materials of sanitary wastes.

Many dangerous chemicals like cadmium, chromium, lead, arsenic, selenium products are likely to be deposited in underground soil. Similarly, underground soil polluted by sanitary wastes generate many harmful chemicals. These can damage the normal activities and ecological balance in the underground soil

Brief causes of soil pollution due to urbanization is given below:

- Polluted water discharged from factories.
- Runoff from pollutants (paint, chemicals, rotting organic material) leaching out of landfill.
- Oil and petroleum leaks from vehicles washed off the road by the rain into the surrounding habitat.
- Chemical fertilizer runoff from farms and crops.
- Acid rain (fumes from factories mixing with rain).
- Sewage discharged into rivers instead of being treated properly.
- Over application of pesticides and fertilizers.
- Purposeful injection into groundwater as a disposal method.
- Interconnections between aquifers during drilling (poor technique).
- Septic tank seepage.
- Lagoon seepage.
- Sanitary/hazardous landfill seepage.
- Cemeteries.
- Scrap yards (waste oil and chemical drainage).
- Leaks from sanitary sewers.

Q.19. What are the main causes of land degradation?

Ans. Land degradation: Land degradation is the temporary or permanent lowering of the productive capacity of the land.

Various reasons of land degradation are listed below:

- **Soil erosion:** Soil erosion means wearing a way of soil. It refers to loss of the superficial layer of the soil by the top layer of the soil is very fertile. Therefore, its removal degrades the quality of land making it less productive.

  Soil erosion is caused due to heavy rainfall, lack of vegetation cover, and soil mismanagement like, uncontrolled grazing by cattle, faulty method of surface drainage, removal of forest cover.
- **Salination:** Salination refers to increase in the concentration of soluble salts in the soil poor irrigation practices, flood water results in accumulation of dissolved salts on the soil surface. In dry areas water evaporate quickly leaving behind a white crust of salts in soil on the surface. The higher concentration of salts severely affects the water absorption process of the plants, resulting into poor production.
- **Water logging:** Water- logging may be due to surface flooding or due to high water table. Excessive irrigation practices may cause water logging due to rise in water table of the area. The productivity of water-logged soil is severely affected due to lesser availability of plants.
- **Shifting cultivation:** Among many tribal communities a very peculiar practice of cutting down a selected portion of the forest and burning. Crops were cultivated for a few years till the fertility of soil is reduced. When the fields become sterile the were abandoned and a fresh patch of forest were burnt and fresh fields were formed for crop cultivation.
- **Urbanization:** Man is to blamed for most of the land degradation. Productive areas is fast reducing because of developmental activities such as human settlement, industries, roads, railways, airports etc.
- **Dumping:** Dumping of solid wastes is one of the most important factors which are responsible for the land pollution. In developing countries like India, the garbage and refuse products are simply dumped leading to land pollution.

Q.20. What are the agricultural, industrial and urban effects of soil pollution?

Ans. Agricultural effects:

- Reduced soil fertility
- Reduced nitrogen fixation
- Increased erodibility
- Larger loss of soil and nutrients
- Deposition of silt in tanks and reservoirs
- Reduced crop yield
- Imbalance in soil fauna and flora

Industrial effects:

- Dangerous chemicals entering underground water
- Ecological imbalance
- Release of pollutant gases
- Release of radioactive rays causing health problems
- Increased salinity
- Reduced vegetation

Urban effects:

- Clogging of drains
- Inundation of areas
- Public health problems
- Pollution of drinking water sources
- Foul smell and release of gases
- Waste management problems

Q.21. What are the environmental long term effects of soil pollution?

Ans. When it comes to the environment itself, the toll of contaminated soil is even more dire. Soil that has been contaminated should no longer be used to grow food, because the chemicals can leach into the food and harm people who eat it. If contaminated soil is used to grow food, the land will usually produce lower yields than it would if it were not contaminated. This, in turn, can cause even more harm because a lack of plants on the soil will cause more erosion, spreading the contaminants onto land that might not have been tainted before.

In addition, the pollutants will change the makeup of the soil and the types of microorganisms that will live in it. If certain organisms die off in the area, the larger predator animals will also have to move away or die because they've lost their food supply. Thus it's possible for soil pollution to change whole ecosystems

Effects of soil pollution in brief:

- pollution runs off into rivers and kills the fish, plants and other aquatic life.
- crops and fodder grown on polluted soil may pass the pollutants on to the consumers.
- polluted soil may no longer grow crops and fodder.
- Soil structure is damaged (clay ionic structure impaired).
- corrosion of foundations and pipelines.
- impairs soil stability.
- may release vapours and hydrocarbon into buildings and cellars.
- may create toxic dusts.
- may poison children playing in the area.

Q.22. What are the health consequences of soil pollution?

Ans. Land Pollution affects humans and wild life in number of ways such as:

- It can cause respiratory problems
- Lead to birth defects
- Cause skin problems
- Causes cancers
- The toxic substances causing soil/land pollution can get into human body directly by skin contacts
- Toxic substances being washed into water sources like rivers, reservoirs, pollutes under ground water courses
- Consumption of vegetables and fruits grown on polluted soils
- Inhalation of polluted dust or particles

Health effects of land pollution are enumerated below:

Cancer: Many of the widely used pesticides on agricultural lands are potentially carcinogenic.

Neurological problems: The health effects of soil pollution, of course, vary with the contaminant. Some toxins, such as lead, are both naturally occurring and come from man made sources. Children are most vulnerable to the health effects of lead in the soil. Their developing neurological systems put them at risk for serious health issues, causing possible brain and nervous system damage.

Tissue damage and irritation: Mercury levels in the soil can increase through deposition from precipitation. Mercury can cause irritation of internal and external tissues, such as your skin and the lining of your gastrointestinal tract. The severity depends upon the toxin and length of exposure.

Reduction of soil health: Soil pollution also degrades soil health, impacting both the environment as well as people in the long term. The same chemicals that can harm people may also affect plants and micro-organisms. Acid rain, for example, can create toxic conditions by mobilizing what might otherwise be harmless chemicals in the soil. These toxins in turn can slow plant growth as well as impact the soil bacteria and other organisms necessary for plant growth.

Q.23. Describe various causes and effects of soil pollution.

Ans. Various causes and effects of soil pollution are listed below:

Discharge of industrial wastes: Most of the industries produce wastes of one kind or the other such as harmful gases and chemicals with polluted water discharged from factories. These wastes are being generated and dumped on the land, mainly include the following:

- Toxic pollutants
- Heavy metals
- Organic compounds
- Inorganic complexes
- Non-biodegradable materials

The industrial pollutants which pollute air and water also pollute the soil.

Urban Wastes: Urban activities generate large quantities of wastes including several biodegradable materials (like vegetables, garbage, waste paper, animal wastes, papers, wooden pieces, Street sweepings, plant twigs, leaves, cloth wastes as well as Packing materials) and many non-biodegradable materials (such as plastic bags, plastic bottles, plastic wastes, glass bottles, glass pieces, Metallic cans, stone/cement pieces). If left uncollected, they are a cause of several problems such as:

- **Clogging of drains:** Causing serious drainage problems leading to health problems.
- **Barrier to movement of water:** Solid wastes have seriously damaged the normal movement of water thus creating problem of inundation, damage to foundation of buildings as well as public health hazards.
- **Foul smell:** Generated by dumping the wastes at a place.
- **Increased microbial activities:** Microbial decomposition of organic wastes generate large quantities of methane besides many chemicals to pollute the soil and water flowing on its surface.
- **When such solid wastes are hospital wastes they create many health problems:** As they may have dangerous pathogen within them besides dangerous medicines, injections etc.

Underground soil in cities is likely to be polluted by:

- **Chemicals released by industrial wastes:** Many dangerous chemicals like cadmium, chromium, lead, arsenic, selenium products are likely to be deposited in underground soil. These can damage the normal activities and ecological balance in the underground soil.
- Decomposed and partially decomposed materials of sanitary wastes may pollute underground soil, since these wastes generate many harmful chemicals.

Urban wastes are also dangerous like industrial wastes because some of these do not degrade easily. Overpopulation and increasing consumption have totally changed the very complexion of urban wastes into a complex mixture of food remains, papers, paints, varnishes and many other toxic chemicals.

Improper agricultural practices:

Indiscriminate use of fertilizers: Soil nutrients are important for plant growth and development. Plants obtain carbon, hydrogen and oxygen from air and water. But other necessary nutrients like nitrogen, phosphorus, potassium, calcium, magnesium, sulfur and more must be obtained from the soil. The three primary soil nutrients often in short supply are potassium, phosphorus and nitrogen compounds. These are called macronutrients. Certain other elements like boron, zinc and manganese are necessary in extremely small amounts and are known as micronutrients. When crops are harvested, a large amount of macronutrients and a small amount of micronutrients are removed with the crops. If the same crop is grown again, depleted levels of these nutrients can result in decreased yields. These necessary nutrients can be returned to the soil through the application of fertilizers. Farmers generally use fertilizers to correct soil deficiencies. Fertilizers contaminate the soil with impurities, which come from the raw materials used for their manufacture.

Mixed fertilizers often contain ammonium nitrate (NH_4NO_3), phosphorus as P_2O_5, and potassium as K_2O. For instance, As, Pb and Cd present in traces in rock phosphate mineral get transferred to super phosphate fertilizer. Since the metals are not degradable, their accumulation in the soil above their toxic levels due to excessive use of phosphate fertilizers, becomes an indestructible poison for crops. The over use of NPK fertilizers reduce quantity of vegetables and crops grown on soil over the years. It also reduces the protein content of wheat, maize, grams, etc., grown on that soil. The carbohydrate quality of such crops also gets degraded. Excess potassium content in soil decreases vitamin C and carotene content in vegetables and fruits. The vegetables and fruits grown on over fertilized soil are more prone to attacks by insects and disease.

Indiscriminate use of pesticides, insecticides and herbicides: Pesticides are a general class of chemicals or biological agents used to control, repel, attract or kill pests. Pests are organisms that include insects, weeds or other plants, birds, mammals, fish and microorganisms that compete with humans for food, destroy property, spread disease or are considered a nuisance. They are categorized according to their target use. The three major groups of pesticides are herbicides (weed control), insecticides (insect control), and fungicides (disease control). Nematicides are pesticides used to control soil, leaf, and stem dwelling nematodes (round worms).

Pesticide can consist of synthetic, organic, inorganic compounds or biological organisms. Organic pesticides contain carbon, while inorganic pesticides contain carbon which can be found only under the form of carbonate or cyanide. These pesticides are derivatives made from arsenic, mercury, fluorine, sulphur and copper, as well as derivatives made from cyanide. Organic pesticides can be divided into three groups: synthetic pesticides (developed in laboratories and manufactured), natural pesticides (from animal, microbial or vegetal origin) and microorganisms. Inorganic pesticides are mostly derived from minerals.

Pesticides are not a modern invention. Elemental sulfur was used by ancient Sumerians to protect their crops from insects. Medieval farmers and scientists experimented with chemicals ranging from arsenic to lead on common crops. In 1938, Dichloro-Diphenyl- Trichloroethane, or DDT was discovered to be extremely effective and rapidly became the most widely used insecticide in the world. The most important pesticides are DDT(Dichloro-Diphenyl- Trichloroethane), BHC(Benzene Hexa Chloride), chlorinated hydrocarbons, organophosphates, aldrin, malathion, dieldrin, furoxan, etc.

Plants on which we depend for food are under attack from insects, fungi, bacteria, viruses, rodents and other animals, and must compete with weeds for nutrients. To kill unwanted populations living in or on their crops, farmers use pesticides. The remnants of such pesticides used on pests may get adsorbed by the soil particles, which then contaminate root crops grown in that soil. The consumption of such crops causes the pesticides remnants to enter human biological systems, affecting them adversely. Pesticides not only bring toxic effect on human and animals but also decrease the fertility of the soil.

DDT, one of the first synthetic organic insecticides to be used, was thought to be the perfect insecticide. DDT has a half-life (the amount of time required for half of the chemical to decompose) of 10-15 years. This means that if 100 kg of DDT were to be sprayed over an area, 50 kg would still be present in the area 10-15 years later. The use of DDT has been banned in some countries. However, India still permits the use of DDT, although only for purposes of mosquito control. Persistent pesticides become attached to small soil particles which are easily moved by wind and water to different parts thus affecting soils elsewhere. Persistent pesticides may also accumulate in the bodies of animals, and over a period of time increase in concentration if the animal is unable to flush them out of its system, thus leading to the phenomenon called bioaccumulation. When an affected animal is eaten by another carnivore, these pesticides are further concentrated in the body of the carnivore. This phenomenon of acquiring increasing levels of a substance in the bodies of higher trophic level organisms is known as 'biomagnification'. This process, especially in the case of insecticides like DDT, has been proved to be disastrous. DDT is a well-known case of biomagnification in ecosystems. The short-term and the long-term health effects to the persons using the pesticide and the public that consumes the food grown by using the pesticides are also major concerns. Exposure to small quantities of pesticides over several years can cause mutations, produce cancers, etc. DDT is still an excellent insecticide, but it has two major disadvantages: (1) Its half-life in the environment is to long (10-15 years), and (2) it is bioaccumulated in the fat of animals, since it is soluble in fat rather than water. Serious concerns about the human safety and biological impacts of DDT has led 100 countries to ban its use.

We should replace chemical fertilizers and pesticides by developing an alternatives. Alternative agriculture is the broadest term that includes all non-traditional agricultural methods and encompasses sustainable agriculture, organic agriculture, alternative uses of traditional crops, alternative methods for raising crops, etc. Sustainable agriculture advocates the use of methods to produce adequate safe food in an economically viable manner, while maintaining the state of the ecosystem. Organic agriculture advocates avoiding the use of chemical fertilizers and pesticides.

Acceptable pesticides must have at least the following characteristics:

- They must be short lived in the environment. They must not exist long enough to bioaccumulation and biomagnified in the food chains. The residues must be gone from foodstuffs. Usually, it must be dissipated within a week or two of application time. To allow workers into fields, it is desirable for many pesticides to be gone within a few days.
- Pesticides must not be carcinogenic, teratogenic, or mutations. But there is no easy nor clear-cut way to determine these characteristics with certainty.
- The pesticide must be effective yet able to be safely handled. Capable of safe handling means that the chemical is retained on the sprayed or treated area, is not overly volatile so that is moves extensively into the air, and can be safely handled by the applicator when reasonable care is used.

Excess salts and water: Irrigated lands can produce higher crop yields than those that only use rainwater. However, this has its own set of ill effects. Irrigation water contains dissolved salts and in dry climates much of the water in the saline solution evaporates leaving its salts, such as sodium chloride in the topsoil. The accumulation of these salts is called salinization, which can stunt plant growth, lower yields and eventually kill the crop and render the land useless for agriculture. These salts can be flushed out of the soil by using more water. This practice, however, increases the cost of crop production and also wastes enormous amounts of water. Flushing out salts can also make the downstream irrigation water saltier.

Another problem with irrigation is waterlogging. This occurs when large amounts of water are used to leach the salts deeper into the soil. However, if the drainage is poor this water accumulates in underground and gradually raising the water table. The roots of the plants then get enveloped in this saline water and eventually die.

Thus, in the long run it is better for us to adopt sustainable farming practices to prevent the degradation of soil.

Dumping of solid wastes: In general, solid waste includes garbage, domestic refuse and discarded solid materials such as those from commercial, industrial and agricultural operations. They contain increasing amounts of paper, cardboards, plastics, glass, old construction material, packaging material and toxic or otherwise hazardous substances. Since a significant amount of urban solid waste tends to be paper and food waste, the majority is recyclable or biodegradable in landfills. Similarly, most agricultural waste and mining waste is left on site. The portion of solid waste that is hazardous such as oils, battery metals, heavy metals from industries have needed to pay particular attention. These can in the long run, get deposited to the soils of the surrounding area and pollute them by altering their chemical and biological properties. They also contaminate drinking water aquifer sources.

Q.24. Describe briefly the overall effects of soil pollution.

Ans. The effects of pollution on soil are quite alarming and can cause huge disturbances in the ecological balance and health of living creatures on earth. Some of the soil pollution effects are described below:

- Excessive use of fertilizers does not allow microbial flora in soil to flourish;
- Excessive use of nitrogen and phosphorus fertilizers makes the soil deficient in other micro-nutrients such as zinc, cobalt, etc;
- Soil fertility is reduced or adversely affected, if pesticides remain in soil for longer duration;
- Pesticides like DDT, dieldrin, etc. are known to seep gradually through soil into groundwater and thus contaminate public drinking water supplies;
- Decrease in soil fertility and therefore decrease in the soil yield;
- Loss of soil and natural nutrients present in it. Plants also would not thrive in such a soil, which would further result in soil erosion;
- Disturbance in the balance of flora and fauna residing in the soil;
- Increase in salinity of the soil, which therefore makes it unfit for vegetation, thus making it useless and barren;
- Generally crops cannot grow and flourish in a polluted soil. Yet if some crops manage to grow, they would be poisonous enough to cause serious health problems in people consuming them;
- Creation of toxic dust leading is another potential effect of soil pollution;
- Foul smell due to industrial chemicals and gases might result in headaches, fatigue, nausea, etc. in many people;
- Soil pollutants would bring in alteration in the soil structure, which would lead to death of many essential organisms in it. This would also affect the larger predators and compel them to move to other places, once they lose their food supply; and
- Soil pollution can lead to water pollution if toxic chemicals leach into groundwater, or if contaminated runoff reaches streams, lakes, or oceans. Soil also naturally contributes to air pollution by releasing volatile compounds into the atmosphere.

Q.25. How can we control soil pollution?

Ans. The following steps have been suggested to control soil pollution.

To help prevent soil erosion, we can limit construction in sensitive area. In general, we would need less fertilizer and fewer pesticides.

If we could all adopt the three R's: Reduce, Reuse, and Recycle ,this would give us less solid waste.

Reducing chemical fertilizer and pesticide use. Applying bio-fertilizers and manures can reduce chemical fertilizer and pesticide use.

Biological methods of pest control can also reduce the use of pesticides and thereby minimize soil pollution.

Reusing of materials such as glass containers, plastic bags, paper, cloth etc. can be reused at domestic levels rather than being disposed, thereby reducing solid waste pollution.

Recycling and recovery of materials-This is a reasonable solution for reducing soil pollution. Materials such as paper, some kinds of plastics and glass can and are being recycled. This decreases the volume of refuse and helps in the conservation of natural resources. For example, recovery of one tonne of paper can save 17 trees.

Reforesting: Control of land loss and soil erosion can be attempted through restoring forest and grass cover to check wastelands, soil erosion and floods. Crop rotation or mixed cropping can improve the fertility of the land.

Solid waste treatment -Proper methods should be adopted for management of solid waste disposal. Industrial wastes can be treated physically, chemically and biologically until they are less hazardous. Acidic and alkaline wastes should be first neutralized; the insoluble material if biodegradable should be allowed to degrade under controlled conditions before being disposed.

As a last resort, new areas for storage of hazardous waste should be investigated such as deep well injection and more secure landfills. Burying the waste in locations situated away from residential areas is the simplest and most widely used technique of solid waste management. Environmental and aesthetic considerations must be taken into consideration before selecting the dumping sites.

Incineration of other wastes is expensive and leaves a huge residue and adds to air pollution. Pyrolysis is a process of combustion in absence of oxygen or the material burnt under controlled atmosphere of oxygen. It is an alternative to incineration. The gas and liquid thus obtained can be used as fuels. Pyrolysis of carbonaceous wastes like firewood, coconut, palm waste, corn combs, cashew shell, rice husk paddy straw and saw dust, yields charcoal along with products like tar, methyl alcohol, acetic acid, acetone and a fuel gas.

Anaerobic/aerobic decomposition of biodegradable municipal and domestic waste is also being done and gives organic manure. Cow dung which releases methane into the atmosphere, should be processed further in 'gobar gas plants' to produce 'gobar gas' and good manure.

Q.26. List out steps to reduce soil pollution.

Ans. Following steps can be taken to stop / reduce /avoid land pollution:

- Adoption of organic farming instead of using synthetic chemicals.
- Adoption of recycle and reuse of a waste materials.
- Proper disposal of industrial wastes into secured landfill sites.
- Educating people not to litter.
- Proper collection, transportation and disposal of municipal wastes.

Q.27. Briefly discuss the facts related to soil pollution.

Ans. Soil pollution refers to an increased level of harmful human-made chemicals in the natural soil environment.

Industrial pollutants and pesticides are two major soil pollution sources.

Soil pollution can lead to serious water contamination and therefore create huge health issues.

Cleaning soil pollution is expensive and (usually) long-term process.

Heavy agricultural machinery used by many farmers makes soil more susceptible to pollution because it negatively affects the density and compactness of soil.

If not treated properly soil pollution could significantly reduce crop yields in years to come which could lead to future food shortages..

Soil pollution is especially big issue in China where contaminated water is often being used to irrigate very large amount of arable land.

If soil is polluted with heavy metals such as chromium and lead this can lead to higher incidence of cancer and leukemia in people.

The excessive soil pollution can also cause major problem for ecosystems by affecting the lower pyramid levels of the food chain and disrupting its proper functioning.

There are several different options used to clean polluted soil. These include excavating soil and taking it to a disposal site, aeration of soils, thermal and bio remediation, etc.

Soil pollution can also lead to soil erosion by decreasing the level of natural nutrients in the soil. Such decrease would cause reduced plant life in given region and without enough plants this would further result in soil erosion.

Even the short term exposure to some harmful chemicals in soil can lead to headaches, nausea, fatigue and skin rashes.

Contaminated soil isn't a safe place to grow food on because the poisonous chemicals may reach into human body once people digest the affected food.

Preventing soil pollution is much easier than to clean it. This is the main reason why soil needs to be constantly tested for dangerous chemicals. Only clean and healthy soil can produce healthy food.

Q.28. What are the types and causes of soil erosion?

Ans. The loss of nutrient rich top soil or disturbances to the soil structure by some natural processes or anthropogenic activities is known as soil erosion. Soil erosion is a part of the soil pollution. Soil erosion is creeping death of the soil. When the rate of removal of soil exceeds the rate of soil formation, the soil erosion occurs.

Soil erosion and deforestation: Soil erosion can be defined as the movement of surface litter and topsoil from one place to another. While erosion is a natural process, often caused by wind and flowing water, it is greatly accelerated by human activities such as farming, construction, overgrazing by livestock, burning of grass cover, and deforestation.

The loss of the topsoil makes a soil less fertile and reduces its water holding capacity. The topsoil, which is washed away, also contributes to water pollution by clogging lakes and increasing the turbidity of the water, ultimately leading to the loss of aquatic life. For one inch of topsoil to be formed it normally requires 200-1000 years, depending upon the climate and soil type. Thus, if the topsoil erodes faster than it is formed, the soil becomes a non-renewable resource. Therefore, it is essential that proper soil conservation measures are, used to minimize the loss of the topsoil.

Forests and grasslands are an excellent binding material that keeps the soil intact and healthy. They support many habitats and ecosystems, which provide innumerable feeding pathways or food chains to all species. Their loss would threaten food chains and the survival of many species. During the past few years quite a lot of vast green land has been converted into deserts. Many scientists believe that a wealth of medicinal substances including a cure for cancer and aids, lie in the forest. Deforestation is slowly destroying the most productive flora and fauna areas in the world, which also form vast tracts of a valuable sink for CO_2.

The soil erosions are of two types:

(*i*) Natural or geological or background soil erosion

(*ii*) Accelerated soil erosion

The natural soil erosion takes place by normal or natural conditions without any interference of man. But the accelerated soil erosion is due to active interference of man and animals in disturbing the equilibrium in between the rate of soil formation and rate of soil loss.

The magnitude of soil erosion is governed by the factors like corrosively of the eroding agent, erodibility of the soil, scope, steepness and length, land use practices and conservation strategies employed.

Causes of soil erosion: The process of soil erosions is caused by the following agents:

(*a*) **Water erosion:** Water is powerful agent causing soil erosion either in the form of rain drops or in the form of surface flow. When the soil is lost like a thin covering from a large surface area, the process is known as sheet erosion. When the sheet erosion takes place with full momentum, leading to cutting of soil surface as thin channels, it is known as rill erosion. Several rills from the steep slope converge forming wider channel known as gully erosion.

(*b*) **Wind erosion:** In dry (arid) regions, the soil is mainly sandy and the vegetation is grossly inadequate. In such places, wind causes loss of soil.

(*c*) **Land slide or slip erosion:** When the rocks at cliffs face heavy rains, their weights undergo sharp rise due to gravitational forces. Such rocks with high weights slip over lower rocks causing land slide and there by slip erosion.

(*d*) **Stream bank erosion:** During flood, the rivers splash their water against banks and cut through them taking away soil from the bank.

(*e*) **Deforestation and overgrazing:** Loss of natural vegetation due to deforestation and overgrazing leads to dry and dust soil which can be taken away easily by wind and water.

(*f*) **Mining erosion:** A lot of soil is also eroded due to mining and industrial activities.

Since soil is very precious for the existence of living organisms, it is the need of the hour to prevent the loss of soil by adopting suitable mechanisms.

The process of soil loss can be checked by the following means:

(*i*) Protection of soil from the severity of rain drops through afforestation.

(*ii*) Preventing water from concentrating and passing through the slopes.

(*iii*) Slowing down the momentum of water flow adopting various scientific techniques.

(*iv*) Growing vegetation covers to minimise wind velocity.

(*v*) Growing soil binders like grasses to prevent direct contact in between soil and eroding agents.

The contribution of different global agent towards total soil degradation (erosion) are as follows:

- Overgrazing accounts for 34.5% of the degraded area
- Deforestation accounts for 29.5%
- Agricultural activities accounts for 28.1 %
- Over exploitation of soil accounts for 7.0%
- Bio-industrial activities accounts for 1.2%

❑❑❑

[illegible] are [illegible] responsible for the [illegible] of soil [illegible] agricultural practices.

(i) Prevention of soil erosion by covering [illegible] through afforestation.

(iv) [illegible] wind velocity.

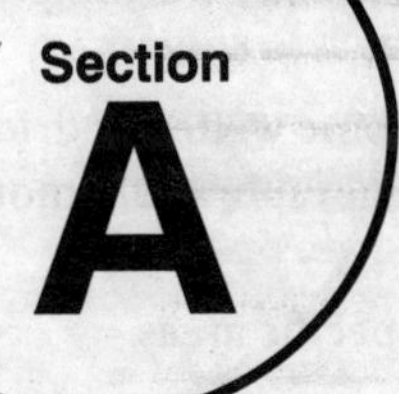

8 Solid Waste Pollution

SHORT QUESTIONS AND ANSWERS ON SOLID WASTE POLLUTION

Q.1. What is solid waste? How do we classify it?

Ans. Any solid material that is thrown away or discarded as useless and unwanted is considered as solid waste.

Classification of solid wastes:

Domestic and municipal wastes: These include garbage and rubbish, like waste paper, plastic, cloth from households, office, hostel and market.

Industrial wastes: The two general categories are process and non-process wastes. The non-process wastes are common to all industries such as packaging, office and cafeteria wastes. Process wastes are more complex and specific to the industrial plants. Their composition depends on type of products produced.

Agricultural wastes: These include cereal and millet straw, paddy husk, sugarcane trash and other crop residues.

Special wastes: The waste materials which endanger public health and welfare and seriously affect environment are: (*a*) **Radioactive wastes** from atomic power stations, labs and hospitals (*b*) **Toxic wastes** such as pesticides, heavy metals, pharmaceuticals (*c*) **Biomedical waste** products such as antibiotics, enzymes, pathogens, hospital waste etc. (*d*) **E-waste** products such as electronics components, computer parts etc.

Q.2. Describe the effects of solid waste, if not managed wisely/properly.

Ans. Effects of improper management of solid waste is listed below:

- Affects our health and our socio-economic conditions.
- Affects our coastal and marine environment.
- Affects our climate.
- Green House Gases (GHGs) accumulate in Earth's atmosphere as a result of human activities, causing global mean surface air temperature and subsurface ocean temperature to rise.
- Rising global temperatures are expected to raise sea levels and change precipitation and other local climate conditions.
- Changing regional climates could alter forests, crop yields, and water supplies.
- This could also affect human health, animals, and many types of ecosystems.
- Deserts might expand into existing range lands, and features of some national parks might be permanently altered.

— Some countries are expected to become warmer, although sulfates might limit warming in some areas.

— Scientists are unable to determine which parts of those countries will become wetter or drier, but there is likely to be an overall trend toward increased precipitation and evaporation, more intense rainstorms, and drier soils.

— Whether rainfall increases or decreases cannot be reliably projected for specific areas.

- **Activities that have altered the chemical composition of the atmosphere:**

  — Buildup of GHGs primarily carbon dioxide (CO_2) methane (CH_4), and nitrous oxide (N_2O).

  — CO_2 is released to the atmosphere by the burning of fossil fuels, wood and wood products, and solid waste.

  — CH_4 is emitted from the decomposition of organic wastes in landfills, the raising of livestock, and the production and transport of coal, natural gas, and oil.

  — NO_2 is emitted during agricultural and industrial activities, as well as during combustion of solid waste and fossil fuels.

Q.3. What are various types of litter we generate and time it takes to degenerate?

Ans. The type of litter we generate and the approximate time it takes to degenerate is given below:

- **Organic waste such as vegetable and fruit peels, leftover foodstuff, etc.:** A week or two
- **Paper:** 10–30 days
- **Cotton cloth:** 2–5 months
- **Wood:** 10–15 years
- **Woollen items:** 1 year
- **Tin, aluminium, and other metal items such as cans:** 100–500 years
- **Plastic bags:** One million years?
- **Glass bottles:** Undetermined

Q.4. Describe the health effects of open dumping of solid waste.

Ans. Health effects of open dumping of waste are listed below:

— The health risks associated with illegal dumping are significant for ragpickers and residents living nearby.

— Areas used for illegal dumping may be easily accessible to people, especially children, who are vulnerable to the physical (protruding nails or sharp edges) and chemical (harmful fluids or dust) hazards posed by wastes.

— Rodents, insects, and other vermin attracted to open dumpsites may also pose health risks.

— Dump sites with scrap tires provide an ideal breeding ground for mosquitoes, which can multiply 100 times faster than normal in the warm stagnant water standing in scrap tire casings.

— Severe illnesses, including encephalitis and dengue fever, have been attributed to disease-carrying mosquitoes originating from scrap tire piles.

— In addition, countless neighbourhoods have evacuated and property damage has been significant because of dumpsites that caught fire, either by spontaneous combustion or, more commonly, by arson. Illegal dumping can impact proper drainage of runoff, making areas more susceptible to flooding when wastes block ravines, creeks, culverts, and drainage basins.

— In rural areas, open burning at dumpsites containing chemicals may contaminate wells and surface water used as sources of drinking water.

— Dioxin emissions from open burning of garbage in one day by four families could equal the emissions from a municipal solid waste incinerator burning 200 tons per day. Dioxins are known to suppress the immune system, disrupt hormonal balances and cause cancer.

— Other emissions resulting from open burning and their associated health risks include:

- benzene (leukemia)
- toluene diisocyanate (asthma)
- nitrogen dioxides (lung damage)
- nitrile compounds (metabolic poisons and carcinogens)
- Burning garbage can also produce emissions of formaldehyde, hydrochloric and sulphuric acid, hydrogen cyanide, polycyclic aromatic hydrocarbons (PAHs), cadmium, lead, mercury and chromium.

Q.5. Describe in detail various types of solid wastes.

Ans. Solid waste includes domestic wastes, municipal wastes, commercial wastes, garbage, rubbish, ashes, construction and demolition wastes, industrial wastes, hazardous wastes, hospital wastes, plastics, e-waste and sewage.

Domestic wastes: These wastes are generated by household activities such as cooking, cleaning, repairs, redecoration, empty containers, packaging, clothing, old books, newspapers, old furnishings, etc.

Commercial wastes: Solid wastes generated in offices, wholesale stores, restaurants, hotels, markets, warehouses and other commercial establishments. These are further classified into garbage and rubbish.

Institutional wastes: Wastes generated from institutions such as schools, colleges, hospitals, research institutions. The waste includes garbage, rubbish and hazardous wastes.

Municipal wastes: Wastes generated due to municipal activities and services such as street waste, dead animals, market waste and abandoned vehicles. Generally, the term is used in a wider sense to incorporate domestic wastes, institutional wastes and commercial wastes.

Garbage: It includes animal and vegetable wastes due to various activities like storage, preparation and sale, cooking and serving. These are biodegradable.

Ashes: Residues from the burning of wood, charcoal and coke for cooking and heating in houses, institutions and small industries. Ashes consist of a fine powdery residue, cinders and clinker often mixed with small pieces of metal and glass.

Rubbish: Apart from garbage and ashes, other solid wastes produced in households, commercial establishments, and institutions are termed as rubbish.

Bulky wastes: Bulky wastes are large household appliances such as cookers, refrigerators and washing machines as well as furniture, crates, vehicle parts, tyres, wood, trees and branches. The bulky metallic wastes are sold as scrap metal but some portion is disposed as sanitary landfills.

Street wastes: Street wastes include paper, cardboard, plastic, dirt, dust, leaves and other vegetable matter collected from streets, walkways, alleys, parks and vacant plots.

Dead animals: It includes animals that die naturally or accidentally killed. It does not include carcass and animal parts from slaughterhouses as these are considered as industrial wastes.

Construction and demolition wastes: Major components of the construction materials are cement, bricks, cement plaster, steel, rubble, stone, timber, plastic and iron pipes. About 50% of the wastes are not currently recycled in India.

Industrial wastes: These are discarded solid material of manufacturing processes and industrial operations and are considered separately from municipal wastes. However, solid wastes from small industries/ plants and ash from power plants are frequently disposed of at municipal landfills. Major producers of industrial wastes are the thermal power plants producing coal ash, integrated iron and steel plants producing blast furnace slag and steel melting slag, non-ferrous industries like aluminium, zinc and copper producing red mud and tailings, sugar industries generating press mud, pulp and paper industries producing lime and fertilizer and allied industries producing gypsum. Management of industrial solid waste is not the responsibility of local bodies. Industries generating solid wastes have to manage by themselves and are required to obtain prior permission from the respective state pollution control boards under relevant rules.

Slaughter house waste: India has the world's largest population of livestock. Slaughter houses slaughter cattle, buffaloes, sheep, goat, pigs and poultry for domestic consumption as well as for export purposes. The waste generated here are liquid and solid in nature. Slaughtering of animals generates waste consisting of non-edible organs, stomach contents, dung, bones and sludge from waste water treatment.

Hospital waste: Hospital waste is generated during the diagnosis, treatment or immunization of human beings or animals. It may include wastes like sharps, soiled waste, disposables, anatomical waste, cultures, discarded medicines, chemical wastes, etc. These are in the form of disposable syringes, swabs, bandages, body fluids, human excreta, etc. This waste is highly infectious and can be a serious threat to human health if not managed in a scientific manner.

Bio-medical waste: Bio-medical waste means "any solid and/or liquid waste including its container and any intermediate product, which is generated during the diagnosis, treatment or immunization of human beings or animals or in research pertaining thereto or in the production or testing thereof."

Bio Medical waste consists of human anatomical waste like tissues, organs, body parts, animal wastes generated during research, from veterinary hospitals, microbiology and biotechnology wastes, waste sharps, hypodermic needles, syringes, scalpels, broken glass, discarded medicines and cyto-toxic drugs, soiled waste, such as dressing, bandages, plaster casts, material contaminated with blood, tubes, catheters, liquid waste from any of the infected areas, incineration ash and other chemical wastes.

Plastics: Plastic, with its exclusive qualities of being light yet strong and economical, has invaded every aspect of our day-to-day life. It has many advantages: it is durable, light, easy to mould and can be adapted to different user requirements. Plastic is now a serious worldwide environmental and health concern, essentially due to its non-biodegradable nature. More than 50% of the plastic waste generated in the country is recycled and used in the manufacture of various plastic products.

Dioxin is a highly carcinogenic and toxic by-product of the manufacturing process of plastics. Burning of plastics, especially PVC, releases this dioxin and also furan into the atmosphere.

Plastics are so versatile in use that their impact on the environment is extremely wide ranging. Careless disposal of plastic bags chokes drains, blocks the porosity of the soil and causes problems for groundwater recharge. Plastic disturbs the soil microbial activity. Plastic bags can also contaminate foodstuffs due to leaching of toxic dyes and transfer of pathogens. In fact, a major portion of the plastic bags *i.e.*, approximately 60-80% of the plastic waste generated in India is collected and segregated for recycling. 20-40% remains strewn on the ground, littered around in open drains.

Designing eco-friendly, biodegradable plastics are the need of the hour. Though partially biodegradable plastics have been developed and used, completely biodegradable plastics based on renewable starch rather than petrochemicals have only recently been developed and are in the early stages of commercialization. The incentives provided for this is not adequate. For example it is both easy and economical to deliver milk in plastic bags rather than in bottles. Plastics used in various applications are listed below.

Household	Health and Medicare
• Carry bags • Pet bottles • Containers • Trash bags	• Disposable syringes • Glucose bottles • Blood and uro bags • Intravenous tubes • Catheters • Surgical gloves

Hotel and Catering	Air/Rail Travel
• Packaging items • Mineral water bottles • Plastic plates, cups, spoons	• Mineral water bottles • Plastic plates, cups, spoons • Plastic bags

Q.6. What are the sources and effects of solid wastes?

Ans. Solid waste, arise from human activities and are discarded as useless or unwanted one.

The types are: Urban solid waste; Industrial solid waste

Urban solid wastes include Residential area wastes, commercial area wastes, open area waste, constructional waste, biomedical waste, e-wastes, etc.

Industrial solid wastes include wastes from Chemical industry, paint industry, petroleum refinery industry, paper industry, metal industry, etc.

The effects of solid wastes include:

- Spreading of fungal diseases.
- Bad odours.
- Burning of waste produce chemicals which are harmful to human life.
- Contaminates water.
- Affects the characteristics of soil.

Q.7. Give examples of various biodegradable and non-biodegradable solid wastes.

Ans. Biodegradable waste includes organic waste, e.g., kitchen waste, vegetables, fruits, flowers, leaves from the garden, and paper.

Non-biodegradable waste can be further segregated into:

(*a*) **Recyclable waste** – plastics, paper, glass, metal, etc.

(*b*) **Toxic waste** – old medicines, paints, chemicals, bulbs, fertilizer, containers, batteries.

(*c*) **Soiled waste** – hospital waste such as cloth soiled with blood and other body fluids.

Q.8. Describe source based classification of solid waste.

Ans. Sources of solid wastes are consistent, dependent on sectors and activities and these include the following:

(*i*) **Residential wastes:** This refers to wastes from dwellings, apartments, etc., and consists of leftover food, vegetable peels, plastic, clothes, ashes, etc.

(*ii*) **Commercial wastes:** This refers to wastes consisting of leftover food, glasses, metals, ashes, etc., generated from stores, restaurants, markets, hotels, motels, auto-repair shops, medical facilities, etc.

(*iii*) **Institutional wastes:** This mainly consists of paper, plastic, glasses, etc., generated from educational, administrative and public buildings such as schools, colleges, offices, prisons, etc.

(*iv*) **Municipal wastes:** This includes dust, leafy matter, building debris, treatment plant residual sludge, etc., generated from various municipal activities like construction and demolition, street cleaning, landscaping, etc.

(*v*) **Industrial wastes:** This mainly consists of process wastes, ashes, demolition and construction wastes, hazardous wastes, etc., due to industrial activities.

(*vi*) **Agricultural wastes:** This mainly consists of spoiled food grains and vegetables, agricultural remains, litter, etc., generated from fields, orchards, vineyards, farms, etc.

(*vii*) **Open areas wastes:** this includes wastes from areas such as Streets, alleys, parks, vacant lots, playgrounds, beaches, highways, recreational areas, etc.

Q.9. Describe the type-based classification of solid wastes.

Ans. Classification of wastes based on types, *i.e.*, physical, chemical, and biological characteristics of wastes, is as follows:

(*i*) **Garbage:** This refers to animal and vegetable wastes resulting from the handling, sale, storage, preparation, cooking and serving of food. Garbage comprising these wastes contains putrescible (rotting) organic matter, which produces an obnoxious odour and attracts rats and other vermin. It, therefore, requires special attention in storage, handling and disposal.

(*ii*) **Ashes and residues:** These are substances remaining from the burning of wood, coal, charcoal, coke and other combustible materials for cooking and heating in houses, institutions and small industrial establishments. When *produced* in large quantities, as in power-generation plants and factories, these are classified as industrial wastes. Ashes consist of fine powdery residue, cinders and clinker often mixed with small pieces of metal and glass. Since ashes and residues are almost entirely inorganic, they are valuable in landfills.

(*iii*) **Combustible and non-combustible wastes:** These consist of wastes generated from households, institutions, commercial activities, etc., excluding food wastes and other highly putrescible material. Typically, while *combustible material* consists of paper, cardboard, textile, rubber, garden trimmings, etc., *non-combustible material* consists of such items as glass, crockery, tin and aluminium cans, ferrous and non-ferrous material and dirt.

(*iv*) **Bulky wastes:** These include large household appliances such as refrigerators, washing machines, furniture, crates, vehicle parts, tyres, wood, trees and branches. Since these household wastes cannot be accommodated in normal storage containers, they require a special collection mechanism.

(*v*) **Street wastes:** These refer to wastes that are collected from streets, walkways, alleys, parks and vacant plots, and include paper, cardboard, plastics, dirt, leaves and other vegetable matter. Littering in public places is indeed a widespread and acute problem in many countries including India, and a solid waste management system must address this menace appropriately.

(*vi*) **Biodegradable and non-biodegradable wastes:** *Biodegradable wastes* mainly refer to substances consisting of organic matter such as leftover food, vegetable and fruit peels, paper, textile, wood, etc., generated from various household and industrial activities. Because of the action of micro-organisms, these wastes are degraded from complex to simpler compounds. *Non-biodegradable* wastes consist of inorganic and recyclable materials such as plastic, glass, cans, metals, etc.

(*vii*) **Dead animals:** With regard to municipal wastes, dead animals are those that die naturally or are accidentally killed on the road. This category does not include carcasses and animal parts from slaughter-houses, which are regarded as industrial wastes. Dead animals are divided into two groups – large and small. Among the large animals are horses, cows, goats, sheep, pigs, etc., and among the small ones are dogs, cats, rabbits, rats, etc. The reason for this differentiation is that large animals require special equipment for lifting and handling when they are removed. If not collected promptly, dead animals pose a threat to public health since they attract flies and other vermin as they decay. Their presence in public places is particularly offensive from the aesthetic point of view as well.

(*viii*) **Abandoned vehicles:** This category includes automobiles, trucks and trailers that are abandoned on streets and other public places. However, abandoned vehicles have significant scrap value for their metal, and their value to collectors is highly variable.

(*ix*) **Construction and demolition wastes:** These are wastes generated as a result of construction, refurbishment, repair and demolition of houses, commercial buildings and other structures. They consist mainly of earth, stones, concrete, bricks, lumber, roofing and plumbing materials, heating systems and electrical wires and parts of the general municipal waste stream.

(*x*) **Farm wastes:** These wastes result from diverse agricultural activities such as planting, harvesting, production of milk, rearing of animals for slaughter and the operation of feedlots. In many areas, the disposal of animal waste has become a critical problem, especially from feedlots, poultry farms and dairies.

(*xi*) **Hazardous wastes:** *Hazardous wastes* are those defined as wastes of industrial, institutional or consumer origin that are potentially dangerous either immediately or over a period of time to human beings and the environment. This is due to their physical, chemical and biological or radioactive characteristics like ignitability, corrosivity, reactivity and toxicity. In some cases, the active agents may be liquid or gaseous hazardous wastes. These are, nevertheless, classified as solid wastes as they are confined in solid containers. Typical examples of hazardous wastes are empty containers of solvents, paints and pesticides, which are frequently mixed with municipal wastes and become part of the urban waste stream. Certain hazardous wastes may cause explosions in incinerators and fires at landfill sites. Others such as pathological wastes from hospitals and radioactive wastes also require special handling. Effective management practices should ensure that hazardous wastes are stored, collected, transported and disposed of separately, preferably after suitable treatment to render them harmless.

(*xii*) **Sewage wastes:** The solid by-products of sewage treatment are classified as sewage wastes. They are mostly organic and derived from the treatment of organic sludge separated from both raw and treated sewages. The inorganic fraction of raw sewage such as grit and eggshells is separated at the preliminary stage of treatment, as it may entrain putrescible organic matter with pathogens and must be buried without delay. The bulk of treated, dewatered sludge is useful as a soil conditioner but is invariably uneconomical. Solid sludge, hence, enters the stream of municipal wastes, unless special arrangements are made for its disposal.

Q.10. Describe the essential features of a solid waste management (SWM) system.

Ans. A SWM system refers to a combination of various functional elements associated with the management of solid wastes. The system, when put in place, facilitates the collection and disposal of solid wastes in the community at minimal costs, while preserving public health and ensuring little or minimal adverse impact on the environment.

The functional elements that constitute the solid waste management system are:

(*i*) **Waste generation:** Wastes are generated at the start of any process, and thereafter, at every stage as raw material are converted into goods for consumption. The source of waste generation, determines quantity, composition and waste characteristics. For example, wastes are generated from households, commercial areas, industries, institutions, street cleaning and other municipal services. The most important aspect of this part of the SWM system is the identification of waste.

(*ii*) **Waste storage:** Storage is a key functional element because collection of wastes never takes place at the source or at the time of their generation. The heterogeneous wastes generated in residential areas must be removed within 8 days due to shortage of storage space and presence of biodegradable material. Onsite storage is of primary importance due to aesthetic consideration, public health and economics involved. Some of the options for storage are plastic containers, conventional dustbins (of households), used oil drums, large storage bins (for institutions and commercial areas or servicing depots), etc. Obviously, these vary greatly in size, form and material.

(*iii*) **Waste collection:** This includes gathering of wastes and hauling them to the location, where the collection vehicle is emptied, which may be a transfer station (*i.e.*, intermediate station where wastes from smaller vehicles are transferred to larger ones and also segregated), a processing plant or a disposal site. Collection depends on the number of containers, frequency of collection, types of collection services and routes. Typically, collection is provided under various management arrangements, ranging from municipal services to franchised services, and under various forms of contracts.

Note that the solution to the problem of hauling is complicated. For instance, vehicles used for long distance hauling may not be suitable or particularly economic for house-to-house collection. Every SWM system, therefore, requires an individual solution to its waste collection problem.

(*iv*) **Transfer and transport:** This functional element involves the transfer of wastes from smaller collection vehicles, where necessary to overcome the problem of narrow access lanes, to larger ones at transfer stations; the subsequent transport of the wastes, usually over long distances, to disposal sites.

The factors that contribute to the designing of a transfer station include the type of transfer operation, capacity, equipment, accessories and environmental requirements.

(*v*) **Processing:** Processing is required to alter the physical and chemical characteristics of wastes for energy and resource recovery and recycling. The important processing techniques include compaction, thermal volume reduction, manual separation of waste components, incineration and composting.

(*vi*) **Recovery and recycling:** This includes various techniques, equipment and facilities used to improve both the efficiency of disposal system and recovery of usable material and energy. Recovery involves the separation of valuable resources from the mixed solid wastes, delivered at transfer stations or processing plants. It also involves size reduction and density separation by air classifier, magnetic device for iron and screens for glass. The selection of any recovery process is a function of economics, *i.e.*, costs of separation versus the recovered-material products. Certain recovered materials like glass, plastics, paper, etc., can be recycled as they have economic value.

A typical SWM system with its functional elements and linkages is shown below in Fig. 8.1.

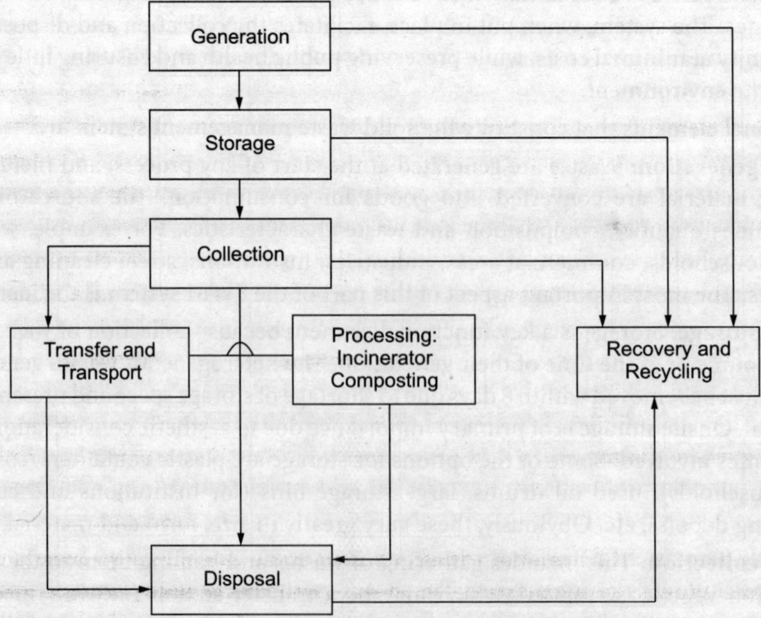

Fig. 8.1: Solid waste management (SWM) system

(*vii*) **Waste disposal:** Disposal is the ultimate fate of all solid wastes, be they residential wastes, semi-solid wastes from municipal and industrial treatment *plants*, incinerator residues, composts or other substances that have no further use to the society. Thus, land use planning becomes a primary determinant in the selection, design and operation of landfill operations. A modern sanitary landfill is a method of disposing solid waste without creating a nuisance and hazard to public health. Generally, engineering principles are followed to confine the wastes to the smallest possible area, reduce them to the lowest particle volume by compaction at the site and cover them after each day's operation to reduce exposure to vermin. One of the most important functional elements of SWM, therefore, relates to the final use of the reclaimed land.

Q.11. Describe the factors affecting solid waste management (SWM)system.

Ans. The factors which influence the decision-making process in the implementation of a SWM system and need to be considered in developing a SWM system are listed below:

(*i*) **Quantities and characteristics of wastes:** The quantities of wastes generated generally depend on the income level of a family, as higher income category tends to generate larger quantity of wastes, compared to low-income category. One of the measures of waste composition (and characteristics) is density, which ranges from 150 to 600 kg/m^3. Proportion of paper and packaging materials in the waste largely account for the differences. When this proportion is high, the density is low and vice versa. The wastes of high density reflect a relatively high proportion of organic matter and moisture and lower levels of recycling.

(*ii*) **Climate and seasonal variations:** There are regions where temperatures are very low for much of the year. In cold climates, drifting snow and frozen ground interfere with landfill operations, and therefore, trenches must be dug in summer and cover material stockpiled for winter use. Tropical climates, on the other hand, are subject to sharp seasonal variations from wet to dry season, which cause significant changes in the moisture content of solid waste, varying from less than 50% in dry season to greater than 65% in wet months. Collection and disposal of wastes in the wet months are often problematic. High temperatures and humidity cause solid wastes to decompose far more rapidly than they do in colder climates. The frequency of waste collection in high temperature and humid climates should, therefore, be higher than that in cold climates. In sub-tropical or desert climate, there is no significant variation in moisture content of wastes (due to low rainfall) and low production of leachate from sanitary landfill. High winds and windblown sand and dust, however, cause special problems at landfill sites. While temperature inversions can cause airborne pollutants to be trapped near ground level, landfill sites can affect ground water by altering the thermal properties of the soil.

(*iii*) **Physical characteristics of an urban area:** In urban areas (*i.e.*, towns and cities), where the layout of streets and houses is such that access by vehicles is possible and door-to-door collection of solid wastes is the accepted norm either by large compaction vehicle or smaller vehicle. The picture is, however, quite different in the inner and older city areas where narrow lanes make service by vehicles difficult. Added to this is the problem of urban sprawl in the outskirts where population is growing at an alarming rate. Access ways are narrow, unpaved and tortuous, and therefore, not accessible to collection vehicles. Problems of solid waste storage and collection are most acute in such areas.

(*iv*) **Financial constraints:** Solid waste management accounts for sizeable proportions of the budgets of municipal corporations. This is allocated for capital resources, which go towards the purchase of equipment, vehicles, and fuel and labour costs. Typically, 10 to 40% of the revenues of municipalities are allocated to solid waste management. In regions where wage rates are low, the aim is to optimise vehicle productivity. The unfavourable financial situation of some countries hinders purchase of equipment and vehicles. This means that the balance between the degree of mechanisation and the size of the labour force becomes a critical issue in arriving at the most cost-effective solution.

(*v*) **Cultural constraints:** In some regions, traditions preclude the intrusion of waste collection on the precincts of households, and therefore, influence the collection system. In others, where the tradition of caste persists, recruits to the labour force for street cleaning and handling of waste must be drawn from certain sections of the population, while others will not consent to placing storage bins in their immediate vicinity. Social norms of a community more often than not over ride what many may consider rational solutions. Waste management should, therefore, be sensitive to such local patterns of living and consider these factors in planning, design and operation.

(*vi*) **Management and technical resources:** Solid waste management, to be successful, requires a wide spectrum of workforce in keeping with the demands of the system. The best system for a region is one which makes full use of indigenous crafts and professional skills and/or ensures that training programmes are in place to provide a self-sustaining supply of trained workforce.

Q.12. What is municipal solid waste and its management?

Ans. Municipal solid waste (MSW) is the trash or garbage that is discarded day to day in a human settlement. *According to MSW rules 2000 MSW includes commercial and residential wastes generated in a municipal or notified areas in either solid or semi-solid form excluding industrial hazardous wastes but including treated bio-medical wastes. Waste generation encompasses activities in which materials are identified as no longer being of value (being in the present form) and are either thrown away or gathered together for disposal.* Municipal solid waste consists of the following kinds of waste:

Horticultural waste, construction and demolition waste, road sweeping waste, general house hold waste (including kitchen waste), drain silt/waste, market and commercial waste, institutional waste, slaughter house waste and dead animals sludge, treated biomedical waste, etc.

Municipal solid waste management (MSWM) refers to a systematic process that comprises of waste segregation and storage at source, primary collection, secondary storage, transportation, secondary segregation, resource recovery, processing, treatment, and final disposal of solid waste.

Q.13. What are the other kinds of waste found in urban settlements?

Ans. These include:

- Industrial or hazardous waste
- Bio-medical or hospital waste
- E-waste.

The Industrial hazardous waste is managed through hazardous waste (management and handling) forth amendments rules 2010. Hazardous waste is identified with properties of *ignitability, corrosivity, reactivity and toxicity*. Hence urban local bodies must ensure that industrial waste does not get mixed with the municipal solid waste stream, failing which will result in economic losses (as hazardous waste treatment costs much more higher than the municipal solid waste) and health and safety hazards (Contaminants like heavy metals, chromium, mercury, etc. when found in the municipal waste stream will contaminate the compost produced by the city. When farmers buy the compost it will indirectly affect the food chain.) while treating such wastes.

According to the hazardous waste management rules 2010, the onus of managing and treating hazardous waste lie with the waste generator, and the urban local body has to ensure that such waste does not contaminate municipal waste stream in their area of authority.

Q.14. What is the significance of municipal solid waste management?

Ans. Efficient municipal solid waste management benefits in maintaining hygienic conditions leading to lesser health issues, better living environment, improved economic prosperity in the area, aesthetically cleaner surroundings with cleaner drains for storm water flow, cleaner water sources and safer neighbourhoods. Some of the key benefits of solid waste management have been discussed below.

- **Environment, health and safety benefits:** Effective solid waste management helps preserve environment in the project area.
  - — As it ***prevents waste*** to contamination of water (in drains) and soil in particular.
  - — ***Reduces waste*** sent to the landfill, which may have negative impacts on ground water and air quality.
  - — ***Reduce emissions*** from energy consumption- as waste when recycled requires less energy than making goods from virgin materials thereby reducing the energy demand and pressure on nonrenewable sources (oil, fossil fuels).
  - — ***Contributes to climate change*** by reducing methane emissions from landfills. Waste prevention and recycling (including composting) divert organic wastes from landfills, reducing the methane released when these materials decompose. Thereby reducing the emission of greenhouse gases and climate change.
  - — Improved waste management reduce chances of spread of diseases
  - — Efficiently designed waste management reduces multiple handling of waste avoiding potentially injurious or dangerous practices.
- **Economic benefits**
  - — Reduction in quantum of waste by diverting it to recycling and other processing unit reduces the landfill costs.
  - — Better managed waste management services indicate better value at same cost.
  - — An effective waste management implies reduce, reuse and recycle waste matter leading to introduction of more and more of waste matter into the value chain leading to economic benefits.
  - — Waste management being a labour intensive activity, it helps in employment generation. Introduction of rag pickers into formalized waste collection and segregation process is a win scenario for implementers.

Q.15. Describe various physical characteristics of solid wastes.

Ans. Information on the physical characteristics of solid wastes are important for the selection and operation of equipment and for the analysis and design of disposal facilities. Important physical characteristics of solid wastes are described below:

(*i*) **Density:** Density of waste, *i.e.*, its mass per unit volume (kg/m^3), is a critical factor in the design of a SWM system, e.g., the design of sanitary landfills, storage, types of collection and transport vehicles, etc. An efficient operation of a landfill demands compaction of wastes to optimum density. Any normal compaction equipment can achieve reduction in volume of wastes by 75%, which increases an initial density of 100 kg/m^3 to 400 kg/m^3. In other words, a waste collection vehicle can haul four times the weight of waste in its compacted state than when it is uncompacted. A high initial density of waste precludes the achievement of a high compaction ratio and the compaction ratio achieved is no greater than 1.5:1. Significant changes in density occur spontaneously as the waste moves from source to disposal, due to scavenging, handling, wetting and drying by the weather, vibration in the collection vehicle and decomposition. Effect of increasing the moisture content of the waste is detrimental in the sense that dry density decreases at higher moisture levels; soil-cover plays an important role in containing the waste, there is an upper limit to the density, and the conservative estimate of in-place density for waste in a sanitary landfill is about 600 kg/m^3.

(*ii*) **Moisture content:** Moisture content is defined as the ratio of the weight of water (wet weight - dry weight) to the total weight of the wet waste. Moisture increases the weight of solid wastes, and thereby, the cost of collection and transport. Besides, moisture content affects the economic feasibility of waste treatment by incineration, because wet waste consumes energy for evaporation of water and in raising the temperature of water vapour. Wastes should be insulated from rainfall or other extraneous water. A typical range of moisture content is 20 to 40%, representing the extremes of wastes in an arid climate and in the wet season of a region of high precipitation. However, values greater than 40% are not uncommon.

(*iii*) **Size:** Measurement of size distribution of particles in waste stream is important because of its significance in the design of mechanical separators and shredders. Generally, the results of size distribution analysis are expressed in the manner used for soil particle analysis. That is to say, they are expressed as a plot of particle size (mm) against percentage, less than a given value.

The physical properties that are essential to analyse wastes disposed at landfills are:

I. **Field capacity:** The field capacity of MSW is the total amount of moisture which can be retained in a waste sample subject to gravitational pull. It is a critical measure because water in excess of field capacity will form leachate, and leachate can be a major problem in landfills. Field capacity varies with the degree of applied pressure and the state of decomposition of the wastes.

II. **Permeability of compacted wastes:** The hydraulic conductivity of compacted wastes is an important physical property because it governs the movement of liquids and gases in a landfill. Permeability depends on the other properties of the solid material include pore size distribution, surface area and porosity.

III. **Porosity:** It represents the amount of voids per unit overall volume of material. The porosity of MSW varies typically from 0.40 to 0.67 depending on the compaction and composition of the waste. Porosity of solid waste, $n = e/(1+e)$ Where, e is void ratio of solid waste

IV. **Compressibility of MSW:** Degree of physical changes of the suspended solids or filter cake when subjected to pressure.

Q.16. What are the chemical characteristics of solid wastes and its importance?

Ans. Knowledge of the chemical characteristics is essential for proper understanding of the behaviour of solid waste, as it moves through the waste management system. The products of decomposition and heating values are two examples of chemical characteristics. If solid wastes are to be used as fuel, or are used for any other purpose, we must know their chemical characteristics. Some of the important chemical characteristics include the following:

(*i*) **Lipids:** This class of compounds includes fats, oils and grease, and the principal sources of lipids are garbage, cooking oils and fats. Lipids have high heating values, about 38,000 kJ/kg, which makes waste with high lipid content suitable for energy recovery. Since lipids become liquid at temperatures slightly above ambient, they add to the liquid content during waste decomposition. Though they are biodegradable, the rate of biodegradation is relatively slow because lipids have a low solubility in water.

(*ii*) **Carbohydrates:** These are found primarily in food and yard wastes, which encompass sugar and polymer of sugars (e.g., starch, cellulose, etc.) with general formula $(CH_2O)_x$. Carbohydrates are readily biodegraded to products such as carbon dioxide, water and methane. Decomposing carbohydrates attract flies and rats, and therefore, should not be left exposed for long duration.

(*iii*) **Proteins:** These are compounds containing carbon, hydrogen, oxygen and nitrogen, and consist of an organic acid with a substituted amine group (NH_2). They are mainly found in food and garden wastes. The partial decomposition of these compounds can result in the production of amines that have unpleasant odours.

(*iv*) **Natural fibres:** These are found in paper products, food and yard wastes and include the natural compounds, cellulose and lignin, that are resistant to biodegradation. (Note that paper is almost 100% cellulose, cotton over 95% and wood products over 40%). Because they are a highly combustible solid waste, having a high proportion of paper and wood products, they are suitable for incineration. Calorific values of oven-dried paper products are in the range of 12,000-18,000 kJ/kg and of wood about 20,000 kJ/kg, *i.e.*, about half that for fuel oil, which is 44,200 kJ/kg.

(*v*) **Synthetic organic material (Plastics):** Accounting for 1-10%, plastics have become a significant component of solid waste in recent years. They are highly resistant to biodegradation and, therefore, are objectionable and of special concern in SWM. Hence the increasing attention is being paid to the recycling of plastics to reduce the proportion of this waste component at disposal sites. Plastics have a high heating value, about 32,000 kJ/kg, which makes them very suitable for incineration. But, polyvinyl chloride (PVC), when burnt, produces dioxin and acid gas. The latter increases corrosion in the combustion system and is responsible for acid rain.

(*vi*) **Non-combustibles:** This includes glass, ceramics, metals, dust and ashes, and accounts for 12-25% of dry solids.

(*vii*) **Heating value:** An evaluation of potential of waste material for use as fuel for incineration requires a determination of its heating value. The heating value is determined experimentally using the *Bomb calorimeter test*, in which the heat generated, from the combustion of a dry sample is measured. Since the test temperature is below the boiling point of water (100°C), the combustion water remains in the liquid state. However, during combustion, the temperature of the combustion gases reaches above 100°C, and the resultant water is in the vapour form.

While evaluating incineration as a means of disposal or energy recovery, we need to consider the heating values of respective constituents. For example: Organic material yields energy only when dry. The moisture content in the waste reduces the dry organic material per kilogram of waste and requires a significant amount of energy for drying. Ash content of the waste reduces the proportion of dry organic material per kilogram of waste and retains some heat when removed from the furnace.

(*viii*) **Ultimate analysis:** This refers to an analysis of waste to determine the proportion of carbon, hydrogen, oxygen, nitrogen and sulphur, and the analysis is done to make mass balance calculation for a chemical or thermal process. Besides, it is necessary to determine ash fraction because of its potentially harmful environmental effects, brought about by the presence of toxic metals such as cadmium, chromium, mercury, nickel, lead, tin and zinc in it. Other metals (e.g., iron, magnesium, etc.) may also be present but they are non-toxic. The result of ultimate analysis of a typical municipal solid waste(dry) is given below:

Table 8.1: Ultimate analysis of a typical municipal solid wasto(dry)

Elements	Range, weight%
Carbon	25-30
Hydrogen	2.5-6.0
Oxygen	15-30
Nitrogen	0.25-1.2
Sulphur	0.02-0.12
Ash	12-30

(*ix*) **Proximate analysis:** This is important in evaluating the combustion properties of wastes or refuse derived fuel. The fractions of interest are: **moisture content**, which adds weight to the waste without increasing its heating value, and the evaporation of water reduces the heat released from the fuel; **ash,** which adds weight without generating any heat during combustion; **volatile matter**, *i.e.*, that portion of the waste that is converted to gases before and during combustion; **fixed carbon**, which represents the carbon remaining on the surface grates as charcoal. A waste or fuel with a high proportion of fixed carbon requires a longer retention time on the furnace grates to achieve complete combustion than a waste or fuel with a low proportion of fixed carbon.

Table 8.2: Proximate analysis of a typical municipal solid waste(dry)

Components	Value, percent	
	Range	Typical
Moisture	15-40	20
Volatile matter	40-60	53
Fixed carbon	5-12	7
Glass, metal, ash	15-30	20

Q.17. What are the public health effect of improper handling and management of solid waste?

Ans. The volume of solid waste is increasing rapidly as a result of increasing population. This increased volume of wastes is posing serious problems due to insufficient workforce and other constraints in disposing of it properly. Public health consequences of improper management and handling of wastes are given below:

(*i*) **Disease vectors and pathways:** Wastes dumped indiscriminately provide the food and environment for thriving populations of vermin, which are the agents of various diseases. The pathways of pathogen transmission from *wastes* to humans are mostly indirect through insects – flies, mosquitoes and roaches and animals – rodents and pigs. Diseases become a public health problem when they are present in the human and animal population of surrounding communities, or if a carrier transmits the etiological agent from host to receptor.

(*ii*) **Flies:** Most common in this category is the housefly, which transmits typhoid, salmonellosis, gastro-enteritis and dysentery. Flies have a flight range of about 10 km, and therefore, they are able to spread their influence over a relatively wide area. The four stages in their life-cycle are egg, larva, pupa and adult. Eggs are deposited in the warm, moist environment of decomposing food wastes. When they hatch, the larvae feed on the organic material, until certain maturity is reached, at which time they migrate from the waste to the soil or other dry loose material before being transformed into pupae. The pupae are inactive until the adult fly emerges. The migration of larvae within 4 to 10 days provides the clue to an effective control measure, necessitating the removal of waste before migration of larvae. Consequently, in warm weather, municipal waste should be collected twice weekly for effective control. In addition, the quality of household and commercial storage containers is very significant. The guiding principle here is to restrict access to flies. Clearly, the use of suitable storage containers and general cleanliness at their location, as well as frequent collection of wastes, greatly reduces the population of flies. Control is also necessary at transfer stations, composting facilities and disposal sites to prevent them from becoming breeding grounds for flies. Covering solid wastes with a layer of earth at landfill sites at the end of every day arrests the problem of fly breeding at the final stage.

(*iii*) **Mosquitoes:** They transmit diseases such as malaria, filaria and dengue fever. Since they breed in stagnant water, control measures should centre on the elimination of breeding places such as tins, cans, tyres, etc. Proper sanitary practices and general cleanliness in the community help eliminate the mosquito problems caused by the mismanagement of solid waste.

(*iv*) **Roaches:** These cause infection by physical contact and can transmit typhoid, cholera and amoebiasis. The problems of roaches are associated with the poor storage of solid waste.

(*v*) **Rodents:** Rodents (rats) proliferate in uncontrolled deposits of solid wastes, which provide a source of food as well as shelter. They are responsible for the spread of diseases such as plague, murine typhus, leptospirosis, histoplasmosis, rat bite fever, dalmonelosis, trichinosis, etc. The fleas, which rats carry, also cause many diseases. This problem is associated not only with open dumping but also poor sanitation.

(*vi*) **Occupational hazards:** Workers handling wastes are at risk of accidents related to the nature of material and lack of safety precautions. The sharp edges of glass and metal and poorly constructed storage containers may inflict injuries to workers. It is, therefore, necessary for waste handlers to wear gloves, masks and be vaccinated. The infections associated with waste handling, include:

- skin and blood infections resulting from direct contact with waste and from infected wounds.
- eye and respiratory infections resulting from exposure to infected dust, especially during landfill operations.
- diseases that result from the bites of animals feeding on the waste.
- intestinal infections that are transmitted by flies feeding on the waste.
- chronic respiratory diseases, including cancers resulting from exposure to dust and hazardous compounds.

In addition, the accidents associated with waste handling include: bone and muscle disorders resulting from the handling of heavy containers and the loading heights of vehicles; infecting wounds resulting from contact with sharp objects; reduced visibility, due to dust along the access routes, creates greater risk of accidents; poisoning and chemical burns resulting from contact with small amounts of hazardous chemical wastes mixed with general wastes such as pesticides, cleaning solutions and solvents in households and commercial establishments; burns and other injuries resulting from occupational accidents at waste disposal sites or from methane gas explosion at landfill sites; serious health hazards, particularly for children, due to careless dumping of lead-acid, nickel-cadmium and mercuric oxide batteries.

(*vii*) **Animals:** Apart from rodents, some animals (e.g., dogs, cats, pigs, etc.) also act as carriers of disease. For example, pigs are involved in the spread of diseases like trichinosis, cysticerosis and toxoplasmosis, which are transmitted through infected pork, eaten either in raw state or improperly cooked. Solid wastes, when fed to pigs, should be properly treated (cooked at 100°C for at least 50 minutes).

Q.18. Describe in detail the environmental effects of improper solid waste handling and management.

Ans. Besides causing health disorders, inadequate and improper waste management causes adverse environmental effects such as the following:

(*i*) **Air pollution:** Burning of solid wastes in open dumps or in improperly designed incinerators emit pollutants (gaseous and particulate matters) to the atmosphere. The environmental consequences of open burning are greater than incinerators, especially with respect to aldehydes and particulates. Emissions from an uncontrolled incinerator system include particulate matter, sulphur oxides, nitrogen oxides, hydrogen chloride, carbon monoxide, lead and mercury. Discharge of arsenic, cadmium and selenium is to be controlled, since they are toxic at relatively low exposure levels. Polychlorinated dibenzofurans (PCDFs), commonly called dioxins and furans, are of concern because of their toxicity, carcinogenicity and possible mutagenicity.

(*ii*) **Water and land pollution:** Water pollution results from dumping in open areas and storm water drains, and improper design, construction and/or operation of a sanitary landfill. Control of infiltration from rainfall and surface runoff is essential in order to minimise the production of leachate. Pollution of groundwater can occur as a result of:

- the flow of groundwater through deposits of solid waste at landfill sites;
- percolation of rainfall or irrigation waters from solid wastes to the water table;
- diffusion and collection of gases generated by the decomposition of solid wastes.

The interaction between leachate contaminants and the soil depends on the characteristics of the soil. Soil bacteria stabilise biochemical oxygen demand (BOD), *i.e.*, the amount of oxygen required by micro-organisms to degrade organic matter, by anaerobic action, if toxic substances are in low concentration. The carbon dioxide produced keeps the pH level low, causing the water to dissolve minerals in the aquifers. Consequently, the change in groundwater quality may take place depending on the characteristics of the aquifer. Contamination can spread over considerable distances from the landfill, if the aquifers are of sand or gravel. In clayey soils, the rate of movement is greatly reduced. The capacity of clay to exchange ions restricts the movement of metal ions by capturing them in the soil matrix. Changes in its chemical characteristics are due to hardness, iron and manganese compounds.

(*iii*) **Visual pollution:** The aesthetic sensibility is offended by the unsightliness of piles of wastes on the roadside. The situation is made worse by the presence of scavengers rummaging in the waste. Waste carelessly discarded in public thoroughfares, along roads and highways and around communal bins (*i.e.*, makeshift containers, without lids, used for the storage of residential, commercial and institutional wastes) gives easy access to animals scavenging for food. The solution to this social problem undoubtedly lies in the implementation of public education at all levels – primary, secondary, tertiary and adult, both short- and long-term, and in raising the status of public health workers and managers in solid waste management.

(*iv*) **Noise pollution:** Undesirable noise is a nuisance associated with operations at landfills, incinerators, transfer stations and sites used for waste recycling. This is due to the movement of vehicles, the operation of large machines and the diverse operations at an incinerator site. The impacts of noise pollution may be reduced by careful siting of SWM operations and by the use of noise barriers.

(*v*) **Odour pollution:** Obnoxious odours due to the presence of decaying organic matter are characteristic of open dumps. They arise from anaerobic decomposition processes and their major constituents are particularly offensive. Proper landfill covering eliminates this nuisance.

(*vi*) **Explosion hazards:** Landfill gas, which is released during anaerobic decomposition processes, contains a high proportion of explosive methane (35-70%). It can migrate through the soil over a considerable distance, leaving the buildings in the vicinity of sanitary landfill sites at risk, even after the closure of landfills. Several methods are available for control of landfill gas, such as venting, flaring and the use of impermeable barriers.

Q.19. What are the occupational hazards associated with waste handling?

Ans. Occupational hazards associated with waste handling are listed below:

Infections

- Skin and blood infections resulting from direct contact with waste, and from infected wounds.
- Eye and respiratory infections resulting from exposure to infected dust, especially during landfill operations.
- Different diseases that results from the bites of animals feeding on the waste.
- Intestinal infections that are transmitted by flies feeding on the waste,

Chronic diseases

- Incineration operators are at risk of chronic respiratory diseases, including cancers resulting from exposure to dust and hazardous compounds.

Accidents

- Bone and muscle disorders resulting from the handling of heavy containers.
- Infecting wounds resulting from contact with sharp objects
- Poisoning and chemical burns resulting from contact with small amounts of hazardous chemical waste mixed with general waste.
- Burns and other injuries resulting from occupational accidents at waste disposal sites or from methane gas explosion at landfill sites.

Q.20. Describe the health impacts of different types of solid waste.

Ans. With increase in the global population and the rising demand for food and other essentials, there has been a rise in the amount of waste being generated daily by each household. This waste is ultimately thrown into municipal waste collection centres from where it is collected by the area municipalities to be further thrown into the landfills and dumps. However, either due to resource crunch or inefficient infrastructure, not all of this waste gets collected and transported to the final dump sites. If at this stage the management and disposal is improperly done, it can cause serious impacts on health and problems to the surrounding environment.

Waste that is not properly managed, especially excreta and other liquid and solid waste from households and the community, are a serious health hazard and lead to the spread of infectious diseases. Unattended waste lying around attracts flies, rats, and other creatures that in turn spread disease. Normally it is the wet waste that decomposes and releases a bad odour. This leads to unhygienic conditions and thereby to a rise in the health problems. The plague outbreak in Surat is good example.

Plastic waste is another cause for ill health. Thus excessive solid waste that is generated should be controlled by taking certain preventive measures. The group at risk from the unscientific disposal of solid waste include – the population in areas where there is no proper waste disposal method, especially the pre-school children; waste workers; and workers in facilities producing toxic and infectious material. Other high-risk group includes population living close to a waste dump and those, whose water supply has become contaminated either due to waste dumping or leakage from landfill sites. Uncollected solid waste also increases risk of injury, and infection.

In particular, *organic domestic waste* poses a serious threat, since they ferment, creating conditions favourable to the survival and growth of microbial pathogens.

Direct handling of solid waste can result in various types of infectious and chronic diseases with the waste workers and the rag pickers being the most vulnerable.

Exposure to hazardous waste can affect human health, children being more vulnerable to these pollutants. In fact, direct exposure can lead to diseases through chemical exposure as the release of chemical waste into the environment leads to chemical poisoning.

Waste from agriculture and industries can also cause serious health risks. Other than this, co-disposal of industrial hazardous waste with municipal waste can expose people to chemical and radioactive hazards.

Uncollected solid waste can also obstruct storm water runoff, resulting in the forming of stagnant water bodies that become the breeding ground of disease. Waste dumped near a water source also causes contamination of the water body or the ground water source. Direct dumping of untreated waste in rivers, seas, and lakes results in the accumulation of toxic substances in the food chain, through the plants and animals that feed on it directly or indirectly.

Disposal of hospital and other medical waste requires special attention since this can create major health hazards. This waste generated from the hospitals, health care centres, medical laboratories, and research centres such as discarded syringe, needles, bandages, swabs, plasters, and other types of infectious waste are often disposed with the regular non-infectious waste.

Waste treatment and disposal sites can create health hazards for the neighborhood. Improperly operated incineration plants cause air pollution and improperly managed and designed landfills attract all types of insects and rodents that spread disease. Ideally these sites should be located at a safe distance from all human settlement. Landfill sites should be well lined and walled to ensure that there is no leakage into the nearby ground water sources.

Recycling too carries health risks if proper precautions are not taken. Workers working with waste containing chemical and metals may experience toxic exposure.

Disposal of health-care wastes require special attention since it can create major health hazards, such as Hepatitis B and C, through wounds caused by discarded syringes. Rag pickers and others who are involved in scavenging in the waste dumps for items that can be recycled, may sustain injuries and come into direct contact with these infectious items.

Q.21. How can we produce less waste by practicing the 3 R's *i.e.*, Reduce, Reuse, and Recycle?

Ans. **Reduce** the amount and toxicity of trash we discard. Source reduction can be a successful method of reducing waste generation. Source reduction has many environmental benefits. It prevents emissions of many greenhouse gases, reduces pollutants, saves energy, conserves resources, and reduces the need for new landfills and combustors.

Reuse containers and products; repair what is broken or give it to someone who can repair it. Reusing items by repairing them, donating them to charity and community groups, or selling them also reduces waste. Use a product more than once, either for the same purpose or for a different purpose.

Recycle as much as possible, which includes buying products with recycled content. Recycling involves the collection of used and discarded materials, processing these materials and making them into new products. Studies have revealed that only 10 to 15% of the waste is recycled. It is being used for composting, making pellets to be used in gasifiers, etc. Plastics are sold to the factories that reuse them.

Q.22. What are the environmental impact of solid waste disposal on land?

Ans. When solid waste is disposed off on land in open dumps or in improperly designed landfills (e.g., in low lying areas), it causes the following impact on the environment:

(*a*) ground water contamination by the leachate generated by the waste dump.
(*b*) surface water contamination by the run-off from the waste dump.
(*c*) bad odour, pests, rodents and wind-blown litter in and around the waste dump.
(*d*) generation of inflammable gas (e.g., methane) within the waste dump.
(*e*) bird menace above the waste dump which affects flight of aircraft.
(*f*) fires within the waste dump.
(*g*) erosion and stability problems relating to slopes of the waste dump.
(*h*) epidemics through stray animals.
(*i*) acidity to surrounding soil.
(*j*) release of greenhouse gas.

Q.23. What are the most common recycled and reuse materials?

Ans. Recyclable materials are as follows:

Paper: Newspaper is almost always recovered in community recycling programs.

Plastics: Most communities recycle plastic items such as detergent bottles, beverage containers (e.g., soda, milk, and juice), and containers for various household products, from shampoo, lotion, and mouth wash containers to plastic peanut butter containers.

Aluminium: Almost all recycling programs include Aluminium beverage cans. Aluminium cans are made into new cans.

Steel: Many steel products manufactured contain a high percentage of recycled steel. Some are even made from 100 percent recycled steel.

Glass: Glass food containers, such as jars and bottles for pickles, juice, jam, or wine.

Reuse items: Reusing items by repairing them, donating them to charity and community groups, or selling them also reduces waste. Use a product more than once, either for the same purpose or for a different purpose. **Ways to reuse include:**

- Refilling bottles and reusing boxes.
- Turning empty jars into containers for leftover food.
- Purchasing refillable pens and pencils.

Q.24. What are the benefits of recycling of materials?

Ans. There are several key benefits to recycling such as:

- Reduces the need for landfilling and incineration.
- Saves energy and prevents pollution caused by the extraction and processing of raw materials and the manufacture of products using raw materials.
- Decreases emissions of greenhouse gases that contribute to global climate change.
- Conserves natural resources such as timber, water, and minerals.
- Helps sustain the environment for future generations.
- **Reduces emissions from incinerators.** Recycling and waste prevention divert materials from incinerators and thus reduce greenhouse gas emissions from waste combustion.
- **Reduces methane emissions from landfills.** Waste prevention and recycling (including composting) divert organic wastes from landfills, reducing the methane that would be released if these materials decomposed in a landfill.

Q.25. List out some items that can be recycled or reused.

Ans. Recyclable items are listed below:

Paper

- Old copies
- Old books
- Paper bags
- Newspapers
- Old greeting cards
- Cardboard box

Plastic

- Containers
- Bags
- Sheets

Glass and ceramics

- Bottles
- Plates
- Cups
- Bowls

Miscellaneous

- Old cans
- Utensils
- Clothes
- Furniture

Q.26. Mention the various activities involved in municipal solid waste (MSW) management.

Ans. Various activities involved are listed below.

- waste generation.
- onsite waste handling, storage and processing.
- waste collection.
- waste transfer and transport.
- waste processing and recovery and disposal.

Q.27. What are the auxiliary operations needed for solid waste treatment?

Ans. Auxiliary operations needed for solid waste treatment are as follows:

(*i*) Transport and handling (*ii*) Pulverization (*iii*) Compaction

Transport and handling: Solid wastes are collected from source, transported in trucks with hydraulic and pneumatic system to a central place and to compact the waste to a high density, for disposal.

Pulverization: Pulverization of solid wastes is carried out prior to loading, land filling, compacting or incineration to facilitate these processes. Taw roll, impact and gyratory crushers and hammer mills are used for pulverization. It makes the solid waste homogenous and helps in greater initial settlement. The land can be more easily reclaimed and built on.

Compaction: Compaction and balling of solid wastes using hydraulic or pneumatic processes lead to reduction in refuse volume, reduction in collection and transport time and cost, lesser storage area and safety hazards and cleaner storage area.

The most effective way to ameliorate the solid waste disposal problem is to reduce the generation and toxicity of waste. But, as people search for better life and higher standard of living they tend to consume more goods and generate more wastes. Consequently, society is searching for improved methods of waste management and ways to reduce the amount of waste management system. This consists of reducing the amount of toxicity of the wastes at the source, recycling, reusing or composting as much of the waste as is economically reasonable. Burning the waste that cannot be economically recycled to generate heat reduces the need for fossils and nuclear fuels.

Q.28. Describe various components of solid waste collection.

Ans. Waste collection does not mean merely the gathering of wastes, but also includes the transporting of wastes to transfer stations and/or disposal sites. The factors and components that influence the waste collection system include the following:

(*i*) **Waste collection points:** These affect such collection system components as crew size and storage, which ultimately control the cost of waste collection. The collection points depend on locality and may be residential, commercial or industrial.

(*ii*) **Waste collection frequency:** Climatic conditions and requirements of a locality as well as containers and costs determine the collection frequency. In hot and humid climates, for example, solid wastes must be collected at least twice a week, as the decomposing solid wastes produce bad odour and leachate. And, as residential wastes usually contain food wastes and other putrescible (rotting) material, frequent collection is desirable for health and aesthetic reasons. Besides climates, the quality of solid waste containers on site also determines the collection frequency. For instance, while sealed or closed containers allow collection frequency up to three days, open and unsealed containers may require daily collection. Collection efficiency largely depends on the demography of the area (such as income groups, community, etc.), where collection takes place. While deciding collection frequency, therefore, we must consider the following:

- cost, e.g., optimal collection frequency reduces the cost as it involves fewer trucks, employees and reduction in total route distance;
- storage space, e.g., less frequent collection may require more storage space in the locality;
- sanitation, e.g., frequent collection reduces concerns about health, safety and nuisance associated with stored refuse.

(*iii*) **Waste storage containers:** Proper container selection can save collection energy, increase the speed of collection and reduce crew size. Containers should be functional for the amount and type of materials and collection vehicles used. Containers should also be durable, easy to handle, and economical, as well as resistant to corrosion, weather and animals. In residential areas, where refuse is collected manually, standardised metal or plastic containers are typically required for waste storage. When mechanised collection systems are used, containers are specifically designed to fit the truck mounted loading mechanisms. While evaluating residential waste containers, consider the following:

- efficiency, *i.e.*, the containers should help maximise the overall collection efficiency.
- convenience, *i.e.*, the containers must be easily manageable both for residents and collection crew.
- compatibility, *i.e.*, the containers must be compatible with collection equipment.
- public health and safety, *i.e.*, the containers should be securely covered and stored.
- ownership, *i.e.*, the municipal ownership must guarantee compatibility with collection equipment.

(*iv*) **Waste collection crew:** The optimum crew size for a community depends on labour and equipment costs, collection methods and route characteristics. The size of the collection crew also depends on the size and type of collection vehicle used, space between the houses, waste generation rate and collection frequency. For example, increase in waste generation rate and quantity of wastes collected per stop due to less frequent collection result in a bigger crew size. The collection vehicle could be a motorised vehicle, a pushcart or a trailer towed by a suitable prime mover (tractor, etc.). It is possible to adjust the ratio of collectors to collection vehicles such that the crew idle time is minimised. However, it is not easy to implement this measure, as it may result in an overlap in the crew collection and truck idle time. An effective collection crew size and proper workforce management can influence the productivity of the collection system. The crew size, in essence, can have a great effect on overall collection costs. However, with increase in collection costs, the trend in recent years is towards:

- decrease in the frequency of collection;
- increase in the dependence on residents to sort waste materials;
- increase in the degree of automation used in collection.

This trend has, in fact, contributed to smaller crews in municipalities.

(*v*) **Waste collection route:** The collection programme must consider the route that is efficient for waste collection. An efficient routing of *collection* vehicles helps decrease costs by reducing the labour expended for collection. Proper planning of collection route also helps conserve energy and minimise working hours and vehicle fuel consumption. It is necessary therefore to develop detailed route configurations and collection schedules for the selected collection system. The size of each route, however, depends on the amount of waste collected per stop, distance between stops, loading time and traffic conditions. Barriers, such as railroad, embankments, rivers and roads with heavy traffic, can be considered to divide route territories. Routing (network) analyses and planning can:

- increase the likelihood of all streets being serviced equally and consistently;
- help supervisors locate or track crews quickly;
- provide optimal routes that can be tested against driver judgement and experience.

(*vi*) **Waste transfer station:** A transfer station is an intermediate station between final disposal option and collection point in order to increase the efficiency of the system, as collection vehicles and crew remain closer to routes. If the disposal site is far from the collection area, it is justifiable to have a transfer station, where smaller collection vehicles transfer their loads to larger vehicles, which then haul the waste long distances. In some instances, the transfer station serves as a pre-processing point, where wastes are dewatered, scooped or compressed. A centralized sorting and recovery of recyclable materials are also carried out at transfer stations. The unit cost of hauling solid wastes from a collection area to a transfer station and then to a disposal site decreases, as the size of the collection vehicle increases. This is due to various reasons such as the following:

- labour costs remain constant;
- the ratio of payload to vehicle load increases with vehicle size;
- the waiting time, unloading time, idle time at traffic lights and driver rest period are constant, regardless of the collection vehicle size.

Q.29. Describe various waste storage containers and waste collection vehicles.

Ans. Waste storage is an important component of a waste management system. Waste storage encompasses proper containers to store wastes and efficient transport of wastes without any spillage to transfer stations/disposal sites.

Waste containers/storage bins: The design of an efficient waste collection system requires careful consideration of the type, size and location of containers at the point of waste generation for storage of wastes until they are collected. While single-family households generally use small containers, residential units, commercial units, institutions and industries require large containers. Smaller containers are usually handled manually *whereas* the larger, heavier ones require mechanical handling. The containers may fall under either of the following two categories:

(*i*) **Stationary containers:** These are used for contents to be transferred to collection vehicles at the site of storage.

(*ii*) **Hauled containers:** These are used for contents to be directly transferred to a processing plant, transfer station or disposal site for emptying before being returned to the storage site.

The desirable characteristics of a well-designed container are low cost, size, weight, shape, resistance to corrosion, water tightness, strength and durability. For example, a container for manual handling by one person should not weigh more than 20 kg, otherwise it may lead to occupational health hazards such as muscular strain, etc. Containers that weigh more than 20 kg, when full, require two or more crew members to manually load and unload the wastes, and which result in low collection efficiency.

Containers should not have rough or sharp edges, and preferably have a handle and a wheel to facilitate mobility. They should be covered to prevent rainwater from entering (which increases the weight and rate of decomposition of organic materials) into the solid wastes. The container body must be strong enough to resist and discourage stray animals and scavengers from ripping it as well as withstand rough handling by the collection crew and mechanical loading equipment. Containers should be provided with a lifting bar, compatible with the hoisting mechanism of the vehicle. The material used should be light, recyclable, easily moulded and the surface must be smooth and resistant to corrosion. On the one hand, steel and ferrous containers are heavy and subject to corrosion; the rust peels off exposing sharp edges, which could be hazardous to the collection crew. On the other, wooden containers (e.g., bamboo, rattan and wooden baskets) readily absorb and retain moisture and their surfaces are generally rough, irregular and difficult to clean.

Communal containers: Generally, the containers used for waste storage are communal/public containers which a compactor collection vehicle can lift and empty mechanically. Communal containers may be fixed on the ground (stationary) or movable (hauled). Movable containers are provided with hoists and tails compatible with lifting mechanism of collection vehicles and such containers have capacities of 1-4 m^3. The waste management authority must monitor, maintain and upgrade the communal containers. In residential and commercial areas in India, the communal containers are often made of concrete. In areas with very high waste generation rates, *i.e.*, rates exceeding two truckloads daily, such as wet markets, large commercial centres and large business establishments, roll-on-roll or hoisted communal containers with capacities of 12-20 m^3 and a strong superstructure with wheels are used. Normally, the collection vehicle keeps an empty container as a replacement before it hauls the filled container. When a truck is used as a collection vehicle, the use of communal containers may be appropriate.

It is advisable to place the containers 100-200 m apart for economic reasons. The communal containers are usually staggered such that the effective distance of 100 m is maintained. This means that the farthest distance the householder will have to walk is 50 meters. However, in narrow streets with low traffic, where the house owner can readily cross the street, a longer distance is advisable. If the collection vehicle has to stop frequently, say, at every 50 m or so, fuel consumption increases, and this must be avoided.

Disadvantages of communal container: The major disadvantage of communal containers is the potential lack of maintenance and upgrading. The residuals and scattered solid wastes emit foul odours, which discourage residents from using the containers properly. In addition, if fixed containers are built below the vehicle level, the collection crew may be held responsible for sweeping and loading the solid wastes into transfer containers before being loaded into the collection vehicle. Sweeping and cleaning the communal containers of residuals obviously impinge on the time of the crew members and take a longer time than if the wastes are placed in smaller containers. As fixed communal containers have higher rates of failure, their use is not advisable

To overcome the problem of maintaining communal containers, individual residents should maintain their own containers and locate them in designated areas. The communal area must have the facilities of water and drains to facilitate the cleaning of the containers. This has the advantage of reducing the number of collections stops and at the same time maintaining the householder's responsibility for cleaning them. The residents must also be properly educated on the importance of good housekeeping as the containers in the communal area are subject to vandalism..

Waste collection vehicles: Almost all waste collections are based on collector and collection crew, which move through the collection service area with a vehicle for collecting the waste material. The collection vehicle selected must be appropriate to the terrain, type and density of waste generation points, the way it travels and type and kind of material. The collection vehicle may be small and simple (e.g., two-wheeled cart pulled by an individual) or large, complex and energy intensive (e.g., rear loading compactor truck). The most commonly used collection vehicle is the dump truck fitted with a hydraulic lifting mechanism. A description of some vehicle types is given below:

(*i*) **Small-scale collection and muscle-powered vehicles:** These are common vehicles used for waste collection in many countries and are generally used in rural hilly areas. These can be small rickshaws, carts or wagons pulled by people or animals, and are less expensive, easier to build and maintain compared to other vehicles..They are suitable for densely populated areas with narrow lanes and squatter settlements, where there is relatively low volume of waste generated. Some drawbacks of these collection vehicles include limited travel range of the vehicles and weather exposure that affect humans and animals.

(*ii*) **Non-compactor trucks:** Non-compactor trucks are efficient and cost effective in small cities and in areas where wastes tend to be very dense and have little potential for compaction. When these trucks are used for waste collection, they need a dumping system to easily discharge the waste. It is generally required to cover the trucks in order to prevent residue flying off or rain soaking the wastes. Trucks with capacities of 10-12 m^3 are effective, if the distance between the disposal site and the collection area is less than 15 km. If the distance is longer, a potential transfer station closer than 10 km from the collection area is required. Non-compactor trucks are generally used, when labour cost is high. Controlling and operating cost is a deciding factor, when collection routes are long and relatively sparsely populated.

(*iii*) **Compactor truck:** Compaction vehicles are more common these days, generally having capacities of 12-15 m^3 due to limitations imposed by narrow roads. Although the capacity of a compaction vehicle, is similar to that of a dump truck, the weight of solid wastes collected per trip is 2 to 2.5 times larger since the wastes are hydraulically compacted. The success of waste management depends on the level of segregation at source. A compactor truck allows waste containers to be emptied into the vehicle from the rear, front or sides and inhibits vectors (of disease) from reaching the waste during collection and transport. It works poorly when waste stream is very dense, wet, collected materials are gritty or abrasive, or when the roads are dusty. The advantages of the compactor collection vehicle include the following:

- containers are uniform, large, covered and relatively visually inoffensive.
- waste is set out in containers so that the crew can pick them up quickly.
- health risk to the collectors and odour on the streets are minimised.
- waste is relatively inaccessible to the waste pickers.

Q.30. Describe the functions and types of waste transfer station.

Ans. Waste transfer station is a centralized facility, where waste is unloaded from smaller collection vehicles and re-loaded into large vehicles for transport to a disposal or processing site. This transfer of waste is frequently accompanied by removal, separation or handling of waste. In areas, where wastes are not already dense, they may be compacted at a transfer station. The technical limitations of smaller collection vehicles and the low hauling cost of solid waste, using larger vehicles, make a transfer station viable. Also, the use of transfer station proves reasonable, when there is a need for vehicles servicing a collection route to travel shorter distances, unload and return quickly to their primary task of collecting the waste.

Limitations in hauling solid wastes are the main factors to be considered, while evaluating the use of transfer stations. These include the additional capital costs of purchasing trailers, building transfer stations and the extra time, labour and energy required for transferring wastes from collection truck to transfer trailer.

The main problem in the establishment of a transfer station, however, is securing a suitable site. Stored solid wastes and recyclable materials, if not properly handled, will attract flies and other insect vectors. Odours from the transferred solid wastes will also be a nuisance, if not properly controlled. Besides, the traffic and noise due to collection vehicles, collectors, drivers, etc., invite the resentment of the communities living in the vicinity of transfer stations.

Types of waste transfer stations

Depending on the size, transfer stations can be either of the following two types:

(*i*) **Small to medium transfer stations:** These are direct-discharge stations that provide no intermediate waste storage area. The capacities are generally small (less than 100 tonnes/day) and medium (100 to 500 tonnes/day). Depending on weather, site aesthetics and environmental concerns, transfer operations of this size may be located either indoor or outdoor. More complex small transfer stations are usually attended during hours of operation and may include some simple waste and materials processing facilities. For example, it includes a recyclable material separation and processing centre. The required overall station capacity (*i.e.*, the number and size of containers) depends on the size and population density of the area served and the frequency of collection.

(*ii*) **Large transfer stations:** These are designed for heavy commercial use by private and municipal collection vehicles. The typical operational procedure for a larger station is as follows:

- when collection vehicles arrive at the site, they are checked in for billing, weighed and directed to the appropriate dumping area;
- collection vehicles travel to the dumping area and empty the wastes into a waiting trailer, a pit or a platform;
- after unloading, the collection vehicle leaves the site, and there is no need to weigh the departing vehicle, if its weight (empty) is known;

Transfer vehicles are weighed either during or after loading. If weighed during loading, trailers can be more consistently loaded to just under maximum legal weights and this maximizes payloads and minimizes weight violations.

Designs for larger transfer operations

Several different designs for larger transfer operations are common, depending on the transfer distance and vehicle type. Most designs, however, fall into one of the following three categories:

(*i*) **Direct-discharge non-compaction station:** In these stations, waste is dumped directly from collection vehicle into waiting transfer trailers and is generally designed with two main operating floors. In the transfer operation, wastes are dumped directly from collection vehicles (on the top floor) through a hopper and into open top trailers on the lower floor. The trailers are often positioned on scales so that dumping can be stopped when the maximum payload is reached. A stationary crane with a bucket is used to distribute the waste in the trailer. After loading, a cover or tarpaulin is placed over the trailer top. Because of the use of little hydraulic equipment, a shutdown is unlikely and this station minimizes handling of waste

(*ii*) **Platform/pit non-compaction station:** In this arrangement, the collection vehicles dump their wastes onto a platform or into a pit using waste handling equipment, where wastes can be temporarily stored, and if desired, picked through for recyclables or unacceptable materials. The waste is then pushed into open-top trailers, usually by front-end loaders. Like direct discharge stations, platform stations have two levels. If a pit is used, however, the station has three levels. A major advantage of these stations is that they provide temporary storage, which allows peak inflow of wastes to be levelled out over a longer period. Construction costs for this type of facility are usually higher because of the increased floor space. This station provides convenient and efficient storage area and due to simplicity of operation and equipment, the potential for station shutdown is less.

(*iii*) **Compaction station:** In this type of station, the mechanical equipment is used to increase the density of wastes before they are transferred. The most common type of compaction station uses a hydraulically powered compactor to compress wastes. Wastes are fed into the compactor through a chute, either directly from collection trucks or after intermediate use of a pit. The hydraulic ram of the compactor pushes waste into the transfer trailer, which is usually mechanically linked to the compactor. Compaction stations are used when:

- wastes must be baled for shipment;
- open-top trailers cannot be used because of size restrictions;
- site topography or layout does not accommodate a multi-level building.

The main disadvantage of a compaction facility is that the facility's ability to process wastes is directly dependent on the operation of the compactor.

Q.31. What are the objectives of solid waste processing?

Ans. The processing of solid wastes helps in achieving the best possible benefit from every functional element of the solid waste management (SWM) system and, therefore, requires proper selection of techniques and equipment for every element. Accordingly, the wastes that are considered suitable for further use need to be paid special attention in terms of processing, in order that we could derive maximum economical value from them. The objectives of solid waste processing are:

(*i*) **Improving efficiency of Solid Waste Management (SWM) system:** Various processing techniques are available to improve the efficiency of SWM system. For example, before waste papers are reused, they are usually baled to reduce transporting and storage volume requirements. In some cases, wastes are baled to reduce the haul costs at disposal site, where solid wastes are compacted to use the available land effectively. If solid wastes are to be transported hydraulically and pneumatically, some form of shredding is also required. Shredding is also used to improve the efficiency of the disposal site.

(*ii*) **Recovering material for reuse:** Usually, materials having a market, when present in wastes in sufficient quantity to justify their separation, are most amenable to recovery and recycling. Materials that can be recovered from solid wastes include paper, cardboard, plastic, glass, ferrous metal, aluminium and other residual metals.

(*iii*) **Recovering conversion products and energy:** Combustible organic materials can be converted to intermediate products and ultimately to usable energy. This can be done either through incineration, pyrolysis, composting or bio-digestion. Initially, the combustible organic matter is separated from the other solid waste components. Once separated, further processing like shredding and drying is necessary before the waste material can be used for power generation.

Q.32. What is the purpose and equipment used for volume reduction (compaction) of solid waste?

Ans. Volume and size reduction is an important factor in the development and operation of any Solid Waste Management (SWM) system. The main purpose is to reduce the volume (amount) and size of the solid waste, as compared to its original form, and produce waste of uniform size. Volume reduction or compaction refers to densifying wastes in order to reduce their volume. The benefits of waste compaction include:

- reduction in the quantity of materials to be handled at the disposal site;
- improved efficiency of collection and disposal of wastes;
- increased life of landfills;
- Economically viable waste management system.

However, following disadvantages are associated with compaction:

- poor quality of recyclable materials sorted out of compaction vehicle.
- difficulty in segregation or sorting (since the various recyclable materials are mixed and compressed in lumps).
- Bio-degradable materials (e.g., leftover food, fruits and vegetables) destroy the value of paper and plastic material.

Equipment used for compaction: Based on their mobility, the compaction equipment used in volume reduction are the following:

(*i*) **Stationary equipment:** This represents the equipment in which wastes are brought to, and loaded into, either manually or mechanically. In fact, the compaction mechanism used to compress waste in a collection vehicle, is a stationary compactor. According to their application, stationary compactors can be described as light duty (e.g., those used for residential areas), commercial or light industrial, heavy industrial and transfer station compactors. Usually, large stationary compactors are necessary, when wastes are to be compressed into:

- steel containers that can be subsequently moved manually or mechanically.

- chambers where the compressed blocks are banded or tied by some means before being removed.
- chambers where they are compressed into a block and then released and hauled away untied.
- transport vehicles directly.

(*ii*) **Movable equipment:** This represents the wheeled and tracked equipment used to place and compact solid wastes, as in a sanitary landfill.

According to their compaction pressure, we can divide the compactors used at transfer stations as follows:

(*i*) **Low-pressure (less than 7 kg/cm^2) compaction:** This includes those used at apartments and commercial establishments, bailing equipment used for waste papers and cardboards and stationary compactors used at transfer stations. In low-pressure compaction, wastes are compacted in large containers. Portable stationary compactors are being used increasingly by a number of industries in conjunction with material recovery options, especially for waste paper and cardboard.

(*ii*) **High-pressure (more than 7 kg/cm^2) compaction:** Compact systems with a capacity up to 350 kg/cm^2 come under this category. In such systems, specialised compaction equipment are used to compress solid wastes into blocks or bales of various sizes. In some cases, pulverised wastes are extruded after compaction in the form of logs. The volume reduction achieved with these high-pressure compaction systems varies with the characteristics of the waste. Typically, the reduction ranges from about 3 to 1 through 8 to 1. When wastes are compressed, their volume is reduced, which is normally expressed in percentage.

Q.33. What are the purpose and equipment used for size reduction of solid waste?

Ans. Size reduction or shredding: This is required to convert large sized wastes (as they are collected) into smaller pieces. Size reduction helps in obtaining the final product in a reasonably uniform and considerably reduced size in comparison to the original form. But size reduction does not necessarily imply volume reduction, and this must be accounted into the design and operation of SWM systems as well as in the recovery of materials for reuse and conversion to energy. In the overall process of waste treatment and disposal, size reduction is implemented ahead of:

- land filling to provide a more homogeneous product. This may require less cover material and less frequent covering than that without shredding. This can be of economic importance, where cover material is scarce or needs to be brought to the landfill site from some distance.
- recovering materials from the waste stream for recycling.
- baling the wastes – a process sometimes used ahead of long distance transport of solid wastes – to achieve a greater density.
- making the waste a better fuel for incineration waste energy recovery facilities. (The size reduction coupled with screening, results in a more homogeneous mixture of relatively uniform size, moisture content and heating value, and thereby improving the steps of incineration and energy recovery reducing moisture, *i.e.*, drying and dewatering of wastes).

Mode of action and applications of various size reduction equipment is tabulated below.

Table 8.3: Mode of action and applications of various size reduction equipments

Size Reduction Equipment Type	Mode of action	Application
Small grinders	**Grinding, mashing**	**Organic residential solid wastes**
Chippers	Cutting, slicing	Paper, cardboard, tree trimmings, yard waste, wood, plastics

Large grinders	Grinding, mashing	Brittle and friable materials, used mostly in industrial operation
Jaw crushers	Crushing, breaking	Large solids
Rasp mills	Shredding, tearing	Moistened solid wastes
Shredders	Shearing, tearing	All types of municipal wastes
Cutters, Clippers	Shearing, tearing	All types of municipal wastes
Hammer mills	Breaking, tearing, cutting, crushing	All types of municipal wastes, most commonly used equipment for reducing size and homogenizing composition of wastes
Hydropulper	Shearing, tearing	Ideally suited for use with pulpable wastes, including paper, wood chips. Used primarily in the paper making industry. Also used to destroy paper records

The most frequently used shredding equipment are the following:

(*i*) **Hammer mill:** These are used most often in large commercial operations for reducing the size of solid wastes. Hammer mill is an impact device consisting of a number of hammers, fastened flexibly to an inner disk, which rotates at a very high speed:

Solid wastes, as they enter the mill are hit by sufficient force, which crush or tear them with a velocity so that they do not adhere to the hammers. Wastes are further reduced in size by being struck between breaker plates and/or cutting bars fixed around the periphery of the inner chamber. This process of cutting and striking action continues, until the required size of material is achieved and after that it falls out of the bottom of the mill.

(*ii*) **Hydropulper:** An alternative method of size reduction involves the use of a hydro pulper. Solid wastes and recycled water are added to the hydro pulper. The high-speed cutting blades, mounted on a rotor in the bottom of the unit, convert pulpable and friable materials into slurry with a solid content varying from 2.5 to 3.5%. Metal, tins, cans and other non-pulpable or non-friable materials are rejected from the side of the hydro pulper tank. The rejected material passes down a chute that is connected to a bucket elevator, while the solid slurry passes out through the bottom of the pulper tank and is pumped to the next processing operation.

Q.34. What is the purpose of component separation of solid wastes?

Ans. Component separation is a necessary operation in which the solid waste components are identified and sorted either manually or mechanically to aid further processing. This is required for the:

- recovery of valuable materials for recycling;
- preparation of solid wastes by removing certain components prior to incineration, energy recovery, composting and biogas production.

The most effective way of separation is manual sorting in households prior to collection. The municipality generally provides separate, easily identifiable containers into which the householder deposits segregated recyclable materials such as paper, glass, metals, etc. Usually, separate collections are carried out for the recyclable material. At curbside, separate areas are set aside for each of the recyclable materials for householders to deliver material – when there is no municipal collection system. In case the separation is not done prior to collection, it could be sorted out through mechanical techniques such as air separation, magnetic separation, etc., to recover the wastes.

Q.35. Describe some important component separation techniques from solid waste.

Ans. Some of the component separation techniques are described below:

Air separation: This technique is in use in industrial operations for segregating various components from dry mixture of solid waste. Air separation is primarily used to separate lighter materials (usually organic) from heavier (usually inorganic) ones. The lighter material may include plastics, paper and paper products and other organic materials. Generally, there is also a need to separate the light fraction of organic material from the conveying air streams, which is usually done in a cyclone separator. In this technique, the heavy fraction is removed from the air classifier (*i.e.*, equipment used for air separation) to the recycling stage or to land disposal, as appropriate. The light fraction may be used, with or without further size reduction, as fuel for incinerators or as compost material. There are various types of air classifiers commonly used, some of which are listed below:

(*i*) **Conventional chute type:** This is one of the simplest types of air classifiers. In this type, when the processed solid wastes are dropped into the vertical chute, the lighter material is carried by the airflow to the top while the heavier materials fall to the bottom of the chute. The control of the percentage split between the light and heavy fraction is accomplished by varying the waste loading rate, air flow rate and the cross section of chute. A rotary air lock feed mechanism is required to introduce the shredded wastes into the classifier.

(*ii*) **Zigzag air classifier:** A zigzag air classifier consists of a continuous vertical column with internal zigzag deflectors through which air is drawn at a high rate. Shredded wastes are introduced at the top of the column at a controlled rate, and air is introduced at the bottom of the column. As the wastes drop into the air stream, the lighter fraction is fluidised and moves upward and out of column, while the heavy fraction falls to the bottom. Best separation can be achieved through proper design of the separation chamber, airflow rate and influent feed rate.

(*iii*) **Open inlet vibrator type:** In this type of air classifier, the separation is accomplished by a combination of the following actions:

Vibration: This helps to stratify the material fed to the separator into heavy and light components. Due to this agitation, the heavier particles tend to settle at the bottom as the shredded waste is conveyed down the length of the separator.

Inertial force: In this action, the air pulled in through the feed inlet imparts an initial acceleration to the lighter particle, while the wastes travel down the separator as they are being agitated.

Air pressure: This action refers to the injection of fluidising air in two or more high velocity and low mass flow curtains across the bed. A final stripping of light particles is accomplished at the point where the heavy fraction discharges from the elutriators. The resulting separation is less sensitive to particle size than a conventional vertical air classifier, be it of straight or zigzag design.

An advantage of this classifier is that an air lock feed mechanism is not required and wastes are fed by gravity directly into the separator inlet.

Magnetic separation: The most common method of recovering ferrous scrap from shredded solid wastes involves the use of magnetic recovery systems. Ferrous materials are usually recovered either after shredding or before air classification. When wastes are mass fired in incinerators, the magnetic separator is used to remove the ferrous material from the incinerator residue. Magnetic recovery systems have also been used at landfill disposal sites. The specific locations, where ferrous materials are recovered will depend on the objectives to be achieved, such as reduction of wear and tear on processing and separation equipment, degree of product purity achieved and the required recovery efficiency.

Equipment used for magnetic separation: Various types of equipment are in use for the magnetic separation of ferrous materials. The most common types are the following:

(*i*) **Suspended magnet:** In this type of separator, a permanent magnet is used to attract the ferrous metal from the waste stream. When the attracted metal reaches the area, where there is no magnetism, it falls away freely. This ferrous metal is then collected in a container. This type of separation device is suitable for processing raw refuse, where separators can remove large pieces of ferrous metal easily from the waste stream.

(*ii*) **Magnetic pulley:** This consists of a drum type device containing permanent magnets or electromagnets over which a conveyor or a similar transfer mechanism carries the waste stream. The conveyor belt conforms to the rounded shape of the magnetic drum and the magnetic force pulls the ferrous material away from the falling stream of solid waste.

Screening: Screening is the most common form of separating solid wastes, depending on their size by the use of one or more screening surfaces. Screening has a number of applications in solid waste resource and energy recovery systems. Screens can be used before or after shredding and after air separation of wastes in various applications dealing with both light and heavy fraction materials. The most commonly used screens are rotary drum screens and various forms of vibrating screens. Rotating wire screens with relatively large openings are used for separation of cardboard and paper products, while vibrating screens and rotating drum screens are typically used for the removal of glass and related materials from the shredded solid wastes.

Besides the mechanical techniques described above for segregating wastes, there are other separation techniques also which is discussed below:

(*i*) **Hand-sorting or previewing:** Previewing of the waste stream and manual removal of large sized materials is necessary, prior to most types of separation or size reduction techniques. This is done to prevent damage or stoppage of equipment such as shredders or screens, due to items such as rugs, pillows, mattresses, large metallic or plastic objects, wood or other construction materials, paint cans, etc.

(*ii*) **Inertial separation:** Inertial separation methods rely on ballistic or gravity separation principles to separate shredded solid wastes into light (*i.e.*, organic) and heavy (*i.e.*, inorganic) particles.

(*iii*) **Flotation:** In the flotation process, glass-rich feedstock, which is produced by screening the heavy fraction of the air-classified wastes after ferrous metal separation, is immersed in water in a tank. Glass chips, rocks, bricks, bones and dense plastic materials that sink to the bottom are removed with belt scrappers for further processing. Light organic and other *materials* that float are skimmed from the surface. These materials are taken to landfill sites or to incinerators for energy recovery. Chemical adhesives (flocculants) are also used to improve the capture of light organic and fine inorganic materials.

(*iv*) **Optical sorting:** Optical sorting is used mostly to separate glass from the waste stream, and this can be accomplished by identification of the transparent properties of glass to sort it from opaque materials (e.g., stones, ceramics, bottle caps, corks, etc.) in the waste stream. Optical sorting involves a compressed air blast that removes or separates the glasses – plain or coloured. An optical sorting machinery is, however, complex and expensive.

Q.36. What is the purpose of drying and dewatering of solid waste?

Ans. Drying and dewatering operations for solid wastes are used primarily for incineration systems, with or without energy recovery systems. These are also used for drying of sludges in waste water treatment plants, prior to their incineration or transport to land disposal. The purpose of drying and dewatering operation is to remove moisture from wastes and thereby make it a better fuel. Sometimes, the light fraction is *pelletised* after drying to make the fuel easier to transport and store, prior to use in an incinerator or energy recovery facility.

Q.37. What are the methods and equipment used for drying and dewatering of solid wastes?

Ans. Drying: The following three methods are used to apply the heat required for drying the wastes:

(*i*) **Convection drying:** In this method, hot air is in direct contact with the wet solid waste stream.

(*ii*) **Conduction drying:** In this method, the wet solid waste stream is in contact with a heated surface.

(*iii*) **Radiation drying:** In this method, heat is transmitted directly to the wet solid waste stream by radiation from the heated body. Of these three methods, convection drying is used most commonly.

A rotary drum dryer is composed of a rotating cylinder, slightly inclined from the horizontal through which the material to be dried and the drying gas are passed simultaneously. The drying of material in a direct rotary dryer occurs in the following stages:

- Heating the wet material and its moisture content to the constant-rate drying temperature.
- Drying the material substantially at this temperature.
- Heating of material to its discharge temperature and evaporation of moisture remaining at the end of the stage.

The retention time in the rotary drum is about 30-45 minutes.

Dewatering: Dewatering is more applicable to the problem of sludge disposal from wastewater treatment of plants, but may also be applicable in some cases to municipal/industrial waste problems. When drying beds, lagoons or spreading on land are not feasible, other mechanical means of dewatering are used. The emphasis in the dewatering operation is often on reducing the liquid volume. Once dewatered, the sludge can be mixed with other solid waste, and the resulting mixture can be:

- incinerated to reduce volume.
- used for the production of recoverable by-products.
- used for production of compost.
- buried in a landfill.

Centrifugation and filtration are the two common methods for the dewatering of sludge. Sludges with solid content of a few percent can be thickened to about 10-5% in centrifugation and about 20-30% in pressure filtration or vacuum filtration.

Q.38. What are the various technologies involved in solid waste management?

Ans. Solid waste management technologies include, technologies associated with the control of generation, storage, collection, transfer and transportation, processing, marketing, incineration and disposal of solid wastes. All of these processes have to be carried out within existing legal and social guidelines that protect the public health and environment and are aesthetically acceptable. The various techniques employed in solid waste management include:

- Composting
- Sanitary land filling (Controlled tipping)
- Thermal process (Incineration and pyrolysis)
- Recycling and reuse

Q.39. Describe in detail various solid waste disposal options and their selection criteria.

Ans. The most common disposal option practised currently in many countries is either uncontrolled dumping or dumping with moderate control. The environmental costs of uncontrolled dumping include breeding of disease-causing vectors (e.g., flies, mosquitoes and rodents), pollution, odour and smoke.

Solid waste disposal options: Some of the options available for waste disposal, are as follows:

(*i*) **Uncontrolled dumping:** This is the most common method being practised in many parts of the world, including India. In this method, wastes are dumped at a designated site without any environmental control. They remain there for a long period of time, pose health risks and cause environmental degradation. Due to the adverse health and environmental impact associated with it, this non-engineered disposal method is not a viable and safe option.

(*ii*) **Sanitary landfill:** Sanitary landfill is a fully engineered disposal option in that the selected location or wasteland is carefully engineered in advance before it is pressed into service. Operators of sanitary landfills can minimise the effects of leachate (*i.e.*, polluted water which flows from a landfill) and gas production through proper site selection, preparation and management. This particular option of waste disposal is suitable when the land is available at an affordable price, and adequate workforce and technical resources are available to operate and manage the site.

(*iii*) **Composting:** This is a biological process of decomposition in which organisms, under controlled conditions of ventilation, temperature and moisture, convert the organic portion of solid waste into humus-like material. If this process is carried out effectively, what we get as the final product is a stable, odour-free soil conditioner. Generally, the option of composting is considered, when a considerable amount of biodegradable waste is available in the waste stream and there is use or market for composts. Composting can be either centralised or small-scale. Centralised composting plants are possible, if adequate skilled workforce and equipment are available. And, small-scale composting practices can be effective at household level, but this needs public awareness.

(*iv*) **Incineration:** This refers to the controlled burning of wastes, at a high temperature (roughly 1200-1400°C), which sterilises and stabilises the waste in addition to reducing its volume. In the process, most of the combustible materials (*i.e.*, self-sustaining combustible matter, which saves the energy needed to maintain the combustion) such as paper or plastics get converted into carbon dioxide and ash. Incineration may be used as a disposal option, when land filling is not possible and the waste composition is highly combustible. An appropriate technology, infrastructure and skilled workforce are required to operate and maintain the plant.Incineration is generally limited to hospital and other biological wastes.

(*v*) **Gasification:** This is the partial combustion of carbonaceous material (through combustion) at high temperature (roughly 1000°C) forming a gas, comprising mainly carbon dioxide, carbon monoxide, nitrogen, hydrogen, water vapour and methane, which can be used as fuel

(*vi*) **Refuse-derived fuel (RDF):** This is the combustible part of raw waste, separated for burning as fuel. Various physical processes such as screening, size reduction, magnetic separation, etc., are used to separate the combustibles.

(*vii*) **Pyrolysis:** This is the thermal degradation of carbonaceous material to gaseous, liquid and solid fraction in the absence of oxygen. This occurs at a temperature between 200 and 900°C. The product of pyrolysis is a gas of relatively high calorific value of 20,000 joules per gram with oils, tars and solid burned residue.

Selection criteria of waste disposal options: We can assess the effectiveness of each of the waste disposal options with the help of proper frameworks which represents an aid to decision-making and helps to ensure the key issues are considered. A sub-framework explains how and why the necessary information should be obtained. A framework contains a list of issues and questions pertaining to the technical, institutional, financial, social and environmental features of a waste disposal system to assess the capacity of a disposal option to meet the requirements.

An appraisal of waste disposal option must include the following factors:

(*i*) **Technical:** This feature, involving efficient and effective operation of the technology being used, evaluates the following components of a SWM system:

- composition of wastes, e.g., type, characteristics and quantity.
- existing practices, e.g., collection, transport, and recycling process.
- siting, e.g., location of disposal site, engineering material, etc.
- technology, e.g., operation, maintenance, technical support, etc.
- impact, e.g., anticipated by-product, requirement for their treatment and disposal, etc.

(*ii*) **Institutional:** This involves the ability and willingness of responsible agencies to operate and manage the system by evaluating the following:

- structures, roles and responsibilities, e.g., current institutional frameworks.

- operational capacity, e.g., municipal capacities, local experience and staff training.
- incentives, e.g., management improvement and waste disposal practices.
- innovation and partnership.

(*iii*) **Financial:** This assesses the ability to finance the implementation, operation and maintenance of the system by evaluating the following:

- financing and cost recovery, e.g., willingness to raise finance for waste management.
- current revenue and expenditure on waste management.
- potential need for external finance for capital cost.

(*iv*) **Social:** This helps in avoiding adverse social impact by evaluating the following:

- waste picking, which has an impact on livelihood and access to waste pickers.
- health and income implication.
- public opinions on the existing and proposed system.

(*v*) **Environmental:** This means setting up an environment friendly disposal system by evaluating the following:

- initial environmental risks, *i.e.*, impact of existing and proposed disposal option.
- long-term environmental risks, *i.e.*, long-term implication (future impacts).

Q.40. Describe briefly the techniques of treatment and disposal of municipal solid waste (MSW).

Ans. Various techniques of MSW treatment and disposal are described below:

- **Composting:** Composting is the controlled biological decomposition of organic matter, such as food and yard wastes, into humus, a soil-like material. Composting is nature's way of recycling organic wastes into new soil used in vegetable and flower gardens, landscaping, and many other applications. Organic matter constitutes 35%–40% of the municipal solid waste generated. It is the natural process of decomposition of organic waste that yields manure or compost, which is very rich in nutrients. Composting is a biological process in which micro-organisms, mainly fungi and bacteria, convert degradable organic waste into humus like substance. This finished product, which looks like soil, is high in carbon and nitrogen and is an excellent medium for growing plants. **Vermi-composting** has become very popular in the last few years. In this method, worms are added to the compost. These help to break the waste and the added excreta of the worms makes the compost very rich in nutrients.
- **Landfills:** Landfills are generally located in urban areas where a large amount of waste is generated and has to be dumped in a common place. Unlike an open dump, it is a pit that is dug in the ground. At the end of each day, a layer of soil is scattered on top of it and usually an earth-moving equipment is used to compress the garbage, which now forms a cell. The contamination of groundwater and soil through landfills is known as leaching.
- **Sanitary landfills:** An alternative to landfills which will solve the problem of leaching to some extent, is a sanitary landfill which is more hygienic and built in a methodical manner. These are lined with materials that are impermeable such as plastics and clay, and are also built over impermeable soil. The major problem is the development of methane gas, which occurs when little oxygen is present, *i.e.*, during anaerobic decomposition. Methane gas produced from sanitary landfills is tapped and sold as fuel.
- **Incineration plants.** To reduce waste volume, a controlled burning process called combustion or incineration is practiced to minimizing volume. Combustors, when properly equipped, can convert water into steam to fuel heating systems or generate electricity. The ash that is buried at the landfills leaches the area and cause severe contamination.

Q.41. What is composting of solid waste?

Ans. Composting is an environment friendly process to convert solid wastes into valuable agricultural inputs. Composts are excellent source of humus and plant nutrients, the application of which improves soil biophysical properties and organic matter status of the soil. Composting is defined as the biological conversion of organic wastes into an amorphous dark brown to black colloidal humus like substance under conditions of optimum temperature, moisture and aeration. Nutrient content of compost depends largely on the nutrient content of the wastes. Composting is a process in which the organic portion of solid waste is allowed to decompose under carefully controlled conditions. It is a biological process of decomposition and transformation of the waste material accomplished by the action of bacteria, fungi, and other microorganisms.

Composting is the aerobic, thermophilic degradation of organic matter present in the refuse by microbes, predominantly by fungi and actinomycetes, which are favoured by semi moist condition that prevail in the process. The control parameters for optimum composting include, temperature (40°C), moisture (40.7%), pH (4.5-9.5), nutrients (C:N ratio 40:1); C:P ratio (100:1), air (0.5-0.8 m^3/d/kg volatile solid) and particle size (6-25 mm).

With proper control of moisture, temperature, and aeration, a composting plant can reduce the volume of raw organic material by as much as 50 per cent. In addition, composting can stabilize the waste and produce an end product that may be recycled for beneficial use. The end product is called **compost** or **humus**. It resembles potting soil in texture and earthy odor, and it may be used as a soil conditioner.

A complete municipal solid waste (MSW) composting operation includes sorting and separating, shredding and pulverizing, digestion, product upgrading, and finally marketing. Sorting and separation operations are required to isolate organic, decomposable waste materials from the plastic, glass, metal, and other non-biodegradable substances. Solid waste sorting and separation methods are a key part of MSW recycling operations.

Shredding and pulverizing serve to reduce the size of the individual pieces of the organic waste, resulting in a relatively uniform mass of material. This facilitates handling, moisture control, and aeration of the decomposing waste. Size reduction also helps optimize bacterial activity and increases the rate of decomposition. After size reduction, the wastes are ready for the actual composting or digestion step. Digestion may take place in open windrow or in an enclosed mechanical facility.

Q.42. What are benefits of composting of municipal solid waste?

Ans. Composting is a controlled biological process that uses natural aerobic processes to increase the rate of biological decomposition of organic materials. It is carried out by successive microbial populations that break down organic materials into carbon dioxide, water, minerals, and stabilized organic matter. Carbon dioxide and water are released into the atmosphere, while minerals and organic matter are converted into a potentially reusable soil-like material called compost. The loss of water and carbon dioxide typically reduces the volume of remaining material by 30 to 60%. Compost can be used as a soil amendment in a variety of agricultural, horticultural or landscaping applications.

Benefits of composting are:

- Increases overall waste diversion from final disposal, especially since as much as 80% of the waste stream in low and middle-income countries is compostable.
- Enhances recycling and incineration operations by removing organic matter from the waste stream.
- Produces a valuable soil amendment-integral to sustainable agriculture.
- Promotes environmentally sound practices, such as the reduction of methane generation at landfills.

- Enhances the effectiveness of fertilizer application.
- Can reduce waste transportation requirements.
- Flexible for implementation at different levels, from household efforts to large-scale centralized facilities.
- Can be started with very little capital and operating costs.
- The climate is optimum for composting.
- Addresses significant health effects resulting from organic waste.
- Provides an excellent opportunity to improve a city's overall waste collection program.
- Accommodates seasonal waste fluctuations, such as leaves and crop residue.
- Can integrate existing informal sectors involved in the collection, separation and recycling of wastes.

Q.43. What is windrow composting?

Ans. A windrow is a long, low pile of the prepared organic waste, usually about 3m wide at the base and about 2 m high. Most windrows are conical in cross section and about 50 m in length. The composting waste is aerated by periodically turning each windrow. Turning frequency varies with moisture content and other factors. When moisture content is maintained at about 50 per cent, windrows are turned two or three times a week and, in some cases, daily.

Generally, open field windrow composting takes about 5 weeks for digestion or stabilization of the waste material. An additional 3 weeks may sometimes be required to ensure complete stabilization. Temperatures in an aerobic compost windrow may reach 65 °C because of the natural metabolic action of thermophilic microbes that thrive at such elevated temperatures. The relatively high temperatures destroy most of the pathogenic or disease-causing organisms that may be present in the waste.

Open-field windrow composting requires relatively large land areas. To reduce land requirements, various types of enclosed mechanical systems can be used in lieu of the open-field method. A variety of mechanical type compost systems are available. Oxygen is supplied to the waste material by forced aeration, stirring, or tumbling. In addition to reducing land requirements, enclosed mechanical compost facilities can reduce the time required for stabilization from about 5 weeks to about 1 week.

The digestion of the waste is carried out naturally in an outside decomposition area in windrows (for five weeks) or in mechanized composting plants (for 4 to 6 days). In natural system, the garbage is mixed with nutrient source (sewage sludge/animal manure) and a filler (wood chips) to provide entry of air. The mixture is turned over twice a week and the process is completed in 4-6 weeks. The darkening of refuse, fall in temperature and a musty odour indicate completion of the process.

Before the stabilized compost or humus can be sold for use as a soil conditioner, it must be processed further to upgrade or improve its quality and appearance. This includes drying, screening, and granulating or pelletizing. Sometimes, the compost is placed in bags, although bulk sale is more efficient and economical. Compost can increase the organic and nutrient content of soil and improve its texture and ability to retain moisture.

Q.44. What is vermi-composting?

Ans. The role of earthworms in improving the soil fertility is well known. Earthworms feed on any organic wastes, consume three to five times their body weight and after using 5 to 10 per cent of the organic wastes for their growth, excrete the mucous coated undigested matter as worm casts. Worm casts consist of organic matter that has undergone physical and chemical breakdown through the activity of the muscular gizzard, that grinds the material to a particle size of 1-2 micron. The nutrients present in the worm casts are readily soluble in water for the uptake of plants. Vermi castings are rich sources of macro and micronutrients, vitamins, enzymes, antibiotics, growth hormones and immobilized micro flora.

Vermicompost refers to organic manure produced by earthworms. It is a mixture of worm castings, including humus, live earthworms, their cocoons and other microorganisms. Vermi composting is an appropriate method for disposal of non-toxic solid and liquid organic wastes. It helps in cost effective and efficient recycling of animal wastes (Poultry droppings, horse, piggery excreta and cattle dung), agricultural residues and industrial wastes using low energy.

Several types of earthworms are found in our soils. Earthworms can be divided into the following two categories:

- Epigeic – the surface living worms
- Endogeic – the burrowing worms

Epigeic: These worms are found on the surface and are reddish brown in colour. They do not process the soil but are efficient in composting of organic wastes. They enhance the rate of organic manure production through biodegradation or mineralization.

E.g., *Lampito mauritii, Octochaetona serrata, Perionyx excavatus*

Endogeic: These species burrow and mix the soil, from different horizons in the profile. They ingest organic and mineral fraction of soil, thus promoting the formation of organo mineral complexes. Organo mineral crumbs are brought from deeper parts of the soil profile to the surface. Different species of earthworms show specificity to soil types, moisture content and temperature.

Method of vermicomposting include:

- **Selection of earthworm:** Earthworm that is native to the local soil may be used.
- **Size of pit:** Any convenient dimension such as 2 m × 1 m × 1 m may be prepared.
- **Preparation of vermin bed:** A layer, 15-20 cm thick of good loamy soil above a thin layer of (5 cms) broken bricks and sand should be made.
- **Inoculation of earthworms:** About one hundred earthworms are introduced as an optimum inoculating density into a compost pit of about 2 m × 1 m × 1 m, provided with vermin bed.
- **Organic layering:** It is done on the vermibed with fresh cattle dung. The compost pit is then layered to about 5 cm with dry leaves or hay or organic wastes. Moisture content of the pit is maintained by the addition of water.
- **Wet organic layering:** It is done after four weeks with moist green organic waste, which can be spread over it to a thickness of 5 cm. This practice can be repeated every 4 days. Mixing of wastes periodically without disturbing the vermin bed ensures proper vermicomposting. Wet layering with organic wastes can be repeated till the compost pit is nearly full.

At maturation (after 120 days), the moisture content is brought down, by stopping the addition of water. This ensures drying of compost and migration of worms in to the vermibed. The mature compost, a fine loose granular mass (about 1500 kg), is removed from the pit, sieved, dried and packed. Matured vermin compost is rich in nutrients and recommended to be used @ 50 t ha^{-1}.

Characteristics of vermicompost

pH — 7.00

EC dsm^{-1} — 1.20

Organic carbon% — 30.50

Macronutrients:

Total nitrogen% — 0.66

Total P_2O_5% — 1.93

Total K_2O% — 0.42

Micro nutrients

Fe (ppm) — 19.8

Zn (ppm) — 0.90

Mn (ppm) — 16.50

Cu (ppm) — 2.30

Q.45. What are thermal treatment processes for solid waste disposal?

Ans. This refers to processes that involve the use of heat to treat waste. Listed below are descriptions of some commonly utilized thermal treatment processes.

Incineration: Incineration is a process of destruction of waste at high temperature. The combustible wastes are converted through controlled combustion to a residue, which contain no combustible matter. If land suitable for solid waste land filling operations is not available within economic haul distances, then incineration is necessary. After incineration, the wastes are converted to carbon dioxide, water vapour and ash. This method may be used as a means of recovering energy to be used in heating or the supply of electricity. In addition to supplying energy incineration technologies have the advantage of reducing the volume of the waste, rendering it harmless, reducing transportation costs and reducing the production of the green house gas methane. The solid waste is reduced in volume (80% - 90%) and height (98-99%). Incinerator can accept toxic and industrial wastes of any size in solid or powdery form. The other special wastes include hospital wastes, putrefiable organic solids from slaughter houses.

Pyrolysis and gasification: Pyrolysis is the process of conversion of biomass into solid, liquid and gaseous energy. Pyrolysis results in the chemical breakdown of organic carbon material into three basic components: (1) gas phase containing mainly H_2, CO_2, CO and CH_4, (2) tar or oil phase containing simple organic acids, methanol and acetone and (3) char phase made up of pure carbon and inert material. Pyrolysis does not cause pollution of the atmosphere and large quantities of potentially hazardous plastics could be treated. Pyrolysis and gasification are similar processes as they both decompose organic waste by exposing it to high temperatures and low amounts of oxygen. Gasification uses a low oxygen environment while pyrolysis allows no oxygen. These techniques use heat and an oxygen starved environment to convert biomass into other forms. A mixture of combustible and non-combustible gases as well as pyroligenous liquid is produced by these processes. All of these products have a high heat value and can be utilised. Gasification is advantageous since it allows for the incineration of waste with energy recovery and without the air pollution that is characteristic of other incineration methods.

Open burning: Open burning is the burning of unwanted materials in a manner that causes smoke and other emissions to be released directly into the air without passing through a chimney or stack. This includes the burning of outdoor piles, burning in a burn barrel and the use of incinerators which have no pollution control devices and as such release the gaseous by products directly into the atmosphere. Open burning has been practiced by a number of urban centres because it reduces the volume of refuse received at the dump and therefore extends the life of their dumpsite. Garbage may be burnt because of the ease and convenience of the method or because of the cheapness of the method. In countries where house holders are required to pay for garbage disposal, burning of waste in the backyard allows the householder to avoid paying the costs associated with collecting, hauling and dumping the waste.

Open burning has many negative effects on both human health and the environment. This uncontrolled burning of garbage releases many pollutants into the atmosphere. These include dioxins, particulate matter, polycyclic aromatic compounds, volatile organic compounds, carbon monoxide, hexachlorobenzene and ash. All of these chemicals pose serious risks to human health. The dioxins have adverse effects on reproduction, development, disrupt the hormonal systems or even cause cancer.

The polycyclic aromatic compounds and the hexachlorobenzene are carcinogenic. The particulate matter can be harmful to persons with respiratory problems such as asthma or bronchitis and carbon monoxide can cause neurological symptoms. The harmful effects of open burning are also felt by the environment. This process releases acidic gases such as the halo-hydrides; it also may release the oxides of nitrogen and carbon. Nitrogen oxides contribute to acid rain, ozone depletion, smog and global warming. The particulate matter creates smoke and haze which contribute to air pollution.

Q.46. Describe in detail the features of sanitary landfills.

Ans. Land filling is the most common and economic method of solid waste disposal. The indiscriminate land filling of solid waste in open dumps without adequate control and consideration of sanitation and public health as generally followed in India is undesirable. It results in water pollution, bad odour, fire and breeding of flies and rats.

It should be replaced by sanitary land filling or controlled tipping. The construction of sanitary land filling includes:

- Deposition of solid waste in such a way to have a working force of minimum area.
- Spreading and compaction of waste in thin layers.
- Covering of the waste with a layer of compacted cover soil daily.
- Final cover of the entire construction with compacted earth layer of 1.0 m thick.

The solid wastes in sanitary land fill are degraded by soil microbes. In comparison with other biological treatment systems such as activated sludge and anaerobic digestion, the microbial degradation of solid waste proceeds at a slow rate.

Sanitary Landfills are designed to greatly reduce or eliminate the risks that waste disposal may pose to the public health and environmental quality. They are usually placed in areas where land features act as natural buffers between the landfill and the environment. For example, the area may be comprised of clay soil which is fairly impermeable due to its tightly packed particles, or the area may be characterised by a low water table and an absence of surface water bodies thus preventing the threat of water contamination.

Besides the strategic placement of the landfill other protective measures are incorporated into its design. The bottom and sides of landfills are lined with layers of clay or plastic to keep the liquid waste, known as leachate, from escaping into the soil. The leachate is collected and pumped to the surface for treatment. Boreholes or monitoring wells are dug in the vicinity of the landfill to monitor groundwater quality.

A landfill is divided into a series of individual cells and only a few cells of the site are filled with trash at any one time. This minimizes exposure to wind and rain. The daily waste is spread and compacted to reduce the volume, a cover is then applied to reduce odours and keep out pests. When the landfill has reached its capacity, it is capped with an impermeable seal which is typically composed of clay soil.

Some sanitary landfills are used to recover energy. The natural anaerobic decomposition of the waste in the landfill produces landfill gases which include Carbon Dioxide, methane and traces of other gases. Methane can be used as an energy source to produce heat or electricity. Thus, some landfills are fitted with landfill gas collection (LFG) systems to capitalize on the methane being produced. The process of generating gas is very slow, for the energy recovery system to be successful there needs to be large volumes of wastes. These landfills present the least environmental and health risk, however, the cost of establishing these sanitary landfills are high when compared to the other land disposal methods of solid waste.

A sanitary landfill is essentially an engineered waste disposal option, where the environmental risk is controlled at an appropriate and acceptable level. But, if it is not properly managed, it can cause the following problems:

(*i*) noise pollution due to vehicles such as earthmovers, compactors, etc.,

(*ii*) scavenger birds, vermins, insects, etc., are attracted to the landfill for breeding and they constitute potential health problems,

(*iii*) leachates, if not treated or controlled, may pollute groundwater and surface water and

(*iv*) gas released, due to degradation or volatilization of waste components, causes problems such as odour, flammability and damage to health and the environment.

Uncontrolled disposal is a non-engineered waste disposal method in which wastes are dumped at a designated point without any environmental control, and it (*i*) causes odour problems due to the putrification of biodegradable wastes, (*ii*) provides a breeding place for disease vectors such as flies, mosquitoes, etc., and (*iii*) causes visual pollution, etc.

When properly managed, a landfill is a better and safer option of disposing wastes than uncontrolled dumping. Also, it is adaptable to all kinds of wastes and there is potential for by-product recovery.

Controlled dumps: Controlled dumps are disposal sites which comply with most of the requirements for a sanitary landfill but usually have one deficiency. They may have a planned capacity but no cell planning, there may be partial leachate management, partial or no gas management, regular cover, compaction in some cases, basic record keeping and they are fenced or enclosed. These dumps have a reduced risk of environmental contamination, the initial costs are low and the operational costs are moderate. While there is controlled access and use, they are still accessible by scavengers and so there is some recovery of materials through this practice.

Main features of a modern landfill is shown below in figure.

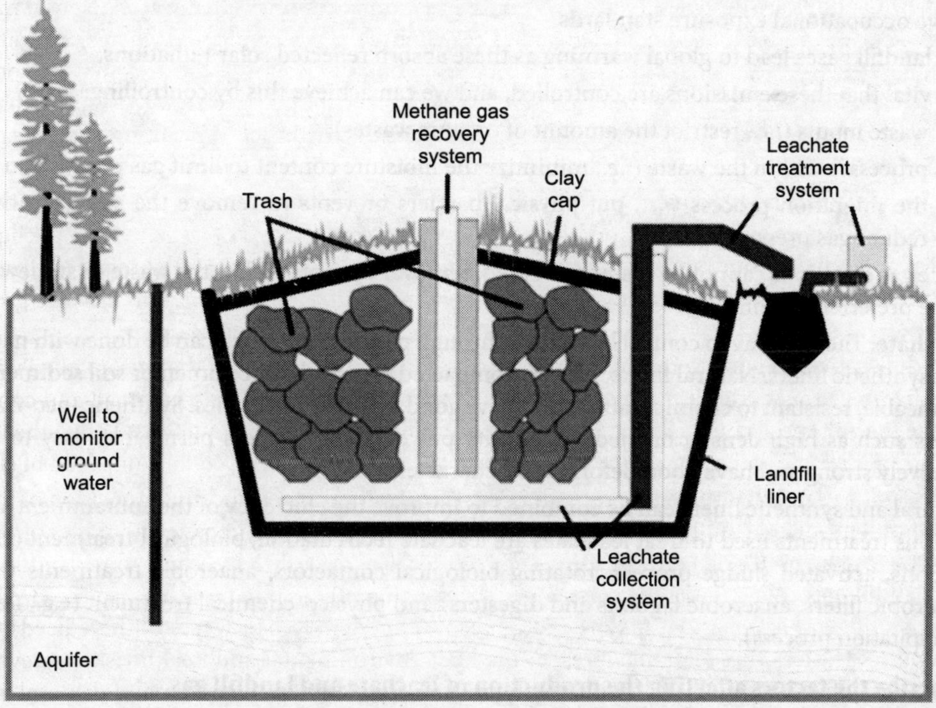

Fig. 8.2: Main features of a modern landfill

Bioreactor Landfills: Recent technological advances have led to the introduction of the Bioreactor Landfill which use enhanced microbiological processes to accelerate the decomposition of waste. The main controlling factor is the constant addition of liquid to maintain optimum moisture for microbial digestion. This liquid is usually added by re- circulating the landfill leachate. In cases where leachate in not enough, water or other liquid waste such as sewage sludge can be used. The landfill may use either anaerobic or aerobic microbial digestion or it may be designed to combine the two. These enhanced microbial processes have the advantage of rapidly reducing the volume of the waste creating more space for additional waste, they also maximize the production and capture of methane for energy recovery systems and they reduce the costs associated with leachate management. For Bioreactor landfills to be successful the waste should be comprised predominantly of organic matter and should be produced in large volumes.

Landfill is a complex reactor where physical, chemical and biological processes transform polluting wastes into environmentally acceptable deposits. Due to the complexity of these processes and their potential environmental effects, monitoring is needed to confirm that the landfill works as expected. Knowledge of the chemical composition of leachate and gas is required, as leachates may contain toxic substances and, if not prevented, may contaminate the groundwater. Landfill gases that may be flammable, asphyxiating and noxious, pose a health hazard. A continued groundwater-monitoring programme is essential for confirming the integrity of the liner system.

Q.47. How can we control landfill gases and leachate?

Ans. Landfill gases: Landfill gases include methane, carbon dioxide, oxygen and nitrogen, and these adversely impact on public health and the environment. Methane is flammable in air and can lead to unpredictable and uncontrolled subsidence and production of smoke and toxic fumes. Trace components such as aldehydes, alcohol and esters are toxicants when present in air at concentrations above occupational exposure standards.

The landfill gases lead to global warming as these absorb reflected solar radiations.

It is vital that these emissions are controlled, and we can achieve this by controlling:

- waste inputs (*i.e.*, restrict the amount of organic wastes);
- processes within the waste (*i.e.*, minimize the moisture content to limit gas production);
- the migration process (*i.e.*, put physical barriers or vents to remove the gas from the site and reduce gas pressure).

Since gas emissions cannot be easily prevented, removal by vents within the waste or stonewalled vents is the preferred option.

Leachate: The best way to control leachate is through prevention, which can be done with natural liners and synthetic liners. Natural liners, such as compacted clay or shale, bitumen or soil sediments, are less permeable, resistant to chemical attack and have good sorption properties. Synthetic (geo-membranes) liners such as high density or medium density polyethylene, are less permeable, easy to install and relatively strong and have good deformation characteristics.

Natural and synthetic liners can be combined to improve the efficiency of the containment system. The various treatments used to treat leachates are leachate recirculation, biological treatment (e.g., aerated lagoons, activated sludge process, rotating biological contactors, anaerobic treatments that include anaerobic filters, anaerobic lagoons and digesters) and physico-chemical treatment (e.g., flocculation-precipitation process).

Q.48. Describe the factors affecting the production of leachate and landfill gas.

Ans. The factors, which affect the production of leachate and landfill gas, are the following:

Nature of solid waste: The decomposition of waste containing biodegradable matter leads to the production of gas and leachate, and the amount depends on the content of biodegradable material in the waste.

Moisture content: Most micro-organisms require a minimum of approximately 12% (by weight) moisture for growth, and thus the moisture content of landfill waste is an important factor in determining the amount and extent of leachate and gas production.

pH: The methanogenic bacteria within a landfill produce methane gas, which will grow only at low pH range around neutrality.

Particle size and density: The size of waste particle affects the density that can be achieved upon compaction and affects the surface area and hence volume. Both affect moisture absorption and therefore are potential for biological degradation.

Temperature: An increase in temperature tends to increase gas production. The temperature affects the microbial activity to the extent that it is possible to segregate bacteria, according to their optimum temperature operating conditions.

The composition of waste, which varies with region and climate (season), determines the variation in pollution potential. Carbohydrates comprise a large percentage of biodegradable matter within municipal waste, the overall breakdown of which can be represented by the following equation:

$$\underset{\text{(Carbohydrate)}}{C_6H_{12}O_6} \xrightarrow{\text{Bacteria}} CH_4 + CO_2 + \text{Biomass} + \text{Heat}$$

Q.49. Describe the composition and properties, hazards,migration and control of landfill gases.

Ans. Landfill gas contains a high percentage of methane due to the anaerobic decomposition of organic matter, which can be utilized as a source of energy. Composition and properties, hazards, migration and control of landfill gases is described below:

Composition and properties: The amount and composition of the landfill gas generated for different substrates can be predicted, depending on the general anaerobic decomposition of solid wastes added. Climatic and environmental conditions also influence gas composition. Due to the heterogeneous nature of the landfill, some acid-phase anaerobic decomposition occurs along with the methanogenic decomposition. Since aerobic and acid-phase degradation give rise to carbon dioxide and not methane, there may be a higher carbon dioxide content in the gas generated than what would otherwise be expected. Depending on the moisture distribution, some carbon dioxide goes into solution. This may appear to increase (artificially) the methane content of the gas measured in the landfill. A typical landfill gas contains a number of components such as the following, which tend to occur within a characteristic range:

Methane: This is a colourless, odourless and flammable gas with a density lighter than air, typically making up 50 – 60% of the landfill gas.

Carbon dioxide: This is a colourless, odourless and non-inflammable gas that is denser than air, typically accounting for 30 – 40%.

Oxygen: The flammability of methane depends on the percentage of oxygen. It is, therefore, important to control oxygen levels, where gas extraction is undertaken.

Nitrogen: This is inert and will have little effect, except to modify the explosive range of methane.

It is difficult to convert the amount of gas measured to the maximum landfill gas production value because gas is withdrawn from a small part of the landfill only, referred to as *zone of influence* during measurement. In other words, it is very difficult to determine this zone and relate it to the whole landfill area.

Hazards of landfill gases: Landfill gas consists of a mixture of flammable, asphyxiating and noxious gases and may be hazardous to health and safety, and hence the need for precautions. Some of the major hazards are listed below:

Explosion and fire hazards: Methane is flammable in air within the range of 5-15% by volume, while hydrogen is flammable within the range of 4.1-7.5% (in the presence of oxygen) and potentially explosive. Fire, occurring within the waste, can be difficult to extinguish and can lead to unpredictable and uncontrolled subsidence as well as production of smoke and toxic fumes.

Trace components: These comprise mostly alkanes and alkenes, and their oxidation products such as aldehydes, alcohols and esters. Many of them are toxicants, when present in air at concentrations above occupational exposure standards.

Global warming: Known also as greenhouse effect, it is the warming of the earth's atmosphere by the accumulation of gases (methane, carbon dioxide and chlorofluorocarbons) that absorbs reflected solar radiation.

Migration of landfill gases: During landfill development, most of the gas produced is vented to the atmosphere, provided the permeable intermediate cover has been used. While biological and chemical processes affect gas composition through methane oxidation, which converts methane to carbon dioxide, physical factors affect gas migration. The physical factors that affect gas migration include:

Environmental conditions: These affect the rate of degradation and gas pressure build up.

Geophysical conditions: These affect migration pathways. In the presence of fractured geological strata or a mineshaft, the gas may travel large distances, unless restricted by the water table.

Climatic conditions: Falling atmospheric pressure, rainfall and water infiltration rate affect landfill gas migration.

The proportion of void space in the ground, rather than permeability, determines the variability of gas emission. If the escape of landfill gas is controlled and proper extraction system is designed, this gas can be utilised as a source of energy. If landfill gas is not utilised, it should be burnt by means of flaring. However, landfill gas utilisation can save on the use of fossil fuels since its heating value is approximately 6 kWh/m^3 and can be utilised in internal combustion engines for production of electricity and heat. Landfill gas is extracted out of the landfill by means of gas wells, which are normally drilled by auger and are driven into the landfill at a spacing of 40-70 m. In addition, horizontal systems can be installed during operation of the landfill. The gas wells consist mainly of perforated plastic pipes surrounded by coarse gravel and are connected with the gas transportation pipe with flexible tubing. The vacuum necessary for gas extraction and transportation is created by means of a blower. The most important factors influencing planning and construction of landfill gas extraction systems are settling of waste, water tables in landfills and gas quality.

Control of landfill gas emission: To control gas emission, it is necessary to control the following:

- waste inputs (*i.e.*, restrict the amount of organic waste).
- processes within the waste (*i.e.*, minimise moisture content to limit gas production).
- migration process (*i.e.*, provide physical barriers or vents to remove the gas from the site and reduce gas pressure).
- Since gas migration cannot be easily prevented, removal is often the preferred option. This is done by using vents (extraction wells) within the waste or stone filled vents, which are placed around the periphery of the landfill site. Some of the gas collection systems include impermeable cap, granular material, collection pipes and treatment systems.

Q.50. Describe the composition and properties, migration, control and treatment of leachate produced in landfill.

Ans. Leachate liquid produced in landfill can pollute both groundwater and surface water supplies. The degree of pollution will depend on local geology and hydrogeology, nature of waste and the proximity of susceptible receptors. Once groundwater is contaminated, it is very costly to clean it up. Landfills, therefore, undergo siting, design and construction procedures that control leachate migration.

Composition and properties: Leachate comprises soluble components of waste and its degradation products enter water, as it percolates through the landfill. The amount of leachate generated depends on:

- water availability
- landfill surface condition
- refuse state
- condition of surrounding strata.

The major factor, *i.e.*, water availability, is affected by precipitation, surface runoff, waste decomposition and liquid waste disposal.

Common toxic components in leachate are ammonia and heavy metals, which can be hazardous even at low levels, if they accumulate in the food chain. The presence of ammoniacal nitrogen means that leachate often has to be treated off-site before being discharged to a sewer, since there is no natural bio-chemical path for its removal. Leachate composition varies with time and location. Table below shows a typical leachate properties and composition at various stages of waste decomposition:

Table 8.4: Typical leachate properties and composition at various stages of waste decomposition

Properties and Composition of Leachate at Various Stages of Decomposition (mg/l except pH) Components	Fresh wastes	Aged wastes	Wastes with high moisture
pH	6.2	7.5	8.0
COD	23800	1160	1500
BOD	11900	260	500
TOC	8000	465	450
Volatile acid (as C)	5688	5	12
NH_3-N	790	370	1000
NO_3-N	3	1	1.0
Ortho-P	0.73	1.4	1.0
Cl	1315	2080	1390
Na	9601	300	1900
Mg	252	185	186
K	780	590	570
Ca	1820	250	158
Mn	27	2.1	0.05
Fe	540	23	2.0
Cu	0.12	0.03	—
Zn	21.5	0.4	0.5
Pb	0.40	0.14	—

Leachate migration: It is difficult to predict the movement of escaped leachate accurately. The main controlling factors are the surrounding geology and hydrogeology. Escape to surface water may be relatively easy to control, but if it escapes to groundwater sources, it can be very difficult both to control and clean up. The degree of ground water contamination is affected by physical, chemical and biological actions. The relative importance of each process may change, however, if the leachate moves from the landfill to the sub-surface region.

Leachate control: The best way to control leachate is through prevention, which should be integral to the landfill site design. In most cases, it is necessary to control liquid access, collection and treatment, all of which can be done using the following landfill liners:

Natural liners: These refer to compacted clay or shale, bitumen or soil sealants, etc., and are generally less permeable, resistant to chemical attack and have good sorption properties. They generally do not act as true containment barriers, because sometimes leachate migrates through them.

Synthetic (geo-membrane) liners: These are typically made up of high or medium density polyethylene and are generally less permeable, easy to install, relatively strong and have good deformation characteristics. They sometimes expand or shrink according to temperature and age.

Natural and geo-membrane liners are often combined to enhance the overall efficiency of the leachate containment system. Some of the leachate collection systems include impermeable liner, granular material, collection piping, leachate storage tank; leachate is trucked to a wastewater treatment facility.

Leachate treatment: Concentrations of various substances occurring in leachate are too high to be discharged to surface water or into a sewer system. These concentrations, therefore, have to be reduced by removal, treatment or both. The various treatments of leachate include:

Leachate recirculation: It is one of the simplest forms of treatment. Recirculation of leachate reduces the hazardous nature of leachate and helps wet the waste, increasing its potential for biological degradation.

Biological treatment: This removes BOD, ammonia and suspended solids. Leachate from land filled waste can be readily degraded by biological means, due to high content of volatile fatty acids (VFAs). The common methods are aerated lagoons (*i.e.*, special devices which enhance the aerobic processes of degradation of organic substances over the entire depth of the tank) and activated sludge process, which differs from aerated lagoons in that discharged sludge is recirculated and is often used for BOD and ammonia removal. While under conditions of low COD, rotating biological contactors (*i.e.*, biomass is brought into contact with circular blades fixed to a common axle which is rotated) are very effective in removing ammonia. In an anaerobic treatment system, complex organic molecules are fermented in filter. The common types are anaerobic filters, anaerobic lagoon and digesters.

Physicochemical treatment: After biological degradation, effluents still contain significant concentrations of different substances. Physico chemical treatment processes could be installed to improve the leachate effluent quality. Some of these processes are flocculation-precipitation. (Note that addition of chemicals to the water attracts the metal by floc formation). Separation of the floc from water takes place by sedimentation, adsorption and reverse osmosis.

Q.51. Describe the environmental effects of landfill operations.

Ans. The environmental effects of a landfill operation include wind-blown litter and dust, noise, obnoxious odour, vermin and insects attracted by the waste, surface runoff and in aesthetic conditions. Gas and leachate problems also arise during the operation phase and require significant environmental controls.

- The major environmental effects of landfill operations are described below:
  - (*i*) Wind-blown litter and dust are continuous problems of the ongoing landfill operation and a nuisance to the neighbourhood. Covering the waste cells with soil and spraying water on dirty roads and waste in dry periods, in combination with fencing and movable screens, may minimize the problem of wind-blown litter and dust. However, the problem will remain at the tipping front of the landfill.
  - (*ii*) Movement of waste collection vehicles, emptying of wastes from them, compactors, earth moving equipment, etc., produce noise. Improving the technical capability of the equipment, surrounding the fill area with soil embankments and plantations, limiting the working hours and appropriately training the workforce will help minimize noise pollution.
  - (*iii*) Birds (e.g., scavengers), vermin, insects and animals are attracted to the landfill for feeding and breeding. Since many of these may act as disease vectors, their presence is a potential health problem.

(*iv*) Surface run-off, which has been in contact with the land filled waste, may be a problem in areas of intense rainfall. If not controlled, heavily polluted run-off may enter directly into creeks and streams. Careful design and maintenance of surface drains and ditches, together with a final soil cover on completed landfill sections, can help eliminate this problem.

(*v*) An operating landfill, where equipment and waste are exposed, appears in aesthetic. This problem may be reduced by careful design of screening soil embankments, plantings, rapid covering and re-vegetation of filled sections.

(*vi*) Gas released, as a result of degradation or volatilisation of waste components, causes odour, flammability, health problems and damage of the vegetation (due to oxygen depletion in the root zone). The measures to control this include liners, soil covers, passive venting or active extraction of gas for treatment before discharge into the atmosphere.

(*vii*) Polluted leachate appears shortly after disposal of the waste. This may cause groundwater pollution and pollution of streams through sub-surface migration. Liners, drainage collection, treatment of leachate, and groundwater and downstream water quality monitoring are necessary to control this problem.

Besides the emissions, incidental events such as flooding, fires, landslides and earthquakes result in severe environmental impacts, and may require preventive measures with respect to landfill site selection, design and operation. In the main, to minimise adverse environmental impacts due to sanitary landfill, proper attention must be paid to the environmental aspects at all stages and phases of landfill management, viz., site selection, design, construction, operation and maintenance.

Regulations for Landfills: Regulations include restrictions on distances from airports, flood plains, and fault areas, as well as limitations on construction in wetlands and others. Prevention of contamination of groundwater and land resources requires synthetic liner and adequate buffer with the restricted activities around the landfill.

Q.52. Describe the criteria for landfill as notified by Ministry of Environment and Forest (MOEF) government of India.

Ans. For the identification of landfill site the following criteria as suggested by MOEF should be followed:

- The landfill site shall be large enough to last for **20-25 years** and preferably within **5 km from** present city limits.
- The site shall be at least **500 mt. away from habitation clusters**, forest areas, monuments, National Parks, Wetlands and places of important cultural, historical or religious interest.
- No landfill should be constructed within **200 mt of any lake or pond**.
- No landfill should be constructed within **100 mt. of a navigable river**.
- No landfill should be constructed within **200 mt. of the right of way of any state or national highway.**
- A site should be at least **500 mt. away from a notified habitated area**.
- A buffer zone of 500 mt. around a landfill boundary should be declared as a No development zone after the landfill location is finalized.
- No landfill site should be constructed within critical habitat areas.
- A landfill should not be constructed in areas where water table is less than 2 mt. below ground surface.
- No landfill should be constructed within 500 mt. of any water supply.
- Topographical and geographical features (contour, slope, soil type) of the site should be carefully analyzed in order to prevent soil contamination and disturbance of drainage pattern.

- Transportation and linkages will be clearly defined for waste movement from the city. The influence of increased heavy vehicle traffic due to land filling should be taken care in order to avoid nuisance.

Q.53. Describe in detail the features of biological waste treatment of solid wastes.

Ans. Features of biological waste treatment methods like composting and anaerobic digestion are described below.

Composting: Composting is the controlled aerobic decomposition of organic matter by the action of micro organisms and small invertebrates. A number of composting techniques being used today include: in vessel composting, windrow composting, vermicomposting and static pile composting. The process is controlled by making the environmental conditions optimum for the waste decomposers to thrive. The rate of compost formation is controlled by the composition and constituents of the materials *i.e.*, their Carbon/Nitrogen (C/N) ratio, the temperature, the moisture content and the amount of air.

The C/N ratio is very important for the process to be efficient. The micro organisms require carbon as an energy source and nitrogen for the synthesis of some proteins. If the correct C/N ration is not achieved, then application of the compost with either a high or low C/N ratio can have adverse effects on both the soil and the plants. A high C/N ratio can be corrected by dehydrated mud and a low ratio is corrected by adding cellulose.

Moisture content greatly influences the composting process. The microbes need the moisture to perform their metabolic functions. If the waste becomes too dry the composting is not favoured. If however there is too much moisture then it is possible that it may displace the air in the compost heap depriving the organisms of oxygen and drowning them.

A high temperature is desirable for the elimination of pathogenic organisms. However, if temperatures are too high, above 75°C then the organisms necessary to complete the composting process are destroyed. Optimum temperatures for the process are in the range of 50-60°C.

Aeration is a very important and the quantity of air needs to be properly controlled when composting. If there is insufficient oxygen the aerobes will begin to die and will be replaced by anaerobes. The anaerobes are undesirable since they will slow the process, produce odours and also produce the highly flammable methane gas. Air can be incorporated by churning the compost.

Although compost has been used for a variety of purposes, contaminants in the compost could be detrimental to the environment and public health as they affect the water (*i.e.*, eutrophication and leaching to surface and groundwater), the land (*i.e.*, acidic soil, nitrogen depletion in soil, impacts of acids, organic and metal contamination), the air (*i.e.*, odour) and health (*i.e.*, drinking of contaminated water, toxins in the finished product and pathogens).

Anaerobic digestion: Anaerobic digestion like composting uses biological processes to decompose organic waste. However, where composting can use a variety of microbes and must have air; anaerobic digestion uses bacteria and an oxygen free environment to decompose the waste. Aerobic respiration, typical of composting, results in the formation of Carbon dioxide and water. While the anaerobic respiration results in the formation of Carbon Dioxide and methane. In addition to generating the humus which is used as a soil enhancer, Anaerobic Digestion is also used as a method of producing biogas which can be used to generate electricity. Optimal conditions for the process require nutrients such as nitrogen, phosphorous and potassium, it requires that the pH be maintained around 7 and the alkalinity be appropriate to buffer pH changes, temperature should also be controlled.

Q.54. Briefly describe the mechanism of composting.

Ans. Biodegradation is a natural, ongoing biological process that is a common occurrence in both human-made and natural environments. Any organic material that can be biologically decomposed is *compostable*. Human beings have used this naturally occurring process for centuries to stabilize and recycle agricultural and human wastes. Today, however, composting is a diverse practice that includes a variety of approaches, depending on the type of organic materials being composted and the desired properties of the final product.

To derive the maximum benefit from the natural, slow decomposition process (e.g., grass clippings left in the lawn or food scraps rotting in dustbins, etc.), it is necessary to control the environmental conditions during the composting process.

The overall composting process is as follows:

Organic matter + O_2 + aerobic bacteria $\longrightarrow$ CO_2 + NH_3 + H_2O + other end products + energy

Compost is the end product of the composting process. The by-products of this process are carbon dioxide and water. Compost is peaty humus, dark in colour and has a crumbly texture, an earthy odour, and resembles rich top soil. Composts will not have any resemblance in the physical form to the original waste from which it was derived. High-quality compost is devoid of weed seeds and organisms that may be pathogenic to humans, animals, or plants. Cured compost is also relatively stable and resistant to further decomposition by microorganisms.

The composting process is an environmentally sound and beneficial means of recycling organic materials and not a means of waste disposal. And, it is important to view compostable materials as usable and not as waste requiring disposal.

A complete composting process is shown below in figure.

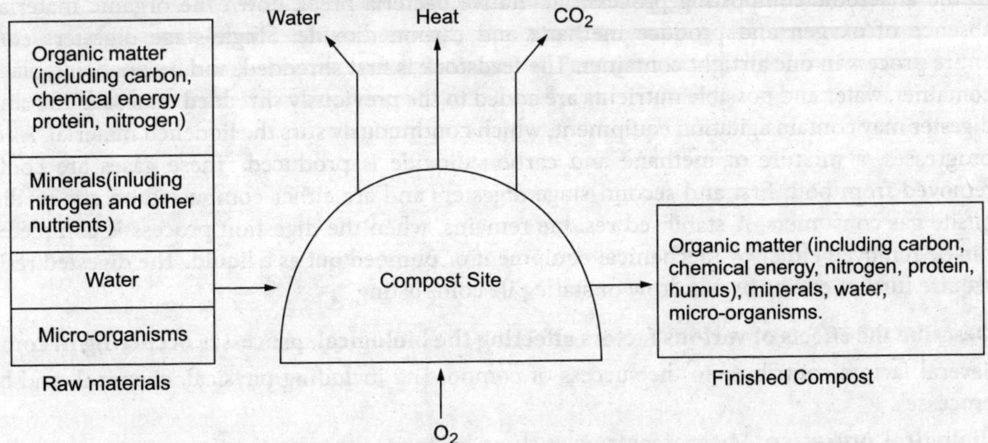

Fig. 8.3: Complete composting process

Q.55. Describe the five basic stages involved in composting and three processes of composting.

Ans. There are five basic stages involved in all composting practices, namely, **preparation, digestion, curing, screening and finishing and storage or disposal. Preparation** consists of sorting the recyclables, removal of non-combustibles, shredding, pulping, grinding and the addition of water sludge. There are five techniques of **digestion**, viz., windrow, aerated, static pile, in-vessel composting and anaerobic processing. These techniques vary in the method of air supply, temperature control, mixing/turning of the material, time required for composting and capital and operating costs. The amount of **curing** necessary at the initial stabilisation depends on the proposed use of the composted material. If the compost is to be pelletised and/or bagged, it must be completely stable before being finished. The amount of **finishing** required is also dependent on the final marketing of the compost. Regrinding and rescreening is done, so as to make the compost uniform in size. The efficiency of the composting process, *i.e.*, the success or failure of the operation, depends on the method of disposal. Even where a good compost market exists, provisions must still be made for storage because the use of composting is seasonal.

There are three processes of composting, viz., **biological, chemical and physical**. In **biological processes**, microorganisms such as bacteria, fungi and actinomycetes break down the organic matter present in the waste. As microorganisms begin to decompose the organic material, the carbon in it is converted to by-products like carbon dioxide and water and a humic end product called compost. Inoculum is added in case of a lack of microbial organisms. If all conditions are ideal for a given microbial population to perform at its maximum potential, composting will occur rapidly. The **chemical environment** is largely determined by the composition of material to be composted. Several factors determine the chemical environment for composting: carbon energy source, the correct amount of water, adequate oxygen, appropriate pH and the absence of toxic constituents that could inhibit microbial activity. A **physical environment** is essential for the composting process to proceed efficiently. This includes factors such as temperature, particle size, mixing and pile size. Smaller particle size is preferred, as it has more surface area per unit weight, and therefore, facilitates more microbial activity, which leads to rapid decomposition. The optimum temperature range is between 32 and 60°C. Temperature above 60°C is not ideal for composting. Mixing is done to distribute moisture and air evenly and promote the breakdown of compost clumps.

Q.56. What is anaerobic composting?

Ans. In the anaerobic composting process, facultative bacteria break down the organic materials in the absence of oxygen and produce methane and carbon dioxide. Single-stage digesters contain the entire process in one airtight container. The feedstock is first shredded, and before being placed in the container, water and possible nutrients are added to the previously shredded material. The single-stage digester may contain agitation equipment, which continuously stirs the liquefied material. As digestion progresses, a mixture of methane and carbon dioxide is produced. These gases are continuously removed from both first and second-stage digesters and are either combusted on site or directed to offsite gas consumers. A stabilised residue remains, when the digestion process is completed, which can be removed either by mechanical equipment or pumped out as a liquid. The digested residue may require further curing by windrow or static pile composting.

Q.57. Describe the effects of various factors affecting the biological, processes occurring in composting.

Ans. Several factors contribute to the success of composting including physical, chemical, and biological processes.

Biological processes: Microorganisms such as bacteria, fungi and actinomycetes as well as larger organisms such as insects and earthworms play an active role in decomposing the organic materials. As microorganisms begin to decompose the organic material, they break down organic matter and produce carbon dioxide, water, *heat* and humus (the relatively stable organic end product). This humus end product is called compost.

Microorganisms consume some of the carbon to form new microbial cells, as they increase their population. They have, preferences on the type of organic material they consume. When the organic molecules they require are not available, they may become dormant or die. The chain of succession of different types of microbes continues, until there is little decomposable organic material left. At this point, the organic material that is remaining is termed *compost*. It is largely made up of microbial cells, microbial skeletons and by-products of microbial decomposition, and undecomposed particles of organic and inorganic origin. Decomposition may proceed slowly at first because of smaller microbial populations, but as populations grow in the first few hours or days, they rapidly consume the organic materials present in the feedstock. The number and kind of microorganisms are generally not a limiting environmental factor in composting non-toxic agricultural materials, yard trimmings, or municipal solid wastes, all of which usually contain an adequate diversity of microorganisms.

Under optimal conditions, composting proceeds through the following four phases

(*i*) **Mesophilic, or moderate-temperature phase:** Compost bacteria combine carbon with oxygen to produce carbon dioxide and energy. The microorganisms for reproduction and growth use some of the energy and the rest is generated as heat. When a pile of organic refuse begins to undergo the composting process, mesophilic bacteria proliferate, raising the temperature of the composting mass up to 44°C. This is the first stage of the composting process. These mesophilic bacteria can include *E. coli* and other bacteria from the human intestinal tract, but these soon become increasingly inhibited by the temperature, as the thermophilic bacteria take over in the transition range of 44 to 52°C.

(*ii*) **Thermophilic, or high-temperature phase:** In the second stage of the process, the thermophilic microorganisms are very active and produce heat. This stage can continue up to about 70°C, although such high temperatures are neither common nor desirable in compost. This heating stage takes place rather quickly and may last only a few days, weeks, or months. It tends to remain localized in the upper portion of a compost pile where the fresh material is being added, whereas in batch compost, the entire composting mass may be thermophilic all at once. After the thermophilic heating period, the manure will appear to have been digested, but the coarser organic material will not be digested. This is when the third stage of composting, *i.e.*, the cooling phase, takes place.

(*iii*) **Cooling phase:** During this phase, the microorganisms that were replaced by the thermophiles migrate back into the compost and digest the more resistant organic materials. Fungi and macroorganisms such as earthworms and sow bugs that break the coarser elements down into humus also move back in.

(*iv*) **Maturation or curing phase:** The final stage of the composting process is called curing, ageing, or maturing stage, and is a long and important one. A long curing period (e.g., a year after the thermophilic stage) adds a safety net for pathogen destruction. Many pathogens have a limited period of viability in the soil, and the longer they are subjected to the microbiological competition of the compost pile the more likely they will die a swift death. Immature compost can be harmful to plants. Uncured compost can, for example, produce phytotoxins (*i.e.*, substances toxic to plants), robbing the soil of oxygen and nitrogen and contain high levels of organic acids.

Different communities of microorganisms predominate during the various composting phases. Initial decomposition is carried out by mesophilic microorganisms, which rapidly break down the soluble, readily degradable compounds. The heat they produce causes the compost temperature to rise rapidly. As the temperature rises above 40°C, the mesophilic microorganisms become less competitive and are replaced by thermophilic (heat loving) ones. At temperatures of 55°C and above, many microorganisms that are pathogenic to humans or plants are destroyed. Temperatures above 65°C kill many forms of microbes and limit the rate of decomposition. Composters use aeration and mixing to keep the temperature below this point. During the thermophilic phase, high temperatures accelerate the breakdown of proteins, fats, and complex carbohydrates like cellulose and hemicellulose, the major structural molecules in plants. As the supply of these high-energy compounds become exhausted, compost temperature gradually decreases and mesophilic microorganisms once again take over the final phase of curing or maturation of the remaining organic matter.

The composting process, therefore, should cater to the needs of the microorganisms and promote conditions that will lead to rapid stabilisation of the organic materials. While several microorganisms are beneficial to the composting process and may be present in the final product, some microbes are potential pathogens (disease causing) to animals, plants and humans. These pathogenic organisms must be destroyed in the composting process before the compost is distributed in the market place. Most of this destruction takes place by controlling the temperature of composting operations.

Q.58. Describe the factors affecting the chemical processes occurring during composting.

Ans. Factors that determine the chemical environment for composting include the presence of an adequate carbon food/energy source, a balanced amount of nutrients, the correct amount of water, adequate oxygen, appropriate pH and the absence of toxic constituents that could inhibit microbial activity The effects of each of these factors is discussed below:

(*i*) **Carbon food/energy source:** For their carbon/energy source, microorganisms in the composting process rely on carbon in the organic material, unlike higher plants that rely on carbon dioxide and sunlight. The carbon contained in natural or human-based organic materials may or may not be biodegradable. The relative ease with which a material is biodegraded depends on the genetic makeup of the microorganisms present and the organic molecules that the organism decomposes. Since most municipal and agricultural organics and yard trimmings contain an adequate amount of biodegradable forms of carbon, it is not a limiting factor in the composting process. As more easily degradable forms of carbon are decomposed, a small portion of the carbon is converted into microbial cells, and a significant portion is converted to carbon dioxide and lost to the atmosphere. As the composting process progresses, the loss of carbon results in a decrease in weight and volume of the feedstock. The less easily decomposed forms of carbon will form the matrix for the physical structure of the final product.

(*ii*) **Nutrients:** Among the plant nutrients (*i.e.*, nitrogen, phosphorus, and potassium), nitrogen is of greatest concern, because it is lacking in some plant materials. The carbon-nitrogen ratio, which is established on the basis of available carbon rather than total carbon, is considered critical in determining the rate of decomposition. Leaves, for example, are a good source for carbon, and fresh grass, manure and slaughter house waste are the sources for nitrogen. To aid the decomposition process, the bulk of the organic matter should be carbon with just enough nitrogen. In general, an initial ratio of 30:1 (Carbon: Nitrogen) is considered ideal. Higher ratios tend to retard the process of decomposition, while ratios below 25:1 may result in odour problems. As the composting process proceeds and carbon is lost to the atmosphere, this ratio decreases.

Finished compost should have ratios of 15 to 20:1. Adding 3-4 kg of nitrogen material for every 100 kg of carbon should be satisfactory for efficient and rapid composting. To lower the carbon to nitrogen ratios, nitrogen-rich materials such as yard trimmings, animal manures, or bio-solids are added. Adding partially decomposed or composted materials (with a lower carbon: nitrogen ratio) as inoculums may also lower the ratio. As the temperature in the compost pile rises and carbon: nitrogen ratio falls below 25:1, the nitrogen in the fertiliser is lost as gas (ammonia) to the atmosphere. The composting process slows, if there is not enough nitrogen, and too much nitrogen may cause the generation of ammonia gas, which can create unpleasant odours.

(*iii*) **Moisture:** Water is an essential part of all forms of life, and the microorganisms living in a compost pile are no exception. Since most compostable materials have lower than ideal water content, *i.e.*, 50 to 60% of total weight, the composting process may be slower than desired, if water is not added. However, it should not be high enough to create excessive free flow of water and movement caused by gravity. Excessive moisture and flowing water form leachate, creates potential liquid management problems including water and air pollution (e.g., odour). For example, excess moisture impedes oxygen transfer to the microbial cells, can increase the possibility of developing anaerobic conditions and may lead to rotting and obnoxious odours. Microbial processes contribute moisture to the compost pile during decomposition. Although moisture is added, it is also being lost through evaporation. Since the amount of water evaporated usually exceeds the input of moisture from the decomposition processes, there is generally a net loss of moisture from the compost pile. In such cases, adding moisture may be necessary to keep the composting process performing at its peak.

Controlling the size of piles can minimise evaporation from compost piles, as piles with larger volumes have less evaporating surface/unit volume than smaller piles. The water added must be thoroughly mixed so that the organic fraction in the bulk of the material is uniformly wetted and composted under ideal conditions. Properly wetted compost has the consistency of a wet sponge.

(*iv*) **Oxygen:** Composting is considered an aerobic process. Decomposition can occur under both aerobic (requiring oxygen) and anaerobic (lacking oxygen) conditions. The compost pile should have enough void space to allow free air movement so that oxygen from the atmosphere can enter the pile and the carbon dioxide and other gases emitted can be exhausted to the atmosphere. To maintain aerobic conditions, in which decomposition occurs at a fast rate, the compost pile is mechanically aerated or turned frequently to expose the microbes to the atmosphere and to create more air spaces by fluffing up the pile. A 10 to 15% oxygen concentration is considered adequate, although a concentration as low as 5% may be sufficient for leaves. While higher concentrations of oxygen will not negatively affect the composting process, circulation of an excessive amount of air can cause problems. For example, excess air removes heat, which cools the pile and also promotes excess evaporation. In other words, excess air slows down the rate of composting. Excess aeration is also an added expense that increases production costs.

(*v*) **pH:** The pH factor affects the amount of nutrients available for the microorganisms, the solubility of heavy metals and the overall metabolic activity of the microorganisms. A pH between 6 and 8 is optimum, and it can be adjusted upward by the addition of lime, or downward with sulphur, although such additions are normally not necessary. The composting process produces carbon dioxide, which, when combined with water, produces carbonic acid, which could lower the pH of the compost. As the composting process progresses, the final pH varies, depending on the specific type of feedstock used and operating conditions. Wide swings in pH are unusual, since organic materials are naturally well buffered with respect to pH changes. Down swings in pH during composting usually do not occur. The composition of material to be composted largely determines the chemical environment. In addition, several modifications can be made during the composting process to create an ideal chemical environment for the rapid decomposition of organic materials.

Q.59. Describe the factors affecting the physical processes occurring during composting.

Ans. The physical environment in the composting process is affected by factors such as temperature, particle size, mixing and pile size. Each of these is essential for the composting process to proceed in an efficient manner. Their effects are described below:

(*i*) **Particle size:** As composting progresses, there is a natural process of size reduction and the particle size of the material being composted is critical. Because smaller particles usually have more surface area per unit weight, they facilitate more microbial activity on their surfaces, which leads to rapid decomposition. The optimum particle size has enough surface area for rapid microbial activity and also enough void space to allow air to circulate for microbial respiration. The feedstock composition can be manipulated to create the desired mix of particle size and void space. For instance, through particle size reduction, we can increase the desired combination of void space and surface area for garden trimmings or municipal solid wastes. To improve the aesthetic appeal of finished composts, we can carry out particle size reduction, after the composting process is completed.

(*ii*) **Temperature:** Composting can occur at a range of temperatures, and the optimum temperature range is between 32 and 60°C. Temperatures above 65°C are not ideal for composting as thermal destruction of cell proteins kill the organisms. Similarly, temperatures below the minimum required for a group of organisms affect the metabolic activity (*i.e.*, regulatory machinery) of the cells. Temperatures can be lowered, if required, by either increasing the frequency of mechanical agitation or using airflow throttling, temperature feedback control or blowers controlled with timers. Mixing or mechanical aeration also provides air for the microbes.

When compost is at a temperature greater than 55°C for at least three days, pathogen destruction occurs. It is important that all portions of the compost material are exposed to such temperatures to ensure pathogen destruction throughout the compost. At these temperatures, weed seeds are also destroyed. After the complete pathogen destruction, temperatures may be lowered and maintained at slightly lower levels (51 to 55°C). Attaining and maintaining 55°C for three days is not difficult for in-vessel composting systems. However, to achieve pathogen destruction with windrow composting systems, 55°C temperature level must be maintained for a minimum of 15 days, during which time the windrows must be turned at least five times. The longer duration and increased turning are necessary to achieve a uniform pathogen destruction in the entire pile. Care should be taken to avoid contact between materials that have achieved these minimum temperatures and materials that have not.

(*iii*) **Mixing:** Mixing of feedstock, water and inoculants is important and is done by turning or mixing the piles after composting has begun. Mixing and agitation distribute moisture and air evenly, and promote the breakdown of compost clumps. Excessive agitation of open vessels or piles, however, can cool them and retard microbial activity.

Q.60. Describe briefly various composting technologies.

Ans. Various composting technologies namely windrow composting, aerated static pile composting, in-vessel composting and anaerobic processing – vary in the method of air supply, temperature control, mixing/turning of the material, time required for composting, and capital and operating costs. Besides these general categories of composting technologies, there are also some supporting technologies, which include sorting, screening, and curing.

Four general categories of composting technologies are described below:

Windrow composting: The windrow system is the least expensive and most common approach. Windrows are defined as regularly turned elongated piles, shaped like a haystack in cross section and up to a hundred meters or more in length. The cross-sectional dimensions vary with feedstock and turning equipment, but most municipal solid waste (MSW) windrows are 1.5 to 3 meters high and 3 to 6 meters wide. Windrows composed of MSW are usually required to be located on an impermeable surface, which greatly improves equipment handling under inclement weather conditions. The optimum size and shape of the windrow depends on particle size, moisture content, pore space and decomposition rate – all of which affect the movement of oxygen towards the centre of the pile.

Process control is normally through pile management, although forced aeration can also be used. Turning the pile re-introduces air into the pile and increases porosity so that efficient passive aeration from atmospheric air continues at all times. The windrow dimensions should allow conservation of the heat generated during the composting process and also allow air to diffuse to the deeper portions of the pile.

Windrows must be placed on a firm surface to turn the piles with ease. They may be turned as frequently as once per week, but more frequent turning may be necessary, if high proportions of bio-solids are present in the feedstock. Turning the piles also moves the materials from the pile surface to the core of the windrow, where they can undergo composting. Machines equipped with augers, paddles or tines are used for turning the piles. Some windrow turners can supplement piles with water, if necessary. When piles are turned, heat is released as steam to the atmosphere. If inner portions of the pile have low levels of oxygen, odours may result when this portion of the pile is exposed to the atmosphere. Piles with initial moisture content within the optimum range have a reduced potential for producing leachate. The addition of moisture from precipitation, however, increases this potential. Any leachate or runoff created must be collected and treated or added to a batch of incoming feedstock to increase its moisture content. To avoid problems with leachate or runoff, piles can be placed under a roof, but doing so adds to the initial costs of the operation.

Aerated static pile composting: Aerated static pile composting technology requires the composting mixture (*i.e.*, a mixture of pre-processed materials and liquids) to be placed in piles that are mechanically aerated. The piles are placed over a network of pipes connected to a blower, which supplies the air for composting.Air can be supplied under positive or negative pressure. That is to say, the air supply blower either forces air into the pile or draws air out of it. Forcing the air into the pile generates a positive pressure system, while drawing air out of the pile creates a negative pressure. When the composting process is nearly complete, the piles are broken up for the first time since their construction. The compost is then taken through a series of post-processing steps. A timer or a temperature feedback system similar to a home thermostat controls the blowers.

Air circulation in the compost piles provides the needed oxygen for the composting microbes and prevents excessive heat build-up in the pile. Removing excess heat and water vapour cools the pile to maintain optimum temperature for microbial activity. A controlled air supply enables construction of large piles, which decreases the need for land. Odours from the exhaust air could be substantial, but traps or filters can be used to control them. The temperatures in the inner portion of a pile are usually adequate to destroy a significant number of the pathogens and weed seeds present. The surface of piles, however, may not reach the desired temperatures for destruction of pathogens because piles are not turned in the aerated static pile technology. This problem can be overcome by placing a layer of finished compost of 15 to 30 cms thick over the compost pile. The outer layer of finished compost acts as an insulating blanket and helps maintain the desired temperature for destruction of pathogens and weed seeds throughout the entire pile.

Aerated static pile composting systems have been used successfully for MSW, yard trimmings, bio-solids and industrial composting. Aerated static pile composting can also be done under a roof or in the open. Producing compost using this technology usually takes about 6 to 12 weeks. The land requirements for this method are lower than that of windrow composting.

In-vessel composting system: In-vessel composting systems enclose the feedstock in a chamber or vessel that provides adequate mixing, aeration and moisture. Drums, digester bins and tunnels are some of the common in vessel type systems. In vessel systems vary in their requirements for pre-processing materials. For example, some require minimal pre-processing, while others require extensive MSW pre-processing. These vessels can be single or multi compartment units. In some cases, the vessel rotates, and in others, it is stationary and a mixing/agitating mechanism moves the material around. Most in-vessel systems are continuous-feed *systems*, although some operate in a batch mode. All in-vessel systems require further composting(curing) after the material has been discharged from the vessel.

A major advantage of in-vessel systems is that all environmental conditions can be carefully controlled to allow rapid composting. The material to be composted is frequently turned and mixed to homogenise the compost and promote rapid oxygen transfer. Retention times range from less than one week to as long as four weeks. These systems, if properly operated, produce minimal odours and little or no leachate. In-vessel systems enable exhaust gases from the vessel to be captured and are subjected to odour control and treatment.

Some of the commonly used in-vessel systems are as follows:

Vertical composting reactor: It is generally over 4 meters high and can be housed in *silos* or other large structures. Organic material, typically fed into the reactor at the top through a distribution mechanism, moves by gravity to an unloading mechanism at the bottom. Process control is usually by pressure-induced aeration, where the airflow is opposite to the downward materials flow. The height of these reactors makes process control difficult due to the high rates of airflow required per unit of distribution surface area. Neither temperature nor oxygen can be maintained at optimal levels throughout the reactors, leading to zones of non-optimal activity. As with static pile composting, a stable porous structure is important in vertical reactors, which usually lack internal mixing. Tall vertical reactors have been used in the sludge composting industry where uniform feedstock and porous amendments can minimise these difficulties in process control, but are rarely used for heterogeneous materials like MSW.

Horizontal composting reactors: These avoid high temperature, oxygen and moisture gradients of vertical reactors by maintaining a short airflow pathway. They come in a wide range of configurations, including static and agitated, pressure and/or vacuum-induced aeration. Agitated systems generally use the turning process to move the material through the system in a continuous mode, while static systems require a loading and unloading mechanism. Material handling equipment may also shred to a certain degree, exposing new surfaces for decomposition, but excessive shredding may also reduce porosity. Aeration systems are usually set in the floor of the reactor and may use temperature and/or oxygen as control variables. Systems with agitation and bed depths less than two to three meters are effective in dealing with the heterogeneity of MSW.

Rotating drum composting reactors: Rotating drum reactors take the trade-off between reactor cost and compost residence time to an even further extreme than the horizontal or vertical in-vessel systems. These reactors, also known as digesters, retain the material for only a few hours or days. While the tumbling action can help homogenise and shred materials, the short residence time usually means the processing is more physical than biological. While rotating drums can play an important role in MSW composting, they are normally followed by other biological processing, which may include in-vessel, static pile and/or windrow systems.

Anaerobic composting: In anaerobic processes, facultative bacteria break down the organic materials in the absence of oxygen and produce methane and carbon dioxide. Anaerobic systems, if configured efficiently, will generate sufficient energy in the form of methane to operate the process and have enough surpluses to either market as gas or convert to electricity. Conventional composting systems, on the other hand, need significant electrical or mechanical energy inputs to aerate or turn the piles. Several approaches are available for anaerobic digestion of feedstock.

Single-stage digesters contain the entire process in one airtight container. In this system, the feedstock is first shredded, and before being placed in the container, water and possibly nutrients are added to the previously shredded material. A single-stage digester may contain agitation equipment, which continuously stirs the liquefied material. The amount of water added and presence or absence of agitation equipment depend on the particular process employed.

The two-stage digestion involves circulating a liquid supernatant from a first stage digester, containing the materials, to a second-stage digester. This circulation eliminates the need for agitation equipment and provides the system operator with more opportunity to carefully control the biological process.

As digestion progresses, a mixture of methane and carbon dioxide is produced. These gases are continuously removed from both first and second stage digesters and are either combusted on-site or directed to off-site gas consumers. A portion of the recovered gas may be converted to thermal energy by combustion, which is then used to heat the digester. A stabilised residue remains, when the digestion process is completed. The residue is either removed from the digester with the mechanical equipment or pumped out as a liquid. It is chemically similar to compost, but contains more moisture. Conventional dewatering equipment can reduce the moisture content enough to handle the residue as a solid. The digested residue may require further curing by windrow or static pile composting. Figure below illustrates an anaerobic digester with an aerobic compost curing.

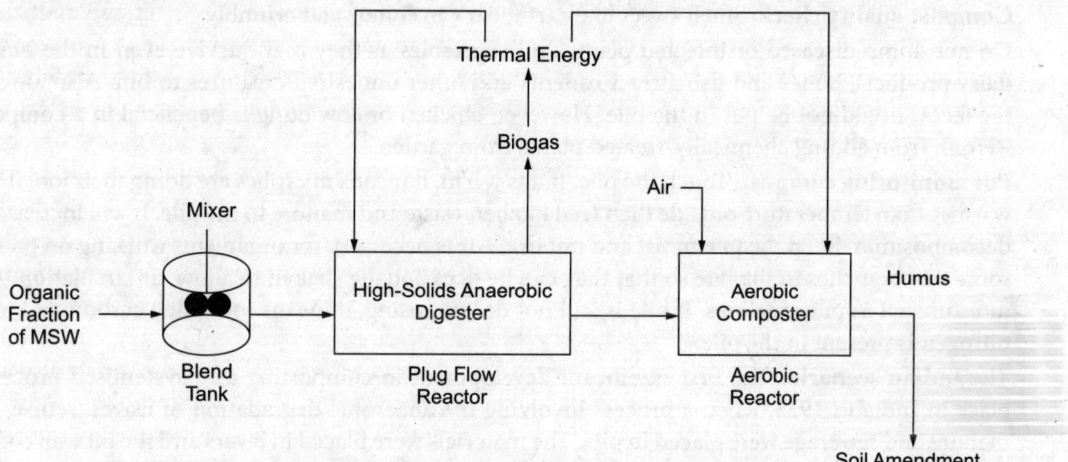

Fig. 8.4: An anaerobic digester with an aerobic compost curing

Q.61. Describe the process and features of backyard composting using dry leaves and kitchen and garden waste.

Ans. Backyard composting using dry leaves and kitchen and garden waste can be done with a little time and effort. Any pile of waste organic matter will rot. To produce good compost for one's own garden following steps are to be followed:

- Choose a compost bin which can be of wood, cement, brick or plastic.
- Place the bin on a leveled, well-drained surface as it does not require base. Earthworm and other micro-organisms enter the bin from the base of bin through the soil and act on compost.

Avoid wood bins that may be chemically treated to keep micro-organisms at bay. Carbon nitrogen ratio in the bin should be 25:1. Carbon-rich organic matters are brown coloured waste products that include hay, saw dust, old leaves, cardboard, newspaper and straw. Nitrogen-rich organic matters are green coloured products that include green grass, manure, coffee, weeds, and vegetable peels. Bottom layer of bin should be around six inches of carbon-rich matter like hay, straw, etc. Then top of it should be covered with a three inch layer of nitrogen-rich organic waste like vegetable scraps, manure, etc.

- Water each layer in the bin. Add these layers alternately till it becomes 1 to 1.5 meters high.
- Shred all the waste into smaller particles that you dump in the pile which makes decomposition process faster. A sample pile composition could be four parts of fruit and vegetable scraps, two parts of cow dung, one part of shredded paper, and one part dry leaves.
- Protect the pile from rain. To avoid sogginess of the contents cover the pile and see to it that contents are not too dry either.
- Ensure only biodegradable organic component are added.
- Proper decomposition occurs in presence of proper air and water. Don't saturate the pile with excess water. Turn the contents of the compost bin thoroughly using a shovel. This will aerate its contents, prevent it from stinking and hasten decomposition. It is advisable to mix the compost

once in one or two weeks. Under proper temperature and moisture conditions, earthworms and other microbes will do the decomposition.

If these steps are properly followed then compost would be ready in six months.

Spread around two inches of compost on the soil until it mixes well. Though the compost is spread on the top layer of the soil, it releases nutrients to the layers lying underneath.

Compost quality check: Smell sweet like earth, dark in colour and crumbly.

Do not dump diseased or infected plants and vegetables as they may survive even in the bin. Meat, dairy products, bones and fish attract rodents and other undesired creatures to bin. Also, dog, cat or pig feces should not be put in the pile. However, chicken or cow dung is beneficial in a compost bin. Refrain from adding chemically-treated plants from garden.

For monitoring compost: Touch the pile. If it is warm, it means microbes are doing their job. If it is not warmer than temperature outside then feed kitchen waste and manure to the pile. It will increase rate of decomposition. Keep the pile moist and not dry. Air is necessary for organisms working on pile. Insert some tree branches in the pile so that they can be occasionally shaken to allow air circulation. Air and moisture act as pile's lifelines. If pile is still not decomposing, it means too much carbon and too little nitrogen is present in the pile.

The Indian scenario: The first significant development in composting as a systemised process took place in India in 1925, when a process involving the anaerobic degradation of leaves, refuse, animal manure and sewerage were placed in pits. The materials were placed in layers and the pit wall conserved some of the heat of degradation, resulting in high temperature than when composting was carried out in the open. This process (often referred as Indore process) took approximately six months to produce usable compost.

Following this, the Indian Council of Agricultural Research (ICAR) improved the method by laying down alternate layers of waste and sewage and this system (referred to as the Bangalore process) is still being used in India. In India, the high humid degradation that occurs in the land requires a large amount of humus for maintaining soil-fertility, and for that reason, composting is an ideal method for recycling organic wastes.

Q.62. Describe the manual composting methods referred as Bangalore method and Indore methods of composting and compare their features.

Ans. Bangalore method: This is an anaerobic composting method conventionally carried out in pits. Formerly the waste was anaerobically stabilised in pits where alternate layers of MSW and night soil were laid. The pit is completely filled and a final soil layer is laid to prevent fly breeding, entry of rain water into the pit and for conservation of the released energy. The material is allowed to decompose for 4 to 6 months after which the stabilised material is taken out and used as compost.

Indore method: This method of composting in pits involves filling of alternate layers of similar thickness as in Bangalore method. However, to ensure aerobic condition the material is turned at specific intervals for which a 60 cm strip on the longitudinal side of the pit is kept vacant. For starting the turning operation, the first turn is manually given using long handled rakes 4 to 7 days after filling. The second turn is given after 5 to 10 more days. Further turning is normally not required and the compost is ready in 2 to 4 weeks.

In the urban areas, due to extensive provision of water carriage system of sanitation, night soil is not available. Composting of MSW alone is hence carried out. Aerobic composting of MSW is commonly carried out in windrows.

Comparison of the two methods: The Bangalore method requires longer time for stablisation of the material and hence needs larger land space, which is in short supply in urban areas. The gases generated in this anaerobic process also pose smell and odour problems.

The Indore method on the other hand stabilises the material in shorter time and needs lesser land space. As no odorous gases are generated in this process, it is environment friendly and hence commonly preferred. While the organic matter is stabilised during the composting process, the moisture content also changes. The non-decomposable are also rejected. Hence the quantity of compost is much lesser than the input and is normally around 50%, and the exact value depends upon the characteristics of the input material.

Q.63. Describe the bio-gasification/bio-methanation process of MSW disposal.

Ans. Bio gasification or Bio methanation is the process of conversion of organic matter in the waste (liquid or solid) to Biomethane (sometimes referred to as "Biogas" with high energy density) and manure (bio compost) by microbial action in the absence of air, known as "anaerobic processing or digestion."

Biogas is a mixture of gases composed of methane (CH_4) 40-70 vol.%, carbon dioxide (CO_2) 30-60 vol.%, other gases 1 – 5 vol.% including, hydrogen (H_2) 0-1 vol.% and hydrogen sulphide (H_2S) 0-3 vol.%. It originates from bacteria in the process of bio-degradation of organic material under anaerobic (without air) conditions. The natural generation of biogas is an important part of the biogeochemical carbon cycle.

Biogasification is particularly suitable for wet substrates, such as sludges or food waste, which present difficulties in composting, as the lack of structural material restricts air circulation. The anaerobic process is used sometimes to digest sewage sludge, and this has been extended to fractions of household solid waste.

In contrast to aerobic processes (*i.e.*, composting), bio gasification is mildly exothermic. Thus, the heat needs to be supplied to maintain the process temperature, especially for thermophilic processes. The advantage of high temperature is that the reaction will occur at a faster rate, and so a shorter residence time is required in the reactor vessel.

The conditions for bio gasification is anaerobic, for which a totally enclosed process vessel is required. Although this necessitates a higher level of technology than compared to composting, it allows a greater control over the process itself and the emission of noxious odours. Greater process control, especially of temperature, allows a reduction in treatment time, when compared to composting. Since a biogas plant is usually vertical, it also requires less area than a composting plant.

Anaerobic processing of organic material is a two-stage process, where large organic polymers are fermented into short-chain volatile fatty acids. These acids are then converted into methane and carbon dioxide. Both the organic polymers fermentation process and acid conversion occur at the same time, in a single-phase system. And, the separation of the acid-producing (acidogenic) bacteria from the methane producing (methanogenic) bacteria results in a two-phase system.

The main feature of anaerobic treatment is the concurrent waste stabilisation and production of methane gas, which is an energy source. The retention time for solid material in an anaerobic process can range from a few days to several weeks, depending upon the chemical characteristics of solid material and the design of the bio gasification system (e.g., single stage, two stage, multi- stage, wet or dry, temperature and pH control).

In the absence of oxygen, anaerobic bacteria decompose organic matter as follows:

Organic matter + anaerobic bacteria $\longrightarrow$ $CH_4 + CO_2 + H_2S + NH_3$ + other end products + energy

Some of the benefits of bio gasification of solid wastes are as follows:

- production of energy (heat, light, electricity);
- transformation of organic waste into high quality fertiliser;
- improvement of hygienic conditions through reduction of pathogens, worm eggs and flies;
- reduction of workload in firewood collection and cooking;

- environmental advantages through protection of soil, water, air and woody vegetation;
- micro-economic benefits through energy and fertiliser substitution, additional income generation and increasing yields of animal husbandry and agriculture;
- macro-economic benefits through decentralised energy generation, import substitution and environmental protection.

Biogas technology can substantially contribute to energy conservation and development, if the economic viability and social acceptance of biogas technology are favourable.

Q.64. Describe various bio-gasification/anaerobic digestion process of organic waste present in MSW.

Ans. According to the solid content of the material digested and the temperature at which the process operates, the various bio gasification processes can be classified as under:

Dry anaerobic digestion: In dry anaerobic digestion, semi-solid wastes are digested to produce biogas in a single stage, either as a batch process or a continuous process. It takes place at a total solid concentration of over 25%, and below this level of solid, the process is described as wet digestion. With regard to temperature, the processes are either described as mesophilic (operating between 30 and 40°C) or thermophilic (operating between 50 and 65°C), and anaerobic microorganisms have optimum growth rates within these temperature ranges.

The dry fermentation process means that only little process water has to be added or heated, which favours thermophilic operation. No mixing equipment is necessary, and crust formation is not possible due to the relatively solid nature of the digester contents. This anaerobic process usually takes between 12 and 18 days, followed by several days in post-digestion for residue stabilisation and maturation.

Wet anaerobic digestion: This process consists of a single stage in a completely mixed mesophilic digester, operating at a total solid content of around 3-8%. To produce this level of dilution, water has to be added and heated, and then removed after the digestion process. This method is used to digest sewage sludge, and animal and house hold wastes. The single-stage wet process suffers from problems such as the formation of a hard scum layer in the digester and difficulty in keeping the content completely mixed.

The major problem with the single-stage process is that the different reactions in the process cannot be separately optimised. The acidogenic microorganisms grow fast and lower the pH of the reaction mixture, whereas the methanogens, which grow slowly, have a pH optimum around 7.0.

The development of the two-stage digestion process solves this problem as hydrolysis and acidification occur in the first reactor vessel, kept at a pH of around 6.0 and methanogenesis occurs in the second vessel, operated at a pH of 7.5-8.2. The whole process can run with a retention time of 5 to 8 days.

Maturing or refining: The residues of both wet and dry bio gasification processes require extensive maturing under aerobic conditions. However, effective aeration can reduce this time considerably. The maturation processes facilitate the release of entrapped methane, elimination of phytotoxins (*i.e.*, substances that are harmful for plant growth, such as volatile organic acids) and reduce the moisture content to an acceptable level. These residues contain a high level of water – even the dry process residue contains around 65% water. Filtering or pressing can reduce excess water, and further drying can be achieved using waste heat from the gas engine, if the biogas is burnt onsite to produce electricity. The digested residue, initially anaerobic, will also contain many volatile organic acids and reduced organic materials. These need to be matured aerobically to oxidize and stabilize the compounds, in the process similar to the maturation of aerobic compost, prior to sale as compost or disposal as residue. Odour production is measured as the total amount of volatile organics produced per ton of bio-waste during composting and the final aerobic maturation after anaerobic digestion.

Q.65. Describe the various factors affecting bio gasification of solid waste.

Ans. As with composting, a number of factors affect bio gasification, which are described below:

Temperature: A temperature range of about 25-40°C (mesophilic) is generally optimal. It can be achieved without additional heating, thus being very economical. In some cases, additional energy input is provided to increase temperature to 50-60°C (thermophilic range) for greater gas production. Often digesters are constructed below ground to conserve heat.

pH and alkalinity: pH close to neutral, *i.e.*, 7, is optimum. At lower pH values (below 5.5), some bacteria carrying out the process are inhibited. Excess loading and presence of toxic materials will lower pH levels to below 6.5 and can cause difficulties. When pH levels are too low, stopping the loading of the digester and/or use of lime is recommended. The presence of alkalinity between 2500 and 5000 mg/l will provide good buffering against excessive pH changes.

Nutrient concentration: An ideal C:N ratio of 25:1 is to be maintained in any digester. It is an important parameter, as anaerobic bacteria need nitrogen compound to grow and multiply. Too much nitrogen, however, can inhibit methanogenic activity. If the C:N ratio is high, then gas production can be enhanced by adding nitrogen, and if the C:N ratio is low, it can be increased by adding carbon, *i.e.*, adding chicken manure, etc., which reduces the possibility of toxicity. Anaerobic digestion not only breaks down plant materials into biogas, but also releases plant nutrients, such as nitrogen (N), potassium (K) and phosphorous (P), and converts them into a form that can be easily absorbed by plants.

Loading: When any digester is designed, the main variable to be defined is the internal volume. The digester volume is related to, feed rate (Q, measured in m^3/day) and hydraulic loading or retention time (R, measured in days). The feed rate (Q) is given by the mass of total solid (m, kg) fed daily, divided by the proportion of total solid (TS) in the mixed slurry. The retention time (R) of any digester is given by the volume of the digester pit (V, m^3), divided by the volume of the daily feed (Q, m^3/day). The loading rate of a digester is defined as the mass of volatile solids added each day per unit volume of digester. The typical values for the loading rate are between 0.2 and 2.0 kg VS/m^3/day.

Effect of toxins: The main cause of biogas plants receiving flak is the presence of toxic substance. Chlorinated hydrocarbons, such as chloroforms and other organic solvents, are particularly toxic to biogas digestion. If the digester has been badly poisoned, it may be difficult to remove the toxins without removing most of the bacteria. In that case, the digester must be emptied, cleaned with water and refilled with fresh slurry.

Q.66. Describe the main components of biogas plant.

Ans. A typical biogas plant has the following components:

- a digester in which the slurry (e.g., dung mixed with water) is fermented.
- an inlet tank used to mix the feed and let it into the digester.
- a gas holder/dome in which the generated gas is collected.
- an outlet tank to remove the spent slurry.
- distribution pipeline(s) to carry the gas into the kitchen.
- a manure pit, where the spent slurry is stored.

In India, the objective of setting up bio gas plant has been:

- to provide fuel for cooking purposes and organic manure to rural households through biogas plants.
- to mitigate the drudgery of rural women, reduce pressure on forest and accentuate social benefits.
- to improve sanitation in villages by linking toilets with biogas plants.

Q.67. Describe the various types of bio gas plants model used in India.

Ans. There are various biogas plant models approved by National Project on Biogas Development (NPDB) initiated by the ministry of non-conventional energy resources for implementation. All these models are based on the basic designs available: floating metal drum type, fixed masonry dome type, and FLEXI, a portable model made of rubberized nylon fabric, which has been approved for promotion in hilly and other terrains.

Some of the NPDB approved models of biogas are described below:

KVIC floating drum: *Khadi* and Village Industries Commission (KVIC) developed this model in the early 1960s. This model has an underground cylindrical digester with inlet and outlet connections at the bottom on either side of a masonry wall. An inverted metal drum, which serves as the gasholder, rests on a wedge type support on top of the digester and, as the gas begins to accumulate, the drum starts rising in height. The weight of the drum applies pressure on the gas to make it pass through the pipeline to the point of use. As the gas flows out, the drum gradually moves down. Due to this smooth two-way motion, the gas remains at constant pressure, which ensures efficient use of gas. Though KVIC has been a very popular model since the beginning, it has two major drawbacks: the plant cost is very high and the metal gasholder has to be painted regularly for protecting it against corrosion, which means high running cost.

Deenbandhu: Action for food production, a voluntary organization based in New Delhi developed this model in 1984. *The* cost of the Deenbandhu bio gas plant is nearly 50% less than that of the KVIC plant, which has brought the biogas technology within the reach of even the poorer sections of the population. The cost reduction was achieved by minimising the surface area through joining the segments of two spheres of different diameters at their bases. This structure acts as a digester, and pressure is exerted on the slurry, which is pushed into a displacement chamber. Once the gas is drawn from the outlet, the slurry again enters the digester. The brick masonry dome, which is fixed, requires skilled workmanship and quality material to ensure no leakage.

Pragati: A Pune-based NGO has developed this model, which is a combination of KVIC and *Deenbandhu* designs. In this model, the lower part of the digester is semi-spherical in shape with a conical bottom. However, instead of a fixed dome, it has a floating drum acting as gas storage chamber. The spread of *Pragati* model has been confined mainly to the state of Maharashtra.

KVIC plant with ferro-cement digester: The Structural Engineering Research Centre, Ghaziabad, has developed the ferro-cement digester design in order to overcome the problems encountered in the construction of traditional models of biogas plants. In this digester, layers of thin steel wire mesh, distributed throughout, are impregnated with rich mortar. Ferro-cement (*i.e.*, concrete made of welded mesh, sand and cement) as a building material offers several advantages like 10 to 15% reduction in cost over KVIC digesters, usage of locally available material, less labour and little or no maintenance. The only limitation though is ferro-cement is vulnerable to damage due to the impact of pointed loads, which, however, can be easily repaired.

Plant with fiber reinforced plastic (FRP) gasholder: FRP is used in place of metal in the floating drum gasholder. Contact Moulding Process, a technique of moulding without the application of external pressure, is adopted to manufacture the FRP gas holder. It employs one of the less expensive types of moulds, resulting in lower cost of the plant. The major advantage of FRP is its resistance to corrosion, which saves the recurring expenditure on painting the drum.

Flexi: Developed by Swastik Rubber Products Ltd., Pune, this is a portable model in which the digester is made of rubberized nylon fabric. The model is particularly suitable for hilly areas, where high transportation cost of construction materials, such as cement and bricks substantially increase the cost of installing the regular type of biogas plants.

Q.68. Describe the environmental effects of composting and biogasification.

Ans. The compost produced and slurry from biogas plants is often used as a soil amendment in a variety of applications. Since the product is used for a variety of purposes, contaminants in the compost could be detrimental to the environment. This is particularly true when one considers that the environmental pollutants are easily minimized through simple treatment procedures (*i.e.*, waste segregation, proper turning and sufficient composting time). The composting facilities of solid waste with manure, sewage sludge and residue from processing create some significant environmental effects as described below:

(*i*) **Water impacts:** Water runoff from garden waste composting facilities could contain large concentration of nutrients (*i.e.*, nitrates and phosphorous), volatile organics and metals that cause algal blooms in the nearby surface water. Retention basins may be used at facilities to limit water runoff. Water impacts are not serious at yard waste composting facilities. Municipal waste composting, sludge composting and co-composting involve a large amount of potential contaminants, and water impact could be greater at these facilities. Leachate from MSW compost facilities can contain high concentration of nutrients, volatile organics and metals. Leachate affects both surface and ground water.

(*ii*) **Land impacts:** At garden waste composting facilities, soil may become more acidic, due to the presence of certain leaves and pine needles in the compost pile. Nitrogen depletion may also occur, which can be limited by proper turning. MSW and co-composting facilities carry the potentially harmful impacts of acids, and organic and metal contamination. Again, careful pre-processing to divert much of the potentially hazardous materials from the compost facility is an important quality control procedure.

(*iii*) **Odour pollution:** Odour is one of the most frequent problems at composting facilities. Frequent turning of compost piles has proven to be effective in limiting odour problems. When in-vessel systems are used, odour control devices (e.g., air scrubber, etc.) can minimise the problem.

(*iv*) **Health impact:** The primary public health concerns associated with composting operation result from:

- drinking contaminated water
- toxins in the finished product (applied on land)
- pathogens.

Nitrate contamination of drinking water can affect the oxygen-carrying capacity of the blood in infants, e.g., blue baby syndrome (methemoglobinemia) and in elderly, but again under proper composting conditions, this risk can be minimised. Insects and vermin can spread pathogens. Note that because of the high temperature achieved during the normal composting operations, pathogens found in manure, sewage sludge or municipal waste are usually destroyed. There are risks to workers as well (e.g., aggravation of respiratory problem). However, proper training and health monitoring as well as proper apparel and equipment can minimise these risks.

Q.69. What are the benefits of composting?

Ans. Composting is one of the important components of solid waste management (SWM). It is a form of source reduction or waste prevention, as the materials are completely diverted from the disposal facilities and require no management or transportation. Community-yard trimming composting programme, source-separated organic composting and mixed municipal solid waste (MSW) composting constitute the various recycling processes. A major portion of municipal solid wastes in India contain up to 70% by weight of organic materials. In addition, certain industrial by-products – those from food processing, agricultural and paper industries – are mostly composed of organic materials.

Composting, being an organic material, can significantly reduce waste stream volume. Diverting such materials from the waste stream frees up landfill space needed for materials that cannot be composted.

Composting owes its current popularity to several factors, including increased landfill tipping fees, shortage of landfill capacity and increasingly restrictive measures imposed by regulatory agencies. In addition, recycling mandates indirectly encourage composting, as they consider it an acceptable strategy for achieving mandatory goals.

Composting may also offer an attractive economic advantage for communities where the costs of using other options are high. However, it is considered a viable option only when the compost can be marketed. In some cases, nevertheless, the benefits of reducing disposal needs through composting may be adequate to justify choosing this option, even if the compost is only used as a landfill cover. Compost, because of its high organic matter content, makes a valuable soil amendment and is used to provide nutrients for plants. When mixed with soil, compost promotes a proper balance between air and water in the resulting mixture, helps reduce soil erosion and serves as a slow release fertiliser.

Q.70. What do you mean by thermal treatment of solid waste?

Ans. Thermal treatment of solid wastes, which is pre-treatment of waste prior to final disposal includes both the burning of mixed MSW (municipal solid waste) in incinerators and the burning of selected parts of the waste stream as fuel for energy recovery. Various incineration technologies such as mass burning system, refuse derived fuel (RDF) system, modular incineration and fluidized bed incineration are in practice. Emission of air pollutants is a major concern of incineration facilities.

Q.71. What is incineration of solid waste?

Ans. Incineration is a chemical reaction in which carbon, hydrogen and other combustible elements in the solid waste mix with oxygen in the combustion zone and generates heat. The air requirements for combustion of solid wastes is approximately 5000 kg of air is required for each tonne of solid wastes burned. Usually, excess air is supplied to the incinerator to ensure complete mixing and combustion and to regulate operating temperature and control emissions. Excess air requirements, however, differ with moisture content of waste, heating values and the type of combustion technology employed. The principal gaseous products of *combustion* are carbon dioxide, carbon monoxide, water, oxygen and oxides of nitrogen.

Many incinerators are designed to operate in the combustion zone of 900°C – 1100°C. This temperature is selected to ensure good combustion, complete elimination of odours and protection of the walls of the incinerator. Incinerator systems are designed to maximise waste burn out and heat output, while minimising emissions by balancing the oxygen (air) and the three "Ts", *i.e.*, time, temperature and turbulence. Complete incineration of solid wastes produces virtually an inert residue, which constitutes about 10% of the initial weight of waste and a larger reduction in volume. The residue is generally landfilled. The incineration facility along with combustion of waste emits air pollutants (*i.e.*, fine particulate and toxic gases), which are an environmental concern, and, therefore, their control is necessary. Other concerns relating to incineration include the disposal of the liquid wastes from floor drainage, quench water, scrubber effluents and the problem of ash disposal in landfills because of heavy metal residues. By optimising the combustion process, we can control the emission of combustible, carbon-containing pollutants. Oxides of nitrogen and sulphur, and other gaseous pollutants are not considered a problem because of their relatively smaller concentration.

Q.72. Describe the main objectives of incineration of solid wastes.

Ans. The purpose of incineration is to combust solid wastes to reduce their volume to about one-tenth, without producing offensive gases and ashes. Key objectives of incineration of solid wastes are described below:

Waste volume reduction: Depending on its composition, incineration reduces the volume of solid wastes to be disposed of by about 90%. The weight of the solid wastes to be dealt with is reduced by about 70 %. This has both environmental and economic advantages since there is less demand for final disposal to landfill, as well as reduced costs and environmental burdens due to transport, if a distant landfill is used.

Waste stabilisation: Incinerator output (*i.e.*, ash) is considerably more inert than incinerator input (*i.e.*, solid wastes), mainly due to the oxidation of the organic components of the waste stream. This leads to a reduction of landfill management problems (since the organic fraction is responsible for landfill gas production) and the organic compounds present in landfill leachate.

Recovery of energy from waste: Energy recovered from burning the wastes is used to generate steam for use in on-site electricity generation. Combined heat and power plants increase the efficiency of energy recovery by producing electricity as well as utilising the residual heat. Solid waste incineration can replace the use of fossil fuels for energy generation. As a large part of the energy content of solid wastes comes from renewable resources (e.g., biomass), there should be a lower overall net carbon dioxide production than that from burning fossil fuels, since carbon dioxide is absorbed in the initial growing phase of the biomass.

Sterilisation of waste: This is of primary importance in the incineration of hospital waste. Incineration of solid wastes will also ensure destruction of pathogens prior to final disposal in a landfill.

Waste volume reduction for both environmental and economic reasons and sterilisation of waste is the main objectives of incineration. Due to the growing concern over the production of landfill gas and the organic compounds in leachate from landfill receiving untreated waste, in future there will be more emphasis on using incineration for stabilising wastes for subsequent landfilling. Landfill gas and leachate arise mainly from the organic fraction of solid wastes, which will be effectively converted to gases and mineralized ash by incineration.

Q.73. Describe the main factors to be considered in site selection for waste incineration facility.

Ans. The important factors to be considered for siting the incineration facility are described below:

Effect on residents: Residents will be concerned with the health effects associated with incinerator plant, decreased property value and increased traffic (e.g., due to truck movement).

Environmental impact: Incineration has the potential to create a variety of environmental concerns like air, water and noise pollution and ash disposal.

Development plans: It is important to evaluate future land use plans at the possible site.

Proximity to waste source: Transportation cost is one of the most significant expenditures in waste management system.

Proximity to energy market: The energy products will have to be delivered to buyers. The location of power line must be considered.

Logistic concerns: Area zoning and access route must be considered.

Residual ash disposal: Access to a secure landfill is necessary.

Q.74. Describe various waste incineration technologies in detail.

Ans. The four important waste incineration technologies mass burning system, refuse derived fuel system, modular incineration and fluidised bed incineration. The two most widely used and technically proven incineration technologies are mass-burning incineration and modular incineration. Fluidised-bed incineration has been employed to a lesser extent and Refuse Derived Fuel (RDF) production and incineration has also been used, with limited success. Some facilities are used in conjunction with pyrolysis, gasification and other related processes that convert solid waste to gaseous, liquid, or solid fuel through thermal processing. These incineration technologies are described below:

Mass-burning system: Mass-burning systems facility consists of a reciprocating grate combustion system and a refractory-lined, water-walled steam generator. Mass-burn systems generally consist of either two or three incineration units ranging in capacity from 50 to 1,000 tonnes per day. That means the facility capacity ranges from about 100 – 150 to 2,000 – 3,000 tonnes per day. These facilities can accept refuse that has undergone little preprocessing other than the removal of oversized items. Although this versatility makes mass-burn facilities convenient and flexible, local programmes to separate household hazardous wastes (e.g., cleaners and pesticides) and recover certain materials (e.g., iron scrap) are necessary to help ensure environmentally viable incineration and resource conservation. To achieve greater combustion and heat recovery efficiency, the larger field-erected incinerators are usually in-line furnaces with a grate system. The steam generator generally consists of refractory-coated water wall systems, *i.e.*, walls comprised of tubes through which water circulates to absorb the heat of combustion.

Figure below shows a typical mass-burn facility.

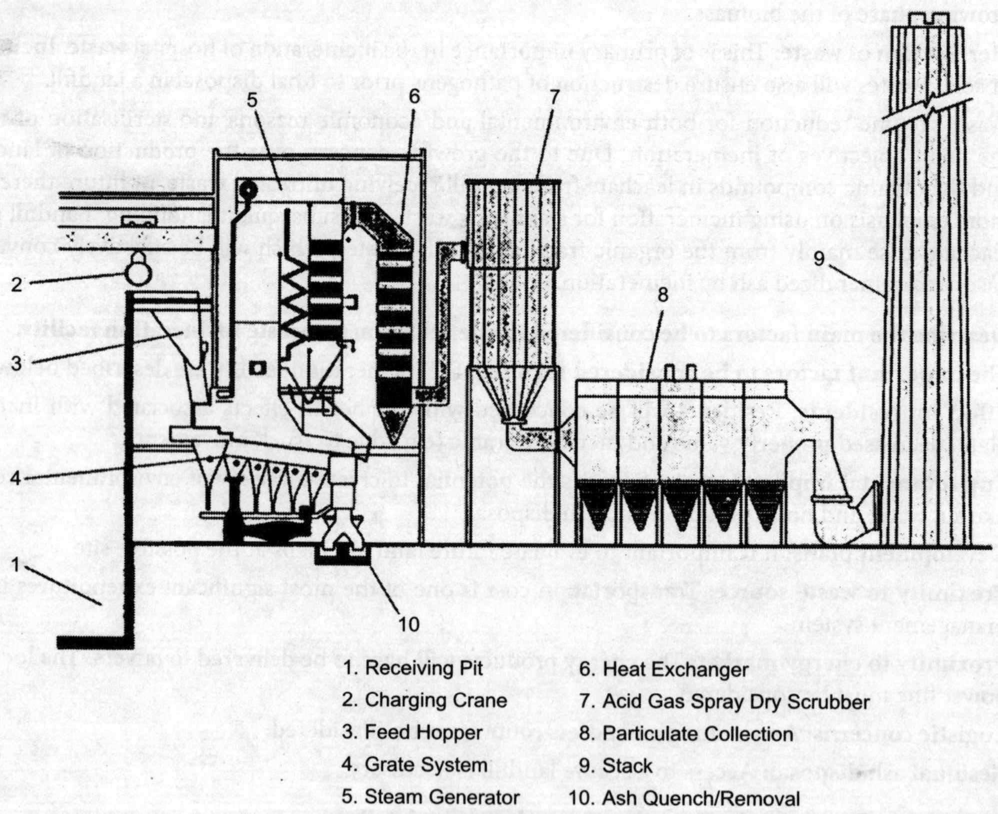

Fig. 8.5: A typical mass-burn facility

Mass-burning of waste can also be done by the use of a rotary kiln. Rotary kilns use a turning cylinder, either refractory or water wall design, to tumble the waste through the system. The kiln is inclined, with waste entering at the high elevation end and ash and non-combustibles leaving at the lower end. The waste intake area includes a tipping floor, pit, crane and sometimes conveyors. Trucks enter the tipping floor and tip their wastes either onto the floor itself, or directly into the pit. When wastes are tipped onto the floor, a front-end loader or a bulldozer is used to push them into the pit or onto a conveyor. MSW is continuously fed to a grate system, from a feed chute, which moves the waste through a combustion chamber using a tumbling motion. A travelling or reciprocating grate may follow rotary combustors to further complete combustion.

At least two combustor units are included to provide a level of redundancy and to allow waste processing at a reduced rate during periods of scheduled and unscheduled maintenance. Mass-burn facilities today generate a higher quality of steam (*i.e.*, pressure and temperature), which is then passed through a turbine generator to produce electricity or through an extraction turbine to generate electricity as well as provide process steam for heating or other purposes.

Refuse derived fuel (RDF) system: Refuse-derived fuel (RDF) refers to solid wastes in any form that is used as fuel. The term RDF, however, is commonly used to refer to solid waste that has been mechanically processed to produce a storable, transportable and more homogeneous fuel for combustion. RDF systems have two basic components: RDF production and RDF incineration.

RDF production facilities make RDF in various forms through material separation, size reduction and pelletizing. Although RDF processing has the advantage of removing recyclables and contaminants from the combustion stream, on an average, capital costs per tonne for incineration units that use RDF are higher than for other incineration options. RDF production plants like mass-burn incinerators have an indoor tipping floor. Instead of being pushed onto a pit, the waste in an RDF plant is typically fed onto a conveyor, which is either below grade or hopper fed. In some plants, the loader doing the feeding will separate corrugated and bulky items, like carpets.

Once on the conveyor, the waste travels through a number of processing stages, usually beginning with magnetic separation. The processing steps typically include one or more screening stages, using trammel or vibrating screens, shredding or hammer milling of waste with additional screening steps, pelletizing or baling of burnable wastes, and, depending on the local recycling markets and the design of the facility, a manual separation line.

A typical RDF facility scheme is shown below in figure.

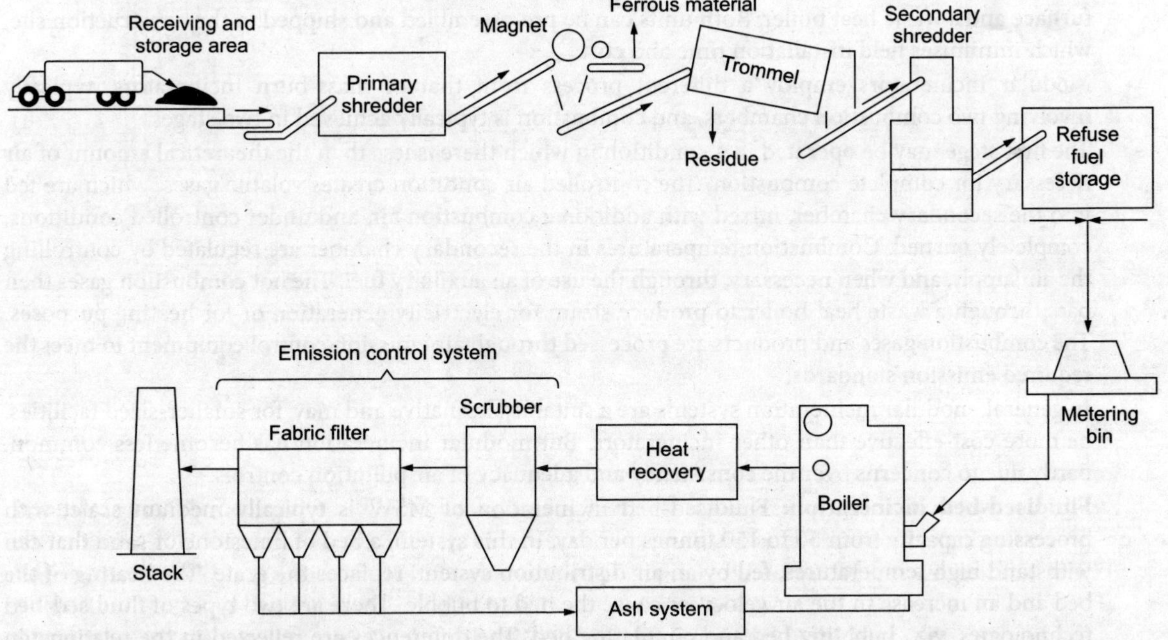

Fig. 8.6: A typical refuse derived fuel (RDF) facility

There are two primary types of RDF systems in operation, and these are:

(*i*) **Shred-and-burn RDF systems:** Shred-and-burn systems are the simplest form of RDF production. The process system typically consists of shredding the MSW to the desired particle size that allows effective feeding to the combustor and magnetic removal of ferrous metal, with the remaining portion delivered to the combustor. There is no attempt to remove other non-combustible materials in the MSW before combustion. This, in essence, is a system with minimal processing and removal of non-combustibles.

(*ii*) **Simplified process RDF systems:** This is a system that removes a significant portion of the non-combustibles. A simplified process system involves processing the MSW to produce an RDF with a significant portion of the non-combustibles removed before combustion. The MSW process removes more than 85% of the ferrous metals, a significant percentage of the remaining non-combustible (*i.e.*, glass, nonferrous metals, dirt, sand, etc.), and shreds the material to a nominal particle top size of 10 to 15 cm to allow effective firing in the combustion unit.

Depending on the type of combustor to be used, a significant degree of separation can be achieved to produce a high-quality RDF (*i.e.*, low ash), which typically results in the loss of a higher percentage of combustibles when compared to systems that can produce a low-quality fuel (*i.e.*, slightly higher ash content) for firing in a specially designed combustor. These types of systems recover over 95% of the combustibles in the fuel fraction.

Modular incineration: Modular incinerator units are usually prefabricated units with relatively small capacities between 5 and 120 tonnes of solid waste per day. Typical facilities have between 1 and 4 units with a total plant capacity of about 15 to 400 tonnes per day. The majority of modular units produce steam as the sole energy product. Due to their small capacity, modular incinerators are generally used in small communities or for commercial and industrial operations.

Their prefabricated design gives modular facilities the advantage of a shorter construction time. Modular combustion systems are usually factory-assembled units consisting of a refractory-lined furnace and a waste heat boiler. Both units can be pre-assembled and shipped to the construction site, which minimises field installation time and cost.

Modular incinerators employ a different process from that of mass-burn incinerators, typically involving two combustion chambers, and combustion is typically achieved in two stages.

The first stage may be operated in a condition in which there is less than the theoretical amount of air necessary for complete combustion. The controlled air condition creates volatile gases, which are fed into the secondary chamber, mixed with additional combustion air, and under controlled conditions, completely burned. Combustion temperatures in the secondary chamber are regulated by controlling the air supply, and when necessary, through the use of an auxiliary fuel. The hot combustion gases then pass through a waste heat boiler to produce steam for electricity generation or for heating purposes. The combustion gases and products are processed through air emission control equipment to meet the required emission standards.

In general, modular incineration systems are a suitable alternative and may, for smaller-sized facilities, be more cost-effective than other incinerators. But modular incineration has become less common, partly due to concerns over the consistency and adequacy of air pollution controls.

Fluidised-bed incineration: Fluidised-bed incineration of MSW is typically medium scale, with processing capacity from 50 to 150 tonnes per day. In this system, a bed of limestone or sand that can withstand high temperatures, fed by an air distribution system, replaces the grate. The heating of the bed and an increase in the air velocities cause the bed to bubble. There are two types of fluidised-bed technologies, viz., bubbling bed and circulating bed. The differences are reflected in the relationship between air flow and bed material, and have implications for the type of wastes that can be burned, as well as the heat transfer to the energy recovery system.

Unlike mass-burn incinerators, fluidised-bed incinerators require fuel preparation. They are generally associated with source separation because glass and metals do not fare well in these systems and also they can successfully burn wastes of widely varying moisture and heat content, so that the inclusion of paper and wood, which are both recyclable and burnable, is not a crucial factor in their operation. Fluidised bed systems are more consistent in their operation than mass burn and can be controlled more effectively to achieve higher energy conversion efficiency, less residual ash and lower air emissions. However, these systems operate efficiently on smaller scales than mass-burn incinerators, which may make them attractive in some situations. For this reason, fluidised-bed technology may be a sound choice for high-recycling cities in developing countries when they first adopt incineration.

Out of most recent Waste to energy technologies adopted in India such as Bio methanation, landfill with gas recovery, gasification/pyrolysis, incineration and composting; incineration is selected as last option.

Q.75. Describe basic types of waste to energy recovery systems following incineration of municipal solid waste (MSW).

Ans. Most of the MSW incineration currently practice energy recovery in the form of steam, which is used either to drive a turbine to generate electricity or directly for heating. In the past, it was common to simply burn MSW in incinerators to reduce its volume and weight, but energy recovery has become more prevalent now.

In waste-to-energy (WTE) plants, heat from the burning waste is absorbed by water in the wall of the furnace chamber, or in separate boilers to produce steam which is used either for heating or to turn turbines to generate electricity. The amount of energy recovered from waste is a function of the amount of waste combusted, energy value of the waste stream and the efficiency of the combustion process.

The three basic types of waste to energy incineration of MSW are described below:

(*i*) **Generation of electricity:** Electricity is the most common form of energy produced and sold from WTE facilities constructed today. By directing the steam produced from a WTE system through a turbine generator, electricity can be produced and sold. A process flow diagram of an incinerator and electrical generation system is shown below in figure.

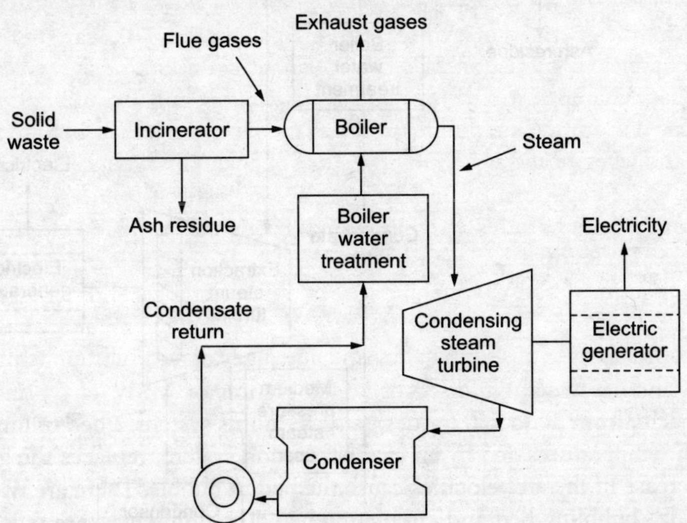

Fig. 8.7: Waste to energy (WTE) recovery system for electrical power generation

Of the electricity produced in incineration facilities, about one-fifth is used at the facility for general operations, and the remainder can be sold.

(*ii*) **Steam generation:** Steam is used widely in a variety of industrial applications. Steam generated in incineration facilities can also be used directly by a customer for manufacturing operations. Steam generated in an incinerator is supplied to a customer through a steam line, and a separate line sometimes returns the condensed steam. It can be used to drive machinery such as compressors, for space heating and generating electricity. Industrial plants, dairies, cheese plants, public utilities, paper mills, tanneries, breweries, public buildings and many other businesses use steam for heating and air conditioning.

When assessing potential markets for steam, it is important to consider a market's proximity to the WTE facility and the quantity of steam produced. Proximity is important because steam cannot usually be economically transported more than 1.5 to 3.0 km.. The WTE facility, therefore, should be as close as possible to the potential market. The advantages of transmitting steam over a long distance to an end user must be weighed against energy losses that will occur in transmission. Installation of a pipeline connecting the facility and the customer can also be prohibitively expensive in certain circumstances. The quantity of steam produced from a given amount of waste will decrease, as the steam temperature and pressure increase, but the equipment using the steam will operate more efficiently. To ensure the continuous availability of high quantity and quality steam, supplementary fuels, may occasionally be used, but this will result in an increase in the operating costs.

(*iii*) **Co-generation:** Co-generation refers to combined production of steam and electricity and can occur in two ways. If the steam conditions (pressure and temperature) required by the customer are less than the incineration plant's design specifications, a turbine-generator is used to produce

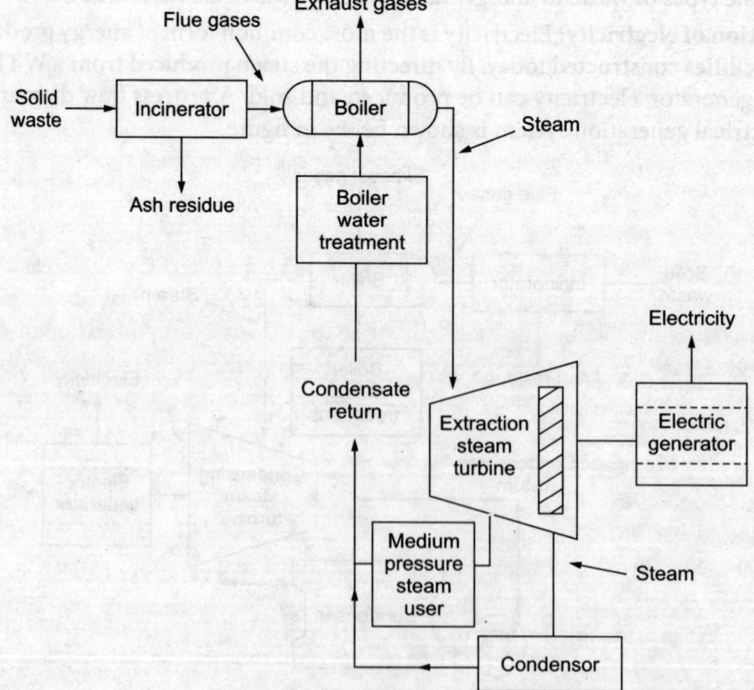

Fig. 8.8: Waste to energy (WTE) recovery system for cogeneration of steam and power

electricity and thus reduce steam conditions to appropriate levels for the customer. If the steam purchaser cannot accept all the steam produced by the facility, the excess can be converted to electricity. In co-generation, high-pressure steam is used first to generate electricity; the steam leaving the turbine is then used to serve the low-pressure steam users for heating purposes.

Co-generation provides greater overall energy efficiency, even though the output of the major energy product – whether electricity or *steam* – may be less than that generated by producing one type of energy alone.

Co-generation allows flexibility, so that seasonal variations in steam demand can be offset by increases in electricity production, and can provide the project a financial base by selling electricity, should the steam customer become unavailable.

Co-generation System for Producing Electricity and Steam *from incineration of MSW is shown in Fig. 8.8.*

The major risks of incinerators are the potential emission of contaminants into the air through exhaust stack or into the water through ash leachate. Proper planning to minimise these problems is essential for a successful incineration programme.

Q.76. What are the various air pollutants emitted from waste incineration plants?

Ans. The operation of the combustion process plays an important role in the formation of pollutants, which are carbon monoxide, NO_x (oxides of nitrogen), hydrocarbons and other volatile organic compounds. It also produces gaseous stream containing dust, acid gases (HCl, SO_x, HF), heavy metals and traces of dioxins. The modern incinerators, however, produce less particulate and gaseous pollutants than the older plants. Also, emissions from incinerators are controlled by a combination of measures that use both the pollution prevention approach and various control equipment.

The main pollutants emitted from incineration plants are described below:

Gaseous pollutants: The various gaseous pollutants formed due to incineration processes are:

Carbon dioxide (CO_2): This is one of the main products of incineration, and the other main product is water. At low concentrations, CO_2 has no short-term toxic or irritating effect, as it is abundant in the atmosphere and necessary for plant life and is not generally considered a pollutant. Nevertheless, due to the high increase in global concentration of CO_2, it has been recognised as one of the greenhouse gases responsible for global warming.

Carbon monoxide (CO): An incomplete combustion of carbon due to the lack of oxygen forms CO. This gas is toxic, as it reacts with the haemoglobin in the blood, causing a decrease of available oxygen to the organisms. This lack of oxygen produces headache, nausea, suffocation and eventually death. Carbon monoxide in the flue gas is used to monitor the incomplete combustion of the other emissions, such as un-burnt hydrocarbons and provide information on the performance of the incinerator.

Sulphur oxides (SO_x): The emission of SO_x is a direct result of the oxidation of sulphur present in solid waste, but other conditions such as the type of incinerator used and its operating conditions also influence its production. Approximately 90% of SO_x emissions are SO_2 and 10% are SO_3. In the atmosphere, most of the SO_2 is transformed into SO_3, which leads to the production of H_2SO_3 (sulphurous acid) and H_2SO_4 (sulphuric acid), increasing the acidity of rain. At high concentrations, it causes eye, nose and throat irritation, and other respiratory problems.

Nitrogen oxides (NO_x): This is predominantly formed during the incineration process. However, they oxidise to NO_2 in the atmosphere. NO_x is formed from two main sources – thermal NO_x and fuel NO_x. In thermal formation, the oxygen and nitrogen react in the air. Fuel NO_x is formed during the reactions between oxygen and nitrogen in the fuel. Nitrogen oxides are important, as they participate in several processes in atmospheric chemistry. They are precursors of the formation of ozone (O_3) and peroxy acetal nitrate (PAN). These photochemical oxidants are responsible for smog formation and cause acid rain.

Particulates: This is formed during the combustion process by several mechanisms. The turbulence in the combustion chambers may carry some ash into the exhaust flow. Other inorganic materials present in the waste volatilise at combustion temperature and later condense downstream to form particles or deposits on ash particles. The main component of fly ash is chemically inert silica; but it may also contain toxic metal and organic substances.

Hydrochloric acid (HCl): Hydrochloric acid results from the high concentration of chlorine containing materials (e.g., some type of plastics like polyvinyl chloride) in solid waste. Chlorine easily dissolves in water to form HCl. Its presence in the gaseous state may increase the acidity of local rain and ground water, which can damage exposed and unprotected metal surfaces, erode buildings and may affect the mobilisation of heavy metals in soil.

Hydrogen fluoride (HF): Hydrogen fluoride is more toxic and corrosive than HCl, although its presence in the emissions from solid waste incinerators occurs in much smaller quantities. It is formed due to the presence of trace amounts of fluorine in the waste.

Heavy metals (Hg, Cd, Pb, Zn, Cu, Ni, Cr): Solid waste contains heavy metals and metallic compounds in the combustible and incombustible fractions. During the incineration process, metals may vaporize directly or form oxides or chlorides at high temperatures in the combustion zone. They condense over other particles and leave the incineration process in the flue gas.

Dioxins and furans: Polychlorinated dibenzo-*p*-dioxins (PCDD) and polychlorinated dibenzofurans (PCDF) have been detected in the emissions from solid waste incinerators. Dioxins can be formed in all combustion processes, where organic carbon, oxygen and chlorine are present. The concern over dioxins and furans has increased after a number of animal studies have shown that for some species, they are carcinogenic and highly toxic, even at very low levels of exposure.

Q.77. Describe the various pollution control flue gas cleaning equipment used in waste incinerators.

Ans. Gas-cleaning equipment employed to carry out the necessary flue gas cleaning in waste incinerator plants are described below:

Electrostatic precipitators (ESP): These are used for particulate control. ESP use electric forces to move the particle flowing out of the gas stream on to the collector electrodes. The particles get a negative charge, when they pass through an ionised field. The electric field that forces the charged particles to the walls comes from discharged electrodes maintained at high voltage in the centre of the flue gas flow lane.

When particles are collected, they must be carefully removed to avoid their reentry into the gaseous stream. This is achieved by knocking them loose from the plates or by intermittent and continuous washing with water. The removal efficiency is more than 99% with a low-pressure drop. Particle size and other physical characteristics such as gas stream temperature, flue gas volume, moisture content, gas stream composition, particle composition and particle surface characteristics affect the performance of the equipment. ESP's come in one or two stage designs and can have different configurations: plate wire, flat plate and tubular. These configurations can be wet or dry, depending on the method of dust collection.

Fabric/bag filters: In fabric filtration, the gas flows through a number of filter bags placed in parallel, leaving the dust captured by the fabric. Extended operation of fabric filters requires periodic cleaning of the cloth surface. After a new fabric goes through a number of cycles of use and cleaning, it forms a residual cake of dust that becomes the filter medium, which is responsible for highly efficient filtering of small particles that characterises fabric filter. They are widely accepted for controlling particulate matter.

The type of cloth fabric limits the temperature of operation of fabric cloth: cotton is least resistant (80°C) while fabric glass is the most resistant (275°C). This temperature requirement makes the use of a cooling system necessary for the gas, before it enters the filter, but it is also necessary that the temperature of the exhaust gas stream be maintained above the dew point because liquid particles block the pores in the fabric very quickly. The major difference between different configurations of fabric filters is the cleaning method used during operating cycles such as shaker, reverse-air and pulse-jet cleaning.

Scrubbers: Scrubbers are used to control particulate matter and acid gases leaving the incinerator. Particulates and gases are removed from the gas stream mainly by absorption and adsorption. The particles are moved to the vicinity of the water droplets and collide with each other. The particles adhere to the liquid media and precipitate to the bottom of the unit containing the dust particle from the gas phase. In addition to removing entrapped particulate matter, scrubbers can also remove gases by absorption and adsorption. Any other type of particulate control equipment does not possess this capability of scrubber. It can remove particles of size 0.1 to 200m efficiently. The various types of scrubbers can be dry, semi-dry or wet, depending on the composition of flue gas.

Q.78. Describe the various environmental concerns related to MSW incineration plants.

Ans. Apart from air pollution, there are other environmental concerns related to waste incineration which are described below:

(*i*) **Water pollution:** Wastewater in an incineration facility can be generated in various forms. These include tipping floor runoff system wash water, ash quench water and water from pollution control systems. For most incineration facilities, wastewater can be recycled in a closed-loop system. In these systems, water from floor drains, ash dewatering, water softener recharge and other process wastewaters are collected and stored in a tank. This water is then reused for ash quenching. Sanitary waste can be directed to municipal sewer systems. In some cases, regulatory authorities may require that the waste stream be pretreated before discharge.

(*ii*) **Land-retained pollution:** Land-retained pollutants originate as stack or fugitive emissions and are of increasing concern. Bioaccumulation and subsequent ingestion from food is an indirect exposure route resulting from land-retained emissions. To provide better understanding of land-retained pollutants, it may be desirable to establish baseline contaminant levels, before incineration plant construction so that changes in those levels throughout the plant's operating lifetime can be monitored.

(*iii*) **Residue disposal:** An incineration facility and its emission control system produce a variety of residues such as large quantity of bottom ash, the unburned and not burnable materials discharged from the incinerator at the end of the burning cycle. The process also produces a lighter emission known as fly ash. Fly ash consists of products in particulate form, which are either produced as a result of the chemical decomposition of burnable materials or are unburned (or partially burned) materials drawn upward by thermal air currents in the incinerator and trapped in pollution control equipment. Fly ash normally comprises only a small proportion (ranging from 10 to 20%) of the total volume of residue from an incineration facility.

Constituents in both ash and scrubber product vary, depending on the materials burned. In systems burning a homogeneous fuel such as coal, oil, or tires, the level of pollutants in residues may be relatively constant. Systems burning a more heterogeneous mixture, such as municipal, industrial, or medical waste may experience wide swings in the chemical composition of residues. The major constituents of concern in municipal waste combustion ash are heavy metals, particularly lead, cadmium and mercury. These metals may impact human health and the environment; if improperly handled, stored, transported, disposed of or reused.

(*iv*) **Noise pollution:** Truck traffic is the greatest source of noise pollution in incineration plant operations. Well-maintained and responsibly operated trucks will help minimise this problem. Local regulations may restrict truck traffic to certain hours of the day and to specified truck corridors. Under these conditions, noise pollution should not be a significant factor. Noise resulting from plant operations and air handling fans associated with the combustion and emission control equipment is also a potential problem. Noise levels are likely to be the highest in front of waste tipping floor doors, ash floor doors and the vicinity of air emission stacks. Walls, fences, trees and landscaped earthen barriers can serve to reduce noise levels.

(*v*) **Aesthetic impact:** Proper site landscaping and building design can help minimise or prevent negative aesthetic impacts. Such impacts are much less problematic, if the facility is sited in an industrial area and not adjacent to residential or commercial areas. Keeping the process building at negative pressure can prevent undesirable odours from escaping outside of *the* building. The use of internal air for combustion in the plant processes will destroy most odours.

Some facilities may emit visible steam or vapour plumes. Smoke, resulting from improper conditions in the combustion chamber, can also be problematic. Air emission stacks and cooling towers may also be unappealing anomalies in the skyline of some areas. If external lights on buildings prove objectionable to neighbours, perimeter lights (on stands) directed towards the plant may be preferable.

Q.79. Describe in detail the solid waste management hierarchy.

Ans. Solid waste is the unwanted or useless solid materials generated from combined residential, industrial and commercial activities. It may be categorized according to its origin (domestic, industrial, commercial, construction or institutional wastes); according to its contents (organic material, glass, metal, plastic paper etc); or according to hazard potential (toxic, non-toxic, flammable, radioactive, infectious etc). Management of solid waste reduces or eliminates adverse impacts on the environment and human health and supports economic development and improved quality of life. A number of processes are involved in effectively managing waste for a municipality. These include monitoring, collection, transport, processing, recycling and disposal of wastes.

Reduce, reuse, recycle: Methods of waste reduction, waste reuse and recycling are the preferred options when managing waste. There are many environmental benefits that can be derived from the use of these methods. They reduce or prevent greenhouse gas emissions, reduce the release of pollutants, conserve resources, save energy and reduce the demand for waste treatment technology and landfill space. Therefore, that these methods should be adopted and incorporated as part of the waste management plan.

Waste reduction and reuse: Waste reduction and reuse of products are both methods of waste prevention. They eliminate the production of waste at the source of generation and reduce the demands for large scale treatment and disposal facilities. Methods of waste reduction include manufacturing products with less packaging, encouraging customers to bring their own reusable bags for packaging, encouraging the public to choose reusable products such as cloth napkins and reusable plastic and glass containers, backyard composting and sharing and donating any unwanted items rather than discarding them. All of the methods of waste prevention require public participation. Also, the government may need to regulate the types and amount of packaging used by manufacturers and make the reuse of shopping bags mandatory.

Recycling: Recycling refers to the removal of items from the waste stream to be used as raw materials in the manufacture of new products. Recycling occurs in three phases: first the waste is sorted and recyclables collected, the recyclables are used to create raw materials. These raw materials are then used in the production of new products.

The sorting of recyclables may be done at the source (*i.e.*, within household or office) for selective collection by the municipality or to be dropped off by the waste producer at a recycling centres. The pre-sorting at the source requires public participation which may not be forthcoming if there are no benefits to be derived. Also, a system of selective collection by the government can be costly. It would require more frequent circulation of trucks within neighborhood to facilitate the waste collection.

Another option is to mix the recyclables with the general waste stream for collection and then sorting and recovery of recyclable materials can be performed by the municipality at a suitable site. The sorting by the municipality has the advantage of eliminating the dependence on the public and ensuring that the recycling does occur. Disadvantage however, is that the value of the recyclable materials is reduced since being mixed in and compacted with other garbage can have adverse effects on the quality of the recyclable material.

Waste collection: Waste from homes is generally collected by local authorities through regular waste collection, or by special collections for recycling. Within hot climates the waste should be collected at least twice a week to control fly breeding, and the harbouring of other pests in the community. Other factors to consider when deciding on frequency of collection are the odours caused by decomposition and the accumulated quantities.

Waste treatment: Waste treatment techniques transform the waste into a form that is more manageable, reduce the volume or reduce the toxicity of the waste thus making the waste easier to dispose of. Treatment methods are selected based on the composition, quantity, and form of the waste material. Some waste treatment methods being used today include subjecting the waste to extremely high temperatures, dumping on land or land filling and use of biological processes to treat the waste. It should be noted that treatment and disposal options are chosen as a last resort to the previously mentioned management strategies of reducing, reusing and recycling of waste.

Solid waste management hierarchy is shown below in figure.

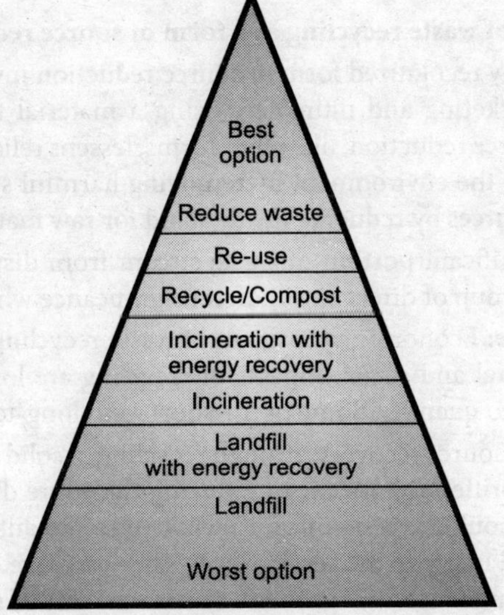

Fig. 8.9: Solid waste management hierarchy

Q.80. What is the purpose and advantages of prevention of waste generation by source reduction?

Ans. Source reduction can serve several purposes, including the following:

(*i*) **Product reuse:** Using reusable products, instead of their disposal equivalents, reduce the amount of materials that are to be managed as wastes. An example of product reuse is the reusable shopping bag.

(*ii*) **Material volume reduction:** Reducing the volume of material used changes the amount of waste entering the waste stream. This helps in controlling the waste generated and its disposal. For example, buying in bulk or using large food containers reduces the amount of packaging waste generated.

(*iii*) **Toxicity reduction:** Source reduction reduces the amount of toxic constituents in products entering the waste stream and reduces the adverse environmental impacts of recycling or other waste management activities. For example, substitution of lead and cadmium in inks (solvent-based to water-based) and paints is a source reduction activity.

(*iv*) **Increased product lifetime:** Source reduction facilitates the use of products with longer lifetime over short-lived alternatives that are designed to be discarded at the end of their useful lives. It encourages a product design that allows for repair and continued use rather than disposal. Manufacturing long-life tyres is a good example of increasing product lifetime.

(*v*) **Decreased consumption:** This refers to the reduced consumption of materials that are not reusable (e.g., using a reusable shopping bag instead of picking up plastic bags from the store). Consumer education about the materials that are difficult to dispose of or are harmful to the environment is essential. Buying practices can thus be altered (e.g., buying in bulk) to reflect environmental consciousness.

Main advantages of source reduction are:

- Reduction in extent of environmental impacts.
- Reduction in resource consumption and generation of pollution.
- It includes producer, consumer, prudent and efficient activities.

Q.81. Describe the significance of waste recycling as a form of source reduction.

Ans. Recycling is the most widely recognized form of source reduction involving the process of separating, collecting, processing, marketing and ultimately using a material that would have otherwise been discarded. This form of source reduction, like other forms, lessens reliance on landfills and incinerators, protects human health and the environment by removing harmful substances from the waste stream and conserves natural resources by reducing the demand for raw materials.

Recycling can divert a significant portion of waste stream from disposal in landfill and combustion facilities. Recycling has a group of direct and indirect significance which is described below:

(*i*) **Economic significance:** Economic assessment of waste recycling is a difficult task as many of the beneficial environmental and social impacts of recycling are long-term and are intangible, and, therefore, are difficult to quantify. Some of the short- and long-term economic benefits are:

- **Cost reduction:** Resource recovery through recycling of solid waste could be of interest to waste management authorities as a means of reducing the waste disposal cost. Any saving in waste management cost could be a significant incentive to the authorities to increase the coverage of service areas and improve the service level. They can save cost from fuel for transportation, operation and maintenance, and generate revenue by sale of recyclables, etc.
- **Employment:** Recycling of waste is a labour intensive activity, and its potential to ease the unemployment problem is high. Enhanced recycling activities, for example, can create an

additional job market for skilled and unskilled workforce, and they can adapt to any of the occupations such as a labourer in recycling business or industry or a dealership.

- **Energy saving:** Use of recyclables in some industrial processes is known to consume less energy than the use of any other raw material. The reduction in energy consumption in one industry could mean its availability for some other industry in need.
- **Reduced health care costs:** Improved health and sanitary conditions in urban areas resulting from indirect benefits of waste recycling can reduce the investment in public health programme.
- **Saving costs for other public utilities:** Enhanced solid waste recycling practices can reduce the frequency of sewer clogging, blocking of natural watercourses and pollution of water bodies. This will benefit the concerned public utilities through reduced cost in cleaning sewers and improved public safety due to blocked sewers and narrowing of natural watercourses.

(*ii*) **Environmental and health significance:** The volume of waste is increasing rapidly because of population growth and economic development. The composition of waste is also changing, leading to waste production with more recyclables. At the same time, polluted waste fractions are increasing because of increasing complex processes being used in industries, and these contribute to environmental degradation. This notwithstanding, recycling helps, among others, in the following ways, to facilitate effective waste management:

- **Improved environment:** The environmental pollution may be due to inadequate SWM as well as due to its effect on other urban infrastructure. Recycling reduces the volume of waste that has to be finally dumped, and thereby causing reduction in pollution at the waste disposal sites. When there is reduction in volume of waste because of its increased reuse, different types of pollution (e.g., water, air and land) will get abated.
- **Natural resource conservation:** Industries with natural products as their raw material for production are depleting natural resources. Use of more and more recyclable solid wastes in industrial production will relieve the tremendous pressure on these precious resources. For example, recycling of waste paper means a lower demand for wood, which means less cutting of trees and an enhanced possibility for sustainable use of the forest. Using recyclable items in the production process would reduce the demand for energy as well.

(*iii*) **Social significance:** People engaged in waste collection activities are normally of low social and economic standing. Different groups of people engaged in waste recycling have a hierarchical social and economic status, in which, processors are at the top of the hierarchy followed by waste dealers and wholesalers, waste buyers and waste collectors in that very order, while scavengers are at the bottom. Although there is this social and economic hierarchy within the waste recycling business, the overall social esteem of waste recycling operators is low.

A formal recycling arrangement will help promote the social esteem of waste workers and facilitate their upward social mobility due to increased earning. In addition, the improved recycling activity will increase the economic value of the waste and will reduce waste scavenging activity providing opportunity for scavengers to switch to a more socially acceptable occupation. In short, institutionalised recycling programmes will help remove the stigma associated with waste scavenging and transform it to an economic enterprise.

Q.82. Describe various elements of a waste recycling programme.

Ans. Recycling includes a mix of strategies, ranging from single material drop off centres to large scale, centralised processing facilities. Elements of a recycling programme include source separation, buy back.curbside (kerbside) collection, storage and collection of recyclables, collection vehicle and processing equipment, material resource facilities and full stream processing which are described below:

Source separation: Source separation refers to the segregation of the recyclable and reusable materials at the point of generation. This requires that several designated materials be separated into their own specific containers, while other programmes use only two containers – one for the storage of mixed recyclables and the other for regular wastes.

Drop-off/buy-back: A *drop-off* programme requires residents to separate the recyclable materials and bring them to a specified drop-off or collection centre. However, we must take into account local conditions when designing a collection programme. For a community that does not provide a curbside pickup, for example, educating and encouraging citizens to deliver materials to a drop-off site may be all that is needed. A recycling centre can be established at the same location where residents deliver waste. Mobile recycling drop-off trailers can also be used. Drop-off recycling, however, is less convenient than curbside pickup.

Buy-back refers to a drop-off programme that provides monetary incentives to participate. In this type of programme, the residents are paid back for their recyclable material directly or indirectly through the reduction in collection and disposal fees. Establishing a buy-back centre (*i.e.*, a place where recyclables are purchased) may help induce citizens to recycle. Some buy back centres purchase some materials depending on current market conditions. Mobile buy-back operations can serve some areas, purchasing recyclables from small communities or from neighborhoods of large metropolitan areas on a regular schedule.

Curbside programme: In a curbside system, source separated recyclables are collected separately from regular refuse from the curbside, alley, or commercial facility. Curbside programmes vary greatly from community to community. Some programmes require residents to separate different materials that are stored in their own containers and collected separately. Other programmes use only one container to store recyclables or two containers, one for paper and the other for heavy recyclables (e.g., glass aluminium, etc.).

Storage and collection of recyclables: Collection of source-separated materials is a necessary component of recycling programme. Establishing a collection system for source-separated materials will require more careful planning than regular trash collection. Some principles of sound recyclables storage and collection programme include:

- **Resident convenience:** The easier it is for residents to separate materials, the higher the participation and recovery rates will be.
- **Collection crew convenience:** The system should be convenient for collection crews. For example, loading and sorting activities should be as simple as possible.
- **Cost effectiveness:** Equipment and procedures must be designed to maximize collection crew and vehicle productivity.
- **Integrity of materials:** The storage and collection system should keep recyclables in the best shape possible. It should be properly handled, dry and contaminant free.

How residents store recyclables in a household and curb has a direct impact on the success of a recycling programme. The storage container should provide a handy way to store materials until collected, and it must be easy for the collector to distinguish recyclables from garbage.

Collection vehicles for recycling: Collection vehicles that are designed specifically for collecting recyclables have several storage bins, which can be easily loaded and often equipped with automatic container-tipping devices. A dedicated, closed-body collection vehicle for recycling with sufficient capacity offers advantages such as easy loading and unloading, flexible compartments and protection from weather. Of course, this warrants a substantial initial investment.

Processing equipment for recycling: Recycling involves a number of processing techniques and these processes require different equipment. Some of the special equipment used in recycling are:

- **Balers:** Balers can be used to densify many types of materials including paper, cardboard, plastics and cans. Balers can improve space utilisation and reduce material transportation costs.
- **Can densifiers:** Can crushers are used to densify aluminium and steel cans prior to transport.
- **Glass crushers:** These are used to process glass fraction separated by colour and break it into small pieces. This crushed material is called cullet, and can be reprocessed into new glass products.
- **Magnetic separators:** These are used to remove ferrous material from a mixture of materials.
- **Wood grinders:** These are chippers and are used to shred large pieces of wood into chips that can be used as mulch or as fuel.
- **Scales:** These are used to measure the quantity of materials recovered or sold.

Material recovery facilities (MRF): MRF is a centralised facility that receives, separates, processes and markets recyclable material. It can be operated with both drop off and curbside programmes. The primary advantage of MRF is that it allows materials directly from the municipalities and processes them uniformly. It is generally designed to handle all type of recyclables. Implementation of MRF in a municipality depends upon a number of factors as given below:

- **Market demand:** When additional processing is required, MRF is more useful as buyers may have certain material specifications.
- **Separate collection:** In systems that require residents to separate their recyclables, intermediate separation and processing is required.
- **Number of different recyclables:** In general, a MRF will be more beneficial when a large number of different recyclables are collected.
- **Quantities of materials:** Because MRF involves substantial capital and operating costs (e.g., buildings, equipment and labour), it is expected to handle a significant amount of materials to justify its operation.

Full stream processing: This is a high technology separation technique, which processes all components of municipal waste. The materials recovered by this process tend to be of lower quality than those recovered or source separated in MRF because the former is a mix of various types of wastes. To achieve a better quality, the materials obtained through the full stream processing must be cleaned, which is a costly process.

However, this technique remains attractive because it does not require source separation, and it is used in the following applications:

(*i*) **Refuse derived fuel (RDF) preparation:** In this application, it is used to extract the combustible portion of municipal waste.

(*ii*) **Municipal waste composting:** In this application, it is used to concentrate the compostable portion of municipal solid waste. This is sometimes performed as part of RDF preparation.

(*iii*) **Material recovery:** In this application, it is used to recoverand resell certain materials, and thereby making material recovery a recycling technology as well.

In full stream processing, depending on the facility design, the materials are separated either mechanically or by hand, and size and weight are the main characteristics used to separate the materials. For example: When the material is dumped, oversized materials such as furniture, etc., are removed; rotating screens are used to separate materials of different sizes (small and large); ferrous material is extracted using a magnet system; air classifier is used to separate the lighter material; light materials including plastic and paper are further processed into RDF; heavy fraction is mechanically or manually sorted to recover saleable materials such as cardboard, etc.

Q.83. Describe the salient features of municipal solid waste (MSW) rules, 2000.

Ans. Main features of Indian MSW rules,2000 are given below:

- **Responsibility of the urban local body**
  - — The rule designates the urban local bodies as sole responsible to manage solid waste in their area and dictates that "*within the territorial area of the municipality, be responsible for the implementation of the provisions of these rules, and for any infrastructure development for collection, storage, segregation, transportation, processing and disposal of municipal solid wastes*"
  - — Prohibits waste to be exposed to open atmosphere.
  - — Prohibits waste disposal by burning (garbage, dry leaves) in open;
- **Collection of MSW**
  - — Mandates collection of waste from slums and open squatter areas, hotels/restaurants/office complexes and commercial areas.
  - — Avoid Manual handling of waste, and ensure that the waste is collected and removed from the municipal area daily.
  - — Vehicles used for transportation of wastes to be covered.
  - — io degradable waste and non bio-degradable waste must be collected in separate bins from source. Waste bins for biodegradable waste shall be painted 'Green', those for storage of recyclable wastes shall be painted 'White' and those for storage of other wastes shall be painted 'Black':
  - — Construction/demolition wastes/debris to be separately collected and disposed off following proper norms.
  - — Stray animals are to be kept out from the waste storage facilities.
- **Processing of MSW:** Recover recyclables from the waste mass before treating for biodegradable portion of the waste.
  - — Treatment of organic waste through biodegradation such as vermi composting, mechanical composting by windrow method or any other suitable methods such as anaerobic digestion etc as approved by Central Pollution Control Board (CPCB) may be adopted.
  - — In case the municipal body is engaged in any other treatment technology, such as incineration, energy recovery from waste etc. it must be duly approved by CPCB.
  - — *Chlorinated plastics should not to be incinerated.*
- **Disposal of MSW:**
  - — Municipal body to develop scientifically designed landfill as disposal facility for residues out of waste processing facilities, as well as pre-processing rejects or unprocessed mixed waste (applicable if the waste is not fit for any treatment) in a scientifically designed sanitary landfill for a long term of 20-25 years.
  - — Site suitability criteria to be ensured for selection of landfill sites.
  - — Land filling of mixed waste must be avoided, unless the waste is found unsuitable for waste processing. Under unavoidable circumstances or till installation of alternate facilities, land-filling shall be done following proper norms.
- **Monitoring of pollution:**
  - — Municipality to take adequate pollution prevention steps for all its waste management and handling units
  - — Measures to reduce air pollution in case the unit has a waste to energy units.
  - — Environment monitoring (ground and surface water, air quality) for waste handling site.

- **Awareness programs:**
  - — The municipality to be engaged in encouraging citizens, by organizing awareness programs for segregation of waste at source and promote recycling or reuse of segregated materials through community participation program involving representatives of local resident welfare associations, community based organizations and non-governmental organizations.

Q.84. What is the role of municipal body towards bio-medical waste management?

Ans. Hospital waste/Bio-medical waste is generated during the diagnosis, treatment, or immunization of human beings or animals or in research activities in these fields or in the production or testing of biological waste. Hospital/Bio-medical waste must be managed through "***Bio-medical waste (Management and Handling) Rules, 1998.***

The biomedical waste management rule directs biomedical waste incineration ash to be disposed into municipal landfills. Also, the municipal body must be a member of the '*Advisory Committee*' in the state to advice on steady implementation of biomedical waste management and handling in their areas.

Q.85. What is the role of municipal body towards e-waste management?

Ans. E-waste or electronic waste means waste from electrical and electronic equipment, whole or in part or rejects from their manufacturing and repair process, which are intended to be discarded. The ***E-waste (Management and Handling) Rules 2011*** has been published by the Ministry of Environment and Forest.

Q.86. Discuss the problems related to plastic waste management.

Ans. Problems related to plastic waste management are listed below:

- The plastic content of the municipal waste is picked up by rag pickers for recycling either at primary collection centers or at dumpsites. Since the rag-picking sector is not formalized, not all the recyclables, particularly plastic bags, get picked up and are found littered everywhere.
- Littering is a very common phenomenon in India. One of the offshoots of littering is the choking of drains, streams, etc.
- Plastic films, bags are not permeable, and so they tend to hold other type of wastes thus blocking the way. This gives rise to flooding of the streets in the urban low-lying areas with wastewater emanating foul smell and causing breakthrough of serious health hazards.
- Recently, cow deaths have been reported due to the consumption of scattered plastic bags along with the organic matter.
- Plastics are recycled mostly in factories, which do not have adequate technologies to process them in a safe manner. This exposes the workers to toxic fumes and unhygienic conditions.
- Dioxin, a highly carcinogenic and toxic by-product of the manufacturing process of plastics, is one of the chemicals believed to be passed on through breast milk of the mother to the nursing infant. Burning of plastics, especially PVC releases this dioxin and also furan into the atmosphere.
- Since toxic dyes and chemicals are used as additives during the recycling, the workers engaged in the recycling of plastic are constantly exposed to various toxic compounds.
- Polybag recycling is carried out in shanties, this problem is compounded due to poor ventilation, as workers find themselves inhaling contaminated air. Indian collection sector employ children below the age of 15 to collect them because of the low wages to be paid to the child and the ease of availability of child labour.
- Backyard smelters and plastic recycling units dot India's suburban/urban sites, taking lead battery scrap and plastic waste imported from developed countries. The dangerous toxins emitting from the smelters affects human, animal and plant life.

- The current situation is that the plastic recycling in the country is creating more problems and with the influx of plastic waste import it is getting aggravated. If imported, India should also import the technology along with the waste.
- Although plastics contribute only about 7% by weight to MSW, they may contribute 15% or more to the total heat content of MSW. Hydrogen chloride gas is emitted during combustion of polyvinyl chloride (or other chlorinated polymers), and may result in corrosion of municipal waste combustor internal surfaces.

Q.87. Describe the environmental issues related to disposal of plastic waste.

Ans. Plastic products have become an integral part in our daily life as a basic need. Its production crosses the 150 million tonnes per year globally. In India approximately 15 Million tonnes plastic products are consumed every year. Its broad range of application is in packaging films, wrapping materials, shopping and garbage bags, fluid containers, clothing, toys, household and industrial products, and building materials. Plastics never degrades and remains on landscape for several years. The recycled plastics are more harmful to the environment than the virgin products due to mixing of colour, additives, stabilizers, flame retardants etc.

The recycling of a virgin plastic material can be done 2-3 time only, because, after every recycling, the strength of plastic material is reduced due to thermal degradation.

Environmental issues on disposal of plastic waste:

Indiscriminate littering of unskilled recycling/reprocessing and non-biodegradability of plastic waste raises the following environmental issues:

- During polymerization process fugitive emissions are released.
- During product manufacturing various types of gases are released.
- Indiscriminate dumping of plastic waste on land makes the land infertile due to its barrier properties.
- Burning of plastics generates toxic emissions such as carbon monoxide, chlorine, hydrochloric acid, dioxin, furans, amines, nitrides, styrene, benzene, 1, 3-butadiene, CCl4, and acetaldehyde.
- Lead and cadmium pigments, commonly used in LDPE, HDPE and PP as additives are toxic and are known to leach out.
- Non-recyclable plastic wastes like multilayer, metalised pouches and other thermoset plastic poses disposal problems.
- Sub-standard plastic carry bags, packaging films (< 40μ) etc. pose problem in collection and recycling.
- Littered plastics give unaesthetic look in the city, choke the drain and may cause flood during monsoon.
- Garbage mixed with plastics interferes in waste processing facilities and also cause problems in landfill operations.
- Recycling industries operating in non-conforming areas are posing threat to environment to unsound recycling practices.

Disposal of plastic waste is a serious concern in India and such no technology has been validated, however, several experiments have been conducted on reuse of plastic waste in road construction, co-processing of plastic waste in cement kilns. Currently, worldwide accepted technology used for the plastic disposal is incineration, however, due to poor maintenance of incinerators, it releases several harmful gases including dioxins and furans in case of chlorinated and brominated plastic waste therefore, raising several environmental issues. In India, for safer disposal of plastic waste, various technologies have been experimented such as utilization of plastic waste in road construction, co-processing of plastic waste in cement kilns.

Q.88. What are activities involved in solid waste management?

Ans. Activities involved are:

- waste generation
- onsite handling, storage and processing
- collection
- transfer and transport
- processing and recovery and disposal

Q.89. What is meant by commercial/industrial waste collection services?

Ans. The use of large movable and stationary containers and also large stationary compactors are done in this method. Compactors are used to compress the material directly into the large containers or in the form of bales which are then placed in the large containers

Q.90. What is hauled container waste collection system?

Ans. Waste collection system in which containers used for storage of waste are hauled to the processing, transfer, emptied and returned to either their original location or some other location are defined as hauled container system.

Q.91. What is stationary container waste collection system?

Ans. The collection system in which the containers are used for storage of waste remains at the point of waste generation except when moved for collection are defined as stationary container system.

Q.92. Mention the advantages of composting.

Ans. The entire process involving both the separation and bacterial conversion of the organic solid waste is called composting. Its advantages include:

- Compost contains nitrogen essential for plant growth.
- Used for agricultural field.
- Used for recycling.
- Increasing water retention.
- Increases ion exchange capacity of soil.

Q.93. Describe briefly the sources, effects and management of solid waste.

Ans. Higher standards of living of ever-increasing population have resulted in an increase in the quantity and variety of solid waste generated. Management of solid waste is very important in order to minimize the adverse effects of solid wastes. Solid waste can be classified as municipal, industrial, agriculture, medical,and mining waste and sewage sludge.

Sources of urban and industrial wastes: Urban waste consists of medical waste from hospitals; municipal solid wastes from homes, offices, markets, small cottage units, and horticulture waste from parks, gardens, orchards etc.

- **Waste from homes** contains a variety of discarded materials like polyethylene bags, empty metal and Aluminium cans, scrap metals, glass bottles, waste paper, diapers, cloth/rags, food waste etc.
- **Waste from shops** mainly consists of waste paper, packaging material, cans, bottles, polyethylene bags, peanut shells, eggshells, tea leaves, etc.
- **Bio medical waste** includes anatomical wastes, pathological wastes, infectious wastes, etc.
- **Construction/demolition waste** includes debris and rubbles, wood, concrete, etc.
- **Horticulture waste and waste from slaughter houses** include vegetable parts, residues and remains of slaughtered animals, respectively.

The urban solid waste materials that can be degraded by micro-organisms are called biodegradable wastes. Examples of this type of waste are vegetable wastes, stale food, tea leaves, egg shells, peanut shells, dry leaves, etc. Wastes that cannot be degraded by micro-organisms are called **non-biodegradable wastes e.g.,** polyethylene bags, scrap metal, glass bottles, etc.

- **Industrial wastes:** Industrial wastes consists of a large number of materials including factory rubbish, packaging material, organic wastes, acids, alkalis and metals etc. During some industrial processing large quantities of hazardous and toxic materials are produced. The main sources of industrial wastes are chemical industries, metal and mineral processing industries. Radioactive wastes are generated by nuclear power plants. Thermal power plants produce fly ash in large quantities. Solid wastes from other types of industries include scrap metal, rubber, plastic, paper, glass, wood, oils, paints, asphalt, tars, dyes, scrap leather, ceramics. Abrasives slag, heavy metals, asbestos, batteries.

Effects of solid wastes: Municipal solid wastes heap up on the roads due to improper disposal system. People clean their own houses and dump in their immediate surroundings which affects the community including themselves. This type of dumping allows biodegradable materials to decompose under uncontrolled and unhygienic conditions. This produces foul smell and breeds various types of insects and infectious organisms besides spoiling the aesthetics of the site.

Industrial solid wastes are sources of toxic metals and hazardous wastes, which may spread on land and can cause changes in physico chemical and biological characteristics thereby affecting productivity of soils. Toxic substances may leach or percolate to contaminate the ground water.

In refuse mixing the hazardous wastes are mixed with garbage and other combustible wastes. This makes segregation and disposal more difficult and riskier. Various types of wastes like cans, pesticides, cleaning solvents, batteries, radioactive materials, and plastics are mixed up with paper, scraps and other non-toxic materials which could be recycled.

Management of solid waste: For waste management we stress on 'three R's' - Reduce, reuse and recycle before destruction and safe storage of wastes.

(*i*) **Reduction in use of raw materials:** Reduction in the use of raw materials will correspondingly decrease the production of waste. Reduced demand for any metallic product will decrease the mining of their metal and cause less production of waste.

(*ii*) **Reuse of waste materials:** The refillable containers which are discarded after use can be reused. Villagers make casseroles and silos from waste paper and other waste materials. Making rubber rings from the discarded cycle tubes which are used by the newspaper vendors, instead of rubber bands, reduces the waste generation during manufacturing of rubber bands. Because of financial constraints poor people reuse their materials to the maximum.

(*iii*) **Recycling of materials:** Recycling is the reprocessing of discarded materials into new useful products.

- Formation of some old type products e.g., old Aluminium cans and glass bottles are melted and into recast new cans and bottles.
- **Formation of new products:** Preparation of cellulose insulation from paper, preparation of fuel pellets from kitchen waste, Preparation of automobiles and construction material from steel cans.

The process of reducing, reusing and recycling saves money, energy, raw materials, land space and also reduces pollution. Recycling of paper will reduce cutting of trees for making fresh paper. Reuse of metals will reduce mining and melting of ores for recovery of metals from ores and prevent pollution.

For **discarding wastes** the following methods can be adopted:

(*i*) **Sanitary landfill:** In a sanitary landfill, garbage is spread out in thin layers, compacted and covered with clay or plastic foam. In the modern landfills the bottom is covered with an impermeable liner, usually several layers of clay, thick plastic and sand. The liner protects the ground water from being contaminated due to percolation of leachate. Leachate from bottom is pumped and sent for treatment. When land is full it is covered with clay, sand, gravel and top soil to prevent seepage of water. Several well are drilled near the landfill site to monitor if any leakage is contaminating ground water.

(*ii*) **Composting:** Due to shortage of space for landfill in bigger cities, the biodegradable yard waste is allowed to degrade or decompose in an oxygen rich medium. A good quality nutrient rich and environmental friendly manure called compost is formed which improves the soil conditions and fertility.

(*iii*) **Incineration:** Incinerators are burning plants capable of burning a large amount of solid waste materials at high temperature. The initial cost is very high. During incineration high levels of toxic dioxins, furans, lead and cadmium may be emitted with the fly ash of incinerator. Dioxin level may reach many times more than in the ambient environment, waste landfill plant is an engineered facility used for disposing of solid wastes on land or within the earth 's mantle without creating nuisance or hazard to public health or safety, such as breeding of rodents and insects and contamination of groundwater.

Q.94. Describe six functional elements of municipal solid waste management.

Ans. The activities associated with the management of municipal solid wastes from the point of generation to final disposal is grouped into the six functional elements as described below:

(*a*) **Waste generation:** Waste generation encompasses activities in which materials are identified as no longer being of value (in their present form) and are either thrown away or gathered together for disposal. Waste generation is, at present, an activity that is not very controllable. In the future, more control is likely to be exercised over generation of wastes. Reduction of waste at source, although not controlled by solid waste managers, is now included in system evaluations as a method of limiting the quantity of waste generated.

(*b*) **Waste handling and sorting, storage, and processing at the source:** Waste handling and sorting involves the activities associated with management of wastes until they are placed in storage containers for collection. Handling also encompasses the movement of loaded containers to the point of collection. Sorting of waste components is an important step in the handling and storage of solid waste at the source. For example, the best place to separate waste materials for reuse and recycling is at the source of generation. Households are becoming more aware of the importance of separating newspaper and cardboard, bottles/glass, kitchen wastes and ferrous and non-ferrous materials. On site storage is of primary importance because of public health concerns and aesthetic consideration. Unsightly makeshift containers and even open ground storage, both of which are undesirable, are seen at many residential and commercial sites. The cost of providing storage for solid wastes at the source is normally borne by the household in the case of individuals, or by the management of commercial and industrial properties. Processing at the source involves activities such as backyard waste composting.

(*c*) **Waste collection:** Functional element of waste collection includes not only the gathering of solid wastes and recyclable materials, but also the transport of these materials, after collection, to the location where the collection vehicle is emptied. This location may be a material processing facility, a transfer station, or a landfill disposal site.

(*d*) **Waste sorting, processing and transformation:** The recovery of sorted materials, processing of solid waste and transformation of solid waste that occurs primarily in locations away from the source of waste generation are encompassed by this functional element. Sorting of mixed wastes usually occurs at a materials recovery facility, transfer stations, combustion facilities, and disposal sites. Sorting includes the separation of bulky items, separation of waste components by size using screens, manual separation of waste components, and separation of ferrous and non-ferrous metals. Waste processing is undertaken to recover conversion products and energy. The organic fraction of Municipal Solid Waste (MSW) can be transformed by a variety of biological and thermal processes. The most commonly used biological transformation process is aerobic composting. The most commonly used thermal transformation process is incineration. Waste transformation is undertaken to reduce the volume, weight, size or toxicity of waste without resource recovery. Transformation may be done by a variety of mechanical (e.g., shredding), thermal (e.g., incineration without energy recovery) or chemical (e.g., encapsulation) techniques.

(*e*) **Waste transfer and transport:** It involves two steps: (i) the transfer of wastes from the smaller collection vehicle to the larger transport equipment and (ii) the subsequent transport of the wastes, usually over long distances, to a processing or disposal site. The transfer usually takes place at a transfer station.

(*f*) **Waste disposal:** The disposal of wastes by landfilling or uncontrolled dumping is the ultimate fate of all solid wastes, whether they are residential wastes collected and transported directly to a landfill site, residual materials from Materials Recovery Facilities residue from the combustion of solid waste, rejects of composting, or other substances from various solid waste processing facilities. A municipal solid waste landfill plant is an engineered facility used for disposing of solid wastes on land or within the earth's mantle without creating nuisance or hazard to public health or safety, such as breeding of rodents and insects and contamination of groundwater.

Q.95. Describe the hierarchy of solid waste management options.

Ans. Current thinking on the best methods to deal with waste is centred on a broadly accepted 'hierarchy of waste management' (arrangement in order of rank) which gives a priority listing of the waste management options available. The hierarchy usually adopted is (*a*) waste minimisation/reduction at source, (*b*) recycling, (*c*) waste processing (with recovery of resources *i.e.*, materials (products) and energy), (*d*) waste transformation (without recovery of resources) and (*e*) disposal on land (landfilling). The highest rank of the ISWM hierarchy is waste minimisation or reduction at source, which involves reducing the amount (and/or toxicity) of the wastes produced. Reduction at source is first in the hierarchy because it is the most effective way to reduce the quantity of waste, the cost associated with its handling, and its environmental impacts.

The second highest rank in the hierarchy is recycling, which involves (*a*) the separation and sorting of waste materials, (*b*) the preparation of these materials for reuse or reprocessing and (*c*) the reuse and reprocessing of these materials. Recycling is an important factor which helps to reduce the demand on resources and the amount of waste requiring disposal by landfilling.

The third rank in the ISWM hierarchy is waste processing which involves alteration of wastes to recover conversion products (e.g., compost) and energy. The processing of waste materials usually results in the reduced use of landfill capacity.

Transformation of waste, without recovery of products or energy, may have to be undertaken to reduce waste volume (e.g., shredding and baling) or to reduce toxicity. This is usually ranked fourth in the ISWM hierarchy.

Ultimately, something must be done with (*a*) the solid wastes that cannot be recycled and are of no further use, (*b*) the residual matter remaining after solid wastes have been pre-sorted at a materials recovery facility and (c) the residual matter remaining after the recovery of conversion products or energy.

Landfilling is the fifth rank of the ISWM hierarchy and involves the controlled disposal of wastes on or in the earth's mantle. It is by far the most common method of ultimate disposal for waste residuals. Landfilling is the lowest rank in the ISWM hierarchy because it represents the least desirable means of dealing with wastes.

Q.96. Describe various methods of waste minimization.

Ans. Waste minimisation or reduction at source is the most desirable activity, since the community does not incur expenditure for waste handling, recycling and disposal of waste that is never created and delivered to the waste management system. To reduce the amount of waste generated at the source, the methods are:

- the adoption of industry standards for product manufacturing and packaging that use less material.
- the passing of laws that minimise the use of virgin materials in consumer products.
- the levying of cess/fees for waste management services that penalise generators in case of increase in waste quantities.

Modifications in product packaging standards can result in reduction of waste packaging material or use of recyclable materials. Minimisation of use of virgin raw materials by the manufacturing industry promotes substitution by recycled materials. Sorting at source, recycling at source and processing at source (e.g., yard composting) help in waste minimisation. One waste management strategy used in some communities in developed countries is to charge a variable rate per can (or ton) of waste, which gives generators a financial incentive to reduce the amount of waste set out for collection. Issues related to the use of variable rates include the ability to generate the revenues required to pay the costs of facilities, the administration of a complex monitoring and reporting network for service, and the extent to which wastes are being put in another place by the generator and not reduced at source.

Q.97. What is the relationship between climate change and solid waste?

Ans. Even before a material or product becomes solid waste, it goes through a long cycle that involves removing and processing raw materials, manufacturing the product, transporting the materials and products to markets, and using energy to operate the product. Each of these activities has the potential to generate greenhouse gas emissions through one or more of the following means:

- **Energy consumption:** Extracting and processing raw materials, manufacturing products, and transporting materials and products to markets all generate greenhouse gas emissions by consuming energy from fossil fuels.
- **Methane emissions:** When organic waste decomposes in landfills, it generates methane, a greenhouse gas.
- **Carbon storage:** Trees absorb carbon dioxide, a greenhouse gas, from the air and store it in wood through carbon sequestration. Waste prevention and recycling of wood and paper products allow more trees to remain standing in the forest, where they can continue to remove carbon dioxide from the air, which helps minimize climate change impacts.

Different wastes and waste management activities have varying impacts on energy consumption, methane emissions, and carbon storage. For example, recycling reduces greenhouse gas emissions by preventing methane emissions from landfills or open dumps and by preventing the consumption of energy for extracting and processing raw materials. Communities looking for ways to help prevent climate change can start by implementing an integrated solid waste management program.

Carbon dioxide is released to the atmosphere when solid waste, fossil fuels (oil, natural gas, and coal), and wood and wood products are burned. **Methane** is emitted during the production and transport of coal, natural gas, and oil; the decomposition of organic wastes in municipal solid waste landfills; and by livestock. **Nitrous oxide** is emitted during agricultural and industrial activities, as well as during the combustion of solid waste and fossil fuels. Each greenhouse gas differs in its ability to trap heat in the atmosphere. Methane traps over 21 times more heat than carbon dioxide, and nitrous oxide absorbs 310 times more than carbon dioxide. The higher the heat trapping potential of the greenhouse gas, the greater the impact on climate change. Efforts to decrease emissions of these gases help reduce climate change impacts.

Q.98. What can you do on personal level to reduce solid waste?

Ans. Following actions can help reduce solid waste:

- Carry your own cloth or jute bag when you go shopping.
- Say no to all plastic bags as far as possible.
- Reduce the use of paper bags also.
- Reuse the soft drinks poly bottles for storing water.
- Segregate the waste in the house – keep two garbage bins and see to it that the biodegradable and the non-biodegradable waste is put into separate bins and dispose off separately.
- Dig a compost pit in your garden and put all the bio degradable into it.
- See to it that all garbage is thrown into the municipal bin as the collection is generally done from there.
- When you go out do not throw paper and other wrappings or even left over food here and there, make sure that it is put in the correct place that is into a dustbin.
- Not to throw the waste/litter on the streets, drains, open spaces, water bodies etc.
- Storage of organic/bio-degradable and recyclable waste separately at source.
- Community storage/collection of waste in flats, multi-storied buildings, societies, commercial complexes, etc.
- Managing excreta of pet dogs and cats appropriately.
- Waste processing/disposal at a community level.
- Pay adequately for the services provided.
- Public education to be organized.

Q.99. Describe the role of rag pickers in solid waste management and occupational hazards associated with it.

Ans. Rag pickers are the people who are actually going through the garbage bins, dumping sites to pick out the 'rags'. These rag pickers: women, children, and men from the lowest rung in the society, are a common sight in most cities and towns around the country. Rag picking is considered the most menial of all activities and it is people who have no other alternative that are generally driven to it. Rag pickers contribute a great deal towards waste management as they scavenge the recyclable matter thereby saving the municipality of the cost and time of collecting and transporting this to the dumps.

They are one of the focal points for the recycling of waste. They are the persons who, in spite of all the dangers that they face, goes on relentlessly picking through the garbage bin, looking for waste that could be useful to them. They sell all the material they picked to the whole sellers and retailers who in turn sell it to the industry that uses this waste matter as raw material. The main items of collection are plastics, paper, bottles, and cans.

While picking through waste, the rag pickers puts themself at a great risk and is always prone to disease as the waste that they handle can be infectious. We can indirectly help the rag pickers by carefully segregating the waste that is generated at our homes, thereby facilitating their search for materials that are useful to them. They will not have to scavenge in the bins/yards for long hours.

Occupational hazards associated with waste handling Infections:

- Skin and blood infections resulting from direct contact with waste, and from infected wounds.
- Eye and respiratory infections resulting from exposure to infected dust.
- Different diseases that results from the bites of animals feeding on the waste.
- Intestinal infections that are transmitted by flies feeding on the waste.

Q.100. Describe the actions to be taken for waste management at source.

Ans. Waste management at source is given below in table.

Table 8.5: Solid waste management at source

Source of waste generation	Action to be taken
Household	• Not to throw any solid waste in the neighborhood, on the streets, open spaces, and vacant lands, into the drains or water bodies. • Keep food waste/biodegradable waste in a non-corrosive container with a cover (lid). • Keep dry, recyclable waste in a bin or bag or a sack. • Keep domestic hazardous waste if and when generated separately for disposal at specially notified locations.
Multi-storeyed buildings, commercial complexes, private societies	• Provide separate community bin or bins large enough to hold food/biodegradable waste and recyclable waste generated in the building or society. • Direct the members of the association to deposit their waste in community bin.
Slums	• Use community bins provided by local body for deposition of food and biodegradable waste.
Shops, offices, institutions, etc	• If situated in a commercial complex, deposit the waste in bins provided by the association.
Hotels and restaurants	• The container used should be strong, not more than 100 litre in size, should have a handle on the top or handles on the sides and a rim at the bottom for easy handling.
Vegetableand Fruit Markets	• Provide large containers, which match with transportation system of the local body. • Shop keepers not to dispose of the waste in front of their shops or open spaces. Deposit the waste as and when generated into the large container placed in the market.
Meat and fish markets	• Not to throw any waste in front of their shops or open spaces around. Keep non-corrosive container/containers not exceeding 100-litre capacity with lid handle and the rim at the bottom and deposit the waste in the said containers as and when generated. • Transfer the contents of this container into a large container provided by the association.
Street food vendors	• Not to throw any waste on the street, pavement or open spaces. Keep bin or bag for the storage of waste that generates during street vending activity. • Preferably have arrangements to affix the bin or bag with the hand cart used for vending.
Marriage halls, community halls, kalyana mandapas	• Not to throw any solid waste in their neighborhood, on the streets, open spaces, and vacant lands, into the drains or water bodies. • Provide a large container with lid which may match with the transportation system of the local body and deposit all the waste generated in the premises in such containers.
Hospitals, nursing homes, etc.	• Not to throw any solid waste in their neighborhood, on the streets, open spaces, and vacant lands, into the drains or water bodies. • Not to dispose off the biomedical waste in the municipal dust bins or other waste collection or storage site meant for municipal solid waste. • Store the waste as per the directions contained in the government of India, Ministry of Environment Biomedical Waste (Management and Handling) Rules, 1998.

Source of waste generation	Action to be taken
Construction/ demolition waste	• Not to deposit construction waste or debris on the streets, footpaths, pavements, open spaces, water bodies etc. • Store the waste within the premises or with permission of the authorities just outside the premises without obstructing the traffic preferably in a container if available through the local body or private contractors.
Garden waste	Compost the waste within the garden, if possible Trim the garden waste once in a week on the days notified by the local body. Store the waste into large bags or bins for handing over to the municipal authorities appointed for the purpose on the day of collection notified.

Q.101. List out the application of some important industrial solid wastes.

Ans. Industrial solid wastes applications are tabulated below.

Table 8.6: Applications of some important industrial solid wastes.

S.No.	Solid Waste	Areas of application
1.	Fly ash	• Cement • Raw material in ordinary portland cement (OPC) manufacture • Manufacture of oil well cement • Making sintered fly ash light-weight aggregates • Cement/silicate bonded fly ash/clay binding bricks and insulating bricks • Cellular concrete bricks and blocks, lime and cement fly ash concrete • Precast fly ash concrete building units • Structural fill for roads, construction on sites, land reclamation, etc. • As filler in mines, in bituminous concrete • As plasticiser • As water reducer in concrete and sulphate resisting concrete
2.	Blast Furnace Slags	• Manufacture of slag cement, super sulphated cement, metallurgical cement • Non-portland cement • Making expansive cement, oil well, coloured cement and high early-strength cement • In refractory and in ceramic as sital • As a structural fill (air-cooled slag) • As aggregate in concrete
3.	Ferro-alloy and other metallurgical slags	• As structural fill • In making pozzolona metallurgical cement
4.	By product gypsum	• In making of gypsum plaster, plaster boards and slotted tiles • As set controller in the manufacture of portland cement • In the manufacture of expensive or non-shrinking cement, super sulphated and anhydrite cement • As mineraliser • Simultaneous manufacture of cement and sulphuric acid
5.	Lime sludge (phos-phochalk paper and sugar sludges)	• As a sweetener for lime in cement manufacture • Manufacture of lime pozzolana bricks/binders • For recycling in parent industry • Manufacture of building lime • Manufacture of masonry cement

S.No.	Solid Waste	Areas of application
6.	Chromium sludge	• As a raw material component in cement manufacture • Manufacture of coloured cement as a chromium – bearing material
7.	Red mud	• As a corrective material • As a binder • Making construction blocks • As a cellular concrete additive· Coloured composition for concrete • Making heavy clay products and red mud bricks • In the formation of aggregate • In making floor and all tiles • Red mud polymer door
8.	Pulp and paper	• Lignin

Q.102. Describe various acts, rules and notifications for solid waste management in India.

Ans. Laws concerning solid waste are passed to improve the solid waste management and to regulate the disposal activity, which causes problems in public health, the environment and economics. Many laws apply to the control of solid waste management problems which are given below:.

Acts, Rules and Notification regarding Solid Waste Management in India

- Constitution of India, 1950
- Code of Criminal Procedure, 1973
- Water (Prevention and Control of Pollution) Act, 1974
- The Air (Prevention and Control of Pollution) Act, 1981
- Environment Protection Act, 1986
- Hazardous waste (Management and Handling) Rules, 1989
- Coastal Regulation Zone Notification, 1991
- Bio-medical wastes (Management and Handling) Rules, 1998.
- Recycled plastics (Manufacture and Usage) Rules, 1999
- Municipal Waste (Management and Handling) Rules 2000

The right to live in a clean and healthy environment is not only a fundamental right guaranteed under Article 21 of our Constitution but also a right recognized and enforced by the courts of law under different laws, like Law of Torts, Indian Penal Code, 1860 Civil Procedure Code, 1908 and Criminal Procedure Code, 1973. **The Constitution of India, 1950 the earliest legislation and which is the supreme law of the land has imposed a fundamental duty on every citizen of India under Article 51-A(g) to protect and improve the environment.**

At the national policy level, the ministry of environment and forests has legislated the Municipal Waste Management and Handling Rules 2000 in exercise of the power conferred under sections 3, 6 and 25 of the Environment Protection Act, 1986. These rules shall apply to every municipal authority responsible for collection, segregation, storage, transportation, processing and disposal of municipal solid wastes.

Composting of wastes is a legal requirement provided under the Municipal Solid Waste Management (MSW) Rules 2000 for all municipal bodies in the country. The MSW Rules 2000 requires that "biodegradable wastes shall be processed by composting, vermi-composting, anaerobic digestion or any other appropriate biological processing for the stabilization of wastes".

Every municipal authority shall, within the territorial area of the municipality, be responsible for the implementation of the provisions of these rules, and for any infrastructure development for collection, storage, segregation, transportation, processing and disposal of municipal solid wastes.

The central government, to perform its functions effectively as contemplated under sections 6, 8, and 25 of the Environment Protection Act, 1986, and has made various Rules, Notifications and Orders including the Bio-medical wastes (Management and Handling) Rules, 1998.

Recycled plastics (Manufacture and Usage) Rules, 1999:

- Prohibit the usage of carry bags or containers made of recycled plastics for storing, carrying, dispensing or packaging of foodstuffs.
- Prescribe that the minimum thickness of carry bags made of recycled plastics shall not be less than 20 microns.
- Directs the manufacturers of carry bags that the carry bags and containers shall be in natural shape or white in colour.
- Stipulate that recycling of plastics shall be undertaken strictly in accordance with the standards prescribed by the Bureau of Indian Standards. The manufacturers of recycled plastics carry bags shall mark their products as "recycled".

This notification also provides that the Plastics Industries Association through member units shall undertake self-regulatory measures. The State Pollution Control Board exercise powers to implement and control the above rules.

Preliminary surveys on municipalities' preparedness in implementing the MSW Rules 2000 show that the majority of the cities have embarked on city-wide implementation of door-to-door collection of waste, source segregation, composting of organics, recycling and creating engineered and safe landfill sites for residual waste disposal.

Segregation of municipal solid waste: In order to encourage the citizens, municipal authority shall organize awareness programmes for segregation of wastes and shall promote recycling or reuse of segregated materials. The municipal authority shall undertake phased programme to ensure community participation in waste segregation. For this purpose, the municipal authorities shall arrange regular meetings at quarterly intervals with representatives of local resident welfare associations and non-governmental organizations.

Storage of municipal solid wastes: Municipal authorities shall establish and maintain storage facilities in such a manner, as they do not create unhygienic and unsanitary conditions around it. Following criteria shall be taken into account while establishing and maintaining storage facilities, namely:

- Storage facilities shall be created and established by taking into account quantities of waste generation in a given area and the population densities. A storage facility shall be so placed that it is accessible to users.
- Storage facilities to be set up by municipal authorities or any other agency shall be so designed that wastes stored are not exposed to open atmosphere and shall be aesthetically acceptable and user-friendly.
- Storage facilities or 'bins' shall have 'easy to operate' design for handling, transfer and transportation of waste. Bins for storage of bio-degradable wastes shall be painted green, those for storage of recyclable wastes shall be painted white and those for storage of other wastes shall be painted black.

Manual handling of waste shall be prohibited. If unavoidable due to constraints, manual handling shall be carried out under proper precaution with due care for safety of workers.

Processing of municipal solid wastes: Municipal authorities shall adopt suitable technology or combination of such technologies to make use of wastes so as to minimize burden on landfill. Following criteria shall be adopted, namely

- The biodegradable wastes shall be processed by composting, vermicomposting, anaerobic digestion or any other appropriate biological processing for stabilization of wastes.

Mixed waste containing recoverable resources shall follow the route of recycling. Incineration with or without energy recovery can also be used for processing wastes in specific cases. Municipal authority or the operator of a facility wishing to use other state-of-the-art technologies shall approach the Central Pollution Control Board to get the standards laid down before applying for grant of authorization.

Disposal of municipal solid wastes Land filling shall be restricted to non-biodegradable, inert waste and other waste that are not suitable either for recycling or for biological processing. Land filling shall also be carried out for residues of waste processing facilities as well as pre-processing rejects from waste processing facilities. Land filling of mixed waste shall be avoided unless the same is found unsuitable for waste processing. Under unavoidable circumstances or till installation of alternate facilities, land filling shall be done following proper norms.

Managing Non-biodegradable solid waste: Non-biodegradable solid waste (NBDSW) or refuse is a carpet word. It covers a variety of materials ranging from asbestos to Zinc batteries. Polythene and its related compounds are the most commonly found solid waste materials in urban environs. Many non-biodegradable solid waste materials are known to cause considerable environmental hazards when released into land, water and atmosphere.

Coastal environment and solid waste management: Solid waste related problems prevail more in megalopolis and the dangers reach great heights in coastal cities. Solid wastes of domestic and industrial units are considered major pollutants of coastal regions of India. Nearly 44000 m^3 of domestic sewage and 440 m^3 of industrial waste are discharged every year into the seas of India.

Q.103. Describe the salient features of the Indian plastic waste management (PWM) rules, 2011.

Ans. Important features of PWM rules, 2011 are given below:

(*i*) The plastic carry bags shall be used for the purpose of carrying or dispensing commodities but don't include these bags which are integral part of packaged products. The thickness of bag shall not be < 40μ.

(*ii*) Carry bags can also be made from compostable plastics conforming IS/ISO:17088:2008.

(*iii*) Prescribed Authority for registration, manufacture and recycling shall be State Pollution Control Board (SPCB) or Pollution Control Committee (PCC). And for enforcement of Rules relating to use, collection, segregation, transportation and disposal of plastic waste, shall be concerned Municipal Authority;

(*iv*) Multilayered pouches or sachets used for packaging of gutkha etc. shall not use plastic material in any form;

(*v*) Every carry bags made from plastic shall bear a label or mark "recycled" as per IS:14534:1998. Each carry bag made from "Compostable material" shall bear a label "Compostable" and shall conform to IS/ISO:17088:2008.

(*vi*) No carry bag shall be made available free of cost by retailers to consumers. The concerned Municipal Authority may be notification determine the minimum price for carry bags in order to encourage re-use so as to minimize plastic waste generation.

(*vii*) Each state government shall for constitute a State Level Advisory (SLA) Body to monitor implementation of Rules. This body shall meet once in a year and may invite experts, if it considers necessary.

(*viii*) **The plastic waste management (PWM) shall be as under:**

(*a*) recycling, recovery or disposal of plastic waste shall be carried out as per the rules, regulations and standards stipulated by the central government from time to time.

(*b*) recycling of plastics shall be carried out in accordance with the Indian Standard IS 14534:1998 titled as Guidelines for Recycling of Plastics, as amended from time to time.

(*c*) the Municipal Authority shall be responsible for setting up, operationalisation and coordination of the waste management system and for performing the associated functions, namely: (*i*) to ensure safe collection, storage, segregation, transportation, processing and disposal of plastic waste. (*ii*) to ensure that no damage is caused to the environment during this process. (*iii*) to ensure setting up of collection centres for plastic waste involving manufacturers. (*iv*) to ensure its channelisation to recyclers. (*v*) to create awareness among all stakeholders about their responsibilities. (*vi*) to engage agencies or groups working in waste management including waste pickers, and (*vii*) to ensure that open burning of plastic waste is not permitted.

(*d*) for setting up plastic waste collection centres, the Municipal Authority may ask the manufacturers, either collectively or individually in line with the principle of Extended Producer's Responsibility (EPR) to provide the required finance to establish such collection centre.

(*e*) recyclers shall ensure that recycling facilities are in accordance with the Indian Standard: IS 14534:194 titled as Guidelines for Recycling of Plastics and in compliance with the rules under the Environment (Protection) Ad, 1986, as amended from time to time.

(*f*) the concerned Municipal Authority shall ensure that the residues generated from recycling processes are disposed of in compliance with Schedule II (Management of Municipal Solid Wastes) and Schedule III (Specifications for Landfill Sites) of the Municipal Solid Wastes (Management and Handling) Rules, 2000 made under the Environment (Protection) Act, 1986, as amended from time to time.

(*g*) the municipal authority shall incorporate the said rules in the municipal bye- laws of all the urban local bodies.

(*h*) the municipal authority shall encourage the use of plastic waste by adopting suitable technology such as in road construction, co-incineration etc. The municipal authority or the operator intending to use such technology shall ensure the compliance with the prescribed standards including pollution norms prescribed by the Competent Authority in this regard.

(*ix*) Each SPCB or PCC shall prepare and submit annual report to CPCB by **30th day of September** each year. The Central Pollution Control Board (CPCB) shall consolidate the report on use of plastic carry bags, sachets/pouches etc. and management of plastic waste. The consolidated report along with recommendations on implementation of the

Plastic waste (Management and Handling) (Amendment) Rules, 2011 will be submitted to MoEF by **30th Day of December.**

Q.104. What are various hazardous wastes we generate?

Ans. Hazardous wastes are those wastes that are highly toxic to humans, animals, and plants, are corrosive, highly inflammable, or explosive, and reactive when exposed to certain things e.g., gases. Industrial and hospital waste is considered hazardous as they may contain toxic substances. Certain types of household waste are also hazardous. India generates around 10 million tonnes of hazardous wastes every year.

Hazardous wastes refer to wastes that may cause adverse health effects on the ecosystem and human beings. These wastes pose risks to human health or living organisms,due to the fact that they: are non-degradable or persistent in nature, can be biologically magnified, are highly toxic and even lethal at very low concentrations. The criteria used to determine the nature of hazard include toxicity, phytotoxicity, genetic activity and bio-concentration. The threat to public health and the environment of a given hazardous waste is dependent on the quantity and characteristics of the waste involved.

Household wastes that can be categorized as hazardous waste include old batteries, shoe polish, paint tins, old medicines, and medicine bottles.

In the industrial sector, the major generators of hazardous waste are the metal, chemical, paper, pesticide, dye, refining, and rubber goods industries. Direct exposure to chemicals in hazardous waste such as mercury and cyanide can be fatal.

Hospital waste is generated during the diagnosis, treatment, or immunization of human beings or animals or in research activities in these fields or in the production or testing of biologicals. It may include wastes like sharps, soiled waste, disposables, anatomical waste, cultures, discarded medicines, chemical wastes, etc. These are in the form of disposable syringes, swabs, bandages, body fluids, human excreta, etc.

This waste is highly infectious and can be a serious threat to human health if not managed in a scientific and discriminate manner. It has been roughly estimated that of the 4 kg of waste generated in a hospital at least 1 kg would be infected.

Hospital waste contaminated by chemicals used in hospitals is considered hazardous. These chemicals include formaldehyde and phenols, which are used as disinfectants, and mercury, which is used in thermometers or equipment that measure blood pressure. Most hospitals in India do not have proper disposal facilities for these hazardous wastes.

Q.105. What are four characteristics of hazardous wastes?

Ans. Four characteristics of hazardous wastes are ignitability, corrosivity, reactivity and toxicity which are described below:

(*i*) **Ignitability:** A waste is an ignitable hazardous waste, if it has a flash point of $< 60°C$, readily catches fire and burns so vigorously as to create a hazard, or is an ignitable compressed gas or an oxidiser. A simple method of determining the flash point of a waste is to review the material safety data sheet, which can be obtained from the manufacturer or distributor of the material. Naphtha, lacquer thinner, epoxy resins, adhesives and oil based paints are all examples of ignitable hazardous wastes.

(*ii*) **Corrosivity:** A liquid waste which has a pH of less than or equal to 2 or greater than or equal to 12.5 is considered to be a corrosive hazardous waste. Sodium hydroxide, a caustic solution with a high pH, is often used by many industries to clean or degrease metal parts. Hydrochloric acid, a solution with a low pH, is used by many industries to clean metal parts prior to painting. When these caustic or acid solutions are disposed of, the waste is a corrosive hazardous waste.

(*iii*) **Reactivity:** A material is considered a reactive hazardous waste, if it is unstable, reacts violently with water, generates toxic gases when exposed to water or corrosive materials, or if it is capable of detonation or explosion when exposed to heat or a flame. Examples of reactive wastes are waste gunpowder, sodium metal or wastes containing cyanides or sulphides.

(*iv*) **Toxicity:** To determine if a waste is a toxic hazardous waste, a sample of the material must be subjected to a test conducted in a certified laboratory. The toxic characteristic identifies wastes that are likely to leach dangerous concentrations of toxic chemicals into ground water.

Q.106. In the regulatory definition of solid waste, which criteria a material must fall in to be classified as hazardous waste?

Ans. Hazardous wastes refer to wastes that may cause adverse health effects on the ecosystem, human being or living organisms, due to the fact that they:

- are non-degradable or persistent in nature.
- can be biologically magnified.
- are highly toxic and even lethal at very low concentrations.

The above list relates only to the intrinsic hazard of the waste, under uncontrolled release, to the environment, regardless of quantity or pathways to humans, plants and animals). These Wastes are secondary materials, which are generally classified into six categories as inherently waste: like materials, spent materials, sludges, by-products, commercial chemical products and scrap metals. Solid wastes form a subset of all secondary materials and hazardous wastes form a subset of solid waste. However, certain secondary materials are not regulated as wastes, as they are recycled and reused.

A material to be classified as a hazardous waste, must also meet the criteria specified in the regulatory definition of solid waste. By using either or both of the following criteria, we can identify as to whether or not a waste is hazardous:

(*i*) The list provided by government agencies declaring that substance as hazardous.

(*ii*) Characteristics such as ignitibility, corrosivity, reactivity and toxicity of the substance.

These two criteria are explained below:

Listed hazardous wastes (priority chemicals):

A specific list showing certain materials as hazardous wastes minimises the need to test wastes as well as simplifies waste determination. In other words, any waste that fits the definition of a listed waste is considered a hazardous waste. Four separate lists cover wastes from generic industrial processes, specific industrial sectors, unused pure chemical products and formulations that are either acutely toxic or toxic, and all hazardous waste regulations apply to these lists of wastes. These wastes, classified in the F, K, P, and U industrial waste codes, respectively are described below:

F-list: The F-list contains hazardous wastes from non-specific sources, that is, various industrial processes that may have generated the waste. The list consists of solvents commonly used in degreasing, metal treatment baths and sludges, wastewaters from metal plating operations and dioxin containing chemicals or their precursors. Examples of solvents that are F-listed hazardous wastes, include benzene, carbon tetrachloride, cresylic acid, methyl ethyl ketone (MEK), methylene chloride, 1, 1, 1, trichloroethane, toluene, and trichloroethylene. Solvent mixtures or blends, which contain greater than 10% of one or more of the solvents listed above are also considered F-listed wastes.

K-list: The K-list contains hazardous wastes generated by specific industrial processes. Examples of industries, which generate K-listed wastes include wood preservation, pigment production, chemical production, petroleum refining, iron and steel production, explosive manufacturing and pesticide production.

P and U lists: The P and U lists contain discarded commercial chemical products, off specification chemicals, container residues and residues from the spillage of materials. These two lists include commercial pure grades of the chemical, any technical grades of the chemical that are produced or marketed, and all formulations in which the chemical is the sole active ingredient. An example of a P or U listed hazardous waste is a pesticide, which is not used during its shelf-life and requires to be disposed in bulk. The primary distinction between the two lists is the quantity at which the chemical is regulated. The P-list consists of acutely toxic wastes that are regulated when the quantity generated per month, or accumulated at any time, exceeds one kilogram, while U-listed hazardous wastes are regulated when the quantity generated per month exceeds 25 kilograms. Examples of businesses that typically generate P or U listed wastes include pesticide applicators, laboratories and chemical formulators.

Characteristics of hazardous wastes: The regulations define characteristic hazardous wastes as wastes that exhibit measurable properties posing sufficient threats to warrant regulation. For a waste to be deemed a characteristic hazardous waste, it must cause, or significantly contribute to, an increased mortality or an increase in serious irreversible or incapacitating reversible illness, or pose a substantial hazard or threat of a hazard to human health or the environment, when it is improperly treated, stored, transported, disposed of, or otherwise mismanaged. In other words, if the wastes generated at a facility are not listed in the F, K, P, or U lists, the final step to determine whether a waste is hazardous is to evaluate it against the four hazardous characteristics *i.e.*, ignitability, corrosivity, reactivity and toxicity.

Q.107. Describe the classification of various hazardous wastes?

Ans. There are many compounds, products and product combinations that fit within the definition of hazardous waste. These groups of hazardous waste are classified into the following five general categories:

(*i*) **Radioactive substance:** Substances that emit ionizing radiation are radioactive. Such substances are hazardous because prolonged exposure to radiation often results in damage to living organisms. Radioactive substances are of special concern because they persist for a long period. The period in which radiation occurs is commonly measured and expressed as *half-life, i.e.*, the time required for the radioactivity of a given amount of the substance to decay to half its initial value. For example, uranium *compounds* have half-lives that range from 72 years for U_{232} to 23,420,000 years for U_{236}. The management of radioactive wastes is controlled by regulatory agencies. Disposal sites that are used for the long-term storage of radioactive wastes are not used for the disposal of any other solid waste.

(*ii*) **Chemicals:** Most hazardous chemical wastes can be classified into four groups: synthetic organics, inorganic metals, salts, acids and bases, and flammables and explosives. Some of the chemicals are hazardous because they are highly toxic to most life forms. When such hazardous compounds are present in a waste stream at levels equal to, or greater than, their threshold levels, the entire waste stream is identified as hazardous.

(*iii*) **Biomedical wastes:** The principal sources of hazardous biological wastes are hospitals and biological research facilities. The ability to infect other living organisms and the ability to produce toxins are the most significant characteristics of hazardous biological wastes. This group mainly includes malignant tissues discarded during surgical procedures and contaminated materials, such as hypodermic needles, bandages and outdated drugs. This waste can also be generated as a by-product of industrial biological conversion processes.

(*iv*) **Flammable wastes:** Most flammable wastes are also identified as hazardous chemical wastes. This dual grouping is necessary because of the high potential hazard in storing, collecting and disposing of flammable wastes. These wastes may be liquid, gaseous or solid, but most often they are liquids. Typical examples include organic solvents, oils, plasticisers and organic sludges.

(*v*) **Explosives:** Explosive hazardous wastes are mainly ordnance (artillery) materials, *i.e.*, the wastes resulting from ordnance manufacturing and some industrial gases. Similar to flammables, these wastes also have a high potential for hazard in storage, collection and disposal, and hence, they should be considered separately in addition to being listed as hazardous chemicals. These wastes may exist in solid, liquid or gaseous form.

(*vi*) **Household hazardous wastes:** Household wastes such as cleaning chemicals, batteries, nail polish, etc. in MSW constitute hazardous waste. Especially batteries contain mercury which are alkaline which is dangerous enough to kill people. Generic household hazardous material include non-chlorinated organic, chlorinated organic, pesticides, latex paint, oil based paints, waste oil, automobile battery and household battery.

Q.108. What are the community sources of generation of various categories of hazardous wastes?

Ans. Hazardous wastes are generated in limited amounts in a community and very little information is available on the quantities of hazardous waste generated within a community and in various industries. Hazardous waste generation outside the industry is irregular and very less in amount

Table below gives a list of hazardous waste generation sources:

Table 8.7: Hazardous waste generation sources

Hazardous waste category	Generation sources
Radioactive substances	Biomedical research facilities, colleges and university laboratories, offices, hospitals, nuclear power plants, etc.
Toxic chemicals	Agricultural chemical companies, battery shops, car washes, chemical shops, college and university laboratories, construction companies, electric utilities, hospitals and clinics, industrial cooling towers, newspaper and photographic solutions, nuclear power plants, pest control agencies, photographic processing facilities, plating shops, service stations, etc.
Biological wastes	Biomedical research facilities, drug companies, hospitals, medical clinics, etc.
Flammable wastes	Dry cleaners, petroleum reclamation plants, petroleum refining and processing facilities, service stations, tanker truck cleaning stations, etc.
Explosives	Construction companies, dry cleaners, ammunition production facilities, etc.

In addition to the sources listed, the spillage of containerized hazardous waste is also considered an important source. The quantities of hazardous wastes that are involved in spillage are usually not known. The effects of spillage are visible to the community. Because the occurrence of spillage cannot be predicted, the potential threat to human health and environment is greater than that from routinely generated hazardous wastes.

Q.109. Describe the methods of storage and collection of hazardous wastes.

Ans. When large quantities of hazardous wastes are generated, special facilities are used that have sufficient capacity to hold wastes accumulated over a period of several days. When only a small amount is generated, the waste can be containerised, and limited quantity may be stored. Containers and facilities used in hazardous waste storage and handling are selected on the basis of waste characteristics. For example, corrosive acids or caustic solutions are stored in fibreglass or glass-lined containers to prevent deterioration of metals in the container. Care is required to avoid storing incompatible wastes in the same container or locations.

The waste generator, or a specialised hauler, generally collects the hazardous waste for delivery to a treatment or disposal site. The loading of collection vehicles is completed in either of the following ways:

(*i*) Wastes stored in large-capacity tanks are either drained or pumped into collection vehicles.

(*ii*) Wastes stored in sealed drums or sealed containers are loaded by hand or by mechanical equipment onto flatbed trucks.

The stored containers are transported unopened to the treatment and disposal facility. To avoid accidents and loss of life, two collectors should be assigned when hazardous wastes are to be collected. The equipment used for collection vary with the waste characteristics, and the typical collection equipment are listed in Table 8.7:

Table 8.8: Hazardous waste collection equipment and accessories

Hazardous waste category	Waste collection equipment and accessories
Radioactive substances	Various types of trucks and railroad equipment depending on characteristics of wastes. Special marking to show safety hazard, heavy loading equipment to handle concrete-encased lead containers.
Toxic chemicals	Flat bed trucks for wastes stored in drums, tractor-trailer tank truck combination for large volumes of wastes, railroad tank cars, special interior linings such as glass, fiberglass or rubber.

Biological wastes	Standard packers' collection truck with some special precautions to prevent contact between wastes and the collector, flatbed trucks for wastes stored in drums.
Flammable wastes	Same as those for toxic chemicals, with special colourings and safety warning printed on vehicles.
Explosives	Same as those for toxic chemicals with some restriction on transport routes, especially through residential areas.

For short-haul distances, drum storage and collection with a flatbed truck is often used. As hauling distances increase, the larger tank trucks, trailers and railroad tank cars are used.

Q.110. Describe the transfer and transport methods for hazardous wastes.

Ans. The economic benefits derived by transferring smaller vehicle loads to larger vehicles, for non-hazardous solid wastes are equally *applicable* to hazardous wastes. However, the facilities of a hazardous waste transfer station are quite different from nonhazardous solid waste transfer station. Typically, hazardous wastes are not compacted (*i.e.*, mechanical volume reduction) or delivered by numerous community residents. Instead, liquid hazardous wastes are generally pumped from collection vehicles and sludge or solids are reloaded without removal from the collection containers for transport to processing and disposal facilities.

It is unusual to find a hazardous waste transfer facility, where wastes are simply transferred to larger transport vehicles. Some processing and storage facilities are often part of the material handling sequence at a transfer station. For example, neutralisation of corrosive wastes might result in the use of a lower-cost holding tank on transport vehicles. As in the case of storage, care is required to avoid the danger of mixing incompatible wastes.

Q.111. What is the purpose of processing of hazardous wastes?

Ans. Processing of hazardous waste is done for purposes of recovering useful materials and preparing the wastes for disposal. Processing can be accomplished on-site or off-site. The variables affecting the selection of processing site include the characteristics of wastes, the quantity of wastes, the technical, economical and environmental aspects of available on-site treatment processes and the availability of the nearest off-site treatment facility (e.g., haul distance, fees, and exclusions). The treatment of hazardous waste can be accomplished by physical, chemical, thermal or biological means.

Table below gives the various individual treatment processes in each category:

Table 8.9: Hazardous waste treatment processes

Operation/Processes	Functions performed$	Types of wastes*	Forms of waste#
	Physical treatment		
Aeration	Se	1, 2, 3, 4	L
Ammonia stripping	VR, Se	1, 2, 3, 4	L
Carbon sorption	VR, Se	1, 2, 4, 5	L, G
Centrifugation	VR, Se	1, 2, 3, 4, 5	L
Dialysis	VR, Se	1, 2, 3, 4	L
Distillation	VR, Se	1, 2, 3, 4, 5	L
Electro dialysis	VR, Se	1, 2, 3, 4, 6	L
Encapsulation	St	1, 2, 3, 4, 6	L, S
Evaporation	VR, Se	1, 2, 5	L
Filtration	VR, Se	1, 2, 3, 4, 5	L, G

Operation/Processes	Functions performed$	Types of wastes*	Forms of waste#
Physical treatment			
Flocculation/Settling	VR, Se	1, 2, 3, 4, 5	L
Flotation	Se	1, 2, 3, 4	L
Reverse osmosis	VR, Se	1, 2, 3, 6	L
Sedimentation	VR, Se	1, 2, 3, 4, 5	L
Thickening	Se	1, 2, 3, 4	L
Vapour scrubbing	VR, Se	1, 2, 3, 4	L
Chemical treatment			
Calcination	CR	1, 2, 5	L
Ion exchange	VR, Se, De	1, 2, 3, 4, 5	L
Neutralisation	De	1, 2, 3, 4	L
Oxidation	De	1, 2, 3, 4	L
Precipitation	VR, Se	1, 2, 3, 4, 5	L
Reduction	De	1, 2	L
Solvent extraction	Se	1, 2, 3, 4, 5	L
Sorption	De	1, 2, 3, 4	L
Thermal treatment			
Incineration	VR, De	3, 5, 6, 7, 8	S, L, G
Pyrolysis	VR, De	3, 4, 6	S, L, G
Biological treatment			
Activated sludges	De	3	L
Aerated lagoons	De	3	L
Anearobic digestion	De	3	L
Anaerobic filters	De	3	L
Trickling filters	De	3	L
Waste stablisation pond	De	3	L

*$ Functions: VR = volume reduction, Se = separation, De = detoxification, St = storage, * Waste types: 1 = inorganic chemical without heavy metals, 2 = inorganic chemical with heavy metal, 3 = organic chemical without heavy metal, 4 = organic chemical with heavy metal, 5 = radiological, 6 = biological, 7 = flammable and 8 = explosive, # Waste forms: S = solid, L = liquid and G = gas*

In practice, the physical, chemical and thermal treatment operations are the most commonly used. (Biological treatment processes are used less often because of their sensitivity.) Depending on the type of wastes being treated, one *or more of these methods may be used.*

Q.112. Describe various disposal methods for hazardous wastes.

Ans. Regardless of their form (*i.e.*, solid, liquid, or gas), most hazardous waste is disposed off either near the surface or by deep burial. Table below shows the various hazardous waste disposal methods:

Table 8.10: Various hazardous waste disposal methods

Operation/Process	Functions performed$	Types of wastes*	Forms of waste#
Deep well injection	Di	1, 2, 3, 4, 5, 6, 7	L

Operation/Process	Functions performed$	Types of wastes*	Forms of waste#
Detonation	Di	6, 8	S, L, G
Engineered storage	St	1, 2, 3, 4, 5, 6, 7, 8	S, L, G
Land burial	Di	1, 2, 3, 4, 5, 6, 7, 8	S, L
Ocean dumping	Di	1, 2, 3, 4, 7, 8	S, L, G

*$ Functions: Di = disposal, St = storage, * Waste types: 1 = inorganic chemical without heavy metals, 2 = inorganic chemical with heavy metal, 3 = organic chemical without heavy metal, 4 = organic chemical with heavy metal, 5 = radiological, 6 = biological, 7 = flammable and 8= explosive.*

 # Waste form: S = solid, L = liquid and G = gas

Although, controlled landfill methods have been proved adequate for disposing of municipal solid waste and limited amounts of hazardous waste, they are not suitable enough for the disposal of a large quantity of hazardous waste, due to the following reasons:

- possible percolation of toxic liquid waste to the ground water;
- dissolution of solids followed by leaching and percolation to the ground water;
- dissolution of solid hazardous wastes by acid leachate from solid waste, followed by leaching and percolation to the ground water;
- potential for undesirable reactions in the landfill that may lead to the development of explosive or toxic gases;
- volatilisation of hazardous waste leading to the release of toxic or explosive vapours to the atmosphere;
- corrosion of containers with hazardous wastes.

Care is therefore required both in the selection of a hazardous waste disposal site and its design. In general, disposal sites for hazardous wastes should be separate from those for municipal solid wastes. As hazardous wastes can exist in the form of liquids, sludges, solids and dusts, a correct approach for co-disposal for each of the hazardous wastes should be determined. To avoid the co-disposal of incompatible wastes, separate storage areas within the total landfill site should be designated for various classes of compatible wastes.

Liquid wastes are usually stored in a tank near the site and can be introduced into the landfill by means of trenches or lagoons, injection or irrigation. Sludges are also placed in trenches. During disposal of lightweight wastes, the disposal area must be kept wet to prevent dust emissions.

Hazardous solid waste characterised by a high degree of impermeability as such must not be disposed of over large areas. When containerised wastes are to be disposed of, precautions must be taken to avoid the rupturing of containers during the unloading operation and the placement of incompatible waste in the same location. To avoid rupturing, the containers are unloaded and placed in position individually. The covering of the containers with earth should be monitored and controlled carefully to ensure that a soil layer exists between each container and the equipment placing the soil does not crush or deform the container.

While designing a landfill site for hazardous waste, provision should be made to prevent any leachate escaping from landfill site. This requires a clay liner, and in some cases, both clay and impermeable membrane liners are used. A layer of limestone is placed at the bottom of the landfill to neutralise the pH of leachate. A final soil cover of 25 cm or more should be placed over the liner. The completed site should be monitored continuously, both visually and with sample wells.

Q.113. Describe the various physical and chemical treatment methods of hazardous wastes?

Ans. Physical and chemical treatments are an essential part of most hazardous waste treatment operations, and the treatments include the following:

(*i*) **Filtration and separation:** Filtration is a method for separating solid particles from a liquid using a porous medium. The driving force in filtration is a pressure gradient, caused by gravity, centrifugal force, vacuum, or pressure greater than atmospheric pressure. The application of filtration for treatment of hazardous waste fall into the following categories:

Clarification, in which suspended solid particles less than 100 ppm (parts per million) concentration are removed from an aqueous stream. This is usually accomplished by depth filtration and cross-flow filtration and the primary aim is to produce a clear aqueous effluent, which can either be discharged directly, or further processed. The suspended solids are concentrated in a reject stream.

Dewatering of slurries of typically 1% to 30 % solids by weight. Here, the aim is to concentrate the solids into a phase or solid form for disposal or further treatment. This is usually accomplished by cake filtration. The filtration treatment, can be used for neutralisation of strong acid with lime or limestone, or precipitation of dissolved heavy metals as carbonates or sulphides followed by settling and thickening of the resulting precipitated solids as slurry. The slurry can be dewatered by cake filtration and the effluent from the settling step can be filtered by depth filtration prior to discharge.

(*ii*) **Chemical precipitation:** This is a process by which the soluble substance is converted to an insoluble form either by a chemical reaction or by change in the composition of the solvent to diminish the solubility of the substance in it. Settling and/or filtration can then remove the precipitated solids. In the treatment of hazardous waste, the process has a wide applicability in the removal of toxic metal from aqueous wastes by converting them to an insoluble form. This includes wastes containing arsenic, barium, cadmium, chromium, copper, lead, mercury, nickel, selenium, silver, thallium and zinc. The sources of wastes containing metals are metal plating and polishing, inorganic pigment, mining and the electronic industries. Hazardous wastes containing metals are also generated from cleanup of uncontrolled hazardous waste sites, e.g., leachate or contaminated ground water.

(*iii*) **Chemical oxidation and reduction (redox):** In these reactions, the oxidation state of one reactant is raised, while that of the other reactant is lowered. When electrons are removed from an ion, atom, or molecule, the substance is oxidised and when electrons are added to a substance, it is reduced.Such reactions are used in treatment of metal bearing wastes, sulphides, cyanides and chromium and in the treatment of many organic wastes such as phenols, pesticides and sulphur containing compounds. Since these treatment processes involve chemical reactions, both reactants are generally in solution. However, in some cases, a solution reacts with a slightly soluble solid or gas.

There are many chemicals, which are oxidising agents; but relatively few of them are used for waste treatment. Some of the commonly used oxidising agents are sodium hypochlorite, hydrogen peroxide, calcium hypochlorite, potassium permanganate and ozone. Reducing agents are used to treat wastes containing hexavalent chromium, mercury, organo metallic compounds and chelated metals. Some of the compounds used as reducing agents are sulphur dioxide, sodium borohydride, etc. In general, chemical treatment costs are highly influenced by the chemical cost. This oxidation and reduction treatment tends to be more suitable for low concentration (*i.e.*, less than 1%) in wastes.

(*iv*) **Solidification and stabilization (S/S):** In hazardous waste management, solidification and stabilisation is a term normally used to designate a technology employing activities to reduce the mobility of pollutants, thereby making the waste acceptable under land disposal requirements. Solidification and stabilisation are designed to improve waste handling and physical characteristics, decrease surface area across which pollutants can transfer or leach, limit the solubility or detoxify the hazardous constituent. To understand this technology, it is important to understand the following terms:

Solidification: This refers to a process in which materials are added to the waste to produce a solid. It may or may not involve a chemical bonding between the toxic contaminant and the additive.

Stabilisation: This refers to a process by which a waste is converted to a more chemically stable form. Subsuming solidification, stabilisation represents the use of a chemical reaction to transform the toxic component to a new, non-toxic compound or substance.

Chemical fixation: This implies the transformation of toxic contaminants to a new non-toxic compound.

Encapsulation: This is a process involving the complete coating or enclosure of a toxic particle or waste agglomerate with a new substance (e.g., S/S additive or binder). The encapsulation of the individual particles is known as micro-encapsulation, while that of an agglomeration of waste particles or micro-encapsulated materials is known as macro-encapsulation.

In S/S method, some wastes can be mixed with filling and binding agents to obtain a dischargeable product. This simple treatment can only be used for waste with chemical properties suitable for landfilling. With regard to wastes with physical properties, it changes only the physical properties, but is unsuitable for landfilling. The most important application of this technology, is the solidification of metal containing waste. S/S technology could potentially be an important alternative technology with a major use being to treat wastes in order to make them acceptable for land disposal. Lower permeability, lower contaminant leaching rate and such similar characteristics may make hazardous wastes acceptable for land disposal after stabilisation.

(*v*) **Evaporation:** Evaporation is defined as the conversion of a liquid from a solution or slurry into vapour. All evaporation systems vaporise the volatile solvent. Evaporation is used in the treatment of hazardous waste and the process equipment is quite flexible and can handle waste in various forms – aqueous, slurries, sludges and tars. Evaporation is commonly used as a pre-treatment method to decrease quantities of material for final treatment. It is also used in cases where no other treatment method is found to be practical, such as in the concentration of trinitrotoluene (TNT) for subsequent incineration.

(*vi*) **Ozonation:** Ozone is one of the strongest oxidising agents known. It can be substituted for conventional oxidants such as chlorine, hydrogen peroxide and potassium permanganate. Ozone and UV radiations have been used to detoxify industrial organic wastes, containing aromatic and aliphatic polychlorinated compounds, ketones and alcohols.

Q.114. Describe various thermal treatment methods of hazardous wastes.

Ans. The two main thermal treatments used for hazardous wastes are:

(*i*) **Incineration:** Incineration can be regarded as either a pre-treatment of hazardous waste, prior to final disposal or as a means of valorising waste by recovering energy. It includes both the burning of mixed solid waste or burning of selected parts of the waste stream as a fuel. The concept of treating hazardous waste is similar to that of municipal solid waste.

(*ii*) **Pyrolysis:** This is defined as the chemical decomposition or change brought about by heating in the absence of oxygen. This is a thermal process for transformation of solid and liquid carbonaceous materials into gaseous components and the solid residue containing fixed carbon and ash.

The application of pyrolysis to hazardous waste treatment leads to a two-step process for disposal. In the first step, wastes are heated separating the volatile contents (e.g., combustible gases, water vapour, etc.) from non-volatile char and ash. In the second step volatile components are burned under proper conditions to assure incineration of all hazardous components.

Pyrolysis is applicable to hazardous waste treatment, as it provides a precise control of the combustion process. The first step of pyrolysis treatment is endothermic and generally done at 425 to 760°C. The heating chamber is called the pyrolyser. Hazardous organic compounds can be volatilised at this low temperature, leaving a clean residue. In the second step, the volatiles are burned in a fume incinerator to achieve destruction efficiency of more than 99%. The pyrolysis process can be applied to solids, sludges and liquid wastes.

Wastes with the following characteristics are especially amenable to pyrolysis:

- Sludge material that is either too viscous, too abrasive or varies too much in consistency to be atomised in an incinerator.
- Wastes such as plastic, which undergo partial or complete phase changes during thermal processing.
- High-residue materials such as high ash liquid and sludges, with light, easily entrained solids that will generally require substantial stack gas clean up.
- Materials containing salts and metals, which melt and volatilise at normal incineration temperatures. Materials like sodium chloride, zinc and lead, when incinerated may cause refractory spalling and fouling of the heat exchanger surface.

Q.115. Describe the biological treatment methods for hazardous waste?

Ans. On the basis of the fact that hazardous materials are toxic to living beings, it is not uncommon assume that biological treatment is not possible for hazardous wastes.

Some of the techniques used for biological treatment of hazardous waste are:

(*i*) **Land treatment:** This is a waste treatment and disposal process, where a waste is mixed with or incorporated into the surface soil and is degraded, transformed or immobilized through proper management. The other terminologies used commonly include land cultivation, land farming, land application and sludge spreading. Compared to other land disposal options (e.g., landfill and surface impoundments), land treatment has lower long term monitoring, maintenance and potential clean up liabilities and because of this, it has received considerable attention as an ultimate disposal method. It is a dynamic, management intensive process involving waste, site, soil, climate and biological activity as a system to degrade and immobilize waste constituents.

In land treatment, the organic fraction must be biodegradable at reasonable rates to minimise environmental problems associated with migration of hazardous waste constituents.

The various factors involved in the operation of the system are as follows:

Waste characteristics: Biodegradable wastes are suitable for land treatment. Radioactive wastes, highly volatile, reactive, flammable liquids and inorganic wastes such as heavy metals, acids and bases, cyanides and ammonia are not considered for land treatment. Land treatability of organic compound often follows a predictable pattern for similar type of compounds. Chemical structure, molecular weight, water solubility and vapour pressure are few of the characteristics that determine the ease of biodegradation.

Soil characteristics: The rate of biodegradation and leaching of waste applied, the availability of nutrients and toxicants to microorganisms and the fate of hazardous waste constituents are determined largely by application rate as well as the soil's chemical and physical characteristics or reaction. Principal soil characteristics affecting land treatment processes are pH, salinity, aeration, moisture holding capacity, soil temperature, etc. Some of the characteristics can be improved through soil amendments (e.g., nutrients, lime, etc.), tillage or through adjustments of loading rate, frequency, etc., at the time of waste application.

Micro organisms: Soil normally contains a large number of diverse microorganisms, consisting of several groups that are predominantly aerobic in well-drained soil. The types and population of microorganisms present in the waste amended soil depend on the soil moisture content, available oxygen, nutrient composition and other characteristics. The key groups of the microorganisms present in the surface soil are bacteria, actinomycetes, fungi, algae and protozoa. In addition to these groups, other micro and macro fauna, such as nematodes and insects are often present.

Waste degradation: Conditions favourable for plant growth are also favourable for the activity of soil microorganisms. The factors affecting waste degradation that (may be adjusted in the design and operation of a land treatment facility) are soil pH (near 7), soil moisture content (usually between 30 to 90 %), soil temperature (activity decreases below 10°C) and nutrients.

(*ii*) **Enzymatic systems:** Enzymes are complex proteins, composed of amino acids, are linked together via peptide bonds. Enzymes capable of transforming hazardous waste chemicals to non-toxic products can be harvested from microorganisms grown in mass culture. Such crude enzyme extracts derived from microorganisms convert pesticides into less toxic and persistent products. The reaction of detoxifying enzymes are not limited to intracellular conditions but have been demonstrated through the use of immobilised enzyme extracts on several liquid waste streams. The factors of moisture, temperature, aeration, soil structure, organic matter content, seasonal variation and the availability of soil nutrients influence the presence and abundance of enzymes.

(*iii*) **Composting:** The principles involved in composting organic hazardous wastes are the same as those in the composting of all organic materials, though with moderate modifications. The microbiology of hazardous wastes differs from that of composting in the use of inoculums. The reaction is that certain types of hazardous waste molecules can be degraded by only one or a very few microbial species, which may not be widely distributed or abundant in nature. The factors important in composting of hazardous wastes are those that govern biological reactions. The principal physical parameters are the shape and dimensions of the particles of the material to be composted and the environmental factors of interest in an operation are temperature, pH, available oxygen, moisture, and nutrient availability.

The compost technology can be divided into two broad classes – windrow (open pile) and in vessel (enclosed), and the former may be further subdivided into turned and forced aeration (static pile). Composting, by no means, is a panacea for the hazardous waste problem. When considering the future of hazardous waste composting needs, attention must be paid to the advantages and disadvantages inherent in composting as compared to those inherent in physical, chemical and thermal method of waste treatment.

(*iv*) **Aerobic and anaerobic treatment:** Hazardous materials are present in low to high concentration in wastewaters, leachate and soil. These wastes are characterised by high organic content (e.g., up to 40,000 mg/l total organic carbon), low and high pH (2 to 12), elevated salt levels (sometimes, over 5%), and presence of heavy metals and hazardous organics. Hazardous wastes can be treated using either aerobic or anaerobic treatment methods.

In aerobic treatment, under proper conditions, microorganisms grow. They need a carbon and energy source, which many hazardous wastes satisfy, nutrients such as nitrogen, phosphorus and trace metals and a source of oxygen. Some organisms can use oxidized inorganic compounds (e.g., nitrate) as a substitute for oxygen. Care is to be taken such that all the required nutrients and substances are supplied in sufficient quantities. Temperature and pH must be controlled as needed and the substances that are toxic to the organisms (e.g., heavy metals) must be removed.

Anaerobic treatment is a sequential biologically destructive process in which hydrocarbons are converted, in the absence of free oxygen, from complex to simpler molecules, and ultimately to carbon dioxide and methane. The process is mediated through enzyme catalysis and depends on maintaining a balance of population within a specific set of environmental conditions. Hazardous waste streams often consist of hydrocarbons leading to higher concentrations of chemical oxygen demand (COD). Depending upon the nature of waste, the organic constituents may be derived from a single process stream or from a mixture of streams.

The treatability of the waste depends upon the susceptibility of the hydrocarbon content to anaerobic biological degradation, and on the ability of the organisms to resist detrimental effect of biologically recalcitrant and toxic organic and inorganic chemicals. The metabolic interactions among the various groups of organisms are essential for the successful and complete mineralisation of the organic molecules. Various parameters such as the influent quality, the biological activity of the reactor and the quality of the reactor environment are monitored to maintain efficient operating conditions within the reactor.

Q.116. Describe various methods and advantages of pollution prevention and waste minimization.

Ans. Pollution prevention is the use of materials, processes, or practices that reduce or eliminate the generation of pollutants or wastes at the source. It includes practices that reduce the use of hazardous and non-hazardous materials, energy, water or other resources as well as those that protect natural resources through conservation or more efficient use. Pollution prevention is the maximum feasible reduction of all wastes generated at production sites. It involves the judicious use of resources through source reduction, energy efficiency, reuse of input materials and reduced water consumption.

Waste minimization means the reduction of hazardous waste that is generated prior to treatment, storage and disposal. It is defined as any source reduction or recycling activity that results in the reduction of the total volume of hazardous waste, or toxicity of hazardous waste, or both. Practices in waste minimization include recycling, source separation, product substitution, manufacturing process changes and the use of less toxic raw materials.

Pollution prevention and waste minimization makes us environmentally responsible. While pollution prevention reduces waste at its source, waste minimization, including recycling and other methods, reduces the amount of waste.

The factors that can contribute to pollution prevention and waste minimization include the following.

(*i*) **Management support and employee participation:** Management commitment through policy, communications and resources for waste minimization and pollution prevention is essential to earn the dedication of all employees. A formal policy statement must be drafted and adopted. for this purpose to reflect commitment and attitude towards protecting the environment, minimizing or eliminating waste and reusing or recycling materials by the laboratories, departments and industries.

(*ii*) **Training:** Employees must be provided with formal and on the job training to increase awareness of operating practices that reduce both solid and hazardous waste generation. The training programme should include the industries' compliance requirements, occupational health and safety requirements, waste minimization and pollution prevention.

(*iii*) **Waste audits:** Waste audits will help industries to identify areas of potential waste reduction. The audit programme includes the identification of hazardous wastes and their sources, prioritization of various waste reduction actions to be undertaken, evaluation of some technically, economically and ecologically feasible approaches to waste minimization and pollution prevention, development of an economic comparison of waste minimization and pollution prevention options and evaluation of their results.

(*iv*) **Good operating practices:** These practices involve the procedural or organisational aspects of industry, changes in operating practices, in order to reduce the amount of waste generated. These practices may include, material handling improvements, scheduling improvements, spill and leak prevention, preventive maintenance, corrective maintenance, material/waste tracking or inventory control and waste stream segregation, according to the toxicity, type of contaminant and their physical state.

(*v*) **Material substitution:** The purpose is to find substitute materials, which are less hazardous than those currently utilised and which result in the generation of waste in smaller quantities and/or of less toxicity.

(*vi*) **Technological modification:** These practices are oriented towards process and equipment modifications to reduce waste generation and can range from changes that can be implemented in a matter of days at low cost to the replacement of process equipment involving large capital expenditures.

(*vii*) **Recycling of materials:** These options are characterised as reclaimation, use/reuse and resource recovery techniques. Which involve the return of a waste material either to the originating process or to another process as a substitute for an input material.

(*viii*) **Surplus chemical waste exchange options:** Inter- and intra-department chemical exchange is to be implemented and encouraged by employers/employees. Material exchanges not only reduce wastes but also save money.

In addition, by auditing each department or section, a knowledge base of chemical purchase and usage can be developed, allowing each department to develop and implement controls on the purchase of chemicals, institute intra departmental chemical sharing/swapping programmes and eliminate excessive purchase and usage.

Facility operations should be examined to determine whether changes in practices and procedures will result in the generation of non-hazardous or less hazardous waste, or waste reduced in toxicity or volume. The specifics in this context include the substitution of non-toxic materials for toxic ones, distillation or evaporation of water-based chemical end-products, reclamation and reuse of common solvents, use of non-chromate cleaners as a standard part of doing business to generate non-hazardous end products. By implementing and adhering to the guidelines for handling and storing wastes at the point of generation, the costs associated with hazardous waste disposal will also be minimised.

Q.117. Describe the provisions and status of hazardous wastes management in India.

Ans. In the USA, about 70% of the hazardous waste generated are produced from chemical and petrochemical industries. As industrialisation proceeds, the management of hazardous wastes is increasingly becoming a serious *problem* in India as well. The Indian chemical industry, which accounts for about 13% of the total industrial production and about 10% of the GNP, employs about 6% of the nation's industrial workforce and is one of the major generators of toxic and hazardous wastes. There are about 15000 industrial units located in 400 districts, out of which 12,000 units have been granted authorization for multiple disposal practices including incineration, storage, land disposal and other disposal options. However, small and medium sized enterprises (SMEs) are the major sources of hazardous wastes. And, the States of Andhra Pradesh, Assam, Gujarat, Karnataka, Kerala, Madhya Pradesh, Maharashtra, Orissa, Rajasthan and Tamil Nadu generate the majority of all hazardous wastes.

India is the first country that has made provisions for the protection and improvement of environment in its Constitution. One of the Fundamental Duties of every citizen of India is to protect and improve the natural environment including forests, lakes, rivers and wild life and to have compassion for living creatures.

India has enacted the following laws, regulations and standards governing the country's environmental protection:

(*i*) The Water (Prevention and Control of Pollution) Act, 1974 as amended in 1988.
(*ii*) Water (Prevention and Control of Pollution) Rules, 1975.
(*iii*) The Water (Prevention and Control of Pollution) Cess Act, 1977, amended in 1991.
(*iv*) The Water (Prevention and Control of Pollution) Cess Rules, 1978.
(*v*) The Air (Prevention and Control of Pollution) Act, 1984, amended in 1987.
(*vi*) The Air (Prevention and Control of Pollution) Rules 1982 and 1983.
(*vii*) The Environment (Protection) Act, 1986.
(*viii*) Hazardous Waste (Management and Handling) Rules, 1989 as amended in 2000.
(*ix*) Management, Storage and Import of Hazardous Chemical Rules, 1989.
(*x*) Manufacture, Use, Import, Export and Storage of Hazardous Microorganisms, Genetically Engineered Microorganisms or Cells Rules, 1989.
(*xi*) The Public Liability Insurance Act, 1991.
(*xii*) The Public Liability Insurance Rules, 1991.
(*xiii*) The Biomedical Wastes (Management and Handling) Rules, 1995.
(*xiv*) Municipal Wastes (Management and Handling) Draft Rules, 1999.
(*xv*) Hazardous Waste (Management and Handling) Amendment Rules 2000.
(*xvi*) E-waste Management and Handling Rules, 2008

Because of these amendments, the legal management of hazardous substances in India will now apply to 50 industrial processes.

The **penal provisions** for non-compliance under Hazardous Waste (Management and Handling) Amended Rules 2000 and Environment (Protection) Act, 1986 are:

The State Pollution Control Board may cancel an authorisation issued under these rules or suspend it for such period as it thinks fit, if, in its opinion, the authorised person has failed to comply with any of the conditions of the authorisation or with any provisions of the Act of these rules, after giving the authorised person an opportunity to show cause and after recording reasons therefore.

The occupier, transporter and operator of a facility shall be liable for damages caused to the environment resulting due to improper disposal of hazardous waste listed in Schedule 1, 2 and 3 of The Hazardous Waste (Management and Handling) Amendment Rules, 2000. The occupier and operator of a facility shall also be liable to reinstate or restore damaged or destroyed elements of the environment. The occupier and operator of a facility shall be liable to pay a fine as levied by the SPCB with the approval of the Central Pollution Control Board (CPCB) for any violation of the provisions under these rules. An appeal shall lie against any order of grantor refusal of an authorisation by the Member Secretary, SPCB, etc., to the Secretary, Department of Environment of the State.

Besides the aforementioned provisions for non-compliance(s), the Penalty Provisions, delineated under Sections 15 (1, 2) and 16 of the Environmental (Protection) Act, 1986 are also applicable.

Furthermore, the Union Ministry of Environment and Forests, through the Gazette Notification of March 24, 1992, introduced Public Liability Insurance Act Policy, which is specially designed to protect any person, firm, association, or company who owns or has control over handling any hazardous substance at the time of accident. These include 179 hazardous substances along with three categories of inflammable substances. The term *handling* means manufacturing, processing, treatment, packaging, storing, transportation by vehicle, use, collection, destruction, conversion, offering for sale, transfer or any other similar form of dealing with hazardous substances.

Hazardous waste (Management, Handling and Transboundary Movement) Rules 2008.

Changes to the Hazardous waste Rules Existing Regulation

The Ministry had notified the Hazardous Wastes (Management and Handling) Rules,1989 as amended in 2000 and 2003 for regulating management and handling of hazardous waste. Based on the experience gained in the implementation of these Rules, the **Hazardous Waste (Management, Handling and Transboundary Movement) Rules, 2008** have been notified repealing the earlier Rules with a view to ensuring effective implementation. The Ministry has also provided financial assistance for strengthening the State Pollution Control Boards (SPCBs) for facilitating implementation of the Rules. Financial assistance has also been provided for setting up Common Treatment, Storage and Disposal Facilities for hazardous waste management. In addition, the Ministry and the Central Pollution Control Board (CPCB) from time to time sponsor training programmes for creation of awareness about the provisions laid down in the Rules. The CPCB has also published guidelines on various aspects of the hazardous waste management for ensuring compliance of the Rules.

In new rules, categories of wastes banned for export and import had also been defined, fulfilling the Basel Convention, ratified by India in 1992. The basic objectives of the Basel Convention are for the control and reduction of transboundary movements of hazardous and other wastes subject to the Convention, prevention and minimization of their generation, environmentally sound management of such wastes and for active promotion of the transfer and use of cleaner technologies.

Current Indian Scenario

The hazardous waste generated in the country per annum currently is around 8 million tonnes out of which 70% is being generated by five states, namely Gujarat, Maharashtra, Tamil Nadu, Karnataka and Andhra Pradesh. Only three States have developed common TSDF (Treatment, Storage, Disposal Facility), which are essential component of proper hazardous waste management activity for ultimate disposal of the hazardous wastes in an environmentally sound manner. These 10 facilities are currently operational only in Gujarat, Andhra Pradesh and Maharashtra.

Though the Hazardous Wastes (Management and Handling) Rules were notified in 1989, the implementation on the ground has left a lot to be desired. Lack of proper infrastructure and strict enforcement mechanism has led to hazardous waste still remaining a grave problem. New emerging wastes and loopholes in the current legislation have also contributed to this. There are still problems of hazardous waste not being managed in sound environmental conditions, improper dumping and lack of proper treatment and disposal facilities.

However, the new draft claims to address sustainable development and also enable the recovery and/or reuse of useful materials from hazardous materials generated from a process, thereby, reducing the hazardous wastes destined for final disposal and to ensure the environmentally sound management of all hazardous materials.

Q.118. Describe the ways and means of properly managing hazardous wastes.

Ans. Management of hazardous waste follows the functional elements of solid waste, viz., generation, storage and collection, transfer and transport, processing and disposal. Special facilities are used that have sufficient capacity to hold wastes accumulated over a period of several days. Small amounts are containerised, and limited quantities may be stored for periods covering months or years. The waste generator or a specialised hauler normally does the collection of hazardous wastes for delivery to a treatment or disposal facility. All storage containers remain unopened till they reach the treatment and disposal facility. Processing of waste can be done to recover useful materials and to prepare the wastes for disposal. Processing can be accomplished either on-site or off-site. Regardless of their forms, most hazardous wastes are disposed of either near the surface or by deep burial. When landfills are used for such wastes, precautions must be taken to avoid the rupturing of containers during the unloading operation and the placement of incompatible wastes in the same location. Also provisions should be made to prevent any leachate from escaping from landfills, which can be done both visually and by monitoring wells.

Q.119. Describe the advantages of waste minimization and pollution prevention.

Ans. Waste minimization means reduction, to the extent feasible, of hazardous waste that is generated prior to treatment storage and disposal. The advantages include reduction of the total volume and toxicity of hazardous waste or reduction in both. By using materials carefully so as to reduce the generation of waste, pollution is reduced, resources are conserved and costs of waste disposal are minimised. Pollution prevention is the use of materials, processes or practices that reduce or eliminate the creation of pollutants or wastes at the source. The advantages include reduction of the use of hazardous and non-hazardous materials, energy, water or other resources as well as those that protect natural resources through conservation or that is more efficient.

Q.120. Describe the three stage hierarchy of options for handling and management of hazardous waste.

Ans. A three stage hierarchy of options for handling and managing hazardous wastes are as follows:

- The top tier includes in plant options such as process manipulation, recycle and reuse options that reduce the production of hazardous waste in the first place. It also contains most desirable options.
- Middle stage highlights processes that convert hazardous waste to less hazardous or non-hazardous substances that include:

  (*a*) Incineration (*b*) Land treatment
  (*c*) Ocean and atmospheric assimilation (*d*) Chemical, physical and biological treatments
  (*e*) Thermal treatments
- Last stage which is the least desirable tier is perpetual storage- cheapest alternative. Few processes include landfill, underground injection, arid region unsaturated zone, surface impoundments, salt formations and waste piles.

Q.121. What is the need and benefits of Integrated Solid Waste Management (ISWM)?

Ans. Integrated solid waste management refers to the strategic approach to sustainable management of solid wastes covering all sources and all aspects, covering generation, segregation, transfer, sorting, treatment, recovery and disposal in an integrated manner, with an emphasis on maximizing resource use efficiency.

Need for ISWM

- Cities are facing an increasing growth in population, and shares in GDP growth, resulting in – among other things – increasing quantities of waste being generated.
- Due to changing lifestyles and consumption patterns, the quantity of waste generated has increased with quality and composition of waste becoming more varied and changing.
- Industrialization and economic growth has produced more amounts of waste, including hazardous and toxic wastes.
- There is a growing realization of the negative impacts that wastes have had on the local environment (air, water, land, human health, etc.)
- Complexity, costs and coordination of waste management has necessitated multi-stakeholder involvement in every stage of the waste stream. This calls for an integrated approach to waste management.

- Local Governments are now looking at waste as a *business opportunity*, (a) to extract valuable resources contained within it that can still be used and (b) to safely process and dispose wastes with a minimum impact on the environment.

Benefits of ISWM:

- Cleaner and safe neighbourhoods
- Higher resource use efficiency
- Resource augmentation
- Savings in waste management costs due to reduced levels of final waste for disposal
- Better business opportunities and economic growth
- Local ownership and responsibilities/Participation.

Q.122. What are the elements of Integrated Solid Waste Management (ISWM)?

Ans. Integrated Solid Waste Management takes an overall approach to creating sustainable systems that are economically affordable, socially acceptable and environmentally effective. An integrated solid waste management system involves the use of a range of different solid waste treatment methods, and key to the functioning of such a system is the collection and sorting of the waste. It is important to note that no one single treatment method can manage all the waste materials in an environmentally effective way. Thus all of the available treatment and disposal options must be evaluated equally and the best combination of the available options suited to particular community chosen. Effective management schemes therefore need to operate in ways which best meet current social, economic, and environmental conditions of the municipality.

Elements of ISWM is shown below in the figure.

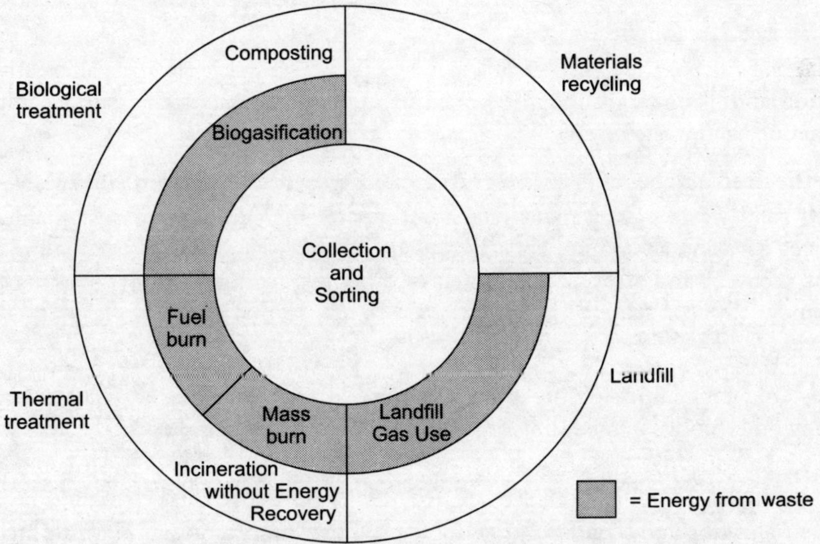

Fig. 8.10: Elements of integrated solid waste management (ISWM)

❑❑❑

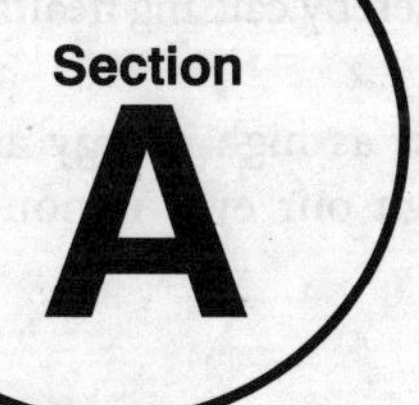

Section A

9

Radioactive/Nuclear Pollution

SHORT QUESTIONS AND ANSWERS ON RADIOACTIVE/NUCLEAR POLLUTION

Q.1. What is radiation, its types and radioactive decay?

Ans. Radiation is energy travelling through space. Energy can be transported either in form of electromagnetic waves (radiations) or a stream of energetic particles, which can be electrically charged or neutral.

These radiations are of two types:

(1) Non-ionizing radiations (2) Ionizing radiations

Non-ionizing radiations are the electromagnetic waves of longer wavelength from near ultraviolet rays to radio waves. These waves have energies enough to excite the atoms and molecules of the medium through which they are moving, causing them to vibrate faster. These do not have enough energy to ionize them.

Ionizing radiations are the electromagnetic radiations having high energy, such as short wavelength ultra violet radiations, *x*-rays and gamma rays. The energetic rays like (α, β and γ etc.) produced in radioactive decay can cause ionization of atoms and molecules of the medium through which they pass and convert them into charged ions.

For example in water molecule, it can induce reaction that can break bonds in proteins, DNA and or other important molecules.

Alpha (α), beta (β) and gamma (γ) radiations are produced by the process called **radioactive decay.** The unstable nuclei decay spontaneously and emit these radiations. These rays (radiations) can affect some other non-radioactive atoms to become radioactive (unstable) and give out radioactive radiations.

Living organisms are continuously exposed to a variety of radiations called background radiations. If the level of the radioactive radiations increases above a certain limit it causes harmful effects to living beings. This harmful level of radiations emitted by radioactive elements is called **radioactive pollution.**

Q.2. What do you understand by the term radiation?

Ans. Radiation is the term used to refer to the emission of any rays, wave motion or particles (e.g., alpha particles, beta particles, neutrons) from a source. Our environment receives the radiation both naturally and from human activities. Radiation is applied to the emission of electromagnetic radiation , which is a form of energy radiated in the form of a *wave* as a result of the changing electric and magnetic fields. There are many different forms of electromagnetic radiation, each with a different wavelength and energy content. Such radiation travels through space at the speed of 300,000 km s^{-1}.

Cosmic rays,gamma rays, X-rays, and ultraviolet radiation are ***ionizing radiation*** as they have high energy to ionize the atoms by knocking out electrons from the atoms. The resulting electrons and positively charge ions are capable of causing damaging effects on living cells and thereby causing health problems such as cancer.

The other forms of energy are ***non-ionizing radiation*** and they do not possess as high energy as ionizing radiation hence, harmless. The visible light which we perceive through our eyes is non-ionizing radiation and does not cause any such adverse effects.

Q.3. What are natural and anthropogenic sources of radiation pollution?

Ans. There are two sources of radiation pollution, namely natural sources and anthropogenic sources.

Natural sources of radiation:

We live in an environment of natural radiations; there is a certain amount of radioactive radiations almost everywhere on the earth. These are only rarely harmful as the radiation level is generally quite low. This small level of radioactive radiation which is present everywhere is called the background radiation.

(*i*) Atomic radioactive minerals are one of natural sources of radioactive pollution. During mining of uranium, radon gas is constantly released into the air. The parent of radon-222 ($t½$ = 3.82 days) is radium 226 which has a half-life of 1602 years. Radium-226 is widely distributed in rocks, sediments and soils along with isotopes of uranium. Radioactive radiations from these natural sources are known as natural or background radiation.

(*ii*) Cosmic rays are high energy ionizing electromagnetic radiation. The cosmic rays originate from the stars in our galaxy by virtue of nuclear reactions primarily in their cores. The cosmic rays are constantly reaching the earth from outer space.

(*iii*) Naturally occurring radioisotopes such as radon-222 found in soil in small quantity is another source of radioactive radiations.

(*iv*) Radioactive elements which like uranium, thorium, radium, isotopes of potassium (K-40) and carbon (C-40) occur in the lithosphere. Potassium-40 contributes radioactivity to all potassium containing systems in the soil. Crops grown on such soil contain radioactive elements like carbon-14 and potassium-40. Water gets contaminated with various radionuclides when it runs through soils and rocks containing radioactive minerals.

Anthropogenic (manmade) sources of radiation: The following human activities add to the source of radioactive pollution:

(*i*) **Diagnostic medical applications:** Radiations are employed for diagnostic and therapeutic applications. *X*-rays are used in general radiology and CT scan. Gamma rays are used in treatment of cancer. In all these procedures we are exposed to varying doses of radiations.

(*ii*) **Nuclear tests:** Nuclear explosion tests especially when carried out in the atmosphere are a major cause of radiation pollution. It is responsible for increasing the background level of radiation throughout the world. During atmospheric nuclear explosion tests, a number of long-lived radionuclides are released into the atmosphere. This **radioactive dust** (also known as **radioactive fall out**) gets suspended in air at a height of 6 to 7 km above the earth's surface and is dispersed over long distances by winds from the test site. These radionuclides often settle down by rain and get mixed with soil and water. From there they can easily enter the food chain and finally get deposited in the human body where they cause serious health hazards. Some of the radioactive isotopes given off during nuclear test affect the human body.

India exploded its nuclear device (equivalent to 12 kilotons of Tri Nitro Toluene) in an underground tunnel at a depth of 107 meters in the Thar desert near Pokhran in 1974. No radioactivity is said to have been released into the surroundings. Again in May, 1998 India conducted nuclear tests without any fall out of radioactivity into the surroundings.

(*iii*) **Nuclear reactors:** Radiations may leak from nuclear reactors and other nuclear facilities even when they are operating normally. It is often feared that even with the best design, proper handling and techniques, some radioactivity is routinely released into the air and water.

However, dangers of radiation leakage are from possibility of accidents that could result in the release of radioactive material which raises the level of radioactive (ionizing) radiations. Such accidents took place at the 'Chernobyl nuclear power plant' in USSR in 1986 and at the 'Three Mile Island Power Plant' in USA in 1979.

The accidents of 'Three Mile Island' plant in Middletown (U.S.A.) in 1979 and at Chernobyl nuclear power plant (U.S.S.R.) in 1986 were the worst disasters in the history of nuclear power industry. In both incidents, a series of mishaps and errors resulted in the overheating of the nuclear core. In both cases radiation was released in the atmosphere. The leakage from the 'Three Mile Island' nuclear reactor has been claimed to be very low with no immediate injuries to workers or people. But the leakage at Chernobyl was very heavy causing death to many workers and radiation was spread over large areas spread all over Europe.

Nuclear fission converts some radioactive fuel in the reactors to some other radioactive fragments. These reactions produce a lot of heat and therefore the core of the reactor containing fuel rods cooled by water circulation to avoid a meltdown of the fuel rods. If a meltdown happens by accident, it will release a large quantity of highly dangerous radioactive materials into the environment. In order to avoid this type of very serious mishap nuclear reactors are designed to have a number of safety features and it is very unlikely that a reactor would blow up.

Although a core meltdown is highly unlikely, but it is possible. For example, the loss of coolant due to some material failure, would allow the reactor core to overheat and eventually meltdown the fuel rods. Another possibility is the build up of some gas or steam inside the reactor vessel which would blow off the top and release large clouds of radioactive materials in the atmosphere resulting in death and injury to thousands of people.

(*iv*) **Nuclear explosions:** Nuclear explosions are a serious source of radiations hazard. The effects of atomic explosions in Nagasaki and Hiroshima are still not forgotten.

(*v*) **Nuclear wastes:** When uranium-235 nuclei split in a nuclear reactor, they break into fission products which are also highly radioactive. Since one uranium atom splits into two radioactive products, there is a doubling of the number of radioactive atoms on the earth.

Furthermore. uranium-235 has a half-life of more than 700 million years. Some of its fission products have shorter half-lives and decompose much faster than uranium and emit higher levels of radiations. There is no method by which we can increase or decrease the rate of decay of these products. The wastes of nuclear reactor emit dangerous radiations for thousands and thousands of years. Since it is not possible to destroy these radionuclides, they must be stored somewhere on this earth in order to ensure least harm to humans.

Plutonium-239 is another nuclear waste. The plutonium-239 isotope is produced as a by-product during uranium fission. It is an alpha particle emitter and has a long half-life of 24000 years. After about 1000 years or more, the main radioactivity from fuel rods of the nuclear reactors will be from plutonium and other such heavy elements, since by then most of the other nuclides produced in fission and having much shorter half-life will have decayed to a large extent. Plutonium is one of the deadliest poisons known. Plutonium does not occur naturally on earth. This element is produced either in nuclear reactors or in nuclear weapons programme. The plutonium produced today will have to be taken care of for thousands of years by future generations. The general approach in dealing with radioactive wastes is to concentrate and contain as much radioactivity as possible. Effluents containing only very low level of radioactivity are allowed to be discharged into the environment.

Radon can diffuse through rocks and soils into the atmosphere. Once the radon reaches the atmosphere it can be breathed in. The transformation into lead is very significant because the solid radioactive particles are trapped in the lungs and are acutely harmful.

(*vi*) **Nuclear material processing:** Many radioactive minerals are processed to extract and processed, for various uses. All those who handle these materials are at a greater risk of man-made radioactive pollution.

Q.4. Describe the features of various natural sources of radiation.

Ans. Four types of natural sources of radiation include:

- Solar rays
- Environmental radiation
- Radionuclide's in earth crust
- Internal radiation

Solar rays: Solar rays coming from the sun keep a steady drizzle of gamma rays, cosmic rays and heavy particles. Solar storms vastly intensify these showers, but the earth's atmosphere shields us from most dangerous radiations. Solar radiations consisting of cosmic rays are highly energetic particles (10^9MeV) which reach the earth's surface from different galaxies.

Environmental radiations: Radioisotopes of naturally occurring radioelement release enormous amount of radiations in the form of alpha, beta and gamma rays. Besides, radioisotopes, additional radiations emanate from soil, rocks, air and ground water. Radioactive elements occurring mainly in lithosphere comprise uranium, thorium, radium, isotopes of potassium and carbon (C-14). These radiations mix and interact with natural particulate materials in atmosphere enhancing the extent of radio pollution. In marine water, sediments have higher concentration of radioactive isotopes. Uranium and thorium extensively occur in nature and are contained in ores, rocks,soils, and river and sea waters and in animal and plant organisms. From soils, the natural radioactive elements are passed to plants and from plants to animal organisms. In addition to uranium, thorium and their decay products, radio active isotopes of K, Ca, Rb, Sn etc are also found in nature.

Radionuclides in earth's crust: Radioactive materials such as uranium (U-238), thorium (Th-232), and potassium (K-40) which are widely distributed in earth's crust give rise to the phenomenon referred to as terrestrial activity. Radio active potassium (K-40) constitutes 0.012 % of natural potassium, while rubidium (Rb-87) constitutes of 28% of natural rubidium. Thus K -40 is considered to be responsible for 20 to 80% radioactivity in thes oil. Rubidium occurs less abundantly in the earth's crust so it is relatively less distributed in the environment. Radon (Rn- 222) and its immediate daughter nuclide radium-A through radium-C are the common radioactive isotopes in radioactive springs.

A gaseous emission of radio-radon is produced during mining of uranium which on decay yields long lived polonium and radio-lead that enter into the soil and ground water. On an average, a man receives about one rad per year from terrestrial radiation and it maybe as high as 2000 m rad per year in areas where uranium containing rocks exist.

Internal radiation: Generally radiations originate within our bodies, particularly during the decay of potassium in our muscles. Besides potassium, radioactive elements such as uranium,thorium, strontium and carbon (C-14) exist in minute quantities in human's body.

Radioactive materials that emit alpha or beta particles are known as **internal emitters** because these rays severely affect the living tissues on absorption or ingestion. The radioactive substances that emit gamma rays called **external emitters** because these are highly penetrating and can produce their effects without being absorbed inside the body. Some metabolic important radio nuclides which release radiations in the body include calcium,cobalt, iodine, phosphorus, carbon, iron, manganese and hydrogen, etc.

Q.5. Describe the features of various anthropogenic/artificial (manmade) sources of radiation.

Ans. Anthropogenic sources of radiation add large doses of radiation to the existing natural radio active pollution to which our bodies have got accustomed with several ill effects.

The major artificial sources of radiation include:

(*a*) Medical X-rays	(*b*) Radio-isotopes	(*c*) Nuclear tests
(*d*) Radioactive fallout	(*e*) Nuclear reactors	(*f*) Nuclear power plants
(*g*) Nuclear installations	(*h*) radioactive ore processing	(*i*) Electric fields
(*j*) Industrial, medical and research use of radioactive materials		(*k*) Miscellaneous sources

(*a*) **Medical X-rays:** Medical X-rays are artificial radiations used in radiotherapy for diagnostic purposes. These rays are highly penetrating like the gamma rays. X-ray exposure is cumulative in the body and creates chronic defects in the internal organs. Radiation from medical sources accounts for diseases resulting from genetic damage.

(*b*) **Radio isotopes:** Radioisotopes administered to patients during radiation therapy are now proving to be a hazardous source of nuclear pollution. The ability of radiations to kill deceased cells leaving mainly the normal ones unaffected had made them as indiscriminate tool in the diagnosis and cure of some lethal diseases like cancer. However the indiscriminate use of radio nuclides, their over doses to patients and improper handlings have resulted in dangerous nuclear pollution.

(*c*) **Nuclear tests:** During atmospheric nuclear explosion tests, large quantities of long lived nucleotides are released to the atmosphere which gets distributed all over the world.

Generally the test including **nuclear fission and fusion** processes uses uranium (235U) and plutonium (239Pu) as fission materials and lighter nuclei such as hydrogen, lithium or beryllium as fusion elements. When an atom bomb is detonated, about 50%, of the energy released goes into the blast, 35% is dissipated as heat and 15% is released as radioactivity. The radioactive dust that falls to the earth after atomic explosion is known as **radioactive fallout.** It becomes suspended to a height of about 7-8 km above the earth's surface that may disperse by air-currents around the world.

Actually during nuclear explosion, the radioactive products are vaporized to hot gases due to greater force of explosion and very high temperature. These radioactive materials are ejected high into the air as extremely fine particles posing atmospheric pollution through radioactive fallout. The radioactive materials formed by nuclear explosions are absorbed by the atmospheric dust as well as rain water and gradually settle to the earth surface at distant places from the explosion site for a long time after the explosion or blast.

(*d*) **Radioactive fallout:** The atomic explosions not only produce the local ionizing radiations but also the radioactive isotopes, which enter the atmosphere and continue to fallout gradually overbroad geographic areas for a very long time. This is known as **nuclear fallout or radioactive fallout.** This fall out is dangerous for life because it also produces ionizing radiations. Radioactive fallout mainly consists of radioactive particles discharged into the air during explosion test. It also includes their decay products along with radio activated dust that rises from the bomb crater. Radio isotopes suspended in air may come down to soil and water in the form of radioactive rain.

Radioactive fallout causing radiation pollution is two types:

- Early fallout
- Delayed fallout

- **Early fallout:** If the nuclear explosion is at a very low altitude, it sucks up large quantities of soil and water affecting severely all the living organisms. The fire ball also condenses in heavy particles falling back to the earth in a short time. The radioactive pollutants are carried by wind in different directions posing several ill effects. It causes lot of damage even at places remote from the site of explosion.

- **Delayed fallout:** If the nuclear explosion occurs at a high altitude, it sucks little dirt and water. The delayed or worldwide fallout may be polluting both troposphere and stratosphere with radioactive materials. The fission products may be injected into the air and spread everywhere to several kilometers.

  About 80% of the fallout from small and big atomic weapons is deposited on the ground, about 5% passes into the lower atmosphere and the debris settles on the land in few weeks. The remaining 15% which is made up of small particles, goes to the upper atmosphere, gets widely dispersed in the air and may comedown in rain at long distances and enter the food chain. This indicates that total amount of radioactivity decreases with distance from the site of the explosion.

(*e*) **Nuclear reactors:** The leakage of nuclear reactions from nuclear reactors, nuclear research laboratories, medicines and industries is also on the increase with the increase in number of such facilities. In nuclear reactors, structural materials and components become radioactive when exposed to radiation and generate active corrosion products in the reactor effluents. Such activation products are formed by fast neutron or thermal/slow neutron.

Operation of nuclear reactors: The processed nuclear fuel is introduced into the reactor, the operation of which is the major contributor of **radiation pollution.** The sequential processes include – **(*a*) Fission process (*b*) Activation process (*c*) Thermal process**

Some nuclear reactors use U-238 ore as the basic raw material. This has to be processed by purification, enrichment with U-235 and then manufactured into fuel elements. This fabrication results in the production of solid, liquid and gaseous wastes as radio pollutants. Even the wastes of nuclear reactors, called **radioactive wastes** contain vast amounts of long lived radionuclides.

Nuclear reactor wastes: Following types of wastes are generated by nuclear reactors;

(*a*) Fission products remaining in both the primary and secondary fuels.

(*b*) Extraneous activation products in the coolant.

(*c*) Gaseous wastes of several nuclides comprising C-14, H-3, Xe-133, I-131, Kr-85, Ar-41, Fe-54 and I-129, etc.

(*d*) Liquid wastes containing H-3, Co-58, Fe-55, Co-59 and other corrosion products are also produced.

(*e*) During chemical treatment to separate the reusable components of U-235, U-238 and Pu-239 from waste fission products, highly active radioactive wastes are produced.

(*f*) **Radiations from nuclear power plants:** Nuclear power plants are more convenient to run. Once fuelled, they can operate for several months. These plants are different from conventional electricity generating plants. In the fuelled plants, fossil fuel is burnt to produce heat. The fuel used in nuclear plants,being radioactive is critically dangerous and the waste materials are equally so. No power plant is perfectively contamination proof. Leakage may occur from several points which may be chronically radioactive.

Radioactive waste generated by nuclear power plants:

- **Low level radioactive liquid waste:** Radioactive wastes in solution coming from power plants contaminate with aquatic life. These radio elements are eventually conveyed to man from water supplies to food chain through soil, vegetation or livestock.
- **Gaseous and particulate radio wastes:** Stack effluents from atomic power plants contain gaseous and particulate radio isotopes such as H-3, C-14, Kr-85 and I-129, etc. Some of these radionuclides have long half-lives and may be distributed in the environment for several years.

- **Fission fragments:** The largest volume of radioactive wastes comes from reprocessing of irradiated fuel. These radionuclides include Sr-90, I-131, Cs-137and Co-58, etc.
- **Release of tritium:** The heavy water reactors contain high tritium (H-3) inventories because of its production through irradiation of deuterium (D_2) in heavy water(D_2O). Tritium is also released from primary coolants in which lithium hydroxide is added to slow down corrosion in Pressurized Water Reactor (PWR).
- **Heat release:** In power plants, atomic pellets of uranium metal are used as fuel in nuclear reactors which contain three million times as much potential energy as fossil fuel. One ton of uranium produces as much energy as 12 million barrels of oil produces.

(*g*) **Nuclear installations:** Dangers posed by power plant installations due to proliferations of plutonium and other isotopes are astounding. In many industries, e.g., those concerned with power generation, radioactive isotopes are used as fuel. After the nuclear fuel has been burned up in the nuclear reactor, the spent fuel so formed is transferred to the processing plant and then to the burial ground or to some other form of container. In reprocessing plant, fission products are removed from the spent fuel

(*h*) **Radioactive ore processing:** Radioactive ores of uranium like pitchblende and uranites as well as thorium are used in nuclear processes. U-235 undergoes natural fission and emits radiations such as alpha, beta and gamma. The half life periods of U-235 is 4.5×10^9 years. The disintegrated products formed constitute

226 Ra and 82 Pb which are extremely dangerous emitting radiations naturally and ^{92}U and ^{90}Th are employed as artificial fissionable radioactive elements.

During milling for recovery of uranium, process effluents are released as slurry with other residues to a tailings pond from which the effluents drain away to join public waters. Of all the radionuclides, radium-226 is considered to be the most dangerous due to its longer half life, its biochemical properties, its highly energetic radiations and its immediate fission to daughter radionuclides.

Besides radioactive elements, chemical contaminants such as chromium,manganese, sulfate and nitrate are also present as effluents in water. Chemical treatment of monazite also forms highly toxic effluents. All these processes, mining, washing, refining,separation and milling, etc. cause nuclear pollution. All the treatments during ore processing result in the release of radioactive gases which subsequently adsorb on the particles present in the atmosphere. Uranium and thorium ores form dusts in air posing deleterious effects on living organisms.

(*i*) **Electric fields:** Electrical gadgets and power transmission lines generate electric fields causing environmental radiation hazard. Scientists believe that radiation from the electric fields might distort molecular aggregations by disturbing the ionic distribution.

(*j*) **Industrial, medical and research use:** Thermal power plants, fertilizer firms and other industries involved in large scale coal combustion could be a major threat of radiological pollution. However, the level of radioactivity released and its exposure to the population through the fly-ash produced is well within the maximum permissible limits as recommended by the International Commission for Radiological Protection.

Radioactive contamination from the productions of plutonium for the nuclear weapons is far higher than was ever believed.

Accidents and deliberate discharges from nuclear power plants have pollutes lakes, over 200 km of county side which makes it the most radioactive contaminated area and anyone going near it for a few hours may suffer from radioactive sickness.

Radionuclides administrated to patients during medical diagnosis used in radiation therapy and scientific research laboratories are the main source of nuclear pollution. X-rays, light gamma rays (ã) are the most dangerous. They have extremely high penetrating power and can destroy the internal body tissues and inflict serious burns quite rapidly. X-rays are also a form of cosmic radiation and can ionize the atoms in the living tissues. Although certain radiations have proved as indispensable tool in curing lethal diseases and have the ability to kill diseased cells, but radiation effects are dangerous even at lowest levels. Even then radiations are frequently used to destroy the malignant cells of tumor, sterilizing medical products and foodstuff by killing bacteria, tracking progress of medicines through the body with radioisotopes and attacking cancer cells and stones from within the cells of body.

The radiations coming from nuclear research work take the form of particle. Some come in the form of alpha, beta rays and others in the form of highly energetic electromagnetic waves i.e., X-rays and gamma rays. There are neutrons too,emitted by devices that humans make.

(*k*) **Miscellaneous sources:** A replaceable type of nuclear pollution also occurs in the environment. In this type of pollution some radioactive nuclides replace other nuclides that are present in the same group in periodic table. For instance, **radium** is present in second group of periodic table along with calcium and strontium. Radium can replace calcium to cause damage to the body tissues. Similarly Sr-90 gets deposited in bones, teeth and tissues in man in place of calcium.

Q.6. Describe various effects of ionizing radiation.

Ans. When ionizing radiations penetrate the living tissue, it wrecks havoc on the atoms and molecules in its path. Actually when a water molecule in the cell is irradiated, an electron is knocked out of its orbit. The ejected electron may then attach to a normal water molecule creating instability. Such unstable water molecules (H_2O) split into hydrogen ions (H^+), hydroxide ions (OH^-) and the free radicals –H• and OH•. Radiation also produces a host of several species like H_2, H_2O^-, H_2O^+, HO_2, H_3O^-, e^-, e^+ and H_2O_2. The free radicals are extremely reactive. They react with protein molecules in the cell, setting a chain of event, that can destroy living cells or make them to function abnormally.

Radiation exposure may damage the cell membranes by making it permeable, while the large doses of ionizing radiations can kill quickly or inflict severe damage. Even the lower doses can initiate cancer throughout the body. Radiation also results in abnormal interchange of materials through an imperfect cell membrane causing temporary or permanent injury in the body. Studies of survivors who have received significant doses such as atom bomb survivors, uranium miners and radium watch dial painters etc. showed that damage depends on how the victims were exposed. Experiments have shown that human organs can repair some radiation damage. The sensitivity to radiation damage appears to be directly proportional to the cell's reproductive capacity and inversely proportional to its degree of differentiation.

The action of ultraviolet (UV) radiations has been extensively studied and their findings are as follows:

(*a*) In body cells, the protein and nucleic acid are mainly responsible for the absorption of radiation. In the region of 240 nm to 280 nm wavelength, the absorption by nucleic acid is 10 to 20 times greater than that by proteins of the same weight.

(*b*) UV radiations are thought to trigger to distinct immunological effects. One is confined to patches of skin that are actually irradiated, while the other damage is caused to the immune system as a whole.

(*c*) UV radiations cause the blood vessels near the skin's epidermis to carry more blood causing the skin hot, swollen or sun burns.

(*d*) Serious skin cancers including the basal cell carcinoma, squamous cell carcinoma and melanoma are rapidly climbing the list of human diseases caused by UV radiations.

(*e*) It has been observed that closer a fair-skinned person lives to the equator, the more likely he is to get non-melanoma cancer by UV rays.

(*f*) Melanoma is caused by intermittently exposing the body;high doses of UV radiations often associated with burning sensation and skin aging.

(*g*) UV radiations cause leukemia and breast cancer, although the reasons are obscure.

(*h*) UV rays can also be absorbed by lens and cornea in the eye leading to photokeratitis and cataracts. Since the radiation is not sensed by the visual receptors of eyes, the damage is done without the individual knowing about its hazards.

(*i*) These radiations are also associated with DNA breakage, inhibition and alterations of its replications and formation of DNA adduct which has been implicated in premature aging and finally death of the cells.

(*j*) UV radiations also affect drastically the micro-phytoplanktons. Increased UV-rays will increase the mortality rate of larvae of zooplanktons in water.

(*k*) Plants absorb strongly the light near 280 nm. So plant proteins are more susceptible to UV injury. In plants 20 to 50% chlorophyll reduction and harmful mutations are reported.

(*l*) It has been reported that UV-B radiations reduce the effectiveness of plant photosynthesis by 70%.

(*m*) Increased solar UV radiation cause green house effect by changing the global energy and radiation balance of the atmosphere at the planet earth.

Q.7. Describe the various effects of non-ionizing radiation.

Ans. The non-ionizing radiations include infrared, radio waves, microwaves, radar etc.

Effects of microwave radiation: Microwaves shorter than 10 cm. are usually absorbed by the skin that can be felt by the heating of the surface tissue. Microwaves between 10 and 30 cm can penetrate the epidermis and fat layer of the skin, while the waves longer than 30 cm can penetrate deep tissues of dermis causing the skin hot. The eyes and other organs that cannot dissipate heat, are most vulnerable to microwave radiations. Russians have set lower exposure limits claiming that microwaves cause skin-burns, fatigue,dizziness, headache, eye-injuries and cataracts.

Effects of radio frequency radiations: Non-ionizing radiations of longer wave length cause a common thermal effect. These radiations induce thermal agitation in molecules of the matter to produce heat. A variety of non-thermal effects are potentially more dangerous as they pose acute physiological effects. Non-thermal effects may be linked to the electric and magnetic fields associated with the electromagnetic radiation. Effects of electric fields of the radiations are more devastating. It has been observed that electrical potentials of nerve fibers of the central or peripheral nervous system get acutely affected by the electric field component of electromagnetic radiations.

Q.8. Describe the effects of fallout nuclear radiations.

Ans. The cloud obtained during nuclear explosion contains a mixture of gases, molten nuclear fuel and some partially melted radioisotopes. As the fireball cools, these materials condense to form the debris which drops down to the earth in the form of radioactive fallout.

The fallout radionuclides either fuse with iron, silica or dust, and from colloidal suspension or insoluble particles or combine with organic compounds to form complexes. The smaller particles of radionuclides form the colloids and adhere tightly to the leaves of plants and produce radiation damage to leaf tissues. These are often ingested by the grazing animals, and as a result, enter into the food chain directly at the primary consumers level. The radionuclides which form complexes with organic substances enter the food chain through producer trophic level. In this way, radionuclides manage to enter the body of all living organisms.

Q.9. What are the biological effects of nuclear radiation?

Ans. The extent of biological damage depends upon various factors such as amount of radiation exposed, its duration, age of the person exposed, and the part of the body affected etc. The hazards of the radiation on human beings may be acute, chronic or genetic damage. The acute radiation damage occurs from relatively large dose of radiation over a short period of time. The **acute radiation damages** include sudden death, death after some weeks, loss of hairs, widespread ulcers, bleeding from the mouth and gums, etc. The **chronic radiation damage** occurs from relatively small continuous dose of radiation over a long period of time. The chronic radiation damages include leukemia, anemia, cancers of skin and other organs, cataracts, reduction in life span and mutations, etc. The genetic radiation damage represents long term effect of radiation and it indicates changes among future generations.

Exposure of the brain and central nervous system to high doses of ionizing radiation causes delirium, convulsions and death within hours or days. The lens of eye is vulnerable to radiation. As its cell die, it becomes opaque, forming cataracts that impair sight. Internal bleeding and blood vessel damage may show up as red spots on the skin.

Unborn children are vulnerable to brain damage or mental retardation, especially if irradiation occurs during formation of the central nervous system in early pregnancy. Damage to bone marrow, the body's blood factory, is especially harmful; it retards the body's ability to fight infections and hemorrhaging.

Q.10. Describe the effects of X-rays exposure.

Ans. Expectant mothers if irradiated during pregnancy give births to malformed babies. X-rays also cause carcinogenicity in women. Scatter is the biggest problem with X-ray machines. Scatter is the term for leaking and subsequent overdose of radiation. This could cause health problems ranging from impotency, dry eyes, low blood count and throat irritation.

Q.11. What are the effects of plutonium?

Ans. Plutonium, a deadly poisonous substance, is bone-seeker. Once deposited in the bone, it can cause serious bone cancer. Insoluble particles of plutonium, if inhaled, can get preferentially lodged in the deep lung, where they can deliver high doses of radiation to the surrounding tissues inducing chronic lung cancer. The damage occurs due to emission of the short range of alpha particles by plutonium.

Q.12. What are the effects of nuclear radiation on plants and polymers?

Ans. Small amounts of radionuclides may lead to an increase in the rate of mutation in the plants also. Trees and shrubs vary in their reactivity and sensitivity towards radioactive substances, mainly due to the difference in chromosome number and size. Plants with less number of chromosomes offer larger 'target' of radiation hit than those with excess of small chromosomes. The small chromosomes possess more duplicate genetic information so that when one or two cells are damaged, a number of others can take up their function.

Polymerization of monomers can be brought about through irradiation. The effect **of** radiation is of great biological and industrial significance and has led to outstanding development in science and technology. Polymers have also been found useful as radiation protecting materials.

Q.13. What are the health effects of ionizing radiation?

Ans. In recent years, number of people exposed to ionizing radiation has increased tremendously, especially people involved in the mining of uranium ores, patients treated with γ-radiations and technical people using *X*-rays and other radioactive isotopes. Before the dangers of radiation on human body were known, workers dealing with radioactive materials were careless and suffered from various types of cancer. Early workers who used phosphorescent radium paint on the dials of watches suffered from bone tumors in 1920s.

Exposure to any type of ionizing radiation (Alpha and Beta-particles, Gaama -rays and X-rays) can proveharmful and even lethal. The two types of effects are:

(*i*) **genetic effect** (*ii*) **nongenetic** or **body damage.**

In **genetic damage,** genes and chromosomes get altered. Its effect may become visible as deformations in the offsprings (children). Alterations or breaks in the genetic material, that is DNA (deoxyribonucleic acid)- the molecule containing genetic information, called **mutation.**

In **nongenetic effects,** the harm is visible immediately in the form of birth defects, burns, some type of leukemia, miscarriages, tumors, cancer of one or more organs and fertility problems.

Alpha (α-) particles: Generally they cannot penetrate the skin. But if their sources is inside the body, they can cause damage to bones or lungs.

β-particles: Can penetrate the skin but cannot damage the tissues. They can cause damage to skin and eyes(cataract).

γ-radiation: Can easily penetrate the body and pass through it. They cause damage to cell structure.

X-rays: Can travel very far and pass through the body tissues except bones. They can cause damage to the cells.

Q.14. What are units of radiation doses and their effects?

Ans. The biological damage caused by the radiation depends upon the following factors:

(i) the time of exposure (*ii*) the intensity of radiation

(*iii*) the type of ionizing radiation (i.e., its penetration power)

(*iv*) whether the radiation is emanating from outside or inside the human body.

On account of these factors the absorbed dose (or simply doses) designated as D, of the radiation is the amount of energy deposited into a region of the body divided by the mass of the portion of the body that absorbed the radiation.

The MKS unit of absorbed dose is 'gray' (Gy), thus one gray is one joule of energy deposited per kg of body mass. More traditional unit of absorbed doses is **radiation absorbed dose** (rad). One rad is equivalent to 0.01 Gy. Biological damage caused by a particle depends not only on the total energy deposited but also on the rate of energy loss per unit distance traversed by the particle. For example, alpha particles do much more damage per unit energy deposited than do electron (β-particles). This effect is represented by the quality factor Q taken 0 for electron and 20 for alpha particle.

The biological impact is specified by the human equivalent dose H, which is the product of the absorbed dose D and the quality factor Q : $H = QD$. The MKS unit of the human equivalent dose is called the sievert, Sv.

A traditional unit of human equivalent dose is the rem, which stands for radiation equivalent in man, 1 rem = 0.01 Sv.

At low doses such as what we receive every day from background radiation (< 1m rem), the cells repairs the damage rapidly, At higher doses (upto 100 rem), the cell might not be able to repair the damage, the cell may either be changed permanently or die. Cells changed permanently may go on to produce abnormal cells when they divide and may become cancerous.

- **Effects of radiation doses:** Alpha and beta particles can cause burns to skin on high level of exposure from outside but they cannot penetrate the skin to cause internal damage. However, if a radioactive isotope which can emit alpha or beta particles is inhaled or ingested in the body the particles can then cause serious damage to nearby tissues. They may affect the replication of cells and induce tumor formation. The damage inside the body from beta particles is lower than that from the alpha particles of the same energy.

Gamma rays and high energy neutrons are so much penetrating that they pass through the body easily and can cause cellular damage both from outside or inside the body. Rapidly growing tissues of embryo are very sensitive and therefore pregnant women should avoid exposure to radioactivity and *X*-rays unless they are very essential. Although the medical use of *X*-rays may involve very low levels of radiation but even such exposures if carried out frequently can result in a significant increase in the dose of ionizing radiation and involves definite risks. The use of radio-isotopes and gamma ray irradiation for cancer treatment can lead to fairly high dosages of radioactivity.

An increase in the altitude, where an individual lives determines the dose received from cosmic radiation. For example, an increase in an altitude of 2000 meters doubles the dose of radiation due to cosmic rays from outer space. The effect of radiation may be immediate or delayed.

Q.15. Discuss the preventive measures required to be taken against exposure to nuclear radiation.

Ans. The following preventive measures may be adopted to reduce the effects due to both natural and artificial radiations:

(*i*) Atomic explosions should not be carried out in the atmosphere.

(*ii*) In nuclear reactors, closed cycle coolant system may be employed, so that no radiation leakage through coolant can take place.

(*iii*) Radioactive wastes generated by nuclear reactors or from nuclear weapons programme must be disposed in a manner that they will do the least harm. First, the wastes may be stored at some place temporarily to allow for the initial, very intense,radioactivity to die down by natural decay. Nuclear wastes should always be sealed in double-walled tanks so that no leaks may take place. In the second stage, some useful isotopes generated during fission in the reactors may be recycled in reprocessing plants. Finally, a permanent storage space for the wastes in geologically stable underground deep mines should be established. It has been suggested that these wastes may be stored till the wastes are reduced to the same radioactivity level as that of a natural uranium mine.

(*iv*) Production and use of radioisotopes should be minimum and only for very essential use because radioisotopes once produced cannot be destroyed by any means except by the passage of time.

(*v*) The number of nuclear installations should be minimised so as to limit the emission of Radioactive pollutants.

(*vi*) Fission reactions should be minimized.

(*vii*) In nuclear mines, wet drilling may be used and tailings properly sealed and protected for radiation leakage.

(*viii*) Industrial wastes contaminated with radio nuclides should be disposed off carefully in specially built tanks.

(*ix*) Working places where radioactive emissions are possible should have high chimneys and good ventilation system.

(*x*) In areas where there is a risk of radon leakage from underground. Radon concentrations should be monitored and protection measures be installed in buildings and homes.

Q.16. What is the biological effect of nuclear radiation on cell?

Ans. Cell is the basic unit of all living organism.

— cell has cell membrane(out side) — Nucleus is in centre

— cytoplasm is out side the nucleus

Nucleus:

— Chromosomes have Deoxyribo nucleous acid(DNA) and protein.

— DNA contains all information of cellular function like growth,development.

— Some sections of Chromosomes having specific function are called genes.

Type of cell:

— Somatic cell and reproductive cell

— Somatic cell-Have 46 chromosomes in 23 pair

— Reproductive cells have 23 chromosomes

— Water content in human tissue is more than 75%.

— Radiation on interaction with water produces free radicals.

— These free radicals react with biomolecule of cell causing damage to the cell.

Nuclear radiation cause:

(*a*) Inhibition of cell division

(*b*) Chromosome aberration

(*c*) Gene mutation

(*d*) Cell death

Q.17. Discuss about sources, effects and control of radioactive pollution.

Ans. Term radioactive pollution refers to the introduction of one or many radioactive substances into our living environment (land, air and water).

Radioactive pollution is mostly a result of human activity, and has its origin in radioactive waste (mostly coming from nuclear power plants).

People that work in uranium mines and nuclear power plants are being continuously exposed to radioactive pollution and therefore require protective clothing to avoid being contaminated by radiation

Nuclear waste remains radioactive for more than 5,000 years meaning that radioactive pollution has long-term effects, and isolating it for such a long period is difficult.

The radioactive pollution has extremely harmful effect on human health because of variety of radioactive gases, liquids or particles.

Radioactive pollutants can enter our body through ingestion, inhalation, absorption, or injection.

Exposure to large quantities of radioactive material always leads to death. This is because high levels of radiation destroy the cells in our body and cause cancer.

Radioactive pollution has been linked with genetic defects in children and grandchildren of people that were at some point of life being exposed to radioactive pollutants.

The worst cases of radioactive pollution in our planet occurred at the end of the World War two, after two atomic bombs were dropped on Japanese cities Hiroshima and Nagasaki. The radiation effect and a resulting cancer took millions of lives in Japan.

In recent times, radioactive pollution is mostly the result of nuclear power plant accidents such as Fukushima and Chernobyl. The levels of radioactive pollution need to be constantly monitored and measured in nuclear power plants in order to avoid nuclear disaster.

World could decrease the global levels of radioactive pollution by switching from nuclear power to other energy sources such as renewable energy.

Reusing old nuclear fuel is also one of the ways to decrease the level of radioactive pollution.

Some scientists have even suggested that the best possible solution to deal with nuclear waste would be its disposal into space. This, however, could easily result in disaster if, for instance, there would to occur the failure of the launch vehicle.

Illegal dumping of nuclear waste could lead to massive increase of radioactive pollution in the world. This issue is certainly in need of very strict regulations.

Q.18. What are various types of radioactive waste?

Ans. Various types of radioactive wastes are discussed below:

Low level Waste (LLW): It is generated from hospitals and industry, as well as the nuclear fuel cycle. It comprises paper, rags, tools, clothing, filters, etc which contain small amounts of mostly short lived radioactivity. It does not require shielding during handling and transport and is suitable for shallow land burial. To reduce its volume, it is often compacted or incinerated before disposal.

Intermediate level Waste (ILW) contains higher amounts of radioactivity and some requires shielding. It typically comprises resins, chemical sludges and metal fuel cladding, as well as contaminated materials from reactor decommissioning. It may be solidified in concrete or bitumen for disposal. Generally short lived waste (mainly from reactors) is buried in a shallow repository, while long lived waste (from fuel reprocessing) will be disposed of deep underground.

Transuranic waste arises mainly from weapons production, and consists of clothing, tools, rags, residues, debris and other such items contaminated with small amounts of radioactive elements -- mostly plutonium. These elements have an atomic number greater than uranium -- thus transuranic (beyond uranium). Because of the long half-lives of these elements, this waste is not disposed of as either low level or intermediate level waste. It does not have the very high radioactivity of high level waste, nor its high heat generation. The United States currently permanently disposes of transuranic waste at the Waste Isolation Pilot Plant.

High level waste (HLW) arises from the use of uranium fuel in a nuclear reactor and nuclear weapons processing. It contains the fission products and transuranic elements generated in the reactor core. It is highly radioactive and hot. It can be considered the "ash" from "burning" uranium. HLW accounts for over 95% of the total radioactivity produced in the process of nuclear electricity generation.

Q.19. What types of fallout result from nuclear detonation?

Ans. Radioactive pollution that is spread through the earth's atmosphere is termed *fallout*. Three types of fallout result from nuclear detonations: local, tropospheric and stratospheric.

- Local fallout is quite intense but short-lived.
- Tropospheric fallout (in the lower atmosphere) is deposited at a later time and covers a larger area, depending on meteorological conditions.
- Stratospheric fallout, which release extremely fine particles into the upper atmosphere, may continue for years after an explosion and attain a worldwide distribution.

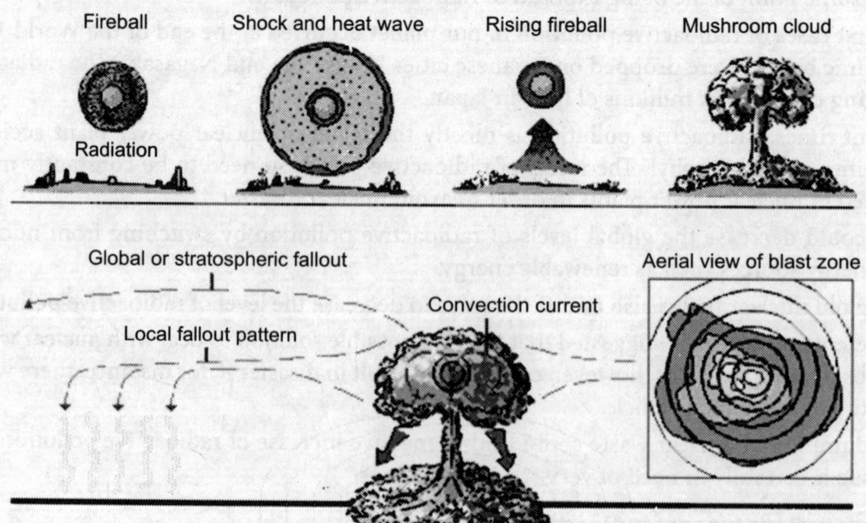

Fig. 9.1: Types of fallout resulting from nuclear detonation

Q.20. What are the factors on which the effects of radioactive wastes depend?

Ans. Radioactive waste causes:

- Soil pollution
- Water pollution

In these two pollutions, pollution hazards finally enter into the food chain of the human who is the final victim of radioactive pollution.

The factors on which the effect of radioactive pollution depends are:

- Half–life
- Energy releasing capacity
- Rate of diffusion
- Rate of deposition of the contaminant.
- Various atmospheric and climatic conditions; such as wind, temperature and rainfall also determine their effects.

The general effects of radioactive wastes are categorized into:

- Somatic effect
- Genetic effect
- Biomagnification

Somatic effect: Affects somatic cells. It appears within individual and disappears with the death of the individual.

Immediate effects: Anemia, Reduced immune response, Hemorrhage , skin burn, mouth ulcers , CNS (Central Nervous System)Damage

Delayed effects: Eye cataract, Leukemia, Cardiovascular disease,Premature ageing,Reduced life span , reduction of fertility

Genetic effects: The radiation affects the genes of the gamete cells. The changes are not apparent in the individual. The effects are exhibited by offspring and in the subsequent generations. They affect the DNA, RNA replication and chromosome. It causes

- Mutation
- Chromosomal aberration
- Chromosomal fragmentation
- Inhibition of RNA,DNA synthesis

Q.21. Describe two major radioactive pollution incidents of the world?

Ans. The two major examples illustrating the effect of radioactive fallout contamination are the bombing of Hiroshima and Nagasaki, Japan in 1945, and Chernobyl Nuclear Power Station (in Russia)disaster in April 1986. Within five years of the American bombing of Japan, as many as 225,000 people had died as a result of long-term exposure to radiation from the bomb blast, chiefly in the form of fallout.

The disaster at the Chernobyl Nuclear Power Station in Ukraine on April 26, 1986 produced a staggering release of radioactivity. In ten days at least 36 million curies spewed across the world. The fallout contaminated approximately 2,590 sq. km of farmland and villages in the Soviet Union. In addition to the hundreds killed at the time of the explosion, scientists predict the eventual Soviet death toll from the Chernobyl accident is around 200,000; the estimated mortality in western Europe may be around 40,000.

Q.22. How does radioactive contamination and radiation exposure occur?

Ans. Radioactive contamination and radiation exposure could occur if radioactive materials are released into the environment as the result of an accident, an event in nature, or an act of terrorism. Such a release could expose people and contaminate their surroundings and personal property.

Radioactive contamination occurs when radioactive material is deposited on or in an object or a person. Radioactive materials released into the environment can cause air, water, surfaces, soil, plants, buildings, people, or animals to become contaminated. A contaminated person has radioactive materials on or inside their body.

Radioactive materials give off a form of energy that travels in waves or particles. This energy is called radiation. When a person is exposed to radiation, the energy penetrates the body. For example, when a person has an X-ray, he or she is exposed to radiation.

A person exposed to radiation is not necessarily contaminated with radioactive material. A person who has been exposed to radiation has had radioactive waves or particles penetrate the body, like having an X-ray. For a person to be contaminated, radioactive material must be on or inside of his or her body. A contaminated person is exposed to radiation released by the radioactive material on or inside the body. An uncontaminated person can be exposed by being too close to radioactive material or a contaminated person, place, or thing.

Q.23. How external and internal contamination occurs from radioactive materials?

Ans. External contamination occurs when radioactive material, in the form of dust, powder, or liquid, comes into contact with a person's skin, hair, or clothing. In other words, the contact is external to a person's body. People who are externally contaminated can become internally contaminated if radioactive material gets into their bodies.

Internal contamination occurs when people swallow or breathe in radioactive materials, or when radioactive materials enter the body through an open wound or are absorbed through the skin. Some types of radioactive materials stay in the body and are deposited in different body organs. Other types are eliminated from the body in blood, sweat, urine, and feces.

Q.24. How radioactive exposure or contamination can happen and spread?

Ans. Radioactive materials could be released into the environment in the following ways:

- A nuclear power plant accident
- An atomic bomb explosion
- An accidental release from a medical or industrial device
- Nuclear weapons testing
- An intentional release of radioactive material as an act of terrorism

People who are externally contaminated with radioactive material can contaminate other people or surfaces that they touch. For example, people who have radioactive dust on their clothing may spread the radioactive dust when they sit in chairs or hug other people.

People who are internally contaminated can expose people near them to radiation from the radioactive material inside their bodies. The body fluids (blood, sweat, urine) of an internally contaminated person can contain radioactive materials. Coming in contact with these body fluids can result in contamination and/or exposure.

Q.25. How our home could become radioactively contaminated?

Ans. People who are externally contaminated can spread the contamination by touching surfaces, sitting in a chair, or even walking through a house. Contaminants can easily fall from clothing and contaminate other surfaces. Homes can also become contaminated with radioactive materials by body fluids from internally contaminated people. Making sure that others do not come in contact with body fluids from a contaminated person will help prevent contamination of other people in the household.

Q.26. How can we limit our radioactive contamination?

Ans. Since radiation cannot be seen, smelled, felt, or tasted, people at the site of an incident will not know whether radioactive materials were involved.

We can take the following steps to limit our level of radioactive contamination;

- **Get out of the immediate area quickly:** Go inside the nearest safe building or to an area to which you are directed by law enforcement or health officials.

- **Remove the outer layer of clothing:** If radioactive material is on your clothes, getting it away from you will reduce the external contamination and decrease the risk of internal contamination. It will also reduce the length of time that you are exposed to radiation.
- If possible, **place the clothing in a plastic bag or leave it in an out-of-the-way area**, such as the corner of a room. Keep people away from it to reduce their exposure to radiation. Keep cuts and abrasions covered when handling contaminated items to avoid getting radioactive material in them.
- **Wash all of the exposed parts of your body** using lots of soap and lukewarm water to remove contamination. This process is called **decontamination**. Try to avoid spreading contamination to parts of the body that may not be contaminated, such as areas that were clothed.
- After authorities determine that **internal contamination may have occurred**, you may be able to take medication to reduce the radioactive material in your body.

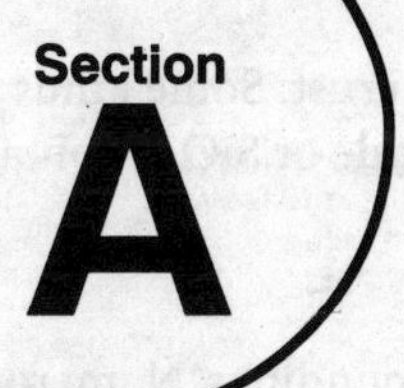

10 Environmental Studies

SHORT QUESTIONS AND ANSWERS ON ENVIRONMENTAL STUDIES

Q.1. Define environment.

Ans. Physical, chemical and biological presence of living and non-living things outside an individual species is called as its environment.

According to ISO 14001, environment can be defined as, "Surroundings in which an organization operates, including air,water,land, natural resources,flora,fauna,humans and inter relation".

Q.2. What are all the categories of environment?

Ans. The main categories of environment are biotic and abiotic environments. The abiotic environment can further be classified into atmosphere (air), lithosphere(soil), and hydrosphere(water). The biotic environment is called as biosphere.

Q.3. Write the components of environment?

Ans. 1. Air (atmosphere) 2. Land (lithosphere) 3. Water (hydrosphere)

Q.4. Mention any two awareness programme of environmental issues to student?

Ans.
1. Participating in seminars and courses related to environment issues.
2. Discussion with people about the environmental problems.

Q.5. Define deforestation?

Ans. Deforestation refers to the removal of plants in the forest.

Q.6. List the causes of deforestation?

Ans.
- Slash and burn farming.
- Commercial agriculture.
- Cattle ranching and livestock grazing.
- Mining and petroleum exploration.
- Infrastructure development.
- Fuel wood collection.
- Tree plantations.

Q.7. Define surface water?

Ans. Precipitation that does not soak into the ground or return to the atmosphere by evaporation or transpiration is called surface water. It forms streams, lakes, wetlands, and artificial reservoirs.

Q.8. What is groundwater?

Ans. The subsurface area where all available soil and rock spaces are filled by water is called the zone of saturation, and the water in these pores is called ground water.

Q.9. What is rock?

Ans. Rock is any material that makes up a large natural, continuous part of the earth's crust. Some kinds of rock, such as limestone (calcium carbonate, or $CaCO_3$) and quartzite(silicon dioxide or SiO_2), contain only one mineral, but most rocks consist of two or more minerals.

Q.10. Define ecosystem.

Ans. Ecosystem has been defined as a system of interaction of organisms with their surroundings. Numerous dynamic interactions are occurring within an ecosystem and these are complex. Always alterations to the biotic and Abiotic components are happening within the ecosystems.

Q.11. Enumerate some characteristics of an ecosystem.

Ans.
- Ecosystem is the major ecological unit.
- It contains both biotic and abiotic components.
- Through the biotic and abiotic components nutrient cycle and energy flow occur.
- The boundary of the ecosystem is not rigidly defined and it is flexible.
- The function of ecosystem is related to the cycling of materials and flow of energy.
- The amount of energy needed to maintain an ecosystem depends on its structure.
- Ecosystems pass from a less to more complex state, which is called as succession.
- Adaptation to local environmental condition is the important feature of the biotic components of an ecosystem, failing which they might perish.

Q.12. What is nitrogen fixation?

Ans. Hydrogen bonding of water molecule is defined as the bonding developed between the positive hydrogen end of one molecule and the negative lone pair of another water molecule. Four hydrogen bonds are formed around each and every water molecules.

Q.13. Define the term nitrification.

Ans. The process of converting the free nitrogen gas available in the atmosphere into compounds of nitrogen is called as nitrogen fixation.

Q.14. What do you mean by denitrification?

Ans. The process of converting ammonia into nitrites with the help of Nitrosomonas bacteria, and nitrites into nitrates with the help of Nitrobacter bacteria is called as nitrification.

Q.15. Define Biome.

Ans. Biome is defined as a major ecological community of organisms occupying in a larger area.

Q.16. Define the term 'anthroposystem'.

Ans. It is defined as the ecosystem developed by the humankind. It is also called as artificial ecosystem.

Q.17. List the major biomes of the world.

Ans. Dessert, Tropical rain forest, Tundra, etc.

Q.18. What are the methods by which nitrogen fixation takes place in the nature?

Ans. Natural nitrogen fixation can be obtained from lightening of clouds and bacteria and fungi present in soil and water. Artificial fixation of nitrogen can be obtained with the help of fertilizer of fertilizer industries, which convert the atmosphere nitrogen into ammonia.

Q.19. Define ecological succession.

Ans. "Ecological succession" is the observed process of change in the species structure of an ecological community over time. Within any community some species may become less abundant over some time interval, or they may even vanish from the ecosystem altogether. Similarly, over some time interval, other species within the community may become more abundant, or new species may even invade into the community from adjacent ecosystems. This observed change over time in what is living in a particular ecosystem is "ecological succession".

Q.20. What are the two types of ecological succession?

Ans. (*i*) Primary ecological succession comprising of Hydrarch (Hydrosere) and Xerarch (Xerosere)

(*ii*) Secondary succession involving establishment of biotic communities.

Q.21. Define primary and secondary succession

Ans. (*i*) Primary ecological succession comprises of Hydrarch(Hydrosere – establishment starting in a watery area like pond and lake) and Xerarch (Xerosere – establishment starting in a dry area like, desert and rock)

(*ii*) Secondary succession involving establishment of biotic communities in an area, where some types of biotic community is already present.

Q.22. What are the forces involved in succession?

Ans. Nutation, Invasion (Migration and Establishment), Competition, Reaction and Stabilization

Q.23. What are the classifications of biotic components of ecosystems?

Ans. Producers, Consumer and Decomposers are the biotic components of an ecosystem

Q.24. How does a biome differ from an ecosystem?

Ans. Biome is defined as a major ecological community of organisms occupying in a larger area.

Q.25. What are the types of grassland ecosystem?

Ans. (*a*) Savannah ecosystem (*b*) Forest ecosystem (*c*) Tundra ecosystem

Q.26. Give the characteristic features of grassland ecosystem.

Ans. Plants(Producers) → Rat (Primary consumer) → Snakes(secondary consumer) → Eagles(Tertiary consumer)

Q.27. What is the structure and function of grassland ecosystem?

Ans. Structure and function of grassland ecosystem. Abiotic: C, H, O, N, P, S, etc supplied by rates, nitrates, phosphates and sulphates. Biotic producers: grasses, forbs and shrubs, consumers: cows, buffaloes, deer, sheep, decomposers: fungi and bacteria.

Q.28. What are the different types of desert ecosystem?

Ans. Different types of desert ecosystem Tropical characterized by only few species, Temperate – characterized by hot summer and cool winter and cold – characterized by cold winters and warm summers

Q.29. Give the characteristic features of desert ecosystem.

Ans. Characteristic of desert ecosystem:

- Desert air is dry and the climate is hot.
- Annual rainfall is less than 2.5 cm
- The soil is very poor in nutrients and organic matter
- Vegetation is very poor.

Q.30. What is the structure and function of desert ecosystem?

Ans. Structure and function of desert ecosystem. Abiotic: temperature, rainfall, sunlight, water, Biotic producers: shrubs, bushes, grasses, consumers: squirrels, mice, foxes, decomposers: fungi and bacteria.

Q.31. What are the different types of aquatic ecosystem?

Ans. Different types of aquatic ecosystem. Fresh water: pond, lake and river and salt water: ocean and estuarine.

Q.32. What are the functional components of ecosystem?

Ans. Biotic and Abiotic are the components of ecosystem

Q.33. Define heterotrophs.

Ans. Those organisms which depend on others (Producers-Autotrophs) for their energy requirements are known as Consumers or Heterotrophs.

Q.34. Name the types of consumers.

Ans.
- Herbivores (or) Primary Consumers
- Carnivores (or) Secondary Consumers
- Omnivores (or) Tertiary Consumers

Q.35. Define food chain and food web.

Ans. A food chain is a picture or model that shows the flow of energy from Autotrophs to a series of organisms in an environment. The energy that it flows can be different for each food chain. The relationships between the components of the food chain are very complex. The network like interaction of organisms is called as food web.

Q.36. What are decomposers?

Ans. The organisms which breakdown the complex compounds into simple products are called as decomposers or reducers.

Q.37. What are the factors to be considered in abiotic ecosystem?

Ans. Climatic factors, Physical factors and Chemical factors.

Q.38. Define biogeochemical cycle.

Ans. Two types of biogeochemical cycles are:
- Gaseous cycle (e.g., N_2 and O_2 cycles);
- Sedimentary cycle (Sulphur and Phosphorus cycles)

Q.39. List some examples for autotrophic succession

Ans. Autotrophs are nothing but producers which synthesize their food through photosynthesis. e.g., All green plants and trees

Q.40. Classify the ecosystem.

Ans. The ecosystem can be generally classified into three types.
- Natural ecosystem
- Artificial ecosystem
- Incomplete ecosystem

Q.41. Describe the structure of an ecosystem.

Ans. An ecosystem has two basic components i) Abiotic components ii) Biotic components The relationship between the abiotic components and the biotic components of the ecosystem is termed 'holocoenosis'.

Abiotic components

Those include the non-living or physico-chemical factors like air, soil, water and the basic compounds and elements of the environment. Abiotic factors are classified broadly under three categories: Climatic factors, including the climatic regime with physical factors in the environment such as light, atmospheric temperature, wind, humidity, etc.; Edaphic factors, which relate to the composition and structure of the soil like its chemical and physical properties – like the soil type, soil profile, organic matter, minerals, soil water, and soil organisms. Inorganic substances like water, carbon, sulphur, nitrogen, phosphorus and so on. Organic substances like proteins, lipids, carbohydrates, humic substances etc.

Biotic components

It consists of the living parts of the environment, including the association of a lot of interrelated populations that belong to different species inhabiting a common environment. The populations are those of the animal community, the plant community and the microbial community. The biotic community is divided into autotrophs, saprotrophs and heterotrophs. Autotrophs (from Greek: auto - self, trophos - feeder) are called producers, transducers or convertors, as well. Those are photosynthetic plants, normally chlorophyll bearing, which synthesize a high-energy complex organic compound (or food) from the inorganic raw materials utilizing the aid of the sun, and this process is called photosynthesis. Autortophs form the core of all biotic systems. In terrestrial ecosystems, autotrophs are usually rooted plants. In the aquatic ecosystems, the floating plants referred to as phytoplankton and the shallow water rooted plants – macrophytes - are the main producers. Heterotrophs (from Greek: heteros - other; trophs - feeder) are the consumers, normally animals that feed on the other organisms. Consumers are also referred to as phagotrophs (phago - to swallow or ingest) while macro consumers are normally herbivores and carnivores. Herbivores are called First order or primary consumers, for they feed directly on green plants. For example, Terrestrial ecosystem consumers are cattle, deer, grass hopper, rabbit, etc. Aquatic ecosystem consumers are protozoans, crustaceans, etc. Carnivores are animals that prey or feed on other animals. Second order consumers or Primary carnivores include those animals that feed on herbivorous animals. For example, fox, frog, smaller fishes, predatory birds, snakes, etc. Third order consumers or Secondary carnivores are the animals that feed on primary carnivores. For example, wolf, owl, peacock, etc. Some larger carnivores prey on Secondary carnivores. Quaternary consumers or Tertiary carnivores include those animals which feed upon secondary carnivores. For example, the lion, the tiger, etc. Those are not eaten by any other animal. The larger carnivores which cannot be preyed on further are also called the top carnivores. Saprotrophs (from Greek again: sapros - rotten; trophos - feeder) are called the reducers or decomposers. They break the complex organic compounds in dead matter down (dead plants and animals). Decomposers don't ingest the food. Instead they secrete a digestive enzyme into the dead, decaying plant or animal remains and digest this organic material. The enzymes act on the complex organic compounds in the dead matter. Decomposers absorb a bit of the decomposition products to provide themselves with nourishment. The remaining substance is added as minerals in the process of mineralization to the substratum. Released minerals are utilized or reused as nutrients by plants - the producers.

Q.42. Describe the function of an ecosystem

Ans. The function of an ecosystem is a broad, vast and often confused topic. The function of an ecosystem can be best studied by understanding the history of ecological studies. The function of an ecosystem can be studied under the three heads.

1. Trophic level interaction 2. Ecological succession 3. Biogeochemistry

Trophic Level Interaction deals with how the members of an ecosystem are connected based on nutritional needs. Ecological Succession deals with the changes in features/members of an ecosystem over a period of time.

Biogeochemistry is focused upon the cycling of essential materials in ecosystem.

Trophic level interaction was developed by zoologist Charles Elton. It deals with who eats who and is eaten by whom in an ecosystem. The study of trophic level interaction in an ecosystem gives us an idea about the energy flow through the ecosystem.

The tropic level interaction involves three concepts namely

1. Food chain 2. Food web 3. Ecological pyramids

Food chain: In an ecosystem one can observe the transfer or flow of energy from one trophic level to other in succession. A trophic level can be defined as the number of links by which it is separated from the producer, or position of the organism in the food chain. The patterns of eating and being eaten forms a linear chain called food chain which can always be traced back to the producers. Thus, primary producers trap radiant energy of sun and transfer that to chemical or potential energy of organic compounds such as carbohydrates, proteins and fats. When a herbivore animal eats a plant (or when bacteria decompose it) and these organic compounds are oxidized, the energy liberated is just equal to the amount of energy used in synthesizing the substances (first law of thermodynamics), but some of the energy is heat and not useful energy (second law of thermodynamics). If this animal, in rum, is eaten by another one, along with transfer of energy from a herbivore to carnivore a further decrease in useful energy occurs as the second animal (carnivore) oxidizes the organic substances of the first (herbivore or omnivore) to liberate energy to synthesize its own cellular constituents. Such transfer of energy from organism to organism sustains the ecosystem and when energy is transferred from individual to individual in a particular community, as in a pond or a lake or a river, we come across the food chains. The numbers of steps in a food chain are always restricted to four or five, since the energy available decreases with each step. Many direct or indirect methods arc employed to study food chain relationships in nature. They include gut content analysis, use of radioactive isotopes, precipitin test, etc.

In nature, basically two types of food chains arc recognized—grazing food chain and detritus food chain.

1. **Grazing food chain:** This type of food chain starts from the living green plants, goes to grazing herbivores and on to the carnivores. Ecosystems with such type of food chain are directly dependent on an influx of solar radiation. Thus, this type of food chain depends on autotrophic energy capture and the movement of this energy to herbivores. Most of the ecosystems in nature follow this type of food chain. These chains are very significant from energy standpoint. The phytoplanktons → zooplanktons → fish sequence or the grasses → rabbit → fox sequence arc the examples of grazing food chain. Further the producer → herbivore → carnivore chain is a predator chain. Parasitic chains also exist wherein smaller organisms consume larger ones without outright killing as the case of the predators.
2. **Detritus food chain:** The organic wastes, exudates and dead matter derived from the grazing food chain are generally termed detritus. The energy contained in this detritus in not lost to the ecosystem as a whole; rather it serves as the source of energy for a group of organisms (detritivores that are separate from the grazing food chain, and generally termed as the detritus food chain. The detritus food chain represents an exceedingly important component in the energy (low of an ecosystem. Indeed in some ecosystems considerably more energy flows through the detritus food chain than through the grazing food chain. In the detritus food chain the energy flow remains as a continuous passage rather than as a stepwise flow between discrete entities. The organisms of the detritus food chain are many and include algae, bacteria, slime molds, actinomycetes, fungi. Protozoa, insects' mites. Crustacea, centipedes, molluscs, rotifers, annelid worms, nematodes and some vertebrates. Some species are highly specific in their food requirements and some can eat almost anything.

   Significance of food chain: The food chain studies/help understand the feeding relationships and the interaction between organisms in an ecosystem. They also help us to appreciate the energy flow mechanism and matter circulation in eco-system, and understand the movement of toxic substances in the eco-system and the problem of biological magnification

Food web: In nature simple food chains occur rarely. The same organism may operate in the ecosystem at more than one trophic level i.e., it may derive its food from more than one source. Even the same organism may be eaten by several organisms of a higher trophic level or an organism may feed upon several different organisms of a lower trophic level. Usually the kind of food changes with the age of the organism and the food availability. Thus in a given ecosystem various food chains are linked together and interested each other to form a complex network called food Web. Generally food webs are not too complex. As more and more species are involved in a web the connectance falls. Expect in insect communities, omnivores are scare and when they occur they usually feed on species in adjacent trophic levels. Within habitats, food webs arc rarely broken up into discrete compartments. The number of species of predators in a food web typically exceeds the number of species of prey by an aver-age of 1.3 predator species per prey species.

Ecological pyramids: In the successive steps of grazing food chain-photosynthetic autotrophs, herbivorous heterotrophs, carnivore's heterotrophy, decay bacteria-the number and mass of the organisms in each step is limited by the amount of energy available. Since some energy is lost as heat, in each transformation the steps become progressively smaller near the top. This relationship is sometimes called ecological pyramid.

The ecological pyramids represent the trophic structure and also trophic function of the ecosystem. In many ecological pyramids, the producer form the base and the successive trophic levels make up the apex. The ecological pyramids may be of following three kinds:

1. **Pyramid of number:** It depicts the number of individual organisms at different trophic levels of food chain. This pyramid was advanced by Charles Elton (1927), who pointed out the great difference in the number of the organisms involved in each step of the food chain. The animals at the lower end (base of pyramid) of the chain are the most abundant. Successive links of carnivores decrease rapidly in number until there are very few carnivores at the top. The pyramid of number ignores the biomass of organisms and it also does not indicate the energy transferred or the use of energy by the groups in volt ved. T he lake ecosystem provides a typical example for pyramid of number.
2. **Pyramid of biomass:** The biomass of the members of the food chain present at any one time forms the pyramid of the biomass. Pyramid of biomass indicates decrease of biomass in each tropical level from base to apex. For example, the total biomass of the producers ingested by herbivores is more than the total biomass of the herbivores in an ecosystem. Likewise, the total biomass of the primary carnivores (or secondary consumer) will be less man the herbivores and so on.
3. **Pyramid of energy:** When production is considered in terms of energy, the pyramid indicates not only the amount of energy flow at each level, but more important, the actual role the various organisms play in the transfer of energy. The base upon which the pyramid of energy is constructed is the quantity of organisms produced per unit lime, or in other words, the rate at which food material passes through the food chain. Some organisms may have a small biomass, but the total energy they dissimilate and pass on, may be considerably greater than that of organisms with a much larger biomass. Energy pyramids are always slopping because less energy is transferred from each level than was paid into it. In cases such as in open water communities the producers have less bulk than consumers but the energy they store and pass on must be greater than that of the next level. Otherwise the biomass that producers support could not be greater than that of the producers themselves. This high energy flow is maintained by a rapid turn over of individual plankton, rather than an increase of total mass.

Q.43. Write notes on values of biodiversity and hotspots of biodiversity.

Ans. Values of biodiversity: Biodiversity is vital to the entire living world since it provides the resources required to sustain life . This includes the air we breathe , water and food as well as the resources we use in our daily lives. The value of particular species will be known only when it is not available. As humans we totally depend on nature for everything. A rich biodiversity is the wealth of a nation. A detailed study shows that we get the following benefits from rich biodiversity.

Food: All of our food comes from other organism. It is estimated that as many as 80,000 edible wild plant species could be utilized by humans. Villages in Indonesia, for instance are known to use some 4,000 native plant and animal species for their food, medicine and other valuable products.

Drugs and medicine: Living organisms provide us many useful drugs and medicine. The United Nations Development Programme estimates the value of pharmaceutical products derived from developing world, plants, animals and microbes to be more than $30 billion per year.

Ecological benefits: We are totally depending on ecological services, which in turn provided by other organisms. The organisms are nothing but value of biodiversity. The following services are generally provided by the organism.

- Soil formatting
- Air and water purification
- Water disposal
- Nutrient cycling
- Food production

It is estimated that the value of these ecological services is at least $33 trillion per year. Thus maintaining biodiversity is essential to preserve these ecological services.

Aesthetic benefits: Millions of people enjoy hunting, fishing, camping, hiking wildlife watching and other natural activities. These activities provide stress relief; good exercise and contact with nature can be psychologically and emotionally restorative.

In many religious natures is considered as God's creation and particular species of plant/animal is being worshipped. Also many plant species are preserved as 'sacred grove'. These can be sustained only because of rich biodiversity.

Social value: The U.S fish and wildlife services estimated that 40% of all adults enjoy wildlife, including 39 million who are hunting or fishing and 76 million, who watch, feed or photograph wildlife. Many communities are finding that local biodiversity can bring cash to remote areas through ecotourism.

Hotspots of biodiversity: The richness of species concentration at a particular region is called biodiversity hotspots. Most of the world's biodiversity concentrations are either near the equator especially tropical rain forests or coral reefs. It was observed that only 10-15% live in North America and Europe, when compared to the distribution species worldwide.

Areas isolated by water, desert or mountains can also have high concentrations of unique species and biodiversity. Madagascar, New Zealand, South Africa and California are all mid latitude areas isolated by barriers that prevent mixing with biological communities from other regions and have a rich population of unusual species.

Q.44. Discuss the threats faced by Indian biodiversity and explain the strategy adopted to conserve biodiversity.

Ans. Threats to biodiversity are numerous and human activity is responsible for most of them.

- **Habitat loss and degradation** affects 86% of all threatened birds, 86% of the threatened mammals assessed and 88% of the threatened amphibians.
- Introductions of **invasive alien species** that establish and spread outside their normal distribution. Some of the most threatening invasive species include cats and rats, green crabs, zebra mussels, the African tulip tree and the brown tree snake. Introductions of alien species can happen deliberately or unintentionally, for example, by organisms "hitch-hiking" in containers, ships, cars or soil.
- **Over-exploitation of natural resources**. Resource extraction, hunting, and fishing for food, pets, and medicine.
- **Pollution and diseases**. For example, excessive fertilizer use leads to excessive levels of nutrients in soil and water.

- **Human-induced climate change**: For example, climate change is altering migratory species patterns, and increasing coral bleaching.

1. **Habitat destruction:** Human disturbances of natural habitat are the largest single cause of loss of biological biodiversity. Over the past 10,000 years billions of hectares of forests, woodlands and grasslands have been converted to commercial forests, croplands or grazing lands.
2. **Fragmentation:** Habitat fragmentation means the reduction of habitat into smaller and smaller, more scattered patches. Fragmentation reduces biodiversity because many species such as bears and large cats require large territories to subsist
3. **Hunting and fishing:** Over harvesting is responsible for depletion or extinction of many species. Hunting of species for human consumption will extinct the entire population. Fish stocks have been seriously depleted by over harvesting in many parts of the world.

Strategy adopted to conserve biodiversity:

India is one of the 12 mega biodiversity countries in the world. The country is divided into 10 biogeographic regions. The diverse physical features and climatic situations have formed ecological habitats like forests, grasslands, wetlands, coastal and marine ecosystems and desert ecosystems, which harbour and sustain immense biodiversity. Biogeographically, India is situated at the tri-junction of three realms - Afro-tropical, Indo-Malayan and Paleo-Arctic realms, and therefore, has characteristic elements from each of them. This assemblage of three distinct realms makes the country rich and unique in biological diversity.

The country is also one of the 12 primary centers of origin of cultivated plants and domesticated animals. It is considered to be the homeland of 167 important plant species of cereals, millets, fruits, condiments, vegetables, pulses, fiber crops and oilseeds, and 114 breeds of domesticated animals.

About 4,900 species of flowering plants are endemic to the country. These are distributed among 141 genera belonging to 47 families. These are concentrated in the floristically rich areas of North-East India, the Western Ghats, North-West Himalayas and the Andaman and Nicobar Islands. These areas constitute two of the 18 hot spots identified in the world. It is estimated that 62 per cent of the known amphibian species are endemic to India of which a majority is found in Western Ghats.

Approximately 65 per cent of the total geographical area has been surveyed so far. Based on this, over 46,000 species of plants and 81,000 species of animals have been described by the Botanical Survey of India (BSI) established in 1890 and Zoological Survey of India (ZSI) established in 1916, respectively. This list is being constantly upgraded, especially in lower plants and invertebrate animals. The Forest Survey of India established in 1981 assess the forest cover with a view to develop an accurate database for planning and monitoring purposes.

Conservation and sustainable use of biological resources based on local knowledge systems and practices is ingrained in Indian ethos. The country has a number of alternative medicines, like Ayurveda, Unani, Siddha and Homeopathic systems which are predominantly based on plant based raw materials in most of their preparations and formulations. Herbal preparations for various purposes including pharmaceutical and cosmetic purposes form part of the traditional biodiversity uses in India.

The strategies for conservation and sustainable utilization of biodiversity have comprised providing special status and protection to biodiversity - rich areas by declaring them as national parks, wildlife sanctuaries, biosphere reserves, ecologically fragile and sensitive areas. Other strategies include off-loading pressure from reserve forests by alternative measures of fuelwood and fodder need satisfaction by afforestation of degraded areas and wastelands and creation of ex-situ conservation facilities such as gene banks. For example, the Tura Range in Garo Hills of Meghalaya is a gene sanctuary for preserving the rich native diversity of wild citrus and musa species.

Approximately, 4.2 per cent of the total geographical area of the country has been earmarked for extensive in-situ conservation of habitats and ecosystems. A protected area network of 85 national parks and 448 wildlife sanctuaries has been created. The results of this network have been significant in restoring viable population of large mammals such as tiger, lion, rhinoceros, crocodiles and elephants.

The Indian Council of Forestry Research and Education (ICFRE) has identified 309 forest preservation plots of representative forest types for conservation of viable and representative areas of biodiversity. Out of these plots, 187 area in natural forests and 112 are in plantations, covering a total area of 8,500 hectares.

A programme "Eco-development" for in-situ conservation of biological diversity involving local communities has been initiated in recent years. The concept of eco-development integrates the ecological and economic parameters for sustained conservation of ecosystems by involving the local communities with the maintenance of earmarked regions surrounding protected areas. The economic needs of the local communities are taken care under this programme through provision of alternative sources of income and a steady availability of forest and related produce.

Programmes have also been launched for scientific management and wise use of wetlands, mangroves and coral reef ecosystems. Twenty one wetlands, and mangrove areas and 4 coral reef areas have been identified for intensive conservation and management purposes. Mangroves conservation is one of the thrust areas of the Ministry of Environment and Forests.

Six significant wetlands of India have been declared as "Ramsar Sites" under the Ramsar Convention. Under the World Heritage Convention, five natural sites have been declared as "World Heritage Sites".

To conserve the representative ecosystems, a Biosphere Reserve Programme is being implemented. Twelve biodiversity rich areas of the country have been designated as Biosphere Reserves applying the UNESCO/MAB criteria. These reserves aim at conserving the biological diversity and genetic integrity of plants, animals and microorganisms in their totality as part of the natural ecosystems, so as to ensure that self-perpetuation and unhindered evolution of the living resources.

The Ministry of Environment and Forests constituted the National Afforestation and Eco-development Board (NAEB) in August 1992. NAEB has evolved specific schemes for promoting afforestation and management strategies which help the states in developing specific afforestation and management strategies and eco-development packages for augmenting biomass production through a participatory planning process of joint forest management and microplanning.

To complement in-situ conservation, attention has been paid to ex-situ conservation measures. According to currently available survey, central government and state governments together run and manage 33 botanical gardens. Universities have their own botanical gardens. There are 275 zoos, deer parks, safari parks and aquaria. A Central Zoo Authority was set up to secure better management of zoos. A scheme 'Assistance to Botanical Gardens' provides one-time assistance to botanical gardens to strengthen and institute measures for ex-situ conservation of threatened and endangered species in their respective regions.

Formal policies and programmes for conservation and sustainable utilisation of biodiversity resources dates back to several decades. The concept of environmental protection is enshrined in the Indian constitution in articles 48a and 51a(g).

Major central acts relevant to biodiversity include Forest Act, 1927, Wildlife (Protection) Act, 1972, Forest (Conservation) Act, 1980, and Environment (Protection) Act, 1986. The various central Acts are supported by a number of state laws and statutes concerning forests and other natural resources. The policies and strategies directly relevant to biodiversity include National Forest Policy amended in 1988, National Conservation Strategy and Policy Statement for Environment and Sustainable Development, National Agricultural Policy, National Land Use Policy, National Fisheries Policy, National Policy and Action Strategy on Biodiversity, National Wildlife Action Plan and Environmental Action Plan.

India is a party to the Convention on Biological Diversity (CBD). The main objectives of this convention are; conservation of biological diversity, sustainable use of the components of biodiversity and fair and equitable sharing of benefits arising out of the utilization of genetic resources.

The main implementation measures for the CBD are through national strategies, legislation, and administrative instruments to be developed in accordance with the country's conditions and capabilities.

Adopting a consultative process with the stakeholders, a national policy and action strategy on biological diversity has been drawn up as a macro-level statement of strategies. This plan aims at consolidating the on-going efforts of conservation and sustainable use of biological diversity, identifying gaps in various sectors and providing a policy and programme regime to ensure attainment of the three objectives of the CBD. This fame work document provides the basis for developing detailed action at micro and regional levels.

The ministry has accessed funding from GEF for preparing detailed National Biodiversity Strategy and Action Plan (NBSAP) using participatory planning approach and is currently being implemented.

Identifying the critically important areas and gaps in taxonomic work, an All India Coordinated Project on Taxonomy (AICOPTAX) has been launched for prioritized purpose-oriented capacity building in taxonomy. To encourage taxonomic work, Janaki Ammal National Award in taxonomy has been instituted.

Q.45. Discuss the biogeographical classification of India.

Ans. Biogeography is the study of the biosphere and of human effects on plants and animals. According to a recent classification done by the Wild-Life Institute of India, the country has ten biogeographical zones:

1. **Trans-Himalayan region:** The Himalayan ranges immediately north of the Great Himalayan range are called the Trans-Himalayas. The Trans-Himalayan region with its sparse vegetation has the richest wild sheep and goat community in the world. The snow leopard is found here, as is the migratory black-necked crane.
2. **Himalayas:** The Himalayas consist of the youngest and loftiest mountain chains in the world. The Himalayas have attained a unique personality owing to their high altitude, steep gradient and rich temperate flora.

   The forests are very dense with extensive growth of grass and evergreen tall trees. Oak, chestnut, conifer, ash, pine, deodar are abundant in Himalayas. There is no vegetation above the snowline. Several interesting animals live in the Himalayan ranges. Chief species include wild sheep, mountain goats, ibex, shrew, and tapir. Panda and snow leopard are also found here.
3. **Semi-arid areas:** Adjoining the desert are the semi-arid areas, a transitional zone between the desert and the denser forests of the western ghats. The natural vegetation is thorn forest. This region is characterized by discontinuous vegetation cover with open areas of bare soil and soil-water deficit throughout the year.

   Thorny scrubs, grasses and some bamboos are present in some regions. A few species of xerophytic herbs and some ephemeral herbs are found in this semi-arid tract. Birds, jackals, leopards, eagles, snakes, fox, buffaloes are found in this region.
4. **Western ghats:** The mountains along the west coast of peninsular India are the western ghats, which constitute one of the unique biological regions of the world. The western ghats extend from the southern tip of the peninsula (8°N) northwards about 1600 km to the mouth of the river Tapti (21°N).

   The mountains rise to average altitudes between 900 and 1500 m above sea level, intercepting monsoon winds from the southwest and creating a rain shadow in the region to their East.

The varied climate and diverse topography create a wide array of habitats that support unique sets of plant and animal species. Apart from biological diversity, the region boasts of high levels of cultural diversity, as many indigenous people inhabit its forests.

The western ghats are amongst the 25 biodiversity hot-spots recognized globally. These hills are known for their high levels of endemism expressed at both higher and lower taxonomic levels. Most of the western ghats endemic plants are associated with evergreen forests.

The region also shares several plant species with Sri Lanka. The higher altitude forests were, if at all, sparsely populated with tribal people. Rice cultivation in the fertile valley proceeded gardens of early commercial crops like areca nut and pepper. The original vegetation of the ill-drained valley bottoms with sluggish streams in elevations below 100 m would be often a special formation, the Myristic swamp.

Expansion of traditional agriculture and the spread of particularly rubber, tea, coffee and forest tree plantations would have wiped out large pockets of primary forests in valleys. The Western Ghats are well-known for harboring 14 endemic species of caecilians (i.e., legless amphibians) out of 15 recorded from the region so far.

5. **North-west desert regions:** This region consists of parts of Rajasthan, Kutch, Delhi and parts of Gujarat. The climate is characterized by very hot and dry summer and cold winter. Rainfall is less than 70 cm. The plants are mostly xerophytic. Babul, Kikar, wild palm grows in areas of moderate rainfall. Indian Bustard, a highly endangered bird is found here. Camels, wild asses, foxes, and snakes are found in hot and arid deserts.
6. **Deccan plateau:** Beyond the ghats is Deccan Plateau, a semi-arid region lying in the rain shadow of the western ghats. This is the largest unit of the Peninsular Plateau of India. The highlands of the plateau are covered with different types of forests, which provide a large variety of forest products.
7. **Gangetic plain:** In the north is the Gangetic plain extending up to the Himalayan foothills. This is the largest unit of the great plain of India. Ganga is the main river after whose name this plain is named. The aggradation Great Plains cover about 72.4 mha area with the Ganga and the Brahmaputra forming the main drainage axes in the major portion.

   The thickness in the alluvial sediments varies considerably with its maximum in the Ganga plains. The physio geographic scenery varies greatly from arid and semi-arid landscapes of the Rajasthan Plains to the humid and per-humid landscapes of the Delta and Assam valley in the east.

   Topographic uniformity, except in the arid western Rajasthan is a common feature throughout these plains. The plain supports some of the highest population densities depending upon purely agro-based economy in some of these areas. The trees belonging to these forests are teak, sal, shisha, mahua, khair etc.
8. **North-east India:** North-east India is one of the richest flora regions in the country. It has several species of orchids, bamboos, ferns and other plants. Here the wild relatives of cultivated plants such as banana, mango, citrus and pepper can be found.
9. **Islands:** The two groups of islands, i.e., the Arabian Sea islands and Bay Islands differ significantly in origin and physical characteristics. The Arabian Sea Islands (Laccadive, Minicoy, etc.) are the foundered remnants of the old land mass and subsequent coral formations. On the other hand, the Bay Islands lay only about 220 km.

   Away from the nearest point on the main land mass and extend about 590 km. With a maximum width of 58 kms the island forests of Lakshadweep in the Bay of Bengal have some of the best-preserved evergreen forests of India. Some of the islands are fringed with coral reefs. Many of them are covered with thick forests and some are highly dissected.

10. **Coasts:** India has a coastline extending over 5,500 km. The Indian coasts vary in their characteristics and structures. The west coast is narrow except around the Gulf of Canbary and the Gulf of Kutch. In the extreme south, however, it is somewhat wider along the south Sahyadri.

    The backwaters are the characteristic features of this coast. The east coast plains, in contrast are broader due to depositional activities of the east-flowing rivers owing to the change in their base levels.

    Extensive deltas of the Mahanadi, Godavari, Krishna and Kaveri are the characteristic features of this coast. Mangrove vegetation is characteristic of estuarine tracts along the coast for instance, at Ratnagiri in Maharashtra.

    Larger parts of the coastal plains are covered by fertile soils on which different crops are grown. Rice is the main crop of these areas. Coconut trees grow all along the coast.

Q.46. Discuss India's mega diverse biodiversity.

Ans. Biodiversity: There is much diversity of flora and fauna in the oceans — India has over 6000 km of coastline with an area of 20,13,410 km^2. It harbors a great variety of marine biota including sea weeds, fish, crustaceans, amphibia, molluscs, corals, reptiles, mammals, etc.

India has 45,000 species of plants (15% of known world plants) and is an important primary centre of origin of about 167 important cultivated plant species like rice, sugarcane, jute, millets, cucurbits, legumes, mango, banana, Dioscoreas, turmeric, spices, many ornamental plants like orchids, medicinal and aromatics. India is also a secondary center of maize, red pepper, soyabean and rubber plants.

There are 15,000 species of flowering plants, 35% of which are endemics and largely concentrated in Western Ghats, North-east and North-west Himalaya, and Andaman and Nicobar islands. About 5000 plants species occur on Western Ghats of Kerala, 235 being endemic. North-east India contains 675 orchid species (1000 in India and 17,000 in the World).

India has about 75,000 species of animals, 80% of which are insects. India has many endemic animal species — 53% fresh water fish, 60% amphibians, 36% reptiles and 10% of mammals. India has wide diversity in domestic animals like cows, buffaloes, sheep, goats, pigs, camels, yaks, horses, poultry etc.

Characteristics of India's biodiversity. Biodiversity is not uniformly distributed in space and time. It is rich in tropics. India's biodiversity is characterized by the following

1. India contains 10 bio-geographic regions which include the Himalayan, Trans Himalayan, the Indian desert, the Semi-arid zone, the Western-Ghats, the Deccan Peninsula, the Gangetic Plain, North-East India and the Islands and Coasts which possess different biodiversity levels.
2. India is one of the world's 12 leading biodiversity centers of the origin of cultivated plants.
3. India has 5 natural world heritage sites, 14 biosphere reserves, 89 national parks, 492 wildlife sanctuaries and 2 hotspots. Heritage sites are the places that attract tourists.
4. About 33 per cent of the country's recorded flora are endemic to India and concentrated in the North-East, Western-Ghats, North-West Himalaya and Andaman and Nicobar islands.

Q.47. Discuss the structure and function of grassland ecosystem.

Ans. A biological community that contains few trees or shrubs, is characterized by mixed herbaceous (non-woody) vegetation cover, and is dominated by grasses or grass-like plants. Mixtures of trees and grasslands occur as savannas at transition zones with forests or where rainfall is marginal for trees. About 1.2×10^8 mi^2 (4.6×107 km^2) of the Earth's surface is covered with grasslands, which make up about 32% of the plant cover of the world. In North America, grasslands include the Great Plains, which extend from southern Texas into Canada. The European meadows cross the subcontinent, and the Eurasian steppe ranges from Hungary eastward through Russia to Mongolia; the pampas cover much of the interior of Argentina and Uruguay. Vast and varied savannas and velds can be found in central and southern Africa and throughout much of Australia.

Grasslands occur in regions that are too dry for forests but that have sufficient soil water to support a closed herbaceous plant canopy that is lacking in deserts. Thus, temperate grasslands usually develop in areas with 10–40 in. (25–100 cm) of annual precipitation, although tropical grasslands may receive up to 60 in. (150 cm). Grasslands are found primarily on plains or rolling topography in the interiors of great land masses, and from sea level to elevations of nearly 16,400 ft (5000 m) in the Andes. Because of their continental location they experience large differences in seasonal climate and wide ranges in diurnal conditions. In general, there is at least one dry season during the year, and drought conditions occur periodically.

Significant portions of the world's grasslands have been modified by grazing or tillage or have been converted to other uses. The most fertile and productive soils in the world have developed under grassland, and in many cases the natural species have been replaced by cultivated grasses (cereals).

Different kinds of grasslands develop within continents, and their classification is based on similarity of dominant vegetation, presence or absence of specific dominant species, or prevailing climate conditions

The climate of grasslands is one of daily and seasonal extremes. Deep winter cold does not preclude grasslands since they occur in some of the coldest regions of the world. However, the success of grasslands in the Mediterranean climate shows that marked summer drought is not prohibitive either. In North America, the rainfall gradient decreases from an annual precipitation of about 40 in. (100 cm) along the eastern border of the tall grass prairie at the deciduous forest to only about 8 in. (20 cm) in the short grass prairies at the foothills of the Rocky Mountains. A similar pattern exists in Europe. Growing-season length is determined by temperature in the north latitudes and by available soil moisture in many regions, especially those adjacent to deserts. Plants are frequently subjected to hot and dry weather conditions, which are often exacerbated by windy conditions that increase transpiration water loss from the plant leaves.

Soils of mesic temperate grasslands are usually deep, about 3 ft (1 m), are neutral to basic, have high amounts of organic matter, contain large amounts of exchangeable bases, and are highly fertile, with well-developed profiles. The soils are rich because rainfall is inadequate for excessive leaching of minerals and because plant roots produce large amounts of organic material. With less rainfall, grassland soils are shallow, contain less organic matter, frequently are lighter coloured, and may be more basic. Tropical and subtropical soils are highly leached, have lower amounts of organic material because of rapid decomposition and more leaching from higher rainfall, and are frequently red to yellow.

Grassland soils are dry throughout the profile for a portion of the year. Because of their dense fibrous root system in the upper layers of the soil, grasses are better adapted than trees to make use of light rainfall showers during the growing season. When compared with forest soils, grassland soils are generally subjected to higher temperatures, greater evaporation, periodic drought, and more transpiration per unit of total plant biomass.

Throughout the year, flowering plants bloom in the grasslands with moderate precipitation, and flowers bloom after rainfall in the drier grasslands. With increasing aridity and temperature, grasslands tend to become less diverse in the number of species; they support more warm-season species; the complexity of the vegetation decreases; the total above-ground and below-ground production decreases; but the ratio of above-ground to below-ground biomass becomes smaller.

There are many more invertebrate species than any other taxonomic group in the grassland ecosystem. Invertebrates play several roles in the ecosystem. For example, many are herbivorous, and eat leaves and stems, whereas others feed on the roots of plants. Earthworms process organic matter into small fragments that decompose rapidly, scarab beetles process animal dung on the soil surface, flies feed on plants and are pests to cattle, and many species of invertebrates are predaceous and feed on other invertebrates. Soil nematodes, small non-arthropod invertebrates, include forms that are herbivorous, predaceous, or saprophagous, feeding on decaying organic matter.

Most of the reptiles and amphibians in grassland ecosystems are predators. Relatively few bird species inhabit the grassland ecosystem, although many more species are found in the flooding pampas of Argentina than in the dry grasslands of the western United States. Their role in the grassland ecosystem involves consumption of seeds, invertebrates, and vertebrates; seed dispersal; and scavenging of dead animals.

Small mammals of the North American grassland include moles, shrews, gophers, ground squirrels, and various species of mice. Among intermediate-size animals are the opossum, fox, coyote, badger, skunk, rabbit, and prairie dog; large animals include various types of deer and elk. The most characteristic large mammal species of the North American grassland is the bison, although many of these animals were eliminated in the late 1800s. Mammals include both ruminant (pronghorns) and non-ruminant (prairie dogs) herbivores, omnivores (opossum), and predators (wolves

Except for large mammals and birds, the animals found in the grassland ecosystem undergo relatively large population variations from year to year. These variations, some of which are cyclical and others more episodic, are not entirely understood and may extend over several years. Many depend upon predator–prey relationships, parasite or disease dynamics, or weather conditions that influence the organisms themselves or the availability of food, water, and shelter.

Within the grassland ecosystem are enormous numbers of very small organisms, including bacteria, fungi, algae, and viruses. From a systems perspective, the hundreds of species of bacteria and fungi are particularly important because they decompose organic material, releasing carbon dioxide and other gases into the atmosphere and making nutrients available for recycling. Bacteria and some algae also capture nitrogen from the atmosphere and fix it into forms available to plants.

Much of the grassland ecosystem has been burned naturally, probably from fires sparked by lightning. Human inhabitants have also routinely started fires intentionally to remove predators and undesirable insects, to improve the condition of the rangeland, and to reduce cover for predators and enemies; or unintentionally. Thus, grasslands have evolved under the influences of grazing and periodic burning, and the species have adapted to withstand these conditions. If burning or grazing is coupled with drought, however, the grassland will sustain damage that may require long periods of time for recovery by succession processes.

Q.48. Discuss the structure and function of aquatic ecosystem.

Ans. An aquatic ecosystem **is an** ecosystem in a body of water. Communities of organisms that are dependent on each other and on their environment live in aquatic ecosystems. The two main types of aquatic ecosystems are marine ecosystems and freshwater ecosystems.

Types:

(*a*) Marine (*b*) Freshwater (*c*) Ponds.

Marine ecosystem: Marine ecosystems cover approximately 71% of the Earth's surface and contain approximately 97% of the planet's water. They generate 32% of the world's net primary production. They are distinguished from freshwater ecosystems by the presence of dissolved compounds, especially salts, in the water. Approximately 85% of the dissolved materials in seawater are sodium and chlorine. Seawater has an average salinity of 35 parts per thousand (ppt) of water. Actual salinity varies among different marine ecosystems.

Marine ecosystems can be divided into the following zones: oceanic (the open part of the ocean where animals such as whales, sharks, and tuna live); profundal (bottom or deep water); benthic (bottom substrates); intertidal (the area between high and low tides); estuaries; salt marshes; coral reefs; and hydrothermal vents (where chemosynthetic sulfur bacteria form the food base).

Classes of organisms found in marine ecosystems include brown algae, dinoflagellates, corals, cephalopods, echinoderms, and sharks. Fishes caught in marine ecosystems are the biggest source of commercial foods obtained from wild populations

Environmental problems concerning marine ecosystems include unsustainable exploitation of marine resources (for example overfishing of certain species), marine pollution, climate change, and building on coastal areas.[1]

Freshwater: Freshwater ecosystems cover 0.80% of the Earth's surface and inhabit 0.009% of its total water. They generate nearly 3% of its net primary production. Freshwater ecosystems contain 41% of the world's known fish species.

There are three basic types of freshwater ecosystems:

- **Lentic:** slow-moving water, including pools, ponds, and lakes.
- **Lotic:** rapi dly-moving water, for example streams and rivers.
- **Wetlands:** areas where the soil is saturated or inundated for at least part of the time.

Lake ecosystems can be divided into zones: pelagic (open offshore waters); profundal; littoral (near shore shallow waters); and riparian (the area of land bordering a body of water).

Wetlands naturally form along most lakeshores, the width of the wetland and littoral zone being dependent upon the slope of the shoreline and the amount of natural change in water levels, within and among years. The production of the lake as a a whole is the result of production from plants growing in wetlands and shorelines, combined with production from plankton growing in the open water.

Two important subclasses of lakes are ponds, which typically are small lakes that intergrade with wetlands, and water reservoirs. Over long periods of time, lakes, or bays within them, may gradually become enriched by nutrients and slowly fill in with organic sediments, a process called succession. When humans use the watershed, the volumes of sediment entering the lake can accelerate this process. The addition of sediments and nutrients to a lake is known as eutrophication.

The major zones in river ecosystems are determined by the river bed's gradient or by the velocity of the current. Faster moving turbulent water typically contains greater concentrations of dissolved oxygen, which supports greater biodiversity than the slow moving water of pools. This distinction forms the basis for the division of rivers into upland and lowland rivers. The food base of streams within riparian forests is mostly derived from the trees, but wider streams and those that lack a canopy derive the majority of their food base from algae. Anadromous fish are also an important source of nutrients. Environmental threats to rivers include loss of water, dams, chemical pollution and introduced species. A dam produces negative effects that continue down the watershed. The most important negative effects are the reduction of spring flooding, which damages wetlands, and the retention of sediment, which leads to loss of deltaic wetlands.

Wetlands are dominated by vascular plants that have adapted to saturated soil. There are four main types of wetlands: swamp, marsh, fen and bog. Wetlands are the most productive natural ecosystems in the world because of the proximity of water and soil. Hence they support large numbers of plant and animal species. Due to their productivity, wetlands are often converted into dry land with dykes and drains and used for agricultural purposes. The construction of dykes, and dams, has negative consequences for individual wetlands and entire watersheds. Their closeness to lakes and rivers means that they are often developed for human settlement. Once settlements are constructed and protected by dykes, the settlements then become vulnerable to land subsidence and ever increasing risk of flooding. The Louisiana coast around New Orleans is a well known example; the Danube Delta in Europe is another.

Ponds: These are a specific type of freshwater ecosystem with shallow water, marsh, and aquatic plants. They can be further divided into four zones: vegetation zone, open water, bottom mud and surface film. The size and depth of ponds often varies greatly with the time of year; many ponds are produced by spring flooding by rivers. Food webs are based both on free-floating algae and upon aquatic plants. There is usually a diverse array of aquatic life, with a few examples including algae, snails, fish, beetles, water bugs, frogs, turtles, otters and muskrats. Top predators may include large fish, herons, or alligators. Since fish are a major predator upon amphibian larvae, ponds that dry up each year, thereby killing resident fish, provide important refugia for amphibian breeding. Ponds that dry up completely each year are often known as vernal pools. Some ponds are produced by animal activity, including alligator holes and beaver ponds, and these add important diversity to landscapes.

Q.49. Discuss the different levels of hierarchical to understand the concept of biodiversity. List out major threats to biodiversity.

Ans. Biodiversity refers to the variety and variability among all groups of living organisms. The three levels of organisms are discussed below:

(*a*) **Genetic diversity:** It is the basic of biodiversity. The genes found in organisms can form enormous number of combinations each of which gives rise to some variability. Genes are the basic units of hereditary information transmitted from one generation to other. When the genes within the same species show different versions due to new combinations, it is called genetic variability. For example, all rice varieties belong to the species Oryza sativa, but there are thousand of wild and cultivated varieties of rice which show variations at the genetic level and differ in their colour, size, shape, aroma and nutrient content of the grain. This is the genetic diversity of rice.

(*b*) **Species diversity:** This is the variability found within the population of a species or between different species of a community. It represents broadly the species richness and their abundance in a community. There are two popular indices of measuring species diversity known as Shannon-Wiener index and Simpson index.

What is the number of species on this biosphere? The estimates of actual number vary widely due to incomplete and indirect data. The current estimates given by Wilson in 1992 put the total number of living species in a range of 20 million to 50 million. Till now only about 1.5 million living and 300,000 fossil species have been actually described and given may become extinct even before they are discovered and enlisted.

(*c*) **Ecosystem diversity:** This is the diversity of ecological complexity showing variations in ecological niches, trophic structure, food-webs, nutrients cycling, etc. The ecosystem also show variations with respect to physical parameters like moisture, temperature, altitude, precipitation, etc. Thus, there occurs tremendous diversity within the ecosystems, along theses gradients. We may consider diversity in forest ecosystem, which is supposed to have mainly a dominance of trees. But, while considering a tropical rainforest, a tropical deciduous forest, a temperature deciduous forest and a boreal due to variations in the above mentioned physical factors. The ecosystem diversity is of great value that must be kept intact. This diversity has it would disrupt the ecological balance. We cannot even replace the diversity of one ecosystem by that of another. Coniferous trees of boreal forests cannot and vice versa, because ecosystem diversity has evolved with respect of the prevailing environmental conditions with well-regulated ecological balance

Threats to biodiversity: Extinction means the elimination of a species .It can be either natural or human caused. Particular species. Some of them are:

- **Habitat destruction:** Human disturbances of natural habitat is the largest single cause of loss of biological biodiversity. Over the past 10,000 years billions of hectares of forests, woodlands and grasslands have been converted to commercial forests, croplands or grazing lands.

- **Fragmentation:** Habitat fragmentation means the reduction of habitat into smaller and smaller, more scattered patches. Fragmentation reduces biodiversity because many species such as bears and large cats require large territories to subsist.
- **Hunting and fishing:** Over harvesting is responsible for depletion or extinction of many species. Hunting of species for human consumption will extinct the entire population. Fish stocks have been seriously depleted by over harvesting in many parts of the world.

Q.50. Write notes on forest ecosystem.

Ans. Forest ecology is the scientific study of the interrelated patterns, processes, flora, fauna and ecosystems in forests. The management of forests is known as forestry, silviculture, and forest management. A forest ecosystem is a natural woodland unit consisting of all plants, animals and micro-organisms (Biotic components) in that area functioning together with all of the non-living physical (abiotic) factors of the environment.

Forest ecosystem: Plants have been dominating the earth for about 3.0 billion years. They have the unique art of manufacturing their own food by photosynthesis from nature and the rest of the living World depends on them for their food and sustenance. Plants constitute 99 percent world for food. If this ratio(99 : 1) is distributed by elimination of plants (i.e., deforestation), then the natural balance will be lost and the entire living world will suffer most.

The dynamic balance is among plants, bacteria and microorganisms and animals including man. Once this dynamic balance is upset, there is ecological crisis whereby the entire biosphere is in danger.

Forests are renewable resources and have a key role in improving the quality of environment by exerting beneficial effect on the life support system. Moreover, forest also contribute much to the economic development of the country by providing goods and services to people and industry. They are the treasure house of valuable plant and animal genes and medicinal plants, most of which are yet to be discovered.

Some of the forest types are:

- Conifer forests
- Deciduous and evergreen forests
- Tropical moist forests

Coniferous forests: These consist of cone-bearing trees. The thin, needle-like leaves of conifers have small surface area and thick, waxy coatings that help trees minimize moisture loss. These occur where moisture limited by sandal soil, low precipitation or a short growing season. These can survive harsh winters or extended droughts. In many areas, frequent fires help to keep the forest open and encourage seedling growth.

Among the dominant conifers are pine, hemlock, spruce, cedar and fir. Warm-climate pine forests occur at lower latitudes. In the United States, southern pine forests grow in a warm, moist climate on sandy soils.

Temperate rainforests, such as those in the Pacific Northwest, are also dominated by conifers. Here, mild temperatures and rainfall as high as 250 cm per year support luxuriant growth and huge trees, such as the California redwood, the largest tree in the world.

Deciduous and evergreen forests: Forests of broad-leaved trees occur throughout the world, where rainfall is plentiful. In temperature regions, the climate supports lush summer plant summer growth when water us scarce or the ground is frozen, deciduous trees shed their leaves to conserve water.

Temperate deciduous forest contains rich and varied association of oak, birch, elm trees. If the climate is warm year-round forest are dominant by broadleaf evergreen trees., which keep their leaves year-round. These include the live oaks, cypresses. Broadleaf, evergreen also dominate tropical rainforests.

These forests provide good habitat for tigers, bears, cranes and other unique and endangered species.

Tropical moist forests: The humid tropical regions of South and Central America, Africa, Southeast Asia, and some of the pacific Islands support one of the most complex and biologically rich biome types in the world. Tropical rainforests occur where rainfall is abundant – more than 200 cm per year and temperatures are warm to hot year-round.

The types of soil in these forests are acidic, and nutrient-poor but very in species concentration. For example, the number of insect species found in the canopy of tropical rainforests has been estimated to be in millions. It has been estimated that one-half to two-thirds of all species of terrestrial plants and insects live in tropical forests. The good growth in tropical rainforests

Depends on rapid decomposition and recycling of dead organic material. Once the forest cleared for logging agriculture, etc. it may not be recovered.

Tropical seasonal forests: Many areas in India, South-east Asia, Australia, West Africa and South America, have tropical regions characterized

By distinct wet dry seasons instead of uniform heavy rainfall throughout the year, although temperatures are hot yet – around.

The resulting forest type of tropical seasonal forests, include semi evergreen, deciduous, and grassy savannas.

Tropical dry forests have typically been more attractive than wet forests for human habitation and have suffered greater degradation. Soils of dry forests often have higher nutrients levels and are more agriculturally productive than those of a rain forest. Destruction of these forests is at faster rate by human interference.

Q.51. What are the main reasons for extinction of biodiversity?

Ans. Reasons for extinction of biodiversity:

- **Habitat loss:** The natural habitat may be destroyed by man for his settlements, grazing grounds, agriculture, mining, industries, dam building, etc. As a consequence of this, the species must adapt to the changes, move elsewhere or may succumb to predation, starvation or disease, eventually dies. In India, several rare butterfly species are facing extinction with the uncannily swift habitat destruction of the Western Ghats.
- **Hunting:** Man has always hunted the animals for its food, wild animals are hunted for their products such as hide and skin, tusk, for meat, pharmaceuticals, perfumes, cosmetics, decoration purposes.

  In our country, rhinoceros is hunted for its horns, tiger for bones and skin, musk deer for musk, elephant for ivory, etc. One of the most published commercial hunts is that of whale.
- **Over exploitation:** This is one of the main causes of the loss of not only economic species but also biological curiosities like the insectivorous, primitive species and other taxa needed for teaching of lab work. Commercial exploitation of wild plants has invariably meant its overuse, eventual destruction.
- **Collection for zoo and research:** Animals and plants are collected throughout the world for zoo and biological lab for study and research in science and medicines.
- **Introduction of exotic species:** Native species are subjected to competition for food and space due to introduction of exotic species. For example, introduction of goats and rabbits in the Pacific and Indian regions has resulted in destruction of habitats of several plants, birds and reptiles.
- **Control of pest and predators:** Predators and pest control measures, generally kill predators that are a component of balanced ecosystem and may also indiscriminately poison non-target species.

- **Pollution:** Pollution alters the natural habitat. Water pollution especially injurious to the biotic components of estuary and coastal ecosystem. Toxic wastes entering the water bodies disturb the food chain and so to the aquatic ecosystems. Insecticides, pesticides sulphur dioxide, nitrogen oxides, acid rain, ozone depletion and global warming too, affect adversely the plant and animal species.

  The impact of coastal pollution is also very important it is seen that coral reefs are being threatened by pollution from industrialization along the coast, oil transport and offshore mining. Noise pollution is also the cause of wildlife extinction.
- **Deforestation:** One of the main causes for the loss of wild life is population explosion and resultant deforestation. Deforestation mainly results from population settlement, shifting cultivation, development projects, demand for fuel wood, demand of wood as a raw material for many industries such as paper and pulp, match, veneer and plywood, furniture, etc.
- **Other factors:** Other ecological factors that may also contribute to the extinction of wildlife are as follows:
  (*i*) **Distribution range:** The smaller the range of distribution, the greater the threat of extinction.
  (*ii*) **Degree of specialization:** The more specialized an organism is, the more vulnerable it is to extinction.
  (*iii*) **Substitution:** During the process of evolution an existing species may be replaced by ecological another one.
  (*iv*) **Position of the organism in the food chain:** The higher the position of the organism is in food chain, the more susceptible it becomes.
  (*v*) **Reproduction rate:** Large organisms tend to produce fewer offspring at widely spaced intervals.
  (*vi*) **Outbreak of diseases:** It is also one of the major causes for the decline in wildlife species.
  (*vii*) **Loss of gene flow:** The individuals of plant and animal life may decline to significant level as a result of loss of gene flow.

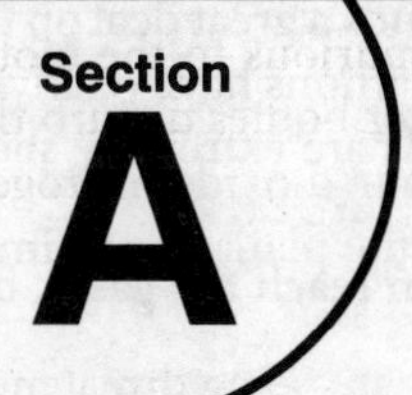

Section A

11 Ozone Layer Depletion and its Health Effects

SHORT QUESTIONS AND ANSWERS ON OZONE LAYER DEPLETION AND ITS HEALTH EFFECTS

Q.1. Describe the process of production and destruction of ozone.

Ans. The ozone molecule (O_3) contains three atoms of oxygen and is mainly formed by the action of the UV rays of the sun on oxygen molecules (O_2) in the upper part of Earth's atmosphere (called the stratosphere). Ozone is also produced locally near Earth's surface from the action of UV radiation on some air pollutants.

About 90% of all ozone molecules are found in the stratosphere, a region that begins about 10-16 kilometers above Earth's surface and extends up to about 50 kilometers. Most of this ozone is found in the lower stratosphere in what is commonly known as the "ozone layer". The stratospheric ozone layer protects life on Earth by absorbing most of the harmful UV radiation from the sun. The remaining 10% of ozone is in the troposphere, between Earth's surface and the stratosphere.

Ozone is a very small fraction of the air. In the stratosphere, near the peak of the ozone layer, there are up to 12 ozone molecules for every million air molecules. In the troposphere near Earth's surface, ozone is less abundant, with a typical range of 0.02 to 0.1 ozone molecules for each million air molecules.

Ozone is destroyed naturally in the upper stratosphere by the UV radiation from the sun. These reactions are most important in the stratosphere of tropical and middle latitudes, where this radiation is most intense. For each ozone molecule that is destroyed an oxygen atom and an oxygen molecule are formed. Some of these recombine to produce ozone again. These naturally occurring reactions of destruction and production of ozone are balanced so that the ozone amount in the stratosphere remains constant.

Other, non-natural, mechanisms for the destruction of ozone involve halogen atoms. Emissions from human activities and natural processes include large sources of chlorine (Cl), bromine (Br) and fluorine (F) containing gases that eventually reach the stratosphere. When exposed to UV radiation from the sun. These halogen-containing gases are converted to more reactive gases, also containing chlorine and bromine monoxide (BrO). These reactive gases participate in "catalytic" reaction cycles that efficiently destroy ozone in the stratosphere.

Atomic oxygen (O) is formed when UV radiation in sunlight interacts with ozone and oxygen molecules. The destruction of ozone involves two separate chemical reactions. The net or overall reaction is that of atomic oxygen (O) with ozone (O_3), forming two oxygen molecules (O_2). The cycle begin and with either ClO or Cl. Cl reacts with (and thereby destroys) ozone and reforms ClO. The cycle then is repeated with another reaction of ClO with O. Because Cl or ClO is reformed each time an ozone molecule is destroyed, chlorine is considered a catalyst for ozone destruction. Similar reaction occur with bromine derivates. The relative potency of the different halogens depends a great deal on the stability of the compounds. Hydrogen fluoride (HF) is so very stable that fluorocarbons have relatively no known impact on ozone. The atmospheric lifetimes of the iodine compounds are extremely short and they do not play an important role in the ozone destruction processes.

Volcanoes can emit some chlorine-containing of the atmosphere before they can reach the gases, but these gases are dissolved in stratosphere.

Rainwater and ice and are usually "washed out".

Q.2. Describe the relationship between ozone and solar ultraviolet radiation?

Ans. There is an inverse relationship between the concentration of ozone and the amount of harmful UV radiation transmitted through the atmosphere since ozone absorbs some of the UV radiation.

A small fraction of the radiation emitted by the sun resides in the ultraviolet range. This range extends from 100 to 400 nm and is divided into three bands: UV-A(315-400 nm), UV-B(280-315 nm) and UV-C (100-280 nm). As the sun's radiation passes vertically through the atmosphere, all the UV-C and approximately 90% of the UV-B is absorbed by ozone and oxygen molecules in the stratosphere. UV-A radiation is less affected by the atmosphere. Therefore, the UV radiation reaching Earth's surface is composed mainly of UV-A with a small UV-B component. The amount and variability of the UV-B component depends on the solar elevation angle which defines the path-length through the atmosphere and also on the amount of ozone. A decrease in the concentration of ozone in the atmosphere results of the Earth. UV-B radiation at surface of the Earth. UV-B radiation is biologically much more active than UV-A radiation and can have either beneficial or detrimental effects on living organisms. Changes in the amount of UV-B radiation (for example due to stratospheric ozone depletion) are very important for humans.

Q.3. What changes have occurred in the ozone layer during past 35 years?

Ans. Stratospheric ozone has decreased over the globe since the 1980s. Averaged over the globe, ozone in the period 1996-2009 is about 4% lower than before 1980. Much larger depletion, up to 40% occurs over the high latitudes of the Southern Hemisphere in October.

The increase in reactive halogen gases in the stratosphere is the primary cause of the average ozone depletion. The lowest ozone values in recent years occurred after the eruption of Mt. Pinatubo volcano in 1991, which increased the number of sulphur-containing particles in the stratosphere. These particles remained in the stratosphere for several years and increased the effectiveness of reactive halogen gasses in destroying ozone. Observed ozone depletion varies significantly with latitude on the globe. The largest losses occurs at the highest southern latitudes as a result of the severe ozone loss over Antarctica that occurs every year during winter and early spring. The next largest ozone losses are observed in the high latitudes of the Northern Hemisphere, caused in part by late winter/early spring losses over the Arctic. Ozone-depleted air over both polar regions is dispersed away from the poles during and after each winter/spring period. Ozone depletion also occurs directly at latitudes between the Equator and Polar regions but is much smaller. Equivalent effective stratospheric chlorine (EESC) concentration based on the findings of the latest Scientific Assessment is used as an indicator of the amount of ozone depleting substances in the atmosphere. The amounts of these substances in the stratosphere have already reached a maximum and now are decreasing. The concentration of the ozone is inversely proportional to the concentration of the EESC.

Q.4. Describe the factors that determines the level of solar UV-B radiation at a specific place.

Ans. The sun is the source of the UV radiation reaching Earth. UV radiation is partly absorbed by the components of Earth's atmosphere. The amount of UV radiation that is absorbed depends mainly on the length of the path of the sunlight through the atmosphere.

The UV-B levels at Earth's surface vary with the time of day, geographic location and season. UV radiation is highest in the tropics, because the sun is higher in the sky, and decreases towards the poles where the sun is lower. The height of the sun above the horizon has an influence on the UV radiation, since the steeper angles mean longer pathways and more opportunity for the radiation to be absorbed. UV radiation is most intense near noon hours and least near sunrise or sunset, and is more in the summer and less in the winter. Clouds, particulate matter, aerosols and air pollutants absorb and scatter some of the UV radiation from the sun is less attenuated. This increase in UV radiation varies between 5 and 20% for each kilometer of height, depending on the specific wavelength, solar angle, atmospheric pollution and other local conditions. Other factors cause even larger differences in UV radiation between different altitudes. Surface reflections, especially from snow, ice and sand, increase the UV radiation levels at a particular location, because the reflected radiation is redirected towards the surface through scattering by particles and molecules in the atmosphere. Snow reflects as much as 90% and dry beach sand and sea foam about 25% of UV radiation. Clouds also reflect an appreciable amount of radiation towards the surface, thus while areas that are not shaded by clouds usually receive more radiation, under certain cloud conditions levels of UV radiation at Earth's surface can actually be higher under these conditions than under clear-sky conditions.

Q.5. What is the importance of solar UV index?

Ans. The solar UV index (UVI) describes the level of solar UV radiation relevant to human sunburn (erythema).

The maximum value of UVI is at midday. In the tropics at sea level the UVI can exceed 16 and peak terrestrial values of 25 can occur at high altitudes. The higher the UVI, the greater the potential for damage, and the less exposure time it takes for harm to occur. For fair-skinned individuals a UVI of more than 10 can cause skin damage (erythema) from an exposure of about 15 minutes. Outside the protective layer of Earth's atmosphere (altitude > 50 km), the UVI can exceed 300.

Information about UV intensities is provided in terms of the internationally adopted UVI colour-scale, along with appropriate health warnings. The colours corresponding to the various ranges are standardized throughout the world.

The UVI can be measured directly with instruments designed specifically to measure sun burning UV radiation. For clear sky conditions, the UVI can be calculated approximately from knowledge of the ozone and the solar elevation angle. The UVI at a specific location and time depends strongly on the cloud cover and on the amount of aerosols. Other influential factors include the seasonally varying Sun-Earth separation, the altitude, atmospheric pollution and surface reflection. When the surface is snow-covered the UVI can be up to 90% greater than for snow-free surfaces. Several countries provide forecasts of UVI that take predicted changes in ozone and cloud cover into account.

The UV index is a next day forecast of the amount of skin damaging UV radiation expected to reach the earth's surface at the time when the sun is highest in the sky (solar noon). The amount of UV radiation reaching the surface in primarily related to the elevation of the sun in the sky, the amount of ozone in the stratosphere, and the amount of clouds present. The UV index can range from 0 (when it is night time) to 16 (in the tropics at high elevations under clear skies). UV radiation is greatest when the sun is highest in the sky and rapidly decreases as the sun approaches the horizon. The higher the UV index, the greater the dose rate of skin damaging (and eye damaging) UV radiation. Consequently, the higher the UV index, the smaller the time it takes before skin damage occurs.

Overexposure to UV radiation causes a severe sun burn following an intense short-term overexposure, and the more serious skin cancers developing after long term overexposure. Melanoma, the deadlier of the two types of skin cancer occurs when the patient has been subjected to several intense short-term overexposures. Non-melanoma skin cancers, which are almost 100% curable, occur in people who are overexposed for very long periods of time, like construction workers, farmers, or fishermen. Long term overexposure to UV radiation has been linked to the formation of cataracts in the eyes as well.

Colours used in the UV index

Table 11.1: Colours used in the UV index

Exposure Category of UV radiation	Colour	UVI Range
Low	Light green	< 2
Moderate	Pale yellow	3-5
High	Deep yellow	6-7
Very High	RED	8-10
Extreme	Light purple	11+

UVI versus Sun angle and ozone amount: A figure indicating the variation of the UVI with the solar elevation. The coloured lines represent different

Ozone concentrations measured in Dobson Units (DU).

100 DU (Dobson unit) = 1 mm.

Q.6. Discuss the effect of the interaction between UV-B radiation, climate change, and human activity on air pollution.

Ans. Pollutants emitted by human activities can reduce UV-B radiation near the earth's surface, while particles may lead to enhancement by scattering. These processes decrease some exposures to UV while enhancing others. Interactions between UV radiation and pollutants resulting from changes in climate and burning of fossil and plant fuels will worsen the effects of ozone on humans and plants in the lower atmosphere.

While most of the atmospheric ozone is formed in the stratosphere, ozone is also produced in the lower atmosphere (troposphere) by the chemical reactions between pollutants such as nitrogen oxides, hydrocarbons and UV radiation. This ground-level ozone is a significant component of photochemical smog and has important adverse effects on human health and the environment.

Increases of this type of air pollution will result in more human deaths than other effects of climate change, Such as increased intensity of storms and flooding.

Climate change will increase circulation of gases in the atmosphere but cause only small increases in ozone in the lower atmosphere in areas where there is little pollution. These increases will be less significant for the health of humans than those caused by tropospheric ozone in polluted areas.

Concentration of ozone at different altitudes in the atmosphere

Q.7. How human activities affects worldwide phenomena such as depletion of the ozone layer and climate change?

Ans. It is not easy to accurately determine the influence of humans on any natural activity. In the case of the ozone layer, the depletion of the ozone over Antarctica cannot be explained by natural cycles alone but is caused by the increase of man-made chemicals (e.g., CFC) in the stratosphere. Relationship between these chemicals (e.g., chlorofluorocarbons). By absorbing the infrared radiation emitted by Earth, some substances influence the natural energy flows through the atmosphere. The greenhouse gases, such as carbon dioxide, (although only a tiny fraction of the atmosphere) play an important role in this process. Carbon dioxide is produced when fossil fuels are used to generate energy an d when forests are burned. Measurements show that its concentration has increased by almost 30% over the past 250 years. In the mid-20th century, the concentration of CO_2 was 280 ppmv, and it is currently 390 ppmv, and is increasing at about 3% per decade. Methane and nitrous oxide emitted from agricultural activities, changes in land use, and other sources are also important.

The increase in greenhouse gases contributes to climate change in the form of increased temperatures (about 0.6°C near Earth's surface over the past 150 years) and a rise in sea level. Models of the climate change induced by emission of greenhouse gases predict that the global temperature will rise between 2 and 6°C in the next century. If this happens, the change would be much larger than any temperature changes experienced over at least the last 10,000 years.

Q.8. What are the effects of exposure to solar UV radiation on the human health (e.g., on eye, skin and immune system)?

Ans. (*a*) **Effect on eyes:** The effects of UV radiation on the eye is almost immediate (acute) occurring several hours after a short, intense exposure. They can also be long-term (chronic), following exposure of the eye to levels of UV radiation below those required for the acute effects but occurring repeatedly over a long period of time. The commonest acute effect, photokeratitis (snow blindness), leaves few or no permanent effects, whereas cataract due to chronic exposure in irreversible and ultimately leads to severe loss of vision requiring surgery.

Avoidance of the sun is an effective but impractical means of avoiding exposure of the eyes to UV radiation. Although the eyes rarely receive direct sunlight, protection is frequently needed under conditions of high ambient UV radiation and/or where there are reflective surfaces.

Appropriate glass and plastic lenses absorb all UV-B and much of the UV-A radiation. Even clear spectacle lenses provide protection from UV-B. However, in the case of non-wrap-around sunglasses, there is potential for ambient UV radiation to enter the eye from the side. This effect can be exacerbated by tinted sunglass lenses which encourage a wider opening of the eyelids. UV radiation-blocking soft contact lenses, that cover the entire cornea, effectively shield the cornea and ocular lens against UV radiation incident from all angles. They offer a UV protection alternative in those situations where the wearing of sunglasses in not practical or convenient.

(*b*) **Effect on skin:** Acute overexposure of the skin to solar UV radiation causes sunburn; chronic sunlight exposure can lead to the development of skin cancers.

Sunburn is the major acute outcome in the skin of overexposure to sunlight. The dose of solar UV-B radiation required to induce sunburn varies considerably from one individual to another, depending on the pigment in their skin.

The major harmful effect of chronic exposure to sunlight, and/or intermittent episodes of sun burning, is the induction of skin cancers and are found at highest frequency in fair-skinned individuals living in sunny climates. They can be readily treated and are rarely fatal. The number of cases of melanoma in dark-skinned individuals is fewer than in fair-skinned individuals and it is also uncommon in people under the age of 20.

(*c*) **Effect on immune system:** The immune system can be suppressed by exposure of the skin and eyes to UV-B radiation leading to reduced immune responses to infectious agents and skin cancers, but a potentially beneficial effect for some autoimmune diseases.

UV radiation is absorbed by specific molecules called chromophores in the skin. These initiate a cascade of events affecting the immune system: decreasing the ability of specialized cells to recognize foreign challenges such as invading microorganisms or tumor proteins, altering the production of a range of immune mediators, and inducing the generation of particular lymphocytes called T regulatory cells. All of these changes lead to suppression of immune responses following the exposure.

UV radiation at a critical time during infection can increase the severity of symptoms and duration of the disease. In addition, UV radiation before (and possibly immediately after) immunization can reduce the immune response generated and therefore the effectiveness of the vaccination. Exposure to UV radiation around the time of vaccination can decrease the immune response generated.

In the case of cold sores, solar UV radiation is a common trigger for the reactivation of latent virus and the reappearance of the vesicles in the skin. The involvement of T regulatory cells in squamous and basal cell carcinomas indicate that UV-induced immunosuppression is an important factor in their development. In contrast, this down regulation in immunity is of potential benefit in protection against the development of some autoimmune diseases, such as multiple sclerosis, where there is an over-activity of T cells directed against specific element of the body's own tissues.

Q.9. What is the beneficial health effect of solar UV-B radiation?

Ans. A major benefit to human health of exposure to UV-B radiation is the production of vitamin D. Although the diet of humans contains some items rich in vitamin D, such as oily fish and eggs, more than 90% of the vitamin D in most people is produced by exposure of the skin to solar UV radiation. Vitamin D is synthesized most effectively when the sun is at its height in the summer months and little or none is synthesized in the winter at mid to high latitudes. Individuals with dark skin require more sun exposure than those with fair skin to make the same amount of vitamin D, and the production is less efficient in older subjects. Vitamin D is required for the growth, development and maintenance of bone. Those individuals deficient in vitamin D can develop bone defects, resulting in an increased risk of osteoporosis and fractures in adults, and of rickets in children. Vitamin D also protects against non-skeletal disorders. Like some internal cancers such as colorectal cancer, autoimmune diseases such as multiple sclerosis and insulin-dependent diabetes, infections such as tuberculosis and influenza, and cardiovascular diseases such as hypertension.

Q.10. What are the main risks to human and the environment caused by the breakdown products of HFCs and HCFCs?

Ans. The hydrofluorocarbons (HFCs) and hydrochlorofluorocarbons (HCFCs) are replacement for the chlorofluorocarbons (CFCs) as they have a smaller effect on the ozone layer. The HFCs and HCFCs are largely degraded before reaching the stratosphere.

HFCs and HCFCs break down relatively rapidly into several products including persistent substances such as trifluoroacetic acid (TFA) and chlorodifluoroacetic acid (CDFA). The compounds are soluble and are washed from the atmosphere by precipitation and reach surface waters, along with other chemicals washed from the soil. Microbiological degradation slowly removes these substances from the water. In locations where there is little or no outflow and high evaporation (seasonal wetlands and salt lakes), these products are expected to increase in concentration over time.

The effects of increased concentrations of naturally occurring mineral salts (from natural sources such as under sea vents and volcanic activity) and other materials would be greater and more biologically significant then those of breakdown products of the HFCs and HCFCs. TFA is very resistant to breakdown, and amounts deposited in flowing surface water will ultimately accumulate in the oceans.

Q.11. How can we protect ourselves from the adverse effects of solar UV-B radiation on the skin?

Ans. The intensity of UV-B radiation from the sun is usually highest during the central hours of the day (about 10 am until 2 pm). If outside, it is important to avoid sunburn. Hats with brims more than 10 cm wide are recommended for head and neck protection, and can reduce exposure of the eyes by up to 50%. The hood of a jacket and headwear with side-flaps can provide protection from side UV-B irradiation. Wrap-around sunglasses are better at protecting the entire eye than conventional sunglasses with open sides.

Protection of the body with suitable clothing is recommended when outside during the central hours. Some textiles are highly effective in this regard while other are less so. Sunscreens are also effective and ones with a sun protection factor of 30 are generally recommended. They need to be applied at the stated concentration and to the frequently re-applied, especially when swimming. It is particularly important to protect children from sunburn, episodes of which could lead to an increased risk of skin cancer development in adulthood. The use of sun protection is recommended when the UV Index is forecast to be 3 or greater.

Q.12. How much time should one spend in the sun in order to produce sufficient vitamin D but avoiding sunburn and minimizing the risk of skin cancer?

Ans. The time will depend on the latitude, season, time of day, weather conditions, the amount of skin exposed and whether your skin is accustomed to exposure. You should also know your skin type and recognize how readily you burn or tan as a result of sun exposure. Most importantly, sunburn should be avoided, particularly in children. One useful guide is to find out the daily forecast for the noon UV Index for your location. In many countries the UV Index is reported in weather forecasts. The UV Index at times other than noon is lower unless there is a big change in cloudiness.

Estimates of the daily variations in sun burning UV radiation and in the UV, radiation needed to make vitamin D at mid-latitudes are shown in Table 11.2.

Some calculations have been made giving an indication of the number of minutes in the sun required to cause sunburn, and the number required for adequate vitamin D production for various UV indices at various latitudes in the summer and winter. The values below should be taken as a rough guide only, and are for times around noon and for (previously unexposed) skin type II. They should be multiplied by 2 for skin type IV and by 5 for skin type VI.

Estimates of daily variations in sun burning UV radiation and in the UV radiation needed to make vitamin D at mid-latitudes.

Table 11.2 Estimates of daily variations in sun burning UV radiation and in the UV radiation needed to make vitamin D at mid-latitudes.

	Minutes to sunburn	Minutes for sufficient vitamin D, full body exposure	Minutes or sufficient vitamin D, 10% body exposure
Mid-latitude, summer UV Index 12	15	1	10
Mid-latitude, winter UV Index 1	180	20	200
Tropics UV Index 16	10	<1	7

Q.13. Will global climate change alter the effects of UV radiation on human health?

Ans. In trying to assess possible interactions between climate change and solar UV radiation, the major uncertainty for human health is whether people will spend more or less time outdoors in sunlight as temperatures rise, but humidity, storms, floods and droughts also increase. The risk of skin cancer is greater for the same level of UV radiation exposure if the ambient temperature increases: Exposure to UV radiation in sunlight can be important for disinfection of waterborne pathogens, which may grow and survive better in warmer conditions. If warmer temperatures encourage people to spend more time outside, it will be important to ensure adequate sun protection. In contrast, if more time is spent indoors, vitamin D may need to be provide as a dietary supplement.

Climate change may after the effects of UV radiation on terrestrial and aquatic ecosystems to change the availability or quality of human food supplies, with resulting effects on human health. In addition, climate- induced changes in air pollution have potential health effects.

Q.14. Discuss the major impacts of UV-B radiation on natural terrestrial ecosystems, crops and forests.

Ans. Vital biomolecules, including DNA, proteins, and lipids, are potentially vulnerable to UV-B radiation. However, only a small portion of the UV-B radiation striking a leaf penetrates into the inner tissues. UV-B radiation induces the synthesis of compounds that act as sunscreens and prevent UV-B radiation, and changes in the protective waxy layer of the leaves. Some protective molecules produced by plants, in response to natural UV-B radiation, are important in our food, enhancing colours, flavours, antioxidant activity and fibre production. The change in biochemical composition can make the leaves less attractive as food for herbivores (including for insect pests) and alter the speed with which leaf litter is broken down in the soil. At an ecosystem level, these effects on palatability and decomposition can be substantial, since they have an impact on the food supply of animals and the recycling of nutrients in the soil.

The negative effects of UV-B radiation on the food supply of plant-eating insects can be substantial. Some of the reduced consumption is due to direct effects or UV-B radiation on insects and some due to changes in plant tissues induced by the UV-B radiation. Under enhanced UV-B radiation, sunscreen compounds both protect the plant from the UV-B radiation directly and deter insects from eating the plant (herbivory). This means that if UV-B radiation is higher, insects generally eat less plant material. By contrast, higher atmospheric CO_2 leads to increased herbivory, so future levels of UV-B irradiance and CO_2 will be important in controlling both plant production and the food supply for insects. The changes that occur at the plant level can influence underground decomposition. Decomposition of dead plant material (leaf litter) is a vital process, since it recycles carbon and nutrients making them available to growing plants. UV radiation affects decomposition indirectly via changes to leaf biochemistry and microbial diversity and directly through light- induced breakdown (photodegradation).

Sunscreen compounds and structural alterations, which allow leaves to withstand UV-B radiation while attached to the plant, can make leaves tougher to break down once they form leaf litter. UV-B radiation changes the composition of the microbes in the soil and this can also influence how easily leaf litter is broken down. When plant litter is directly exposed to sunlight, it is degraded photochemically (photodegradation).

It is possible to breed and genetically engineer UV-B tolerant crops. While many forest tree species appear to be UV-B tolerant, there is some evidence that detrimental UV-B effects accumulate slowly from year to year in certain species.

The present rate of global change is so rapid that evolution may not keep up with it, particularly in high latitudes where temperature and UV-B radiation have increased dramatically over recent decades. In Antarctica and the southern tip of South America, plants adapted to environments with relatively low levels of UV-B radiation have been affected by the increased levels of UV-B radiation due to ozone depletion. Although the negative impact of UV-B radiation on plant growth is usually relatively small (about 6%), some species are more affected than others. Over time, these difference between species may lead to changes in terrestrial ecosystems, especially in regions like Antarctica where UV-B radiation is likely to remain elevated for many more decades.

Q.15. How does exposure to UV-B radiation affect aquatic life?

Ans. In clear ocean and lake waters. UV-B radiation can penetrate to tens of meters. In turbid river and wetlands, however, UV-B radiation may be completely absorbed within the top few centimeters. In aquatic ecosystems, most organisms, such as phytoplankton, live in the illuminated upper layer of a body of water that allows the penetration of enough light (and UV-B radiation) to support photosynthetic, or green, plants. In particular, UV-B radiation may damage those organisms that live at the surface of the water during their early life stages.

Detrimental effects of UV-B radiation have been shown for phytoplankton, fish eggs and larvae, zooplankton and other primary and secondary consumers. Most adult fish are well protected from excessive solar UV radiation, since they inhabit deep waters. Some shallow-water fish can develop skin cancer and other UV-related diseases.

UV-B radiation reaches different depths in ocean water depending on water chemistry, the density of phytoplankton, and the presence of dissolved and particulate matter. UV radiation penetrates to considerable depth where the embryos and larvae of the Atlantic cod develop. Exposure to UV radiation equivalent tot heat present at 10 m depth results in s significant mortality of developing embryos and a significant decrease in the length of the larvae. Such radiation levels occur at many temperate latitudes, these ecologically and freshwater lakes and ponds, amphibian embryos are protected from UV-B radiation by the enzyme photolyase, melanin pigmentation of eggs, jelly surrounding the eggs, the water depth and dissolved organic matter in the water. Larvae can seek shelter from sunlight by swimming into shaded areas and vegetation in ponds. In contrast, lobster larvae seem to be tolerant to UV radiation even though they develop in the surface layers of the water column.

Q.16. How does climate change alter the effect of UV radiation on aquatic ecosystems?

Ans. Climate change will influence various aspects of how UV-B radiation affects ecosystems, such as through changes in temperature and sea-level, shifts in the timing and extent of sea-ice cover, changes in the wave climate, ocean circulation and salinity and alterations in the stratification of the water column.

These complex changes are likely to have significant effects on ecosystems, including biological production as well as changes in the global hydrological cycle, vertical mixing and efficiency of carbon dioxide uptake by the ocean. Such effects will vary over time and between different locations. Changes in temperature and the intensity and frequency of rainfall may alter the input of terrestrially-derived coloured dissolved organic matter (CDOM) to inland and coastal aquatic ecosystems. For example, decreased rainfall and increased temperature cause reductions in CDOM inputs and consequent increases in the depth to which UV radiation penetrates. In addition, the dissolved and suspended material in the water column changes the ratio of UV-A to UV-B to photosynthetically active radiation in the penetrating UV radiation. Climate change influences the amount of ice and snow cover in polar and sub polar areas. Ice and snow strongly attenuate the penetration of solar radiation into the water column. Any substantial decrease in ice and snow cover will alter the exposure of aquatic ecosystems to solar UV radiation.

Shifts in atmospheric circulation will change wind fields, influencing mixing and the depth of the upper mixed layer of the ocean. Such changes, and increases in temperature, affect the stratification of the surface layer of the ocean and the potential impact of UV-B radiation on near-surface organisms. Changing winds will also influence coastal upwelling systems and the potential for possible influence of these systems on UV radiation.

Q.17. How does increasing atmospheric CO_2 concentrations influence the sensitivity of aquatic organisms to solar UV-B radiation?

Ans. Acidification of lakes and oceans is a major stress factor closely related to climate change and solar UV-B radiation. Increasing acidification of marine waters due to increased atmospheric CO_2 concentrations hampers carbonate incorporation in many calcified marine organisms and makes them more vulnerable to solar UV-B radiation. Before industrialization, the concentration of CO_2 in the atmosphere was about 280 ppmv (parts per million by volume). This corresponded to an average PH in the oceans of about 8.2. The current level of 390 ppmv CO_2 has lowered the pH of the water by about 0.1 units, which corresponds to 26% increase in acidity. A further pH decrease of 0.3-0.4 units is expected by the year 2100. This acidification is expected to adversely affect the uptake and incorporation of calcium carbonate in mollusks, phytoplankton and some seaweeds. Encrustations of calcium carbonate efficiently absorb UV radiation. At lower pH values, increased UV transmission will expose the vulnerable organisms to higher levels of solar UV-B radiation. Experiments have shown that UV radiation significantly inhibits growth, photosynthetic oxygen evolution and calcification rates in the seaweed Corallina at high CO_2 concentrations (1000 ppmv as compared to 390 ppmv). In one of the major biomass producers in the ocean. Calcification rates are diminished by increased acidity, resulting in thinner protective scales on the surface. The transmission of UV radiation is significantly higher because of the acidification.

Q.18. Discuss the effects of the depletion of ozone on environmental processes and cycles.

Ans. UV radiation influences the biological productivity of oceans, including the production of gases at their surfaces and their subsequent transfer to the atmosphere. Once in the atmosphere, gases such as carbon dioxide (CO_2) interact with the physical climate system resulting in alterations to climate and feedbacks in the global biogeochemical system. Since atmospheric CO_2 plays a central role in the distribution of heat in the atmosphere, its increasing concentrations may affect many components of the physical climate system, such as wind, precipitation and the exchange of heat and energy between the air and the oceans.

There are also similarly complex interactions between biogeochemical cycles on land and the integrated climate system that may have important implications for organisms on Earth. At this stage, it is not possible to predict the overall environmental effects of these complex interactions between changes in climate and UV radiation.

Q.19. What is the deleterious effects of UV-B radiation on plastics and wood products used outdoors due to global warming?

Ans. Useful lifetimes of plastic and wood products are determined to a large extent by degradation due to the action of solar UV radiation. A partial depletion of the stratospheric ozone layer results in increased terrestrial UV-B flux, affecting organic materials used in construction. Plastics materials undergo oxidation and chemical breakdown when exposed to UV-B radiation in the atmosphere. This results in discolouration, surface cracking and loss of strength or stiffness in the product, limiting its useful lifetime. Additives called light stabilizers are used to partly mitigate the deleterious effects of UV-B radiation in plastics used outdoors. Similar degradation and breakdown occur in wood products, although here biodegradation also contributes to the deterioration very significantly. The action of UV-B radiation on the surface of wood also renders it hydrophilic; rainwater is better absorbed by the surface, leading to the growth of wood-rot fungi that biodegrade the wood. Photodamage by UV-B radiation facilitating fungal biodegradation is seen in wood- plastic composites (or plastic lumber) that are essentially plastics mixed with large amounts of wood flour. The rates of chemical reactions that cause the UV-initiated oxidative degradation of plastics or wood are increased at higher temperatures. Thus, the useful lifetimes of these materials outdoors will shorten as the ambient air temperature increases. Plastics with a deeper colour exposed to sunlight reach bulk temperatures higher than those of light-coloured products of the same plastic. Which the oxidative breakdown will be faster, the available light-stabilizer technologies are likely to control this, thus maintaining the useful lifetimes at the present levels. Because of the need to use higher levels of conventional stabilizers or coating, the cost of preservation may be higher in some locations.

Several environmental factors affect the outdoor degradation of plastics and wood as shown below:

Material	UV-B radiation	Temperature	Humidity	Pollutants
Plastics	++++	+++	+	+
Wood	+++	++	++++	+

++++ Very susceptible, +++ Moderately Susceptible, ++ Susceptible, + Likely to be susceptible.

Q.20. How does ozone depleting substances (ODS) and their substitutes affect climate?

Ans. Ozone depleting substances (ODSs) such as the CFCs have contributed to increase in global average surface temperature. On the other hand, ozone depletion itself has a cooling effect. Warming due to ODSs and cooling associated with ozone depletion are two distinct climate forcing mechanisms that do not simply offset one another. Bromine-containing gases currently contribute much less to warming than to cooling. HFCs and PFCs contribute only to warming.

Actions taken under the Montreal Protocol have led to the replacement of CFCs with HCFCs, HFCs, and other substances. Because replacement species (with the exception of HFCs) generally have lower global warming potentials (GWPs) and because total halocarbon emission have decreased, their contribution to climate change has been reduced. Ammonia and hydrocarbons used as halocarbon substitutes will have a negligible effect on global climate. The relative future warming and cooling effects of emissions of CFCs, HCFCs, HFCs, PFCs and halons vary. The indirect cooling effect of ODSs is projected to cease upon ozone layer recovery, so that GWPs associated with the indirect cooling effect depend on the year of emission, compliance with the Montreal protocol and gas lifetimes.

Substitution for ODSs in air conditioning, refrigeration, and foam blowing by HFCs, PFCs, and other gases such as hydrocarbons are not expected to have a significant effect on global tropospheric chemistry.

Q.21. Is ozone depletion affected by climate change?

Ans. Climate change affects ozone depletion through changes in atmospheric conditions that affect the chemical production and loss of stratospheric ozone. Climate change is expected to decrease temperatures and water vapour abundances in the stratosphere. This will tend to speed up ozone recovery outside polar Regions, but slow down the recovery in Polar Regions.

Ozone, the CFCs and their substitutes are minor greenhouse gases with a relatively small (+ 13%) contribution to climate change. Several other gases involved in the chemistry of ozone depletion are also active greenhouse gases. They include water vapour, methane, and nitrous oxide. Increases in those will ultimately lead to increases in stratospheric gases that destroy ozone. Change in solar output and future volcanic eruptions will influence both climate change and ozone depletion.

While current ozone depletion compared with period prior to the 1980s is dominated by chlorine and bromine in the stratosphere, in the longer term (~ 100 years) the impact of climate change will dominate through the effect of changes in atmospheric circulation and chemistry. The result is that over the first half of the current century, increases in greenhouse gases may contribute to a colder stratosphere. This will lead to a decrease in the rate of destruction of ozone outside Polar Regions. In Polar Regions, however, the lower temperatures may lead to increased polar stratospheric clouds, thus exacerbating ozone depletion. The temperature changes will also lead to changes in atmospheric circulation which may aid the mixing of long-lived CFCs from the troposphere to the stratosphere, which will increase their rate of photochemical destruction leading to more sever ozone depletion in the short term but will contribute to a faster ultimate recovery of ozone. Changes in circulation pattern in the lower atmosphere, which in turn affect surface climate. The effects of climate change on UV radiation total ozone directly (and thus indirectly affect UV radiation), and those that depend on changes in other variables (such as clouds, aerosols or snow cover that influence solar UV radiation directly).

Q.22. What is the effect of control measures envisaged in the Montreal Protocol on (*a*) phase out of ODS(ozone depleting substances) (*b*) UV radiation changes (*c*) climate?

Ans. (*a*) **Phase-out of ODC:** The Montreal Protocol for the protection of the Ozone Layer is the most successful environmental international agreement to date. It has been ratified by all of the 196 countries of the world. All the CFCs have been phased out since January 2010. The phase-out of the HCFCs is on schedule and has been advanced. Stratospheric ozone is no longer decreasing and is predicted to return to pre-1980 values before 2050 at mid-latitudes and a few years later at high latitudes. Concentrations of ozone depleting substances have been decreasing for over ten years, and are expected to continue to decrease in the future.

(*b*) **UV radiation changes:** The Montreal Protocol has a huge influence on UV-B radiation. Prediction of the UV Index indicating what could have happened in the absence of a Montreal Protocol.

(*c*) **Effect on the climate:** As a result of the phase-out schedules of the Montreal Protocol, the global production and use of chlorofluorocarbons (CFCs) and halons has decreased significantly. However, the sustained growth in demand for refrigeration, air-conditioning and insulating foam products in developing countries has led to an increase in the consumption and emissions of hydrofluorocarbons (HFCs). Consequently, the use of HCFCs and HFCs as replacements for CFCs and halons has increased. The HCFCs are low-ozone-depletion-potential substitutes for high-ozone-depletion-potential substances, particularly CFCs and halos, and were classified under the protocol as "transitional substitutes" during the time it took to commercialize new ozone-safe alternatives and replacements. Ultimately, HCFCs will be phased out globally under the Montreal Protocol leaving much of the application demand for refrigeration, air conditioning, heating and thermal-insulating foam production to be met by HCFs. The demand for HCFCs and/or HFCs in many applications is expected to increase. HFCs do not deplete the ozone layer but, along with CFCs and HCFCs are greenhouse gases that contribute to the radiative forcing of climate. Thus, the transition away from ozone depleting substances (ODSs) has implications for future climate. HFCs are in the "basket of gases" regulated under the 1997 Kyoto Protocol, a global treaty to reduce developed-country emissions of greenhouse gases.

Figure below shows the GWP-weighted emission of CFC, HCFC, HFC, and CO_2 for the period 1960-2050 and the corresponding emission for CFCs from 1987-2020 following a scenario in which there is no Montreal Protocol except HCFCs. The emissions of individual compounds are multiplied by their respective Global Warming Potentials (GWPs) to obtain aggregate emissions expressed as equivalent $GtCO_2\ yr^{-1}$ (as used in the 2007 IPCC report). The colour-shaded regions show ranges of emissions of CFCs, HCFCs, HFCs, and CO_2 as indicated in the panel legends. The high and low labels identify the upper and lower limits in the global baseline scenarios. Shown for reference are emission for the range of one of the scenarios (SRES CO_2) from the IPCC 2007 report and the 450- and 550-ppm CO_2 stabilization scenarios. The reduction of CFCs has been very beneficial in mitigating climate change due to increasing CO_2, but if the projected growth rate of the HFC replacement chemicals continues, the rate of warming due to their increase could approach that due to increasing CO_2.

Background: CO_2^- equivalence is the contribution to climate forcing from different greenhouse-gas emissions. It can be compared by scaling the emissions with the associated global warming potentials (GWPs) to form 'CO_2^- equivalent' emissions. Typically, the 100 GWP is used. HFC emissions in 2050 will largely offset the climate benefits already achieved by the Montreal Protocol in terms of annual emission reductions. HFC emissions are projected to reach 5.5-8.8 GtCO - eq yr^{-1} in 2050. In comparison, ODS emission peaked at 9.4 GtCO - eq yr^{-1} in 1988 and could have reached 15-18 GtCO - eq yr^{-1} in 2010 in the absence of Montreal Protocol regulations.

Q.23. Why ozone is classified as "good ozone" and "bad ozone?

Ans. Ozone in the stratosphere absorbs some of the Sun's biological harmful ultraviolet radiation. Because of this beneficial role, stratosphere ozone is considered "good "ozone. In contrast, excess ozone on Earth's surface that is formed from pollutants is considered "bad" ozone because it can be harmful to humans, plants and animals.

Good ozone: Stratospheric ozone is considered good for humans and other life forms because it absorbs ultraviolet (UV)-B would reach Earth's surface and harm to a variety of life forms. In humans, increased exposure to UV-B increases the risk of skin cancer, cataracts, and a suppressed immune system. UV-B exposure also can damage terrestrial plant life. Single-cell organisms, and aquatic ecosystems. Other UV radiation, UV-A, which is not absorbed significant by ozone, causes premature aging of the skin.

The absorption of UV-B radiation by ozone is a source of heat in the stratosphere. This help to maintain the stratosphere as a stable region of the atmosphere, with temperatures increasing with altitude. As a result, ozone plays a key role in controlling the temperature structure of Earth's atmosphere.

In the mid-1970s, it was discovered that halogen source gases released in human activities could cause stratospheric ozone depletion Ozone depletion increases harmful UV-B amounts at Earth's surface. Global efforts have been undertaken to protect the ozone layer through regulation of ozone depleting gases.

Bad ozone: Excess ozone formed near Earth's surface in reactions caused by the presence of the human- made pollutant gases is considered bad ozone. Increased ozone amounts are harmful to humans, plant, and other living systems because ozone reacts strongly to destroy or alter many other molecules. Excessive ozone exposure reduces crop yields and forest growth. In human, exposure to high levels of ozone can reduce lung capacity; cause chest pains, throat irritation, and coughing; and worsen preexisting health, conditions related to the heart and lungs. In additions, increases in tropospheric ozone lead to a warming of Earth's surface. The negative effects of increasing tropospheric ozone contrast sharply with the positive effects of stratospheric ozone as an absorber of harmful UV-B radiation from the sun.

Reducing the emissions of pollutants can reduce bad ozone in the air surrounding humans, plants, and animals. Major sources of pollutants include large cities where fossil fuel consumption and industrial activities are greatest.

The ozone layer resides in the stratosphere and surrounds the entire Earth. UV-B radiation (280- 315) nanometer (nm) wavelength) from the Sun is partially absorbed in this layer. As a result, the amount of UV-B reaching Earth's surface is greatly reduced. UV-A (315-400 nm wavelength) and other solar radiation are not strongly absorbed by the ozone layer._

Q.24. Is total ozone uniformly distributed over the globe?

Ans. NO, the total amount of ozone above the surface of Earth varies with location on time scales the range from daily to seasonal and longer. The variations are caused by stratospheric winds and the chemical production and destruction of ozone. Total ozone is generally lowest at the equator and higher near the poles because of the seasonal wind patterns in the stratosphere.

Total ozone at any location on the globe is found by measuring all the ozone in the atmosphere directly by measuring all the ozone includes that present in the stratospheric ozone layer and that present throughout the troposphere. The contribution from the troposphere is generally only about 10% of total ozone. Total ozone values are often reported in Dobson units, denoted "DU". Typical values vary between 200 and 500 DU , for example , is equivalent to a layer of pure ozone gas on Earth's surface having a thickness of only 1.5 centimeters (0.2 inches).

Total ozone varies strongly with latitude over the globe, with the largest values occurring at middle and high latitudes. This is a result of winds that circulate air in the stratosphere, moving tropical air rich in ozone toward the poles. Since about 1980s , region of low total ozone have occurred at polar latitudes in winter and springs as a result of the chemical destruction of ozone by chlorine and bromine gases. The smallest values of total ozone (other than in the Antarctic in spring) occur in the tropics in all seasons, in part because of the troposphere extends to a higher altitude in the tropics, and consequently, the thickness of the ozone layer is smallest there.

The variations of total ozone with latitude and longitude come about for two reasons. First, natural air motions mix air between regions of the stratosphere that have high ozone values and those that have low ozone values. Air motions also increase the vertical thickness of the ozone layer near the poles, which increases the value of total ozone in those regions. Tropospheric weather systems can temporarily reduce the thickness of the stratospospheric ozone layer in region, lowering total ozone at the same time. Second, variations occur as a result of changes in the balance of chemical production and loss processes as air moves to different locations over the globe. Reductions in ultraviolet radiation from the sun in the its 11–year cycle, for example, reduce the production of ozone.

Q.25. What are the variations methods of measurements of ozone is the atmosphere?

Ans. The amount of ozone in the atmosphere is measured by instruments on the ground and carried aloft contains a system for detecting ozone. Other measurements are based on ozone's unique absorption of light in the atmosphere. In the case, sunlight or laser light is carefully measured after passing through a portion of atmosphere containing ozone.

The abundance of ozone in the atmosphere is measured techniques that make use of ozone's unique optical and chemical properties. There are two principal categories of measurement techniques: local and remote.

Local measurement: Local measurement of atmospheric ozone abundance are those that require air to be drawn directly into an instrument . once inside an instruments, ozone can be measured by its absorbtion of ultraviolet (UV) light or by the electrical current produced in an ozone chemical reactions. The balloons ascend far enough in the atmosphere to measure ozone in the stratospheric ozone layer. Local ozone- measuring instrument using optical or chemical detection schemes are also used routinely on board research aircraft to measure the distribution of ozone in the troposphere and lower stratosphere. High-altitude research aircraft can reach the ozone layer at most locations over the globe and can reach farthest into the layer at high latitudes in polar regions.

Remote measurements: Most remote measurement of ozone rely on its unique absorption of UV radiation. Sources of UV radiation that can be used are the Sun and lasers. For example, satellites use the absorption of sunlight scattered from the surface of Earth to measure ozone over nearly the entire globe on a daily basis. A network of ground –based detectors measurement ozone by the amount of the Suns's UV light that reaches Earth's surface. Other instruments measure ozone using its absorption of infrared or visible radiation or its emission of microwave on infrared radiation. Total zone amounts and the altitude distribution of ozone can be obtained with remote measurement techniques. Lasers are deployed at ground sites or on board aircraft to detect ozone over a distance of many kilometers along the laser light path.

Q.26. Discuss the various steps involved in the stratospheric ozone depletion.

Ans. The initial step in the depletion of stratospheric ozone by human activities is the emission, at Emissions, at Earth's surface, of ozone-depleting gases containing chlorine and bromine. Most of these gases accumulate in the lower atmosphere because they are unreactive and do not dissolve readily in rain or snow. Eventually, these emitted source gases are transported to the stratosphere, where they are converted to more reactive gases containing chlorine and bromine. These more reactive gases then participate in reactions that destroy ozone. Finally, when air returns to the lower atmosphere, these reactive chlorine and bromine gases are removed from Earth's atmosphere by rain and snow.

The process of ozone deletion begins with the emission, of source gases containing the halogens chlorine and bromine. The halogen source gases include manufactured chemicals released to the atmosphere by a variety of human activities. Chloroflurocarbons (CFCs) are an important example of chlorine-containing gases. Emitted source gases accumulate in the lower atmosphere (troposphere) and are eventually transported to the stratosphere. The accumulation occurs because most source gases are unreactive in the lower atmosphere. However, small amounts of these gases dissolve or are taken up in ocean waters.

Some emissions of halogen gases come from natural sources. These emissions also accumulate in the troposphere and are transported to the stratosphere.

Halogen source gases do not react directly with ozone. Once in the stratosphere, halogen source gases are chemically converted to reactive halogen gases by ultraviolet radiation from the Sun. The rate of conversion is related to the atmospheric lifetime greater than a few years may circulate between the troposphere and stratosphere multiple times before full conversion occurs.

The reactive gases formed in the eventual conversion of the halogen source gases react chemically to destroy ozone in the stratosphere.

After a few years, air in the stratosphere returns to the troposphere, bringing along reactive halogen gases. These gases are then removed from the atmosphere by rain and other precipitation and deposited on Earth's surface. This removal brings to an end the destruction of ozone by chlorine and bromine atoms that were first released to the atmosphere as components of halogen source gas molecules.

Halogen source gases with short lifetimes undergo significant chemical conversion in the troposphere, producing reactive halogen gases and other compounds. Source gas molecules that are not converted accumulate in the troposphere and are transported to the stratosphere. Because of removal by precipitation, only small portions of the reactive halogen gases produced in the troposphere are also transported to the stratosphere. Important examples of gases tat undergo some tropospheric removal are the HCFCs, which are used as substitute gases for other halogen source gases, bromoform, and gases containing iodine.

Principal steps in the depletion of stratospheric ozone is as follows:

- **Emissions:** Halogen source gases are emitted at Earth's surface by human activities and natural processes.
- **Accumulation:** Halogen source gases accumulate in the atmosphere and are distributed throughout the lower atmosphere by winds and other air motions.
- **Transport:** Halogen source gases are transported to the stratosphere by air motions.
- **Conversion:** Most halogen source gases are converted in the stratosphere to reactive halogen gases in chemical reactions involving ultraviolet radiation from the Sun.
- **Chemical reaction:** Reactive halogen gases cause chemical depletion of stratospheric total ozone over the globe except at tropical latitudes.

  Polar stratospheric clouds increase ozone depletion by reactive halogen gases, causing severe ozone loss in polar regions in winter and spring.
- **Removal:** Air containing reactive halogen gases returns to the troposphere and these gases are removed from the air by moisture in clouds and rain.

Q.27. Discuss about the anthropogenic emissions leading to ozone depletion.

Ans. Certain industrial processes and consumer products result in the emission of "halogen source gases" to the atmosphere. These gases bring chlorine and bromine to the stratosphere, which cause depletion of the ozone layer, for example, Chlorofluorocarbons (CFCs), once used in almost all refrigeration and air conditioning systems, eventually reach the stratosphere, where they are broken apart to release ozone-depleting chlorine atoms. Other examples of human- produced ozone-depleting bromine atoms. The production and consumption of all principal halogen source gases by human activities are regulated worldwide under the Montreal Protocol.

The source gases that contain only carbon, chlorine, and fluorine are called "chlorofluorocarbons, abbreviated as CFCs. CFCs, along with carbon tetrachloride (CCl_4) and methyl chloroform historically have been the most important chlorine-containing gases that are emitted by human activities and destroy stratospheric ozone. These and other chlorine-containing gases have been used in many applications, including refrigeration, air conditioning, foam blowing, aerosol propellants, and cleaning of metals and electronic components.

Another category of halogen source gas contains bromine. The most important of these are the "halons" and methyl bromide (CH_3Br). Halons are halogenated hydrocarbon gases originally developed to extinguish fires. Halons are widely used to protect large computers, military hardware, and commercial aircraft engines. Because of these uses, halons are often -1301 are the most abundant halons emitted by human activities. Methyl bromide, used primarily as an agricultural fumigant, is also a significant source of bromine to the atmosphere.

Other chlorine – and bromine – containing gases are released regularly in human activities. Common examples are the use of chlorine gases to disinfect swimming pools and wastewater, fossil fuel burning, and various industrial processes.

There are a few halogen source gases present in the stratosphere that have large natural sources. These include methyl chloride (CH_3Cl) and methyl bromide (CH_3Br), both of which are emitted by oceanic and terrestrial ecosystems. Very short-lived source gases containing bromine, such as bromoform are also released to the atmosphere primarily from the oceans. Only a small fraction of these emissions reaches the stratosphere, because these gases are rapidly removed in the lower atmosphere.

After emission, halogen source gases are either naturally removed from the atmosphere or undergo chemical conversion. The time to remove or convert about 60% of a gas is often called its atmospheric "lifetime." Lifetime vary from less than 1 year to 100 years for the principal chlorine-and bromine-containing gases. Gases with the shortest lifetime (e.g., the HCFCs, methyl bromide, methyl chloride, and the very short-lived gases) are substantially destroyed in the troposphere, and therefore only a fraction of each emitted gas contributes to ozone depletion in the stratosphere.

The halogen source gases are also known as "ozone-depleting substances" because they are converted in the stratosphere to reactive gases containing chlorine in the stratosphere to reactive gases containing chlorine and bromine. Some of these reactive gases participate in reactions that destroy ozone. Ozone-depleting substances (ODS) are compared in their effectiveness to destroy stratospheric ozone using the "Ozone Depletion Potential" (ODP). A gas with a larger ODP has a greater potential to destroy ozone over its lifetime in the atmosphere. The ODP is calculated on a "per mass" basis for each gas relative to CFC-11, which has an ODP defined to be 1. Halon-1211 and halon-1301 have ODPs significantly larger than CFC-11 and most other emitted gases, because bromine is much more effective overall (about 60 times) on a per-atom basis than chlorine in chemical reactions that destroy ozone in the stratosphere. The gases with small ODP values generally have short atmospheric lifetimes or fewer chlorine and bromine atoms.

Fluorine and iodine are also halogen atoms. Many of the source gases also contain fluorine atoms in addition to chlorine or bromine. After the source gases undergo conversion in the stratosphere, the fluorine content of these gases is left in chemical forms that do not cause ozone depletion. Iodine is a component of several gases that are naturally emitted from the oceans. Although iodine can participate in ozone destruction reactions, this iodine containing source gases generally have very short life-times and, as a result, most are removed in the troposphere before they reach the stratosphere.

Other gases that influences stratospheric ozone abundances also have increased in the stratosphere as a result of human activities. Important examples are methane (CH_4) and nitrous oxide (N_2O), which react in the stratosphere to form water vapour and reactive hydrogen, and nitrogen oxides, respectively. These reactive products also participate in the production and loss balance of stratospheric ozone. The overall effect of increases in these other gases on ozone is much smaller than that caused by increases in chlorine-and bromine-containing gases from human activities.

Atmospheric lifetimes, emissions, and Ozone Depletion Potentials of halogen source gases have been tabulated below:

Table 11.3: Atmospheric lifetimes, emissions, and ozone depletion potentials of halogen source gases

Halogen source gas	Atmospheric lifetime (years)	Global emissions in 2003[b]	Ozone depletion potential (ODP)[d]
Chlorine:			
CFC-12	100	101-144	1
CFC-113	85	1-15	1
CFC-11	45	60-126	1
Carbon tetrachloride (CCL_4)	26	58-131	0.73
HCFCs	1-26	312-403	0.02-0.12
Methyl chloroform (CH_3CCl_3)	5	20	0.12
Methyl chloride	1.0	1700-13600	0.02
Bromine:			
Halon-1301	65	3	16
Halon-1211	16	7-10	7.1
Methyl bromide (CH_3Br)	0.7	160-200	0.51
Very short-lived gases (e.g., $CHBr_3$)	< 0.5	c	c

(*a*) Includes both human activities and natural sources.

(*b*) Emission in gigagrams per year (1 gigagram = 10^9 grams = 1000 metric tons.)

(*c*) Estimates are uncertain for most species.

(*d*) Values are calculated for emissions of equal mass for each gas.

When exposed to ultraviolet radiation from the Sun, these halogen source gases are converted to more reactive gases also containing chlorine and bromine. Important examples of the reactive gases that destroy stratospheric ozone are chlorine monoxide (CIO)and bromine monoxide (BrO). thee reactive gases participate in "catalytic" reaction cycles that efficiently destroy ozone. Volcanoes can emit some chlorine- containing gases, but these gases are ones that readily dissolve in rainwater and ice and are usually "washed out" of the atmosphere before they can reach the stratosphere.

Reactive gases containing the halogen chlorine and bromine lead to the chemical destruction of stratospheric ozone. Halogen-containing gases present in the stratosphere can be divided into two groups: halogen source gases and reactive halogen gases. The source gases are emitted at Earth's surface by natural processes and by human activities. Once in the stratosphere, the halogen source gases chemically convert at different rates to form the reactive halogen gases. The conversion occurs in the stratosphere instead of the troposphere because solar UV radiation is more intense in the stratosphere. The chemical conversion of halogen source gases, which involves ultraviolet sun-light and other chemical reactions, produces a number of reactive halogen gases. These reactive gases contain all of the chlorine and bromine atoms originally present in the source gases.

The abundances of chlorine source gases and reactive chlorine gases as measured from space are displayed with altitude for a mid-latitude location. In the troposphere (below about 11 kilometers), all chlorine is contained in the source gases. In the stratosphere, reactive chlorine gases increase with altitude as chlorine source gases decrease. This is a consequence of chemical reactions involving ultraviolet sunlight. The principal reactive gases formed are HCI, $CIONO_2$, and CIO. Summing the source gases with the reactive gases gives *total available chlorine,* which is nearly constant with altitude up to 47 km. In the ozone layer, HCI and $CIONO_2$ are the most abundant reactive chlorine gases. The most widely observed bromine gas is bromine monoxide (BrO).

Reactive chlorine present in the atmosphere as sea salt (sodium chloride) produced by evaporation as sea ocean spray dissolves in water and is removed and hence does not reach the stratosphere in appreciable quantities. Another ground-level source is emission of chlorine gases from swimming pools, household bleach, and other uses.

When released to the atmosphere, this chlorine is rapidly converted to forms that are soluble in water and removed. The space Shuttle and other rocket motors release reactive chlorine gases directly in the stratosphere: in this case, the quantities are very small in comparison with other tropospheric sources. Volcanic plumes generally contain large quantities of chlorine in the form of hydrogen chloride (HCI). Because the plumes also contain a considerable amount of water vapor, the HCI is efficiently scavenged by rainwater and ice and removed from the atmosphere. As a result, most of the HCI in the plume does not enter the stratosphere.

Reactive gases containing chlorine and bromine destroy stratosphere ozone in "catalytic" cycles made up of two or more separate reactions. As a result, a single chlorine or bromine atom can destroy many hundreds of ozone molecules before it reacts with another gas, breaking the cycle. In this way, a small amount of reactive chlorine or bromine has a large impact on the ozone layer. Certain ozone destruction reactions become most effective in polar regions because the reactive gas chlorine monoxide reaches very high levels there in the late winter/early spring season.

Ozone destruction cycle is made up of two basic reactions: CIO + O and CI + O_3. The net result of cycle is to convert one ozone molecule and one oxygen atom into two oxygen molecules. In each cycle, chlorine acts as a *catalyst* because CIO and CI react and are reformed. In this way, one CI atom participates in many cycles, destroying many ozone molecules. For typical stratospheric conditions at middle or low latitudes, a single chlorine atom can destroy hundreds of ozone molecules before it happens to react with another gas, breaking the catalytic cycle.

Q.28. How the changes in solar radiation and volcanic eruptions affect the ozone layer?

Ans. Factors such as changes in solar radiation, as well as the formation of stratospheric particles after volcanic eruptions, do influence the ozone layer. However, neither factor can explain the average decreases observed in global total ozone over the last two decades. If large volcanic eruptions occur in the coming decades, ozone depletion will increase for several years after the eruption.

Solar changes: The formation of stratospheric ozone is initiated by ultraviolet (UV) radiation coming from the Sun. As a result, an increase in the Sun's radiation output increases the amount of ozone in Earth's atmosphere. The Sun's radiation output and sunspot number vary over the well-known 11 year solar cycle.

Over the last two decades, average total ozone has decreased over the globe. Average values in recent years show about 4% depletion from pre-1980 values. The lòng-term decreases in global ozone cannot result from changes in solar output alone.

Past volcanoes: Large volcanic eruptions inject sulfur gases directly into the stratosphere, causing new sulfate particles to be formed. The particles initially form in the stratosphere above and downwind of the volcano location and then spread throughout the hemisphere or globally as air is transported by stratospheric winds. When large amounts of particles are present in the stratosphere, transmission of solar radiation is significantly reduced.

Chemical reactions on the surface of volcanically produced particles increase ozone destruction by increasing the amount of the highly reactions chlorine gas, chlorine monoxide (CIO). The amount of CIO produced is proportional to the total abundance of reactive chlorine in the stratosphere. Ozone depletion increases as a consequence of increased CIO. After a few years, however, the effect of volcanic particles on ozone is diminished by their gradual removal from the stratosphere by natural air circulation.

If large volcanic eruptions occur in early decades, ozone depletion will increase for several years. Only later in the 21st century when halogen gas abundances have declined close to pre-1980 values will the effect of volcanic eruptions on ozone be lessened.

Q.29. Is depletion of the ozone layer the principal cause of climate change?

Ans. No ozone depletion itself is not the principal cause of climate change. However, because ozone absorbs solar radiation and is a greenhouse gas, ozone changes and climate change are linked. Stratospheric ozone depletion and increases in global tropospheric ozone that have occurred in recent decades both contribute to climate change. These contributions to climate change are significant but small compared with the total contribution from all other greenhouse gases. Ozone and climate change are indirectly linked because both ozone depleting gases and substitute gases contribute to climate change.

Radiative forcing of climate change. Human activities and natural processes have led to the accumulation in the atmosphere of several long-lived and radiatively active gases known as "greenhouse gases." Ozone is a greenhouse gas, along with carbon dioxide (CO_2), methane (CH_4), nitrous oxide (N_2O), and halogen source gases. The accumulation of these gases in Earth's atmosphere changes the balance between incoming solar radiation and outgoing infrared radiation. Greenhouse gases generally change the balance by absorbing outgoing radiation, leading to a warming at Earth's surface. This change in Earth's radiative balance is called a *radiative forcing of climate change.*

All forcings shown relate to human activities. Positive forcings generally lead to *warming* and negative forcings lead to *cooling* of Earth's surface. The accumulation of carbon dioxide represents the largest forcing term. Carbon dioxide concentrations are increasing in the atmosphere primarily as the result of burning coal, oil, and natural gas for energy and transportation; and from cement manufacturing. Much of the observed surface warming over the last 50 years has been linked to increases in carbon dioxide and other greenhouse gas concentrations caused by human activities.

Stratospheric and tropospheric ozone both absorb infrared radiation emitted by Earth's surface, effectively trapping heat in the atmosphere. Stratospheric ozone also significantly absorbs solar radiation. As a result, increases or decreases in stratospheric or tropospheric ozone cause radiative forcings and represent direct links of ozone to climate change, in recent decades, stratospheric ozone has decreased due to rising chlorine and bromine amounts in the atmosphere, while troposphere ozone in the industrial era has increased due to pollution from human activities. Stratospheric ozone depletion causes a negative radiative forcing, while increases in tropospheric ozone cause a positive radiative forcing. The radiative forcing due to tropospheric ozone increases is currently larger than that associated with stratospheric ozone depletion. The negative forcing from ozone depletion represents an offset to the positive forcing from the halogen source gases, which cause ozone depletion.

Halogen source gases and HFCs. An important link between ozone depletion and climate change is the radiative forcing from halogen source gases and hydrofluorocarbons (HFCs). Halogen source gases are the cause of ozone depletion and HFCs are substitute gases. Both groups of gases cause radiative forcing in the atmosphere, but with a wide range of effectiveness. The ODPs of CFC-11 and CFC-12, and the GWP of CO_2 all are assigned a value of 1.0. For ozone depletion, the halons are the most effective gases (for equal mass amounts) and HFCs cause no ozone depletion For climate change, all gases make a contribution, with CFC- 12 and HFC-23 having the largest effect (for equal mass amounts). Montreal Protocol actions that have led to reductions in CFC concentrations and increases in HCFC and HFC concentration have also reduced the total radiative forcing from these gases. It is important to note that, despite a GWP that is small in comparison to many other greenhouse gases, CO_2 is the most important greenhouse gas related to human activities because its atmospheric abundance is so much greater than the abundance of other gases.

As a group, the principal halogen source gases represent a positive direct radiative forcing in the Industrial Era that is comparable to the forcing from methane, the second most important greenhouse gas. In the coming decades, the abundances of these ozone-depleting gases and their associated positive radiative forcings are expected to decrease. Future growth in HFC emissions, while uncertain, will contribute a positive forcing that will counter the decrease from ozone-depleting gases. Finally, reductions in ozone-depleting gases will be followed by reductions in stratospheric ozone depletion and its associated *negative* radiative forcing.

Impact of climate change on ozone. Certain changes in Earth's climate could affect the future of the ozone layer. Stratospheric ozone is influenced by changes in temperatures and winds in the stratosphere. For example, lower temperatures and stronger polar winds could both affect the extent and severity of winter polar ozone depletion. While the Earth's surface is expected to warm in response to the net positive radiative forcing from greenhouse gas increases, the stratosphere is expected to cool. A cooler stratospheric clouds (PSCs) are present in polar regions and, as a result, might increase winter ozone depletion. In the upper stratosphere at altitudes above PSC formation regions, a cooler stratosphere is expected to increase ozone amounts and hence, hasten recovery, because lower temperatures favor ozone production over loss. Similarly, changes in atmospheric composition that lead to a warmer climate may also alter ozone amounts.

Ozone-depleting gases (halogen source gases) and their substitutes can be compared via their ozone depletion potentials (ODPs) and global warming potentials (GWPs). The GWPs are evaluated for a 100-yr time interval after emission. The CFCs, halons, and HCFCs are ozone-depleting gases and HFCs, used as substitute or replacement gases, do not destroy ozone. The ODPs of CFC-11 and CFC-12, and the GWP of CO_2 have values of 1.0 by definition. Larger ODPs or GWPs indicate greater potential for ozone depletion or climate change, respectively. The top panel compares ODPs and GWPs for emissions of equal mass amounts of each gas. The ODPs of the halons far exceed those of the CFCs. HFCs have zero ODPs. All gases have non-zero GWPs that span a wide range of values. The bottom panel compares the contributions of the 2004 emissions of each gas, using CFC-11 as the reference gas. Each bar represents the product of a global emission value and the respective ODP or GWP factor. The comparison shows that 2004 emissions of ozone- depleting gases currently contribute more than substitute gas emissions to both ozone depletion and climate change. Future projections guided by Montreal Protocol provisions suggest that the contributions of ozone- depleting gases to climate change will decrease, while those of the substitute gases will increase.

Q.30. Discuss the regulations on the production of ozone depleting gases under the provisions in the "Montreal Protocol".

Ans. The production of ozone-depleting gases is regulated under a 1987 international agreement known as the "Montreal Protocol a Substances that Deplete the Ozone Layer" and its subsequent Amendments and Adjustments. The Protocol now ratified by 196 nations, establishes legally binding controls on the national production and consumption of ozone-depleting gases. Production and consumption of all principal halogen- containing gases by developed and developing nations will be significantly phases out before the middle of the 21st century.

Montreal Protocol: In 1985, a treaty called the *Vienna Convention for the Protection of the Ozone Layer* was signed by 20 nations in Vienna. Who agreed to take appropriate measures to protect the ozone layer from human activities. In response to growing concern, the *Montreal Protocol on Substances that Deplete the Ozone Layer* was signed in 1987 and, following country ratification, entered into force in 1989. The Protocol established legally binding controls for developed and developing nations on the production and consumption of halogen source gases known to cause ozone depletion. National consumption of a halogen gas is defined as the amount that production and imports of a gas exceed its export to other nations.

Effects of the Montreal Protocol: The purpose of the Montreal Protocol is to achieve reductions in stratospheric abundances of chlorine and bromine. The reductions follow from restrictions on the production and consumption of manufactured halogen source gases. Projections of the future abundance of *effective stratospheric chlorine* are shown in the top panel assuming (1) no Protocol regulations. (2) only the regulations in the original 1987 Montreal Protocol, and (3) additional regulations from the subsequent Amendments and Adjustments. The city names and years indicate where and when changes to the original 1987 Protocol provisions were agreed upon. Effective stratospheric chlorine as used here accounts for the combined effect of chlorine and bromine gases. Without the Protocol, stratospheric halogen gases are projected to increase significantly in the 21st century. The "zero emissions" line shows a hypothetical case of stratospheric abundance if all emissions were reduced to zero beginning in 2007. The lower panel shows how excess skin cancer cases might increase with no regulation and how they might be reduced under the Protocol provisions.

Effect of the Montreal Protocol: The initial Protocol called for only a slowing of chlorofluorocarbon (CFC) and halon production. The 1990 London Amendments to the Protocol called for a phaseout of the production and consumption of the most damaging ozone-depleting substances in developed nations by 2000 and in developing nations by 2010. The 1992 Copenhagen Amendments accelerated the date of the phaseout to 1996 in developed nations. Further controls on ozone-depleting substances were agreed upon in later meetings in Vienna (1995), Montreal (1997), and Beijing (1999).

Without the Montreal Protocol and with continued production and use of CFCs and other ozone-depleting gases, effective stratospheric chlorine is projected to have increased tenfold by the mid-2050s compared with the 1980 value. Such high values likely would have increased global ozone depletion far beyond that currently observed. As a result, harmful UV-B radiation would have also increased substantially at Earth's surface, causing a rise in excess skin cancer cases.

The 1987 provisions of the Montreal Protocol alone would have only slowed the approach to high effective chlorine values by one or more decades in the 21st century. Not until the 1992 Copenhagen Amendments and Adjustments did the Protocol projections show a *decrease* in the future effective stratospheric chlorine values. Now, with full compliance to the Montreal Protocol and its Amendments and Adjustments, use of the major human-produced ozone-depleting gases will ultimately be phased out and effective stratospheric chlorine will slowly decay, reaching pre-1980 values in the mid-21st century.

The Montreal Protocol provides for the transitional use of hydrochlorofluorocarbons (HCFCs) as substitute compounds for principal halogen source gases such as CFC-12. HCFCs differ chemically from most other halogen source gases in that they contain hydrogen (H) atoms in addition to chlorine and fluorine atoms. HCFCs are used for refrigeration, for blowing foams, and as solvents, which were primary uses of CFCs. HCFCs are 88 to 98% less effective than CFC-12 in depleting stratospheric ozone because they are chemically removed primarily in the troposphere. This removal partially protects stratospheric ozone from the halogens contained in HCFCs. In contrast, CFCs and many other halogen source gases are chemically inert in the troposphere and, hence, reach the stratosphere without beings significantly removed. Because HCFCs still contribute to the chlorine abundance in the stratosphere, the Montreal Protocol requires a gradual phaseout of HCFC consumption in developed and developing nations that will be complete in 2040.

Hydrofluorocarbons (HFCs) are also used as substitute compounds for CFCs and other halogen source gases. HFCs contain only hydrogen, fluorine, and carbon atoms. Because HFCs contain no chlorine or bromine, they do not contribute to ozone depletion. As a consequence, the Montreal Protocol does not regulate the HCFCs. However, HFCs (as well as all halogen source gases) are radiatively active gases that contribute to human-induced climate change as they accumulate in the atmosphere. HFCs are included in the group of gases listed in the Kyoto Protocol of the United Nations Framework Convention on Climate Change (UNFCCC).

The rise in effective stratospheric chlorine values in the 20th century has slowed and reversed in the last decade (top left panel). Effective stratospheric chlorine levels as shown here for midlatitudes will return to 1980 values around 2050. The return to 1980 values will occur around 2065 in polar regions. In 1980, ozone was not significantly depleted by the chlcrine and bromine then present in the stratosphere. A decrease in effective stratospheric chlorine abundance follows reductions in emissions of individual halogen source gases. Overall emissions and atmospheric concentrations have decreased and will continue to decrease given international compliance with the Montreal Protocol provisions. The past increases of CFCs, along with those of CCI_4 and CH_3CCI_3 have slowed significantly and most have reversed in the last decade. HCFCs, which are used as CFC substituted, will continue to increase in the coming decades. Some halon abundances will also continue to grow in the future while current halon reserves are depleted. Smaller relative decreases are expected for CH_3Br in response to production and use restrictions because it has substantial natural sources. CH_3CI has large natural sources and is not regulated under the Montreal Protocol.

The Montreal Protocol allows for the use of hydrochlorofluorocarbons (HCFCs) as short-term substitutes for CFCs. As a result, the abundances of HCFC-22, HCFC-141b, and HCFC-142b continue to grow in the atmosphere. HCFCs pose a lesser threat to the ozone layer than CFCs because they are partially destroyed in the troposphere by chemical processes, thus reducing the overall effective-ness of their emissions in destroying stratospheric ozone. Under the Montreal Protocol, HCFC consumption will reach zero in developed nations by 2030 and in developing nations by 2040. Thus, the future projections show HCFC abundances reaching a peak in the first decades of the 21st century and steadily decreasing thereafter.

The atmospheric abundances of halon-1211 and halon-1301 account for a significant fraction of bromine from all source gases and continue to grow despite the elimination of production in developed nations in 1994. The growth in abundance continues because substantial reserves are held in fire-extinguishing equipment and are gradually being released, and production and consumption are still allowed in developing nations. Atmospheric halon abundances can be expected to remain high well into the 21st century because of their long lifetimes and continued release.

Both methyl chloride (CH_3CI) and methyl bromide (CH_3Br) are distinct among principal halogen source gases because a substantial fraction of their emissions is associated with natural processes. The average atmospheric abundance of methyl chloride, which is not regulated under the Montreal Protocol, will remain fairly constant throughout this century, if natural sources remain unchanged. At century's end, methyl chloride is expected to account for a large fraction of remaining effective stratospheric chlorine because the abundances of other gases, such as the CFCs, are expected to be greatly reduced. The abundance of methyl bromide, which is regulated under the Montreal Protocol, has already decreased in recent years and is projected to decrease further as a result of production phaseouts in developed and developing countries. For the later decades of the century, methyl bromide abundances are shown as nearly constant. However, these abundances are uncertain because the amounts of exempted used of methyl bromide under the Montreal Protocol are not known for future years.

Q.31. Why has an "ozone hole" appeared over Antarctica when ozone-depleting gases are present throughout the stratosphere?

Ans. Ozone-depleting gases are present throughout the stratospheric ozone layer because they are transported great distances by atmospheric air motions. The severe depletion of the Antarctic ozone layer known as the "ozone hole" occurs because of the special whether conditions that exist there and nowhere else on the globe. The very low temperatures of the Antarctic stratosphere create ice clouds polar stratospheric clouds (PSCs). Special reactions that occur on PSCs and the relative isolation of polar stratospheric air allow chlorine and bromine reactions to produce the ozone hole in Antarctic springtime.

The severe depletion of stratospheric ozone in Antarctic winter is known as the "ozone hole". Which first appeared over Antarctica because atmospheric conditions there increased the effectiveness of ozone destruction by reactive halogen gases. The formation of the Antarctic ozone hole requires abundant reactive halogen gases, temperatures low enough to form polar stratospheric clouds (PSCs), isolation of air form other stratospheric regions, and sunlight.

Distributing halogen gases. Halogen source gases emitted at Earth's surface are present in comparable abundances throughout the stratosphere in both hemispheres even though most of the emissions occur in the Northern Hemisphere. The abundances are comparable because most source gases have no important natural removal processes in the lower atmosphere and because winds and warm-air convection redistribute and mix air efficiently throughout the troposphere. Halogen gases enter the stratosphere primarily from the tropical upper troposphere, Atmospheric air motions then transport them upward and toward the poles in both hemispheres.

The severe ozone destruction represented by the ozone hole requires that low temperature. Stratospheric air temperature in both polar regions reach minimum values in the lower stratosphere in the winter season. Average minimum values over Antarctica are as low as – 90°C in July and August in a typical year. Over the Arctic, average minimum values are near – 80°C in January and February. Polar stratospheric clouds (PSCs) are formed when winter minimum temperatures fall below the formation temperature (about – 78°C). This occurs on average for 1 to 2 months over the Arctic and 5 to 6 months over Antarctica. Reactions on PSCs cause the highly reactive chlorine gas CIO to be formed, which increases the destruction of ozone. The range of winter minimum temperatures found in the Arctic is much greater than in the Antarctic. In some years, PSC formation temperatures are not reached in the Arctic, and significant ozone depletion does not occur. In the Antarctic, PSCs are present for many months, and severe ozone depletion now occurs in each winter season.

In the Antarctic winter, minimum temperatures are generally lower and less variable than in the Arctic winter. Antarctic temperatures also remain below the PSC formation temperature for much longer periods during winter. The winter temperatures are low enough for PSCs to form for nearly the entire Antarctic winter but usually only for part of every Arctic winter.

Air in the polar stratospheric regions is relatively isolated from other stratospheric regions for long periods in the winter months. The isolation comes about because of strong winds that encircle the poles, preventing substantial motion of air in or out of the polar stratospheres. The isolation is much more effective in the Antarctic than the Arctic. Once chemical changes occur in the cold air as a result of the presence of PSCs, the changes remain for many weeks to months.

Polar stratospheric clouds changes in the relative abundances of reactive chlorine gases. Reactions occur on the surfaces of PSC particles that convert the reservoir forms of reactive chlorine gases, CIONO, and HCI, to the most reactive form, CIO. CIO increases from a small fraction of available reactive chlorine gases to nearly all that is available. With increased CIO, additional catalytic cycles involving CIO and BrO become active in the chemical destruction of ozone when sunlight is available.

PSCs form when stratospheric temperatures fall below about – 78°C in polar regions. as a result, PSC are often found over large areas of the winter polar regions and over a significant altitude range. At low polar temperatures, nitric acid (HNO_3) and water condense on preexisting sulfur-containing particles to form solid and liquid PSC particles. At even lower temperatures, ice particles also form. PSCs are often found near mountain ranges in polar regions because the motion of air over the mountains can cause local cooling of stratospheric air.

When temperatures increase by early spring, PSCs no longer form and the production of CIO ends. Without continued CIO production, CIO amounts decrease as other chemical reactions reform $CIONO_2$ and HCI. As a result, the intense period of ozone depletion ends.

Once formed, PSC particles move downward because of gravity. The largest particles move down several kilometers or more in the stratosphere during the low-temperatures winter/spring period. Because most PSCs contain nitric acid, their downward motion removes nitric acid from regions of the ozone layer. That process is called *denitrification.* With less nitric acid, the highly reactive chlorine gas CIO remains chemically active for a longer period, thereby increasing chemical ozone destruction. Denitrification occurs each winter in the Antarctic and in some, but not all, Arctic winters, because PSC formation temperatures are required over an extensive time period.

The role of PSCs in converting reactive chlorine gases to CIO was not understood until after the discovery of the Antarctic ozone hole in 1985.

The first decreases in Antarctic total ozone were observed in the early 1980s over research stations located on the Antarctic continent. The observations showed unusually low total overhead ozone during the late winter/early spring months of September, October, and November. Total ozone was lower in these months compared with previous observations made as early as 1957. Soon, after, satellite measurements confirmed the spring ozone depletion and further showed that in each late winter/early spring season starting in the early 1980s, the depletion extended over a large region centered near the South Pole. The team "ozone hole" came about from satellite images of total ozone that showed very low values encircling the Antarctic continent each spring. Currently, the formation and severity of the Antarctic "ozone hole" are documented each year by a combination of satellite, ground-bases, and balloon observations of ozone.

Q.32. Discuss the severity of the depletion of Antarctic ozone layer.

Ans. Antarctic ozone depletion is seasonal, occurring primarily in late winter and early spring (August November). Peak depletion occurs in early October when ozone is often completely destroyed over a range of altitudes, reducing overhead total ozone by as much as two-thirds at some locations. This severe depletion creates the "ozone hole" in images of Antarctic total ozone made from space. In most years the maximum area of the ozone hole far exceeds the size of the Antarctic continent.

The severe depletion of Antarctic ozone, known as the "ozone hole," was first observed in the early 1980s. The depletion is attributable to chemical destruction by reactive halogen gases, which increased in the stratosphere in the latter half of the 20th century. Conditions in the Antarctic winter stratosphere are highly suitable for ozone depletion because of (1) the long periods of extremely low temperatures, which promote polar stratospheric cloud (PSC) formation; (2) the abundance of reactive halogen gases, which chemically destroy ozone; and (3) the isolation of stratospheric air during the winter, which allows time for chemical destruction to occur.

Satellite images made during Antarctic winter and spring show a large region centered near the South Pole in which total ozone is highly depleted. This region has come to be called the "ozone hold" because of the near-circular contours hole is defined here as the area contained within the 220-Dobson unit (DU) contour in total ozone maps. The maximum area has reached 25 million square kilometers in recent years, which is nearly twice the area of the Antarctic continent. Minimum values of total ozone inside the ozone hole averaged in late September have reached below 100 DU, which is well below normal springtime values of about 200 DU Normal value is 300 DU.

Ozone within the "ozone hole" is also measured using balloon house instruments. Balloon measurements show changes within the ozone layer, the vertical region that contains the highest ozone abundances in the stratosphere. Balloon measurements show that the chemical destruction of ozone is complete over a vertical region of several kilometers.

Low winter temperatures and isolated conditions occur each year in the Antarctic stratosphere, but significant spring ozone depletion has been observed every year only since the early 1980s. In prior years, the amounts of reactive halogen gases in the stratosphere were insufficient to cause significant depletion.

Average values decreased steadily through the 1980s and 1990s, reaching minimum values that were 37% less than in pre-ozone-hole years (1970-1982). The year-to-year changes in the average values reflect variations in the meteorological conditions, which affect the extent of low polar temperatures and the transport of air into and out of the Antarctic winter stratosphere. However, essentially all of the ozone depletion in the Antarctic in most years is attribute to chemical loss from reactive halogen gases.

The depletion of Antarctic ozone occurs primarily in the late winter/early spring season. In spring, temperatures in the lower polar stratosphere eventually warm, thereby ending PSC formation as well as the most effective chemical cycles that destroy ozone. The transport of air between the polar stratosphere and lower latitudes also increases during this time, ending winter isolation. This allows ozone-rich air to be transported to polar regions, displacing air in which ozone has been severely depleted. This displaced air is mixed at lower latitudes with more abundant ozone-rich air. As a result, the ozone hole disappears by December and Antarctic ozone amounts remain near normal until the next winter season.

The stratospheric ozone layer resides between about 10 and 50 kilometers above Earth's surface over the globe. In the Antarctic at the South Pole, halogen gases have destroyed ozone in the ozone layer beginning in the 1980s. Before that period, the ozone layer was clearly present. In more recent years, (for 2 October 2001) ozone is destroyed completely between 14 and 20 kilometers in the Antarctic in spring. Average October values in the ozone layer now are reduced by 90% from pre-1980 values. The Arctic ozone layer is still present in spring. However, March Arctic ozone values in some years are often below normal average values. In such years, winter minimum temperatures are generally below PSC formation temperatures for long periods.

Q.33. Discuss the nature of UV radiation.

Ans. Ultraviolet (UV) radiation is a specific part of the sun's entire spectrum of wavelengths. UV radiation can be further sub-divided into three parts, UV-C is characterized by wavelengths less than 280 nm (1 nanometre = -9 10 metre) Although highly dangerous to plants and animals, UV-spectrum is completely absorbed by stratospheric ozone and does not reach the earth's surface. UV-B ranges in wavelengths between 280 and 320 nm. Ozone absorbs much of the shorter wavelength radiation, but this absorption weakens as 320 nm is approached. Plants and animals are particularly affected by this part of the UV spectrum. UV-B effects to humans are reddening of the skin (erythema) and reduction of vitamin-D, synthesis in the short term, development of skin cancer, cataracts and suppression of the immune system in the long term. The wavelengths of UV-A radiation range between 320 and 400 nm. Ozone absorbs very little of UV-A spectrum. UV-A radiation is needed by humans for the synthesis of vitamin-D; however, too much UV-A causes photoaging (toughening of the skin), suppression of the immune system and, to a lesser degree, reddening of the skin, and cataract formation.

The shorter wavelengths UV radiations are highly sensitive in comparison to the longer wavelengths. The erythemal action spectrum is a composite of several investigators' measurements of the response of many different human skin types to UV radiation and is not representative of any one skin type which is meant to depict not only the natural pigmentation of the skin, but also the likelihood of one's skin to either burn or tan.

Q.34. Discuss the diurnal variability of ultraviolet (UV) radiation.

Ans. The variation throughout the course of the day (diurnal) of UV radiation is much like that of visible light. UV radiation is much more attenuated than the global at high solar zenith angles in the early morning and late afternoon. This is because the two components which make up UV and all other forms of radiation; the *direct* and the *diffuse,* are both greatly affected at these low sun angles. Direct UV radiation is greatly reduced by the increased absorption by stratospheric ozone during its increased path length through the atmosphere (about 6 times more than when the sun is directly overhead). Also, radiation at the UV wavelengths is scattered much more than visible light. This further decreases the direct component and increases the diffuse component. As the Sun rises above the horizon, the amount of absorption in the stratosphere and scattering in the troposphere is reduced. The result is a drastic increase in UV radiation reaching the surface.

The variation of surface temperature differs significantly from both UV and global radiation. Whereas the diurnal variation of incoming infrared radiation is similar to that of global radiation, there exists a delay between the time when the earth-atmosphere system is irradiated to when the temperature begins to increase. This is known as the thermal response. Depending upon the time of the year the lag in time between the peak radiation flux reaching the surface (solar noon) to when the surface temperature reaches its maximum can be as great as 3 to 4 hours. For example, it is hottest in the mid afternoon not at noon. Summer surface temperature is at the afternoon. By the time of the temperature maximum, the amount of UV radiation reaching the surface has decreased almost by half of that at solar noon.

Q.35. What are the effects of clouds, elevation, and surface pollution on the UV radiation reaching the earth?

Ans. Clouds, air pollution, haze and elevation all have effects on the amount of ultraviolet (UV) radiation reaching the earth's surface. UV radiation reaches the surface as a sum of its direct component (normal to the sun) and its diffuse component (from all directions). UV radiation reaches the top of the troposphere in mostly its direct component. This is because there are few molecules to scatter the radiation. Decreases in UV radiation intensity has resulted due to absorption by ozone. Once the UV radiation reaches the troposphere it encounters much greater numbers of scattering air molecules and dust. Below are discusses how UV radiation is affected by:

Elevation: In the troposphere, air molecules and dust increase as the UV radiation travels from the stratosphere to the troposphere. The further down in to the atmosphere UV radiation travels, the more the direct component is reduced and the more the diffuse component is increased. As more UV radiation is scattered, the smaller the amount the reaches the surface. As a result, there is more UV radiation at high elevations than at lower elevations.

Clouds: Made up of millions of water droplets, clouds can transmit, reflect and scatter UV radiation. The amount of each is dependent upon the thickness of the cloud and its morphology. Generally, the larger and thicker the cloud is the lesser amount of UV radiation that is transmitted. UV radiation can and does reflect off the sides of towering cumulus clouds. Such conditions result in actual enhancements of surface UV radiation.

Dust/Haze: These two conditions act on UV radiation the same way. They both scatter UV radiation. Enough UV radiation is scattered that on hazy or dusty days there is less UV radiation reaching the surface than would otherwise be there on a clear day.

Air Pollution/Smog: This encompasses many greenhouse gases. Emission from traffic and industrial manufacturing plants form smog as UV radiation and solar heat cause the necessary chemical reactions to take place. As a result, the amounts of UV radiation reaching the surface is smaller under these conditions.

❑❑❑

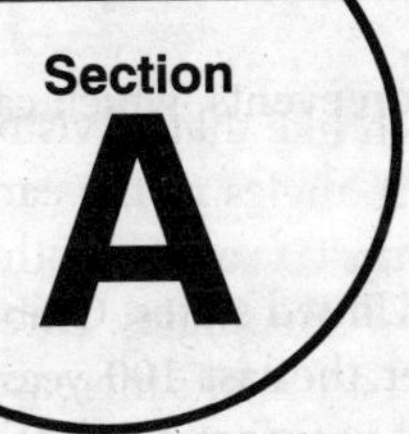

Section A

12 Climate Change

SHORT QUESTIONS AND ANSWERS ON CLIMATE CHANGE

Q.1. What is the evidence that proves that climate is changing?

Ans. The global average temperature increased by more than 1.4°F over the last century The decade from 2000 to 2010 was the warmest on record, and 2010 was tied with 2005 as the warmest year on record. Rising global temperatures have also been accompanied by other changes in weather and climate. Many places have experienced changes in rainfall resulting in more intense rain, as well as more frequent and severe heat waves The planet's oceans and glaciers have also experienced changes: oceans are warning and becoming more acidic, ice caps are melting, and sea levels are rising All of these changes are evidence that our climate is changing and the world is getting warmer.

Q.2. Are human activities or natural variations in climate responsible for the climate change being observed today?

Ans. The Earth does go through natural cycles of warming and cooling, caused by factors such as changes in the sun or volcanic activity. The warming we have seen in the past 50 years cannot be explained by natural factors alone. The recent global warming is primarily a result of greenhouse gas emissions from human activities.

Q.3. Is climate change influenced more by human activities and excess greenhouse gases or changes in the sun's energy?

Ans. The sun has natural periods of warming and cooling. With satellites, scientists have measures fluctuations in the sun's energy and found that these recent variations have been small in comparison to human influences in the last several centuries, with no increase in solar energy in the past 50 years. Thus, changes in the sun's energy cannot explain the warming we have seen over the past several decades. In contrast, the warming we are observing is consistent with the warming properties of carbon dioxide and other heat-trapping gases that we are adding to the atmosphere.

Q.4. How can carbon dioxide affect/hurt us?

Ans. Carbon dioxide is a necessary ingredient for plants to perform photosynthesis, and a critical component of our atmosphere. However, we cannot have too much of a good thing. The excess carbon dioxide we are adding to the atmosphere increases global temperatures, leading to climate changes that can harm plants, animals, and humans.

Q.5. How can a change of one or two of in global average temperatures have an impact on our lives?

Ans. Changing the average global temperature by even a degree or two can lead to serious consequences around the globe. For about every 2°F of warming, we can expect to see:

- 5-15% reductions in the yields of crops as currently grown.
- 3-10% increases in the amount of rain falling during the heaviest precipitation events, which can increase flooding risks.
- 5-10% decreases in stream flow in some river basins,
- 200%-400% increases in the area burned by wildfire in parts of the western United states. Global average temperatures have increases more than 1.4 degrees Fahrenheit over the last 100 years. Many of the extreme precipitation and heat events that we have seen in recent years are consistent with what we would expect give this amount of warming. Scientists project that Earth's average temperatures will rise between 2 and 12 degrees Fahrenheit by 2100

Q.6. Do a few extra cold or snowy winters mean that global warming and climate change is not happening?

Ans. A few extra cold or snowy winters doesn't mean that global warming isn't happening. We now that global average temperatures are rising However, even with this global warming, at the local or regional level, we can expect to have some colder-than-average seasons or even colder-than-average years For example, in the Eastern United States, the winters of 2010 and 2011 were colder than the average winters from the previous decades. In fact, extra snowy winters can be expected. In a warmer climate, more water vapour is held in the atmosphere causing more intense rain and snow storms. As the climate warms, we do expect the duration of the snow seasons to decrease – however, as long as it is still cold enough to snow, a warming climate can lead to bigger snowstorms.

Q.7. How does water vapour in our atmosphere contribute to global warming?

Ans. Higher concentrations of carbon dioxide and other greenhouse gases in the atmosphere cause earth to warm. Warmer temperatures increase the amount of water vapour in the atmosphere. Because water vapour also is a greenhouse gas, the this leads to even further warming. In this way, water vapour actually magnifies the warming caused by excess carbon dioxide and other greenhouse gases.

Q.8. Do emissions of carbon dioxide from human activities have a big impact on Earth's climate?

Ans. Plants, oceans, and soils release and absorb large quantities of carbon dioxide as a part of the Earth's natural carbon cycle. These natural emissions and absorptions of carbon dioxide on average balance out over time However, the carbon dioxide from human activities is not part of this natural balance. Ice core measurements reveal that carbon dioxide levels in the atmosphere are higher than they have been for at least 800,000 years. The global warming that has been observed in recent decades was caused by elevated levels of carbon dioxide and other greenhouse gases in the atmosphere, due primarily to human activities.

Q.9. In the past, has Earth been warmer than it is today? If so, does that mean we shouldn't worry about global warming?

Ans. There were times in the distant past when Earth was warmer than it is now. However, human societies have developed and thrived during the relatively stable climate that has existed since the last ice age. Due to excess carbon dioxide pollution, the climate is no longer stable and is instead projected to change faster than at any other time in human history. This rapid climate change will expose people to serious risks, sea level, rise, increasing droughts and wildfires in some regions and increasing flooding in others, more heat waves, and other effects of climate change all pose risks to human health infrastructure critical to our homes, roads, and cities, and the ecosystems that support us.

Q.10. Is the hole in the ozone layer related to the climate change?

Ans. The ozone hole and climate change are essentially two separate issues The "ozone hold" refers to the destruction of a layer of ozone molecules found high is Earth's atmosphere (stratosphere) when healthy, this ozone layer helps to shield Earth from the sun's harmful ultraviolet rays The ozone layer has become thinner because of chemicals called chlorofluorocabons that were once commonly used in products ranging from spray cans to foam furniture cushions, A thinner ozone layer allows more ultraviolet rays to reach Earth, increasing the risk to humans from skin cancer, cataracts, and other health impacts. This, however, has only minor effects on climate change.

Q.11. Will a small rise in sea level affect people?

Ans. A small rise in sea level will affect many people. The amount of sea level rise expected to occur as a result of climate change will increase the risk of coastal flooding for hundreds of millions of people around the world, many of whom would have to permanently leave their homes. Global sea level has risen by 23 cms, on average, in the last 140 years This has already put some coastal homes, beaches, roads, bridges and wildlife at risk. By the year 2100, sea level is expected to rise another 45 to 90 cms. Rising seas will make coastal storms and the associated storm surges more frequent and destructive.

Q.12. Are the temperature records showing global warming is happening reliable?

Ans. Multiple temperature records from all over the world have all shown a warming trend, and these records have been deemed reliable by the National Aeronautics and Space Administration (NASA), and the National Oceanic and Atmospheric Administration (NOAA), among others. Other observations that point to higher global temperature includes: warmer oceans, melting arctic sea ice and glaciers, sea level rise, increasing precipitation, and changing wind patterns.

Q.13. Is it too late to do anything about climate change?

Ans. It is not too late to have a significant impact on future climate change and its effects on us. With appropriate actions by, we can reduce the amount of greenhouse gas pollution we release and lower the risk of much greater warming and severe consequences. Many of the actions that we can take to address climate change will have other benefits, such as cleaners healthier air. In addition, communities can take action to prepare for the changes we know are coming.

Q.14. How the oceans influence climate?

Ans. The oceans influence climate over long and short time-scales. On the longest time-scale of geologic time, the shape and location of the continents helps to determine the oceans circulation patterns. Since continental plates drift at about 5 cm per year and mountain ranges rise by about 1 mm, it usually takes millions of years for new land formations to change the oceans. Patterns of ocean circulation and up-welling can also change much more rapidly, resulting in climate variations and fluctuations on a human time-scale. Records of global and, in particular, regional climate show periods lasting from years to centuries during which the climate was systematically different from earlier and later periods This behaviour is related to changes in the way the oceans store and transport heat, although the precise causes of these changes are not clear.

The oceans and the atmosphere are tightly linked and together form the most dynamic component of the climate system. Changes in external factors such the sun's energy, the distribution of various plant species, or the emission of greenhouse gases into the atmosphere can alter the temperatures and circulation patterns of the atmosphere-ocean system. Because the atmosphere and oceans are turbulent, they can also generate their own internal fluctuations. Short-term fluctuations in wind or temperature (that is, weather) can directly influence the current and temperature of the underlying ocean, while oceanic fluctuations can magnify, diminish, or modify atmospheric fluctuations.

The oceans play a critical role in storing heat and carbon. When the earth's surface cools or is heated by the sun, the temperature change is greater – and faster – over the land than over the oceans. Because it is a fluid, the ocean diffuses the effects of a temperature change for great distances via vertical mixing and convective movements. The solid land cannot, so the sun's heat penetrates only the thin, upper crust. One consequence of the ocean's ability to absorb more heat is that when an area of ocean becomes warmer or cooler than usual, it takes much longer for that area to revert to "normal" than it would for a land area. This also explains why "maritime" climates tend to be less extreme than "continental" ones, with smaller day- night and winter-summer differences.

The ocean's waters are constantly being moved about by powerful currents. Surface currents are largely wind-driven, although the rotation of the earth, the presence of continents, and the oceans internal dynamics also have a strong influence. Deep-ocean flow (and, to a lesser extent, surface flow) is driven by density differences produced by heating and cooling and by precipitation and evaporation (cool saline water is denser than warm fresh-water). The behaviour of the atmosphere strongly affects these density differences. For example, clouds can cool the sea the blocking the warming rays of the sun or reduce surface salinity by bringing rain. The wind can influence evaporation rates by blowing more strongly or more weakly.

These currents influence the climate by transporting heat. Horizontal currents, particularly those moving north or south, can carry warmed or cooled water as far as several thousand kilometers. The displaced water can then warm or cool the air and, indirectly, the land over which this air blow. For example, water from the tropical and subtropical Atlantic moves north through the Atlantic in a current popularly called the "Gulf Stream". There it bathes the shores of Western Europe, producing a climate that is surprisingly mild for that latitude. In addition to currents, up-wellings of cold water in places where the wind blow surface water away can also affect climate. Thus San Francisco, influenced by coastal up-welling, is barely warmer than Dublin, which is influenced by the Gulf Stream, despite being over 1,600 km further south.

Currents involved in "deep-water formation" are particularly important for climate. In winter, surface cooling causes water to become denser. (While fresh-water that is cooled starts to expand at temperatures below 4 C, salt-water continues to compress all the way down to its freezing point of – 2°C). In areas where evaporation exceeds precipitation, the resulting rise in salinity also increases density. When the surface water becomes denser than the underlying water, "convective overturning" occurs and the dense surface water mixes downwards. In certain places this downward mixing can occasionally extend all the way to the bottom, even in deep oceans. The dense, deep water thus formed spreads out over the whole ocean. As a result, when downward mixing takes place at high latitudes it creates a circulation pattern in which warm water from tropical and subtropical regions moves poleward, surrenders heat to the atmosphere, cools and sinks, and flows back towards the equator. The net result is a transport of heat poleward.

An apparently small change in just one aspect of the ocean's behaviour can produce major climate variations over large areas of the earth. The areas of cold-water formation are one known example of this possibly wide-spread phenomenon. Although more research is needed, there is some agreement among oceanographers that, for the entire area north of 3 N latitude, the ocean's poleward transport of heat is the equivalent of about 15 watts per square metre of the earth's surface. This can be compared with some 200 W/m^2 from direct sunshine, and about 6 W/m^2 for what climate change models predict will happen if the atmospheric concentration of carbon dioxide doubles. Recent observations, ocean core records, and some modelling results indicate that North Atlantic deep-water formation and its associated ocean heat flow fluctuate substantially over time-scales ranging from years to millennia. The system is vulnerable because even a relatively small decrease in surface salinity prevents water – no matter how cold it is – from sinking. This could occur if there is a flood of fresh-water run-off from the Arctic due to global warming.

❑❑❑

Section

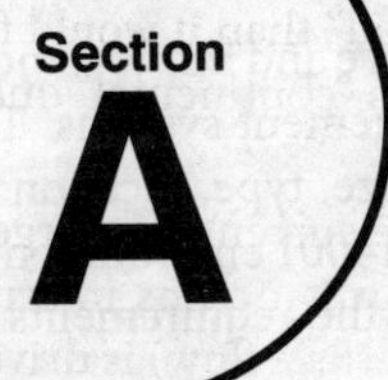

13

ISO-14000

SHORT QUESTIONS AND ANSWERS ON ISO-14000

Q.1. What is ISO?

Ans. ISO (International Organization for Standardization) is a worldwide federation of national standards bodies from each country. The object of ISO is to promote the development of standardization and related activities in the world with a view to facilitating international exchange of goods and services, and to developing cooperation in the spheres of intellectual, scientific, technological and economic activity. The results of ISO technical work are published as International Standards.

Q.2. What is ISO 14000?

Ans. ISO 14000 is the general term used for ISO 14000 family of standards published by International Organization for Standardization. (ISO) which addresses various aspects of environmental managements. It provides a framework for the development of environmental management system.

Q.3. What is the role of BIS (Bureau of Indian Standards) in ISO 14000?

Ans. BIS is the National Standards Body of India and is a founder member of ISO. BIS represents India, in ISO. The Technical Committee (TC) number 207 (ISO/TC 207), and its sub-committees of ISO are responsible for the development of ISO 14000 standards. Industry experts from India including BIS officers nominated by BIS participate in the meetings of the Technical Committee ISO/TC 207 and its sub-committees.

Q.4. What is IS/ISO 14001?

Ans. ISO-140001 'Environmental Management systems – Specifications with Guidance for use' is a standard and adopted by BIS as IS/ISO 14001. IS/ISO 14001 is the exact replica of ISO 14001. It is requirements standard. It contains a set of requirements to define the operation of the Environmental management system. Since the requirements are expressed in a general form, it has the flexibility to be applied to any organization.

It helps an organization to meet its Environmental obligations reliably and consistently.

IS/ISO 14001 is the only certification standard in the IS/ISO 14000 family.

Q.5. What is the difference between ISO 14000 standards and IS/ISO 14000 standards?

Ans. There is no difference. They are exactly the same. BIS had adopted the above-mentioned ISO 14000 series of standards and these are numbered as IS/ISO 14000, IS/ISO 14001, IS/ISO 14004. These standards published by BIS are exact replica at ISO 14000 series of standards. BIS also provides certifications against IS/ISO 14001 under its Management Systems certification activity.

Q.6. Which IS/ISO 14000 standard is meant for certification?

Ans. Any organization can apply for certification against IS/ISO 14001. The other standards *i.e.*, IS/ISO 14004 is guidance standard and is not meant for certification.

Q.7. What is IS/ISO 14004?

Ans. IS/ISO 14004 provides guidance on the establishment, implementation, maintenance and improvement of an environmental management system and its coordination with other management systems. The guidelines in IS/ISO 14004 are applicable to any organization, regardless of its size, type, location or level of maturity. While the guidelines in ISO 14004 are consistent with the IS/ISO 14001 environmental management system model, they are not intended to provide interpretations of the requirements of ISO 14001. This standard is not amenable to certification.

Q.8. How are the IS/ISO 14001 standard applicable to an organization providing services?

Ans. The ISO 14000 environmental management standards exist to help organizations minimize how their operations negatively affect the environment (cause adverse changes to air, water or land) and comply with applicable laws and regulations.

Q.9. Is there any guidance standard for implementation of IS/ISO 14001?

Ans. ISO 14004 has been adopted by BIS as IS/ISO 14004 which is exact replica of ISO 14004 and is the companion standard to IS/ISO 14001 For those organizations willing to implement the Environmental Management System specified in IS/ISO 14001, IS/ISO 14004 provides guidance and illustrations. As this is guidance standard, it is not meant for certification.

Q.10. What are the standards in ISO 14000 family of standards?

Ans. The other standards of ISO 14000 family have also been adopted by BIS as IS/ISO 14000 and are exact replica of ISO 14000 standards. These standards are:

IS/ISO 14004: Environmental management systems – General guidelines or principles, system and supporting techniques

IS/ISO 14020: Environmental labels and declarations – General principles

IS/ISO 14021: Environmental labels and declarations – Self declared environmental claims (Type II Environmental labelling)

IS/ISO 14024: Environmental labels and declaration – Type I Environmental labelling principles and procedures.

IS/ISO 14031: Environmental management – Environmental performance evaluation – Guidelines.

IS/ISO 14040: Environmental Management – Life cycle assessment – principles and framework

IS/ISO 14041: Environmental Management – Life cycle assessment = Goal and scope definition and inventory analysis

IS/ISO 14042: Environmental management – Life cycle assessment – life cycle impact assessment

IS/ISO 14043: Environmental management – Life Cycle Assessment – Life cycle interpretation.

IS/ISO 14050: Environmental management – Vocabulary

Q.11. How to implement the requirements of ISO 14001 in any organization?

Ans. As a minimum to organization familiarize itself not only with the requirements of IS/ISO 14001 but also with the content and philosophies given in:

IS/ISO 14050 'Environmental Management – Vocabulary'

IS/ISO 14004 'Environmental Management Systems – General Guidelines on principles, System and Supporting Techniques'

IS/ISO 19011 'Guidelines for quality and/or environmental management systems auditing' in addition it should also refer to the other standards of ISO 14000 family of standards.

Q.12. What is the difference between certification and accreditation?

Ans. Accreditation is like certification of the certification body. Accreditation should not be used as an interchangeable alternative for certification or registration.

❑❑❑

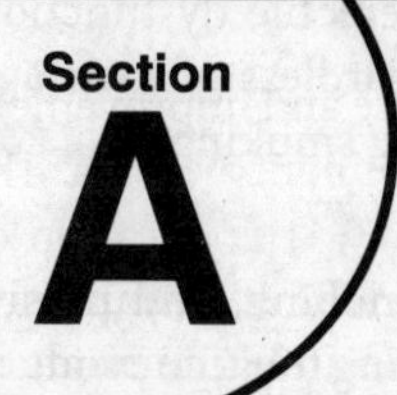

14

Smoking and its Health Effects

SHORT QUESTIONS AND ANSWERS ON SMOKING AND ITS HEALTH EFFECTS

Q.1. What are the key points related to smoking?

Ans.
- Tobacco smoke is harmful to smokers and nonsmokers.
- Cigarette smoking causes many types of cancer, including cancers of the lung, esophagus, larynx (voice box), mouth, throat, kidney, bladder, pancreas, stomach, and cervix, as well as acute myeloid leukemia.
- Quitting smoking reduces the health risks caused by exposure to tobacco smoke.

Q.2. What harmful chemicals are present in tobacco smoke?

Ans. Tobacco smoke contains chemicals that are harmful to both smokers and nonsmokers. Breathing even a little tobacco smoke can be harmful.

Of the more than 4,000 chemicals in tobacco smoke, at least 250 are known to be harmful, including hydrogen cyanide, carbon monoxide, and ammonia.

Among the 250 known harmful chemicals in tobacco smoke, at least 69 can cause cancer. These confirmed cancer-causing chemicals include the following.

- Arsenic
- Benzene
- Beryllium (a toxic metal)
- 1,3-Butadiene (a hazardous gas)
- Cadmium (a toxic metal)
- Chromium (a metallic element)
- Ethylene oxide
- Nickel (a metallic element)
- Polonium-210 (a radioactive chemical element)
- Vinyl chloride

Other toxic chemicals in tobacco smoke are suspected to cause cancer, including the following

- Formaldehyde
- Benzo [α] pyrene
- Toluene

Q.3. Discuss some of the health problems caused by smoking?

Ans. Smoking harms nearly every organ of the body and diminishes a person's overall health. Smoking is a leading cause of cancer and death from cancer. It causes cancers of the lung, esophagus, larynx, mouth, throat, kidney, bladder, pancreas, stomach, and cervix, as well as acute myeloid leukemia. Smoking also causes heart disease, stroke, aortic aneurysm (a balloon-like bulge in an artery in the chest), chronic obstructive pulmonary disease (COPD) (chronic bronchitis and emphysema), asthma, hip fractures, and cataracts. Smokers are at higher risk of developing pneumonia and other airway infections.

A pregnant smoker is at higher risk of having her baby born too early and with an abnormally low birth weight. A woman who smokes during or after pregnancy increases her infant's risk of death from

Sudden Infant Death Syndrome (SIDS). Men who smoke are at greater risk of erectile dysfunction. Cigarette smoking and exposure to tobacco smoke cause premature deaths. Regardless of their age, smokers can substantially reduce their risk of disease, including cancer, by quitting smoking.

Q.4. What are the risks of tobacco smoke to non-smokers?

Ans. Second hand smoke (also called environmental tobacco smoke, involuntary smoking, and passive smoking) is the combination of "side stream" smoke (the smoke given off by a burning tobacco product) and "mainstream" smoke (the smoke exhaled by a smoker). Second hand smoke is a known human carcinogen (cancer-causing agent). Inhaling second hand smoke causes lung cancer in nonsmoking adults. Living with a smoker increases a nonsmoker's chances of developing lung cancer by 20 to 30 percent.

Second hand smoke causes disease and premature death in nonsmoking adults and children. Exposure to second hand smoke may increase the risk of heart disease by an estimated 25 to 30 percent. Pregnant women exposed to second hand smoke are at risk of having a baby with low birth weight. Children exposed to second hand smoke are at an increased risk of SIDS, ear infections, colds, pneumonia, bronchitis, and more severe asthma. Being exposed to second hand smoke slows the growth of children's lungs and can cause them to cough, wheeze, and feel breathless.

Q.5. Is smoking addictive?

Ans. Yes. Nicotine is a drug that is naturally present in the tobacco plant and is primarily responsible for a person's addiction to tobacco products, including cigarettes. During smoking, nicotine enters the lungs and is absorbed quickly into the bloodstream and travels to the brain in a matter of seconds. Nicotine causes addiction to cigarettes and other tobacco products that is similar to the addiction produced by using drugs such as heroin and cocaine.

Q.6. How much nicotine is in cigarettes and cigars?

Ans. Cigarettes, cigars, and other tobacco products vary widely in their content of nicotine, cancer- causing substances, and other toxicants. In a cigarette (which contains less than 1 gram of tobacco), the nicotine content can vary between 13.7 and 23.2 milligrams per gram of dry tobacco. In a cigar (which can contain as many as 20 grams of tobacco), the nicotine content can vary between 5.9 and 335.2 milligrams per gram of tobacco.

The way a person smokes a tobacco product is more important than the nicotine content of the product in determining how much nicotine gets into the body. Nicotine is absorbed in the lungs and through the lining of the mouth. Increased levels of nicotine are absorbed by inhaling the smoke into the lungs and taking frequent and deep puffs.

Q.7. Are other tobacco products, such as smokeless tobacco or pipe tobacco, harmful and addictive?

Ans. Yes. All forms of tobacco are harmful and addictive. There is no safe tobacco product.

In addition to regular cigarettes and cigars, other forms of tobacco include smokeless tobacco (also called chewing tobacco, snuff, and snus), pipes, hookahs (waterpipes) and bidis. Although most research has focused on the harms of cigarette smoking, all forms of tobacco are harmful.

All tobacco products contain nicotine and cancer-causing substances. Both smokeless tobacco and smoking tobacco are known to cause cancer in humans. These products may also cause heart attacks, mouth problems, and other diseases.

Pipe smoking causes lung cancer and increases the risk of cancers of the mouth, throat, larynx, and esophagus.

Hooka or waterpipe is a device used to smoke tobacco. The smoke passes through a partially filled water bowl before being inhaled by the smoker. Some people think hookah smoking is less harmful and addictive than smoking regular cigarettes but the fact is that all forms of tobacco smoking are harmful and addictive. Tobacco smoke, including the smoke produced by a hookah, contains harmful chemicals such as carbon monoxide and cancer- causing substances. A bidi is a flavored cigarette made by rolling tobacco in a dried leaf from the tendu tree, which is native to India. Bidi use is associated with heart attacks and cancers of the mouth, throat, larynx, esophagus, and lung. A kretek is a cigarette made with a mixture of tobacco and cloves. Smoking kreteks is associated with lung cancer and other lung diseases.

Q.8. Is there a tobacco product that is less hazardous than cigarettes?

Ans. All tobacco products are harmful and cause cancer, and the use of these products is strongly discouraged. There is no safe level of tobacco use. People who use any type of tobacco product should quit.

Q.9. What are the immediate benefits of quitting smoking?

Ans. The immediate health benefits of quitting smoking are as follows:

- Heart rate and blood pressure, which are abnormally high while smoking, begin to return to normal.
- Within a few hours, the level of carbon monoxide in the blood begins to decline. (Carbon monoxide reduces the blood's ability to carry oxygen.)
- Within a few weeks, people who quit smoking have improved circulation, produce less phlegm, and don't cough or wheeze as often.
- Within several months of quitting, people can expect substantial improvements in lung function.
- In addition, people who quit smoking will have an improved sense of smell, and food will taste better.

Q.10. What are the long-term benefits of quitting smoking?

Ans. Quitting smoking reduces the risk of cancer and other diseases, such as heart disease and COPD, caused by smoking.

People who quit smoking, regardless of their age, are less likely than those who continue to smoke to die from smoking-related illness:

- **Quitting at age 30:** Studies have shown that smokers who quit at about age 30 reduce their chance of dying prematurely from smoking-related diseases by more than 90 percent.
- **Quitting at age 50:** People who quit at about age 50 reduce their risk of dying prematurely by 50 percent compared with those who continue to smoke.
- **Quitting at age 60:** Even people who quit at about age 60 or older live longer than those who continue to smoke.

Q.11. Does quitting smoking lower the risk of cancer?

Ans. Yes. Quitting smoking reduces the risk of developing and dying from cancer. However, it takes a number of years after quitting for the risk of cancer to start to decline. This benefit increases the longer a person remains smoke free. The risk of premature death and the chance of developing cancer from smoking cigarettes depend on many factors, including the number of years a person smokes, the number of cigarettes he/she smokes per day, the age at which he/she began smoking, and whether or not he/she was already ill at the time of quitting. For people who have already developed cancer, quitting smoking reduces the risk of developing a second cancer.

Q.12. Should someone already diagnosed with cancer bother to quit smoking?

Ans. Yes. There are many reasons that people diagnosed with cancer should quit smoking. For those having surgery, chemotherapy, or other treatments, quitting smoking helps improve the body's ability to heal and respond to therapy. It also lowers the risk of pneumonia and respiratory failure. Moreover, quitting smoking may lower the risk of the cancer returning or a second cancer developing.

Q.13. What's in a cigarette?

Ans. Cigarettes are made from the dried leaves of the tobacco plant. The leaves of the tobacco plant are dried by. When a person smokes, they contribute to damaging the environment and they are also damaging their health. After the leaves of the tobacco plant have been dried, they are treated with many different chemicals. Cigarette smoke contains over 4000 chemicals and many of them cause cancer.

When someone smokes a cigarette, they breathe in:

(*a*) **Tar,** a black, sticky substance that contains many poisonous chemicals such as: ammonia (found in floor and window cleaner), toluene (found in industrial solvents) and acetone (found in paint stripper and nail polish remover). Tar is the main cause of throat and lung cancer. Tar also causes the yellowish-brown stains on smoker's fingers teeth and lung tissue and on the ceilings in rooms where people smoke heavily.

(*b*) **Nicotine:** The drug in tobacco which contributes to addiction to cigarette smoking. Nicotine is poisonous and has a number of effects on the body. These include: stimulating the nervous system, increasing heart rate raising blood pressure and making the small blood vessels under the skin shrink. Which can cause wrinkles.

(*c*) **Carbon monoxide:** A poisonous gas that reduces the amount of oxygen taken up by a person's red blood cells. This means less oxygen goes to organs of the body and. the heart has to work harder.

(*d*) **Hydrogen cyanide:** It damages the tiny hairs which act as natural lungs cleaners in our bodies. As a result, toxic substances can build up in the lungs.

(*e*) **Metals:** Lead, nickel, arsenic (white ant poison) and cadmium (used in car batteries) are among the many metals found in tobacco smoke.

(*f*) **Radioactive compounds:** Radioactive compounds are found in cigarettes and cause cancer.

(*g*) **Pesticides:** Pesticides such as DDT, methoprene (Found in flea powder) are used in growing tobacco. Other chemicals such as benzene (found in petrol) and naphthalene (found in mothballs) are added when the cigarettes are being made.

Q.14. Discuss the effects of smoking on various parts of the body?

Ans. Smoking affects many parts of the body, both inside and outside. Some of the effects happen straight away and others take longer to occur. The diagram below shows some of the harmful consequences of smoking:

We can greatly reduce the risks of these effects by choosing to be smoke free. Some of the benefits of being smoke free are:

healthier skirt	better sense of taste and smell	saving money
fresh-smelling clothes and hair	improved fitness	helping save the environment
fresher breath	less sickness	most other people are smoke free
whiter teeth		

Many people who start smoking think they will be able to stop whenever they want. Unfortunately, when the younger people start smoking, the harder it can be to quit.

Q.15. Discuss in detail, the effects of smoking on various types of diseases?

Ans. Smoking causes more deaths each year than all of these combined:

- Human immunodeficiency virus (HIV)
- Illegal drug use
- Alcohol use
- Motor vehicle injuries
- Firearm-related incidents

— Many citizens have died prematurely from cigarette smoking.

— Smoking causes about 90% of all lung cancer deaths in men and women. More women die from lung cancer each year than from breast cancer.

— About 80% of all deaths from chronic obstructive pulmonary disease (COPD) are caused by smoking.

— Cigarette smoking increases risk for death from all causes in men and women.

Smokers are more likely than nonsmokers to develop heart disease, stroke, and lung cancer.

- Smoking is estimated to increase the risk:
  - — For coronary heart disease by 2 to 4 times
  - — For stroke by 2 to 4 times
  - — Of men women developing lung cancer by 25 times
- Smoking causes diminished overall heath, such as self-reported poor health, increased absenteeism from work, and increased health care utilization and cost.

(*a*) **Smoking and cardiovascular disease:** Smokers are at greater risk for diseases that affect the heart and blood vessels (cardiovascular disease).

- Smoking causes stroke and coronary heart disease— one of the leading causes of death.
- Even people who smoke fewer than five cigarettes a day can have early signs of cardiovascular disease.
- Smoking damages blood vessels and can make them thicken and grow narrower. This makes heart beat faster and blood pressure goes up. Clots can also form.
- A heart attack occurs when a clot blocks the blood flow to heart. When this happens, heart cannot get enough oxygen. This damages the heart muscle, and part of the heart muscle can die.
- A stroke occurs when a clot blocks the blood flow to part of brain or when a blood vessel in or around brain bursts.
- Blockages caused by smoking can also reduce blood flow to legs and skin.

(*b*) **Smoking and respiratory disease:** Smoking can cause lung disease by damaging your airways and the small air sacs (alveoli) found in lungs.

- Lung diseases caused by smoking include COPD, which includes emphysema and chronic bronchitis.
- Cigarette smoking causes most cases of lung cancer.
- If you have asthma, tobacco smoke can trigger an attack or make an attack worse.
- Smokers are 12 times more likely to die from COPD than nonsmokers.

(*c*) **Smoking and cancer:** Smoking can cause cancer almost anywhere in body e.g.,

- Bladder
- Blood (acute myeloid leukemia)
- Cervix
- Colon and rectum (colorectal)
- Esophagus
- Kidney and ureter
- Larynx
- Liver
- Oropharynx (includes parts of the throat, tongue, soft palate, and the tonsils)
- Pancreas
- Stomach
- Trachea, bronchus, and lung

Smoking increases the risk of dying from cancer and other diseases in cancer patients and survivors.

(*d*) **Smoking and other health risks:** Smoking harms nearly every organ of the body and affects a person's overall health.

- Smoking can make it harder for a woman to become pregnant and can affect her baby's health before and after birth. Smoking increases risks for:
  - — Preterm (early) delivery
  - — Stillbirth (death of the baby before birth)
  - — Low birth weight
  - — Sudden infant death syndrome (known as SIDS or crib death)
  - — Ectopic pregnancy
  - — Orofacial clefts in infants
- Smoking can also affect men's sperm, which can reduce fertility and also increase risks for birth defects and miscarriage (loss of the pregnancy).
- Smoking can affect bone health.
  - — Women past childbearing years who smoke have lower bone density (weaker bones) than women who never smoked and are at greater risk for broken bones.
- Smoking affects the health of your teeth and gums and can cause tooth loss.
- Smoking can increase your risk for cataracts (clouding of the eye's lens that makes it hard for you to see) and age-related macular degeneration (damage to a small spot near the center of the retina, the part of the eye needed for central vision).
- Smoking is a cause of type 2 diabetes mellitus and can make it harder to control. The risk of developing diabetes is 30-40% higher for active smokers than nonsmokers.
- Smoking causes general adverse effects on the body. It can cause inflammation and adverse effects on immune function.
- Smoking is a cause of rheumatoid arthritis.

Benefits of quitting smoking:

- Quitting smoking cuts cardiovascular risks. Just 1 year after quitting smoking, risk for a heart attack drops sharply.
- Within 2 to 5 years after quitting smoking, risk for stroke could fall to about the same as a nonsmoker's.
- If you quit smoking, your risks for cancers of the mouth, throat, esophagus, and bladder drop by half within 5 years.
- Ten years after you quit smoking, your risk for lung cancer drops by half.

Q.16. Discuss about the causing chemicals and toxic chemical present in mainstream, sidestream and second hand cigarette smoke?

Ans. Cigarette smoke is a mixture of over 4000 chemicals, many of which are harmful to the human body. All currently available tobacco products that are smoked deliver substantial amounts of toxic chemicals to their users and those who breathe their smoke.

Cigarette smoke is a combination of:

- mainstream smoke - the smoke inhaled by a smoker.
- sidestream smoke - the smoke from the end of a lit cigarette.
- second hand smoke - the smoke exhaled by a smoker plus sidestream smoke.

Of the more than 4000 chemicals present in cigarette smoke, more than 60 have been identified as cancer causing chemicals, 11 of which are known to cause cancer in humans and 8 that probably cause cancer in humans.

Cancer causing chemicals in tobacco smoke include:

- Benzene
- Ethylene oxide
- 2-naphthylamine
- Arsenic
- 4-aminobiphenyl
- Beryllium
- Chromium
- Nickel
- Cadmium
- Polonium-210
- Vinyl chloride

Toxic chemicals in tobacco smoke include:

- **Nicotine:** the addictive agent in tobacco smoke.
- **Formaldehyde:** used in preservation of laboratory specimens.
- **Ammonia:** used in toilet cleaner.
- **Hydrogen Cyanide:** used in rat poison.
- **Acetone:** used in nail polish remover.
- **Carbon monoxide:** found in car exhaust.
- **Tar:** particulate matter in cigarette smoke.
- **Toluene:** found in paint thinners.
- **Phenol:** used in fertilizers 1,2

These chemicals are considered toxic because they have serious health impacts on the human body. For example:

- Hydrogen cyanide, carbon monoxide and tar cause, or are associated with, cardiovascular disease and chronic obstructive, lung disease.
- Ammonia and formaldehyde cause eye, nose and throat irritations and other breathing problems.

The chemicals present in mainstream, sidestream and second hand smoke are similar; however the *quantities* of the various chemicals present differ. Sidestream smoke is generated at lower temperatures and under different conditions than mainstream smoke, and consequently it contains higher concentrations of many of the toxins found in mainstream smoke.

With approximately one non-smoker dying due to second hand smoke exposure for every eight smokers dying of smoking-related disease, it is no surprise that second hand smoke has been designated a known human carcinogen (cancer-causing agent). Further, about half of regular smokers will die of a smoking-related disease and have a reduced life expectancy of about 13 to 16 years as compared with non-smokers.

Q.17. What is electronic cigarette (e-cigarette)? Is it safer or less harmful than the conventional/cigar/ pipes? Does it help in quitting cigarette smoking? Discuss in detail.

Ans. A relatively new nicotine delivery product on the market is the electronic cigarette, commonly referred to as an e-cigarette. Although not a tobacco product, they are designed to look and feel in every way like a conventional cigarette. E-cigarettes are actively marketed as an alternative to regular cigarettes for use in places where smoking is not permitted. However, since e-cigarettes have not been scientifically evaluated, there is no credible evidence that these products are safe for human consumption or that they are- effective at helping smokers or other tobacco users quit, as some manufacturers claim. Unlike other nicotine replace ment products such as gums or patches, the U.S. Food and Drug Administration (FDA) has not approved e-cigarettes as smoking cessation aids. As a result, e-cigarettes cannot be recommended as a safe alternative to smoking or tobacco use.

Working of e-cigartte: A typical e-cigarette consists of a battery, a heating element and a cartridge containing nicotine, propylene glycol and water. The levels of nicotine in the cartridges can vary drastically and many also contain candy- like flavourings. They operate by electronically vaporizing the nicotine solution, creating a mist that is inhaled into the lungs. After the user inhales, the residual aerosol, or vapor. is exhaled into the surrounding air.

Health effects of nicotine

- Nicotine is a highly addictive drug, even more so than cocaine or heroin,

- Nicotine raises blood pressure and cholesterol, increasing the likelihood of a heart attack or stroke.
- Exposure to high doses of nicotine can be fatal for adults; the lethal dose for children is just 10 mg.
- Nicotine promotes insulin resistance, also called prediabetes, which is a risk factor for cardiovascular disease.
- Acute nicotine toxicity is associated with over-stimulation of nicotinic receptors; burning in the throat with nausea and vomiting occurs quickly after ingestion. More toxic manifestations include cardiac tachyarrhythmia's (abnormal electrical activity in the heart, i.e. heartbeat is too fast, too slow, or irregular), seizures, and hypertensive crisis.
- Nicotine is used in a number of pesticides because of its toxic properties.

Electronic cigarettes and second hand smoke:

- After the user inhales, the residual aerosol, or vapour, is exhaled into the surrounding air.
- E-cigarettes may emit nicotine, propylene glycol, carcinogens, and other substances.
- Nicotine deposits on indoor surfaces and lasts for weeks to months.

e-cigarette's public health concerns (untested and unproven):

- Preliminary testing of e-cigarette samples indicate that they contain carcinogens and toxic chemicals such as diethylene glycol, an ingredient used in anti-freeze.
- E-cigarettes produce a vapour of potentially harmful substances, which may appear similar to the smoke emitted by traditional tobacco products. Their use in workplaces and public places where smoking of traditional tobacco products is prohibited
  — creates confusion and leads to difficulties in enforcing existing smoking prohibitions.
- Research shows that reduced risk product claims undermine adult cessation and youth prevention. Lulled by a false sense of security, consumers may not quit and instead continue their addiction.
- Misleading advertisements misrepresent e-cigarettes as harmless and a healthy way to quit. E-cigarettes have not been adequately evaluated for safety or approved as a cessation aid.
- A wide array of flavoured nicotine cartridges are available, many of which are candy/fruit flavours, including vanilla, chocolate, strawberry, grape, bubble gum, cherry, mint, peach cobbler, etc. This increases the appeal of e-cigarettes to youth and can become a "gateway" to tobacco product use.
- When smokers use vaporized nicotine as a smoking alternative or to quit, the vaporized nicotine produces similarly addictive and psychological effects as cigarettes.
- Claims of reduced exposure to toxins does not necessarily mean reduced risk. Allowing smoking indoors puts businesses at risk which can occur when people with respiratory problems have limited access or cannot work in a business due to exposure to indoor air pollutants.

Safety concerns (no regulation, no oversight)

- FDA has not evaluated e-cigarette to verify safety or effectiveness claims.
- Some of the poor manufacturing practices include
- Poor cartridge labelling that lacks any indication of cartridge content, expiration date, or health warnings.
- Once removed from packaging, cartridges claiming high nicotine content are indistinguishable from those claiming to have low or no nicotine content.
- No instructions for proper disposal of used cartridges.
- Sold with vague amounts of nicotine on the label. It is unclear if a value (e.g., 24 mg), is mg/cartridge or mg/ml.
- Because e-cigarettes have not been submitted to the FDA for evaluation or approval for use as nicotine delivery devices, the agency has no way of knowing the levels of nicotine or the amounts or kinds of other chemicals that these products contain.

- The nicotine solutions for e-cigarettes are offered in different strengths and vary by company, each with their own concentration methods. Solutions are very concentrated and can contain 500 to over 1,000 mg of nicotine in one 30 ml bottle. This creates a risk of overdosing or poisoning, as the lethal dose of nicotine for adults is 30-60 mg if swallowed, and for children is just 10 mg. Solutions come in a small bottle, are often fruit-flavoured and without a child safety cap.
- More research is needs around what inhaling water vapour, propylene glycol, pure nicotine (and any other undisclosed substance contained in the vapour) will do to human lungs, heart or cardiovascular system.

Q.18. What are the various implications of cigarette smoking?

Ans. (*i*) Cigarette smoking is addictive, it fulfills these three criteria:

- Smokers develop a tolerance to nicotine (they need to smoke more and more for an effect).
- Smokers become dependent on it (they need it to feel comfortable).
- Smokers suffer withdrawal symptoms (physical and psychological discomfort, when they try to stop smoking).

(*ii*) There are thousand of chemical substances in cigarette smoke. Three of the most damaging are:

- **Tans:** damage delicate lung tissue and are considered the main cancer- causing agent in cigarette smoke.
- **Nicotine:** a poison found only in tobacco leaves. It can be extracted as a colourless, oily transparent liquid and use of pure nicotine can be fatal to humans.
  (*a*) It is a powerful stimulant to the brain and central nervous system that 'hits" the brain within four seconds. Like the stimulants, it has a depressant effect.
  (*b*) Nicotine constricts (narrows) the blood vessels, cutting down the flow of blood and oxygen throughout the body increasing the chance of heart disease. It raises the blood pressure and also narrows bronchioles (air passageways)
- Carbon monoxide-replaces needed oxygen in our red blood cells. Even after one stops smoking, carbon monoxide continues depriving the body of oxygen until the oxygen level in your blood returns to normal. Carbon monoxide is a waste product of gasoline engines.

(*iii*) Some of the diseases caused by cigarette smoking are:

- Chronic bronchitis (inflammation of the bronchi which are the breathing tubes in the lungs)
- Laryngitis (inflammation of the throat)
- Emphysema (a degenerative lung disease that destroys breathing capacity)
- Cancer of the lungs, mouth, and esophagus.

(*iv*) Smoking cigarettes stains the teeth, reduces the efficiency of the body's ability to taste and smell, and increases the prose around the eyes).

(*v*) The effects of some medication taken by a person may be increased, decreased, or cancelled out by smoking.

(*vi*) Cigarette smoking by pregnant women may cause harm to the fetus.

(*viii*) Cigarette smoke pollutes the air in enclosed places, which also affects the nonsmokers present.

(*ix*) Smoke from an idle cigarette contains at least as much tar and nicotine as inhaled smoke.

(*x*) Chewing tobacco may lead to cancer of the mouth and to an addiction because of the nicotine that is absorbed through the mouth's lining.

(*xi*) It is generally accepted that "peer pressure" encourages many young people between the ages of 10 and 18 to begin experimenting with smoking.

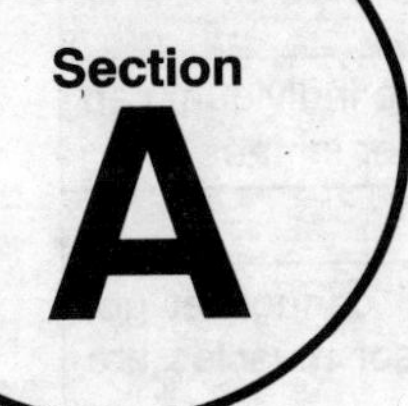

15 Electrical Hazards

SHORT QUESTIONS AND ANSWERS ON ELECTRICAL HAZARDS

Q.1. How does water affect the flow of electricity through the skin of human body?

Ans. Pure water is a poor conductor. But small amount of impurities in water like salt, acid, solvents or other materials can turn water itself and substances that generally act as insulators into conductors or better conductors. Dry wood, for example, generally slows or stops the flow of electricity. But when saturated with water, wood turns into a conductor. The same is true of human skin. Dry skin has a fairly high resistance to electric current. But when skin is moist or wet, it acts as a conductor. Hence anyone working with electricity in a damp or wet environment needs to exercise extra caution to prevent electrical hazards.

Q.2. What are the causes of electric shocks?

Ans. Electricity travels in closed circuits, through a conductor. But sometimes a person's body – an efficient conductor of electricity – mistakenly becomes part of the electric circuit. This can cause an electrical shock. Shocks occur when a person's body completes the current path with:

- Both wires of an electric circuit.
- One wire of an energized circuit and the ground.
- A metal part that accidentally becomes energized due, for example, to a break in its insulation.
- Another "conductor" that is carrying a current.

When a person receives a shock, electricity flows between parts of the body or through the body to a ground or the earth.

Q.3. What effect do electric shocks have on the body?

Ans. An electric shock can result in anything from a slight tingling sensation to immediate cardiac arrest. The severity depends on the following:

- The amount of current flowing through the body.
- The current's path through the body.
- The length of time the body remains in the circuit.
- The current's frequency.

The table below shows the relationship between the amount of current received and the reaction when current flows from the hand to the foot for just 1 second.

Table 15.1: Effects of electric current in the human body

Current, milliampere (mA)	Reaction on human body
<1	Generally, not perceptible
1	Faint tingle

Current, milliampere (mA)	Reaction on human body
5	Slight shock felt, not painful but disturbing. Average individual can let go. Strong involuntary reactions can lead to other injuries.
6-25 (women)	Painful shock, loss of muscular control
9-30 (men)	The freezing current or "let-go" range. Individual cannot let go, but can be thrown away from the circuit if extensor muscles are stimulated.
50-150	Extreme pain, respiratory arrest, severe muscular contractions, Death is possible
1,000-4,300	Rhythmic pumping action of the heart ceases. Muscular contraction and nerve damage occur; death likely.
10,000	Cardiac arrest, severe burns; death probable

* if the extensor muscles are excited by the shock, the person may be thrown away from the power source.

Q.4. What type of burns can an electric shock cause to the human body?

Ans. Burns are the most common shock-related injury. An electrical accident can result in an electrical burn, are burn, thermal contact burn, or a combination of burns.

Electrical burns are the most serious burns and they occur when electric current flow through issues or bone, generating heat that causes tissue damage.

Arc or flash burns result from high temperatures caused by an electric arc or explosion near the body. These burns should be treated promptly.

Thermal contact burns are caused when the skin touches hot surfaces of overheated electric conductors, conducts, or other energized equipment. Thermal burns also can be caused when clothing catches on fire, may occur when an electric arc is produced.

In addition to shock and burn hazards, electricity poses other damages. For example, arcs that result from short circuits can cause injury or start a fire. Extremely high-energy arcs can damage equipment, causing fragmented metal to fly in all directions. Even low-energy arcs can cause violent explosions in atmospheres that contain flammable gases, vapours, or combustible dusts.

Q.5. When do people sometimes "freeze" when they are shocked by electricity?

Ans. When a person receives an electrical shock, sometimes the electrical stimulation causes the muscles to contract. This "freezing" effect makes the person unable to pull free of the circuit. It is extremely dangerous because it increases the length of exposure to electricity and because the current causes blisters, which reduce the body's resistance and increases the current.

The longer the exposure, the greater the risk of serious injury. Longer exposures at even relatively low voltages can be just as dangerous as short exposures at higher voltages. Low voltage does not imply low electrical hazard.

In addition to muscle contractions that cause "freezing" electrical shock also can cause involuntary muscle reactions. These reactions can result in other injuries from collisions or falls, including bruises, bone fractures, and even death. If a person's "frozen" to a live electrical contact, shut off the current immediately. If this is not possible, use brands, poles, or sticks made of wood or any other nonconducting materials and safely push or pull the person away from the contact. It's important to act quickly, but remember to protect yourself as well from electrocution or shock. A severe shock can cause considerably more damage than meets the eye. A victim may suffer internal hemorrhages and destruction of tissues, nerves, and muscles that aren't readily visible. Renal damage also can occur.

Q.6. What is the danger of static electricity?

Ans. Static electricity also can cause a shock, but generally not as severe as the type of shock described previously. Static electricity can build up on the surface of an object and, can discharge to a person, causing a shock. The most familiar example of this is when a person reaches for a door knob or other metal object on a cold, relatively dry day and receives a shock.

However, static electricity also can cause shocks or can just discharge to an object with much more serious consequences, as when friction causes a high level of static electricity to build up at a specific spot on an object. This can happen simply through handling plastic pipes and materials or during normal operation of rubberized drive or machine belts. In these cases, static electricity can potentially discharge when sufficient amounts of flammable or combustible substances are located nearby and cause an explosion. Grounding or other measures may be necessary to prevent this static electricity buildup.

Q.7. What are the different ways to protect against electrical hazards?

Ans. Most electrical accidents result from

- Unsafe equipment or installation
- Unsafe environment, or
- Unsafe work practices,

Some ways to prevent these accidents are through the use of insulation, guarding grounding, electrical protective devices, and safe work practices.

Insulators such as glass, mica, rubber, or plastic used to coat metals and other conductors help stop or reduce the flow of electrical current. This helps prevent shock, fires, and short circuits. The insulation must be suitable for the voltage used and conditions such as temperature and other environmental factors like moisture, oil, gasoline, corrosive fumes, or other substances that could cause the insulator to fail. Insulation on conductors is often colour coded. Insulated equipment grounding conductors usually are either solid green or green with yellow stripes. Insulation covering grounded conductors is generally white or gray. Ungrounded conductors, or "hot wires," often are black or red, although they may be any colour than green, white, or gray.

Before connecting electrical equipment to a power source, It's a good idea to check the insulation for any exposed wires for possible defects. Insulation covering flexible cords such as extension cords is particularly vulnerable to damage.

Guarding involves locating or enclosing electric equipment to make sure people don't accidently come into contact with its live parts. Effective guarding requires equipment with exposed parts operating at 50 volts or more to be placed where it is accessible only to authorized people qualified to work with it. Recommended locations are a room, vault, or similar enclosure; a balcony, gallery, or elevated platform; or a site elevated 12.44 meters or more above the floor, Study, permanent screens also can serve as effective guards.

Conspicuous signs must be posted at the entrances to electrical rooms and similarly guarded locations to alert people to the electrical hazard and to forbid entry to unauthorized people. Signs may contain the word "Danger," "Warning," or "Caution,"

"Grounding" a tool or electrical system means intentionally creating a low-resistance path that connects to the earth. This prevents the buildup of voltages that could cause an electrical accident.

Grounding is a secondary protective measure to protect against electric shock. It does not guarantee that you won't get a shock or be injured or killed by an electrical current. It will, however, substantially reduce the risk, especially when used in combination with other safety measures.

A service or system ground is designed primarily to protect machines, tools, and insulation against damage. One wire, called the "neutral" or "grounded" conductor, is grounded. In an ordinary low-voltage circuit, the white or gray wire is grounded at the generator of transformer and at the building's service entrance.

An equipment ground helps protect the equipment operator. It furnishes a second path for the current to pass through from the tool or machine to the ground. This additional safeguard the operator if a malfunction causes the tool's metal frame to become energized. The resulting flow of current may activate the circuit protection devices.

Electrical Circuit protection devices limit or stop the flow of current automatically in the event of a ground fault. Overload, or short circuit in the wiring system. Examples of these devices are fuses, circuit breakers, ground-fault circuit interrupters, and arc-fault circuit interrupters.

Fuses, and circuit breakers open or break the circuit automatically when too much current flows through them. When that happens, fuses melt and circuit breakers trip the circuit open. Fuses and circuit breakers are designed to protect conductors and equipment. They prevent wires and other components from overheating and open the circuit when there is a risk of a ground fault.

Ground-fault circuit interrupters, are used in wet locations, construction sites, and other high-risk areas. These devices interrupt the flow of electricity within as little as 1/40 of a second to prevent electrocution They compare the amount of current going into electric equipment with the amount of current returning from it along the circuit conductors. If the difference exceeds 5 milliamperes, the device automatically shuts off the electric power.

Arc-fault devices provide protection from the effects of arc-faults by recognizing characteristics unique to arcing and by functioning to deenergize the circuit when an arc-fault is detected.

Electrical accidents are largely preventable through safe work practices. Examples of these practices include the following:

- Deenergizing electric equipment before inspection or repair.
- Keeping electric tools properly maintained.
- Exercising caution when working near energized lines, and
- Using appropriate protective equipment.

Q.8. How can you protect yourself against metal parts that become energized?

Ans. A break in an electric tool's or machine's insulation can cause its metal parts to become "hot" or energized, meaning that they conduct electricity. Touching these energized parts can result in an electrical shock, burn, or electrocution. The best way to protect yourself when using electrical tools or machines is to establish a low-resistance path from the device's metallic case to be ground. This requires an equipment groundling conductor, a low-resistance wire that directs unwanted current directly to the ground. A properly installed groundling conductor has a low resistance to ground and greatly reduces the amount of current that passes through your body. Cord and plug equipment with a three-prong plug is a common example of equipment incorporating this ground conductor.

Another form of protection is to use portable tools and appliances protected by system of double insulation.

Q.9. How can you protect yourself from overhead power lines?

Ans. Before working under or near overhead power lines, ensure that you maintain a safe distance to the lines and, for very high-voltage lines, ground any equipment such as cranes that can become energized. If working on power lines, ensure that the lines have been deenergized and grounded. Other protective measure like guarding or insulating the lines help prevent accidental contact.

Employees unqualified to work with electricity, as well as mechanical equipment, should remain at least 3 meters away from overhead power lines. If the voltage is more than 50,000 volts, the clearance by 10 centimeters for each additional 10,000 volts.

When mechanical equipment is operated near overhead lines, employees standing on the ground should avoid contact with the equipment unless it is located outside the danger zone.

Section B

1. **Multiple Choice Question and Answers on Environmental Engineering**
2. **Immediately Dangerous to Life or Health (IDLH) Concentrations of Various Substances/Chemicals (as per NIOSH as on 03/01/1995)**

Section

B 16

MCQ on Environmental Engineering

Multiple Choice Question and Answer on Environmental Engineering

1. The earth's atmosphere is an envelope of gases present upto a height of about ______ kms.

(*a*) 10 (*b*) 200 (*c*) 1000 (*d*) 2000

2. The density of gas (present in air) decreases with increasing altitude to such an extent, that about 70% of the mass of atmospheric air is found in the lower 5 km. of the atmosphere. This lower region of atmosphere is called the

(*a*) ionosphere (*b*) troposphere

(*c*) stratosphere (*d*) none of these

3. The upper layer of atmosphere is call the

(*a*) stratosphere (*b*) troposphere

(*c*) ionosphere (*d*) none of these

4. Which of the following is not a natural sources of air pollution?

(*a*) Volcanic eruptions and lightening discharges.

(*b*) Biological decay of vegetable matter.

(*c*) Photochemical oxidation of organic matter.

(*d*) None of these

5. Which of these following is a manmade source of air pollution?

(*a*) Automobile exhaust.

(*b*) Forest fire.

(*c*) Bacterial action in soil and swamp areas.

(*d*) All (*a*), (*b*) and (*c*)

6. Which of the following sources is responsible for maximum air pollution?

(*a*) Industrial chimney exhaust.

(*b*) Forest fire

(*c*) Automobile exhaust.

(*d*) Photochemical oxidation of organic matter.

7. Lagooning process is mainly a means of the

(*a*) sludge disposal.

(*b*) reduction of excessive flow in sewers.

(*c*) biological treatment of wastes.

(*d*) none of these.

8. For existence of aquatic life in water, the dissolved oxygen content in it, should not be less than ______ ppm.

(*a*) 10000 (*b*) 5

(*c*) 500 (*d*) 1000

9. Which of the following is the most major constituents of air pollutants?

(*a*) Oxides of sulphur

(*b*) Oxides of nitrogen

(*c*) Carbon monoxide

(*d*) Hydrogen sulphide

Answers

1. (*b*) 2. (*b*) 3. (*a*) 4. (*d*) 5. (*a*) 6. (*c*) 7. (*a*) 8. (*b*) 9. (*a*)

10. Sulphur dioxide present in the industrial chimney exhaust gas causes

(*a*) respiratory and lung disease.
(*b*) reduction in plant's productivity owing to acid rain.
(*c*) corrosion of building materials.
(*d*) all (*a*), (*b*) and (*c*)

11. Exposure to SO_2 containing chimney gases results in the

(*a*) reduction in strength of leather and cloth.
(*b*) acceleration of corrosion rates of metals.
(*c*) increased drying and hardening time of paints.
(*d*) all (*a*), (*b*) and (*c*)

12. Limestone powder is injected during pulversed coal burning in boilers to .... the flue gases.

(*a*) reduce SO_2 content in
(*b*) catalytically convert SO_2 to SO_3 in
(*c*) increase the dew point of
(*d*) none of these

13. In water chemical treatment plant, use of chloramines ensures

(*a*) disinfection.
(*b*) taste and odour control.
(*c*) weed control in reservoirs.
(*d*) removal of permanent hardness.

14. Carbon monoxide is a pollutant, which causes

(*a*) respiratory disease (e.g. asthma),
(*b*) asphyxiation (suffocation) leading to death.
(*c*) retardation in crop growth.
(*d*) damage to building materials like marble.

15. On prolonged exposure to high concentration of carbon monoxide (> 5000 ppm), man dies because

(*a*) of clotting of blood.
(*b*) of jamming of respiratory tract.
(*c*) it forms carboxyhaemoglobin by combining with haemoglobin of blood, thereby making it incapable of absorbing oxygen.
(*d*) it forms CO_2 by combining with oxygen present in the blood.

16. Four major constituents of atmospheric air are

A. nitrogen B. argon
C. oxygen D. carbon dioxide

Arrange them in decreasing order of their percentage in air.

(*a*) A, C, B, D (*b*) A, D, B, C
(*c*) C, D, A, B (*d*) A, C, D, B

17. The effect of increase in carbon dioxide level of the atmosphere and its profound effect on our climate is called the

(*a*) catalytic conversion
(*b*) green house effect (*c*) global warming
(*d*) both (*b*) and (*c*)

18. Pick out the wrong statement pertaining to 'green house' for the plants.

(*a*) 'Green house' (made of glass) allows visible sunlight (i.e., short wavelength solar radiation like ultra-violet rays) to pass through the glass and heat up the soil thereby warming up plants inside it.
(*b*) The emitted longer wavelength radiation (e.g., infrared) is partly reflected and partly absorbed by the glass of the green house.
(*c*) 'Green house effect' in earth's atmosphere is due to increasing CO_2 level in atmosphere, where CO_2 acts like the glass of 'green house' thereby warming up the earth's surface.
(*d*) 'Green house' is colder than the outside atmosphere.

19. A 'body' which allows the short wavelength incoming solar radiation to enter in, but does not allow long wave length outgoing infra red radiation to escape out is called the

(*a*) global warming (*b*) green house
(*c*) atmospheric effect
(*d*) ionosphere

Answers

10. (*d*) 11. (*d*) 12. (*a*) 13. (*a*) 14. (*b*) 15. (*c*) 16. (*a*) 17. (*d*) 18. (*d*) 19. (*b*)

20. Which of the following fine dust removal equipment is the most efficient?
(*a*) Bag filter (*b*) Scrubber
(*c*) Electrostatic precipitator
(*d*) Cyclone separator

21. Higher concentration of CO_2 in atmosphere
(*a*) allows visible solar radiation (ultraviolet) of short wave-length to pass through.
(*b*) reflects and absorbs the longer wavelength (infra-red) radiations.
(*c*) prevents solar heat being radiated out completely, resulting in 'heat trap' i.e., global warming.
(*d*) all (*a*), (*b*) and (*c*).

22. Green house effect is accentuated by
(*a*) deforestation.
(*b*) rapid industrialization.
(e) increased transportation activity.
(*c*) all (*a*), (*b*) and (*c*).

23. Global warming may result in
(*a*) flood (*b*) cyclone
(*c*) decrease in food productivity
(*d*) all (*a*), (*b*) and (e)

24. Ozone is _______
(*a*) a primary pollutant.
(*b*) a secondary pollutant.
(*c*) impervious to ultra-violet rays.
(*d*) (*d*) both (*b*) and (e)

25. Smog is _______
(*a*) nothing but black smoke.
(*b*) a combination of smoke and fog.
(c) a liquid particle resulting from vapor condensation.
(*d*) a solid particle e.g., flyash.

26. Particulates(<1μm size) remaining suspended in air indefinitely and transported by wind currents are called
(*a*) fumes (*b*) mists
(*c*) smoke (*d*) aerosols

27. Which of the following is not a secondary air pollutant?
(*a*) Ozone
(*b*) Photochemical smog
(*c*) Sulphur dioxide
(*d*) All (*a*), (*b*) and (*c*)

28. 'Particulate' air pollutants are finely divided solids and liquids. Which of the following is not a 'particulate' ?
(*a*) Dust and mists (*b*) Smoke and fumes
(*c*) Photochemical smog and soot
(*d*) None of these

29. 'Safe limit' called Threshold Limit Value (TLV) of carbon monoxide in atmospheric air is < 50 ppm. The main source of carbon monoxide pollutant in air is the
(*a*) industrial chimney exhaust.
(*b*) automobiles exhaust.
(*c*) photochemical reaction in polluted atmosphere.
(*d*) burning of domestic fuel.

30. High concentration of carcinogenic hydrocarbon pollutants in atmospheric air
(*a*) cancer (*b*) silicosis
(*c*) respiratory disease (e.g., asthma)
(*d*) reduced crop yield

31. Persons working in cement plants and limestone quarries are more prone to disease like
(*a*) cancer (*b*) asthma
(*c*) silicosis
(*d*) flourosis (bone disease)

32. Higher concentration of nitrogen dioxide in atmospheric air causes
(*a*) cancer (*b*) bronchitis
(*c*) asphyxiation (*d*) corrosion

33. Fluorosis (a bone disease) is caused by the presence of high concentration of _______ in atmospheric air.

Answers

20. (*c*)	21. (*d*)	22. (*d*)	23. (*d*)	24. (*d*)	25. (*b*)	26. (*d*)	27. (*c*)	28. (*d*)
29. (*b*)	30. (*a*)	31. (*c*)	32. (*b*)	33. (*b*)				

(*a*) hydrocarbons (*b*) hydrogen flouride
(*c*) hydrogen sulphides
(*d*) nitrogen dioxide

34. Presence of high concentration of ozone and smog in atmospheric air causes the
(*a*) embrittlement and decrease of folding resistance of paper.
(*b*) cracking of rubber products.
(*c*) fading of dye on textiles.
(*d*) damage of electrical insulator on high tension power line.

35. Moist atmospheric air at high temperature (e.g., in summer) having high concentration of sulphur dioxide causes
(*a*) fading of dyes on textiles.
(*b*) corrosion, tarnishing and soiling of metals.
(*c*) reduced strength of textiles.
(*c*) all (*a*), (*b*) and (*c*).

36. Which of the following air pollutants is not toxic to vegetation i.e., doesn't cause vegetation damage?
(*a*) Smog and ozone
(*b*) Hydrogen fluoride and nitrogen oxides
(*c*) Sulphur dioxide and spray of weed killers
(*d*) Carbon monoxide

37. Lead exhausted in the atmosphere by automobiles using leaded petrol (i.e., tetraethyl lead for improving octane number) is a lethal air pollutant which causes
(*a*) paralysis of muscles and loss of appetite.
(*b*) nervous depression.
(*c*) gastritis and diarrhea.
(*d*) all (*c*), (*b*) and (*c*).

38. Which of the following causes death by asphyxiation, if its presence in atmospheric air exceeds maximum allowable concentration i.e., > 50 ppm) ?
(*a*) Benzopyrene (*b*) Peroxyacyl nitrate
(*c*) Carbon monoxide (*d*) Sulphur dioxide

39. During which of the following operating conditions of an automobile, carbon monoxide content in the exhaust gas is maximum?
(*a*) Idle running (*b*) Acceleration
(*c*) Cruising (*d*) Deaccelaration

40. Dust collection efficiency of a cyclone separator depends upon its
(*a*) diameter. (*b*) inlet gas velocity.
(*c*) overall height.
(*d*) all (*a*), (*b*) and (*c*).

41. Most efficient and suitable dust removal equipment for removal of flyash from flue gas in a thermal power plant is the
(*a*) gravity settling chamber
(*b*) cyclone separator
(*c*) electrostatic precipitator
(*d*) bag filter

42. Ethanolamine is an absorb ant used for the removal of ________ from air/gas.
(*a*) HF (*b*) SO_2
(*c*) H_2S (*d*) both (*b*) and (*c*)

43. Which of the following is an adsorbant for removal of nitrogen oxides from gas/air?
(*a*) Active carbon (*b*) Silica gel
(*c*) Bog iron (iron oxide)
(*d*) Pulverised limestone

44. Which of the following is an adsorbant used for the removal of SO_2 from gas/air?
(*a*) Bog iron
(*b*) Limestone powder or alkalised alumina
(*c*) Silica gel (*d*) Active carbon

45. H_2S present in gaseous stream can be removed by adsorption on
(*a*) silica gel (*b*) active carbon
(*c*) bog iron (*d*) limestone powder

46. Particles having diameter greater than 75 μm (micrometer = 10^{-6} mm) are called
(*a*) grit (*b*) dust
(*c*) powder (*d*) smoke

Answers

34. (*b*) 35. (*d*) 36. (*d*) 37. (*d*) 38. (*c*) 39. (*a*) 40. (*d*) 41. (*c*) 42. (*d*)
43. (*b*) 44. (*b*) 45. (*c*) 46. (*a*)

47. Carbonaceous particles having size less than 1 μm are called
(*a*) grit (*b*) aggregates
(*c*) aerosols (*d*) smoke

48. 5-200 μm size particles are called
(*a*) colloids or aerosols
(*b*) powder (*c*) dust
(*d*) smoke

49. 80% less than 200 mesh size particles are called
(*a*) smoke (*b*) powder
(*c*) grit (*d*) aggregates

50. Sooty and sulphurous smog formed due to combustion of fossil fuels particularly in winter, continues throughout the day and night. When the moist atmospheric air's water vapour condenses on the solid particles of smoke thereby forming smog (smoke + fog), it causes
(*a*) poor visibility due to sky darkening.
(*b*) irritation in eyes.
(*c*) respiratory trouble.
(*d*) all (*a*), (*b*) and (*c*).

51. Operating principle of cyclone separator is based on the action of _______ dust particles.
(*a*) diffusion of
(*b*) centrifugal force on
(*c*) gravitational force on
(*d*) electrostatic force on

52. Which is the most efficient dust removal equipment for removal of sub-micronic dust particles from blast furnace gas?
(*a*) Packed scrubber
(*b*) Gravity settling chamber
(*c*) Electrostatic precipitator
(*d*) Hydrocyclone

53. Aerosols present in atmospheric air may be
(*a*) positively charged.
(*b*) negatively charged.
(*c*) neutral.
(*d*) combination of all (*a*), (*b*) and (*c*).

54. Ionisation potential employed in the industrial electrostatic precipitator is of the order of
(*a*) 30 to 70 kV DC (*b*) 30 to 70 kV AC
(*c*) 230 V AC (*d*) 230 V DC

55. Which of the following pollutants is not emitted during volcanic eruptions ?
(*a*) SO_2 (*b*) H_2S
(*c*) CO (*d*) hydrocarbons

56. Main pollutant present in automobile exhaust is
(*a*) CO (*b*) CO_2
(*c*) NO (*d*) hydrocarbons

57. Main pollutants released from iron and steel industry is
(*a*) CO, CO_2 and SO_2
(*b*) H_2S, NO and SO_3
(*c*) CO_2, H_2S and NO_2
(*d*) SO_3, NO_2 and CO_2

58. Main pollutants released from petroleum refineries is
(*a*) CO, SO_2 and H_2S
(*b*) CO_2, NO and SO_2
(*c*) CO_2, H_2S and NO_2
(*d*) SO_3, NO_2 and CO_2

59. _______ plant emits large amount of SO_2 as an air pollutant.
(*a*) Nitric acid (*b*) Sulphuric acid
(*c*) Chloralkali (*d*) Iron and Steel

60. Which of the following plants does not emit appreciable amount of SO_2 in atmosphere?
(*a*) Thermal power plant
(*b*) Petroleum refinery
(*c*) Nitric acid plant
(*d*) Sulphuric acid

61. Which of the following is a green house gas other than CO_2?

Answers

47. (*d*)	48. (*a*)	49. (*b*)	50. (*d*)	51. (*b*)	52. (*c*)	53. (*d*)	54. (*a*)	55. (*d*)
56. (*a*)	57. (*a*)	58. (*a*)	59. (*b*)	60. (*c*)	61. (*d*)			

(*a*) Methane (*b*) Nitrous oxide
(*c*) Chlorofluro carbons (CFC)
(*d*) All (*a*), (*b*) and (*c*)

62. The main industrial source of emission of hydrogen sulphide air pollutant is
(*a*) petroleum refineries.
(*b*) coal based thermal power plants.
(*c*) pulp and paper plant.
(*d*) metallurgical roasting and smelting plant.

63. Pick out the wrong statement.
(*a*) Catalytic converter is fitted in automobiles to reduce carbon monoxide concentration in exhaust emissions.
(*b*) Inhalation of pollutant carbon monoxide results in death by asphyxiation.
(*c*) Sulphur dioxide is the main pollutant emitted from the exhaust of petrol driven automobiles.
(*d*) Decomposition of plants containing chlorophyll is a natural source of carbon monoxide in atmosphere.

64. The progressive warming up of the earth's surface is mainly due to the
(*a*) automobile exhaust.
(*b*) blanketing effect of CO_2 in the atmosphere.
(*c*) de-forestation.
(*d*) thickening of ozone layer.

65. Pick out the correct statement.
(*a*) Deforestation helps in controlling the green house effect.
(*b*) Global warming is detrimental for increase in food productivity and may cause flood and cyclone.
(*c*) Lightening discharges are natural source of production of SO_2 and H_2S pollutant.
(*d*) Sulhur dioxide causes death by asphyxiation.

66. Which of the following dust collection equipment is the least efficient (for sub-micronic particles)?
(*a*) Dust catcher (gravity type)
(*b*) Cyclone separator
(*c*) Bag filter
(*d*) Hollow wet scrubber

67. Photochemical smog is formed from automobile exhaust
(*a*) by reaction of hydrocarbon and nitric oxide in presence of sunlight.
(*b*) appears only on sunny days.
(*c*) is harmful for crops and trees also besides causing eye irritation and asthma.
(*d*) all (*a*), (*b*) and (*c*)

68. Threshold Limit Value (TLV) is the maximum allowable concentration (i.e., safe limit) of pollutants in air. Safe limit for SO_2 in air is ________ ppm.
(*a*) 5 (*b*) 1000 (*c*) 500 (*d*) 2000

69. Maximum allowable concentration of CO_2 in air for safe working is ________ ppm (parts per million).
(*a*) 50 (*b*) 1000 (*c*) 2000 (*d*) 5000

70. Threshold limit value (TLV) of CO in air is ________ ppm.
(*a*) 5 (*b*) 50 (*c*) 2000 (*d*) 5000

71. TLV of ammonia, nitrogen dioxide and phenol vapour in air is ________ ppm.
(*a*) 5 (*b*) 100 (*c*) 1000 (*d*) 2000

72. Acute danger to human life (i.e. death) exists, if the concentration of CO_2 in atmospheric air exceeds ________ percent (by volume).
(*a*) 1 (*b*) 3 (*c*) 7 (*d*) 20

73. Death may occur, when SO_2 concentration in atmospheric air exceeds ________ ppm.
(*a*) 20 (*b*) 100 (*c*) 400 (*d*) 200

74. If carbon monoxide content in atmospheric air exceeds ________ ppm, death is bound to occur.
(*a*) 50 (*b*) 500
(*c*) 1000 (*d*) 3000 (i.e., 0.3%)

Answers

62. (*b*) 63. (*c*) 64. (*d*) 65. (*d*) 66. (*a*) 67. (*d*) 68. (*a*) 69. (*d*) 70. (*b*)
71. (*a*) 72. (*d*) 73. (*c*) 74. (*d*)

75. Beyond what concentration of H_2S in air, acute danger to human life exists?

(*a*) 50 ppm (*b*) 100 ppm
(*c*) 300 ppm (*d*) 700 ppm

76. TLV of ozone (O_3) and phosgene ($COCl_2$) in air is _______ ppm.

(*a*) 0.1 (*b*) 25 (*c*) 100 (*d*) 1000

77. Black smoke coming out of the chimney of a furnace is an indication of the use of....... in the furnace.

(*a*) low amount of excess combustion air
(*b*) large quantity of excess combustion air
(*c*) hydrocarbon fuel
(*d*) pulverised coal as fuel

78. White smoke coming out of the chimney of a furnace indicates the use of

(*a*) low excess air.
(*b*) very high excess air.
(*c*) gaseous fuel in the furnace.
(*d*) liquid fuel in the furnace.

79. Pick out the wrong statement.

(*a*) A slight haze at the top of chimney indicates good combustion in the furnace.
(*b*) A bag filter incurs very small pressure drop and is very efficient for removal of sub-micronic dust particles from flue gases at very high temperature.
(*c*) Electrostatic precipitator is the most efficient dust collection equipment for removal of sub-micronic dust particles present in flue gas.
(*d*) none of these.

80. Smoke is produced due to

(*a*) insufficient supply of combustion air and insufficient time for combustion.
(*b*) poor quality of fuel and improper mixing of fuel and combustion air.
(*c*) poor design and overloading of furnace.
(*d*) all (*a*), (*b*) and (*c*)

81. Reinglemann chart is used for the measurement of the

(*a*) combustibles present in automobile exhaust.
(*b*) smoke density from a chimey.
(*c*) exhaust gas density.
(*d*) flue gas temperature.

82. When the concentration of SO_2 in air is greater than _______ ppm, it gives a pungent smell.

(*a*) 0.01 (*b*) 0.1 (*c*) 1 (*d*) 4

83. Peroxyacyl nitrate (PAN), a pollutant is found in the

(*a*) automobile exhaust.
(*b*) flue gas of coal based power plant.
(*c*) exhaust of nitric acid plant.
(*d*) exhaust of sulphuric acid plant.

84. Which of the following is the most detrimental for water used in high pressure boiler?

(*a*) Silica (*b*) Turbidity
(*c*) Phenol (*d*) Dissolved oxygen

85. Pick out the wrong statement.

(*a*) Caustic embrittlement of boiler's metallic parts is caused by high concentration of caustic soda in boiler feed water.
(*b*) Cooling and freezing of water kills the bacteria present in it.
(*c*) With increasing boiler operating pressure of steam, the maximum allowable concentration of silica in feed water goes on decreasing.
(*d*) Dissolved oxygen content in high pressure boiler feed water should be nil.

86. Foaming and priming in boiler operation can be reduced by reduction in _______ in feed water.

(*a*) turbidity (*a*) color (Hazen)
(*b*) total solids (*c*) all (*a*), (*b*) and (*c*)

87. Scale formation in boiler is controlled by

(*a*) preheating of feed water.
(*b*) reduction in hardness, silica and alumina in feed water.

Answers

75. (*d*) 76. (*a*) 77. (*a*) 78. (*b*) 79. (*b*) 80. (*d*) 81. (*b*) 82. (*d*) 83. (*a*)
84. (*a*) 85. (*b*) 86. (*d*) 87. (*b*)

(*c*) keeping the pH value of feed water just below 7.
(*d*) eliminating H_2S in feed water.

88. Which is the most widely used coagulant for the treatment of turbid water?
(*a*) Alum (aluminium sulphate)
(*b*) Lime (*c*) Ferric chloride
(*d*) Sodium aluminate

89. Presence of iron and manganese in water causes
(*a*) reduction in its dissolved oxygen content.
(*b*) discoloration of bathroom fixtures.
(*c*) temporary hardness.
(*d*) none of these.

90. Soluble silica present in boiler feed water can be removed by
(*a*) coagulation (*b*) filtration
(*c*) anion exchanger (*d*) preheating it

91. Corrosion in boilers can be prevented by
(*a*) deaeration of feed water.
(*b*) enhancing pH of feed water by adding alkali.
(*c*) Feeding sodium sulphite or hydrazine phosphate to the boilers, which combines with oxygen and prevents corrosion.
(*d*) all (*a*), (*b*) and (*c*).

92. Tri-sodium phosphate is used in boiler water treatment to reduce
(*a*) turbidity
(*b*) caustic embrittlement
(*c*) suspended silica (*d*) dissolved oxygen

93. Which of the following is the most widely used disinfectant in water treatment?
(*a*) Chlorine
(*b*) Irradiation of water by ultraviolet light
(*c*) Cation exchanger (*d*) Coagulation

94. Replenishment of dissolved oxygen in water stream polluted with industrial waste occurs by
(*a*) natural aeration of water stream.
(*b*) photosynthetic action of algae.
(*c*) both (*a*) and (*b*).
(*d*) neither (*a*) nor (*b*).

95. Pick out the wrong statement.
(*a*) Biological oxygen demand (BOD) is a characteristic and not a constituent of water.
(*b*) BOD is a measure of the amount of oxygen which will be demanded and used in 5 days by the biological decomposition of the organic matter present in water stream present as food for the living organism.
(*c*) BOD is expressed in mg/litre (typically, BOD = 2.5 mg/litre for potable water).
(*d*) none of these

96. Pick out the correct statement.
(*a*) Chemical oxygen demand (COD) is a measure of chemically oxidisable organic matter present in water.
(*b*) COD is determined by oxidising the organic matter present in water with potassium dichromate in cone. sulphuric acid solution at boiling temperature for specified time.
(*c*) COD is related to BOD of a given waste in water but the relationship varies for different wastes. Typically COD of potable water may be 1-2 mg/litre.
(*d*) all (*a*), (*b*) and (*e*).

97. Which of the following is the most lethal water pollutant?
(*a*) Phenol and cyanide
(*b*) Chlorine (*c*) Alkalis
(*d*) Suspended solids

98. Presence of soluble organics In polluted water causes
(*a*) undesirable plants growth.
(*b*) depletion of oxygen.

Answers

88. (*a*) 89. (*b*) 90. (*c*) 91. (*d*) 92. (*b*) 93. (*a*) 94. (*c*) 95. (*d*) 96. (*d*)
97. (*a*) 98. (*b*)

(*c*) fire hazards.
(*d*) explosion hazards.

99. Presence of non-biodegradable substances like alkyl benzene sulphonate (ABS) from the detergents in polluted water stream causes
(*a*) fire hazards. (*b*) explosion hazards.
(*c*) persistent foam.
(*d*) depletion of dissolved oxygen.

100. BOD of raw municipal sewage may be in the range of about ______ mg/litre.
(*a*) 1-2 (*b*) 5-10
(*c*) 150-300 (*d*) 2000-3000

101. COD of raw municipal sewage may be in the range of about ______ mg/litre.
(*a*) 1-2 (*b*) 5-10
(*c*) 90-120 (*d*) 1500-2500

102. Disinfection of water is done to destroy pathogenic bacteria and thus prevent water-borne diseases. Disinfection of water may be done by the use of
(*a*) ozone and iodine.
(*b*) chlorine or its compounds.
(*c*) ultraviolet light for irradiation of water.
(*d*) all (*a*), (*b*) and (e).

103. Irradiation of water by ultraviolet light of suitable wavelength is commonly used for disinfection of water in
(*a*) food industry.
(*b*) municipal sewage treatment.
(*c*) petroleum refinery.
(*d*) iron and steel plant.

104. Deaeration of high pressure boiler feed water is done to reduce
(*a*) foaming from boilers.
(*b*) its dissolved oxygen content.
(*c*) its silica content.
(*d*) caustic embrittlement.

105. Presence of ______ in water stream are deleterious to aquatic life.
(*a*) soluble and toxic organics.
(*b*) suspended solids.
(*c*) heavy metals and cynides.
(*d*) all (*a*), (*b*) and (*c*).

106. Presence of nitrogen and phosphorous in waste water discharged into lakes and ponds causes
(*a*) foaming (*b*) odour nuisances
(*c*) undesirable plant growth
(*d*) turbidity

107. Presence of volatile compounds like gasoline, oil, alcohol, ether etc. in municipal sewers may cause
(*a*) explosion
(*b*) non biodegradable foam
(*c*) undesirable plant growth
(*d*) corrosion

108. Oil and grease present in an emulsified state in waste water discharged from industries can be removed by
(*a*) biological oxidation.
(*b*) skimming off.
(*c*) settling out using chemical reagents.
(*d*) chlorination.

109. Which is the most practical and economical method for removal of suspended solid matter from polluted water?
(*a*) Sedimentation (*b*) Skimming off
(*c*) Chlorination
(*d*) Biological oxidation

110. Which of the following is the most efficient for removal of very finely divided suspended solids and colloidal matter from the polluted water stream?
(*a*) Sedimentation tank
(*b*) Circular clarifier
(*c*) Mechanical flocculation
(*d*) Chemical coagulation

111. Pick out the one which is not a chemical coagulant.

Answers

99. (*c*) 100. (*c*) 101. (*c*) 102. (*d*) 103. (*a*) 104. (*b*) 105. (*d*) 106. (*c*) 107. (*a*)
108. (*c*) 109. (*a*) 110. (*d*) 111. (*b*)

(*a*) Aluminium sulphate
(*b*) Ferrous sulphate (*c*) Hydrated lime
(*d*) Chloramine

112. Biological oxidation ponds remove organic matters present in the polluted water by
(*a*) using the activities of bacteria and other micro organisms.
(*b*) aerobic oxidation.
(*c*) both (*a*) and (*b*). (*d*) neither (*a*) nor (*b*).

113. Bacterial aerobic oxidation of polluted water in biological oxidation ponds is done to purify it. Presence of bacteria helps in
(*a*) coagulation and flocculation of colloids.
(*b*) oxidation of carbonaceous matter to CO_2.
(*c*) nitrification or oxidation of ammonia derived from breakdown of nitrogeneous organic matter to the nitrite and eventually to the nitrate.
(*d*) all (*a*), (*b*) and (*c*).

114. Polluted water having low BOD are most economically treated in
(*a*) sedimentation tanks
(*b*) oxidation ponds . (*c*) sludge digester
(*d*) clarifier

115. Lagoons used for purification of polluted water
(*a*) are large shallow artificial lakes also known as clarification lakes, maturation ponds or oxidation ponds.
(*b*) use micro-organisms/bacteria in presence of dissolved oxygen.
(*c*) gives an excellent final effluent (with 3 to 4 lagoons arranged in series) having suspended solid < 1 mg/litre and BOD = 3.8 mg/litre.
(*d*) all (*a*), (*b*) and (*c*).

116. Aerobic biological oxidation ponds used for the purification of polluted water
(*a*) destroys/removes pathogen from the sewage.
(*b*) is not very effective for non-biodegradable substances (e.g., ABS) containing effluents.
(*c*) destroys/removes pathogen much more effectively if the sewage is chlorinated.
(*d*) all (*a*), (*b*) and (*c*).

117. Which is the best and the most effective method for the removal of organic contaminant present in the polluted water in very small quantity (say < 200 mg/litre)?
(*a*) Lagooning
(*b*) Activated carbon adsorption
(*c*) Biological oxidation pond
(*d*) Chemical coagulation

118. Water generated in coke ovens and by-product plant of a steel plant are disposed off by
(*a*) quenching of hot coke.
(*b*) discharging in the river stream.
(*c*) filtration and recycling for cooling coke oven gas.
(*d*) none of these.

119. Iron and manganese present in the polluted water is removed by
(*a*) simple filtration.
(*b*) oxidation followed by settling and filtration.
(*c*) chemical coagulation.
(*d*) chlorination only.

120. The commonest form of iron and manganese found in ground water as pollutant is in the form of their
(*a*) carbonates (*b*) bi-carbonate
(*c*) chlorides (*d*) sulphides

121. Iron and manganese present as pollutant in water can not be removed by
(*a*) ion exchange process.
(*b*) oxidation followed by settling and filtration.
(*c*) lime soda process or manganese zeolite process.
(*d*) chlorination.

Answers

112. (*c*) 113. (*d*) 114. (*b*) 115. (*d*) 116. (*d*) 117. (*b*) 118. (*a*) 119. (*b*) 120. (*d*)
121. (*d*)

122. Salt content in sea water is about_______ percent.
(*a*) 0.5 (*b*) 1 (*c*) 3.5 (*d*) 10

123. Dose of chlorine for disinfection of water is about_______mg/litre of water.
(*a*) 0.01 (*b*) 0.1 (*c*) 0.3 (*d*) 1

124. What is the major constituents of waste/polluted water discharged from textile, pulp and paper, tanning, distillery, dairy and meat packing industries?
(*a*) Radioactive substances
(*b*) Natural organic products
(*c*) Inorganic pollutants
(*d*) None of these

125. Water generated in coke ovens and by-product plant attached to an integrated steel plant containing phenol in concentration of less than 100 mg/litre can be removed by
(*a*) chlorination
(*b*) treating in biological oxygen pond
(*c*) chemical coagulation
(*d*) none of these

126. Waste/polluted water discharged from electroplating, blast furnace and coal mining industries contain mainly_______substances.
(*a*) radioactive (*b*) organic
(*c*) inorganic (*d*) none of these

127. Radioactive substances present in the polluted water stream can be removed by
(*a*) biological oxygen treatment.
(*b*) coagulation and filtration.
(*c*) adsorption in ion exchange materials.
(*d*) none of these.

128. The main pollutant in waste water discharged from a petroleum refinery is oil (both in free and emulsified form). Free oil is removed by
(*a*) biological oxygen pond.
(*b*) aerated lagoons. (*c*) trickling filters.
(*d*) gravity separator having oil skimming devices.

129. Suspended solid present in the waste water generated in blast furnace gas cooling and cleaning plant is removed by
(*a*) biological oxygen pond.
(*b*) radial settling tank (thickener) using coagulant (lime and ferrous sulphate).
(*c*) lagoons.
(*d*) filtration.

130. Which of the following radioactive wastes emits all α, β and γ rays and hence is the most hazardous of all radioactive emitters?
(*a*) I-131 (*b*) Sr-90
(*c*) Au-198 (*d*) Ra-226

131. What are the methods of treatment and disposal of radioactive wastes?
(*a*) Evaporation and chemical precipitation.
(*b*) Biological methods and adsorption in ion exchange materials.
(*c*) Fixing into a solid mass with cement and sinking deep in the sea.
(*d*) All (*a*), (*b*) and (*c*).

132. The amount of chemical coagulant added for treatment of polluted water _______ with increase in temperature of the polluted water to be treated.
(*a*) decreases. (*b*) increases.
(*c*) remains constant.
(*d*) may increase or decrease; depends on the chemical characteristics of polluted water.

133. Presence of bacteria in potable (drinking) water causes
(*a*) turbidity (*b*) disease
(*c*) bad odour
(*d*) bad taste and colour

134. Coal washing waste water containing about 3% suspended solids (comprising of clay, slate, stone, etc.) is treated for solid particles removal
(*a*) by chemical coagulation.
(*b*) in sedimentation tanks equipped with mechanical scrapper.
(*c*) in vacuum filter. (*d*) in clarifiers.

Answers

122. (*c*) 123. (*d*) 124. (*b*) 125. (*b*) 126. (*c*) 127. (*c*) 128. (*d*) 129. (*b*) 130. (*d*)
131. (*d*) 132. (*a*) 133. (*b*) 134. (*b*)

135. Coal mines drainage waste water (acidic in nature) results from the earth's water percolating through the voids created in coal bed during mining. This polluted water which either drains out naturally to water courses or are removed before starting the mining is

(*a*) neutralised by alkali treatment.

(*b*) left as such without any treatment.

(*c*) diluted with fresh water to reduce its acidity.

(*d*) none of these.

136. Which of the following pollutants is absent in the emissions from a fertiliser plant?

(*a*) SO_2 and SO_3 (*b*) NO_2

(*c*) NH_3 (*d*) CO

137. Noise pollution level in a chemical plant is expressed in

(*a*) roentgen (*b*) decibel

(*c*) hertz (*d*) none of these

138. Noise level during normal conversation among men is about ______ decibels.

(*a*) 10 (*b*) 45 (*c*) 90 (*d*) 115

139. A man exposed to excessive noise level in the working environment may suffer from

(*a*) hearing loss

(*b*) rupture of ear drum

(*c*) nervousness and fatigue

(*d*) all (*a*), (*b*) and (*c*)

140. Shouting by a man at his full voice corresponds to a voice level of about ______ decibels.

(*a*) 25 (*b*) 50 (*c*) 80 (*d*) 120

141. Maximum allowable noise exposure limits for a man working for 8 hours a day in a noisy chemical plant is about ______ decibels.

(*a*) 20 (*b*) 60 (*c*) 90 (*d*) 120

142. Major sources of noise in furnace operation arises from air inspirator, combustion, side wall vibrations and fluctuation in the process heat load. A 50% increase in fuel firing rate may increase the furnace noise level by about ______ decibels.

(*a*) 3 (*b*) 30 (*c*) 70 (*d*) 100

143. The major reason of hydrodynamic noise (i.e., noise resulting from liquid flow) is

(*a*) pipe vibrations.

(*b*) cavitation.

(*c*) boundary layer separation.

(*d*) fluctuation in liquid f1ow.

144. High noise levels produced during operation of fans and compressors can be reduced by using

(*a*) mufflers (silencers).

(*b*) accoustical absorbent.

(*c*) lagging of noisy duct.

(*d*) none of these.

145. Thermal pollution due to excessive heat and temperature in the working place causes

(*a*) reduction in working efficiency of man-power.

(*b*) fatigue.

(*c*) high breathing rate.

(*d*) all (*a*), (*b*) and (*c*).

146. Radioactive solid nuclear wastes are disposed off by

(*a*) high temperature incineration.

(*b*) pathological incineration.

(*c*) pyrolysis.

(*d*) underground burial in concrete containers.

147. High noise level in a chemical plant can be controlled by the

(*a*) suppression of noise at the source itself.

(*b*) path control of noise.

(*c*) protection of operating personnel.

(*d*) all (*a*), (*b*) and (*c*).

148. Noise produced by cooling fans (employed in air cooled heat exchangers or cooling tower) is mainly caused due to the turbulence created

Answers

135. (*a*) 136. (*d*) 137. (*b*) 138. (*b*) 139. (*d*) 140. (*c*) 141. (*c*) 142. (*a*) 143. (*b*)
144. (*a*) 145. (*d*) 146. (*d*) 147. (*d*) 148. (*c*)

by blade passage through air. It can be reduced by use of a slower fan
(*a*) with greater number of blades.
(*b*) of increased diameter.
(*c*) both (*a*) and (*b*),
(*d*) neither (*a*) nor (*b*).

149. Aerodynamic noise resulting from turbulent gas flow is the most prevalent source of valve noise in fluid How control. It is caused due to
(*a*) Reynold stresses (*b*) shear forces
(*c*) both (*a*) and (*b*) (*d*) neither (*a*) nor (*b*)

150. Pick out the correct statement,
(*a*) Noise from ball mills can be dealt with by fiber glass lined enclosures.
(*b*) Noise from vibrating chutes can be reduced by lining the metallic chute with rubber.
(*c*) Noise of conveyor belt system is reduced by using urethane coated cloth conveyor belts and rubber/plastic covered metallic rollers.
(*d*) all (*a*), (*b*) and (*c*).

151. The ratio of oxygen available to the oxygen required for stabilisation of sewage is called the
(*a*) bacterial stability factor.
(*b*) relative stability.
(*c*) biological oxygen demand (BOD).
(*d*) oxygen ion concentration.

152. In sewage treatment, its sedimentation is speeded up by commonly adding
(*a*) hydrochloric acid.
(*b*) lime. (*c*) copper sulphate.
(*d*) sodium sulphate.

153. The lowest layer of atmosphere is known as the
(*a*) stratosphere (*b*) troposphere
(*c*) ionosphere (*d*) none of these

154. Septic tanks are used for the ______ of the deposited solids.
(*a*) separation
(*b*) anaerobic decomposition
(*c*) aerobic decomposition
(*d*) none of these

155. The principal gas evolved from sludge digestion tank is
(*a*) CO (*b*) CO_2 (*c*) CH_4 (*d*) N_2

156. Which of the following processes is involved in the biochemical treatment of sewage effluents?
(*a*) Oxidation (*b*) Reduction
(*c*) Dehydration (*d*) Fermentation

157. Most of the bacteria in sewage are
(*a*) parasitic (*b*) saprophytic
(*c*) anaerobic (*d*) none of these

158. Solid content in most of the domestic sewage is about ______ percent.
(*a*) 0.001 (*b*) 0.1 (*c*) 5 (*d*) 10

159. Siderosis is a disease caused by the inhalation of ______ dust.
(*a*) coal (*b*) silica
(*c*) iron (*d*) none of these

160. Average human body contains about ______ percent water by weight.
(*a*) 20 (*b*) 40 (*c*) 70 (*d*) 85

161. Threshold limit value (TLV) means maxi-. mum permissible/acceptable concentration. TLV of phosgene in air is about ______ ppm (parts per million).
(*a*) 0.002 (*b*) 0.2 (*c*) 1.2 (*d*) 4.8

162. The maximum permissible noise level to which a man working in a chemical plant can be exposed for eight hours per day is about ______ decibels.
(*a*) 60 (*b*) 90 (*c*) 105 (*d*) 120

163. Inhalation of silica dust causes a disease called
(*a*) bronchitis (*b*) silicosis
(*c*) pneumoconiosis
(*d*) none of these

Answers

149. (*c*) 150. (*d*) 151. (*b*) 152. (*b*) 153. (*b*) 154. (*b*) 155. (*c*) 156. (*a*) 157. (*b*)
158. (*b*) 159. (*c*) 160. (*c*) 161. (*b*) 162. (*b*) 163. (*b*)

164. From pollution control point of view, the maximum permissible concentration of sulphur dioxide in 'atmospheric air is about ________ ppm.
(*a*) 1 (*b*) 5 (*c*) 50 (*d*) 500

165. Tolerable limit of nitrogen oxides in air is ________ ppm.
(*a*) 0.1 (*b*) 1 (*c*) 5 (*d*) 25

166. Which of the following industries discharge mercury as a pollutant?
(*a*) Chloro-alkali industry
(*b*) Tanneries (*c*) Beverage plant
(*d*) Phosphoric acid plant

167. TLV of mercury in potable (drinking) water is about ________ ppm.
(*a*) 0.001 (*b*) 0.1 (*c*) 1 (*d*) 5

168. Industrial workers working in leather tanning and manufacturing units are prone to suffer from
(*a*) respiratory ailments (e.g., bronchitis).
(*b*) skin diseases (e.g., dermatities).
(*c*) silicosis. (*d*) blurred vision.

169. There are thirteen metals which are treated as pollutants. Which of the following metals is not a pollutant?
(*a*) Mercury (*b*) Arsenic
(*c*) Aluminium (*d*) Lead

170. Maximum permissible limit of mercury in human blood is ________ micrograms/l00 c.c.
(*a*) 1 (*b*) 7 (*c*) 19 (*d*) 82

171. TLV of lead for public sewer/waste water is about ________ ppm.
(*a*) 1 (*b*) 25 (*c*) 150 (*d*) 650

172. Fluorosis is caused due to the presence of excessive amount of .... in drinking water.
(*a*) mercury (*b*) lead
(*c*) fluoride (*d*) arsenic

173. The destruction of water-borne pathogens is termed as disinfection of water. Which of the following is a water disinfectant?
(*a*) Chlorine (*b*) Alkalis
(*c*) Benzene hexachloride
(*d*) Alkyl benzene sulphonate (ABS)

174. 'Pneumoconiosis' is a disease caused by the inhalation of ________ dust.
(*a*) coal (*b*) uranium ore
(*c*) iron ore (*d*) lime

175. Polyvinyl chloride containers are not suitable for storing
(*a*) foodstuffs (*b*) woolen clothes
(*c*) metallic powder (*d*) none of these

176. Out of the following, TLV of ________ is the minimum (about 0.02 ppm).
(*a*) phosgene (*b*) bromine
(*c*) MIC (methyl iso-cynate)
(*d*) ozone

177. The killer gas which caused Bhopal gas tragedy in 1984 was
(*a*) phosgene
(*b*) methyl iso-cynate (MIC)
(*c*) carbon monoxide
(*d*) sulphur dioxide

178. Out of the following, TLV of ________ is maximum (about 500 ppm).
(*a*) carbon dioxide (*b*) toluene
(*c*) carbon disulphide (*d*) acetaldehyde

179. Reingleman chart is used for the evaluation of pollution.
(*a*) air (*b*) water
(*c*) noise (*d*) radioactive

180. Which of the following gases is having the widest explosion limit (about 2 to 81% gas in gas-air mixture), rendering it the property of the most explosive gas?
(*a*) Hydrogen (*b*) Acetylene
(*c*) Carbon monoxide
(*d*) Ammonia

181. Exposure to chemicals having carcinogenic properties cause

Answers

164. (*b*) 165. (*c*) 166. (*a*) 167. (*a*) 168. (*b*) 169. (*c*) 170. (*b*) 171. (*a*) 172. (*c*)
173. (*a*) 174. (*a*) 175. (*a*) 176. (*c*) 177. (*b*) 178. (*a*) 179. (*a*) 180. (*b*) 181. (*b*)

(*a*) dermatitis (skin disorder)
(*b*) cancer (*c*) asphyxiation
(*d*) asthma

182. Inhalation of silica dust by human being during hand drilling in mica mining, lead and zinc mining, silica refractory manufacture and in foundaries causes
(*a*) asphyxiation (suffocation).
(*b*) shortness of breath.
(*c*) tuberculosis. (*d*) all (*a*), (*b*) and (*c*).

183. Inhalation of lead compounds present in the automobile exhausts (using leaded petrol) causes -
(*a*) blood poisoning. (*b*) anaemia.
(*c*) nervous system disorder.
(*d*) all (*a*), (*b*) and (*c*).

184. Which is a secondary air pollutant?
(*a*) Photochemical smog
(*b*) Sulphur dioxide (*c*) Nitrogen dioxide
(*d*) Dust particles

185 Content of the phosphate rock is the pollutant of primary interest in a phosphatic fertiliser plant.
(*a*) Calcium (*b*) Fluorine
(*c*) Phosphorous (*d*) Sulphur

186. A gas is termed as non-toxic, if its maximum permissible concentration (TLV) ranges from ______ ppm.
(*a*) 1000 to 2000 (*b*) 3000 to 6000
(*c*) 6000 to 9000 (*d*) 10000 to 100000

187. The permissible colour for domestic water supply is ______ ppm.
(*a*) 1 (*b*) 20 (*c*) 100 (*d*) 1000

188. Presence of nitrates in water in excess of 50 ppm causes
(*a*) mathenoglobenemia
(*b*) gastroentetitis (*c*) asphyxiation
(*d*) tooth decay

189. Presence of a certain minimum quantity of flourine is desirable in potable water to prevent
(*a*) dental cavities (*b*) scale formation
(*c*) water-borne disease
(*d*) corrosion

190. Presence of ______ hardness is responsible for the temporary hardness in water.
(*a*) carbonate (*b*) calcium
(*c*) chloride (*d*) sulphate.

191. ______ is the process of killing organism in water.
(*a*) Coagulation (*b*) Sterilisation
(*c*) Disinfection (*d*) Sedimentation

192. Growth of ______ is promoted by the presence of manganese in water.
(*a*) files (*b*) algae
(*c*) micro-organisms (*d*) mosquitoes

193. Presence of ______ bacteria in water causes disease like typhoid.
(*a*) aerobic (*b*) pathogenic
(*c*) anaerobic (*d*) non-pathogenic

194. Maximum permissible turbidity in potable water is ______ ppm.
(*a*) 1 (*b*) 10 (*c*) 250 (*d*) 1000

195. Turbidity of water is an indication of the presence of
(*a*) suspended inorganic matter
(*b*) dissolved solids (*c*) floating solids
(*d*) dissolved gases

196. Presence of dissolved impurities of ______ is responsible for the red brownish colour of water.
(*a*) carbonates (*b*) bi-carbonates
(*c*) iron and manganese
(*d*) arsemic.

197. Presence of excess flourine in water causes
(*a*) dental cavity (*b*) tooth decay
(*c*) fluorosis (*d*) respiratory disease

198. Infective bacteria in water is killed by the ______ process.

Answers

182. (*d*) 183. (*d*) 184. (*a*) 185. (*b*) 186. (*d*) 187. (*b*) 188. (*a*) 189. (*a*) 190. (*b*)
191. (*c*) 192. (*c*) 193. (*b*) 194. (*b*) 195. (*a*) 196. (*c*) 197. (*c*) 198. (*a*)

(*a*) sterilisation (*b*) aeration
(*c*) disinfection (*d*) none of these.

199. Removal of _______ is accomplished by aeration of water.
(*a*) dissolved gases (*b*) suspended solids
(*c*) dissolved solids (*d*) none of these.

200. The pH value of potable water should be between
(*a*) 1 to 1.5 (*b*) 6.5 to 8
(*c*) 13 to 14 (*d*) 4 to 5

201. A standard test for determination of hardness in water is termed as _______ test.
(*a*) EDTA (*b*) electometric
(*c*) total count (*d*) presumptive

201. In a sedimentation tank, the detention period for water ranges from _______ hours.
(*a*) 2 to 4 (*b*) 8 to 12
(*c*) 16 to 20 (*d*) 24 to 32

203. In water treatment, alum $[Al_2(SO_4)^3]$ is used for the process of
(*a*) filtration (*b*) coagulation
(*c*) sedimentation (*d*) disinfection.

204. Color test of water is done with an instrument called
(*a*) tintometer (*b*) colorimeter
(*c*) electro-chemical cell
(*d*) turbidimeter.

205. Hydrogen ion concentration in distilled water is
(*a*) 10^{-7} (*b*) 7×10^{-7}
(*c*) 10^7 (*d*) 7×10^7

206. Removal of _______ results from the disinfection of water.
(*a*) turbidity (*b*) odour
(*c*) hardness (*d*) bacteria.

207. The resistance of water to the passage of light through it is a measure of the
(*a*) turbidity (*b*) color
(*c*) hardness (*d*) dissolved gases.

208. Water filtration rate in a rapid sand filter ranges from _______ kilolitres/m²/hr.
(*a*) 0.1 to 1 (*b*) 3 to 6
(*c*) 10 to 15 (*d*) 15 to 20

209. Disinfection of water is done to remove
(*a*) color (*b*) bad taste
(*c*) foul odour (*d*) bacteria

210. Algae growth in water is controlled by
(*a*) deoxidation (*b*) chlorination
(*c*) bleaching (*d*) aeration.

211. Foul odour and bad taste of water is removed by treating with
(*a*) alum (*b*) bleaching powder
(*c*) activated carbon (*d*) copper sulphate.

212. In water treatment plant, zeolite process is used to remove the _______ of water.
(*a*) acidity (*b*) alkalinity
(*c*) hardness (*d*) iron and zinc.

213. _______ is removed from water by lime-soda process.
(*a*) Foul smell and taste
(*b*) Iron and manganese
(*c*) Temporary hardness
(*d*) Permanent hardness

214. Maximum permissible residual chlorine in treated water should be _______ mg/litre.
(*a*) 0.001 to 0.01 (*b*) 02 to 0.3
(*c*) 2 to 3 (*d*) 5 to 10

215. Fresh domestic sewage is _______ in colour.
(*a*) grey (*b*) dark brown
(*c*) red (*d*) black.

216. Brown spots in fabrics will be caused by washing with water containing large amount of
(*a*) iron (*b*) zinc
(*c*) iodine (*d*) bromine.

217. Water filtration rate in a slow sand filter ranges from _______ litres/m²/hr.

Answers

199. (*a*) 200. (*b*) 201. (*a*) 202. (*a*) 203. (*b*) 204. (*a*) 205. (*c*) 206. (*d*) 207. (*a*)
208. (*b*) 209. (*d*) 210. (*b*) 211. (*c*) 212. (*c*) 213. (*c*) 214. (*b*) 215. (*a*) 216. (*a*)
217. (*b*)

(*a*) 10 to 20 (*b*) 100 to 200
(*c*) 1500 to 2500 (*d*) 4000 to 5000.

218. Insufficient washing of sand grains in a rapid sand filter causes
(*a*) air binding
(*b*) shrinkage of filtering media
(*c*) mud balls
(*d*) expansion of filtering media.

219. Fresh sewage is ______ in nature.
(*a*) acidic (*b*) neutral
(*c*) alkaline (*d*) highly acidic.

220. The pH value of oxidised sewage is about
(*a*) 1.8 (*b*) 6.2 (*c*) 7.3 (*d*) 13.4

221. ______ substances present in sewage are removed in grit chamber during sewage treatment.
(*a*) Organic (*b*) Fatty
(*c*) Inorganic (*d*) Dissolved

222. The type of bacteria which is active in trickling filter during biological treatment of sewage is the ______ bacteria.
(*a*) anaerobic (*b*) saphrophytic
(*c*) anaerobic (*d*) parasitic.

223. During sewage treatment, the sewage is subjected to ______ treatment in Imhoff tank.
(*a*) filtration (*b*) digestion
(*c*) sedimentation (*d*) both (*b*) and (*c*).

224. A shallow pond in which the sewage is retained and biologically treated is called
(*a*) oxidation (*b*) Imhoff tank
(*c*) lagoon (*d*) skimming tank

225. In sewage treatment, the detention period allowed for oxidation ponds ranges from ______ weeks.
(*a*) 1 to 2 (*b*) 4 to 5
(*c*) 9 to 10 (*d*) 15 to 20.

226. The function of skimming tank in sewage treatment is to remove ______ substances.
(*a*) dissolved solid (*b*) suspended solid
(*c*) oil and fatty
(*d*) gritty and inorganic.

227. Fine grit present in sewage is removed in the ______ during sewage treatment.
(*a*) grit chamber (*b*) detritus tank
(*c*) trickling filter (*d*) skimming tank.

228. The main type of sludge gas evolved during sewage treatment in Imhoff tank is
(*a*) CO_2 (*b*) CH_4 (*c*) CO (*d*) H_2

229. The biological decomposition of organic substances in wastes controlled conditions is called
(*a*) incineration
(*b*) biological oxidation
(*c*) composting (*d*) none of these.

230. The most commonly used chemical coagulant in water treatment is
(*a*) ferrous sulphate (*b*) alum
(*c*) lime (*d*) hydrazine

231. The weakest sound that can be heard by a person in a quiet environment is equivalent to ______ decibel.
(*a*) 1 (*b*) 5 (*c*) 10 (*d*) 20

232. Sound produced by an automobile horn heard at a distance of 1.5 metres corresponds to about ______ decibels.
(*a*) 90 (*b*) 120 (*c*) 150 (*d*) 180

233. Noise emitted by a ventilation fan at a distance of 3 metres is about ______ decibels.
(*a*) 85 (*b*) 105 (*c*) 125 (*d*) 145

234. Noise level audible to audience sitting in the 5th row from the stage during a large orchestra show corresponds to about ______ decibels.
(*a*) 105 (*b*) 135 (*c*) 160 (*d*) 185

235. Noise level heard inside a bus in busy city traffic or inside a sub-way train corresponds to about ______ decibels.
(*a*) 75 (*b*) 95 (*c*) 120 (*d*) 140

Answers

218. (*c*) 219. (*c*) 220. (*c*) 221. (*c*) 222. (*c*) 223. (*d*) 224. (*a*) 225. (*b*) 226. (*c*)
227. (*b*) 228. (*b*) 229. (*c*) 230. (*b*) 231. (*a*) 232. (*b*) 233. (*b*) 234. (*b*) 235. (*b*)

236. Noise level inside a jet air liner in normal flight is about _______ decibels.
(*a*) 80 (*b*) 100 (*c*) 125 (*d*) 145

237. Noise level heard at a distance of about 100 metres from a jet engine with after burner is about _______ decibels.
(*a*) 120 (*b*) 140 (*c*) 170 (*d*) 200

238. Noise level in a quiet private business office is about _______ decibels.
(*a*) 25 (*b*) 50 (*c*) 70 (*d*) 85

239. Reingleman chart No. 2 corresponds to _______ percent black smoke.
(*a*) 10 (*b*) 20 (*c*) 40 (*d*) 80

240. Maximum permissible concentration (i.e., TLV) of DDT in public water supply system is _______ micro gram (μg)/litre.
(*a*) 2 (*b*) 42 (*c*) 332 (*d*) 1050

241. Which of the following is the common pollutant emitted from metallurgical smelters, thermal power plant and cement plants?
(*a*) NO_x (*b*) Hg (*c*) SO_2 (*d*) F

242. Arsenic pollutant is not generated in _______ industries.
(*a*) tanneries (*b*) glass and ceramic
(*c*) beverages (*d*) any of these.

243. The widest explosive limit is of _______ thereby making it the most explosive gas.
(*a*) acetylene (*b*) petrol vapor
(*c*) hydrogen (*d*) carbon monoxide

244. Nitrogen oxides (NO_x) is not produced in the _______ industry.
(*a*) nitric acid making
(*b*) nitrogenous fertiliser
(*c*) detergent (*d*) any of these.

245. TLV of aldrin in public water supply system is about _______ μg/litre.
(*a*) 0.5 (*b*) 17 (*c*) 357 (*d*) 1097

246. The common pollutant generated in chlor-alkali industry and battery manufacture is
(*a*) mercury (*b*) brine
(*c*) phosphate (*d*) none of these.

247. Permissible safe limit (TLV) of _______ toxic gas is 100-1000 parts per million (ppm).
(*a*) highly (*b*) moderately
(*c*) extremely (*d*) very extremely.

248. Hazardous/polluting chemical industries should have an 'exclusion zone' with a green belt and general public access prohibited around it, covering a radius of.......... metres.
(*a*) 100 (*b*) 1000 (*c*) 400 (*d*) 4000

249. Methyl iso-cynate (MIC) gas (TLV < 1 ppm), which caused Bhopal gas tragedy in 1984 falls under the category of _______ toxic gas.
(*a*) moderately (*b*) highly
(*c*) extremely (*d*) very extremely.

250. Threshold limit value (TLV) i.e., the maximum permissible safe limit of phosgene gas which Hitler used to use to kill his enemies in 'gas chamber' is about _______ ppm.
(*a*) < 1 (*b*) 10-100
(*c*) 100-200 (*d*) 100-1000

251. There are five concentric layers within the atmosphere which is differentiated on the basis of temperature. The atmospheric layer which lies close to the earth's surface in which human being along with other organisms live is called troposphere. The rate at which air temperature in the troposphere gradually decreases with height is about _______ °C/km.
(*a*) 0.05 (*b*) 1 (*c*) 6.5 (*d*) 15

252. Pick out the wrong statement
(*a*) The concentric atmosphere layer just above troposphere is called stratosphere, which is rich in ozone.
(*b*) Mesosphere is characterised by very low atmospheric pressure and low temperature.

Answers

236. (*b*) 237. (*c*) 238. (*b*) 239. (*c*) 240. (*b*) 241. (*c*) 242. (*c*) 243. (*a*) 244. (*c*)
245. (*b*) 246. (*a*) 247. (*b*) 248. (*b*) 249. (*d*) 250. (*a*) 251. (*c*) 252. (*d*)

(*c*) Troposphere is a dusty zone containing water vapor and clouds.
(*d*) The radio waves used in the long distance radio communication are reflected back to earth by stratosphere.

253. Most of the atmospheric air pollutants are present in large quantity in
(*a*) stratosphere (*b*) thermosphere
(*c*) troposphere (*d*) mesosphere

254. In large thermal power stations very fine particulates present in flue gas are removed by
(*a*) wet scrubber (*b*) bag filter
(*c*) electrostatic precipitators
(*d*) dust catcher

255. Direct reaction of unsaturated hydrocarbons with either NO or NO_2 produces an eye irritating pollutant compound known as
(*a*) photochemical smog.
(*b*) peroxy acetyl nitrate. (PAN) or methyl nitrile.
(*c*) benzopyrene. (*d*) poly acrylonitrile.

256. The average thickness of ozone layer in stratosphere is about _______ dobson unit (DU).
(*a*) 20 (*b*) 230 (*c*) 750 (*d*) 1500

257. Inorganic impurities causing water pollution is
(*a*) fats (*b*) carbohydrates
(*c*) salts of metals (*d*) protein.

258. The detrimental effect on organism and water quality with temperature rise of aquatic system is the reduction of _______ of water.
(*a*) dissolved oxygen content
(*b*) biological oxygen demand
(*c*) vapor pressure (*d*) all (*a*), (*b*) and (*c*)

259. Water effluent generated in printing industry is decolourised by
(*a*) ion exchange technique.
(*b*) reverse osmosis.
(*c*) electrolytic decomposition.
(*d*) adsorption.

260. Workers working in _______ industry are most prone to white lung cancer.
(*a*) coal mining (*b*) limestone mining
(*c*) textile (*d*) asbestos.

261. Automobile exhaust is passed through two compartments catalytic converter employing platinum as catalyst for
(*a*) conversion of CO into CO_2 in the second compartment.
(*b*) conversion of NO_x into N_2 and NH_3 in the first compartment.
(*c*) oxidation of unburnt hydrocarbon fuel in the second compartment.
(*d*) all (*a*), (*b*) and (*c*).

262. Presence of carbon monoxide in atmosphere produced by decomposition of chlorophyll and haemoglobin breakdown of some animals, beyond TL V (> 50 ppm)
(*a*) acts as a green house gas thereby raising earth's temperature.
(*b*) causes asphyxia.
(*c*) causes increase in sea level.
(*d*) inhances the green house effect.

263. Oceans act as sinks for atmospheric gases including carbon dioxide whose concentration in the atmosphere is increased by the
(*a*) forestation (*b*) rain
(*c*) green house effect (*d*) vegetation

264. Bag filter design is predominantly dependent on gas temperature, as it affects the gas density and viscosity and the selection of filtering material. The pressure drop in a bag filter is
(*a*) inversely proportional to viscosity of gas.
(*b*) proportional to the viscosity and density of the gas.
(*c*) proportional to the pressure of the gas.
(*d*) both (*b*) and (*c*).

265. Maximum permissible limit of industrial noise as recommended by World Health Organisation (WHO) is _______ decibels.
(*a*) 35 (*b*) 75 (*c*) 95 (*d*) 105

Answers

253. (*c*) 254. (*c*) 255. (*b*) 256. (*b*) 257. (*c*) 258. (*a*) 259. (*c*) 260. (*c*) 261. (*d*)
262. (*b*) 263. (*c*) 264. (*d*) 265. (*b*)

266. Match the pollutants with their source of generation.

List I		List II	
(*a*)	Ammonia	I.	Automobile exhaust
(*b*)	Hydrogen sulphide	II.	Coal combustion in boiler
(*c*)	Sulphur dioxide	III.	Heavy water plant
(*d*)	Nitrogen oxides	IV.	Nitrogeneous fertiliser plant

267. Match the ill effects of various pollutants.

List I		List II	
(*a*)	Carbon monoxide	I.	Bone disease
(*b*)	Sulphur dioxide	II.	Asphyxiation
(*c*)	Chromium and arsenic	III.	Respiratory problems
(*d*)	Flourides	IV.	Cancer

268. Match the major pollutant emitted from various chemical industries.

List I		List II	
(*a*)	Nitric acid plant.	I.	Arsenic
(*b*)	Glass and Ceramic industry and tanneries.	II.	Hydroflouric acid
(*c*)	Phosphoric acid plant.	III.	Mercury
(*d*)	Caustic-chlorine industry.	IV.	Nitrogen oxides

269. Match the tolerate limit (in ppm) of various atmospheric pollutants.

List I		List II	
(*a*)	Benzene	I.	500 ppm
(*b*)	Methyl iso-cynate (MIC)	II.	10 ppm
(*c*)	NO_x and SO_x	III.	5 ppm
(*d*)	Carbon dioxide	IV.	0.02 ppm

270. Match the deleterious effect of various metallic water pollutants on human health.

List I		List II	
(*a*)	Cadmium	I.	Lung irritation.
(*b*)	Cobalt	II.	Bone deformation and growth retardation
(*c*)	Zinc	III.	Liver and kidney damage.
(*d*)	Selenium	IV.	Renal damage.

271. Exposure to chemicals having carcinogenic properties cause

(*a*) dermatitis (skin disorder)
(*b*) cancer
(*c*) asphyxiation (suffocation)
(*d*) asthma

272. Inhalation of silica dust by human being causes

(*a*) asphyxiation (*b*) shortness of breath
(*c*) tuberculosis (*d*) both (*b*) and (*c*)

273. Inhalation of lead compounds present in automobile exhaust (using leaded petrol) causes

(*a*) blood poisoning. (*b*) anaemia.
(*c*) nervous system disorder.
(*d*) all (*a*), (*b*) and (*c*).

274. Which of the following is a secondary air pollutant?

(*a*) Photochemical smog
(*b*) Sulphur dioxide (*c*) Nitrogen dioxide
(*d*) Dust particles

275. A considerable part of the harmful ultraviolet radiation of the sum does not reach the earth surface, because of the fact that, there is a layer of ______ high above earth's atmosphere, which absorbs it,

(*a*) hydrogen (*b*) carbon dioxide
(*c*) ozone (*d*) none of these

276. As per the world health organisation (WHO) specification, the maximum permissible

Answers

266. (*a*)-IV(*b*)-III (*c*)-II (*d*)-I 267. (*a*)-II (*b*)-III (*c*)-IV (*d*)-I 268. (*a*)-IV (*b*)-I (*c*)-II (*d*)-III
269. (*a*)-II (*b*)-IV (*c*)-IV (*d*)-I 270. (*a*)-II (*b*)-I (*c*)-IV (*d*)-III 271. (*b*) 272. (*d*) 273. (*d*)
274. (*a*) 275. (*c*) 276. (*b*)

concentration (i.e., TLV) of particulate matter in air is ______ μg/m^3. (μg-microgram)
(*a*) 10 (*b*) 90 (*c*) 300 (*d*) 750

277. World environment day is observed every year on the 5th of
(*a*) June (*b*) December
(*c*) July (*d*) September

278. Pick out the wrong statement.
(*a*) The concentric layer in atmosphere which contains about 70% of the total mass of atmosphere and characterised by a steady decrease in temperature is called stratosphere.
(*b*) Stratosphere is rich in ozone and is located just above the troposphere.
(*c*) Troposphere is a turbulent dusty zone containing much of water vapor and clouds.
(*d*) The earth's atmosphere is an envelope of gases extending upto a height of about 200 kms.

279. Which of the following pollutants, if present in atmosphere is detectable by its odour?
(*a*) CO (*b*) SO_2 (*c*) NO_2 (*d*) CO_2

280. Thermal pollution of water increases its toxicity and oxidation of oxygen demanding waste besides favouring bacterial growth. A rise in water temperature by 10°C, doubles the toxic effects of ______ present in it.
(*a*) coal ash (*b*) potassium cyanide
(*c*) ortho-xylene (*d*) none of these.

281. Pick out the wrong statement.
(*a*) Biological oxygen demand (B.O.D.) value of a sewerage sample is always lower than its chemical oxygen demand (C.O.D.) value.
(*b*) Environmental pollution by NO_x emission is much higher by four stroke petrol engines as compared to the two stroke engines.
(*c*) Temperature in stratosphere rises with increasing altitude.
(*d*) The characteristic of a green house body is that it allows the long wavelength incoming solar radiation to come in but does not allow the short wavelength infra red radiation to escape out of the earth's atmosphere.

282. The concentration of water vapour in troposphere, which depends upon the altitude and temperature varies in the range of zero to ______ percent.
(*a*) 1 (*b*) 4 (*c*) 8 (*d*) 12

283. Oilish impurities present the effluent discharged from the electroplating industry is normally not removed by
(*a*) chemical coagulation
(*b*) floatation and skimming
(*c*) centrifugation
(*d*) ultrafiltration

284. ______ can not control the noise pollution.
(*a*) Use of silencers (*b*) Green house gases
(*c*) Vibration damping
(*d*) Tree plantation

285. Which of the following is the most active zone of atmosphere in which weathering events like rain, storm and lightning occur?
(*a*) Thermosphere (*b*) Troposphere
(*c*) Stratosphere (*d*) None of these

286. Very small amount of air pollutants are present in stratosphere also; though most of the atmospheric pollutants are present in the troposphere. Which of the following atmospheric pollutants does not cause the ozone layer depletion in atmosphere at tremendous rate?
(*a*) CO (*b*) SO_2
(*c*) NO_x
(*d*) CFC (chloro fluoro carbons)

287. CFC (chloro fluoro carbon) is very highly reactive in causing depletion of ozone layer in the atmosphere. Each atom of chlorine

Answers

277. (*a*) 278. (*a*) 279. (*b*) 280. (*b*) 281. (*d*) 282. (*b*) 283. (*a*) 284. (*b*) 285. (*b*)
286. (*a*) 287. (*b*)

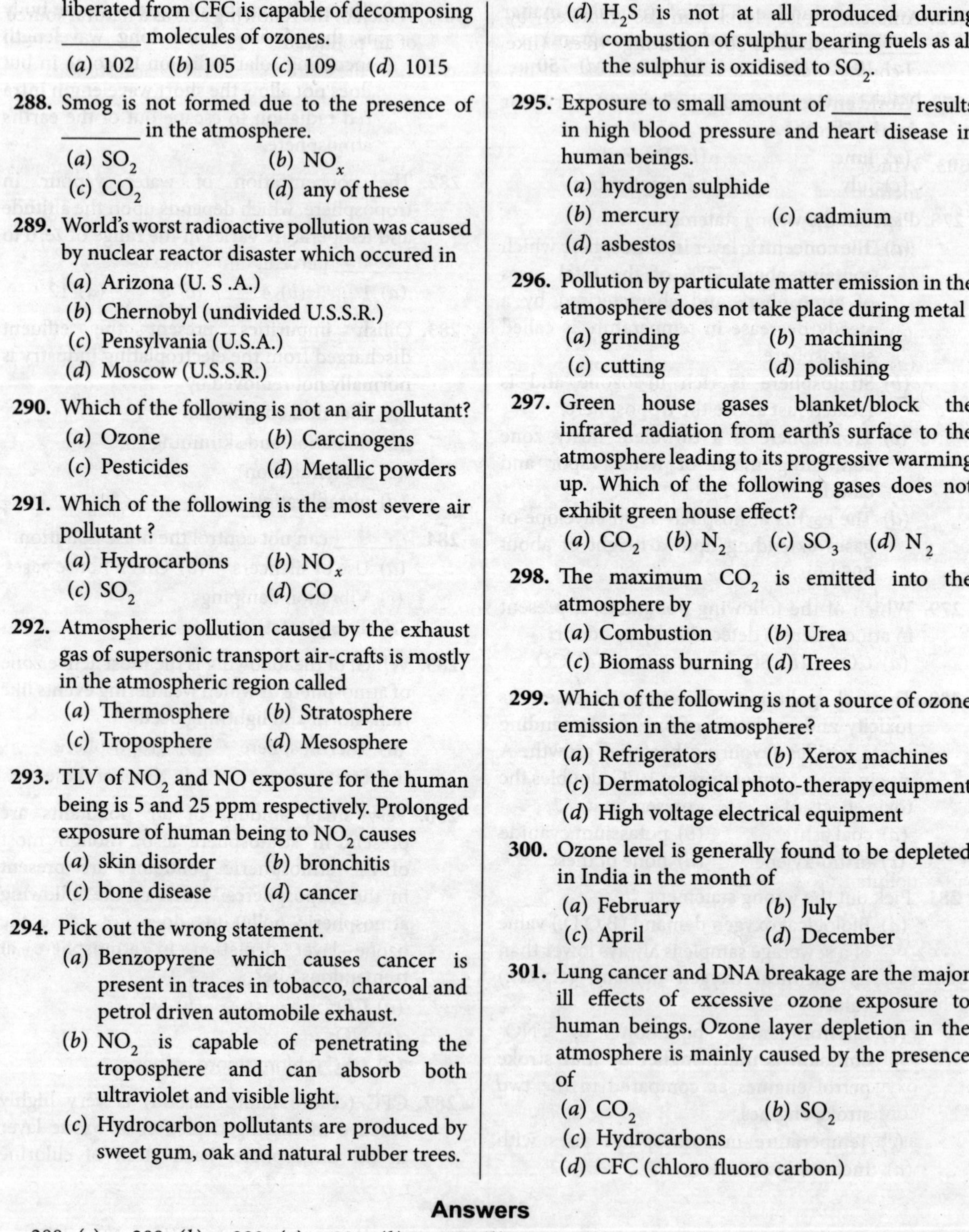

liberated from CFC is capable of decomposing ______ molecules of ozones.

(a) 102 (b) 105 (c) 109 (d) 1015

288. Smog is not formed due to the presence of ______ in the atmosphere.

(a) SO_2 (b) NO_x
(c) CO_2 (d) any of these

289. World's worst radioactive pollution was caused by nuclear reactor disaster which occured in

(a) Arizona (U. S .A.)
(b) Chernobyl (undivided U.S.S.R.)
(c) Pensylvania (U.S.A.)
(d) Moscow (U.S.S.R.)

290. Which of the following is not an air pollutant?

(a) Ozone (b) Carcinogens
(c) Pesticides (d) Metallic powders

291. Which of the following is the most severe air pollutant ?

(a) Hydrocarbons (b) NO_x
(c) SO_2 (d) CO

292. Atmospheric pollution caused by the exhaust gas of supersonic transport air-crafts is mostly in the atmospheric region called

(a) Thermosphere (b) Stratosphere
(c) Troposphere (d) Mesosphere

293. TLV of NO_2 and NO exposure for the human being is 5 and 25 ppm respectively. Prolonged exposure of human being to NO_2 causes

(a) skin disorder (b) bronchitis
(c) bone disease (d) cancer

294. Pick out the wrong statement.

(a) Benzopyrene which causes cancer is present in traces in tobacco, charcoal and petrol driven automobile exhaust.
(b) NO_2 is capable of penetrating the troposphere and can absorb both ultraviolet and visible light.
(c) Hydrocarbon pollutants are produced by sweet gum, oak and natural rubber trees.
(d) H_2S is not at all produced during combustion of sulphur bearing fuels as all the sulphur is oxidised to SO_2.

295. Exposure to small amount of ______ results in high blood pressure and heart disease in human beings.

(a) hydrogen sulphide
(b) mercury (c) cadmium
(d) asbestos

296. Pollution by particulate matter emission in the atmosphere does not take place during metal

(a) grinding (b) machining
(c) cutting (d) polishing

297. Green house gases blanket/block the infrared radiation from earth's surface to the atmosphere leading to its progressive warming up. Which of the following gases does not exhibit green house effect?

(a) CO_2 (b) N_2 (c) SO_3 (d) N_2

298. The maximum CO_2 is emitted into the atmosphere by

(a) Combustion (b) Urea
(c) Biomass burning (d) Trees

299. Which of the following is not a source of ozone emission in the atmosphere?

(a) Refrigerators (b) Xerox machines
(c) Dermatological photo-therapy equipment
(d) High voltage electrical equipment

300. Ozone level is generally found to be depleted in India in the month of

(a) February (b) July
(c) April (d) December

301. Lung cancer and DNA breakage are the major ill effects of excessive ozone exposure to human beings. Ozone layer depletion in the atmosphere is mainly caused by the presence of

(a) CO_2 (b) SO_2
(c) Hydrocarbons
(d) CFC (chloro fluoro carbon)

Answers

288. (c) 289. (b) 290. (a) 291. (b) 292. (b) 293. (b) 294. (d) 295. (c) 296. (d)
297. (c) 298. (a) 299. (a) 300. (d) 301. (d)

302. Ambient noise level can be reduced by _______ decibels by planting trees (like coconut, neem etc.) near public utility buildings (like hospitals and schools).

(*a*) 1-2 (*b*) 5-10 (*c*) 15-20 (*d*) 25-30

303. Which of the following is not a practical method of low level radioactive waste disposal? ‘

(*a*) Dilution with inert material.

(*b*) Discharging to atmosphere through tall stacks after dilution.

(*c*) Disposing off in rivers and oceans.

(*d*) Filling in steel crates and shooting it off out of earth's gravity.

304. Pick out the wrong statement

(*a*) Low ozone layer thickness in polar regions is due to cold climatic conditions

(*b*) Amount of unburnt hydrocarbons emitted by two stroke petrol engine is more as compared to that emitted by a four stroke engine.

(*c*) Carbon monoxide present in the two stroke petrol engine exhaust is much less as compared to that emitted from a four stroke engine.

(*d*) Mercury as a pollutant can enter the blood stream and the digestive system! lungs and is responsible for causing minamate disease.

305. Which of the following is not a weightless pollutant?

(*a*) SPM (*b*) Thermal pollution

(*c*) Radioactive rays (*d*) Noise pollution

306. Presence of nitrogen in high concentration in contaminated air reduces partial pressure of oxygen in lungs, thereby causing asphyxia (suffocation) leading to death from oxygen deficiency. Concentration of N_2 in contaminated air at which it acts as a natural asphyxant is ≥ _______ percent.

(*a*) 84 (*b*) 88 (*c*) 80 (*d*) 92

307. Which of the following acts as a natural source of air pollution?

(*a*) Forest fire (*b*) Deforestation

(*c*) Volcanic eruption

(*d*) None of these

308. Dissolved oxygen content in river water is around _______ ppm.

(*a*) 5 (*b*) 100 (*c*) 250 (*d*) 500

309. Carcinogenic air pollutants cause

(*a*) bone decay (*b*) cancer

(*c*) asphyxiation (suffocation)

(*d*) anemia

310. The term Biological Oxygen Demand (BOD) is used in relation to

(*a*) potable water (*b*) cooling water

(*c*) distilled water (*d*) industrial effluents

311. A masonry structure built below ground level, where biochemical reaction takes place due to anaerobic bacteria is called

(*a*) cesspool (*b*) lagoon

(*c*) skimming mill (*d*) septic tank

312. In the context of the chemical process industries, the term BOD is normally associated with the

(*a*) characterisation of solid wastes.

(*b*) organic concentration in gaseous effluents.

(*c*) characterisation of liquid effluents.

(*d*) characterisation of boiler feed water.

[GATE ‘98]

313. In troposphere (the weather domain), the temperature ‘*t*’ at height ‘*h*’ above the sea level in metres is given by

(*a*) $t = 15 - 0.0065\,h$

(*b*) $t = 15 + 0.0065\,h$

(*c*) $t = 0.0035\,h - 15$

(*d*) $t = 15 - 0.0035\,h$

where, temperature at sea level is 15°C and *t* is in 0°C.

Answers

302. (*b*) 303. (*d*) 304. (*c*) 305. (*a*) 306. (*a*) 307. (*c*) 308. (*a*) 309. (*b*) 310. (*d*)
311. (*d*) 312. (*c*) 313. (*a*)

314. Match the major sources of various pollutants.

	List I		List II
(*a*)	Coal based thermal power plants	I.	NH_3
(*b*)	Oil refineries	II.	NO_x
(*c*)	Automobiles exhaust	III.	H_2S
(*d*)	Dye and explosive manufacture	IV.	SO_2

315. Match the ill effects of various gaseous pollutants on human health.

	List I		List II
(*a*)	Hydrogen sulphide	I.	Causes coughing, irritation and fatal pulmonary edema.
(*b*)	Phosgene	II.	Irritates eyes and throat and causes nausea.
(*c*)	Hydrogen cyanide	III.	Dries throat, interferes with nerve cells.
(*d*)	Ammonia	IV.	Causes inflammation of upper respiratory passage.

316. Match the characteristics of various components of NO_x pollutants produced

	List I		List II
(*a*)	NO	I.	Produces aerosol and damages nylon fibre.
(*b*)	NO_2	II.	Main component of NO_x produced in petrol driven automobile exhaust.
(*c*)	N_2O	III.	Main component of NO_x that reduces the pH value of rain water.
(*d*)	N_2O_5	IV.	Main component of NO_x generated by biomass burning.

317. Match the item in the left hand column to the appropriate item in the right hand column.

	Column I		Column II
P.	Global warming	1.	Unleaded gasoline
Q.	Methyl tertiary-butyl ether (MTBE)	2.	Nitrous oxides
		3.	Chlorofluoro carbons
		4.	Leaded gasoline

[GATE '97]

(*a*) P-2, Q-4 (*b*) P-3, Q-1
(*c*) P-1, Q-3 (*d*) P-2, Q-3

318. LD_{50} stands for

(*a*) legal dose of a specified toxic chemical in which 50 test animals die.

(*b*) lethal dose of a toxic chemical in which 50% of test animals survive and remaining die.

(*c*) lethal dose of toxic chemical in which 50 test animals out of 500 die.

(*d*) lethal dose of a toxic chemical in which 50 test animals out of 500 survive.

319. PAN which stands for ________ is a secondary pollutant.

(*a*) peroxyacetyl nitrate
(*b*) peroxyacetate nitrate
(*c*) peroacetate nitrile
(*d*) peroacetate nitrate.

320. ________ method of waste treatment will be recommended, if the ratio of biological oxygen demand (BOD) to chemical oxygen demand (COD) is 1.

(*a*) Thermal (*b*) Biological
(*c*) Physical (*d*) Chemical

321. What is BOD (ppm) of a water sample having initial oxygen concentration of 10 ppm and final oxygen concentration of 4 ppm after keeping it in a closed bottle for 6 days at 25°C?

(*a*) 2.5 (*b*) 6 (*c*) 11 (*d*) 1.5

322. Main pollutant in the manufacturing of various super phosphates are the compounds.

Answers

314. (*a*)-IV (*b*)-III (*c*)-II (*d*)-I 315. (*a*)-II (*b*)-I (*c*)-III (*d*)-IV 316. (*a*)-II (*b*)-I (*c*)-IV (*d*)-III
317. (*b*) 318. (*b*) 319. (*a*) 320. (*b*) 321. (*b*) 322. (*a*)

(*a*) Fluorine (*b*) Sulphur
(*c*) Chlorine (*d*) Phosphorous

323. Uppermost region of atmosphere is called
(*a*) ionosphere (*b*) exosphere
(*c*) stratosphere (*d*) troposphere

324. The ozone layer depletion is caused by NO and CFC (chloroflouro carbon). Depletion of ozone layer causes ________ cancer.
(*a*) Lung (*b*) Blood
(*c*) Skin (*d*) Breast

325. In which region of atmosphere, the temperature does not decrease with the attitude?
(*a*) Troposphere (*b*) Thermosphere
(*c*) Mesosphere (*d*) None of these

326. Which of the following is the coldest region of atmosphere?
(*a*) Thermosphere (*b*) Troposphere
(*c*) Stratosphere (*d*) Mesosphere

327. Ozone layer is present in
(*a*) Exosphere (*b*) Troposphere
(*c*) Stratosphere (*d*) Mesosphere

328. Pick out the wrong statement about carbon dioxide (CO_2).
(*a*) CO_2 can absorb infrared radiation but does not allow them to pass through.
(*b*) CO_2 is involved in the formation of photochemical fog.
(*c*) CO_2 is responsible for green house effect.
(*d*) CO_2 is not considered as an atmospheric pollutant.

329. White lung cancer is caused by inhalation of
(*a*) paper dust (*b*) silica dust
(*c*) asbestos dust (*d*) textiles dust

330. Which is not a green house gas?
(*a*) O_2 (*b*) CFC
(*c*) CH_4 (*d*) CO_2

331. Pick out the wrong statement about photochemical smog (also called Los Angeles smog). Photochemical smog
(*a*) causes irritation in eyes
(*b*) is formed in summer during day time
(*c*) is reducing in nature
(*d*) formation involves NO, O_3 and hydrocarbons.

332. Pick out the wrong statement about London smog (also called sulphurous smog or classical smog). London smog
(*a*) is formed during winter in morning time.
(*b*) causes bronchitis.
(*c*) contains H_2SO_4 droplets.
(*d*) is oxidising in nature.

333. Which of the following is present in maximum amount in acid rain?
(*a*) H_2SO_3 (*b*) H_2SO_4
(*c*) HCl (*d*) HNO_3

334. Which of the following has affinity for hemoglobin?
(*a*) CO (*b*) CO_2
(*c*) O_2 (*d*) NO the greatest

335. Pick out the wrong statement
(*a*) NO is more harmful than NO_2
(*b*) Mesosphere and thermosphere are collectively called ionosphere
(*c*) Gases involved in the formation of photo chemical smog arc NO and hydrocarbons
(*d*) Slowing down of formation of chlorophyll in presence of SO_2 is called chlorosis which causes loss of green color in leaves

336. Pick out the wrong statement.
(*a*) In troposphere, the fan in temperature with increasing altitude is due to fall of pressure and density of air.
(*b*) In mesosphere, the fall in temperature with increasing altitude is due very low absorption of ultra violet radiations by ozone.
(*c*) In stratosphere, the rise in temperature with increasing altitude is due to absorption of solar ultraviolet radiation by ozone.

Answers

323. (*b*) 324. (*c*) 325. (*b*) 326. (*d*) 327. (*c*) 328. (*b*) 329. (*d*) 330. (*a*) 331. (*c*)
332. (*d*) 333. (*b*) 334. (*d*) 335. (*a*) 336. (*d*)

(*d*) Ozone layer is being depleted because it is dissociated into O_2 and O-atoms by ultraviolet rays.

337. Which of the following oxide of Nitrogen is not common air pollutant?
(*a*) NO (*b*) NO_2 (*c*) N_2O (*d*) N_2O

338. Which region of the atmosphere is closset to the earth's surface?
(*a*) Thermosphere (*b*) Troposphere
(*c*) Stratosphere (*d*) Mesosphere

339. Which of the following is the most toxic particulates?
(*a*) Soot (*b*) Fly ash
(*c*) Smog
(*d*) Inoraganic compounds

340. Pick out the wrong statement
(*a*) SO_2 is a primary pollutant.
(*b*) Marable acts as a sink for acidic pollutant.
(*c*) NO_2 causes maximum air pollution.
(*d*) Ozone absorbs the solar radiation of wavelength 200-300 nm.

341. Global warming is mostly caused by the emission of
(*a*) SO_2 (*b*) CO_2 (*c*) CO (*d*) CH_4

342. The ozone depletion process is due to
(*a*) ultra violet light breaking the ozone molecules.
(*b*) carbon dioxide.
(*c*) chlorine atoms destroying ozone molecules.
(*d*) carbon monoxide.

343. Which of the following is a green house gas?
(*a*) Carbon monoxide
(*b*) Nitrogen (*c*) Nitrous oxide
(*d*) Sulphur dioxide

344. Which of the following gases has the least impact on global warming?
(*a*) Nitrous oxide (*b*) Carbon monoxide
(*c*) Methane (*d*) Carbon dioxide

345. Air layer from 11 kms to 20 kms above earth's surface is called stratosphere. The temperature in stratosphere is approximately constant at ______ °C.
(*a*) – 56 (*b*) + 56 (*c*) 0 (*d*) 600

346. Which of the following is absent in acid rain?
(*a*) Sulphurous acid. (*b*) Sulphuric acid.
(*c*) Acetic acid. (*d*) Nitric acid.

347. Pick out the wrong statement about the effect of green house gases.
(*a*) Major gas responsible for green house effect is carbon dioxide.
(*b*) Methane and ozone also contribute to the green house effect.
(*c*) Green house effect result in lowering of sea levels over a period of time.
(*d*) Green house effect result in global warming.

348. Which of the following does not cause pollution?
(*a*) Nuclear power plant.
(*b*) Hydroelectric power plant.
(*c*) CNG powered automobiles.
(*d*) Underground coal mining.

349. Which of the following is the most harmful for life on the globe?
(*a*) Soil erosion. (*b*) Deforestation.
(*c*) Increasing desert. (*d*) Nuclear fallout.

350. Organic wastes which can be readily decomposed by micro-organisms are called biodegradable. Pick out the biodegradable pollutant out of the following:
(*a*) Pesticides (*b*) Mercuric salts
(*c*) Domestic wastes (*d*) Lead compounds

351. ______ docs not fall into the category of chemical pollutant.
(*a*) Solid wastes (*b*) Liquid wastes
(*c*) Pesticides (*d*) Noise

352. Which of the following is not a component of biosphere?

Answers

337. (*c*) 338. (*b*) 339. (*a*) 340. (*c*) 341. (*b*) 342. (*c*) 343. (*c*) 344. (*b*) 345. (*a*)
346. (*c*) 347. (*c*) 348. (*b*) 349. (*d*) 350. (*c*) 351. (*d*) 352. (*a*)

(*a*) Thermosphere (*b*) Atmosphere
(*c*) Lithosphere (*d*) Hydrosphere

353. Water pollution is not caused by
(*a*) oxygen demanding wastes.
(*b*) automobile exhaust.
(*c*) plant nutrients.
(*d*) disease causing agents.

354. Air pollution is not caused by
(*a*) ozone (*b*) nitric oxide
(*c*) hydrogen (*d*) hydrogen sulphide

355. Pick out the wrong statement.
(*a*) Acid rain is a primary pollutant.
(*b*) Pollutants emitted directly from the source are termed as primary pollutants.
(*c*) Pollutants formed in the atmosphere by chemical interaction among primary pollutants and atmospheric constituents are termed as secondary pollutants.
(*d*) Particulates refer to all atmospheric substances, which are not gases.

356. Predominantly very large cases of lung diseases in urban areas as compared to rural areas is attributed to the presence of ________ in the atmosphere.
(*a*) CO (*b*) CO_2 (*c*) SO_2 (*d*) N_2

357. Bone disease, fluorosis, is caused by the presence of ________ in drinking water.
(*a*) fluorides (*b*) pesticides
(*c*) phenol (*d*) cyanide

358. Reaction with ultraviolet light is responsible for the production of
(*a*) smog (*b*) ozone
(*c*) nitrogen dioxide (*d*) sulphur trioxide

359. Photochemical oxidant, PAN (peroxy acyl-nitrates) is formed by the action of ________ on hydrocarbons in the presence of sunlight.
(*a*) oxides of nitrogen
(*b*) carbon dioxide (*c*) hydrogen sulphide
(*d*) sulphur dioxide

360. Which of the following air pollutants is absent in vehicular emissions?
(*a*) Hydrocarbons (*b*) Sulphur dioxide
(*c*) Carbon monoxide (*d*) Flyash

361. If BOD (biological oxygen demand) of river water is high, then it implies that the river is
(*a*) polluted with pesticides.
(*b*) not at an polluted.
(*c*) highly polluted with inoraganic chemicals.
(*d*) highly polluted with biodegradable organic chemicals.

362. Earth is protected from harmful effects of ultraviolet radiations by
(*a*) carbon dioxide layer
(*b*) ozone layer (*c*) troposphere
(*d*) oxygen layer

363. The pollutants released by jet aeroplanes in the atmosphere as fluorocarbons are called
(*a*) aerosols
(*b*) photochemical oxidants
(*c*) photochemical reductants
(*d*) physical pollutants

364. pH value of acid rain lies below
(*a*) 6.8 (*b*) 5.6 (*c*) 1.0 (*d*) 2.8

365. The part of atmosphere, where temperature increase with increasing attitude is
(*a*) mesosphere (*b*) stratosphere
(*c*) thermosphere (*d*) both (*b*) and (*c*)

366. The oxide of nitrogen not present in atmosphere is
(*a*) NO (*b*) NO_2 (*c*) N_2O (*d*) N_2O_5

367. Which of the following pollutants present in atmosphere plays no role in the formation of photochemical smog?
(*a*) Hydrocarbons (*b*) Nitrogen dioxide
(*c*) Carbon monoxide (*d*) None of these

368. Pick out the correct statement.
(*a*) Conventional sources of energy cause less pollution as compared to non-conventional sources of energy.

Answers

353. (*b*) 354. (*c*) 355. (*a*) 356. (*c*) 357. (*a*) 358. (*b*) 359. (*a*) 360. (*d*) 361. (*d*)
362. (*b*) 363. (*a*) 364. (*b*) 365. (*d*) 366. (*d*) 367. (*c*) 368. (*d*)

(*b*) Oxides of nitrogen involved in the atmospheric pollution is represented by NO_x, where 'X' varies from 1 to 4.
(*c*) CFC, which is a common abbreviation for Freons, cause serious type of water pollution.
(*d*) Ocean and green plants act as the sink for carbon dioxide in atmosphere.

369. India ratified the Kyoto Protocol in the year
(*a*) 1998 (*b*) 2000 (*c*) 2002 (*d*) 2001

370. The main constituent of green house gases (GHG) are
(*a*) SO_x (*b*) NO_x
(*c*) CO (*d*) CO_2 and CH_4

371. Greenhouse gases (GHG) approximately make up about percent of the atmosphere.
(*a*) 1 (*b*) 3 (*c*) 5 (*d*) 10

372. Which of the following gas damages stratospheric ozone layer?
(*a*) Ar (*b*) CFC's (*c*) N_2 (*d*) CO_2

373. The ozone layer in the stratosphere acts as an efficient filter for
(*a*) solar UV-B rays. (*b*) X-rays.
(*c*) gamma rays. (*d*) solar UA-A rays.

374. The world average per capita CO_2 emission is ______ tons/year.
(*a*) 0.87 (*b*) 2 (*c*) 3.87 (*d*) 1.53

375. Which of the following is a secondary air pollutant?
(*a*) Pollen grains (*b*) Flyash
(*c*) Sulphur dioxide (*d*) Ozone

376. Which of the following is primary air pollutant?
(*a*) Particulates
(*b*) Photochemical smog
(*c*) Peroxy acetyl nitrate (PAN)
(*d*) Formaldehyde

377. Which of the following air pollutants is not responsible for the formation of Photochemical smog?
(*a*) NO_x (*b*) CO and HC
(*c*) O_3 (*d*) SO_2

378. If pH value of rain water is less than ______, it is termed as acid rain.
(*a*) 7 (*b*) 6.5 (*c*) 6 (*d*) 5.6

379. Concentration of ozone in stratopheric ozone layer is about ______ ppm as compared to about 0.05 ppm in the troposphere.
(*a*) 1 (*b*) 5 (*c*) 10 (*d*) 50

380. Combustion of fossil fuels emits pollutants like SO_x, CO, NO_x, particulates etc., in the atmosphere. Which of the following is not a fossil fuel?
(*a*) Coal (*b*) Fuel oil
(*c*) Natural gas (*d*) Wood

381. Which of the following is not a flexible mechanism under Kyoto Protocol?
(*a*) Joint implementation
(*b*) Emission trading
(*c*) Clean development mechanism (CDM)
(*d*) Sectoral approaches

382. Ozone layer is found in stratosphere between ______ km above the ground level.
(*a*) 10-50 (*b*) 20-50
(*c*) 50-100 (*d*) 250

383. The global warming potential (GWP) of nitrous oxide (N_2O) is (for CO_2, it is assumed as 1)
(*a*) 21 (*b*) 270
(*c*) 13700 (*d*) 25200

384. According to Kyoto Protocols, Annex II parties
(*a*) are developing countries.
(*b*) have some obligation as of Annex I countries.
(*c*) arc also annex I countries but have a special obligation to provide new and additional financial sources to developing countries.
(*d*) none of these.

Answers

369. (*c*) 370. (*d*) 371. (*a*) 372. (*c*) 373. (*a*) 374. (*c*) 375. (*d*) 376. (*a*) 377. (*d*)
378. (*d*) 379. (*c*) 380. (*d*) 381. (*d*) 382. (*b*) 383. (*b*) 384. (*c*)

385. Reduction of which of the following does not qualify for CDM (clean development mechanism)?
(*a*) CO_2 (*b*) SF_6 (*c*) SO_2
(*d*) PFC (Perflurocarbons)

386. Pick out the wrong statement pertaining to green house gases (GHG).
(*a*) Carbon dioxide CO_2 is responsible for 60% of the enhanced green house effect.
(*b*) The escape of slow moving re-emitted infrared radiation from the earth is delayed by green house gases.
(*c*) Green house gases make only 10% of the atmosphere.
(*d*) Nitrous oxide (N_2O) and methane (CH_4) are green house gases.

387. Pick out the wrong statement pertaining to nitrogen oxides (NO_x).
(*a*) It combines with water vapor to form nitric acid
(*b*) It reacts with hydrocarbons to form ozone in the presence of sunlight.
(*c*) It is formed during low temperature combustion process from the oxidation of nitrogen in the air and/or fuel.
(*d*) It irritates the respiratory system and may worsen asthma.

388. Global warming potential of methane is
(*a*) 21 (*b*) 270
(*c*) 6400 (*d*) 13700

389. One of the flexible instrument as stated in the Kyoto Protocol is
(*a*) clean development mechanism (CDM)
(*b*) CO_2 sequestration mechanism
(*c*) CO_2 adaptation
(*d*) none of these

390. The prototype carbon fund has been established by
(*a*) World bank (*b*) Japan
(*c*) U.S.A (*d*) United nations

391. Pick out the wrong statement about carbon monoxide (CO).
(*a*) CO is emitted into atmosphere from combustion processes and from oxidation of hydrocarbon and other organic compounds.
(*b*) In urban areas, CO is produced almost entirely (90%) from road traffic emission.
(*c*) CO survives in the atmosphere for a period of approximately 1 year and finally gets oxidised to CO_2.
(*d*) High concentration of CO in ambient air reduces the oxygen carrying capacity of blood.

392. Green house effect refers to increase in
(*a*) atmosphere pressure.
(*b*) greenery.
(*c*) global temperature.
(*d*) carbon monoxide.

393. Which is not a green house gas?
(*a*) SO_2 (*b*) SF_6 (*c*) N_2O (*d*) CH_4

394. Which of the following docs not damage the ozone layer?
(*a*) GFC (*b*) HCFC
(*c*) HFC (*d*) CCl_3

395. Green house gas (GHG) emission are represented in ________ units.
(*a*) million metric tons of oil equivalent (MMTOE)
(*b*) million metric tons of carbon equivalent (MMTGE)
(*c*) million metric tons of metal equivalent (MMTME)
(*d*) million metric tons (MMT)

396. Which one of the following is not considered as a pollutant?
(*a*) CO (*b*) CO_2 (*c*) SO_3
(*d*) VOC's (volatile organic chemicals)

397. The major contribution (upto 90%) of CO pollution in urban area is due to

Answers

385. (*c*) 386. (*c*) 387. (*c*) 388. (*a*) 389. (*a*) 390. (*a*) 391. (*c*) 392. (*c*) 393. (*a*)
394. (*c*) 395. (*b*) 396. (*b*) 397. (*a*)

(*a*) road transport. (*b*) waste incineration.
(*c*) industries.
(*d*) fossil fuel combustion.

398. Size of the market for CO_2 emission reduction in the world is one _______ ton of carbon reduction by/during the commitment period 2008-2012.
(*a*) thousand (*b*) million
(*c*) billion (*d*) trillion

399. Which of the following is not a fast track CDM (clean development mechanism) project?
(*a*) Biomass power plant upto 15 MW
(*b*) Photovoltaic power plant upto 15 MW
(*c*) Plantation project up to 15 kilo ton of CO_2 equivalent reduction annually
(*d*) Energy efficiency improvement projects upto 15 MW power reduction annually

400. Projected temperature increase by the year 2100 due to climate change is _______ °C.
(*a*) 2 (*b*) 4 (*c*) 6 (*d*) 8

401. Which country emits maximum CO_2?
(*a*) China (*b*) U.S.A
(*c*) Russia (*d*) Australia

402. The predicted rise in mean sea level (MSL) due to global warming by the year 2100 is _______ cms.
(*a*) 5-6 (*b*) 9-88 (*c*) 100 (*d*) 200

403. COP stands for
(*a*) co-efficient of pollution.
(*b*) conference of parties,
(*c*) coalition of parties,
(*d*) convection of people.

404. The global agency responsible to look after the climate change and [or actions to cut green house gas (GHG) emission is
(*a*) UNFCCC (*b*) WHO
(*c*) USEPA (*d*) CPCB

405. Which of the following compounds can easily break the ozone molecules in stratosphere?
(*a*) Bromine (*b*) Chlorine
(*c*) both (*a*) and (*b*) (*d*) neither (*a*) nor (*b*)

406. The main source of CFC (chloroflurocarbons) in household sector is
(*a*) refrigerators. (*b*) televisions.
(*c*) washing machines
(*d*) none of these.

407. HCFC's (Hydrochloro fluorocarbons) have _______ per cent ozone depletion potential as compared to CFC's.
(*a*) 0.1-1 (*b*) 2-10
(*c*) 10-20 (*d*) 100-150

408. Which of the following refrigerants is completely chlorine free?
(*a*) CFCs (*b*) HCFCs
(*c*) HFCs (*d*) R_{22}

409. Match the residence time of various air pollutants in the atmosphere.

	List I		List II
A.	CO, CO_2 and hydro-carbons (non-methane)	I.	10 years
B.	NO_x and SO_x	II.	200 years
C.	Chloroflurocarbolls (CFC)	III.	4-5 year
D.	CH_4	IV.	3 years

(*a*) I-A, II-B, III-C, IV-D
(*b*) I-D, II-C, III-B, IV-A
(*c*) I-A, II-13, III-C, IV-D
(*d*) I-A, II-B, III-D, IV-C

410. Pick out the correct statement about green house gases.
(*a*) They constitute about 10% of the atmosphere.
(*b*) CO_2 alone is responsible for about 55% of the enhanced green house effect.
(*c*) CH_4 and N_2O are not green house gases.
(*d*) Escape of slow moving re-emitted infrared radiation from the earth is accelerated by the green house gases.

Answers

398. (*c*) 399. (*d*) 400. (*a*) 401. (*b*) 402. (*b*) 403. (*b*) 404. (*a*) 405. (*c*) 406. (*a*)
407. (*b*) 408. (*c*) 409. (*b*) 410. (*b*)

411. Stratospheric ozone depletion process is due to

(*a*) ultraviolet light rays breaking the ozone molecule.

(*b*) carbon dioxide.

(*c*) CFC molecules destroying the ozone molecules.

(*d*) chlorine atoms destroying the ozone molecules.

412. The commitment period for CO_2/green house gases emissions reductions of industrialised countries as per Kyoto Protocol is

(*a*) 2004-2008 (*b*) 2008-2012

(*c*) 2012-2020 (*d*) 2025-2030

413. India comes under which of the following categories as per Kyoto Protocol?

(*a*) Annex I parties. (*b*) Annex II parties.

(*c*) Annex I and II parties.

(*d*) Non- annex parties.

414. In CDM (Glean development mechanism) terminology, CER means

(*a*) clean environment rating.

(*b*) carbon emission reduction.

(*c*) certified emission reduction.

(*d*) carbon emission rating.

415. The zone of earth, where alone life exists, is called

(*a*) hydrosphere (*b*) biosphere

(*c*) lithosphere (*d*) troposphere

416. The average daily consumption rate of respiratory air by an adult human being is about ______ kg/day.

(*a*) 6 (*b*) 12 (*c*) 24 (*d*) 36

417. Which of the following is not an anthropogenic source/cause of air pollution?

(*a*) Agriculture activities.

(*b*) Burning of firewood.

(*c*) Burning of forests due to lightening.

(*d*) Burning of fossil fuels.

418. A major environmental disaster known as "London smog" (caused primarily by burning of coal) occurred in London (U.K) in the year

(*a*) 1912 (*b*) 1952 (*c*) 1964 (*d*) 1978

419. Most major source of all pollution in modern days is

(*a*) volcanic eruptions.

(*b*) forest fires. (*c*) dust storms.

(*d*) automobiles and industries.

420. About 90%of the world's air pollution is caused by primary air pollutants. Which of the following is a secondary air pollutant?

(*a*) Hydrocarbons (*b*) Sulfur dioxide

(*c*) Ozone (*d*) None of these

421. Which of the following is not a secondary air pollutant?

(*a*) Formaldehyde

(*b*) Peroxyacetyl nitrate

(*c*) Sulfuric acid (*d*) None of these

422. Which of the following is not a primary air pollutant?

(*a*) NO_x (oxides of nitrogen).

(*b*) VOC's (volatile organic compounds).

(*c*) SPM (suspended particulate matter).

(*d*) PAN (peroxyacyl nitrate).

423. The poisonous gas responsible for causing the catastrophic Bhopal gas tragedy in 1984 was

(*a*) laughing gas (N_2O).

(*b*) methyl isocynate (MIC).

(*c*) phosphine. (*d*) phosgene.

424. Photochemical smog formation is caused by reaction of hydrocarbon and oxides of nitrogen (NO_x) in presence of sunlight. Which of the following is not a measure constituent of photochemical smog?

(*a*) Ozone (*b*) Carbon monoxide

(*c*) Hydrocarbons (*d*) PAN

425. The particulates present in cigarette smoke are usually smaller than 1 micron. Hair present

Answers

411. (*d*) 412. (*d*) 413. (*b*) 414. (*c*) 415. (*b*) 416. (*b*) 417. (*c*) 418. (*b*) 419. (*d*)
420. (*c*) 421. (*d*) 422. (*d*) 423. (*b*) 424. (*b*) 425. (*c*)

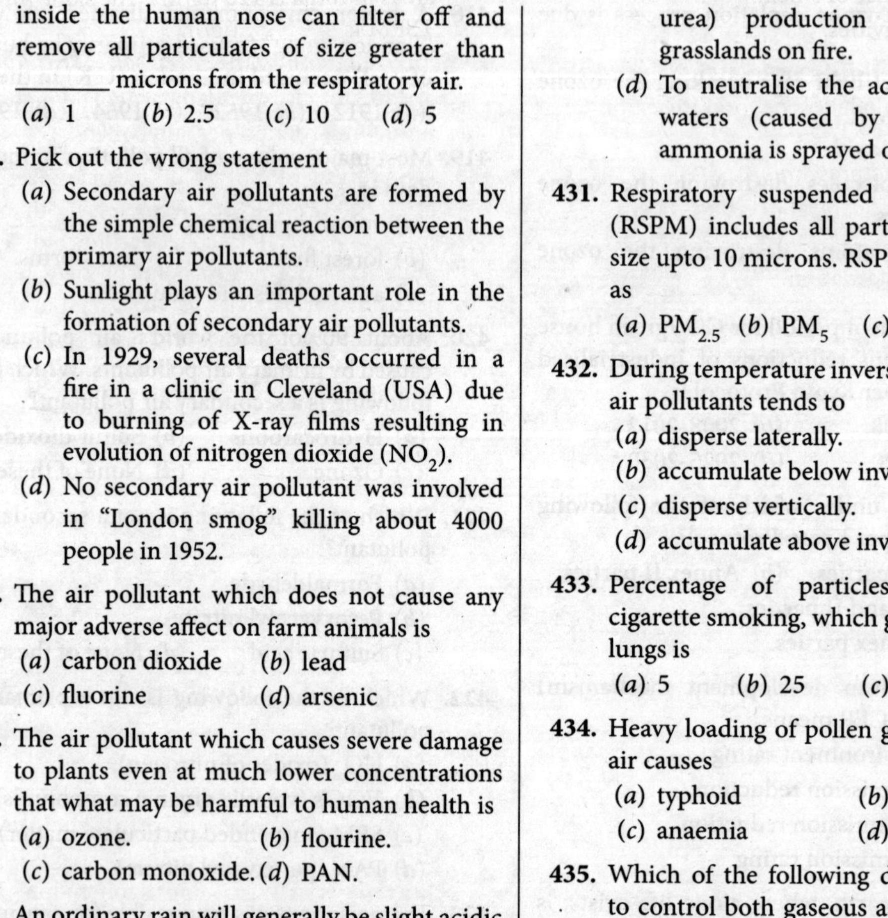

inside the human nose can filter off and remove all particulates of size greater than ______ microns from the respiratory air.

(*a*) 1 (*b*) 2.5 (*c*) 10 (*d*) 5

426. Pick out the wrong statement

(*a*) Secondary air pollutants are formed by the simple chemical reaction between the primary air pollutants.

(*b*) Sunlight plays an important role in the formation of secondary air pollutants.

(*c*) In 1929, several deaths occurred in a fire in a clinic in Cleveland (USA) due to burning of X-ray films resulting in evolution of nitrogen dioxide (NO_2).

(*d*) No secondary air pollutant was involved in "London smog" killing about 4000 people in 1952.

427. The air pollutant which does not cause any major adverse affect on farm animals is

(*a*) carbon dioxide (*b*) lead

(*c*) fluorine (*d*) arsenic

428. The air pollutant which causes severe damage to plants even at much lower concentrations that what may be harmful to human health is

(*a*) ozone. (*b*) flourine.

(*c*) carbon monoxide. (*d*) PAN.

429. An ordinary rain will generally be slight acidic in nature. The important gaseous pollutants contributing to the formation of acid rain are

(*a*) NO_x and SO_x (*b*) NO_x and O_3

(*c*) SO_x and CO_2 (*d*) CO_2 and H_2S

430. Pick out the wrong statement.

(*a*) Two third of acid rains are caused by SO_2 gas which is mainly produced by the burning of coal and oil.

(*b*) A rainfall is generally classified as acidic, if its pH value is less than 5.6.

(*c*) One third of acid rains is caused by nitrogen oxides (NO_x) which are mainly produced by automobiles and industrial emissions, nitrogenous fortiliser (e.g., urea) production and forests and grasslands on fire.

(*d*) To neutralise the acidic nature of lake waters (caused by acid rain) liquid ammonia is sprayed on lake water.

431. Respiratory suspended particulate matter (RSPM) includes all particles in air mass of size upto 10 microns. RSPM is also designated as

(*a*) $PM_{2.5}$ (*b*) PM_5 (*c*) PM_{10} (*d*) PM_{25}

432. During temperature inversions in atmosphere, air pollutants tends to

(*a*) disperse laterally.

(*b*) accumulate below inversions layer.

(*c*) disperse vertically.

(*d*) accumulate above inversions layer.

433. Percentage of particles eminating from cigarette smoking, which gets deposited in the lungs is

(*a*) 5 (*b*) 25 (*c*) 55 (*d*) 75

434. Heavy loading of pollen grains in respiratory air causes

(*a*) typhoid (*b*) influenza

(*c*) anaemia (*d*) hayfever

435. Which of the following devices can be used to control both gaseous as well as particulate pollutants in the industrial emissions?

(*a*) Cyclone separator

(*b*) Bag filter (*c*) Spray scrubber

(*d*) Electrostatic precipitator (dry type)

436. The most significant primary gaseous pollutant found in vehicular emissions, is

(*a*) SO_2 (*b*) NO_x (*c*) CO (*d*) O_3

437. Presence of excessive amount of carbon monoxide in atmospheric air is hazardous to human health, because it causes

(*a*) reduction in oxygen carrying capacity of blood.

(*b*) high blood pressure.

Answers

426. (*d*) 427. (*a*) 428. (*b*) 429. (*a*) 430. (*d*) 431. (*c*) 432. (*b*) 433. (*b*) 434. (*d*)
435. (*c*) 436. (*c*) 437. (*a*)

(*c*) loss of sense of smell.
(*d*) conjunctivities.

438. High lead exposures in human beings either through air inhalation or food ingestion does not cause
(*a*) high blood pressure.
(*b*) disorder of central nervous system.
(*c*) cancer.
(*d*) mental retardation.

439. Longer exposures of human beings to nitrogen dioxide (NO_2) to even in small concentration may cause ________ disease.
(*a*) heart (*b*) lung
(*c*) kidney (*d*) lever

440. Higher exposure of nitrogen oxides (NO_x) does not cause
(*a*) allergies and viral attacks
(*b*) pneumonia (*c*) bronchitis
(*d*) conjunctivities

441. Pick out the wrong statement about acid rains.
(*a*) Europe and north east America is the most affected region of the world by acid rain.
(*b*) Lime powder is sprayed on lake waters to neutralise the acidic nature of lake water caused by acid rain.
(*c*) Acid rains are more prominent in temperate regions.
(*d*) ESP (electrostatic precipitators) are installed in fossil fuel fired thermal power plants to check emission of SO_x and NO_x gases to reduce acid rains.

442. The permissible annual average ambient air quality standard for residential area in India is not equal to 60 μg/m³ in respect of the following pollutants (It is 140 μg/m³). Which is that pollutant?
(*a*) NO_2 (*b*) SO_2 (*c*) RSPM (*d*) TSPM

443. The permissible RSPM (annual average) ambient air quality standard for industrial areas in India is 120 μg/m³, The same limit for TSPM is ________ μg/m³.
(*a*) 140 (*b*) 210 (*c*) 360 (*d*) 120
TSPM: Total suspended particulate matter, μg/m³-microgram per cubic meter

444. Monitoring of RSPM of particle size < 10 microns (PM_{10}) or even lesser than 2.5 microns ($PM_{2.5}$) is a better indicator of air pollution from health point of view, because such smaller particles get deposited in the
(*a*) trachco bronchial and pulmonary region causing bronchitis and other lung disease.
(*b*) nasopharyngeal region causing serious health problems.
(*c*) both (*a*) and (*b*). (*d*) neither (*a*) nor (*b*).

445. The most significant environmental issue threatening human health is
(*a*) acid rain. (*b*) global warming.
(*c*) ozone layer depletion.
(*d*) noise pollution.

416. Excessive emission of CFC's (chlorofluro-carbons) in atmosphere causes depletion of ozone layer in the stratosphere (thereby reducing the thickness of ozone layer causing ozone hole) leading to increase in the ________ level in the atmosphere.
(*a*) radioactivity (*b*) temperature
(*c*) sound
(*d*) ultraviolet radiation

447. How many times more acidic an acid rain having, pH = 5.6 will be in comparison to a neutral rain having pH = 7 ?
(*a*) $\sqrt{7/56} = 1.11$ (*b*) $7^2 5.6^2 = 1.56$
(*c*) $7/5.6 = 1.25$ (*d*) $10^7/10^{5.6} = 20$

448. Ozone layer thickness is measured in Dobson units (DU). One DU is equal to ________ mm.
(*a*) 0.001 (*b*) 0.01 (*c*) 0.1 (*d*) 1

449. The normal average thickness of stratospheric ozone across the globe is ________ DU.
(*a*) 100 (*b*) 350 (*c*) 650 (*d*) 1000

Answers

438. (*c*) 439. (*b*) 440. (*d*) 441. (*d*) 442. (*d*) 443. (*c*) 444. (*c*) 445. (*c*) 446. (*d*)
447. (*d*) 448. (*b*) 449. (*b*)

450. Depletion of ozone layer in stratosphere is caused due to the reaction of ozone with

(*a*) SO_2 (*b*) Cl_2 (*c*) N_2O (*d*) CO_2

451. Reduction in thickness of ozone layer to less than 150 DU (1.5 mm) is termed as ozone hole, which is likely to increase the incidence of ______ in human beings.

(*a*) skin cancer
(*b*) respiratory disease (asthma)
(*c*) kidney failure (*d*) lung disease

452. Which of the following is not an effect of excessive noise on human beings?

(*a*) General annoyance and sleeplessness
(*b*) High blood pressure and gastric secretion
(*c*) Gradual hearing loss
(*d*) Asthamic attacks.

453. Maximum sound level beyond which it is regarded as an ambient pollutant is ______ decibels.

(*a*) 50 (*b*) 80 (*c*) 95 (*d*) 105

454. The taste threshold and odour threshold of SO_2 are ______ ppm respectively.

(*a*) 0.1 and 0.5 (*b*) 0.3 and 0.5
(*c*) 0.5 and 0.1 (*d*) 1 and 3

455. The average NO_2 concentration in tobacco smoke is about ______ ppm.

(*a*) 1 (*b*) 5 (*c*) 25 (*d*) 50

456. The study of earth's atmosphere and its changes is called "meteorology". The effect of earth's rotation on wind velocity and direction is termed as the ______ effect.

(*a*) gravitational (*b*) centrifugal
(*c*) corriolis (*d*) Vanderwaal's

457. The temperature gradient of ambient air is, change of ambient temperature with height is termed as the

(*a*) environmental lapse rate (ELR).
(*b*) adiabatic lapse rate (ALR).
(*c*) dry adiabatic lapse rate (DALR).
(*d*) super adiabatic lapse rate (SALR).

458. When the environmental lapse rate exceeds the adiabatic lapse rate, the ambient lapse rate is termed as the ______ adiabatic lapse rate.

(*a*) dry (*b*) sub
(*c*) super (*d*) saturated.

459. Match the cause of various environmental pollutant's effects.

	Causes		Effects
A.	CO	I.	acid rain
B.	CO_2	II.	explosion
C.	CH_4	III.	asphyxiation
D.	SO_2	IV.	green house effect

Codes:	A	B	C	D
(*a*)	II	III	I	IV
(*b*)	III	IV	II	I
(*c*)	I	III	IV	II
(*d*)	IV	II	I	III

460. The atmosphere extends up to a height of 10000 kMs. It is divided into 4 thermal layers namely:

I. mesosphere II. stratosphere
III. thermosphere IV. troposphere

The correct sequence of these layers starting from the surface of the earth upwards is

(*a*) II, IV, I, III (*b*) IV, II, I, III
(*c*) IV, II, III, I (*d*) II, IV, III, 1.

461. Which of the following solid waste disposal methods is ecologically most acceptable?

(*a*) Pyrolysis (*b*) Incineration
(*c*) Composting (*d*) Sanitary landfill.

462. Non disposal of solid waste may cause the spread of

(*a*) dysentery (*b*) malaria.
(*c*) typhoid.
(*d*) rodents related plague.

463. Which of the following air pollutants are responsible for the formation of photochemical smog?

Answers

450. (*b*) 451. (*a*) 452. (*d*) 453. (*b*) 454. (*b*) 455. (*b*) 456. (*c*) 457. (*a*) 458. (*c*)
459. (*b*) 460. (*b*) 461. (*c*) 462. (*d*) 463. (*d*)

I. NO_x II. O_3
III. unburant HC IV. CO
(*a*) I alone (*b*) II, III, and IV
(*c*) I,III, and IV (*d*) I, andIII.

464. Match the pollution control equipment with the pollutants removed:

	Equipment		Pollutant removed
A.	Electrostatic precipitators	I.	coarse particles
B.	Cyclone separator	II.	fine dust
C.	Wet scrubbers	III.	gas
D.	Absorbers	IV.	sulfur dioxide

Codes:	A	B	C	D
(*a*)	I	II	III	IV
(*b*)	II	I	III	IV
(*c*)	II	I	IV	III
(*d*)	I	II	IV	III

465. Aerosol is
(*a*) finely divided ash particles.
(*b*) diffused liquid particles.
(*c*) dispersion of small solid/liquid particles in gaseous media.
(*d*) carbon particles of microscopic size.

466. Which of the following methods can be employed for plastic and rubber waste disposal?
(*a*) Incineration (*b*) Composting
(*c*) Sanitary landfill (*d*) Pyrolysis

467. Match the environmental effects of various air pollutants.

	Air pollutants		Environmental effects
A.	Carbon monoxide	I.	respiratory problems for living beings
B.	Particulate matter	II.	chemical reaction with hemoglobin in the blood.
C.	Nitrogen oxides	III.	reduction in visibility and aero allergens.
D.	Sulfur dioxide.	IV.	photochemical smog in atmosphere.

Codes:	A	B	C	D
(*a*)	II	III	I	IV
(*b*)	III	II	IV	I
(*c*)	II	III	IV	I
(*d*)	III	II	I	IV

468. Match the features of various solid waste disposal methods

	Solid waste disposal methods		Features
A.	Incineration	I.	Requires presorting, grinding and turning.
B.	Sanitary landfill	II.	Limited to special wastes and selected materials.
C.	Composting	III.	High operational maintenance cost
D.	Salvage by sorting	IV.	Tractor
		V.	Rat and fly breeding

Codes:	A	B	C	D
(*a*)	II	V	IV	III
(*b*)	I	IV	II	III
(*c*)	III	IV	I	II
(*d*)	III	V	IV	II

469. Which of the following air pollutants is formed due to photo chemical reactions?
(*a*) O_3 and PAN (*b*) CO
(*c*) CO and NH_3 (*d*) NH_3 and PAN

470. Which of the fallowing plume behaviour occurs, when atmospheric inversion begins from the ground level and continues?
(*a*) Looping (*b*) Fuming
(*c*) Fanning (*d*) Fumigation

471. Organisms that mineralise organic matter in an ecosystem are called decomposers. Aerobic method of composting is called ______ method.
(*a*) Nagpur (*b*) Indore
(*c*) Delhi (*d*) Bangalore

Answers

464. (*b*) 465. (*c*) 466. (*a*) 467. (*c*) 468. (*c*) 469. (*a*) 470. (*c*) 471. (*b*)

472. The term 'refuse' generally does not include
(*a*) excreta
(*b*) putrescible (decomposable) solid waste
(*c*) ashes
(*d*) non-putrescible solid waste.

473. Which of the following environmental acts/rules has provision for "No right to appeal"?
(*a*) Environment (protection) Acts, 1986.
(*b*) Environment (protection) Rules, 1992.
(*c*) Manufacture, storage and import of hazardous chemicals rules, 1989.
(*d*) The hazardous waste (management and handling) rules, 1989.

474. Which type of light energy is effectively absorbed by atmospheric CO_2 in the lower boundary of the troposphere?
(*a*) Ultraviolet rays (*b*) Infrared rays
(*c*) X-rays (*d*) Visible light.

475. Which of the following types of noise can be controlled/abated by providing lining walls and ceiling with sound absorbing materials?
(*a*) Reflection noise. (*b*) Source noise.
(*c*) Structural noise.
(*d*) Direct air-borne noise.

476. Most major pollutant present in "photochemical smog" is
(*a*) NO_2 (*b*) SO_2 (*c*) PAN (*d*) HC

477. Match the plume behaviour/pattern observed with various atmospheric conditions.

Plume pattern/ behaviour		Atmospheric conditions	
A.	Looping	I.	Unstable
B.	Fumigation	II.	Inversion above and lapse below the stack
C.	Fanning	III.	Inversion
D.	Trapping	IV.	Inversion above and below the stack with lapse in between

Codes:	A	B	C	D
(*a*)	I	II	III	IV
(*b*)	IV	III	II	I
(*c*)	I	II	IV	III
(*d*)	IV	III	I	II

478. Which one of the following is the range of ozonosphere in the atmosphere?
(*a*) Stratosphere to mesosphere.
(*b*) Tropopause to stratopause.
(*c*) Tropopause to mesosphere.
(*d*) Troposphere to stratosphere.

479. The intensification of green house effect is attributed to the increased level of ________ in the atmosphere.
(*a*) CO (*b*) CO_2 (*c*) CFC's (*d*) SO_2

480. Ozone layer in the upper atmosphere is getting destroyed due to its reaction with
(*a*) CO_2 (*b*) H_2O_2
(*c*) SO_x and NO_x (*d*) CFC's

481. For rural areas, most suitable solid waste disposal method is
(*a*) land filling (*b*) composting
(*c*) pyrolysis (*d*) incineration

482. On mass basis, the largest fraction of the municipal solid waste (MSW) from a typical Indian city is made up of the ________ substances.
(*a*) salvagable (*b*) composting
(*c*) combustible (*d*) toxic and hazardous

483. The most popular method of solid disposal in India is
(*a*) pyrolysis (*b*) incineration
(*c*) landfill
(*d*) fluidisod bed combustion.

484. Global warming is caused mainly by ________ present in the atmosphere.
(*a*) CFC's (*b*) O_3
(*c*) CH_4 and CO_2 (*d*) NO_x

Answers

472. (*a*) 473. (*d*) 474. (*d*) 475. (*a*) 476. (*c*) 477. (*a*) 478. (*b*) 479. (*b*) 480. (*d*)
481. (*b*) 482. (*b*) 483. (*c*) 484. (*c*)

485. Major air pollutants from a coal fired thermal power station are fly ash and
(*a*) ozone. (*b*) hydrogen fluoride.
(*c*) hydrogen sulfide. (*d*) sulfur dioxide.

486. Biomedical waste (comprising of both solid and liquid wastes) generated in Indian hospitals is about 1.45 kg/day/patient. Which of the following methods is generally not adopted for safe disposal of biomedical wastes?
(*a*) Incineration. (*b*) Landfilling.
(*c*) Hydroclaying.
(*d*) Shredding after disinfection.

487. Pick out the wrong statement. Autoclaving of biomedical wastes is
(*a*) followed by its shredding off and suitable disposal.
(*b*) a high heat thermal process.
(*c*) meant for disposal of solid wastes only.
(*d*) a low heat thermal process.

488. The aerobic and anaerobic methods or mechanical composting for refuse; disposal as practiced in India, are caned ________ respectively.
(*a*) Indore method and Bangalore method
(*b*) Bangalore method and Indore method
(*c*) Mangalore method and Nagpurmethod
(*d*) Indore method and Mangalore method

489. The method of' refuse disposal burial in trenches is called
(*a*) composting (*b*) incineration
(*c*) pulverisation (*d*) pyrolysis

490. Sanitary landfills may cause troubles during peak monsoons. Leachate is coloured liquid that comes out of
(*a*) septic tanks. (*b*) compost plants.
(*c*) sanitary landfills. (*d*) aerated lagoons.

491. Pick out the wrong statement.
(*a*) Incineration can be adopted for disposal of both solid and liquid biomedical wastes.
(*b*) Liquid biomedical wastes can be disposed off by disinfection and discharge into sewers.
(*c*) According to biomedical waste management and handling rules various biomedical waste types are segregated by collecting the wastes into yellow, red and black colored containers.
(*d*) Biomedical waste (management and handling rules) were enacted in India under Environment Protection Act, 1986 in the year 1986.

492. Match the effects of number of rems of radiation exposure (radiation dose) on human beings

	Effects on human beings		Magnitude of radiation exposure, rem
A.	Produces radiation sickness.	I.	100
B.	Hair loss and loss of body's inability to fight infections, usually resulting in death.	II.	300
C.	Severe changes in blood cells and haemorrhage.	III.	600

(*a*) I-A, II-C, III-B (*b*) I-A, II-B, III-C
(*c*) I-C, II-B, III-A (*d*) I-B, II-A, III-C

493. Which of the following is not a natural source of radioactive radiations in the environment?
(*a*) Radioactive rocks and minerals.
(*b*) Nuclear power plants and nuclear explosions.
(*c*) Radioactive nuclides present in the bodies of organisms.
(*d*) Cosmic rays.

494. Pick out the wrong statement about ionising radiation.
(*a*) That dose of ionising radiation that imparts 1 Joule of energy to 1 kg of radiation absorbing material is termed as "rad".

Answers

485. (*d*) 486. (*b*) 487. (*b*) 488. (*a*) 489. (*a*) 490. (*c*) 491. (*d*) 492. (*a*) 493. (*b*)
494. (*d*)

(*b*) Fish is the most badly affected by radioactive radiation among aquatic animals.

(*c*) Radioactive radiations may cause mutagenic effects in both humans and animals.

(*d*) Trees/plants are very badly affected by longer exposure to α and β-rays.

495. Pick out the wrong statement about RBE (relative biological effectiveness) pertaining to nuclear radiation pollution.

(*a*) The cell damaging capacity of different types of radioactive radiations is compared by relative biological effectiveness (RBE).

(*b*) RBE for α-particles is 20 while that for electrons is 1.

(*c*) Statement in (*b*) implies that the dense track of ions formed by α -particles causes 20 times as much damage to biological body as is caused by a thin ion frail generated by electrons.

(*d*) None of these.

496. Pick out the correct relationship pertaining to measurement of nuclear radiation.

(*a*) 1 rem = 1/100 sievert.

(*b*) 1 rad = 1/100 gray.

(*c*) rem = rad*RBE.

(*d*) 1 rem = 100 sievert.

497. _______ is not having a permanent site for disposal of high level radioactive wastes generated in nuclear power plants

(*a*) India (*b*) Russia

(*c*) U.S.A (*d*) Canada.

498. Spent fuels generated in nuclear power plants contain plutonium-239 (a long life hazardous radioactive substance), which has a very long half life period of the order of _______ years.

(*a*) 1000 (*b*) 5000

(*c*) 10000 (*d*) 24000

499. Air changes for ventilation of occupied rooms/ halls in a building is mainly required to

(*a*) reduce body odours to control indoor air pollution.

(*b*) dissipate body heat.

(*c*) keep CO_2 content under control.

(*d*) increase oxygen content.

500. Fresh air is required to be supplied to an occupied hall at the rate of _______ m^3/hr.

(*a*) 1-5 (*b*) 15-30

(*c*) 75-100 (*d*) 300-500

501. Match the number of air changes required per hour to ensure good ventilation of various occupied halls/rooms in a building.

	Occupied room/halls		No. of air changes required/hr
A.	Cinema halls	I.	3
B.	Offices and libraries	II.	10-40
C.	Kitchens	III.	6-10

(*a*) A-I, B-II, C-III (*b*) A-III, B-II, C-I

(*c*) A-III, B- I, C-II (*d*) A -I, B-III, C-II

502. More than 4000 people died in December 1952 in London due to 'London smog', which was caused by _______ in the city.

(*a*) evolution of smoke containing SO_2

(*b*) arrival of anticyclone

(*c*) presence of fog (*d*) all (*a*), (*b*), and (*c*).

503. The ozone depletion in the stratosphere is mainly caused by the release of chloronurocarbons (CFC's), which are released into the atmosphere by humans, as they are largely used

(*a*) in refrigeration and air-conditioning equipment.

(*b*) in aerosol sprays and as cleaning solvent in factories.

(*c*) as blowing agents to produce certain kinds of plastic foams.

(*d*) all (*a*), (*b*) and (*c*)

504. Pick out the wrong statement.

(*a*) Harmful ultra-violet (UV) radiations eminating from the sun are prevented

Answers

495. (*d*) 496. (*d*) 497. (*a*) 498. (*d*) 499. (*b*) 500. (*b*) 501. (*c*) 502. (*d*) 503. (*d*)
504. (*a*)

from reaching the earth by the presence of ozone in the troposphere.

(*b*) Presence of chlorides, bromides, solar radiation and thin ice crystals in the stratosphere is crucial for the formation of anthropogenic (man-made) ozone hole.

(*c*) The chances of development of on ozone hole are more at arctic and Antarctic regions.

(*d*) The ozone hole appears in Antarctica during late winter.

505. The possible substitutes of CFC's are

(*a*) Halons (*b*) HCF's
(*c*) ODS's (*d*) None of these.

506. High altitude flying air-crafts lead to

(*a*) formation of ozone.
(*b*) destruction of ozone.
(*c*) destruction of oxygen.
(*d*) increased ozone layer thickness.

507. World ozone day is celebrated on 16th September. Ozone strongly absorbs UV light radiation in the region having wavelength in the range of _______ nm (nanometer).

(*a*) 20-60 (*b*) 80-125
(*c*) 220-330 (*d*) 660-820

508. Increased CO_2 concentration in earth's atmosphere leads to warming of the earth due to green house effect. Had there been no or little CO_2 in the earth's atmosphere, the temperature on earth would have been below the freezing point as that existing on

(*a*) Mars (*b*) Moon
(*c*) Venus (*d*) None of these

509. 80% the total CO_2 annually emitted into atmosphere comes from

(*a*) combustion of fossil fuels (coal, oil and gas).
(*b*) landfills.
(*c*) methane release from termites and cattle sheep
(*d*) rotting vegetation.

510. Pick out the wrong statement.

(*a*) Green house effects are produced due to the absorption of long range infrared radiation in the atmosphere.

(*b*) Had there been too much CO_2 in the earth's atmosphere, the temperature on the earth would have been much higher to make life impossible as that existing on Venus.

(*c*) The international treaty which aims at reducing global warming is 'Montreal Protocol'.

(*d*) Carbon emission trading and clean development mechanism (CDM) are an inbuilt mechanism of an international treaty for controlling global warming

511. USA release about _______ percent of the total green house gases (GHG's) emission into the world's atmosphere.

(*a*) 5 (*b*) 10 (*c*) 30 (*d*) 50

512. Present (year 2010) level of CO_2 in the atmosphere is about 390 ppm (by volume), which has during the last 50 years (compared to year 1960) increased by about _______ per cent.

(*a*) 5 (*b*) 10 (*c*) 20 (*d*) 50

513. Water level rise in oceans due to global warming result from melting of glacier and thermal expansion of water volume contained in the ocean. What percentage of total land area is likely to be submerged, if all the world's ice caps and glaciers melt due to global warming?

(*a*) 1 (*b*) 5 (*c*) 10 (*d*) 20

514. Pick out the wrong statement.

(*a*) Global warming due to green house effect is not likely to cause severe climate changes

(*b*) Earth's temperature is likely to increase by 2°C during 21st century due to increasing concentration of GHG's at its present rate.

Answers

505. (*b*) 506. (*b*) 507. (*c*) 508. (*a*) 509. (*a*) 510. (*c*) 511. (*c*) 512. (*c*) 513. (*d*)
514. (*a*)

(*c*) Global warming may spoil the ground water quality by the ingress of salt sea water towards the presently habited land areas.

(*d*) Ozone hole formation means reduction of stratospheric ozone layer. Thickness reduction from 350 DU (35 mm) to ≤ 150 DU (1.5 mm).

515. Stoneware pipes arc generally not used for sewer mains, because they are.

(*a*) less resistant to organic corrosion.

(*b*) hydraulically less efficient.

(*c*) weak in tension.

(*d*) weak in compression.

516. Which of the following gas is generally not present in sewers?

(*a*) H_2S (*b*) CH_4 (*c*) CO_2 (*d*) CO

517. When a sheet of paper moistened with lead acetate, held for 5 minutes in a manhole, turns black; then the sewer certainly contains

(*a*) H_2S (*b*) CO_2 (*c*) CH_4 (*d*) CO

518. Pick out the wrong statement.

(*a*) Sewage contain about 99.9% water and its specific gravity is slightly more than 1.

(*b*) Temperature variations affect the viscosity, gas solubility and biological activity of bacteria in sewage.

(*c*) pH value of fresh sewage is usually more than 7.

(*d*) Well oxidised sewage contains nitrogen in the form of nitrites and sulfur in the arm of sulphites,

519. Which of the following gases is generally not evolved during decomposition of sewage?

(*a*) H_2S (*b*) CO_2 (*c*) CH_4 (*d*) NH_3

520. Solubility of oxygen in sewage when compared to its solubility in distilled water is about ________ percent.

(*a*) 50 (*b*) 60 (*c*) 75 (*d*) 90

521. Minimum dissolved oxygen (D.O) content prescribed for a river stream to avoid fish kill is ________ ppm.

(*a*) 0.1 (*b*) 1 (*c*) 2 (*d*) 4

522. Pick out the wrong statement.

(*a*) Pathogenic (disease causing) bacteria enters waste waters primarily from domestic wastes.

(*b*) Biodegradable organic enters waste waters from both domestic as well as industrial wastes.

(*c*) Soluble organics in domestic waste waters include carbohydrates, proteins and lipids.

(*d*) Biological oxygen demand (BOD) of waste water is the oxygen required to oxidise the biologically inactive organic matter.

523. Most of the bacteria in sewage are

(*a*) anaerobic (*b*) parasitic

(*c*) saprophytic (*d*) pathogenic

524. As compared to fresh river water, sea water contains ________ oxygen.

(*a*) 5% more (*b*) 5% less

(*c*) 20% more (*d*) 20% less

525. Dissolved oxygen content in river/water streams is

(*a*) minimum at noon.

(*b*) maximum at noon.

(*c*) maximum at midnight.

(*d*) same throughout the day.

526. The natural process under which flowing river water gets cleaned is known as 'self purification', Dissolved oxygen content may fall to zero. causing anaerobic bio-condition in a river reach called zone of

(*a*) recovery (*b*) degradation

(*c*) active decomposition

(*d*) none of these

Answers

515. (*c*) 516. (*d*) 517. (*a*) 518. (*d*) 519. (*b*) 520. (*d*) 521. (*d*) 522. (*d*) 523. (*c*)
524. (*d*) 525. (*b*) 526. (*c*)

527. BOD_5 represents 5-days bio-chemical oxygen demand of sewage at a temperature of 20°C; Standard 5 days BOD at 20°C when compared to ultimate BOD, is about _______ per cent.
(*a*) 21 (*b*) 38 (*c*) 68 (*d*) 98

528. Pick out the wrong statement
(*a*) Microbial utilisation of organics in sewage is first order chemical reaction.
(*b*) While testing for COD (chemical oxygen demand) of sewage, organic matter is oxidised by $K_2Cr_2O_7$ in presence of H_2SO_4.
(*c*) Between DOD and COD of sewage, the greater value of the two is BOD.
(*d*) The average BOD_5 of domestic sewage is about 0.08 kg/day/person.

529. Sewage disposal causes formation of sludge banks in sea waters. The most common method of waste water disposal is its
(*a*) utilisation for irrigational purposes.
(*b*) rapid infiltration.
(*c*) dilution in surface water.
(*d*) natural evaporation.

530. Complete through treatment should generally be given to sewage, before its disposal in a stream, if the dilution factor available is less than
(*a*) 500 (*b*) 350 (*c*) 1250 (*d*) 150

531. Sewage disposal by dilution not preferred, where
(*a*) the sewage arriving at outfall points is fresh and non-septic.
(*b*) there are strong forward currents.
(*c*) rivers run with very low flows during summer.
(*d*) discharge in water body is riot low.

532. Primary treatment of sewage is meant for the removal of
(*a*) fine suspended organic matter
(*b*) larger suspended matter
(*c*) pathogenic (disease causing) bacteria
(*d*) dissolved organic matter

533. Match the equipment/techniques used for the removal of unwanted materials during preliminary treatment of sewage

	List I		List II
I.	Skimining tanks	A.	Oil At grease
II.	Grit chambers	B.	Floating materials
III.	Screening	C.	Dissolved organic matter
IV.	Reverse osmosis	D.	Sand

(*a*) I-A, II-D, III-B, IV-C
(*b*) I-A, II-B, III- C, IV-D
(*c*) I-D, II-C, III-13, IV-A
(*d*) I-D, 1I-A, III-B, IV-C

534. The sewage treatment units, which work on anaerobic bio-decomposition or organic mutter is
(*a*) detritus tank. (*b*) oxidation ponds.
(*c*) septic tanks.
(*d*) activated sludge plants.

535. The BOD_5 of the effluents obtained from septic tanks is of the order of about _______ mg/litre.
(*a*) 1-5 (*b*) 20-50
(*c*) 100-200 (*d*) 1000-2000

536. The sewage treatment units, which work on aerobic decomposition of organic matter is
(*a*) Imhoff tanks.
(*b*) sludge digestion tank.
(*c*) grit channel. (*d*) trickling filter.

537. The gas which is evolved in sludge digestion tank is mainly composed by
(*a*) NH_3 (*b*) H_2S (*c*) CO_2 (*d*) CH_4

538. Pick out the wrong statement.
(*a*) Sludge digestion tank works on the principle of aerobic decomposition.

Answers

527. (*c*) 528. (*c*) 529. (*c*) 530. (*d*) 531. (*c*) 532. (*b*) 533. (*a*) 534. (*c*) 535. (*c*)
536. (*d*) 537. (*d*) 538. (*a*)

(*b*) A grit chamber having a higher detention period (about 3-4 times) is called detritus tank.
(*c*) Chlorine is sometimes used in sewage treatment to help in grease separation.
(*d*) A septic tank is both a settling tank as well as digestion tank.

539. Maximum efficiency of BOD removal is achieved in
(*a*) aerated lagoon (*b*) oxidation ditch
(*c*) oxidation pond (*d*) trickling filter

540. High COD to BOD ratio of an organic water pollutant indicates
(*a*) high biodegradability of the pollutant.
(*b*) presence of free toxic material in the pollutant.
(*c*) low biodegradability of the pollutant.
(*d*) presence of free oxygen for aerobic decomposition.

541. In a conventional activated sludge plant, the oxygen demand is the highest near the inlet end of the aeration tank, Activated sludge is the
(*a*) aerated sludge in the aeration unit.
(*b*) sludge settled in the humus tank
(*c*) sludge in the secondary tank after aeration and rich in microbial mass.
(*d*) sludge in the secondary tank after aeration and rich in nutrients.

542. Composting and lagooning are the methods of sewage disposal. For satisfactory working of a sludge digestion unit, the pH range of digested sludge should be maintained at
(*a*) 3 to 5 (*b*) 6.5 to 8
(*c*) 8.5 to 10.5 (*d*) 11 to 13.5

543. The secondary treatment of sewage is caused by
(*a*) chlorine (*b*) bacteria
(*c*) algae (*d*) coagulants

544. Match the detention period adopted for various sewage treatment units.

	Sewage treatment units		Detention period
I.	Grit chambers	A.	1 minute
II.	Sedimentation tanks	B.	1-2 minute
III.	Oxidation ponds	C.	2-4 hours
IV.	Septic tanks	D.	12-36 hours
V.	Imhoff tanks	E.	2-6 weeks

(*a*) I-A, II-B, III-C, IV-D, V-E
(*b*) I- A, II-B, III-E, IV-D, V-C
(*c*) I-E, II-D, III-C, IV-B, V-A
(*d*) I-A, II-B, III-C, IV-D, V-E

545. The phenomenon by virtue of which soil is clogged with sewage matter is called sewage
(*a*) sulking (*b*) bulking
(*c*) sickness (*d*) farming

546. Algae dies out, though fish life may survive, in a river zone, known as the zone of
(*a*) degradation. (*b*) aeration.
(*c*) active decomposition.
(*d*) recovery.

547. Imhoff cone is used to measure the total ________ content in sewage.
(*a*) solids (*b*) settleable solids
(*c*) organic solids (*d*) inorganic solids

548. Criteria pollutants contributing to respiratory illness is
(*a*) tobacco smoke/random.
(*b*) lead. (*c*) carbon monoxide.
(*d*) carbohydrates.

549. Bad ozone (ground level ozone) is formed when
(*a*) O_2 mixes with H_2O.
(*b*) NO_x and VOC's interact with sunlight.
(*c*) VOC'S interact with good ozone.
(*d*) none of these.

550. Best known effect of acid precipitation is the ________ of lakes and streams,
(*a*) rising pH (*b*) lowering sediment
(*c*) decreasing water clarity
(*d*) rising temperature

Answers

539. (*b*) 540. (*c*) 541. (*c*) 542. (*b*) 543. (*b*) 544. (*b*) 545. (*c*) 546. (*a*) 547. (*b*)
548. (*a*) 549. (*b*) 550. (*a*)

551. Primary force responsible for dust particle collection in settling chamber is
(*a*) impaction. (*b*) gravity.
(*c*) brownion diffusion.
(*d*) electrostatic.

552. Gravity settling chambers are normally effective for removing dust particle in the size range of
(*a*) 10 microns (*b*) 5-40 microns
(*c*) 70 microns
(*d*) submicron particle only

553. Increasing the dusty gas volumetric feed rate to an existing gravity settling chamber would result in ______ in dust collection efficiency.
(*a*) decrease (*b*) increase
(*c*) nochange (*d*) unpredictable

554. Pick out the wrong statement.
(*a*) Collection efficiency of a cyclonic dust collector increases proportional to the square of the particle diameter.
(*b*) Collection efficiency of particulate control device using inertial impaction (e.g. wet scrubber) increases proportional to the square of the particle diameter.
(*c*) Effectiveness of Brownian motion/ diffusion increases proportional to the inverse of the particle diameter.
(*d*) Particle that are collected in wet scrubbers with great difficulty arc in the size range 1 to 10 μm diameter.

555. Which of the following air pollutants is not a colourless gas?
(*a*) SO_2 (*b*) NO
(*c*) O_3 (*d*) None of these

556. Which of the following is the best particulate control device for the flue gas containing an explosive gas with aerodynamic diameter greater than 0.5 microns having moderate electrical resistivity but no sulfuric acid?
(*a*) Bag filter.
(*b*) Mechanical dust collector.
(*c*) Wet scrubber.
(*d*) Electrostatic precipitator.

557. Which of the following substances stores more heat energy than all other atmospheric constituents combined?
(*a*) Nitrogen (*b*) Ozone
(*c*) Water vapor (*d*) Carbon dioxide

558. What particle characteristics influences the resistivity of particulate matter in a dusty gas stream being by an electrostatic precipitator?
(*a*) Particle size (*b*) Particle density
(*c*) Particle composition
(*d*) Particle surface area

559. Which of the following air pollutants is not responsible for the formation of photochemical smog?
(*a*) Nitrogen oxides (NO_x).
(*b*) Volatile organic compounds (VOC's).
(*c*) Carbonmonoxide (CO).
(*d*) Sulfur dioxide (SO_2).

560. Which of the following is generally not used as an adsorbant for removal of air pollutants?
(*a*) Zeolites
(*b*) Activated carbon
(*c*) Carbon fibre
(*d*) Polymeric materials

561. What percentage of sulfur present in a fossil fuel (e.g., coal, oil) is converted into sulfur dioxide in a utility or industrial boiler?
(*a*) 25 (*b*) 50
(*c*) 75 (*d*) 95

562. Which factor does not influence the formation of the NO_x in a boiler?
(*a*) Nitrogen content of the fuel being fired.
(*b*) Volume of flue gas.
(*c*) Temperature of combustion zone.
(*d*) Oxygen concentration.

563. Which category of source has the highest NO_x emission?
(*a*) Fuel handling and utilisation.

Answers

551. (*b*) 552. (*c*) 553. (*a*) 554. (*d*) 555. (*d*) 556. (*c*) 557. (*c*) 558. (*c*) 559. (*d*)
560. (*d*) 561. (*d*) 562. (*b*) 563. (*d*)

(*b*) Fuel combustion (electric utilities).
(*c*) Solvent utilisation.
(*d*) Transportation (automobiles, planes).

564. The emission of which air pollutant is likely to increase when the combustion zone temperature of a boiler increases?
(*a*) Volatile organic compounds (VOC's).
(*b*) Carbon monoxide (CO).
(*c*) Dioxins and fur an compounds.
(*d*) Nitrogen oxides NO_x.

565. A plant is planning to instal a selective non-catalytic reduction (SNCR) system to control emissions. Which pollutant will this control system help reduce?
(*a*) PM_{10} particulate matter
(*b*) Carbon monoxide
(*c*) Ozone
(*d*) Volatile organic compound (VOCS)

566. The burning of fuel containing trace amounts of chlorine can not emit which pollutant?
(*a*) Hydrogen chloride
(*b*) Dioxins (*c*) Furans
(*d*) Carbon monoxide

567. The most appropriate particulate control device for flue gas consisting of non-explosive gas, a moderate concentration of H_2SO_4 and particles > 0.5 micron (without the liability to treat waste water) is
(*a*) electrostatic precipitator.
(*b*) wet scrubber.
(*c*) mechanical collector.
(*d*) biological oxidation.

568. Which category of sources is most responsible for sulfur dioxide emissions?
(*a*) Industrial processes.
(*b*) Utility and industrial boilers.
(*c*) Transportation (automobiles, planes).
(*d*) Solvent utilisation.

569. An industrial facility has exceeded the emission limits of SO_2, HCl and HF. Which of the following is the most appropriate add-on control device to reduce these three pollutants?
(*a*) Low NO_x burner.
(*b*) Spray dryer type dry scrubber.
(*c*) Thermal oxidizer.
(*d*) Selective non-catalytic reduction (SNCR)

570. Pick out the wrong statement.
(*a*) The source of energy responsible for atmospheric and oceanic motion is the sun.
(*b*) The fraction of energy reflected by a surface compared to the amount incident on the surface is its "albedo".
(*c*) Oblique sun-rays produce more heating per unit area than vertical rays do.
(*d*) When the air is cloudy or heavily polluted, less direct insolation will be received at the earth's surface.

571. Which force is not used to remove particles in woven and felted bags in fabric filter?
(*a*) Electrostatic attraction
(*b*) Brownion diffusion
(*c*) Inertial impaction
(*d*) Seiving

572. The environmental or prevailing lapse rate can be determined from the actual temperature profile of the atmosphere. The dry adiabatic lapse rate (DALR) isminus ______ °C/km.
(*a*) 6.5 (*b*) 9.8 (*c*) 14.5 (*d*) 1.5

573. An infinitesimally small well defined body of air that does not readily mix with the surrounding air is termed as "air parcel". The maximum level to which a parcel of air will ascend under a given set of conditions is known as the
(*a*) mixing trough. (*b*) mixing layer.
(*c*) mixing height.
(*d*) ascend/descend level.

Answers

564. (*d*) 565. (*c*) 566. (*d*) 567. (*a*) 568. (*b*) 569. (*b*) 570. (*c*) 571. (*d*) 572. (*b*)
573. (*c*)

574. Pick out the wrong statement.

(*a*) Buoyancy of an air parcel is influenced by atmospheric temperature and pressure.

(*b*) If the temperature of an air parcel is cooler than the surrounding air, it will usually descend.

(*c*) The temperature of air decreases as atmospheric pressure increases.

(*d*) The changes in the temperature of an air parcel that are due to changes in atmospheric pressure are called adiabatic.

575. At the wet adiabatic lapse rate, the cooling rate of air parcel is usually ______ at the dry rate.

(*a*) slower than (*b*) faster then

(*c*) the same as (*d*) unpredictable

576. The adiabatic lapse rate for a given air parcel will intersect the environmental lapse rate at the

(*a*) mixing trough (*b*) mixing height

(*c*) mixing layer (*d*) mixing station

577. Unstable atmospheric conditions most commonly develop on

(*a*) cloudy days (*b*) cloudy nights

(*c*) clear night (*d*) sunny days

578. On cloudy days with no strong surface heating, atmospheric conditions are likely to be

(*a*) stable. (*b*) unstable.

(*c*) extremely stable. (*d*) neutral.

579. Vertical mixing of air pollutants due to buoyancy is increased when atmospheric conditions are

(*a*) unstable (*b*) neutral

(*c*) stable (*d*) extremely stable

580. Pick out the correct statement.

(*a*) The dry adiabatic lapse rate (DALR) is fixed and entirely independent of ambient air temperature.

(*b*) A displaced air parcel cools at the dry adiabatic lapse rate. once it becomes saturated.

(*c*) A large mixing layer implies that air pollutants have a lesser volume of air for dilution.

(*d*) An unstable atmosphere resists vertical motion of air pollutants.

581. Which type of bag/fabric filter systems collect the dust cake on the exterior surface of the filter medium?

(*a*) Pulse jet (*b*) Cartridge

(*c*) Reverse air (*d*) None of these

582. Which of the following fabric materials of a fabric filter can not work satisfactorily for longer duration, if the dusty gas temperature is more than 205°C?

(*a*) Fiberglass (*b*) Stainless steel

(*c*) Ceramic (*d*) Tenon

583. What problem occurs at excessive air to cloth ratios in case of a bag/fabric filter?

(*a*) Increased particulate emission through the filter media.

(*b*) Accelerated bag failure.

(*c*) Decreased static pressure drop.

(*d*) all (*a*), (*b*) and (*c*).

584. Which of the following action should be taken to lower the emission of incompletely oxidised organic compounds leaving the municipal waste incinerator?

(*a*) Decrease the excess air levels

(*b*) Increase the moisture level in the combustion zone

(*c*) Adjust the fuel to air distribution to increase the combustion temperature.

(*d*) Decrease the load to the incinerator

585. Purpose of using offline cleaning in a multi-compartment pulse jet fabric filter collector is not to minimise

(*a*) solids build up problems in hoppers.

(*b*) high static pressure drop problems.

(*c*) gravity settling problems during cleaning of the bags.

(*d*) none of these

Answers

574. (*c*) 575. (*a*) 576. (*b*) 577. (*d*) 578. (*d*) 579. (*a*) 580. (*a*) 581. (*c*) 582. (*d*)
583. (*a*) 584. (*c*) 585. (*a*)

586. Which of the following forces used to remove particles in a membrane bag is the most prominent?
(*a*) Electrostatic precipitation
(*b*) Inertial impaction
(*c*) Seiving (*d*) Brownian diffusion

587. Pick out the wrong statement.
(*a*) The environmental lapse rate influences the extent to which a parcel of air can rise or descend.
(*b*) When the earth's surface cools rapidly, such as between late night and early morning under clear skies, a radiation inversion is likely to occur.
(*c*) When vigrous vertical mixing follows a radiation inversion, pollutant plumes will be trapped near the surface.
(*d*) A high pressure system can cause an elevated temperature inversion to form.

588. Surface based inversions associated with horizontal air flow, such as when warm air moves over a cold surface are called ________ inversions,
(*a*) frontal (*b*) subsidence
(*c*) advection (*d*) adiabatic

589. The ________ plume is characteristic of neutral or slightly stable atmospheric conditions.
(*a*) fanning (*b*) coning
(*c*) lofting (*d*) looping

590. A fanning plume will occur, when atmospheric conditions are generally
(*a*) neutral (*b*) stable
(*c*) unstable (*d*) highly unstable

591. Pick out the wrong statement.
(*a*) The looping plume can cause high ground level concentrations of air pollutants.
(*b*) If a plume is released just under an inversion layer, a serious air pollution situation could develop.
(*c*) A subsidence inversion generally tends to last for relatively shorter period of time compared to a radiation inversion.
(*d*) The subsidence inversion is associated with anticyclones because it usually forms high above the surface during the day.

592. Which of the following will be the most suitable particulate control device for the flue gas consisting of an explosive gas containing particles > 5 microns with moderate electrical resistivity and no H_2SO_4?
(*a*) Wet scrubber
(*b*) Mechanical collector
(*c*) Fabric filter
(*d*) Electrostatic precipitator

593. Match the appropriate control device to reduce volatile organic compounds (VOC's) from the flue gas of different characteristics emitted from separate processes.

	Flue gas characteristics		VOC's controlling method/device
P.	Flue gas exiting from a small diameter pipe at a slow velocity containing a high concentration of a variety of VOC's.	I.	Condensation or refrigeration
Q.	Flue gas (containing acetone and methylene chloride) How rate and level of VOC concentration is moderate. Plant management wants to collect and sell these solvents.	II.	Catalytic oxidiser
R.	The flue gas has a moderate flow rate and contains a variety of VOC's (of lower molecular weight = 16 to 32). The gas temperature exiting the control device should not exceed 600°C due to process constraints.	III.	Carbon bed adsorber

Answers

586. (*c*) 587. (*c*) 588. (*c*) 589. (*b*) 590. (*b*) 591. (*c*) 592. (*b*) 593. (*a*)

(*a*) P-I, Q-III, R-II (*b*) P-I, Q-II, R-III
(*c*) P-III, Q-II, R-I (*d*) P-III, Q-I, R-II

591. Which of the following contaminants can not be present in untreated compressed air used to clean pulse jet fabric filters and cartridge fabric filters?

(*a*) Carbon monoxide.
(*b*) Condensed oil droplets.
(*c*) Condensed water droplets.
(*d*) all (*a*), (*b*) and (*c*).

595. What problem is created, if a pulse jet bag does not seal properly to the tube sheet in a pulse jet fabric filter?

(*a*) The bag could develop holes due to its vulnerability to flex failure.
(*b*) Unfiltered gas could leak around the bag into the clean gas plenum.
(*c*) The bags could fall into the hoppers due to inadequate support.
(*d*) The filter media static pressure drop would decrease.

596. A reverse air bag filter constructed of fibre glass has very low tension and is sagging severely at the connection to the tube sheet thimble. Which of the following problems can not be created by this condition?

(*a*) The sagging bag could choke off flow of unfiltered air into the bag.
(*b*) The bag could develop holes due to its vulnerability to flex failure.
(*c*) The filter media static pressure drop would decrease.
(*d*) The bag could be abraded by the high velocity inlet gas stream.

597. Which of the following particulate control device is the most appropriate to use for flue gas containing particles with aerodynamic diameter > 0.1 micron having high electrical resistivity? The Due gas composition (except for the presence of particles) is identical to ambient air and contains no H_2SO_4. Plant management wants a highly efficient particulate control device for the entire particle size range with no waste water generation.

(*a*) Electrostatic precipitator.
(*b*) Mechanical collector.
(*c*) Wet scrubber. (*d*) Fabric filter.

598. Which of the following factors does not govern the amount of insolation received by the earth from the sun?

(*a*) Daily sunlight duration.
(*b*) Angle at which the sun's rays strike the earth.
(*c*) Atmosphere's transparency.
(*d*) Environmental lapse rate.

600. Since the earth's atmosphere depletes solar radiation passing through it, how much radiation received at the outermost boundary of the atmosphere reaches the surface of the earth?

(*a*) All of it. (*b*) None of it.
(*c*) Half of it. (*d*) One quarter of it.

601. Pick out the wrong statement.

(*a*) Oceanic and atmospheric circulations redistribute energy received by the sun.
(*b*) The heating process that causes the vertical mixing of air above earth's surface is called convection.
(*c*) When air touches the heated earth and in turn becomes warm, convection has just occurred.
(*d*) The fraction of energy reflected by a surface compared to the amount incident on the surface is its 'albedo'.

602. Which categories of sources is most responsible for volatile organic compounds (VOC's) emissions?

(*a*) Fuel handling and distribution.
(*b*) Fuel combustion.
(*c*) Transportation (automobiles, planes).
(*d*) Solvent utilisation.

Answers

594. (*a*) 595. (*b*) 596. (*c*) 597. (*d*) 598. (*d*) 599. (*b*) 600. (*c*) 601. (*c*) 602. (*d*)

603. Volatile organic compounds (VOC's) emissions are controlled to achieve National Ambient Air quality standards (NAAQS) in respect of

(*a*) NO_x (*b*) SO_x (*c*) O_3
(*d*) HC (hydrocarbons)

604. Impaction, direct interception and diffusion can all be responsible for particulate collection in case of

(*a*) adsorbers (*b*) incinerators
(*c*) fabric filter (*d*) condensers

605. The typical concentration limits for Ignitable volatile organic compounds (VOC's) in the inlet gas stream to an air pollution control system is ______ per cent of the lower explosion limit (LEL).

(*a*) 10 (*b*) 25 (*c*) 50 (*d*) 75

606. Pick out the wrong statement.

(*a*) Both physical adsorption and chemisorption are used for volatile organic compounds (VOC's) removal in regenerative system.

(*b*) Thermal oxidisers usually operate at gas temperatures that ate 90 to 150°C above the auto ignition temperature of the most difficult to oxidise major constituents.

(*c*) A catalytic oxidiser used for volatile organic compounds removal operates with a catalyst bed inlet temperature that is between 200 to 425°C.

(*d*) Physical adsorption of VOC's is more effective when the contaminated gas temperature is very high,

607. The capacity of an adsorbent for a gaseous contaminant (gms/100 gm of adsorbent) increases with the

(*a*) decreased gas pressure.
(*b*) reduced gas temperature.
(*c*) decreased contaminant concentration.
(*d*) presence of multiple containinants.

608. What materials and/or techniques are not used to desorb VOCs (volatile organic compounds) from regenerative type adsorption system?

(*a*) Hot air (*b*) Vacuum
(*c*) Low pressure steam
(*d*) Hot water flush

609. What is the normal maximum gas temperature °C used in VOC adsorption systems?

(*a*) 30 (*b*) 55 (*c*) 100 (*d*) 150

610. That is typical VOC removal efficiency (%) of a properly designed and operated adsorption system?

(*a*) 40-50 (*b*) 60-75
(*c*) 90-98 (*d*) 99.5

611. Why are high molecular weight VOC's (mol wt. > 200) difficult to handle in a regenerative adsorption system?

(*a*) They are not adsorbed efficiently.
(*b*) They decompose at elevated adsorption temperature.
(*c*) They can not be separated in decanters, condensers and distillation columns.
(*d*) They have too high an affinity and hence can not be desorbed efficiently.

612. Which of the following is not a combustion modification technique used to reduce nitrogen oxides (NO_x) emissions from boilers?

(*a*) Staged combustion of fuel.
(*b*) Tangential firing of coal through burners.
(*c*) Use of high excess air for combustion.
(*d*) Flue gas recirculation.

613. The plume in the following diagram is

(*a*) coning (*b*) looping
(*c*) lofting (*d*) fumigation

614. Which plume is represented by the lapse rate and stack shown below in the diagram?

(*a*) Lofting (*b*) Coning
(*c*) Fanning (*d*) Looping

615. What is the name of the plume shown below in the diagram?

Answers

603. (*c*) 604. (*c*) 605. (*b*) 606. (*d*) 607. (*b*) 608. (*d*) 609. (*b*) 610. (*c*) 611. (*d*)
612. (*c*) 613. (*d*) 614. (*a*) 615. (*a*)

(*a*) Looping (*b*) Coning
(*c*) Fanning (*d*) Trapping

616. In the diagram shown below, a displaced air parcel becomes saturated at an elevation of 2 kms. Which of the following stability conditions does the diagram depict?
(*a*) Stable below 1 km.
(*b*) Conditional stability above 1 km.
(*c*) Neutral from 0 to 2 km.
(*d*) Conditional instability above 2 kms.

617. The total dissolved solids in drinking water should not exceed ______ mg/litre.
(*a*) 1 (*b*) 100 (*c*) 500 (*d*) 1000

618. Methamoglobanumia (blue baby disease) is due to presence of excess of ______ in the drinking water.
(*a*) chlorides (*b*) nitrates
(*c*) dissolved oxygen (*d*) PH

619. Flame ionisation detector (FID) is used for the determination of present in the polluted air.
(*a*) carbon monoxide (CO)
(*b*) nitrogen oxides (NO_x)
(*c*) hydrocarbons (HC)
(*d*) chlorine

620. Orthotolidine test is used to determine ______ present in water.
(*a*) residual ammonia
(*b*) nitrates (*c*) sulfates
(*d*) residual chlorine

621. Ether soluble solids in waste water represents its content.
(*a*) dissolved solids (*b*) total solids
(*c*) grease (*d*) none of these

622. Enrichment of biogas is done by scrubbing out its ______ content.
(*a*) SO_2 (*b*) CO_2 (*c*) CH_4 (*d*) NO_x

623. Contact stabilisation process is a form of
(*a*) stabilisation ponds.
(*b*) activated sludge process.
(*c*) trickling fillers. (*d*) anerobic lagoons.

624. All particulate control devices operate using
(*a*) water
(*b*) electrostatic attraction
(*c*) an applied force (*d*) none of these

625. In a simple particulate control device such as settling chamber, large particles moving slow enough in a gas stream can be overcome by
(*a*) centrifugal force (*b*) centripetel force
(*c*) impaction (*d*) gravity

626. Carbon bed adsorber is a measure of the resistance to gas movement through a particulate control device
(*a*) Collection efficiency
(*b*) Particle to particle density
(*c*) Pressure drop (*d*) None of these

627. Aerodynamic diameter describes h ow a particle moves in a gas stream. It has a direct bearing on the
(*a*) ability of particle to be collected
(*b*) collection efficiency of a specific particulate control device
(*c*) both (*a*) and (*b*) (*d*) neither (*a*) nor (*b*)

628. If carbon bed adsorber is large or constantly fluctuating, some particulate control devices may not function efficiently.
(*a*) gas velocity
(*b*) particle concentration
(*c*) grain loading (*d*) all (*a*), (*b*) and (*c*)

629. In a boiler, SO_2 is emitted when the sulfur in the fuel is oxidised while NO_x is emitted, when the nitrogen in the ______ is oxidised.
(*a*) fuel (*b*) air
(*c*) both (*a*) and (*b*) (*d*) neither (*a*) nor (*b*)

630. Boilers burning low sulfur coal produce fly ash having high resistivity which is difficult to collect using a
(*a*) dry scrubber (*b*) wet scrubber
(*c*) electrostatic precipitator
(*d*) bag filter

Answers

616. (*d*) 617. (*a*) 618. (*b*) 619. (*c*) 620. (*d*) 621. (*c*) 622. (*b*) 623. (*b*) 624. (*c*)
625. (*d*) 626. (*c*) 627. (*c*) 628. (*d*) 629. (*c*) 630. (*c*)

631. Pick out the wrong statement about cyclone separator

(*a*) Purpose of using a solid discharge valve on the hoppers of cyclone collectors is to minimise air infiltration into the cyclone.
(*b*) Tangential gas inlet initiates the spinning gas flow in a large diameter cyclone.
(*c*) Normal range of inlet gas stream velocity for large diameter cyclone ranges from 5 to 15 metres/second.
(*d*) Typical number of complete turns (360°) achieved in a large diameter cyclone operating with normal inlet gas velocity is 20 to 30.

632. The most popular particulate control device used to reduce electric furnace emissions in steel mills is

(*a*) bag houses (*b*) wet scrubber
(*c*) electrostatic precipitator
(*d*) settling chamber

633. Cyclones are used pre-cleaners to remove particles that are larger than ________ in diameter at more than ________ efficiency.

(*a*) 0.1 µm, 80% (*b*) 20 µm, 80%
(*c*) 1 µm, 90% (*d*) 5 µm, 100%

634. Primary pollutant produced in the manufacture of cement is the particulate matter. Which of the following control devices is usually the primary dust controller at a cement plant.

(*a*) Mechanical collector and carbon adsorber
(*b*) Cyclone and twin bed adsorber
(*c*) Baghouse and electrostatic precipitator
(*d*) Wet scrubber and cyclone

635. Which of the following processes generates most of the pollutant emissions in the production of nitric acid?

(*a*) Adsorption of NO_2 in H_2O
(*b*) Conversion of NO_2 to NO
(*c*) Oxidation of NO to NO_2
(*d*) Absorption of NO into H_2SO_4 form HNO_3

636. The major potential source of carbon monoxide at a petroleum refinery is from the

(*a*) storage tanks
(*b*) oil-water separators
(*c*) vacuum systems
(*d*) catalyst regenerator on the FCC unit

637. In large boilers, the majority of NO_x emitted (approximately 95%) is in what form?

(*a*) NO_3 (*b*) NO_2 (*c*) NO (*d*) N_2O

638. Which of the following devices at a copper smelter will not emit SO_2?

(*a*) Roaster (*b*) Converter furnace
(*c*) Reverberatory furnace
(*d*) Froth notation tank

639. The amount of particulate matter generated from the combustion of fuel oil depends on the

(*a*) grade of fuel oil (*b*) type of boiler
(*c*) firing practices (*d*) all (*a*), (*b*) and (*c*)

640. Performance of a cyclone separator is related to the ________ of the radial gas velocity.

(*a*) first power (*b*) second power
(*c*) third power (*d*) none of these

641. Thermal incinerators generally operate between ________ °C and with a residence time of 0.1 to 0.5 seconds.

(*a*) 150 and 260 (*b*) 370 and 480
(*c*) 700 and 820 (*d*) 1150 and 1370

642. Multicyclone collectors are capable of effectively removing particles down to approximately ________ micrometers.

(*a*) 0.5 (*b*) 3 (*c*) 1 (*d*) 20

643. Which of the following materials can be used as a solid absorbant?

(*a*) Silica gel (*b*) Molecular seive
(*c*) Activated carbon (*d*) All (*a*), (*b*) and (*c*)

644. Typical static pressure drop in a multicyclone collector is 100 to 150 mm water column. Static pressure drop across a cyclone collector is related to the ________ of the gas flow rate.

Answers

631. (*d*) 632. (*a*) 633. (*b*) 634. (*c*) 635. (*a*) 636. (*d*) 637. (*c*) 638. (*d*) 639. (*d*)
640. (*b*) 641. (*c*) 642. (*b*) 643. (*d*) 644. (*c*)

(*a*) first power (*b*) second power
(*c*) third power (*d*) none of these

645. Pick out the wrong statement about multi-cyclone separators.
(*a*) Clean side tube sheet in a multicyclone collector separates the inlet gas stream from the outlet gas stream.
(*b*) Multi cyclone collectors have higher radial gas velocity than large diameter cyclones.
(*c*) Typical range of diameters of multicyclone tubes is 150 to 300 mm.
(*d*) Most multicyclone tubes should be oriented vertically (inlet at the top. cyclone discharge at the bottom) in order to operate properly.

646. Performance of a cyclone separator is related to the _______ of the particle diameter.
(*a*) first power (*b*) second power
(*c*) third power (*d*) none of these

647. In order to ensure complete combustion in an incinerator, the waste gas must be heated to the
(*a*) vapor pressure (*b*) boiling point
(*c*) ignition temperature
(*d*) critical temperature

648. Fixed bed carbon adsorption system have been used to control _______ emissions.
(*a*) dry cleaning (*b*) solvent degreasing
(*c*) organic (*d*) all (*a*), (*b*) and (*c*)

649. Dust particles are held into the collection electrodes/plates in an electrostatic precipitator by
(*a*) intermolecular cohesive and adhesive force
(*b*) a strong electric field force
(*c*) a high pulsating direct current
(*d*) electric sponsors

650. For a bag filter, air to cloth ratio is defined as the
(*a*) amount of dust filtered by the bags
(*b*) amount of air used to clean reverse air bags
(*c*) amount of air passing through a given area of cloth in a given time
(*d*) highest setting allowed on the reverse air fan.

651. _______ baghouse filters are composed of randomly placed fibres that are compressed into a mat and attached to a loosely woven backing material.
(*a*) Felted (*b*) Woven
(*c*) Synthetic (*d*) Membrane

652. The furnace of a municipal incinerator consists of a grate system and combustion chambers. Because of the high temperature and high moisture content of the exhaust stream, _______ have not been used to control emission from municipal incinerators.
(*a*) baghouses
(*b*) electrostatic precipitators
(*c*) cyclones (*d*) wet scrubbers

653. Pick out the correct statement about soil
(*a*) Most soils have pH value between 1 and 5.
(*b*) Lime is used to treat soil, that is too alkaline.
(*c*) Loam is a poor soil which drains poorly and is low in nutrients.
(*d*) Organisms like worms help circulate air in soil.

654. Which of the following is used to treat soil which is too alkaline?
(*a*) Sand (*b*) Lime
(*c*) Peat (*d*) Marble

655. Which of the following most likely provides evidence of global warming?
(*a*) Increased drought in parts of Africa
(*b*) Increased ice sheet area at the south pole

Answers

645. (*b*) 646. (*b*) 647. (*c*) 648. (*d*) 649. (*c*) 650. (*a*) 651. (*a*) 652. (*a*) 653. (*d*)
654. (*c*) 655. (*a*)

(*c*) Thinner growth rings in trees in Northern Europe
(*d*) Decreased fish stock of the Atlantic ocean.

656. Which is believed to be the current main cause of the so called 'Green house effect?
(*a*) Increased volcanic activity
(*b*) Burning fossil fuels
(*c*) Reduction of ozone layer
(*d*) Increase in cloud cover above the earth.

657. Which of the following is a dark solid pollution product from the incomplete combustion of a fossil fuel?
(*a*) Carbon (*b*) Carbon dioxide
(*c*) Water (*d*) Nitrogen

658. Which of the following word equation corresponds to 'photosynthesis'?
(*a*) Nitrogen + oxygen ⇄ Nitrogen oxide
(*b*) Glucose + oxygen ⇄ water + carbon dioxide
(*c*) water + oxygen ⇄ glucose + oxygen
(*d*) hydrocarbon + oxygen ⇄ water + carbon dioxide

659. Which of the following is not one of the major environment problems resulting from human interference in the nitrogen cycle?
(*a*) Stratospheric ozone depletion
(*b*) Increased acid rain
(*c*) Eutrophication
(*d*) Nitrous oxide release increases global warming

660. Which of the following is not a major green house gas?
(*a*) Carbon dioxide (*b*) Methane
(*c*) Calcium carbonate
(*d*) Water vapor

661. In 1987, Montreal Protocol was signed for which of the following reasons?
(*a*) To ban nuclear weapon testing in tropical oceans.
(*b*) To stop the global trade in products made from endangered species.
(*c*) To phase out the use of CFC's found to be causing depletion of the .ozone layer.
(*d*) To begin converting from fossil fuels use to more renewable energy sources to reduce the anthropogenic greenhouse effect.

662. Approximately what proportion of the global land surface is used for agriculture and cattle grazing by the world's over 6 million people.
(*a*) one third (*b*) three quarter
(*c*) one eightieth (*d*) one tenth

663. Which of the following is not a major positive feedback mechanism in which the activity of humans to increase global climate temperatures leads to even further increase?
(*a*) Global warming causes increased CO_2 release from bio-mass decomposition.
(*b*) Global warming causes increased rainfall, plant growth and photosynthesis.
(*c*) Tropical deforestation causes warming arid drying so that remaining forests begin to decline.
(*d*) Global warming causes snow to melt in polar regions and therefore increases global albedo.

664. One water ways are not pollutant by
(*a*) fish farming.
(*b*) leaking of old landfills.
(*c*) leaking underground gasoline storage tanks.
(*d*) dumping used oil on the ground.

665. What rubbish material do we throw away the most?
(*a*) Plastic (*b*) Aluminium
(*c*) Paper (*d*) Glass

666. Used oil should be disposed of by
(*a*) collecting it in a sealed container and taking it to an oil changing business

Answers

656. (*b*) 657. (*a*) 658. (*c*) 659. (*a*) 660. (*c*) 661. (*d*) 662. (*a*) 663. (*b*) 664. (*a*)
665. (*c*) 666. (*a*)

(b) keeping it in the garage for 10 years before throwing it away in the trash
(c) watering it down and spreading it as a fertiliser
(d) dumping it on your driver way.

667. How long does it take styrofoam to break down in a landfill?
(a) 10 years (b) 50 years
(c) 100 years (d) 500 years

668. What is the environmentally friendly way to dispose off used tyros?
(a) Recycle them to a playground matting.
(b) Dump them in a ditch along the roadway
(c) Collect them and put them in a big pile
(d) Send them to a landfill.

669. Mercury is a hazardous chemical. Which item does not contain mercury?
(a) Clinical thermometers
(b) Fluorescent light bulbs
(c) Computer discs
(d) Blinking light in sneakers

670. Which of the following items are recyclable?
(a) Plastic milk jugs (b) Cereal boxes
(c) Aluminium soda cans
(d) all (a), (b) and (c)

671. Most of the domestic trash thrown away each day ends up getting
(a) recycled (b) burned
(c) eaten by cattle (d) landfilled

672. What is ground water?
(a) Water that is spilled onto the ground
(b) Water beneath the ground surface
(c) The bottom part of a lake/river
(d) Rain water that lands on the ground.

673. Particulate matter emissions from coke quenching tower in a steel plant is reduced by
(a) baghouses
(b) electrostatic precipitator
(c) baffles
(d) scyclone separators

674. Energy for particulate collection in a venturi scrubber comes primarily from the
(a) water jet velocity (b) gas stream
(c) convergent section
(d) divergent section.

675. Which of the following bag fabrics are not used in high temperature (> 200°C) industrial gas cleaning?
(a) Teflon (b) Fibre glass
(c) Nylon (d) None of these

676. Which of the following is not one of the three T's of combustion?
(a) Thermal barrier (b) Time
(c) Temperature (d) Turbulence

677. Pick out the wrong statement about indoor air pollution
(a) Smoking is only dangerous to the person who is smoking and not others around.
(b) Formaldehyde is a gas that can come from building materials and furnishings.
(c) Printers and fax machines are the source of ozone in an official building.
(d) Pesticides used in the buildings only hurt the pests as they are designed to kill them.

678. The maximum level to which a parcel of air will ascend under a given set of conditions is known as the
(a) ascend level (b) mixing trough
(c) mixing height (d) mixing layer

679. _______ pollutants are emitted directly from identifiable sources.
(a) Toxic (b) Hazardous
(c) Primary (d) Secondary

680. The major component of photochemical smog is
(a) carbon dioxide (b) ozone
(c) nitrogen dioxide (d) hydrocarbons;

681. Ozone layer depletion problems arises due to the use of

Answers

667. (d) 668. (a) 669. (c) 670. (d) 671. (d) 672. (b) 673. (c) 674. (b) 675. (c)
676. (a) 677. (d) 678. (c) 679. (b) 680. (c) 681. (d)

(*a*) CO_2 (*b*) CH_4
(*c*) H_2O (*d*) none of these

682. Respirable particle of size PM2.5 (microns) will affect the ________ system on inhalation.
(*a*) nervous (*b*) cardiovascular
(*c*) respiratory (*d*) urinary

683. Which of the following pollutants affects the composition of leather and paper causing significant deterioration?
(*a*) NO_2 (*b*) SO_2 (*c*) CO (*d*) CO_2

684. Increasing the gas volumetric feed rate to an existing settling chamber would expectedly result in ________ in collection efficiency.
(*a*) a decrease (*b*) an increase
(*c*) no change (*d*) unpredictable

685. What forces are used to remove particles in a membrane bag?
(*a*) Inertial impaction
(*b*) Brownion diffusion
(*c*) Electrostatic attraction
(*d*) Seiving

686. Inertial impaction is the main particle collection mechanism used in particulate matter wet scrubber. What is the most important factor affecting the ability of a particulate matter wet scrubber to achieve high dust removal efficiency?
(*a*) Particle size
(*b*) Droplet size distribution
(*c*) Static pressure drop
(*d*) Droplet surface tension

687. Pick out the wrong statement about metrology of air pollutants dispersion:
(*a*) Vertical mixing due to buoyancy is increased when atmospheric conditions are 'extremely stable.
(*b*) If the temperature of an air parcel is cooler than surrounding air, it will usually descend.
(*c*) At the wet adiabatic lapse rate, the cooling rate of air parcel is usually slower than at the dry rate.
(*d*) Changes in temperature of air parcel that are due to changes in atmospheric pressure are called adiabatic.

688. The heating process that causes the vertical mixing of the air above the earth's surface is termed as
(*a*) conduction (*b*) convection
(*c*) advection (*d*) none of these

689. Surface based inversions associated with horizontal air flow, such as when warm air moves over a cold surface, are called ________ inversions.
(*a*) adiabatic (*b*) advective
(*c*) frontal (*d*) subsidence

690. Plume rise from a stack is due to
(*a*) momentum and buoyancy
(*b*) the composition of the stack gas
(*c*) heat and type of pollutant
(*d*) none of these

691. The environmental or prevailing lapse rate can be determined from the
(*a*) temperature profile of the atmosphere
(*b*) rate of pressure change in the atmosphere
(*c*) rate of frontal system passage
(*d*) rate of wet air lapse rate Vs. pressure change.

692. If DALR < ELR, then the atmospheric condition will be
(*a*) neutral (*b*) inversion
(*c*) super-adiabatic (*d*) sub-adiabatic
where, DALR = Dry Adiabatic Lapse rate;
ELR = Environmental Lapse rate

693. The atmospheric condition is super adiabatic, if
(*a*) ELR > DALR (*b*) ELR < DALR
(*c*) ELR = DALR (*d*) ELR < DALR

Answers

682. (*c*) 683. (*b*) 684. (*a*) 685. (*c*) 686. (*a*) 687. (*a*) 688. (*b*) 689. (*c*) 690. (*a*)
691. (*a*) 692. (*d*) 693. (*a*)

694. Ground level concentration is minimum, if plume shape is
(*a*) looping (*b*) coning
(*c*) fanning (*d*) fumigating

695. Suppression of turbulence occurs in atmosphere.
(*a*) neutral (*b*) sub-adiabatic
(*c*) super adiabatic (*d*) none of these

696. In Pascal stability class, slightly unstable condition is
(*a*) B (*b*) C (*c*) D (*d*) E

697. On cloudy day with no strong surface heating, atmospheric conditions are likely to be
(*a*) stable (*b*) unstable
(*c*) neutral (*d*) extremely stable

698. Pick out the wrong statement about air pollutants dispersion:
(*a*) The momentum term in plume rise equations generally involves ambient air temperature wind speed and stack gas temperature.
(*b*) If stack height is lower than the building height and is located downwind of the building, the smoke will form wake downwind.
(*c*) Particulates from plume gets deposited on the ground both by wet deposition and dry deposition.
(*d*) Fate of air pollutants incorporates source, transport, receptor re-entrainment.

699. Atmospheric turbulence is dominantly responsible for its
(*a*) transport (*b*) dispersion
(*c*) dilution (*d*) advection

700. Lofting plume occurs, when unstable condition
(*a*) is below inversion
(*b*) above inversion
(*c*) an inversion coincide
(*d*) none of these.

701. In elevated inversion condition, the fumigation plume occurs if stack height is ________ mixing height.
(*a*) less than (*b*) more than
(*c*) equal to (*d*) none of these

702. In the afternoon house, NO_x concentration decreases due to
(*a*) global warming (*b*) acid rain
(*c*) chemical reactions
(*d*) photochemical reactions

703. Diesel engine exhaust is the main source of three highly toxic pollutants that have a widespread impact on the urban air quality and human health. Main pollutants in the diesel exhaust
(*a*) NO_x, SO_2, SPM (*b*) SPM, C_6H_6
(*c*) SO_2, C_6H_{65}, NH_3 (*d*) Pb, NO_x, CO_2

704. Which of the following devices is fitted to motor vehicles to chemically reduce some gases produced by internal combustion engines (Petrol and diesel engines) like NO_x, CO and HC into less harmful products?
(*a*) Tail pipe (*b*) Carburettor
(*c*) Catalytic converter
(*d*) Supercharger

705. Which of the following is not a primary contributor to the greenhouse effect?
(*a*) N_2O (*b*) CH_4 (*c*) CFC's (*d*) CO_2

706. Depletion in the stratospheric ozone layer is not caused due to
(*a*) CO_2 (*b*) N_2O (*c*) CH_4
(*d*) all (*a*), (*b*) and (*c*)

707. What is the ill effect and harm caused from the depletion of earth's ozone layer?
(*a*) The oxygen content of the atmosphere will decrease.
(*b*) Increased amount of ultravoilet rays will reach the earth's surface.
(*c*) Sea levels will rise as the polar ice caps win gradually melt.

Answers

694. (*c*) 695. (*c*) 696. (*b*) 697. (*c*) 698. (*d*) 699. (*b*) 700. (*b*) 701. (*a*) 702. (*d*)
703. (*a*) 704. (*c*) 705. (*c*) 706. (*d*) 707. (*b*)

(*d*) The average temperature of earth's surface will increase gradually.

708. Which of the following is a prime health risk associated with greater Ultravoilet radiation through the atmosphere due to depletion of stratospheric ozone?
(*a*) Damage to digestive system
(*b*) Increased liver cancer
(*c*) Increased skin cancer
(*d*) Neurological disorder

709. Which of the following is not a direct consequence of global warming?
(*a*) Increased storm frequency and intensity
(*b*) Rising sea level
(*c*) Depletion of stratospheric ozone layer thickness
(*d*) Worsening health effects

710. The increase in concentration of CO_2 in our environment in last fifty years since 1960 is about ________ per cent. Current level of CO_2 is 385 ppm in the year 2010.
(*a*) 2.5 (*b*) 7.5 (*c*) 15 (*d*) 25

711. Which of the following is not a major green house gas?
(*a*) CH_4 (*b*) O_3 (*c*) N_2O (*d*) H_2O

712. Depletion of stratospheric of ozone layer in earth's atmosphere does not result in high rates of incidence of
(*a*) lung cancer (*b*) skin cancer
(*c*) premature skin ageing
(*d*) severe sunburns

713. In order to address the problem of stratospheric ozone depletion, the United Nations convened a meeting in 1987 in Canada in which all the member nations agreed to reduce the chloroflurocarbons (CFC's) production and use. What was the name this agreement?
(*a*) Kyoto agreement (*b*) Montreal protocol
(*c*) Vancouver convection
(*d*) UNFFC convention

714. Out of the following greenhouse gases, which one is entirely anthropogenic in nature?
(*a*) CO_2 (*b*) CFC's (*c*) N_2O (*d*) CH_4

715. Acid rain formation is due to contribution from the following pair of gases?
(*a*) CH_4 and O_3 (*b*) O_2 and N_2O
(*c*) CH_n4 and SO_2 (*d*) CO_2 and SO_2

716. The stratospheric layer of ozone normally absorbs what proportion of ultravoilet radiation from the sun?
(*a*) > 99% (*b*) about 25%
(*c*) < 10% (*d*) about 50%

717. Which of the following environmental problems is not caused by human interference in the nitrogen cycle?
(*a*) Eutrophication
(*b*) Increased global warming due to release of nitrous oxide
(*c*) Increased acid rain
(*d*) Ozone depletion in stratosphere

718. Which of the following is not a major use of CFC's?
(*a*) Buoyancy gas for blimps and balloons
(*b*) Pressurising agent in aerosol cans
(*c*) Production of plastic foam
(*d*) Cleaning of computer parts

719. Among the given choices below regarding the greenhouse effect, which one is having the least certainty?
(*a*) The heat energy in the atmosphere increases due to the addition of greenhouse gases.
(*b*) An increase in the number of tropical storms will be observed due to global warming.
(*c*) Sea levels will rise due to global warming.
(*d*) Various human activities arc resulting into higher concentration of greenhouse gases.

Answers

708. (*c*) 709. (*c*) 710. (*c*) 711. (*b*) 712. (*a*) 713. (*b*) 714. (*b*) 715. (*b*) 716. (*a*)
717. (*d*) 718. (*a*) 719. (*b*)

720. Which of the following is non-biodegradable?
(*a*) Tea leaves (*b*) Wool
(*c*) Animal bones (*d*) Nylon

721. The most serious environmental effects posed by hazardous waste is
(*a*) increased use of land for landfills
(*b*) contamination of ground water
(*c*) destruction of habitat
(*d*) air pollution.

722. The presence of high caliform counts in water indicates
(*a*) decreased biological oxygen demand (B.O.D)
(*b*) contamination by human waste
(*c*) hydrocarbon contamination
(*d*) phosphorous contamination.

723. Pick out the wrong statement:
(*a*) Biotic environment includes decomposers like bacteria and fungi.
(*b*) Abiotic environment includes air, water and soil but does not include plants.
(*c*) The group of organisms which convert light into food are called autotrophs.
(*d*) Ground water mining in coastal area can result into decrease in its toxicity .

724. Vermicomposting of solid waste is done by
(*a*) bacteria (*b*) fungus
(*c*) animals (*d*) worms

725. Which of the following in non-biodegrable?
(*a*) Fruit (*b*) Vegetables
(*c*) Earthworm (*d*) Aluminium foil

726. If waste materials contaminate the source of drinking water, which of the following disease will spread?
(*a*) Typhoid (*b*) Scurvy
(*c*) Anemia (*d*) Malaria

727. Which of the following solid waste materials can not be decomposed by bacteria to form compost?
(*a*) Dead plants (*b*) Plastic bags
(*c*) Kitchen wastes
(*d*) Bodies of insects living in the soil.

728. Which of the following problems is not created by noise pollution?
(*a*) Irritation (*b*) Hypertension
(*c*) Diarrhoea (*d*) Deafness

729. Pick out the wrong statement:
(*a*) Plants are green because presence of a pigment chlorophyll.
(*b*) When trees are cut, the amount of oxygen in atmosphere decreases.
(*c*) Three primary soil nutrients are carbon, oxygen and water.
(*d*) Process through which plants reproduce is called pollination,

730. Which of the following is not an environmental issue of global significance?
(*a*) Global warming
(*b*) Suspended particulate matter (SPM)
(*c*) Ozone layer depletion
(*d*) Loss of biodiversity

731. What is the global warming potential (GWP) of nitrous oxide (N_2O) gas w.r.t., CO_2?
(*a*) 21 (*b*) 58 (*c*) 270 (*d*) 11300

732. Ozone layer is found in the stratosphere between ______ km above the earth's surface.
(*a*) 1 to 10 (*b*) 10 to 50
(*c*) 50 to 80 (*d*) 80 to 400

733. The commitment period for GHG emission reductions of Annex-I countries as per Kyoto protocol is
(*a*) since 2005 (*b*) 2008-2012
(*c*) 2012-2016 (*d*) 2016-2020.

734. Which of t he following air pollutants is not emitted from furnace oil fired boilers and furnaces?
(*a*) NO_x (*b*) SO_2 (*c*) CFC's (*d*) CO

735. Carbon emission reduction per year by replacement of 60 Watts incandescent lamp

Answers

720. (*d*) 721. (*b*) 722. (*d*) 723. (*d*) 724. (*d*) 725. (*d*) 726. (*b*) 727. (*b*) 728. (*c*)
729. (*c*) 730. (*b*) 731. (*c*) 732. (*b*) 733. (*b*) 734. (*c*) 735. (*a*)

with 15 watts CFL (Compact Fluorescent Lamp), if carbon emission per unit is 1 kg CO_2/kWh and annual burning/usage is 3000 hours, will be ______ ton.
(*a*) 0.135 (*b*) 2.565 (*c*) 179.5 (*d*) 29.30

736. Which of the following green house gases has the least impact on global warming?
(*a*) O_3 (*b*) CO (*c*) CH_4 (*d*) N_2O

737. The process of which Annex-I countries can invest in greenhouse gas mitigation projects in developing countries is called
(*a*) Green Technology
(*b*) Certified Emission Reduction (CER)
(*c*) Conference Of Party (COP)
(*d*) Clean Development Mechanism (CDM)

738. One Certified Emission Reduction (CER) in equivalent of CO_2 emission is ______ ton of CO_2.
(*a*) 0.01 (*b*) 1 (*c*) 0.1 (*d*) 10

739. Which of the following is used to control SO_x in Fludised Bed Combustion (FBC) and fired boilers?
(*a*) Ammonia (*b*) Quartzite
(*c*) Limestone (*d*) Lime

740. Green house effect refers to increase in
(*a*) greenery
(*b*) atmospheric pressure
(*c*) atmospheric temperature
(*d*) carbon monoxide

741. Which of the following gases causes global warming?
(*a*) SO_2 (*b*) O_2 (*c*) CFC
(*d*) None of these

742. Which of the following does not damage ozone layer?
(*a*) HCFC (*b*) HFC (*c*) $CC1_4$ (*d*) CFC

743. The ozone layer in the stratosphere acts as an efficient filter for
(*a*) Infrared rays (*b*) UV-B rays
(*c*) X-rays (*d*) UV-A rays

744. Green house gas emissions are represented in ______ units.
(*a*) MMTOE (*b*) MMTCE
(*c*) MMTME (*d*) MMT

745. Ozone layer is destroyed when it absorbs UV (Ultravoilet) radiation with wavelength greater than ______ nm (nanometer; 1 nm = 1^{-9} m)
(*a*) 55 (*b*) 100 (*c*) 240 (*d*) 290

746. What is the approximate % share of CO_2 in total green house gas emission?
(*a*) 25% (*b*) 55% (*c*) 75% (*d*) 95%

747. Prototype carbon fund has been established by
(*a*) U.S.A (*b*) world bank
(*c*) united nations
(*d*) international monetary fund

748. Which out of the following is the greatest contribution to the air pollution?
(*a*) Hydrocarbons (*b*) Sulfur oxides
(*c*) Particulates (*d*) Carbon monoxide

749. Pick out the wrong statement:
(*a*) The spread of stratosphere above the earth's surface is between 10 to 50 km.
(*b*) The pungent smelling and light bluish gas in stratosphere is due to ozone molecules.
(*c*) The ozone molecules are easily broken by chlorine and bromine compounds.
(*d*) CFC stands for Compact Fluro Carbon.

750. The main source of CFC (responsible for ozone layer depletion) in household sector is
(*a*) washing machines (*b*) televisions
(*c*) refrigerators (*d*) vacuum cleaners.

751. Maximum CO_2 (GHG) is emitted in the atmosphere by
(*a*) U.S.A (*b*) China
(*c*) India (*d*) Japan

752. The raise in present temperature of earth compared to 100 years ago is about ______ °C.

Answers

736. (*b*) 737. (*d*) 738. (*b*) 739. (*c*) 740. (*c*) 741. (*c*) 742. (*b*) 743. (*b*) 744. (*b*)
745. (*d*) 746. (*b*) 747. (*b*) 748. (*d*) 749. (*d*) 750. (*c*) 751. (*a*) 752. (*a*)

(a) 0.3 to 0.6 (b) 2 to 4
(c) 0.3 to 0.04 (d) 5 to 6

753. Pick out the wrong statement:
(a) The agency to look after the climate change and fraction to cut GUG emission is World Health Organisation.
(b) India ratified the Kyoto protocol in year 2002.
(c) The number of major green house covered for their reduction by the Kyoto protocol are six.
(d) The fund intended to invest in projects that will reduce green house gas emissions is Prototype Carbon Funds

754. Which of the following refrigerant has a zero ozone depletion potential?
(a) HCFC 22 (b) HCFC 123
(c) HCFC 124 (d) HCFC 134a

755. Which of the following is not an oxidant causing air pollutant?
(a) H_2O_2 (b) PAN
(c) O_2 (d) HCl fumes

756. Most odourous sulfur compound as air pollutant is
(a) methyl mercaptans
(b) hydrogen sulfide (c) sulfur dioxide
(d) sulfuric acid fumes

757. Match the occupational diseases caused by various causative agents

	Causative agents/ pollutants		Occupational disease
A.	Coal dust particles	I.	Byssinosis
B.	Cotton dust particles	II.	Siderosis
C.	Iron dust particles	III.	Chalcosis
D.	Copper dust particles	IV.	Pneumo-coniosis

(a) A-IV, B-I, C-II, D-IV
(b) A-IV, B-III, C-II, D-I
(c) A-II, B-III, C-I, D-IV
(d) A-I, B-III, C-II, D-IV

758. Which of the following materials is fairly resistant to sulfur dioxide attack?
(a) Nylon (b) Leather
(c) Aluminium (d) Paper

759. Air pollution by Beryllium is caused from following industry?
(a) Coal mining industry
(b) Fertiliser industry
(c) Rocket fuel (d) Pesticide industry.

760. Leprous stone (*e.g.*, yellowing of marble of Taj Mahal) is formed due to air pollution by
(a) O_3 (b) NO_2 (c) SO_2 (d) CO

761. H_2S as an air pollutant is not produced by
(a) petroleum industry
(b) sewage treatment plant
(c) pesticide industry
(d) automobile exhaust.

762. Plant/Vegetation suffer from a disease called ______ due to exposure to polluted air.
(a) necrosis (b) silicosis
(c) tobacosis (d) byssinosis

763. Which of the following air pollutants is emitted from pesticide industry?
(a) Methyl isocynate (MIC)
(b) HF (c) NO (d) SO_2.

764. Arsenic pollution does not take place from the ______ industry.
(a) coal mining (b) paper
(c) petroleum (d) iron ore mining

765. Nickel as an air pollutant is not emitted from
(a) coal mining (b) diesel oil
(c) tobacco smoke
(d) viscose rayon plant

766. SO_2 present is pollutant air will not cause
(a) rapid corrosion of aluminium
(b) acute and chronic vegetable leaf injury
(c) irritation of upper respiratory tract
(d) destruction of paint pigment

Answers

753. (a) 754. (d) 755. (d) 756. (a) 757. (a) 758. (c) 759. (c) 760. (c) 761. (b)
762. (a) 763. (a) 764. (b) 765. (d) 766. (a)

767. A three way catalytic converter does not convert
(*a*) CO to CO_2 (b) HC to CO_2 & H_2O
(c) NO_x to N_2 (d) none of these

768. Total hydrocarbon in polluted air is commonly determined by
(*a*) Lead peroxide candle method
(*b*) West and Gaeke method
(*c*) Flame ionisation method
(*d*) Dust fall jar method,

769. If chimney of an industrial plant is emitting particulate matter @Q ton/hr, the stack height in meter is given by (subject to not less than 30 metres).
(*a*) $74.Q^{0.27}$ (*b*) $14Q^{0.3}$ (*c*) $14Q^{0.27}$ (*d*) $74Q^2$

770. Recommended maximum noise level in a residential building is _______ decibels.
(*a*) 35 (*b*) 55 (*c*) 75 (*d*) 85

771. Pick out the wrong statement about meterology of air pollutant dispersion
(*a*) Classical smog (London smog) is formed by interaction of SO_2 with particulate matter at temperature of 2 to 5°C.
(*b*) Environmental lapse rate (ELR) and dry adiabatic lapse rate (DALR) are equal in case of neutral atmospheric condition.
(*c*) When the temperature increases with altitude, atmospheric condition becomes very unstable.
(*d*) Normally subsidence inversion occurs at high altitude while radiation inversion occurs near the earth surface.

772. When the environmental lapse rate is higher than adiabatic lapse rate
(*a*) inversion layer will not occur
(*b*) atmospheric condition is termed as stable
(*c*) air pollutants will be dispersed quickly
(*d*) all (*a*), (*b*) and (*c*).

773. Industrial chimney located near tall buildings should be at least _______ times the height of the building.
(*a*) 0.5 (*b*) 2.5 (*c*) 1 (*d*) 5.5

774. Air borne disease caused due to hide handling is
(*a*) anthrax (*b*) lung cancer
(*c*) hypertension (*d*) asthma

775. Workers working is _______ industry are liable to suffer from silicosis disease.
(*a*) stone crushing (*b*) textile
(*c*) sugar (*d*) storage battery

776. Gaseous fluorine is not emitted from _______ industry.
(*a*) fertiliser (*b*) sugar
(*c*) aluminium (*d*) brick and tile

777. Bexyl1ium fumes is emitted from
(*a*) auto exhaust
(*b*) fluorescent lamp industry
(*c*) thermal power plant
(*d*) fertiliser industry

778. Cadmium is not emitted to atmosphere from
(*a*) tobacco smoke (*b*) coal mining
(*c*) zinc mining (*d*) viscose rayon plant

779. Which of the following equipment is not used for dust control in industry?
(*a*) Multicyclones (*b*) Bag filter
(*c*) Trickling filter (*d*) Scrubber.

780. Microscopic solid or liquid particles disappeared in atmospheric air is termed as
(*a*) fume (*b*) smog
(*c*) mist (*d*) aerosol

781. Of the total air pollution of a city, generally automobile exhaust alone contributes about _______ per cent of the total
(*a*) 10-20 (*b*) 25-35
(*c*) 50-60 (*d*) 85-90

782. Stack height in metre of an industrial plant emitting SO_2 @ Q kg/hr is given by (subject to a minimum height of 30 metres)
(*a*) $14Q^{0.3}$ (*b*) $Q^{2.5}$ (*c*) $Q^{0.3}$ (*d*) $Q^{2.7}$

783. Carbon monoxide present in polluted air does not cause

Answers

767. (*d*)	768. (*c*)	769. (*a*)	770. (*b*)	771. (*c*)	772. (*d*)	773. (*b*)	774. (*a*)	775. (*a*)
776. (*b*)	777. (*b*)	778. (*d*)	779. (*c*)	780. (*d*)	781. (*c*)	782. (*a*)	783. (*d*)	

(a) respiratory problem
(b) typhoid (c) dysentery
(d) all (a), (b), and (c)

784. Carbon monoxide reacts with hemoglobin of red blood cells forming carbonxyhemoglobin causing asphyxiation as per the following reaction:
(a) $HbO_2 + 2CO = Hb + 2CO_2$
(b) $HbO_2 + CO = COHb + O_2$
(c) $HbO_2 + 2CO = Hb + 2C + 2O_2$
(d) none of these

785. Fly ash emitted from chimney of thermal power plant can be used
(a) in land as fertiliser
(b) for disinfection of water
(c) for brick manufacturing
(d) all (a), (b), and (c)

786. Pick out the wrong statement about solid water management
(a) Sanitary land fill method of solid waste disposal is suitable for garbage disposal.
(b) Nuisance from rodent and fly is controlled in the disposal of solid waste by sanitary landfill method.
(c) Disposal of solid waste by composting is efficient with temperature and pH control.
(d) The effective measure of solid waste disposal employs the principle of "4 R's" which includes reduction, recycling, reuse and recovery.

787. Windrow is a method of solid waste disposal by
(a) sanitary landfill (b) composting
(c) incineration (d) none of these

788. Which of the following disease is not caused by virus?
(a) Polio (b) Influenza
(c) Dengu (d) Typhoid

789. Presence of which of the following organism is an indicator of sewage pollution?
(a) Volvox (b) Escherichia Coli
(c) Aerobacter aerogene
(d) Spirogyra

790. Property of earth to allow water to pass through it is termed as 'perviousness'. What is the capacity of soil to absorb moisture is generally known as?
(a) Permeability (b) Porosity
(c) Transmissibility
(d) Infiltration capacity

791. Water borne amoebic dysentery is caused by
(a) virus (b) protozoa
(c) bacteria (d) all (a), (b) and (c)

792. E. Coli bacterias die in water having pH greater than
(a) 5.5 (b) 6.5 (c) 7.5 (d) 9.5.

793. Presence of pathogenic bacteria in drinking water may cause
(a) cholera (b) dysentery
(c) typhoid (d) all (a), (b) and (c)

794. Match the maximum permissible concentration of various contaminants in treated water of public water supply for domestic use.

Water contaminants		Maximum permissible concentration, ppm	
I.	Chloride content	A.	0.35
II.	Nitride content	B.	45
III.	Nitrate content	C.	Nil
IV.	Fe and Mn content	D.	250

(a) I-D, II-C, III-B, IV-A
(b) I-A, II-B, III-C, IV-D
(c) I-D, II-C, III-A, IV-B
(d) I-C, II-D, III-A, IV-B

795. Methanoglobinemia or blue baby disease is caused due to the presence of _______ air drinking water.

Answers

784. (b) 785. (c) 786. (a) 787. (b) 788. (d) 789. (b) 790. (d) 791. (b) 792. (d)
793. (d) 794. (a) 795. (c)

(*a*) sulfides (*b*) nitrites
(*c*) nitrates (*d*) chlorides

796. Presence of high algal content in water indicates that water is
(*a*) hard (*b*) soft
(*c*) turbidity free (*d*) acidic.

797. Biochemical Oxygen Demand (BOD) of safe drinking water must be _______ mg/litre.
(*a*) 0 (*b*) 3-5 (*c*) 7-9 (*d*) < 10

798. To ensure proper growth of teeth of children, the quantity of fluoride used in water supply is _______ mg/litre.
(*a*) 0.1 (*b*) 1 (*c*) 5 (*d*) 10

799. Disinfection of water is done by chlorination for removal of
(*a*) color (*b*) turbidity
(*c*) bacteria (*d*) odour

800. Pick out the wrong statement about water treatment:
(*a*) Aeration of water is done for removal of odour.
(*b*) Treatment of water with alum $[(NH_4)SO_4]$ to control algae increases its hardness and alkalinity.
(*c*) Dissolved carbon dioxide can be removed from water supply main by aeration.
(*d*) Use of hard water causes scaling of boiler tubes and corrosion and incrustation of pipes.

801. Hard water for public water supply is discarded because, it
(*a*) contains pathogenic (disease causing) bacteria
(*b*) consumes more soap
(*c*) contains lot of turbidity
(*d*) makes detergents non-biodegradable.

802. Most commonly used coagulant for water treatment is
(*a*) lime (*b*) bleaching powder
(*c*) alum (*d*) chlorine

803. Pick out the wrong statement about water treatment:
(*a*) Efficiency of removing bacteria from water by a slow sand filter is about 98-99%.
(*b*) Wash water is required for slow sand filter.
(*c*) Ozone removes bacteria as well as color I and odour from water.
(*d*) Most ideal disinfectant for drinking water is chlorine.

804. Which of the following is not a chemical used for dechlorination of water?
(*a*) Chloramine (*b*) Sodium thiosulfate
(*c*) Sulfur dioxide (*d*) Sodium sulfate

805. Pick out the wrong statement about sewerage system:
(*a*) The sewer which receives the discharge from a collecting system and delivers it to the treatment plant is known as trunk sewer.
(*b*) The rate of flow of sewage is generally assumed less than the rate of water supply.
(*c*) Fresh sewage is generally highly acidic in nature.
(*d*) The sewer which resists sulfide corrosion is stone ware pipes.

806. The gas which may cause explosion in sewer pipes is
(*a*) CO (*b*) CO_2 (*c*) H_2S (*d*) CH_4

807. A circular sewer section is preferred because it provides maximum
(*a*) economy in construction cost
(*b*) hydraulic main depth
(*c*) area for a given perimeter
(*d*) all (*a*), (*b*) and (*c*).

808. Clogging of sewer pipes is not caused due to
(*a*) sitting and low discharge
(*b*) domestic wastes thrown in manholes
(*c*) greasy and oily matters
(*d*) alkalinity of sewage.

Answers

796. (*d*) 797. (*a*) 798. (*b*) 799. (*c*) 800. (*b*) 801. (*b*) 802. (*c*) 803. (*b*) 804. (*a*)
805. (*c*) 806. (*d*) 807. (*d*) 808. (*d*)

809. The non-clog pump which permits solid mater to pass out with the liquid sewage is
(*a*) reciprocating pump
(*b*) centrifugal pump (*c*) pneumatic ejector
(*d*) gear pump

810. Which of the following gases is generally not found in sewers?
(*a*) CO (*b*) CO_2 (*c*) H_2S (*d*) CH_4

811. Pick out the wrong statement about sewage:
(*a*) If the flame of a miner's safety lamp in a manhole extinguishes wit h in five minutes, the sewer certainly contains CO_2.
(*b*) If paper moistened with lead acetate when placed in a manhole for five minutes. turns black. the sewer certainly contains H_2S.
(*c*) Large sewers are cleaned by flushing.
(*d*) Asbestors cement sewers are susceptible to corrosion by sulfuric acid.

812. Match the pumping duty for which various types of pumps are suitable

Type of pumps	Suitable pumping duty
I. Axial flow pump	A. Pumping muddy water and sewage sludge through low height
II. Turbine Pump	B. Lifting activated sludge
III. Screw pump	C. Pumping clean water to higher level
IV. Diaphragm pump	D. Pumping of storm water

(*a*) I-D, II-C, III-B, IV-A
(*b*) I-A, II-B, III-C, IV-D
(*c*) I-D, II-C, III-A, IV-B
(*d*) I-C, II-B, III-D, IV-A

813. In sewer pipes, a manhole is generally provided at each
(*a*) junction (*b*) bend
(*c*) change of gradient
(*d*) all (*a*), (*b*) and (*c*).

814. Dried sewage after treatment is used as a
(*a*) base material for paints
(*b*) building material
(*c*) chemical for lowering B.O.D
(*d*) fertiliser

815. Temperature does not affect the ________ in sewage.
(*a*) biological activity of bacteria
(*b*) solubility of gases (*c*) viscosity
(*d*) alkalinity

816. As compared to sea water fresh river water contains ________ oxygen.
(*a*) 20% more (*b*) 20% less
(*c*) 5% less (*d*) 50% more

817. Sewage treatment is not required if dilution factor is
(*a*) 50-100 (*b*) 150-200
(*c*) 300-350 (*d*) 500

818. Which of the following crop should not be grown by sewage irrigation?
(*a*) Banana (*b*) Cotton
(*c*) Potato (*d*) Coconut

819. The coagulant which is generally not used for treating the sewage is
(*a*) chlorinated copperas
(*b*) ferric chloride (*c*) ferric sulfate
(*d*) alum.

820. Presence of high algae content in water indicates that water is
(*a*) having high B.O.D.
(*b*) having high C.O.D.
(*c*) free from oxygen.
(*d*) saturated with oxygen.

821. In sewage having fully oxidised organic matter, nitrogen exists in the form of nitrates. The presence of free ammonia in sewage is detected by
(*a*) adding H_2SO_4 (*b*) boiling
(*c*) adding $KMnO_4$
(*d*) phenol-disulfuric acid.

Answers

809. (*b*) 810. (*a*) 811. (*c*) 812. (*a*) 813. (*d*) 814. (*d*) 815. (*d*) 816. (*a*) 817. (*d*)
818. (*c*) 819. (*d*) 820. (*d*) 821. (*b*)

822. Dilution method of disposing off sewage is not preferred when the diluting water has
(*a*) to be added to fresh sewage
(*b*) high dissolved oxygen content
(*c*) very high C.O.D.
(*d*) low dissolved oxygen content

823. If dissolved oxygen concentration falls down to zero in any drainage, it indicates zone of ______ in which algae dies out completely.
(*a*) recovery (*b*) degradation
(*c*) active decomposition
(*d*) cleaner water

824. Generally, the activity of enzymes is not affected by
(*a*) temperature pH value
(*b*) pH value
(*c*) substrate concentration
(*d*) none of these,

825. Pick out the wrong statement about sewage:
(*a*) In sewage, generally B.O.D. is higher than C.O.D.
(*b*) Fresh sewage becomes stale in 3-4 hours.
(*c*) Water content of sewage is about 99.9%.
(*d*) Sewage in oxidation ponds is treated by the action of bacteria and algae both.

826. The digested sludge from septic tanks is removed after a maximum period of ______ years.
(*a*) 1 (*b*) 3 (*c*) 5 (*d*) 7

827. Liquid waste from kitchen, bathrooms and wash basins excluding urinals and water closets is known as 'sullage'. Dry weather flow is the liquid waste generated from
(*a*) storm run off (*b*) industrial effluent
(*c*) domestic discharge
(*d*) all (*a*), (*b*) and (*c*) together.

828. Decomposition of organic matter is carried out by ______ bacteria in septic tanks.
(*a*) aerobic (*b*) anaerobic
(*c*) facultative (*d*) none of these.

829. The gas mainly produced in a sludge digestor is
(*a*) nitrogen (*b*) oxygen
(*c*) hydrogen (*d*) methane

830. Most suitable method of solid waste disposal in rural area is
(*a*) composting (*b*) land filling
(*c*) incineration (*d*) pyrolysis.

831. Aerobic activity is maximum in
(*a*) fresh sewage (*b*) stale sewage
(*c*) sewer pipe (*d*) sewage

832. Fertilising element applied by sewage farming is
(*a*) potassium (*b*) phosphorus
(*c*) nitrogen (*d*) all (*a*), (*b*) and (*c*)

833. The term 'elutriation' is related to
(*a*) forced sedimentation
(*b*) land disposal of sewage
(*c*) washing of sludge
(*d*) none of these

834. Particle settlement in a sedimentation tank is affected by the
(*a*) viscosity of water (*b*) flow velocity
(*c*) size, shape and specific gravity of solid
(*d*) all (*a*), (*b*) and (*c*).

835. Pick out the wrong statement:
(*a*) The percentage ratio of oxygen available to the oxygen required for complete stabilisation of a sewage sample is known as relative stability.
(*b*) The effluent of a septic tank is fit for discharge into open drain.
(*c*) The process of decomposing the organic matter under controlled anaerobic conditions is called sludge digestion which produces 75% CH_4 containing gases.
(*d*) In normal condition, period of sludge digestion is 30 days.

Answers

822. (*c*) 823. (*c*) 824. (*d*) 825. (*a*) 826. (*b*) 827. (*c*) 828. (*b*) 829. (*d*) 830. (*a*)
831. (*a*) 832. (*d*) 833. (*c*) 834. (*d*) 835. (*b*)

836. For the design of sewers in India, % of sewage discharge is assumed to be ______ % of water supplied from water works
(*a*) 50-60 (*b*) 75-80 (*c*) 90-95 (*d*) 100

837. The most suitable section of a sewer in a combined or separate sewerage system is circulate. However, an e.g., shaped sewer, when compared to a circular sewer.
(*a*) is easier to construct
(*b*) is more stable (*c*) is economical
(*d*) provides better self cleaning velocity at

838. Branch and main sewers are widely made of
(*a*) reinforced cement concrete (RCC)
(*b*) prestressed cement concrete
(*c*) cast iron (*d*) glazed stoneware

839. Match the type of joint used for joining various sewer pipes

Types sewer pipes to be joined		Preferred joint used	
I.	Joining smaller sewer pipes of diameter 0.6 metre	A.	Collar joint
II.	Joining bigger sewer pipes of diameter 0.6 metre	B.	Simplex joint
III.	Joining asbestos cement pipes	C.	Spigot and socket joint

(*a*) I-A, II-B, III-C (*b*) I-C, II-B, III-A
(*c*) I-A, II-C, III -B (*d*) I-C, II-A, III-B

840. A pipe installed in house drainage for ventilation purposes is called
(*a*) waste pipe (*b*) soil pipe
(*c*) vent pipe (*d*) antisiphonage pipe

841. Which of the following is not a man made source of radio activity in the environment?
(*a*) Cosmic ray
(*b*) Radioisotope production
(*c*) Nuclear explosion
(*d*) Nuclear power plants

842. Which of the following radioisotope is capable of contaminating the milk of cows and other animals due to their exposure to the radiation caused by that isotope?
(*a*) Cesium-137 (*b*) Cobalt-60
(*c*) Strontium-90 (*d*) Iodine-131

843. A circular sewer section looses most of its merits, when the sewage depth becomes less than ______ full.
(*a*) 1/2 (*b*) 3/4 (*c*) 2/3
(*d*) none of these.

844. B.O.D removal in a oxidation pond may be upto Spigot and socket joint percent.
(*a*) 70 (*b*) 80 (*c*) 90 (*d*) 99

845. When chlorine is added to sewage both at the beginning as well as at the end of the treatment process, the phenomenon is called ______ chlorination.
(*a*) break point (*b*) post
(*c*) super (*d*) split

846. The end product formed after separation and anaerobic bacterial digestion of organic municipal solid wastes is called
(*a*) leachate (*b*) humus
(*c*) sewage (*d*) refuse.

817. Particles of around one micron (10^{-6} metre) size are best removed by
(*a*) slow sand filtration
(*b*) chemical precipitation
(*c*) chemical coagulation
(*d*) plain sedimentation

848. Match the detention time in various treatment units

	Treatment units		Detention time
A.	Primary sedimentation	I.	2 hours
B.	Sludge digestion	II.	6 hours
C.	Grit chamber	III.	2 minutes
D.	Activated sludge	IV.	20 days

Answers

836. (*b*) 837. (*d*) 838. (*a*) 839. (*d*) 840. (*c*) 841. (*a*) 842. (*c*) 843. (*a*) 844. (*c*)
845. (*d*) 846. (*b*) 847. (*c*) 848. (*a*)

(*a*) A-I, B-IV, C-III, D-II
(*b*) A-IV, B-III, C-II, D-I
(*c*) A-I, B-II, C-IV, D-III
(*d*) A-III, B-IV, C-II, D-I

849. Sewage sickness occurs, when the sewage
(*a*) clogs the voids of soil due to continuous application of sewage on a piece of land
(*b*) contains pathogenic organism
(*c*) enters the water supply system
(*d*) clogs the sewers due to accumulation of solids

850. Match the property of various sewer pipe materials of construction

	Sewer pipe material		Property of the pipe material
A.	Steel sewer pipe	I.	Corrosion resistant to most acids.
B.	Concrete sewer pipe	II.	Corrosion resistance is most natural soils.
C.	Stoneware sewer pipe	III.	Unable to withstand high external load.
D.	Cast iron sewer pipe	IV.	Unsuitable where soil contains excessive sulfates.

(*a*) A-III, B-IV, C-I, D-II
(*b*) A-IV, B-III, C-II, D-I
(*c*) A-IV, B-III, C-I, D-II
(*d*) A-III, B-IV, C-II, D-I

851. Which of the following' materials is used as landfill sealant for the control of gas and leachate movement?
(*a*) Flyash (*b*) Sand
(*c*) Bentonite (*d*) Lime

852. In the oxidation ditch, the excess sludge is taken to
(*a*) aerobic digestor (*b*) incinerator
(*c*) anaerobic digestor
(*d*) drying beds

853. Two main gases liberated from an anaerobic sludge digestion tank are
(*a*) CH_4 and H_2S (*b*) CH_4 and CO_2
(*c*) NH_3 and CO_s (*d*) NH_3 and CH_4

854. Sludge bulking can be controlled by
(*a*) denitrification (*b*) chlorination
(*c*) aeration (*d*) coagulation

855. With rise in temperature, ultimate B.O.D. value of waste water
(*a*) Increases (*b*) decreases
(*c*) remain constant
(*d*) doubles for every 10°C rise

856. A sewer is mainly designed to attain self-cleaning velocity at
(*a*) average hourly rate of flow
(*b*) minimum hourly rate of flow
(*c*) peak hourly rate of flow
(*d*) sewer running half full

857. Which of the following sewage treatment method requires the largest area for the same discharge?
(*a*) Oxidation pond (*b*) Oxidation ditch
(*c*) Trickling filter (*d*) Anaerobic pond

858. The daily cover of municipal solid waste landfills consists of
(*a*) bentonite (*b*) compacted soil
(*c*) lime (*d*) sand

859. Eutrophication of lakes is primarily due to
(*a*) multiplication of bacteria
(*b*) excessive inflow of nutrients
(*c*) increase in benthic organisms
(*d*) thermal and density currents.

860. Presence of nitrogen in a waste water sample is due to the decomposition of
(*a*) fats (*b*) vitamins
(*c*) protein (*d*) carbohydrates.

861. In context of water polluted with sewage BOD signifies
(*a*) biology of degradation
(*b*) bacteriological oxygen demand

Answers

849. (*a*) 850. (*a*) 851. (*c*) 852. (*d*) 853. (*b*) 854. (*b*) 855. (*c*) 856. (*b*) 857. (*a*)
858. (*b*) 859. (*b*) 860. (*c*) 861. (*d*)

(*c*) biological oxygen demand
(*d*) bio-chemical oxygen demand.

862. Iron may be present in significant quality in
(*a*) irrigation return water
(*b*) septic tank effluent
(*c*) fresh water from deep tubewells
(*d*) rain water.

863. The grown algae is useful in the operation of
(*a*) septic tanks (*b*) oxidation ponds
(*c*) slow and filters
(*d*) sludge digestion lagoons.

864. Secondary treatment of sewage removes
(*a*) suspended organic matter
(*b*) dissolved and colloidal organic matter
(*c*) fats and grease
(*d*) pathogenic (disease causing) bacteria

865. Pick out the wrong statement:
(*a*) Biochemical oxygen demand of safe drinking water is zero.
(*b*) The amount of oxygen consumed by sewage from potassium dichromate is termed as C.O.D.
(*c*) The color of fresh domestic sewage is slightly more than l.
(*d*) The specific gravity of sewage is slightly more than 1.

866. Most commonly used chemical for controlling the algae is
(*a*) alum (*b*) copper sulfate
(*c*) bleaching powder (*d*) lime.

867. The stabilisation of sewage in oxidation pond is due to symbiotic action of
(*a*) inorganic and organic compounds
(*b*) anaerobic and aerobic bacteria
(*c*) protozoa and algae
(*d*) bacteria and algae

868. Eutrophic condition in lakes are due to excess of
(*a*) turbidity (*b*) color
(*c*) suspended solid (*d*) algae.

869. Bacteria which use CO_2 as a source of carbon are known as
(*a*) heterotrophic (*b*) autotrophic
(*c*) anaerobic (*d*) microbacteria.

870. Match the contaminants to their environmental degradation significance

	Water pollutant		Environmental significance
A.	Suspended solids	I.	Causes sludge deposits
B.	Dissolved inorganic solids	II.	Interferes with effluent reuse
C.	Heavy metals	III.	Toxic, may interfere with effluent reuse
D.	Nutrients e.g., N and P	IV.	Causes eutrophication of lakes

(*a*) A-IV, B-III, C-II, D-I
(*b*) A-IV, B-II, C-I, D-III
(*c*) A-III, B-I, C-II, D-IV
(*d*) A-I, B-II, C-III, D-IV

871. Velocity control device is an essential component of
(*a*) rapid sand filter
(*b*) flow equalisation tank
(*c*) activated sludge process
(*d*) grit chamber

872. Pick out the wrong statement:
(*a*) Algal growth is not desirable in waste stabilisation ponds.
(*b*) The sewage strength in stabilisation pond is reduced by the principle of algal-bacterial symbiosis.
(*c*) About 80% of suspended solids are 1 removed in plain sedimentation of sewage.
(*d*) The sewage in a waste stabilisation pond is treated by algae.

873. Municipal sewer is separated from house drainage system by ________ trap.

Answers

862. (*c*) 863. (*b*) 864. (*b*) 865. (*c*) 866. (*d*) 867. (*c*) 868. (*d*) 869. (*b*) 870. (*d*)
871. (*d*) 872. (*c*) 873. (*b*)

(*a*) gully (*b*) intercepting
(*c*) S- (*d*) P-

874. Septic tank is the most suitable unit for
(*a*) hotels (*b*) hostels
(*c*) hospitals (*d*) all (*a*), (*b*) and (*c*).

875. B.O.D is preferred to C.O.D as an index of sewage concentration, because
(*a*) B.O.D represents both nitrogeneous and carbonaceous organic matter, while C.O.D may indicate only carbonaceous matter.
(*b*) B.O.D relates specifically to putrescible(decomposable) organic matter, which is the most objectionable sewage constituent.
(*c*) C.O.D relates to the impurities which can only be removed by chemical treatment which is expensive.
(*d*) B.O.D test is easier to perform and gives more reliable results.

876. Re-aeration rate of river does not depend upon the _______ of river water.
(*a*) depth (*b*) velocity
(*c*) temperature (*d*) none of these

877. Match the following:

	Column I		Column II
A.	Vent pipe	I.	Carries liquid waste that does not contain human excreta.
B.	Soil pipe	II.	Carries liquid waste including human excreta.
C.	Antisi-phonage pipe	III.	Preserves the water seal of trap through access to atmospheric air.
D.	Waste pipe	IV.	Provides flow of air to or from drainage system in order to prevent vacuum pressures and excessive pressure and provides escape for fuel gases.

(*a*) A-IV, B-III, C-II, D-I
(*b*) A-I, B-II, C-IV, D-III
(*c*) A-IV, B-II, C-III, D-I
(*d*) A-III, B-IV, C-II, D-I

878. Under normal condition of now, an unpolluted river will contain
(*a*) more dissolved oxygen in summer than in winter
(*b*) less dissolved oxygen in summer than in winter
(*c*) same amount of dissolved oxygen in winter and summer
(*d*) the least amount of dissolved oxygen during the floods.

879. Correct sequence of unit operations in a sewage treatment plant is
I. Screening II. Grit removal
III. Secondary sedimentation
IV. Aeration
V. Primary sedimentation
(*a*) I, II, V, IV, III (*b*) II, I, III, V, IV
(*c*) I, II, III, IV, V (*d*) II, I, IV, V, III

880. Pick out the wrong statement:
(*a*) The normal temperature of sewage is higher than supply water.
(*b*) Strength of waste water is measured by its pH valve.
(*c*) The pipe installed in the house drainage to preserve the water seal is called antisiphonage pipe.
(*d*) The leakage of sewage from sewers into the surrounding is known as infiltration.

881. Match the various sewage treatment processes with the biological agent used

	Processes		Biological agents
A.	Oxidation ditch	I.	Facultative bacteria
B.	Waste stabilisation pond	II.	Anaerobic bacteria
C.	Imholf tank	III.	Aerobic bacteria (suspended culture)

Answers

874. (*d*) 875. (*b*) 876. (*d*) 877. (*c*) 878. (*b*) 879. (*a*) 880. (*b*) 881. (*c*)

D.	Rotating Biological Contractor (RBC)	IV.	Aerobic bacteria (attached culture)

(*a*) A-III, B-II, C-I, D-IV
(*b*) A-IV, B-III, C-II, D-I
(*c*) A-III, B-I, C-II, D-IV
(*d*) A-II, B-IV, C-I, D-III

882. The waste stabilisation pond can be
(*a*) anaerobic (*b*) aerobic
(*c*) facultative (*d*) any of these.

883. Which of the following pairs is not correctly matched about waste water characteristics?
(*a*) C.O.D: Biodegradability of waste water
(*b*) B.O.D: strength of sewage
(*c*) Nitrate: Methemoglobinemia (blue baby disease)
(*d*) Methane: Product of anaerobic decomposition.

884. The entry of foul smelling gases into the house coming from the sewers can be prevented by providing
(*a*) water seals for all the fixtures
(*b*) exhaust fans
(*c*) water seals for all the fixtures and a vent pipe in the plumbing system
(*d*) efficient vent pipes in the pumpling system.

885. Which of the following will help to prevent the escape of foul sewer gases from a water closet?
(*a*) Vent pipe (*b*) Air gap
(*c*) Gully trap (*d*) none of these

886. Match the sewage impurities removed by various sewage treatment units

Impurities in sewage		Sewage treatment units	
A.	Large floating matter	I.	Trickling filter
B.	Suspended inorganic matter	II.	Primary clarifier
C.	Suspended organic matter	III.	Grit chamber
D.	Dissolved organic matter	IV.	Screens

(*a*) A-IV, B-III, C-II, D-I
(*b*) A-IV, B-III, C-I, D-II
(*c*) A-III, B-IV, C-II, D-I
(*d*) A-III, B-IV, C-I, D-II

887. Which of the following processes is not involved in self purification of running water streams?
(*a*) Dilution (*b*) Sedimentation
(*c*) Oxidation (*d*) Coagulation

888. Waste water solid digestion is done to
(*a*) separate liquids from solids to facilitate drying
(*b*) decompose enough of the organic matter to avoid creating a nuisance
(*c*) decrease the bulk of the waste water solids to facilitate handling
(*d*) all (*a*), (*b*) and (*c*).

889. Corrosion of concrete sewers are generally caused due to
(*a*) higher concentration of grit in the sewage
(*b*) high pH value of the sewage
(*c*) high velocity of flow of sewage
(*d*) septic condition of the sewage

890. Pick out the wrong statement:
(*a*) Sludge digestion in a gobar gas plant is done by anaerobic bacteria.
(*b*) Permissible B.O.D of sewage effluents when discharged into river is 200 mg/litre
(*c*) BOD value of medium strength domestic sewage could be above 200 mg/litre.
(*d*) The optimum temperature [or sludge digestion is 25°C.

891. % of activated sludge mixed with the sewage in primary settling tanks is
(*a*) 5-10 (*b*) 15-20 (*c*) 20-40 (*d*) 40-60

Answers

882. (*d*) 883. (*a*) 884. (*c*) 885. (*c*) 886. (*a*) 887. (*d*) 888. (*d*) 889. (*d*) 890. (*b*)
891. (*c*)

892. The first stage of sludge digestion process is
(*a*) alkaline fermentation
(*b*) acidic fermentation
(*c*) acid regression (*d*) none of these

893. The sludge does not contain waste water from
(*a*) kitchen sinks (*b*) toilets
(*c*) bath rooms (*d*) wash basins.

894. Activated sludge process is used to remove
(*a*) settleable organic matter
(*b*) soluble organic matter
(*c*) inorganic matter and toxic chemicals
(*d*) pathogenic bacteria.

895. Pick out the wrong statement
(*a*) High C.O.D to B.O.D ratio of waste water indicates its high biodegradability.
(*b*) Trickling filter removes dissolved organics from waste water.
(*c*) The hospital waste may be disposed off with municipal wastes.
(*d*) The most critical characteristics of sewage is Biochemical Oxygen Demand (B.O.D), when it is disposed off in river.

896. Indices of water pollution in case of water streams are dissolved oxygen (D.O) and biochemical oxygen demand (B.O.D). The D.O content in self purification of water steams is nearly equal to its saturation valve in the zone of
(*a*) recovery
(*b*) active decomposition
(*c*) degradation (*d*) none of these

897. pH valve range to be maintained for proper functioning of a sludge digestion tank may be
(*a*) 3.5-4.5 (*b*) 6.5-8.5
(*c*) 8.5-10.5 (*d*) < 5

898. Laying of sewer pipes is usually done with the help of
(*a*) slight rails and boning rods
(*b*) a compass (*c*) a plane table
(*d*) atheodolite

899. Match the following:

	Column I		Column II
A.	Water closet	I.	Prevents gases entering a house from public sewers.
B.	Gully trap	II.	Prevents nesting of birds etc.
C.	Cowl	III.	Discharges into soil pipe through floor trap.
D.	Intercepting trap	IV.	Receives human excreta directly.

(*a*) A-IV, B-III, C-II, D-I
(*b*) A-III, B-II, C-IV, D-I
(*c*) A-IV, B-III, C-I, D-II
(*d*) A-III, B-IV, C-II, D-I

900. In combined sewer system, storm water are admitted through
(*a*) leaping weirs (*b*) intercepting weirs
(*c*) inlets (*d*) storm regulators

901. Anaerobic biological treatment of high B.O.D wastes are suggested to
(*a*) obtain end products having a fuel value
(*b*) yield least amount of sludge
(*c*) avoid high requirement of compressed air
(*d*) all (*a*), (*b*) and (*c*).

902. Recirculation in activated sludge process is done to
(*a*) supply seed materials to the aeration tank
(*b*) dilute the incoming sewage
(*c*) dampen the effect of flow variation
(*d*) operate the plant continuously.

903. Which of the following solid waste disposal methods conserves energy most efficiently?
(*a*) Fludised bed combustion
(*b*) Incineration with heat recovery
(*c*) Pyrolysis (*d*) Composting.

904. Which of the following terms describes the addition of water to the atmosphere by plants?

Answers

892. (*b*) 893. (*b*) 894. (*b*) 895. (*d*) 896. (*d*) 897. (*b*) 898. (*a*) 899. (*a*) 900. (*c*)
901. (*d*) 902. (*a*) 903. (*b*) 904. (*a*)

(*a*) Transpiration (*b*) Infiltration
(*c*) Evaporation (*d*) Condensation

905. Pick out the wrong statement:
(*a*) A rock that has been changed due to heat and pressure is an example of metamorphic rock.
(*b*) Carbon cycle is important. to humans because it is a major factor is distributing heat and energy around the earth.
(*c*) The force responsible for driving the water cycle is the earth's rotation.
(*d*) Role played by nitrogen fixing bacteria in the nitrogen cycle is that they convert ammonia to nitrates.

906. Problem encountered by water polluted by excess nutrients is
(*a*) availability of too much oxygen for fish
(*b*) decrease in photosynthesis for aquatic plants
(*c*) increase in aerobic bacteria
(*d*) reduction in water's ability to hold dissolved oxygen.

907. Impact of influx of agricultural waste or manure on an aquatic ecosystem is
(*a*) increase in dissolved oxygen
(*b*) reduction in photosynthesis
(*c*) excessive growth of algae
(*d*) lower biological oxygen demand.

908. Buildings/statue made of ______ are particularly more susceptible to damage by acid rain.
(*a*) Granite (*b*) Wood
(*c*) Marble (*d*) Concrete

909. Which of the following is a hazard of thermal pollution of water?
(*a*) Disruption of aquatic life
(*b*) Pollution of public drinking water
(*c*) Eutrophication
(*d*) Increase in algal content.

810. Thermal pollution of aquatic ecosystem
(*a*) reduces the ability of water to hold dissolved oxygen.
(*b*) reduces the ability of water to hold dissolved CO_2.
(*c*) increases the ability of water to hold dissolved CO_2.
(*d*) does hot have any impact on dissolved CO_2 or O_2.

811. In which of the following methods of solid wastes disposal, methane recovery is a positive trend?
(*a*) Open dumping
(*b*) Sanitary land filling
(*c*) Ocean dumping (*d*) Incineration

912. Which of the following indoor air pollutants has the most significant impact on human health?
(*a*) Cigarette smoke (*b*) Formal dehyde
(*c*) Radon (*d*) Nitrogen oxides.

913. Long term exposure of radon gas can lead to
(*a*) headache
(*b*) skin building syndrome
(*c*) lung cancer (*d*) dizziness.

914. Chronic condition caused by exposure to polluted air is exemplified by
(*a*) skin cancer
(*b*) decrease in O_2 level
(*c*) lung cancer (*d*) emphysema

915. Average residence time of a molecule of CO_2 in atmosphere is ______ years.
(*a*) 1 (*b*) 10 (*c*) 100 (*d*) 1000

916. Sedimentary rocks are a major sink for
(*a*) oxygen (*b*) methane
(*c*) sulfur (*d*) carbon

917. Pick out the wrong statement:
(*a*) Parasites are consumers that feed at all trophic levels.
(*b*) Atmospheric deposition is a non- point source of water pollution.

Answers

905. (*c*) 906. (*b*) 907. (*c*) 908. (*c*) 909. (*a*) 910. (*a*) 911. (*b*) 912. (*a*) 913. (*c*)
914. (*d*) 915. (*c*) 916. (*d*) 917. (*b*)

(*c*) Geothermal spring is a source of thermal pollution.
(*d*) Water temperature is more stable than air temperature.

918. Average residence time of CH_4 in the atmosphere is _______ years.
(*a*) 3 (*b*) 12 (*c*) 125 (*d*) 500

919. Most significant source of methane currently is
(*a*) coal mining (*b*) landfills
(*c*) waste water treatment
(*d*) cultivation of rice.

920. CFC's (chlorofluorocarbons) in the atmosphere decompose from exposure to
(*a*) atomic chlorine (*b*) sunlight
(*c*) ultraviolet radiation
(*d*) stratospheric ozone.

921. Which of the following found in fire extinguishers and naturally released by marine phytoplankton, has more of an impact on the depletion of ozone in the stratosphere than chlorine?
(*a*) Methane (*b*) Bromine
(*c*) Benzene (*d*) Hydrogen

922. What would have been the average temperature (OF) on earth, if the green house effect did not exist?
(*a*) 0 (*b*) 50 (*c*) 150 (*d*) 200

923. Most of the available fresh water worldwide is used for which of the following purposes?
(*a*) Power plant cooling
(*b*) Agriculture (*c*) Public drinking
(*d*) Industrial processes.

924. Pick out the wrong statement:
(*a*) Soil forms as a result of weathering of bed rock.
(*b*) North Atlantic ocean is the warmest of all oceans.
(*c*) Over the past 40 years, the upper 300 metres of ocean water has warmed by an average of 0.56°F.
(*d*) Forest area will be the most affected area by a rise in sea level (out of wetland, grassland, deserts and forests).

925. Which of the following fuels burns cleaner and hotter than others?
(*a*) Oil shale (*b*) Oil
(*c*) Natural gas (*d*) Tar sand

926. Phosphorous is a primary ingradient in fertiliser. Which of the following is a significant phosphorous sink?
(*a*) Ocean water (*b*) Atmosphere
(*c*) Vegetation
(*d*) Deep ocean sediments.

927. _________ cycle is most influenced by non-living processes.
(*a*) Nitrogen (*b*) Water
(*c*) Carbon (*d*) Phosphorous

928. Which of the following is not a factor that influences temperature on earth?
(*a*) Ocean currents (*b*) Latitude
(*c*) Cloud cover (*d*) Altitude

929. Virtually all the molecular oxygen in the atmosphere today comes from
(*a*) respiration of animals
(*b*) photosynthesis of plants
(*c*) volcanic eruptions
(*d*) outgassing

930. Which of the following is a trace element that is cycled through an ecosystem?
(*a*) Hydrogen (*b*) Phosphorous
(*c*) Nitrogen (*d*) Iron

931. An aluminium beverage can takes about _______ to decompose.
(*a*) 40 weeks (*b*) 15 days
(*c*) 10 years (*d*) 100 to 500 years

932. Which of the following contributes to the formation of ozone?
(*a*) SO_2 (*b*) NO (*c*) N_2O (*d*) CO

933. Most of the municipal solid waste (MSW) generated is

Answers

918. (*c*) 919. (*b*) 920. (*c*) 921. (*b*) 922. (*a*) 923. (*b*) 924. (*d*) 925. (*c*) 926. (*a*)
927. (*b*) 928. (*c*) 929. (*b*) 930. (*d*) 931. (*d*) 932. (*b*) 933. (*c*)

(*a*) burned (*b*) recycled
(*c*) placed in landfills
(*d*) recovered and reuse.

934. Which of the following routes of transmission of biological agents to humans has the highest risk?
(*a*) Occupational exposure
(*b*) Skin contact (*c*) Aerosol routes
(*d*) Ingestion

935. Pick out the wrong statement about acid rain and its effects:
(*a*) Acid rain may leach nutrients from the soil leading to decrease in plant growth.
(*b*) Acid rain causes increase in plankton and moss in aquatic ecosystems.
(*c*) Acid rain causes increases in sulfur and nitrogen in the soil.
(*d*) Acid rain causes nitrogen saturation in aquatic ecosystem.

936. Which of the following pH level of water is harmless to most fish?
(*a*) 4 (*b*) 5 (*c*) 6 (*d*) 8

937. At the mouth of a river, there is an area that is depleted in oxygen. Most likely cause of I this may be
(*a*) acid precipitation falling in the ocean
(*b*) thermal pollution of water
(*c*) fertiliser being carried by river
(*d*) overfishing in water

938. Which of the following reactions becomes slower as atmospheric altitude increases?
(*a*) $O_2 + h\nu \longrightarrow O + O$
(*b*) $O + O_3 \longrightarrow O_2 + O_2$
(*c*) $O + O_2 \longrightarrow O_3$
(*d*) $O_3 + h\nu \longrightarrow O_2 + O$

939. As a result of global warming, temperature will rise faster and higher in which of the following areas?
(*a*) Deserts (*b*) Polar regions
(*c*) Equatorial regions
(*d*) Mid-latitudes.

940. Which of the following areas would be most affected by a rise in sea level as a consequence of global warming?
(*a*) Grassland (*b*) Wetland
(*c*) Forests (*d*) Deserts

941. Most major source of ocean pollution is
(*a*) run off from pesticides, fertilisers and sewer discharges
(*b*) oil spills from ships
(*c*) oil exploration and production
(*d*) acid rain.

942. Which of the following sources of CH_4 emission is getting caused?
(*a*) Burning of biomass
(*b*) Anthropogenic activities
(*c*) Wetland
(*d*) Extraction of natural gas and oil.

943. Which of the following is a major contributor to declining fish catches?
(*a*) Overfishing (*b*) Water pollution
(*c*) Lower reproductive rate
(*d*) Global warming

944. Using pesticides and practises such as monoculture decreases biodiversity.
Which of the following will lead to an increase in biodiversity?
(*a*) Decreasing geographic isolation
(*b*) Increased capture rates for fishing
(*c*) Agriculture and logging
(*d*) Increase in global temperature.

945. Which ecosystem produces the most biomass?
(*a*) Desert (*b*) Tropical rain forest
(*c*) Temperature deciduous forest
(*d*) Tundra

946. Which of the following is a major component in acid rain?
(*a*) SO_2 (*b*) SO_3 (*c*) H_2SO_4 (*d*) NO_2

Answers

934. (*c*) 935. (*d*) 936. (*d*) 937. (*c*) 938. (*c*) 939. (*b*) 940. (*b*) 941. (*a*) 942. (*c*)
943. (*b*) 944. (*a*) 945. (*b*) 946. (*c*)

947. Particulate matter
(*a*) includes dust, ash, soot, smoke, pollen and spores
(*b*) solid or liquid particles suspended in gas
(*c*) can directly enter lungs damaging lung tissue and can cause or promote cancer
(*d*) all (*a*), (*b*) and (*c*)

948. Nearly of the cancer caused by' outdoor air pollution are the result of
(*a*) mobile sources (cars. trucks etc.)
(*b*) acid rain (*c*) smoking
(*d*) industrial sources.

949. A rapid increase in CO_2 in the atmosphere coincided with which of the following events?
(*a*) Invention of automobiles
(*b*) The industrial revolution
(*c*) Thinning of ozone layer
(*d*) The Kyoto protocol.

950. Which organism created the methane found in the early atmosphere?
(*a*) Blue green algae
(*b*) Methanogenic bacteria
(*c*) Early fish (*d*) Plants.

951. Global temperature might be more extreme during the night than during the day, because
(*a*) earth is closer to sun at night
(*b*) evaporation proceeds faster at night
(*c*) heat is trapped near the surface at night
(*d*) more solar radiation is absorbed during the day.

952. Most effective way of reducing global warming is to
(*a*) curb emission of green house gases
(*b*) reduce the use of CFC's
(*c*) consume energy more efficiently
(*d*) use cleaner fuels and practice recycling.

953. How does global warming impact the spread of infectious disease?
(*a*) Rising temperatures allow disease carrying mosquitoes to move into new areas.
(*b*) Global warming increases moisture, which drowns disease spreading mosquitoes.
(*c*) Rising temperatures kill the larvae of disease carrying mosquitoes.
(*d*) Global warming decrease the amount of infectious disease.

954. Which of the following factors does not influence a soil's ability to hold water nutrients and air?
(*a*) Soil texture and structure
(*b*) Soil porosity (*c*) Soil pH value
(*d*) Soil composition.

955. What consequence of overusing ground water would affect areas that are distant from the point of use?
(*a*) Saltwater intrusion
(*b*) Ground water contamination
(*c*) Aquifer replenishment
(*d*) Depletion of stream flow.

956. A body of water is considered safe for swimming, if there are no more than 200 calories of coliform bacteria per 100 ______ of water.
(*a*) litre (*b*) gallon
(*c*) milliliters (*d*) ppm

957. Air pollutants returning to the earth surface may return as
(*a*) a solid and liquid
(*b*) a gas (*c*) acid rain only
(*d*) all solid, liquid or gas

958. The eggs and small fry of some aquatic organisms are damaged when a rapidly melting snow pack contains acidic particles, resulting in acidic concentration of 5 to 10 times higher than acidic rainfall. This condition is known as 'acid shock'. Fish may die of acid shock due to too much.
(*a*) iron (*b*) aluminium
(*c*) oxygen (*d*) lead.

Answers

947. (*d*) 948. (*a*) 949. (*b*) 950. (*b*) 951. (*c*) 952. (*a*) 953. (*a*) 954. (*c*) 955. (*d*)
956. (*c*) 957. (*d*) 958. (*b*)

959. The process by which a body of water becomes enriched in dissolved nutrients, stimulating the growth of aquatic plant life and usually resulting in the depletion of dissolved oxygen, is called

(*a*) algal blooms (*b*) eutrophication
(*c*) ultraviolet oxidation
(*d*) acid deposition.

960. What impacts do algal blooms have on plants living deeper beneath the water surface?

(*a*) They supply extra nutrients to deeper growing plants.
(*b*) They cause deeper growing plants to seek other forms of food.
(*c*) They kill them
(*d*) They increase the amount of sunlight to deeper growing plants.

961. The lenses of our eyes absorb UVB radiations. An increase of ultraviolet B radiation has a link to

(*a*) cataracts (*b*) retina problems
(*c*) nearsightedness and farsightedness
(*d*) increased pupil dilation.

962. An increase in evaporation at lower latitudes could lead to

(*a*) a decrease in precipitation
(*b*) an increased number of hurricanes
(*c*) unpredictable hurrican seasons
(*d*) lower temperature.

963. Residence time of nitrogen in the environment is about _______ years.

(*a*) 4000 (*b*) 4×10^6
(*c*) 40×10^6 (*d*) 400×10^6

964. _______ waste does not come under the category of hazardous waste.

(*a*) Reactive (*b*) Non-toxic
(*c*) Inflammable (*d*) Corrosive

965. The middle layer separating the lower stratosphere from thermosphere is called the

(*a*) mesosphere (*b*) ionosphere
(*c*) outer space (*d*) troposphere

966. About _______ % of chlorine in the stratosphere comes from anthropogenic (manmade) sources

(*a*) 25 (*b*) 45 (*c*) 70 (*d*) 85

967. Which atmospheric layer can reach a temperature of nearly 2000°C?

(*a*) Lithosphere (*b*) Mesosphere
(*c*) Thermosphere (*d*) Stratosphere.

968. Which of the following does not cause water turbidity?

(*a*) Clay (*b*) Silt
(*c*) Plankton (*d*) None of these

969. In which atmospheric layer does all the weather we experience take place?

(*a*) Thermosphere (*b*) Stratosphere
(*c*) Mesosphere (*d*) Troposphere

970. Atmospheric pollutants from volcanic eruptions affect

(*a*) only insects and plants
(*b*) only human beings living at the base of the mountain
(*c*) all living organisms worldwide
(*d*) polar climates only.

971. For hurricans to begin forming, meteorologists have found that ocean water must be warmer than _______ °C.

(*a*) 16.5 (*b*) 20.5 (*c*) 26.5 (*d*) 36.5

972. If the amount of human produced CO_2 being absorbed into oceans gets high enough, the ocean's top layer may become increasingly

(*a*) basic (*b*) acidic
(*c*) murky (*d*) neutral.

973. Which pair of phrases best describes the effect of ozone molecules in the environment?

(*a*) Green house effect; lung tissue irritant
(*b*) Ozone depletion; green house effect
(*c*) Green house effect; cause nerve damage
(*d*) Blocks UV rays; lung tissue irritant.

Answers

959. (*b*) 960. (*c*) 961. (*a*) 962. (*b*) 963. (*d*) 964. (*b*) 965. (*a*) 966. (*d*) 967. (*c*)
968. (*d*) 969. (*d*) 970. (*c*) 971. (*c*) 972. (*b*) 973. (*d*)

974. Most ozone depletion occurs under conditions that create a catalytic effect. One of the essential condition is the presence of an ice crystal. These ice crystals can be found in clouds that occur in

(*a*) Stratosphere (*b*) Troposphere
(*c*) Ionosphere (*d*) Mesosphere

975. Large water projects that slow the flow of rivers may show many economic benefits; but such projects in warm climates can also lead to

(*a*) valley fever (*b*) schistosomiasis
(*c*) river blindness (*d*) malaria.

976. Which of the following units are equivalent to the unit, ppm (parts per million)?

(*a*) mg/litre (*b*) mg
(*c*) mg/kg (*d*) gm/litre

977. Which sequence of responses towards solid waste leads to the least amount of energy used?

(*a*) Reduce, recycle reuse
(*b*) Reduce, reuse, recycle
(*c*) Recycle, reduce, reuse
(*d*) Reuse, recycle, reduce.

978. Which of the following water compartments contains the greatest amount of fresh water?

(*a*) Rivers and streams
(*b*) Lakes and reservoirs
(*c*) Ice and snow (*d*) Oceans.

979. A temperature inversion is different from a normal weather pattern, because

(*a*) pollutants get trapped in warm air
(*b*) warm air is trapped under a blanket of cool air
(*c*) cool air is trapped under a blanket of warm air
(*d*) a done of warm air holds in pollutants and diverts weather.

980. Which of the following is a source of asbestos and other fibrous particles?

(*a*) Photocopier (*b*) Carpet
(*c*) Foam insulation
(*d*) Vinyl floor and cement products.

981. Match the sources of various indoor air pollutants:

	Air pollutants		Indoor sources
A.	Ozone	I.	Ceiling tiles, plywood, foam insulation, carpets
B.	Volatile organics	II.	Kerosene space heater
C.	Sulfur dioxide	III.	Sprays, paints, varnishes, solvents
D.	Formaldehyde	IV.	Electrostatic air cleaners

(*a*) A-IV, B-III, C-II, D-I
(*b*) A-IV, B-III, C-I, D-II
(*c*) A-I, B-II, C-III, D-IV
(*d*) A-III, B-IV, C-II, D-I

982. Match the causal sources with the name of disease

	Causal sources		Disease caused
A.	Cotton fibre	I.	Byssinosis
B.	Barium	II.	Silicosis
C.	Silicon	III.	Black lung disease
D.	Coal dust	IV.	Baritosis

(*a*) A-IV, B-III, C-II, D-I
(*b*) A-I, B-III, C-II, D-IV
(*c*) A-I, B-IV, C-II, D-III
(*d*) A-IV, B-I, C-II, D-III

983. Match the year of occurrence of various environmental disaster:

	Environment disaster		Year of occurrence
A.	Bhopal gas tragedy	I.	1991
B.	Indian ocean tsunami	II.	1986
C.	Chernobyl nuclear accident	III.	1984
D.	Super cyclone (Orissa)	IV.	2001

Answers

974. (*a*) 975. (*b*) 976. (*c*) 977. (*b*) 978. (*c*) 979. (*d*) 980. (*d*) 981. (*a*) 982. (*c*)
983. (*b*)

(*a*) A-III, B-IV, C-I, D-II
(*b*) A-III, B-IV, C-II, D-I
(*c*) A-I, B-IV, C-II, D-III
(*d*) A-IV, B-III, C-II, D-I

984. Match the water pollutants generated by various industries:

	Industries		Pollutants generated
A.	Tannaries	I.	Oil, grease, suspended solid, low pH
B.	Sugar factory	II.	High pH and BOD, high chromium concentration
C.	Distillery	III.	High pH, suspended solids, colour, toxicity
D.	Dairy farm	IV.	High pH, High BOD and COD, grease.

(*a*) A-IV, B-III, C-II, D-I
(*b*) A-II, B-III, C-IV, D-I
(*c*) A-II, B-III, C-I, D-IV
(*d*) A-I, B-II, C-III, D-IV

985. Match the name of treatment processes with the mechanism involved in it

	Treatment processes		Mechanisms involved
A.	Electrodialysis	I.	Separation of macro molecular solution
B.	Ultrafiltration	II.	Territory process of treatment
C.	Ion exchange	III.	Desalting of ionic solutions
D.	Reverse osmosis	IV.	Separation from gas mixture

(*a*) A-III, B-I, C-II, D-IV
(*b*) A-IV, B-III, C-I, D-II
(*c*) A-I, B-II, C-III, D-IV
(*d*) A-IV, B-III, C-II, D-I

986. Match the pollutants generated by various industries:

	Industries		Pollutants produced
A.	Mining industries	I.	Solids, surface wash offs, heavy metals
B.	Soap and detergent industries	II.	Acids, alkalis, sulfur, metal nitrates
C.	Pulp and paper industries	III.	Tertiary ammonium compounds
D.	Chemical industries	IV.	Wood, bark, cellulose fibres

(*a*) A-IV, B-III, C-II, D- I
(*b*) A-II, B-III, C-IV, D-I
(*c*) A-I, B-II, C-III, D-IV
(*d*) A- I, B-III, C-IV, D-II

987. Match the disease caused by various causative agents:

	Causative agents		Disease caused
A.	Constant radiation exposure	I.	Insomnia
B.	Polluted drinking water	II.	Shigellosis
C.	Pesticide ingestion	III.	Leukemia
D.	Noise	IV.	Neuralgic paralysis

(*a*) A-III, B-II, C-IV, D-I
(*b*) A-II, B-III, C-IV, D-I
(*c*) A-III, B-II, C-I, D-IV
(*d*) A-I, B-IV, C-III, D-II

988. Match the diseases caused by various pollutants:

	Pollutants		Disease caused
A.	Lead	I.	Black lung disease
B.	Arsenic	II.	Tonic effect
C.	Coal dust	III.	Plumbism
D.	Bauxite	IV.	Aluminosis

(*a*) A-IV, B-III, C-II, D-I
(*b*) A-III, B-II, C-IV, D-I

Answers

984. (*c*) 985. (*a*) 986. (*d*) 987. (*a*) 988. (*b*)

(*c*) A-I, B-II, C-III, D-IV
(*d*) A-III, B-II, C-IV, D-I

989. Match the chemical detection methods used for detections various environmental chemicals in the laboratory

Environment chemicals		Detection method used	
A.	Pesticides	I.	Flameless atomic adsorption
B.	Arsenic	II.	AAS through hydride generation
C.	Oil and grease	III.	Gas chromatography
D.	Mercury	IV.	Gravimetry

(*a*) A-III, B-II, C-IV, D-I
(*b*) A-II, B-III, C-IV, D-I
(*c*) A-II, B-III, C-I, D-IV
(*d*) A-IV, B-III, C-II, D-I

990. Match the detecting instruments used for detecting various environmental chemicals:

Environment chemicals		Detecting instrumental techniques	
A.	Nitrates	I.	Potentiometry
B.	Ozone	II.	Ion specific electrodes
C.	Metals	III.	UV-vis-spectroscopy
D.	Fluorides	IV.	Emission spectroscopy

(*a*) A-IV, B-III, C-II, D-I
(*b*) A-III, B-II, C-I, D-IV
(*c*) A-I, B-II, C-III, D-IV
(*d*) A-II, B-III, C-IV, D-I

991. Match the symptoms of carbon monoxide (CO) poisoning at various % of carbonxyhemoglobin (COHb) in the blood:

% of COHb in blood		Symptoms of CO poisoning	
A.	20	I.	Increased respiration and pulse syncope, dizziness of vision severe headache
B.	40	II.	Respiratory failure, weak pulse, death
C.	60	III.	Comma with intermittent convulsions, depressed cardiac function
D.	75	IV.	Tightness across forehead, headache, throbbing in temple.

(*a*) A-IV, B-I, C-II, D-III
(*b*) A-I, B-IV, C-II, D-III
(*c*) A-IV, B-I, C-III, D-II
(*d*) A-I, B-IV, C-III, D-II

992. Pick out the wrong statement:
(*a*) Most non-polluted fresh water contain total phosphorous concentrations between 10 to 50 mg/litre.
(*b*) A plume formed in the absence of wind velocity is called neutral plume.
(*c*) Decayed organic matter of soil is called humus.
(*d*) Excessive use of fossil fuel resulting in production of CO_2 causes stratospheric ozone depletion.

993. Fluids such as gasoline, paint solvents, and organic cleaning solutions release _________.
(*a*) volatile organic compounds
(*b*) sulfur dioxides (*c*) carbon monoxide
(*d*) nitrogen oxides

994. Which of the following is a pollutant in one part of the atmosphere and valuable filter in another part of the atmosphere?
(*a*) oxygen (*b*) sulfur dioxide
(*c*) ozone (*d*) nitrous oxide

995. Which of the following locations is most likely to have a temperature inversion that results in large amounts of trapped air pollution?
(*a*) a city in the plains with large numbers of people who use personal vehicles to commute
(*b*) a city in the mountains with many large industries

Answers

989. (*a*) 990. (*d*) 991. (*c*) 992. (*d*) 993. (*a*) 994. (*c*) 995. (*c*)

(c) a city in a valley with large numbers of people who use personal vehicles to commute
(d) a city on a lake with many large industries

996. Your community is home to a large coal-fired power plant. What pollutants are most likely to be found in the air?
(a) particulate matter, nitrous oxides, and sulfur dioxide.
(b) lead, nitrogen oxides, and sulfur dioxide.
(c) particulate matter, radon, and sulfur dioxide,
(d) lead radon, and particulate matter.

997. Which of the following air pollutants has shown the greatest decline in recent years?
(a) Particulates (b) Lead
(c) Carbon monoxide
(d) Sulfur dioxide

998. If all transportation were suddenly stopped, which three pollutants would decrease the most?
(a) sulfur dioxide, particulate matter, and volatile organic compounds
(b) carbon monoxide, particulate matter, and nitrous oxides
(c) sulfur dioxide, nitrous oxides, and volatile organic compounds
(d) nitrous oxides, volatile organic compounds, and carbon monoxide.

999. Tomato juice has a pH of 4.0. How many times more acidic is lemon juice, which has a pH of 2.0?
(a) 2 times (b) 4 times
(c) 10 times (d) 100 times

1000. Children in your community are suffering from brain damage and respiratory problems What two pollutants are the likely causes?
(a) carbon monoxide and particulate matter
(b) lead and sulfur dioxide
(c) carbon monoxide and sulfur dioxide
(d) carbon monoxide and lead.

1001. If all volatile organic compounds (VOCs) suddenly disappeared, what would happen to ozone concentrations in the troposphere?
(a) Ozone concentrations in the troposphere would decrease.
(b) Ozone concentrations in the troposphere would stay the same.
(c) It is unknown what would happen to ozone concentrations in the troposphere.
(d) Ozone concentrations in the troposphere would increase.

1002. Which of the following can result in the formation of acids in the atmosphere?
(a) ozone and asbestos
(b) carbon monoxide and lead
(c) nitrogen oxides and sulfur oxides
(d) volatile organic compounds and radon

1003. Which of the following crops would be best to grow in areas with high levels of ozone pollution?
(a) Cotton and corn
(b) Soya beans and wheat
(c) Corn (d) Peanuts and wheat

1004. Which of the following industries would be most impacted if the pollution control board mandated the elimination of sulfur dioxide and nitrous oxides from the air?
(a) car manufacturers
(b) agriculture (c) electrical utilities
(d) paper manufacturers

1005. Over time, if all primary pollutants were eliminated, what would happen to concentrations of secondary pollutants?
(a) Concentrations of secondary pollutants would increase.
(b) It is unknown what would happen to concentrations of secondary pollutants
(c) Concentrations of secondary pollutants would drop to zero.
(d) Concentrations of secondary pollutants would stay the same.

Answers

996. (a) 997. (b) 998. (d) 999. (d) 1000. (b) 1001. (a) 1002. (c) 1003. (c) 1004. (c) 1005. (c)

1006. Photochemical smog is produced when sunlight acts on ________ and ________.
(a) sulfur oxides/carbon dioxide
(b) nitrogen oxides/volatile organic carbon compounds
(c) ground-level ozone/nitrogen oxides
(d) airborne particulates / volatile organic carbon compounds

1007. Which of the following is true regarding catalytic converters employed in automobiles?
(a) They convert a primary pollutant into a secondary pollutant.
(b) They convert a greenhouse gas into a carcinogen
(c) They convert an air pollutant into a greenhouse gas.
(d) They have increased air pollution

1008. If chlorine were not a catalyst, how would CFCs have affected ozone depletion?
(a) Ozone depletion would have been faster
(b) Ozone depletion would have been slower.
(c) It is unknown what would have happened with ozone depletion.
(d) Ozone depletion would have occurred at the same rate .

1009. Which of the following is true?
(a) As the concentration of ozone In the stratosphere over the Antarctic decreased, the size of the ozone hole also decreased.
(b) As the concentration of ozone in the stratosphere over the Antarctic increased, the size of the ozone hole decreased.
(c) As the concentration of ozone in the stratosphere over the Antarctic increased, the size of the ozone hole also increased.
(d) As the concentration of ozone in the stratosphere over the Antarctic increased, the size of the ozone hole remained the same.

1010. HCFCs have replaced CFCs for many uses. However, HCFCs still damage ozone. Why?
(a) They both contain oxygen.
(b) They both contain carbon.
(c) They both contain chlorine.
(d) They both contain hydrogen.

1011. Which of the following best describes a command-and-control approach to decreasing air pollution.
(a) paying a utility company to emit less sulfur dioxide
(b) allowing utility companies to buy emissions from each other
(c) fining a utility company for emitting too much sulfur dioxide
(d) allowing consumers to pay more to ensure that their electricity comes from a renewable resource

1012. Eliminating which of the following would help alleviate the problem of acid deposition?
(a) oxygen gas (b) hydroxide ions
(c) carbon (d) hydrogen ions

1013. What would be best way to eliminate photochemical smog.
(a) Banning coal fired power plants
(b) Banning petrol powered vehicles
(c) Banning HCFs
(d) Banning nuclear power plants.

1014. Emission of ________ cause acid rain
(a) NO_3 (b) SO_2 and NO_3
(c) SO_2 (d) Green house gases

1015. Bhopal gas tragedy was due to
(a) CO (b) SO_2
(c) MIC (d) H_2S

1016. Photochemical smog is a pollutant formed in
(a) Excessive SO_2 Atmosphere.
(b) High temperature (c) Low temperature
(d) Excessive Nitrogen

1017. ________ electrostatic precitators
(a) Heavy metals
(b) Particulate pollutants

Answers

1006. (b) 1007. (c) 1008. (b) 1009. (b) 1010. (c) 1011. (a) 1012. (b) 1013. (b) 1014. (b)
1015. (c) 1016. (a) 1017. (b)

(c) Sulfur compounds
(d) Dioxins

1018. Byssinosis is an Occupational disease found in workers of
(a) Coal mine (b) Textile industry
(c) Slaughter house
(d) Galvanizing Industries

1019. PAN is a ________ pollutant.
(a) Primary (b) Secondary
(c) Water (d) Soil

1020. Smog is ________
(a) Fog + Smoke. (b) Smoke + Rain
(c) Smoke + CO_2 (d) Fog + CO_2

1021. Photo chemical smog is formed during
(a) Early Morning (b) Afternoon
(c) Rainfall (d) Early afternoon.

1022. Peak SO_2 concentration was
(a) 13 ppm and smoke 40 mg/M^3
(b) 0.1 ppm and smoke 400 mg/M^3
(c) 0.3 ppm and smoke 0.4 mg/M^3
(d) 1.3 ppm and smoke 4 mg/M^3

1023. Which of the following pollutant reacts with marble stone ($CaCO_3$) to produce calcium sulphate, causing darkening and disfigurement of Taj Mahal?
(a) Acid Rain (b) Smog
(c) CO_2 (d) O_3

1024. Planet earth along with the atmosphere that sustains life is known as
(a) Hydrosphere (b) Biosphere
(c) Atmosphere (d) Lithosphere

1025. Stratosphere is rich in ________
(a) Ozone (b) Chlorine
(c) Helium (d) Freon

1026. Lead poisoning affects the function of
(a) Liver (b) Kidney
(c) Lungs (d) Heart

1027. Microbes involved in methane oxidation are ________.
(a) Acidophiles (b) Methanotrophs
(c) Thermophiles. (d) Methanogens

1028. ________ the silent killer gas, which has more affinity towards heamoglobin of the blood.
(a) NO_2 (b) CO_2 (c) SO_2 (d) CO

1029. Time taken by sunlight to reach the earth is
(a) 20 seconds (b) 45 minutes
(c) 8 seconds (d) 8 minutes

1030. The unit of measurement of ozone thickness is
(a) Decibel (b) Dobson
(c) Richter (d) Decimal

1031. Which satellite recorded the presence of ozone hole?
(a) Landsat-3 (b) Nimbus
(c) GOES (d) TRIPS-N

1032. The part of human body first to be affected due to vehicular pollution is
(a) Kidney (b) Liver
(c) Lungs (d) Brain

1033. Black lung disease occurs among
(a) Quarry workers (b) Refinery workers
(c) Coal miners (d) Smokers

1034. One per cent reduction in O_3 increases the UV radiation on earth by ________ %.
(a) 1 (b) 2 (c) 10 (d) 5

1035. Plumbism is nothing but ________ poisoning.
(a) Cd (b) Hg (c) Pb (d) Cr

1036. The most prominent inorganic oxidant in atmosphere is
(a) O_3 (b) O_2 (c) OH (d) NO_2

1037. Mesothelima is a type of cancer which occur due to
(a) Asbestos (b) Copper
(c) Silver (d) Iron

1038. White lung disease is common among
(a) Coal miners (b) Farmers
(c) Traffic police (d) Textile workers

Answers

1018. (b) 1019. (b) 1020. (a) 1021. (d) 1022. (d) 1023. (a) 1024. (b) 1025. (a) 1026. (c)
1027. (c) 1028. (d) 1029. (d) 1030. (b) 1031. (b) 1032. (c) 1033. (c) 1034. (b) 1035. (c)
1036. (a) 1037. (a) 1038. (d)

1039. Heavy duty diesel powered vehicles contribute more of
(*a*) NO_x and particulate matter
(*b*) CO and HC (*c*) CO and CO_2
(*d*) SO_2 and NO_2

1040. Ozone at ground surface is responsible for the formation of a hazardous pollutant namely,
(*a*) Smoke (*b*) Fog
(*c*) Smog (*d*) SPM

1041. The major element responsible for photochemical smog is
(*a*) SO_2 (*b*) CO (*c*) CO_2 (*d*) NO_2

1042. The most obvious toxic product of combustion of plastic is
(*a*) SO_2 (*b*) CO
(*c*) CO_2 (*d*) Dioxins

1043. The more toxic gaseous component released due to burning of fossil fuel
(*a*) CO (*b*) CO_2 (*c*) SO_2 (*d*) NO_2

1044. Human Ear can detect sound in the frequency range of about
(*a*) 20 Hz (*b*) 20-20,000 Hz
(*c*) 40,000 Hz (*d*) 24,000 Hz

1045. Unit of measuring noise is ________
(*a*) Dopson (*b*) Decimel
(*c*) Decibel (*d*) Dobson

1046. The permitted noise level in a silent zone during day time is ________
(*a*) 65 (*b*) 55 (*c*) 50 (*d*) 70

1047. Human generated noise in the ocean may interfere with marine organisms in
(*a*) Communication (*b*) Mobility
(*c*) Food finds ability (*d*) Migration

1048. When a person is exposed to a voice level of 160 decibels for long time it causes
(*a*) Annoyance (*b*) Restlessness
(*c*) Tiredness
(*d*) Rapture of tympanic membrane

1049. For one megawatt energy generation, nuclear power plant requires ________ Million m^3 cooling water each day.
(*a*) 1.3 (*b*) 0.3 (*c*) 1 (*d*) 5

1050. Thermal pollution in a river ________
(*a*) Reduces the O_2 Dissolution
(*b*) Increases O_2 solubility
(*c*) Stabilizes O_2 level
(*d*) None of these

1051. Urban heat island is nothing but ________
(*a*) A dome of cool air over the urban area
(*b*) A dome of warm air over the urban area
(*c*) A dome of noise pollution over the urban area
(*d*) Reduced wind speed over the urban area

1052. Chondrococcus columnaris is harmless to fish at 10°C but cases diseases when the temperature of water is
(*a*) More than 12°C (*b*) More than 21°C
(*c*) Less than 10°C (*d*) At 10°C

1053. Tinnitus is due to
(*a*) Water pollution (*b*) Noise pollution
(*c*) Soil pollution (*d*) Xenobiotics

1054. The major effect of discharging heated water from power plants to aquatic ecosystems is
(*a*) Reduction in DO concentration
(*b*) Reduction in Biodiversity of the ecosystem
(*c*) Increase in metabolic of organism
(*d*) All of above

1055. The noise level during normal conversation is ________ decibel (dB).
(*a*) 30 (*b*) 35-60
(*c*) 20 (*d*) 90

1056. Persistent sound in one or both ears, often experienced as a high pitch ringing sound causes
(*a*) Annoyance (*b*) Headache
(*c*) Deafness (*d*) Tinnitus

Answers

1039. (*a*) 1040. (*c*) 1041. (*d*) 1042. (*d*) 1043. (*c*) 1044. (*b*) 1045. (*c*) 1046. (*c*) 1047. (*a*)
1048. (*d*) 1049. (*d*) 1050. (*a*) 1051. (*b*) 1052. (*b*) 1053. (*b*) 1054. (*d*) 1055. (*b*) 1056. (*d*)

1057. The amount of any gas that can be dissolved in water is
(*a*) Inverse with temperature
(*b*) Direct with temperature
(*c*) No relation with temperature
(*d*) None of above

1058. The smoke in London smogs came primarily from:
(*a*) exhaust from diesel engines
(*b*) trash fires
(*c*) factories in Eastern Europe
(*d*) coal combustion

1059. Collectively, particles of soot, smoke, dust and pollen are called:
(*a*) hydrocarbons (*b*) aerosols
(*c*) carcinogens (*d*) haze

1060. Which of the following gases will replace oxygen in blood hemoglobin and thereby reduce the transport of oxygen to the brain?
(*a*) sulfur dioxide (SO_2)
(*b*) carbon monoxide (CO)
(*c*) carbon dioxide (CO_2)
(*d*) methane (CH_4)

1061. A primary component of photochemical smog is:
(*a*) ozone (*b*) carbon monoxide
(*c*) sulfur dioxide
(*d*) chlorofluorocarbons

1062. A mixing layer is characterized by:
(*a*) enhanced vertical air motions
(*b*) suppressed vertical air motions
(*c*) strong horizontal winds
(*d*) high concentrations of pollutants

1063. Which of the following conditions would act to prevent a buildup of pollutants near the surface?
(*a*) light surface winds
(*b*) a strong subsidence inversion
(*c*) a large, slow-moving anticyclone
(*d*) a deep mixing layer

1064. Atmospheric stagnation is a condition normally brought on by:
(*a*) thunderstorms
(*b*) slow-moving anticyclones
(*c*) overcast skies
(*d*) tall buildings in a city
(e) movement of an upper level trough overhead

1065. Sunshine in a city is typically less intense than in surrounding rural areas because:
(*a*) of the higher pollution levels in cities
(*b*) of the air temperature in cities
(*c*) tall structures in cities block our view of the horizon
(*d*) cement and asphalt in cities absorb a larger percentage of the incident sunshine

1066. On clear, cold winter nights, cities tend to cool ________ than rural areas and have ________ temperatures.
(*a*) more slowly, higher
(*b*) more quickly, higher
(*c*) more slowly, lower
(*d*) more quickly, lower

1067. Rain with a pH of 5.6 would be considered:
(*a*) acidic (*b*) alkaline
(*c*) neutral (*d*) polluted

1068. The ozone hole is found in this atmospheric layer:
(*a*) thermosphere (*b*) troposphere
(*c*) stratosphere (*d*) ionosphere

1069. If the concentration of ozone were to decrease significantly, which of the following might also occur?
(*a*) less absorption of ultraviolet radiation in the stratosphere, resulting in its cooling
(*b*) an increase in the number of cases of skin cancer
(*c*) more ultraviolet radiation would be absorbed at the earth's surface of
(*d*) all of the above

Answers

1057. (*a*) 1058. (*a*) 1059. (*b*) 1060. (*b*) 1061. (*a*) 1062. (*a*) 1063. (*d*) 1064. (*c*) 1065. (*a*)
1066. (*c*) 1067. (*a*) 1068. (*c*) 1069. (*d*)

1070. Which of the following catastrophic air pollution events was not an air pollution episode that resulted from a severe temperature inversion?
(*a*) Belgium (Meuse River Valley)
(*b*) Bhopal, India (*c*) Poza Rica, Mexico
(*d*) Donora, Pennsylvania

1071. Each of the following factors has directly influenced the policy objectives of air pollution control programs, except
(*a*) Air pollution episodes
(*b*) Improvements in science
(*c*) Atmospheric dispersion of pollutants
(*d*) Environmental activism

1072. Match the following environmental disasters/ episodes

Column I		Column II	
I.	(1930) Meuse River Valley	A.	Twenty people died and almost 7,000 became sick due to a four-day fog of SO that surrounded a small mill town
II.	(1948) Donora, ennsylvania	B.	Two hundred people died due to pneumonia, influenza, and cardiac arrest. Excess deaths continued several weeks after the episode caused by an epidemic of influenza.
III.	(1952) London, England	C.	Sixty-three people died due to high concentrations of SO_2 in the atmosphere during a severe temperature inversion
IV.	(1953) New York City	D.	The most acute air pollution episode in recorded history in which over 4,000 people died of bronchitis.

(*a*) I-C, II-A, III-D, IV-B
(*b*) I-D, II-C, III-A, IV-B
(*c*) I-A, II-B, III-D, IV-C
(*d*) I-C, II-D, III-B, IV-A

1073. The human health effect that best describes carbon monoxide (CO) poisoning is
(*a*) Accumulation of toxins in the bones and soft tissue resulting in renal failure
(*b*) Increased airway resistance resulting in bronchial constrictions
(*c*) Interference with the body's ability to supply oxygen to tissue resulting in cardiovascular disease.
(*d*) Inflammation of the lungs resulting in emphysema or fibrosis

1074. Each of the following is a direct effect of air pollution on the respiratory system except
(*a*) Pulmonary emphysema
(*b*) Lung cancer (*c*) Pneumoconiosis
(*d*) Blood poisoning

1075. Which of the following is a part of the risk assessment process?
(*a*) Hazard identification
(*b*) Dose-Response Data
(*c*) Exposure Assessment
(*d*) All of the above.

1076. ________ accounts for most of the total SO_2 emissions
(*a*) Petroleum refining
(*b*) Sulfuric Acid Manufacturing
(*c*) Fuel combustion (*d*) Smelting

1077. Match the following pertaining to environmental effects of air pollution.

Column I		Column II	
1.	Acidification of takes and streams	A.	Ecosystem damage
2.	Pristine vistas and national parks irreparably altered	B.	Global Warming

Answers

1070. (*b*) 1071. (*c*) 1072. (*a*) 1073. (*c*) 1074. (*d*) 1075. (*d*) 1076. (*c*) 1077. (*a*)

3.	Soiled buildings, clothing, and automobiles	C.	Property Damage
4.	Melting snow and ice at both poles	D.	Quality of Life
5.	Cracked and split rubber products, nylon, and polymer plastics		
6.	Damage to vegetation, forests, and crops		
7.	Variations in annual and seasonal precipitation patterns		
8.	Reduction in open-air visibility		

(*a*) A(1, 6); B(4, 7); C(3, 5); D(2, 8)
(*b*) A(2, 8); B(3, 5); C(4, 7); D(1, 6)
(*c*) C(2, 5); B(3, 6); C(2, 8); D(1, 8)
(*d*) A(1, 3); B(2, 4); C(3, 6); D(4, 8)

1078. Match the following pertaining to human health effects of air pollution.

	Column I		Column II
1.	Respiratory system	A.	Particulate Matter (PM)
2.	Reduced Lung Elasticity	B.	Lead (Pb)
3.	Lower Respiratory system	C.	Sulfur Dioxide (SO_2)
4.	Circulatory system	D.	Carbon Monoxide (CO)
5.	Lowered Resistance	E.	Nitrogen Dioxide (NO_2) to infection
6.	Organs and Soft Tissue	F.	Ozone

(*a*) A-1, B-2, C-3, D-4, E-6, F-5
(*b*) A-2, B-3, C-5, D-4, E-1, F-6
(*c*) A-3, B-6, C-1, D-4, E-5, F-2
(*d*) A-6, B-5, C-4, D-2, E-1, F-3

1079. Match the various organs of human upper and lower respiratory system as shown in diagram.

Upper Respiratory System		Lower Respiratory System
A.	1.	Bronchi
B.	2.	Alveoli
C.	3.	Larynx
D.	4.	Bronchiole
E.	5.	Trachea
F.	6.	Tongue

(*a*) A-6, B-5, C-3, D-4, E-1, F-2
(*b*) A-6, B-3, C-5, D-1, E-4, F-2
(*c*) A-1, B-2, C-4, D-3, E-6, F-5
(*d*) A-6, B-2, C-4, D-3, E-5, F-1

1080. As wind speed increases, the volume of air dispersed from a pollutant source over a specified period of time
(*a*) increases (*b*) decreases
(*c*) results in super adiabatic conditions
(*d*) remains unchanged

1081. The three dimensional movement of the wind created by an uneven earth surface is call wind
(*a*) Direction (*b*) Turbulence
(*c*) Circulation (*d*) Mixing

1082. Which topographical feature stores the most heat?
(*a*) Lakes (*b*) Cultivated Fields
(*c*) Mountains (*d*) Valleys

1083. A ________ occurs in stable atmospheric conditions; is generally produced on clear nights with little or light wind; and may spread horizontally but has little vertical air mixing.
(*a*) Looping plume (*b*) Coning plume
(*c*) Fanning plume (*d*) Lofting plume

1084. A ________ plume occurs in unstable atmospheric conditions; is generally produced on clear sunny days with light winds; and can draw pollution to ground level near the pollutant source.
(*a*) Looping (*b*) Coning
(*c*) Fanning (*d*) Lofting

Answers

1078. (*c*) 1079. (*b*) 1080. (*a*) 1081. (*b*) 1082. (*a*) 1083. (*c*) 1084. (*a*)

1085. Match the diagram of the various plume types observed during air pollutant dispersion from chimney

1. Looping plume
2. Coning plume
3. Fanning plume
4. Lofting plume
5. Fumigating plume
6. Trapping plume

(*a*) A-1, B-4, C-3, D-2, E-6, F-5
(*b*) A-4, B-3, C-2, D-5, E-6, F-1
(*c*) A-2, B-1, C-3, D-6, E-5, F-4
(*d*) A-1, B-2, C-4, D-6, E-3, F-3

1086. Indoor air pollution has become a matter of great concern because the average person spends ________ of their time indoors.

(*a*) 90% (22 hrs./day)
(*b*) 70% (17 hrs./day)
(*c*) 50% (12 hrs./day)
(*d*) 30% (7 hrs./day)

1087. A radioactive gas, that is a by-product of uranium decay and is found in nearly all soils is ________.

(*a*) potassium (K) (*b*) plutonium (Pu)
(*c*) radium (Ra) (*d*) radon (Rn)

1088. Each of the following are ways to control indoor air pollution except ________.

(*a*) eliminate the-source
(*b*) install carbon monoxide (CO) detectors
(*c*) modify the source
(*d*) ventilate and dilute the air

1089. The most common method of controlling indoor air pollution today is to ________.

(*a*) identify and eliminate the source
(*b*) install carbon monoxide (CO) detectors
(*c*) modify the source

1090. The leading cause of "Sick Building" Syndrome is ________.

(*a*) inadequate ventilation
(*b*) poor maintenance
(*c*) spread of biological contaminants
(*d*) ingestion of respirable particles

1091. Sources of indoor organic chemicals may be

(*a*) paints, varnishes and solvents
(*b*) cleaning, degreasing and hobby products
(*c*) gasoline
(*d*) all of the above

1092. Biological contaminants include all of the following except ________.

(*a*) bacteria (*b*) molds
(*c*) animal danger and pollen
(*d*) carbon monoxide

1093. Indoor levels of many pollutants may be ________ times higher indoors than outdoors.

(*a*) 5 (*b*) 10 (*c*) 15 (*d*) 25

1094. Pontiac fever, Legionnaire's disease, and humidifier fever are all examples of

(*a*) "Sick Building" Syndrome
(*b*) Asbestosis (*c*) Pulmonary edema
(*d*) Mesotherlioma

1095. Match the following pertaining to selected sources of indoor air pollutants

	Column I		Column II
1.	Insulation	A.	Biological contaminants
2.	Glue	B.	CO, SO_2, NO_2
3.	Arachind Excretia	C.	Formaldehyde
4.	Tobacco Products	D.	Asbestos
5.	Office Equipment	E.	Respirable particulates
6.	Oil Furnance	F.	Organic chemical
7.	Interior Grade Plywood		
8.	Kerosene Heaters		
9.	House Dust		

Answers

1085. (*a*) 1086. (*a*) 1087. (*d*) 1088. (*b*) 1089. (*a*) 1090. (*a*) 1091. (*d*) 1092. (*d*) 1093. (*d*) 1094. (*a*) 1095. (*d*)

10.	Foam Insulation		
11.	Siding		
12.	Dry Cleaned Clothing		
13.	Furniture		
14.	Furniture Polish		

(*a*) A(1, 5); B(3, 4, 5); C(4, 6, 5); D(2, 14); E(1, 13); F(5, 9, 11)
(*b*) A(2, 7); B(6, 8); C(7, 13); D(10); E(5, 12); F(7, 13, 14)
(*c*) A(2, 9); B(7, 13); C(4, 13); D(4, 7); E(5, 12); F(7, 5, 13)
(*d*) A(3, 4, 9); B(6, 8); C(7, 10, 13); D(1, 11); E(4, 5, 8); F(2, 5, 12, 14)

1096. The purpose of an ambient air quality monitoring program includes all of the following except ______.
(*a*) to determine compliance with standards and to observe pollution trends
(*b*) to activate emergency control procedures
(*c*) to develop long-term control strategies
(*d*) to establish air quality control regions

1097. Which of the following is not an element of an effective air quality assurances and quality control program?
(*a*) sample collection (*b*) sample analysis
(*c*) data reporting
(*d*) compliance and enforcement

1098. The most commonly used non-continuous method of gas collection and sampling is through the use of
(*a*) bubblers or impingers
(*b*) electrostatic precipitation
(*c*) impaction (*d*) filtration

1099. A technique used to collect and sample particulate matter is ______.
(*a*) absorption (*b*) adsorption
(*c*) condensation (*d*) filtration

1100. A device used to sample particulate matter in which pollutants are drawn through a glass fiber filter using a vacuum motor to draw the air is a
(*a*) electrostatic precipitator
(*b*) Hi-vol. sampler
(*c*) paper tape sampler
(*d*) surface condenser

1101. A sampling technique used to collect ambient air gases with a collection tube containing solid surfaces such as carbon, silica gel or molecular sieves is called ______.
(*a*) filtration (*b*) adsorption
(*c*) condensation (*d*) impaction

1102. A principle of particulate matter collection and sampling in which particles are drawn through a device, but deflected from their original path is called ______.
(*a*) electrostatic precipitation
(*b*) gravitational settling
(*c*) impaction
(*d*) thermostatic precipitation

1103. Match each pollutants sampling technique with the proper description

	Sampling Techniques		Descriptions:
A.	Static sampling	1.	The preferred method of sampling particulate matter and acid rain in which contaminants are collected on a periodic basis, typically over longer averaging time (i.e., 24 hours).
B.	Grab sampling	2.	Collection of contaminants by diffusion of gas, sedimentation of particles, or impaction on a sticky surface.
C.	Inter-mittent sampling	3.	Time preferred method of sampling used by monitoring programs to obtain real-time sampling averages of contaminants.

Answers

1096. (*d*) 1097. (*d*) 1098. (*a*) 1099. (*d*) 1100. (*b*) 1101. (*b*) 1102. (*c*) 1103. (*b*)

D.	Continuous sampling	4.	A sampling method used to collect organic gases on a solid surface such as carbon, silica gel, or a molecular sieve.
E.	Absorption	5.	A method of collection in which a small volume of air quickly captured in an evacuated bottle, gas syringe, or synthetic bag.
F.	Adsorption	6.	A sampling technique in which gas is drawn through a liquid reagent and removed from the air stream as it becomes dissolved into the liquid.

(*a*) A-4, B-3, C-2, D-1, E-5, F-6
(*b*) A-2, B-5, C-1, D-3, E-4, 6, F-4
(*c*) A-1, B-6, C-5, D-4, E-2, F-3
(*d*) A-1, B-2, C-3, D-4, E-6, F-5

1104. The most effective means of capturing condensable gaseous pollutants during stationary source sampling is through ________.
(*a*) electrostatic precipitation
(*b*) filtration (*c*) condensation
(*d*) centrifugal force

1105. In order to obtain a representative collection an isokinetic sample must be drawn into the sampling device at a
(*a*) velocity greater than the gas stream
(*b*) velocity less than the gas stream
(*c*) volume greater than the gas stream
(*d*) velocity equal to the gas stream

1106. Sampling velocities that are greater than the isokinetic rate result in the collection of
(*a*) a smaller number of small particles than those found in the gas stream
(*b*) smaller volumes than those found in the gas stream
(*c*) a larger number of small particles than those found in the sampled gas stream
(*d*) the same proportional volumes as those found in the gas stream

1107. Sampling velocities that are less than the isokinetic rate results in the collection of
(*a*) a smaller number of small particles than those found in the gas stream
(*b*) smaller volumes than those found in the gas stream
(*c*) a larger number of large particles than those found in the gas stream
(*d*) larger volumes than those found in the gas stream

1108. Misalignment of the sample probe results in a measurement error that could cause
(*a*) smaller particle to not be included in the collected sample
(*b*) smaller volumes of gas
(*c*) some larger particles to not be included in the collected sample
(*d*) larger volumes of gas to be collected

1109. The ________ nozzle tapered from the outside inward results in the least disruption of the gas flow streamline.
(*a*) square-edged (*b*) blunt-edged
(*c*) curved-edged (*d*) sharp-edged

1110. Nozzles whose design not include sharp and tapered edges, or are dented
(*a*) collect all particles better than other nozzles
(*b*) collect small particles better than other nozzles
(*c*) collect smaller volumes of sample gas than other nozzles
(*d*) should not be used for particle sampling

1111. The following are all principles used to control particulate matter, except
(*a*) gravity (*b*) centrifugal force
(*c*) infrared analysis
(*d*) electrostatic attraction

Answers

1104. (*c*) 1105. (*d*) 1106. (*c*) 1107. (*c*) 1108. (*c*) 1109. (*d*) 1110. (*d*) 1111. (*c*)

1112. Which of the following control devices might be effectively used to control coarse particulate matter (> 50 microns)?
(*a*) Electrostatic Precipitation
(*b*) Baffle Chambers (*c*) Baghouse
(*d*) Venturi Scrubber

1113. Settling Chambers use ________ to remove pollutants from gas exhaust.
(*a*) diffusion (*b*) fabric filters
(*c*) centrifugal force (*d*) gravity

1114. Match each of the following advantages and disadvantages with the proper particulate control technology.

Control Technology		Advantages and Disadvantages	
A.	Electrostatic Precipitators	1.	Is highly efficient (> 99%) for collecting all types of dust and high to low flow rates.
B.	Fabric Filters	2.	Expensive to purchase and maintain and limited by space.
		3.	Ability to treat large volumes of gas economically.
		4.	Limited to dry environments.

(*a*) A(1, 2); B(3, 4) (*b*) A(2, 3); B(1, 4)
(*c*) A(3, 4); B(1, 2) (*d*) A(1, 4); B(2, 3)

1115. The following are all principles used to control gaseous emissions, except
(*a*) impaction (*b*) condensation
(*c*) absorption (*d*) adsorption

1116. Which of the following control devices might be effectively used to control gaseous emissions?
(*a*) Fabric filters
(*b*) Absorption spray tower
(*c*) Impingement plate scrubber
(*d*) Process boiler

1117. Absorbers collect gaseous pollutants by
(*a*) trapping them in a carbon reagent
(*b*) transferring them from the gas stream into the liquid reagent
(*c*) freezing or condensing the pollutant into a more solid substance
(*d*) heating the pollutant to hasten disposal through combustion

1118. Contact condensers are
(*a*) used to control nitrogen oxide emissions
(*b*) control devices in which coolants come in direct contact with vapors
(*c*) sometimes referred to as shell and tube heat exchangers
(*d*) allow exhaust gases and coolant to exit separately

1119. Match each of the following advantages and disadvantages with the proper pollutant control technology.

Control Technology		Advantages and Disadvantages	
A.	Scrubbers	1.	Can be used for condensable gases and vapours.
B.	Surface and Shell condensers	2.	Collects both gas and particles while neutralizing corrosive gases.
C.	Adsorbers	3.	For collecting a mixture of volatile organic gases.
		4.	Effectively collects various organic gases found in the electronic and petrochemical operations
			Effluents can create water pollution problems.
			Effectively collects a wide variety of organic gases and is typically used by the computer industry.

Answers

1112. (*b*) 1113. (*d*) 1114. (*b*) 1115. (*a*) 1116. (*d*) 1117. (*b*) 1118. (*b*) 1119. (*a*)

(*a*) A(2); B(1, 2, 4); C(3, 5, 6)
(*b*) A(1, 2); B(2, 3); C(5, 6)
(*c*) A(2, 5); B(1, 6); C(3, 4)
(*d*) A(2, 4); B(4, 5); C(6, 1)

1120. Which of the following accounts approximately 90% of all motor vehicle emissions?
(*a*) Exhaust from combustion
(*b*) Crankcase blowby gases
(*c*) Carburetor emissions
(*d*) Evaporative fuel loss (fuel tank)

1121. The most significant pollutant contained in the emissions of automobile vehicle exhaust are
(*a*) carbon monoxide (CO), lead (Pb), particulate matter (PM_{10})
(*b*) carbon monoxide (CO), sulfur dioxide (SO_2), acid deposition
(*c*) carbon monoxide (CO), lead (Pb), nitrogen oxides (NO_x)
(*d*) carbon monoxide (CO), nitrogen oxides (NO_x), non-methane VOCs

1122. The following are effective emission control devices, except
(*a*) Exhaust Gas Recycle (EGR)
(*b*) Evaporative Emissions Canister
(*c*) Camshaft and Piston Valves
(*d*) Catalytic Converter

1123. Ozone pollution control continues to be a problem despite extraordinary improvements in vehicle emissions and cleaner fuel sources because
(*a*) NMHCs are an ozone-depleting compound
(*b*) fewer people use high occupancy vehicle (HOV) lanes or public transportation
(*c*) significant increases in passenger miles offset gains
(*d*) regulatory measures have been relaxed by recent legislation (NMHCs-Non-Methane hydrocarbons)

1124. What engine operating factor is most responsible for the emission of CO?
(*a*) Incomplete fuel combustion
(*b*) Crankcase emissions
(*c*) Evaporative fosses
(*d*) Lean air/fuel mixture

1125. The following gases are produced during the combustion cycle, except
(*a*) unburned hydrocarbons
(*b*) nitrogen oxides (*c*) carbon dioxide
(*d*) lead

1126. Lead alkalis were added to gasoline in order to
(*a*) avoid the additional cost of refining high-octane fuels
(*b*) significantly improve fuel volatility, anti-knock and carbon deposit-forming tendencies
(*c*) boost octane ratings to enhance ignition and burn rates
(*d*) increase the amount of fuel produced by a given quantity of crude oil

1127. Match each of the following mobile source pollution control measure with the proper description.

Control Measures for vehicle emissions		Description	
1.	Thermactor system	A.	A method in which chamber temperatures and NO_x Emissions are reduced.
2.	Engine modifications	B.	A method of changing the ignition process so that ignition occurs in the compression stroke, reducing the temperatures produced and decreasing the production of NO_x.
3.	Lower compression ratios	C.	A method of reducing NOX emissions by relocating the HC-CO oxidation catalyst downstream from the NO_x reduction catalyst.

Answers

1120. (*a*) 1121. (*d*) 1122. (*c*) 1123. (*c*) 1124. (*a*) 1125. (*d*) 1126. (*d*) 1127. (*a*)

4.	Recir-culate exhaust gas	D.	A method of controlling NMHCs and CO in which air is pumped into the exhaust manifold. The air rich environment permits a more complete combustion of volatile gases.
5.	Delayed spark timing	E.	Electric cells, oxygenated additives, and compressed or liquefied gases are a few of the means used to reduce vehicle emissions.
6.	Alter-native fuels	F.	Changes air-fuel mixture, ignition timing, and catalytic converters are some of the ways to effectively reduce vehicle emissions.
7.	Dual or three-way catalytic systems	G.	The most effective way to reduce NO_x emissions is to recirculate controlled amounts of exhaust gas through the intake manifold.

(*a*) A-3, B-5, C-7, D-1, E-4,6, F-2, G-4
(*b*) A-3, B-7, C -5, D-1, E-4, F-6, G-2,5
(*c*) A-4, B-6, C-5,7, D-3,5, E-2,4, F-1,7, G-3,4
(*d*) A-2,3, B-5,6, C-1,4, D-2,5, E-3,7, F-1,7, G-5,6

1128. An example of pollution prevention techniques includes all of the following, except
(*a*) changes in industrial processing
(*b*) improved add-on control technology
(*c*) additional automation
(*d*) changes to operational settings

1129. The least desirable pollution prevention strategy is
(*a*) a source reduction
(*b*) recycling (*c*) waste treatment
(*d*) disposal

1130. The method or equipment best used to dispose of solid waste is
(*a*) anaerobic digestion
(*b*) rotary kiln incinerator
(*c*) oxidation (*d*) wet scrubbers

1131. Which of the following gases is not a green house gas?
(*a*) CO (*b*) O_3
(*c*) CH_4 (*d*) H_2O vapour

1132. Photochemical smog occurs in warm, dry and sunny climate. One of the following is not amongst the components of photochemical smog, identify it.
(*a*) NO_2 (*b*) O_3 (*c*) SO_2
(*d*) Unsaturated hydrocarbon

1133. Which of the following statement is not rue about classical smog?
(*a*) Its main components are produced by the action of sunlight on emissions of automobiles and factories
(*b*) produced in cold and humid climate.
(*c*) it contains compounds of reducing nature.
(*d*) it contains smoke, fog and sulphur dioxide.

1134. Biochemical Oxygen Demand, (BOD) is a measure of organic material present in water, BOD value less than 5 ppm indicates as water sample to be
(*a*) rich in dissolved oxygen.
(*b*) poor in dissolved oxygen.
(*c*) highly polluted.
(*d*) not suitable for aquatic life.

1135. Which of the following statements is wrong?
(*a*) Ozone is not responsible for green house effect.
(*b*) Ozone can oxidise sulphur dioxide present in the atmosphere to sulphur trioxide.
(*c*) Ozone hole is thinning of ozone layer present in stratosphere.
(*d*) Ozone is produced in upper stratosphere by the action of UV rays on oxygen.

Answers

1128. (*b*) 1129. (*d*) 1130. (*b*) 1131. (*a*) 1132. (*c*) 1133. (*a*) 1134. (*a*) 1135. (*a*)

1136. Sewage containing organic waste should not be disposed in water bodies because it cause major water pollution. Fishes in such a polluted water die because of

(*a*) Large number of mosquitoes.
(*b*) Increase in the amount of dissolved oxygen.
(*c*) Decrease in the amount of dissolved oxygen in water.
(*d*) Clogging of gills by mud.

1137. Which of the following statements about photochemical smog is wrong?

(*a*) It has high concentration of oxidising agents.
(*b*) It has low concentration of oxidising agent.
(*c*) It can be controlled by controlling the release of NO_2, hydrocarbons, ozone etc.
(*d*) Plantation of some plants like pinus helps in controlling photochemical smog.

1138. The gaseous envelope around the earth is known as atmosphere. The lowest layer of this is extended upto 11 km from sea level, this layer is

(*a*) stratosphere (*b*) troposphere
(*c*) mesosphere (*d*) hydrosphere

1139. Dinitrogen and dioxygen are main constituents of air but these do not react with each other to from oxides of nitrogen because

(*a*) the reaction is endothermic and requires very high temperature.
(*b*) the reaction can be initiated only in presence of a catalyst.
(*c*) Oxides of nitrogen are unstable.
(*d*) N_2 and O_2 are unreactive.

1140. The pollutants which come directly in the air from sources are called primary pollutants. Primary pollutants are sometimes converted into secondary pollutants. Which of the following belongs to secondary air pollutants?

(*a*) CO (*b*) Hydrocarbon
(*c*) Peroxyacetyl nitrate
(*d*) NO

1141. Which of the following statements is correct?

(*a*) Ozone hole is a hole formed in stratosphere from which ozone oozes out.
(*b*) Ozone hole is a hole formed in the troposphere from which ozone oozes out.
(*c*) Ozone hole is thinning of ozone layer of stratosphere at some places.
(*d*) Ozone hole means vanishing of ozone layer around the earth completely.

1142. Which of the following particles will not come under green chemistry?

(*a*) If possible, making use of soap made of vegetable oils instead of using synthetic detergents.
(*b*) Using H_2O_2 for bleaching purpose instead of using chlorine based bleaching agents.
(*c*) Using bicycle for travelling small distances instead of using petrol/diesel based vehicles.
(*d*) Using plastic cans for neatly storing substances.

1143. Which of the following conditions shows the polluted environment.

(*a*) pH of rain water is 5.6.
(*b*) amount of carbondioxide in the atmosphere is 0.03%.
(*c*) biochemical oxygen demand 10 ppm.
(*d*) eutrophication.

1144. Phosphate containing fertilisers cause water pollution. Addition of such compounds in water bodies causes

(*a*) enhanced growth of algae.
(*b*) decreases in amount of dissolved oxygen in water.
(*c*) deposition of calcium phosphate.
(*d*) increase in fish population

1145. The acid not present in acid rain is

(*a*) Peroxyacetylnitrate

Answers

1136. (*c*) 1137. (*b*) 1138. (*b*) 1139. (*a*) 1140. (*c*) 1141. (*c*) 1142. (*d*) 1143. (*c*) 1144. (*a*)
1145. (*a*)

(*b*) H_2CO_3 (*c*) HNO_3
(*d*) H_2SO_4

1146. The consequences of global warming may be
(*a*) increase in average temperature of the earth.
(*b*) melting of Himalayan Glaciers.
(*c*) increased biochemical oxygen demand
(*d*) eutrophication.

1147. Match the terms given in Column I with the compounds given in Column II.

Column I	Column II
I. Acid rain	A. $CHCl_2$-CHF_2
II. Photochemical smog	B. CO
III. Combination with haemoglobin	C. CO_2
IV. Depletion of ozone layer	D. SO_2

(*a*) I-(C, D); II-(C, D); III-(D); IV-(A)
(*b*) I-(C); II-(D); III-(A); IV-(B)
(*c*) I-(D); II-(C); III-(A); IV-(B)
(*d*) I-(D); II-(C); III-(B); IV-(A)

1148. Match the pollutants (s) in Column I with the effects (s) in Column II.

Column I	Column II
I. Oxides of sulphur	A. Global warming
II. Nitrogen dioxide	B. Damage to kidney
III. Carbon dioxide	C. 'Blue baby' syndrome
IV. Nitrate in drinking water	D. Respiratory diseases
V. Lead	E. Red haze in traffic and congested areas

(*a*) I-(D); II-(A); III-(E); IV-(C); V-(B)
(*b*) I-(A); II-(C); III-(D); IV-(E); V-(B)
(*c*) I-(A); II-(B); (III)-(C); IV-(D); V-(E)
(*d*) I-(D); II-(E); III-(A); IV-(C); V-(B)

1149. Match the activity given in Column I with the type of pollution created by it given in Column II.

Column I (Activity)	Column II (Effect)
I. Burning waste material containing sulphur.	A. Water pollution photochemical smog, damage to plant life, corrosion to
II. Using carbamates as pesticides	B. Building material, induce breathing problems, water pollution
III. Using synthetic detergents for washing clothes	C. Damaging ozone layer
IV. Releasing gases produced by automobiles and factories in the atmosphere.	D. May cause nerve diseases in human classical smog, acid rain, water pollution, induce breathing problems
V. Using chlorofluoro-carbon compounds for cleaning computer parts	E. Damage to buildings, corrosion of metals

(*a*) I-(A); II-(B); III-(C); IV-(D); V-(E)
(*b*) I-(E); II-(D); III-(C); IV-(B); V-(E)
(*c*) I-(E); II-(D); III-(A); IV-(C); V-(B)
(*d*) I-(A); II-(D); III-(B); IV-(C); V-(E)

1150. Match the pollutants given in Column I with their effects given in Column II.

Column I	Column II
I. Phosphate fertilisers in water	A. BOD level of water increases
II. Methane in air	B. Acid rain
III. Synthetic detergents in water	C. Global warming
IV. Nitrogen oxides in air	D. Eutorphication

(*a*) I-(A); II-(C); III-(A); IV-(B)

Answers

1146. (*a*) 1147. (*a*) 1148. (*d*) 1149. (*c*) 1150. (*b*)

(*b*) I-(A, d); II-(C); III-(A); IV-(B)
(*c*) I-(A); III-(B); III-(C); IV-(D)
(*d*) I-(C); II-(B); III-(D); IV-(A, D)

In the following questions a statement of Assertion (A) followed by a statement of Reason (R) is given. Choose the correct option out of the choices given below each question.

1151. Assertion (A): Green house effect was observed in houses used to grow plants and these are made of green glass.
Reason (R): Green house name has been given because glass houses are made of green glass.
(*a*) Both A and R are correct and R is the correct explanation of A.
(*b*) Both A and R are correct but R is not the correct explanation of A.
(*c*) Both A and R are not correct.
(*d*) A is not correct but R is correct.

1152. Assertion (A): The pH of acid rain is less than 5.6.
Reason (R): carbon dioxide present in the atmosphere dissolves in rain water and forms carbonic acid.
(*a*) Both A and R are correct and R is the correct explanation of A.
(*b*) Both A and R are correct but R is not the correct explanation of A.
(*c*) Both A and R are not correct.
(*d*) A is not correct but R is correct.

1153. Assertion (A): Photochemical smog is oxidising in nature.
Reason (R): Photochemical smog contains NO_2 and O_3, which are formed during the sequence of reactions.
(*a*) Both A and R are correct and R is the correct explanation of A.
(*b*) Both A and R are correct but R is not the correct explanation of A.
(*c*) Both A and R are not correct.
(*d*) A is not correct but R is correct.

1154. Assertion (A): carbon dioxide is one of the important greenhouse gases.
Reason (R): It is largely produced by respiratory function of animals and plants.
(*a*) Both A and R are correct and R is the correct explanation of A.
(*b*) Both A and R are correct but R is not the correct explanation of A.
(*c*) Both A and R are not correct.
(*d*) A is not correct but R is correct.

1155. Assertion (A): Ozone is destroyed by solar radiation in upper stratosphere.
Reason (R): Thinning of the ozone layer allows excessive UV radiations to reach the surface of earth.
(*a*) Both A and R are correct and R is the correct explanation of A.
(*b*) Both A and R are correct but R is not the correct explanation of A.
(*c*) Both A and R are not correct.
(*d*) A is not correct but R is correct.

1156. Assertion (A): Excessive use of chlorinated synthetic pesticides causes soil and water pollution.
Reason (R): Such pesticides are non-biodegradable.
(*a*) Both A and R are correct and R is the correct explanation of A.
(*b*) Both A and R are correct but R is not the correct explanation of A.
(*c*) Both A and R are not correct.
(*d*) A is not correct but R is correct.

1157. Assertion (A): If BOD level of water in a reservoir is less than 5 ppm it is highly polluted.
Reason (R): High biological oxygen demand means low activity of bacteria in water.
(*a*) Both A and R are correct and R is the correct explanation of A.
(*b*) Both A and R are correct but R is not the correct explanation of A.

Answers

1151. (*c*) 1152. (*b*) 1153. (*a*) 1154. (*b*) 1155. (*d*) 1156. (*a*) 1157. (*c*)

(*c*) Both A and R are not correct.
(*d*) A is not correct but R is correct.

1158. Which of the following are biodegradable waste products?
(*a*) rubber, plastic, manure
(*b*) metal, glass, plastic
(*c*) vegetable peels, garden refuse, paper
(*d*) glass, wood cuttings, batteries

1159. The main cause of water pollution is:
(*a*) untreated sewage, glass bottles and plastic bags
(*b*) chemical effluent, leaching of pesticides and algae
(*c*) pesticides, treated sewage and litter
(*d*) oil spills untreated sewage and chemical effluent

1160. Typhoid, cholera and dysentery are diseases that are spread by:
(*a*) air pollution (*b*) water pollution
(*c*) land/soil pollution
(*d*) The use of pesticides

1161. Non-biodegradable means a substance that
(*a*) cannot decompose naturally
(*b*) is thrown away as kitchen refuse
(*c*) costs too much to recycle
(*d*) will not dissolve in water

1162. Deforestation is the
(*a*) removal of all plants and trees from a large area of land
(*b*) complete removal of a forest to clear a large area for agriculture
(*c*) result of a veld fire
(*d*) result of concentrated soil erosion

1163. One solution to reduce air pollution may be ______
(*a*) the use of leaded petrol
(*b*) the burning of wood and coal
(*c*) the use of public transport
(*d*) causing wild fires

1164. One of the following has a positive human influence on the environment:
(*a*) Deforestation (*b*) Overpopulation
(*c*) Population (*d*) Conservation

1165. Biological control is useful because it ______
(*a*) usually attacks all plant species
(*b*) can destroy an alien plant infestation or reduce its population numbers
(*c*) usually originates in the country where the alien destroys indigenous vegetation
(*d*) usually attacks all useful micro-organisms

1166. Pollutant gases have an adverse effect on the environment and on our health. The amounts of these gases have been steadily increasing over the years.
The table below shows the source and the amount of some pollutant gases produced by human activities:

Pollutant gases	Sources	Amount produced per year (millions of tonnes)
Carbon dioxide	Vehicle exhaust fumes	350
Sulphur dioxide	Burning coal and oil, industries	200
Nitrogen oxide	Vehicle exhaust fumes	55
Hydro-carbons	Vehicle exhaust fumes, industry	90

Which one of the following statements is correct
(*a*) The total sulphur dioxide and nitrogen dioxide produced per year is 145 million tones.
(*b*) Carbon dioxide is produced by industries.
(*c*) The combined total pollutant gases produced by nitrogen oxide and hydrocarbons, is less than the total produced by sulphur dioxide.

Answers

1158. (*c*) 1159. (*d*) 1160. (*b*) 1161. (*a*) 1162. (*a*) 1163. (*c*) 1164. (*d*) 1165. (*b*) 1166. (*c*)

(*d*) Hydrocarbons are produced by vehicle exhaust gases only.

1167. The following are all sources of pollution:

(*i*) Solid waste (*ii*) Pesticides

(*iii*) Sewage (*iv*) Fertilisers

Which of the following combination contains sources of soil pollution?

(*a*) (*i*), (*ii*) and (*iii*) (*b*) (*ii*), (*iii*), (*iv*)

(*c*) (*i*), (ii), (*iii*) and (*iv*)

(*d*) (*ii*) and (*iv*)

1168. ________ is used to convert organic pollutants into water vapour CO_2.

(*a*) Electrostatic precipitator

(*b*) Flame combustion

(*c*) Absorption (*d*) Wet-scrubber

1169. ________ are used in electrostatic precipitators.

(*a*) Catalysts (*b*) Absorption

(*c*) Electrodes (*d*) Chemicals

1170. Biological oxygen demand (Bod) of ________ is the least.

(*a*) sewage (*b*) sea water

(*c*) pure water (*d*) polluted water

1171. Compound of ________ cause Minamata disease.

(*a*) mercury (*b*) cadmium

(*c*) cobalt (*d*) fertilizers

1172. Due to eutrophication ________.

(*a*) BOD increases.

(*b*) BOD decreases.

(*c*) algae are destroyed.

(*d*) water becomes less harmful

1173. ________ is the first step of sewage treatment.

(*a*) Precipitation (*b*) Chlorination

(*c*) Sedimentation (*d*) Aeration

1174. We and our surrounding together are called ________

(*a*) environment (*b*) atmosphere

(*c*) lithosphere (*d*) hydrosphere

1175. Which of the following is not an environmental problem?

(*a*) Wastage of water

(*b*) Conservation of water

(*c*) Deforestation (*d*) Land erosion

1176. Which harmful rays strike on earth due to depletion of ozone layer?

(*a*) Ultraviolet (*b*) Infrared

(*c*) Microwaves

1177. Which method is preferentially used for the removal of sulphur dioxide and ammonia from the polluted air?

(*a*) Wet scrubbers (*b*) Absorption

(*c*) Gravitational method

(*d*) Electrostatic precipitator

1178. BOD is ________ in polluted water and ________ in potable water.

(*a*) more, less (*b*) less, medium

(*c*) medium, more (*d*) less, more

1179. Which of the following is responsible for acid rain?

(*a*) Hydrogen (*b*) Oxygen

(*c*) Sulphur dioxide (*d*) Methane

1180. Which pollutants are responsible for bronchitis?

(*a*) O_2, CO_2 (*b*) CO, CO_2

(*c*) SO_2, NO_2 (*d*) Cl_2, H_2S

1181. What is the harm from the depletion of Earth's ozone layer

(*a*) The average temperature of earth's surface will increase gradually.

(*b*) The oxygen content of the atmosphere will decrease

(*c*) Increased amount of Ultra violet radiation will reach earth's surface

(*d*) Sea levels will rise as the polar ice caps will gradually melt

1182. Acid rain is formed due to reaction from the following pair of gases

Answers

1167. (*c*) 1168. (*b*) 1169. (*c*) 1170. (*c*) 1171. (*a*) 1172. (*a*) 1173. (*c*) 1174. (*a*) 1175. (*b*)
1176. (*a*) 1177. (*a*) 1178. (*a*) 1179. (*c*) 1180. (*c*) 1181. (*c*) 1182. (*b*)

(*a*) Methane and ozone
(*b*) Oxygen and nitrous oxide
(*c*) methane and sulphur dioxide
(*d*) Carbon dioxide and sulphur dioxide

1183. Which of the following is a prime health risks associated with greater UV (ultra-violet) radiation through the atmosphere due to depletion of stratospheric ozone?
(*a*) Damage to digestive system
(*b*) Increased liver cancer
(*c*) Neurological disorder
(*d*) Increased skin cancer

1184. The most serious environmental effect posed by hazardous wastes is
(*a*) air pollution.
(*b*) contamination of groundwater.
(*c*) increased use of land for landfills.
(*d*) destruction of habitat.

1185. The concentration of which gas is highest in our environment?
(*a*) Oxygen (*b*) Hydrogen
(*c*) Nitrogen (*d*) Carbon dioxide

1186. Which of the following is not a consequence of global warming?
(*a*) rising sea level
(*b*) increased agricultural productivity
(*c*) worsening health effect
(*d*) increased storm frequency and intensity

1187. Which of the following is not a primary contributor to the greenhouse effect?
(*a*) carbon dioxide
(*b*) chlorofluorocarbons
(*c*) methane gas (*d*) nitrogen oxide

1188. The increase in the concentration of CO_2 in our environment in last fifty years; since 1960 is about ________ percent
(*a*) 20 (*b*) 10 (*c*) 15 (*d*) 40

1189. The depletion in the Ozone layer is caused by
(*a*) nitrous oxide. (*b*) carbon dioxide.
(*c*) chlorofluorocarbons.
(*d*) methane.

1190. A major in-stream use of water is for
(*a*) producing hydroelectric power.
(*b*) dissolving industrial wastes.
(*c*) agricultural irrigation.
(*d*) domestic use.

1191. Which of the following are example of Municipal and industrial discharge pipes?
(*a*) non-point sources of pollution.
(*b*) violations of the Clean Water Act.
(*c*) point sources of pollution.
(*d*) irrigation.

1192. The presence of high coliform counts in water indicate
(*a*) contamination by human wastes.
(*b*) phosphorus contamination.
(*c*) decreased biological oxygen demand.
(*d*) hydrocarbon contamination.

1193. How the biological oxygen demand gets affected with the increased presence of organic mater in water?
(*a*) the oxygen demand increases
(*b*) the oxygen demand decreases
(*c*) the oxygen demand remains unchanged
(*d*) None of the above

1194. Which of the following is not a major source of groundwater contamination?
(*a*) agricultural products
(*b*) landfills (*c*) septic tanks
(*d*) all of the above are major sources of groundwater contamination

1195. Which of the following is not considered as a part of water use planning?
(*a*) waster treatment
(*b*) storm sewer drainage
(*c*) salinization
(*d*) Water use planning considers all of the above issue

Answers

1183. (*d*) 1184. (*b*) 1185. (*c*) 1186. (*b*) 1187. (*b*) 1188. (*c*) 1189. (*c*) 1190. (*a*) 1191. (*c*)
1192. (*a*) 1193. (*d*) 1194. (*d*) 1195. (*d*)

1196. The stage in which the biological processes is used to purify water in a waste water treatment plants is called
(*a*) secondary sewage treatment
(*b*) primary sewage treatment
(*c*) waste water reduction
(*d*) biochemical reduction

1197. Groundwater mining in coastal areas can result into
(*a*) increase in the salinity of groundwater.
(*b*) decrease in the toxicity of groundwater.
(*c*) decrease in the salinity of groundwater.
(*d*) increase in the water table.

1198. Which of the following is to an important characteristic of the Green Revolution?
(*a*) mechanized agriculture
(*b*) hybrid seeds (*c*) slash and burn
(*d*) monoculture

1199. The three primary soil macronutrients are
(*a*) carbon, oxygen, and water.
(*b*) copper, cadmium and carbon.
(*c*) potassium, phosphorus, and nitrogen.
(*d*) boron, zinc, and manganese.

1200. Which of the following has negative effects on the soil and water due to conventional, mechanized farming practices?
(*a*) soil compaction and soil erosion
(*b*) reduction in soil organic matter
(*c*) leaching of pesticides and fertilizers into the groundwater
(*d*) all of the above

1201. Standard EDTA (ethylene diamine tetra acetic acid) solution is used to determine the
(*a*) Hardness in water
(*b*) turbidity in water
(*c*) dissolved oxygen in water
(*d*) residual chlorine in water

1202. The water distribution mains are designed for
(*a*) maximum daily demand
(*b*) maximum hourly demand
(*c*) average daily demand
(*d*) maximum hourly demand on maximum day

1203. The maximum permissible limit for fluoride in drinking water is ______ mg/litre.
(*a*) 0.1 (*b*) 1.5 (*c*) 5 (*d*) 10

1204. The polluted water is one which
(*a*) contains pathogenic bacteria
(*b*) consists of undesirable substances rendering unfit for drinking and domestic use
(*c*) is safe and suitable for drinking and domestic use
(*d*) is contaminated

1205. The suitable method of forecasting population for a young and rapidly increasing city is
(*a*) arithmetical increase method
(*b*) geometrical increase method
(*c*) incremental increase method
(*d*) graphical method

1206. If the coliform bacteria is present in a sample of water, then the coliform test to be conducted is (*i*) presumptive coliform test (*ii*) confirmed coliform test (*iii*) completed coliform test. correct answer is
(*a*) only (*i*) (*b*) both (*i*) and (*ii*)
(*c*) both (*i*) and (*iii*)
(*d*) all (*i*), (*ii*) and (*iii*)

1207. Alkalinity in water is expressed as milli-grams per litre in terms of equivalent
(*a*) calcium carbonate
(*b*) magnesium carbonate
(*c*) sodium carbonate
(*d*) calcium hydroxide

1208. The most common cause of acidity in water is
(*a*) carbon dioxide (*b*) oxygen
(*c*) hydrogen (*d*) nitrogen

1209. Ground water is usually free from
(*a*) suspended impurities

Answers

1196. (*a*) 1197. (*a*) 1198. (*c*) 1199. (*c*) 1200. (*d*) 1201. (*a*) 1202. (*d*) 1203. (*b*) 1204. (*b*)
1205. (*b*) 1206. (*d*) 1207. (*a*) 1208. (*a*) 1209. (*a*)

(b) dissolved impurities
(c) both suspended and dissolved impurities
(d) none of the above

1210. The phenolic compounds in public water supply should not be more then _______ ppm.
(a) 0.1 (b) 0.01 (c) 0.001 (d) 0.0001

1211. If the average daily consumption of a city is 100,000 m^3, the maximum daily consumption on peak hourly demand will be m^3
(a) 100000 (b) 150000
(c) 180000 (d) 270000

1212. As compared to shallow-wells, deep wells have
(a) more depth (b) less depth
(c) more discharge (d) less discharge

1213. Which of the following caused a decrease in per capita water consumption?
(a) use of metering system
(b) good quality of water
(c) better standard of living of the people
(d) hotter climate

1214. The devices which are installed for drawing water from the sources are called
(a) aquifers (b) aquiclude
(c) filters (d) intakes

1215. Which of the following is not a water borne disease?
(a) dysentery (b) cholera
(c) typhoid (d) malaria

1216. Alum as a coagulant is found to be most effective when pH range of water is
(a) 2 to 4 (b) 4 to 6
(c) 6 to 8 (d) 8 to 10

1217. The detention period in coagulation tanks is usually kept as
(a) 1 to 2 minutes (b) 30 to 45 minutes
(c) 2 to 6 hours (d) 2 to 6 days

1218. Air binding phenomena in rapid sand filters occur due to
(a) excessive negative head
(b) mud ball formation
(c) higher turbidity in the effluent
(d) low temperature

1219. The chemical most commonly used to increase The speed of sedimentation of sewage is
(a) sulphuric acid (b) copper sulphate
(c) lime
(d) sodium permanganate

1220. The detention period and overflow rate respectively for plain sedimentation as compared to sedimentation with coagulation are generally
(a) less and more (b) less and less
(c) more and less (d) more and more

1221. The alum, when added as a coagulant in water
(a) does not affect pH value of water
(b) decreases pH value of water

1222. If the total hardness of water is greater than its total alkalinity the alkalinity the carbonate hardness will be equal to
(a) total alkalinity (b) total hardness
(c) total hardness-total alkalinity
(d) non carbonate hardness

1223. The effective size of sand particles used in slow sand filters is _______ mm.
(a) 0.25 to 0.35 (b) 0.35 to 0.60
(c) 0.60 to 1.00 (d) 1.00 to 1.80

1224. The amount of coagulant needed for coagulation of water increases with (i) increase in turbidity if water (ii) decrease in turbidity of water (iii) increase in temperature of water (iv) decrease in temperature of water. The correct answer is
(a) (i) and (ii) (b) (i) and (iv)
(c) (ii) and (iii) (d) (ii) and (iv)

1225. In water treatment, rapid gravity filters are adopted to remove
(a) dissolved organic substances

Answers

1210. (c) 1211. (d) 1212. (c) 1213. (a) 1214. (d) 1215. (d) 1216. (c) 1217. (c) 1218. (a)
1219. (c) 1220. (c) 1221. (d) 1222. (a) 1223. (a) 1224. (b) 1225. (d)

(*b*) dissolved solids and dissolved gases
(*c*) floating solids and dissolved inorganic solids
(*d*) bacteria and colloidal solids

1226. The settling velocity of a particle in a sedimentation tank increases, it
(*a*) particle size is decreased
(*b*) the surface area of ank is increased
(*c*) the depth of tank is decrease
(*d*) none of these

1227. Percentage of bacterial load that can be removed from water by the process of plain sedimentation is about
(*a*) 10 to 20 (*b*) 50
(*c*) 75 (*d*) 95

1228. The dissolved oxygen level in natural unpolluted waters at normal temperature is of the order of ______ mg/litre.
(*a*) 1 (*b*) 10 (*c*) 100 (*d*) 1000

1229. For a given discharge, the efficiency of sedimentation tank can be increases by
(*a*) increasing the depth of tank
(*b*) decreasing the depth of tank
(*c*) increasing the surface area of tank
(*d*) decreasing the surface area of tank

1230. As compared to rapid sand filters, slow sand filters give (*i*) slower filtration rate (*ii*) higher filtration rate (*iii*) lesser efficiency in removal of bacteria (*iv*) higher efficiency in removal of bacteria. The correct answer is
(*a*) (*i*) and (*ii*) (*b*) (*ii*) and (*iii*)
(*c*) (*i*) and (*iv*) (*d*) (*ii*) and (*iv*)

1231. Residual chlorine in water is determined by
(*a*) starch iodide method
(*b*) orthotolidine method
(*c*) both (*a*) and (*b*)
(*d*) none of the above

1232. The setting velocity of a particle in a sedimentation tank depends on
(*a*) depth of tank
(*b*) surface area of tank
(*c*) both depth and surface tank
(*d*) none of the above

1233. Assertion A: Slow sand filters and more efficient in removal of a bacteria than rapid filters.

Reason R: The sand used in slow sand filters is finer than that in rapid sand filters. Select your answer based on the coding system given below:
(*a*) Both A nd R are true and R is the correct explanation of A.
(*b*) Both A and R are true but R is not the correct explanation of A.
(*c*) A is true but R is false.
(*d*) A is false but R is true.

1234. The amount of residual chlorine left in public water supply for safety against pathogenic bacteria is about ______ ppm.
(*a*) 0.01 to 0.05 (*b*) 0.05 to 0.5
(*c*) 0.5 to 1.0 (*d*) 1.0 to 5.0

1235. The rate of filtration in slow sand filters in million litres per day per hectare is bout
(*a*) 50 to 60 (*b*) 100 to 150
(*c*) 500 to 600 (*d*) 1400 to 1500

1236. The overflow rate for the plan sedimentation tanks is about ______ lit res/hr/m^2.
(*a*) 500 to 750 (*b*) 100 to 1250
(*c*) 1250 to 1500 (*d*) 1500 to 2000

1237. Orthotolidine test in used for determination of
(*a*) dissolved oxygen (*b*) residual chlorine
(*c*) biochemical oxygen demand
(*d*) does of coagulant

1238. The velocity of flow water in a sedimentation tank is about
(*a*) 5 to cm/sec. (*b*) 15 to 30 cm/sec.
(*c*) 15 to 30 cm/ minute
(*d*) 15 to 30 cm/hour

Answers

1226. (*d*) 1226. (*a*) 1227. (*c*) 1228. (*b*) 1229. (*c*) 1230. (*d*) 1231. (*c*) 1232. (*b*) 1233. (*a*)
1234. (*c*) 1235. (*a*) 1236. (*a*) 1237. (*c*) 1238. (*c*)

1239. On standard silica, the turbidity in drinking water should be limited to ________ ppm
(*a*) 10 (*b*) 20 (*c*) 30 (*d*) 50

1240. The major disadvantage of lime soda process of water softening is that
(*a*) It is unsuitable for turbid and acidic water.
(*b*) huge amount of precipitate is formed which creates a disposal problem
(*c*) the effluent cannot be reduced to zero hardness
(*d*) it is unsuitable for softening the water of excessive hardness

1241. In chlorination. With the rise in temperature of water. Death rate of bacteria
(*a*) increases (*b*) decreases
(*c*) remains unaffected
(*d*) none of the above

1242. The process in which the chlorination is done beyond the break point is known as
(*a*) prechlorination
(*b*) post chlorination
(*c*) super chlorination
(*d*) break point chlorination

1243. Disinfection of water results in
(*a*) removal of turbidity
(*b*) removal hardness
(*c*) killing of disease causing bacteria
(*d*) complete sterilization

1244. Chlorine demand of water is equal to
(*a*) applied chlorine (*b*) residual chloride
(*c*) sum of applied and residual chlorine
(*d*) difference of applied and residual chlorine

1245. Which of the following chemical compounds can be used for dechlorination of water?
(*a*) carbon dioxide (*b*) bleaching powder
(*c*) sulphur dioxide (*d*) chloramines

1246. The treatment of water with bleaching powder is known as
(*a*) prechlorination
(*b*) super chlorination
(*c*) dechlorination (*d*) hypochlorination

1247. The suitable layout of a water distribution system for irregularly growing town is
(*a*) dead end system (*b*) grid iron system
(*c*) radial system (*d*) ring system

1248. The suitable method for disinfection of swimming pool water is
(*a*) ultra violet rays treatment
(*b*) lime treatment
(*c*) by using potassium permanganate
(*d*) clorination

1249. The percentage of chlorine in fresh bleaching powder is about
(*a*) 10 to 15 (*b*) 20 to 25
(*c*) 30 to 35 (*d*) 40 to 50

1250. Disinfection efficiency is
(*a*) reduced at higher pH value of water
(*b*) unaffected by pH value of water
(*c*) increased at higher pH value of water
(*d*) highest at pH value equal to 7

1251. Activate carbon is used ________ of water
(*a*) disinfection
(*b*) removing hardness
(*c*) removing odours
(*d*) removing corrosiveness

1252. The layout of water distribution system in which water flows towards the outer periphery is
(*a*) ring system (*b*) dead end system
(*c*) radial system (*d*) grid iron system

1253. The disinfection efficiency of chlorine increases by (*i*) decreasing the time of contact (*ii*) decreasing the temperature of water (*iii*) increasing the temperature of water. The correct answer is
(*a*) only (*i*) (*b*) both (*i*) and (*ii*)
(*c*) only (*iii*)

Answers

1239. (*a*) 1240. (*b*) 1241. (*a*) 1242. (*c*) 1243. (*c*) 1244. (*d*) 1245. (*c*) 1246. (*d*) 1247. (*a*)
1248. (*a*) 1249. (*c*) 1250. (*a*) 1251. (*c*) 1252. (*c*) 1253. (*d*)

1254. The suitable layout of water distribution system for a city with roads of rectangular pattern is
(*a*) grid iron system (*b*) dead end system
(*c*) ring system (*d*) radial system

1255. Which of the following compounds is widely used for algae control?
(*a*) sodium sulphate (*b*) copper sulphate
(*c*) sodium chloride (*d*) calcium chloride

1256. The commonly used material for water supply pipes, which has the properties of being strong not easily corroded and long life but is heavy and brittle is
(*a*) steel (*b*) cast iron
(*c*) copper
(*d*) reinforced cement concrete

1257. Double filtration is used
(*a*) to increase the filtration capacity of show send filters.
(*b*) to increase the filtration capacity of rapid send filters.
(*c*) for swimming pods, hotels etc.
(*d*) all of the above

1258. In lime-soda process
(*a*) only carbonate hardness is removed
(*b*) only non-carbonate hardness is re-moved
(*c*) lime reduces the carbonate hardness and soda-ash removes the non-carbonate hardness
(*d*) lime reduces the non-carbonate hardness and soda-ash removes the carbonate hardness

1259. The main disadvantage of cement concrete sewers is
(*a*) less strength
(*b*) difficulty in construction
(*c*) difficulty in transportation due to heavy weight
(*d*) less life

1260. An egg shaped section of sewer
(*a*) is economical than circular section.
(*b*) provides self cleansing velocity at low discharges.
(*c*) is more stable than circular section.
(*d*) is easy to construct

1261. A sewer that receives the discharge of a number of house sewer is called
(*a*) house sewer (*b*) lateral sewer
(*c*) intercepting (*d*) submain sewer

1262. Which of the following sewer is preferred for combined system of sewage?
(*a*) circular sewer (*b*) egg shaped sewer
(*c*) rectangular (*d*) none of the above

1263. Average rate of water consumption perheap per day as per Indian Standard is
(*a*) 100 litres (*b*) 135 litres
(*c*) 165 litres (*d*) 200 litres

1264. A pipe conveying sewage from plumbing system of a single building to common sewer or point of immediate disposal is called
(*a*) house sewer (*b*) lateral sewer
(*c*) main sewer (*d*) submain sewer

1265 The minimum and maximum diameters (in cms) of sewer should preferably be
(*a*) 15 and 100 (*b*) 15 and 300
(*c*) 30 and 450 (*d*) 60 and 300

1266. For a country like India, where rainfall is mainly confined to one season, the suitable sewerage system will be
(*a*) Separate system (*b*) Combined system
(*c*) Partially combined system
(*d*) Partially separate system

1267. Sewerage system is usually designed for ________ years.
(*a*) 10 (*b*) 25 (*c*) 50 (*d*) 75

1268. Most suitable section of sewerin separate sewage system is
(*a*) rectangular section

Answers

1254. (*a*) 1255. (*b*) 1256. (*b*) 1257. (*a*) 1258. (*c*) 1259. (*c*) 1260. (*b*) 1261. (*b*) 1262. (*b*)
1263. (*b*) 1264. (*a*) 1265. (*b*) 1266. (*a*) 1267. (*b*) 1268. (*b*)

(*b*) circular section
(*c*) standard form of egg shaped sewer
(*d*) modified egg shaped section

1269. The specific gravity of sewage is
(*a*) much greater than I
(*b*) slightly less than I
(*c*) equal to I
(*d*) slightly greater than I

1270. The suitable system of sanitation for area of distributed rainfall throughout the year less intensity is
(*a*) separate system (*b*) combined system
(*c*) partially separate system
(*d*) partially combined system

1271. The slope of sewer should be
(*a*) given in the direction of natural slope of ground.
(*b*) given in the direction opposite to natural slope of ground
(*c*) zero.
(*d*) steeper than 1 in 20.

1272. The self cleansing velocity for all sewers in India is usually ______ m/sec.
(*a*) less than 1.0 (*b*) 1.0 to 1.2
(*c*) 1.5 to 2.0 (*d*) 3.0 to 3.5

1273. Setting velocity increases with
(*a*) specific gravity of solid particles
(*b*) size of particles (*c*) depth of tank
(*d*) temperature of liquid

1274. The type of sewer which is suitable for both combined and separate system is
(*a*) circular sewer (*b*) egg shaped sewer
(*c*) horseshoe type sewer
(*d*) semi-elliptical sewer

1275. Laying of sewers is usually done with the help of
(*a*) atheodolite (*b*) a compass
(*c*) sight rails and boning rods
(*d*) a plane table

1276. In a BOD test 0 ml of raw sewage was diluted to 100 ml and the dissolved oxygen concentration diluted sample at the beginning was 6 ppm and it was 4 ppm at the end of 5 day incubation at 20°C. The BOD of raw sewage will be ______ ppm.
(*a*) 100 (*b*) 200 (*c*) 300 (*d*) 400

1277. Dissolved oxygen in strams is
(*a*) maximum at noon
(*b*) minimum at noon
(*c*) maximum at midnight
(*d*) same throughout the day

1278. The correct relation between theoretical oxygen demand (TOD), Biochemical oxygen demand (BOD) and chemical oxygen demand (COD) is given by
(*a*) TOD > BOD > COD
(*b*) TOD > COD > BOD
(*c*) BOD > COD > TOD
(*d*) COD > BOD > TOD

1279. The ratio of 5 day BOD to ultimate BOD is about
(*a*) 1/3 (*b*) 2/3 (*c*) 3/4 (*d*) 1.0

1280. The effect of increasing diameter of seweron the self cleansing velocity is
(*a*) to decrease it (*b*) to increase it
(*c*) fluctuation (*d*) nil

1281. The pathogens in sewage
(*a*) nitrification (*b*) chlorination
(*c*) oxidation (*d*) none of the above

1282. The minimum dissolved oxygen which should always be present in water in order to save the aquatic life is ______ ppm.
(*a*) 1 (*b*) 4 (*c*) 10 (*d*) 40

1283. The characteristics of fresh and septic sewage respectively are
(*a*) acidic and alkaline
(*b*) alkaline and acidic
(*c*) both acidic (*d*) both alkaline

Answers

1269. (*d*) 1270. (*b*) 1271. (*a*) 1272. (*b*) 1273. (*c*) 1274. (*b*) 1275. (*c*) 1276. (*b*) 1277. (*a*)
1278. (*b*) 1279. (*b*) 1280. (*b*) 1281. (*b*) 1282. (*b*) 1283. (*b*)

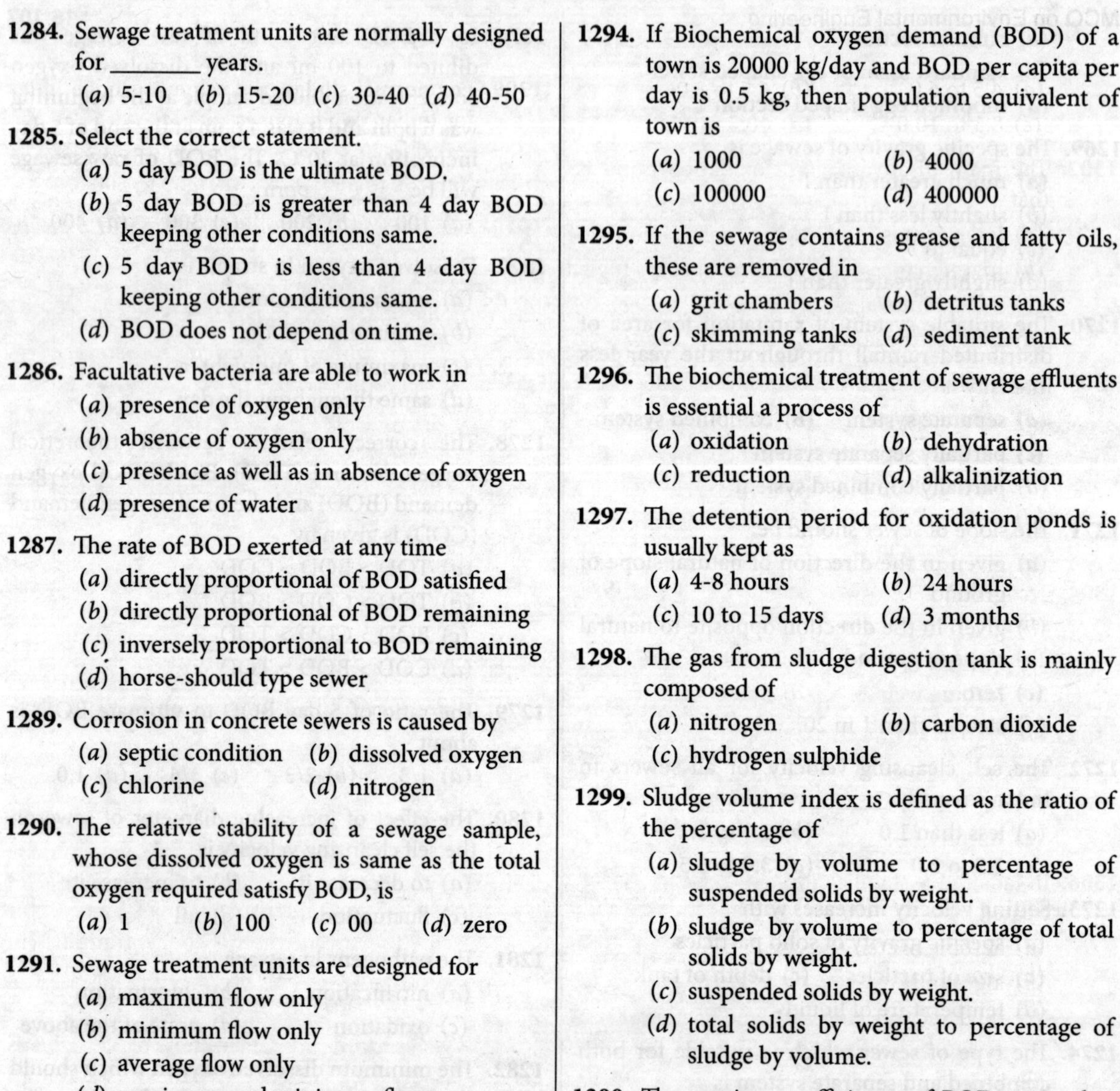

1284. Sewage treatment units are normally designed for ______ years.
(*a*) 5-10 (*b*) 15-20 (*c*) 30-40 (*d*) 40-50

1285. Select the correct statement.
(*a*) 5 day BOD is the ultimate BOD.
(*b*) 5 day BOD is greater than 4 day BOD keeping other conditions same.
(*c*) 5 day BOD is less than 4 day BOD keeping other conditions same.
(*d*) BOD does not depend on time.

1286. Facultative bacteria are able to work in
(*a*) presence of oxygen only
(*b*) absence of oxygen only
(*c*) presence as well as in absence of oxygen
(*d*) presence of water

1287. The rate of BOD exerted at any time
(*a*) directly proportional of BOD satisfied
(*b*) directly proportional of BOD remaining
(*c*) inversely proportional to BOD remaining
(*d*) horse-should type sewer

1289. Corrosion in concrete sewers is caused by
(*a*) septic condition (*b*) dissolved oxygen
(*c*) chlorine (*d*) nitrogen

1290. The relative stability of a sewage sample, whose dissolved oxygen is same as the total oxygen required satisfy BOD, is
(*a*) 1 (*b*) 100 (*c*) 00 (*d*) zero

1291. Sewage treatment units are designed for
(*a*) maximum flow only
(*b*) minimum flow only
(*c*) average flow only
(*d*) maximum and minimum flow

1292. Standard BOD is measured at
(*a*) 20°C – 1 day (*b*) 25°C – 3 day
(*c*) 20°C – 5 day (*d*) 30°C – 5 day

1293. Generally the detention period for grid chambers is kept as
(*a*) 1 minute (*b*) 5 minutes
(*c*) 2-4 hours (*d*) 12 hours

1294. If Biochemical oxygen demand (BOD) of a town is 20000 kg/day and BOD per capita per day is 0.5 kg, then population equivalent of town is
(*a*) 1000 (*b*) 4000
(*c*) 100000 (*d*) 400000

1295. If the sewage contains grease and fatty oils, these are removed in
(*a*) grit chambers (*b*) detritus tanks
(*c*) skimming tanks (*d*) sediment tank

1296. The biochemical treatment of sewage effluents is essential a process of
(*a*) oxidation (*b*) dehydration
(*c*) reduction (*d*) alkalinization

1297. The detention period for oxidation ponds is usually kept as
(*a*) 4-8 hours (*b*) 24 hours
(*c*) 10 to 15 days (*d*) 3 months

1298. The gas from sludge digestion tank is mainly composed of
(*a*) nitrogen (*b*) carbon dioxide
(*c*) hydrogen sulphide

1299. Sludge volume index is defined as the ratio of the percentage of
(*a*) sludge by volume to percentage of suspended solids by weight.
(*b*) sludge by volume to percentage of total solids by weight.
(*c*) suspended solids by weight.
(*d*) total solids by weight to percentage of sludge by volume.

1300. The maximum efficiency of BOD removal is achieved in
(*a*) oxidation (*b*) oxidation ditch
(*c*) aerated lagoons
(*d*) trickling filters

1301. For satisfactory working of a sludge digestion unit, the pH range of digested sludge should be maintained at

Answers

1284. (*b*) 1285. (*c*) 1286. (*c*) 1287. (*b*) 1288. (*a*) 1289. (*a*) 1290. (*b*) 1291. (*c*) 1292. (*c*)
1293. (*a*) 1294. (*d*) 1295. (*c*) 1296. (*a*) 1297. (*c*) 1298. (*d*) 1299. (*a*) 1300. (*b*) 1301. (*b*)

(*a*) 4.5 to 6.0 (*b*) 6.5 to 8.0
(*c*) 8.5 to 10.0 (*d*) 10.5 to 12.0

1302. The main disadvantage of oxidation pond is that
(*a*) large area is required for construction
(*b*) maintenance and operation cost are high
(*c*) BOD removal is very low
(*d*) none of the above

1303. The working conditions in Imhoff tanks are
(*a*) aerobic only (*b*) anaerobic only
(*c*) aerobic in lower compartment and anaerobic in upper compartment
(*d*) anaerobic in lower compartment and aerobic in upper compartment

1304. To divert excessive flow from combined sewers, the most commonly used weir is a
(*a*) cipoletti weir (*b*) broadcrested weir
(*c*) leaping weir (*d*) submerged weir

1305. When there is no recirculation of sewage, then recirculation factor is
(*a*) 0 (*b*) 1
(*c*) infinity (*d*) none of these

1306. In facultative stabilization pond, the sewage is treated by
(*a*) aerobic bacteria only.
(*b*) algae only.
(*c*) dual action of aerobic bacteria and anaerobic bacteria.
(*d*) sedimentation.

1307. Septic tank is a (*i*) setting tank (*ii*) digestion tank (*iii*) aeration tank. The correct answer is
(*a*) only (*i*) (*b*) (*i*) and (*ii*)
(*c*) (*i*) and (*iii*) (*d*) only (*iii*)

1308. For the same solid content, if the quantity of sludge with moisture content of 98% is *x*, then the quantity of sludge with moisture content of 96% will be
(*a*) *x*/4 (*b*) *x*/2 (*c*) *x* (*d*) 2*x*

1309. For normal sludge, the value of sludge index for Indian conditions is
(*a*) 0 to 50 (*b*) 50 to 150
(*c*) 150 to 350 (*d*) 350 to 500

1310. The process of lagooning is primarily a means of
(*a*) reducing the excessive flow in sewers
(*b*) disposing of sludge
(*c*) increasing the capacity of storage reservoirs
(*d*) increasing flow of sewage through imhoff tanks

1311. If the average water consumption of a city is 200 liters per capita day, and the population of city is 5,00,000. The maximum hourly demand of water for the maximum day will be ______ million litres
(*a*) 150 (*b*) 180 (*c*) 270 (*d*) 330

1312. The minimum width of a sewer French in mm (millimeter) must be greater than; when *D* is the external diameter of sewer in mm:
(*a*) 1.5 *D* + 100 (*b*) 1.5 *D* + 300
(*c*) 1.5 *D* + 400 (*d*) *D* + 1000

1313. Turbidity in water is a physical characteristic of water. It can be easily detected in a glass of water and are usually objectionable for aesthetic reason, when it concentration is in excess of ______ turbidity unit.
(*a*) 0.1 (*b*) 1.0 (*c*) 4.0 (*d*) 5.0

1314. If water having a total hardness of 375 mg/l as $CaCO_3$, the Carbonate Hardness. Is 200 mg/l as $CaCO_3$, will be ______ mg/l as $CaCO_3$.
(*a*) 375 (*b*) 200
(*c*) 175 (*d*) none of these

1315. Ground water has to be treated with impurities like hardness, micrograms and low dissolved Oxygen concentration along with different gases. What will be the proper treatment process from the following options:
(*a*) Sedimentation; Filtration: Coagulation Flocculation Disinfection

Answers

1302. (*a*) 1303. (*d*) 1304. (*c*) 1305. (*b*) 1306. (*c*) 1307. (*b*) 1308. (*b*) 1309. (*c*) 1310. (*b*)
1311. (*c*) 1312. (*b*) 1313. (*d*) 1314. (*b*) 1315. (*b*)

(*b*) Aeration; Sedimentation; Softening; Disinfection

(*c*) Aeration; Sedimentation; lime soda process; process; Filtration

(*d*) Screen; Filtration; softening; Disinfection

1316. The normal length of a sedimentation tank to remove suspended particles from water should not be more than ______ time of its width.

(*a*) 2 (*b*) 4 (*c*) 6 (*d*) 8

1317. If the length of setting tank is 5 m, width is 3 m; depth is m of a rectangular tank with discharge (*Q*) of 1.0 m^3/second; the setting velocity will be ______ m/sec.

(*a*) 0.033 (*b*) 0.067 (*c*) 0.167 (*d*) 0.200

1318. The total weight of solid waste generated in a town is 72,850 ibs/week, whose population is 3500 persons. The unit generation rate of solid waste (kg/capita/day) is:

(*a*) 1.0 (*b*) 1.5 (*c*) 1.3 (*d*) 2.0

1319. The major gases produced from sanitary landfill are due to:

(*a*) Anaerobic decomposition of biodegradable organic matter.

(*b*) Aerobic decomposition of biodegradable organic matter.

(Anaerobic decomposition of non-biodegradable organic matter.

1320. The minimum are of land desirable for sanitary landfill that is sufficient to operate for at least:

(*a*) One year (*b*) Six months

(*c*) None of these

1321. In general, the organic waste material decomposes rapidly (3 to 5 years) is:

(*a*) Food waste (*b*) News paper

(*c*) Laves (*d*) All of these.

1322. Sainitary land filling is an operation in which waste to be disposed off is compacted and covered with a layer of soil at the end of ______ day's operation.

(*a*) two (*b*) one

(*c*) three (*d*) none of these.

1323. Screening stage in EIA means:

(*a*) Determining whether or not an environment impact assessment study would be required for a project.

(*b*) Determining the size and nature of the project.

(*c*) Preparing a list of activities to be performed during EIA.

(*d*) Screening the location of the project.

1324. During EIA, flora and fauna are areas covered under ______ environment.

(*a*) Biological (*b*) Physical-chemical

(*c*) Social (*d*) cultural

1325. Paper accounts for approximately what percent of the solid waste stream in the USA

(*a*) 35 (*b*) 45 (*c*) 55 (*d*) 65

1326. Improvements to landfill design include

(*a*) Large size

(*b*) Reducing the amount of yard clippings allowed

(*c*) Leachate collection systems

(*d*) All of the above

1327. In Japan and Switzerland the most common waste disposal methods are

(*a*) Landfills and incineration

(*b*) Landfills and recycling

(*c*) Recycling and incineration

(*d*) None of the above

1328. Incinerator ash usually contains which of the following chemicals

(*a*) Heavy metals

(*b*) Organic compounds

(*c*) VOCs (*d*) All of the above

1329. The cheapest method of municipal solid waste disposal the current time is

(*a*) Incineration (*b*) Land disposal

(*c*) Recycling (*d*) None of the above

Answers

1316. (*b*) 1317. (*c*) 1318. (*c*) 1319. (*a*) 1320. (*a*) 1321. (*d*) 1322. (*b*) 1323. (*a*) 1324. (*a*)
1325. (*a*) 1326. (*c*) 1327. (*c*) 1328. (*a*) 1329. (*b*)

1330. Composting uses ________ processes to transform organic material into compost
(*a*) Industrial (*b*) Chemical
(*c*) Natural (*d*) None of these

1331. Banning the sale of one-time-use beverage containers would
(*a*) Reduce litter (*b*) Save energy
(*c*) Save money (*d*) All of the above

1332. One technological problem with plastic recycling is that
(*a*) People do not like to recycle
(*b*) Different types of plastics cannot be recycled together
(*c*) The technology to recycle plastics does not exist
(*d*) All of the above

1334. The United States produces one-half of the world's e-waste and recycles
(*a*) none of it (*b*) all of it
(*c*) one-half of it (*d*) 10-15% of it

1335. Which of the following is *not* a type of e-waste?
(*a*) discarded TV sets
(*b*) discarded computer printer paper
(*c*) discarded cell pones
(*d*) discarded computer monitors

1336. Waste that includes paper, food waste, cans, bottles, yard, glass, wood, and similar items is called
(*a*) industrial solid waste
(*b*) hazardous waste
(*c*) municipal solid waste
(*d*) toxic solid waste

1337. Hazardous waste includes radio active waste, which will have to be stored safely for as long ________ years
(*a*) 15,00 (*b*) 50,000
(*c*) 100,000 (*d*) 240,000

1338. How much of the waste we mix, crush, and bury could actually be recycled?
(*a*) 30% (*b*) 50% (*c*) 60% (*d*) 75%

1339. With 5% of the world's population, the United States produces about ________ of the world's solid waste
(*a*) one-tenth (*b*) one-fifth
(*c*) one-forth (*d*) one-third

1340. For each on kg. of electronics in a computer, how much solid and liquid wastes are created?
(*a*) 9 tons (*b*) 2 tons
(*c*) 1 tons (*d*) 0.5 ton

1341. Garbage produced directly by households and business account for ________ % of the solid waste produced in the United States.
(*a*) 2 (*b*) 5 (*c*) 10 (*d*) 15

1342. The LD_{50} refers to the
(*a*) Last dose administered before a test organism is killed.
(*b*) Lethal dose that will kill 50% of a test population.
(*c*) Probability of killing a test organism 50% of the time.
(*d*) None of these

1343. The term threshold level describes the concentration of a chemical at which
(*a*) Industrial exposure is expected to occur
(*b*) Industrial exposure is allowed by law
(*c*) Exposure to the chemical does not affect the test animals
(*d*) None of the above

1344. DDT is an example of a
(*a*) Non-persistent chemical
(*b*) Persistent chemical
(*c*) Biodegradable chemical
(*d*) None of the above

1345. Which of the following produces the greatest percent of toxic waste material released to the environment in the US?
(*a*) Chemical industry
(*b*) Paper industry (*c*) Electric utilities
(*d*) metal mining

Answers

1330. (*c*) 1331. (*d*) 1332. (*d*) 1333. (*c*) 1134. (*d*) 1335. (*b*) 1336. (*c*) 1337. (*d*) 1338. (*d*)
1339. (*d*) 1340. (*a*) 1341. (*a*) 1342. (*b*) 1343. (*c*) 1344. (*b*) 1345. (*d*)

1346. Each discarded computer contains approximately 2 kg of which metal?
(*a*) Cadmium (*b*) Lead
(*c*) Mercury (*d*) Arsenic

1347. The primary emphasis in the US EPA's pollution prevention hierarchy is to
(*a*) Recycle wastes
(*b*) Treat wastes to reduce their hazard
(*c*) Reduce the amount of population generated at the source
(*d*) Sell pollution emission permits to other companies

1348. Compared with 25 years ago, water quality
(*a*) has improved (*b*) has worsened
(*c*) has stayed about the same
(*d*) is impossible to determine

1349. Which of the following statements is true about Love Canal situated near Niagra Falls?
(*a*) A Love Canal was an abandoned water canal in Nw York State
(*b*) An elementary school disposed of industrial chemical in Love Canal
(*c*) Hooker Chemical disposed of industrial chemical in Love Canal
(*d*) All of the above

1350. Concerning the mineral fractio of soil, clays is important, because it:
(*a*) has negative charges.
(*b*) has a large surface area.
(*c*) is organic. (*d*) both (*a*) and (*b*).

1351. The three soil primary essential elements are:
(*a*) N, P < K (*b*) C, H, O
(*c*) N, P, S (*d*) C, N, K

1352. Which of the following is incorrect?
(*a*) A current hazardous waste regulation control the transportation of hazardous waste.
(*b*) Current hazardous waste regulation control the storage of hazardous waste.
(*c*) Current hazardous waste regulation control the treatment and disposal of hazardous waste.
(*d*) Current hazardous waste regulation limit the amount of hazardous waste that can be generated

1353. Excluding heat, ______ is/are the leading cause of pollution in streams.
(*a*) Agricultural chemicals
(*b*) Municipal sewage (waste water)
(*c*) Industrial waste waste water
(*d*) Solid waste

1354. The mineral fraction of soil contains.
(*a*) clay (*b*) Organic matter
(*c*) Microorganisms (*d*) All of the above

1355. What is the most desirable method of dealing with municipal solid waste?
(*a*) Land filling (*b*) Incineration
(*c*) Energy recovery (*d*) Recycling

1357. Pesticides include:
(*a*) Herbicides (*b*) Insecticides
(*c*) Fungicides (*d*) All of the above

1358. The US EPA defines hazardous waste by
(*a*) toxicity and reactivity
(*b*) corrosivity (*c*) ignitability
(*d*) all of the above

1359. Which of the following statement does *not* describe a problem associated with modern landfills?
(*a*) Chemicals leaking out of the waste stored there may pollute wells or groundwater supplies.
(*b*) Materials buried in landfills decompose in three years.
(*c*) Explosive gases can be produced by landfills and may build up in the basements of nearby buildings.
(*d*) All of the above

1360. What is the most helpful thing that a consume can do to make sure that **more** materials are recycled?

Answers

1346. (*b*) 1347. (*c*) 1348. (*a*) 1349. (*d*) 1350. (*d*) 1351. (*a*) 1352. (*d*) 1353. (*a*) 1354. (*a*)
1355. (*d*) 1356. (*a*) 1357. (*d*) 1358. (*d*) 1359. (*b*) 1360. (*d*)

(a) Encourage friends and neighbors to recycle their newspaper.
(b) Carefully wash and sort all materials put out for recycling.
(c) Write letters to legislators asking that recycling be made mandatory.
(d) Make an effort to buy products made from recycled materials whenever

1361. Which of the following produces the maximum solid waste?
(a) manufacturing (b) plastic packaging
(c) agriculture
(d) nuclear power plants

1362. Which items make up the largest percentage of waste produced by households and business
(a) aluminum cans
(b) plastic bottles and packaging
(c) paper products
(d) yard waste, such as grass clippings and tree limbs

1363. What percentage of the energy needed to make aluminium cans from ore is needed to make recycled aluminium cans from used aluminium cans?
(a) 95% (b) 5% (c) 60%
(d) None of the above

1364. Which of following correctly describes the disposal of waste by composting?
(a) Composting is a good way for individuals to reduce the amount of garbage they send to landfills, but large-scale composting is impractical.
(b) Composting could reduce the amount of waste that restaurants, food-processing plants, and animal feedlots send to landfills.
(c) Compost made from glass and plastics can be recycled and used as building materials.
(d) yard wastes, such as grass clippings, are not suitable for composting because they contain low level of nutrients.

1365. Which of the following correctly describes biodegradable plastics?
(a) Some plastics are designed to deteriorate when exposed to sunlight.
(b) Plastic that contains sugars can be broken down by soil bacteria.
(c) Degradable plastics usually break down into smaller pieces of plastic.
(d) All of the above

1366. Which of the following is *not* considered hazardous waste?
(a) dead batteries such as those used in flashlights and radios
(b) rotting meat and other food waste
(c) metals such as lead and mercury
(d) Both (a) and (c)

1367. Some hazardous waste is disposed of by deep-well injection. This procedure
(a) pollutes the wells and water supplies of nearby residents.
(b) Stores wastes in layers of dry rock below the water table.
(c) violates federal law.
(d) All of the above

1368. The methane found in landfills is a gas that
(a) is produced when wastes react with oxygen.
(b) must be pumped to the surface in leachate pipes.
(c) is a major contaminant of groundwater supplies.
(d) may be captured and used to produce energy.

1369. Which of the following is a true statement about recycling?
(a) It takes more energy to make steel from scrap than from newly mined ore.
(b) Cities have never had trouble finding buyers for their recycled materials.
(c) After they are collected, recyclable materials are cleaned and processed.
(d) All of the above

Answers

1361. (a) 1362. (c) 1363. (c). 1364. (b) 1365. (d) 1366. (b) 1367. (b) 1368. (d) 1369. (c)

1370. The process of composting
(*a*) uses synthetic materials.
(*b*) has little potential to conserve landfill space.
(*c*) is always carried out in city-run facilities .
(*d*) could effectively handle a variety of wastes.

1371. The main problem with so-called biodegradable plastic is that
(*a*) only its organic portion actually decomposes.
(*b*) animals often choke on the plastic litter that forms.
(*c*) it releases toxic compounds as it break apart.
(*d*) it includes a chemical made by a rare bacterium.

1372. Which of the following is *not* a method of reducing hazardous waste?
(*a*) allowing fungi to break down petroleum sludge.
(*b*) burning toxic substance in specially designed incinerators.
(*c*) using nontoxic beads instead of solvents to clean metal parts.
(*d*) All of the above.

1373. A common characteristic of all hazardous-waste land-disposal facilities is that they
(*a*) are long-term storage sites.
(*b*) make use of deep-well injection.
(*c*) pose no threat to the environment.
(*d*) are not regulated by state or federal laws.

1374. From 1970 to the present, the percentage of waste that is recycled has increased from 5% to
(*a*) 30% (*b*) 15% (*c*) 50% (*d*) 70%

1375. A example of source reduction is
(*a*) composting. (*b*) recycling.
(*c*) using canvas shopping bags.
(*d*) All of the above

1376. Making products from recycled materials
(*a*) is limited to metals.
(*b*) Saves energy and other resources during the manufacturing
(*c*) uses of much energy that is not economically workable.
(*d*) None of the above

1377. An example of something that is biodegradable is
(*a*) a leaf in a compost pile.
(*b*) a plastic milk jug in a landfill.
(*c*) a polystyrene burger container in a trash can.
(*d*) All of the above

1378. Which of the following is *not* a requirement for a landfill built today?
(*a*) It must be lined with clay and plastic.
(*b*) It must be specified distance from a city or other populated area.
(*c*) It must have system for collecting and treating leachate.
(*d*) Vent pipes must be installed for removing methane.

1379. Which of the following are problems associated with landfills?
(*a*) source reduction and composting
(*b*) leachate and methane production
(*c*) surface impoundment and leachate
(*d*) All of the above

1380. A material is biodegradable if, it
(*a*) is made of synthetic material.
(*b*) is made from petroleum or natural gas.
(*c*) can be broken down by biological processes
(*d*) can be disposed of I a landfill.

1381. Materials that are formed naturally are
(*a*) bio-degradable
(*b*) non-biodegradable
(*c*) non-compostable (*d*) non combustible

Answers

1370. (*d*) 1371. (*a*) 1372. (*a*) 1373. (*a*) 1374. (*a*) 1375. (*c*) 1376. (*b*) 1377. (*a*) 1378. (*b*)
1379. (*b*) 1380. (*c*) 1381. (*a*)

1382. An advantage of a solid waste ______ is that it reduces the amount of trush destined for landfills.
(*a*) sanitary land filing
(*b*) vicinerator (*c*) composting
(*d*) bio-treatment

1383. More than half of the solid produced in the United States comes from
(*a*) municipality (*b*) construction
(*c*) slaughter house
(*d*) manufacturing sector

1384. The total amount of solid waste generated in the United States in the last 40 years
(*a*) double (*b*) trebled
(*c*) quadrupled (*d*) halved

1385. Problems associated waste include leachate, methane gas, and inadequate space.
(*a*) composting (*b*) landfills
(*c*) incineration (*d*) pyrolysis

1386. Disposing of hazardous wastes by pumping them into the ground below groundwater level is called
(*a*) burial (*b*) encapsulation
(*c*) deep well injection
(*d*) composting

1387. You can consumer buying power to reduce solid waste by using
(*a*) disposable (*b*) recycled
(*c*) rechargeable (*d*) solar

1388. The steps involved in the recycling of materials are ______ cleaning, and crushing
(*a*) strong (*b*) transferring
(*c*) sorting (*d*) disposing

1389. An advantage of ______ plastic is that its production uses 20 to 50 percent less fossil fuel than does the production of regular plastic.
(*a*) green (*b*) recycled
(*c*) biodegradable
(*d*) non-biodegradable

1390. Waste containing toxic, corrosive, or materials are considered
(*a*) radioactive
(*b*) non-biodegradable
(*c*) biodegradable (*d*) hazardous

1391. Polyester is a (n) material ______, and therefore non-biodegradable, material commonly used clothing.
(*a*) natural (*b*) synthetic
(*c*) radioactive (*d*) non-hazardous

1392. A modern landfill has a plastic liner to prevent ______ from seeping into the soil and contaminating the groundwater.
(*a*) methane gas (*b*) moisture
(*c*) leachate (*d*) carbon dioxide gas

1393. Some landfills generate electricity by burning the ______ that is produced by decomposing garbage
(*a*) carbon monoxide (*b*) acetylene
(*c*) sulfur dioxide (*d*) methane

1394. Manufactures could reduce waste ad conserve resources by making products that are more ______.
(*a*) environment friendly
(*b*) non-biodegradable
(*c*) durable (*d*) flexible

1395. Reducing solid waste by using less material is called ______ reduction.
(*a*) source
(*b*) environment friendly
(*c*) straight (*d*) indirect

1396. More companies will use recycled products, if consumers buy recycled products and create an increased ______ for such products.
(*a*) awareness (*b*) acceptability
(*c*) demand (*d*) none of these

1397. Creating steel from scrap instead of from ore uses about ______ percent less energy.
(*a*) 20 (*b*) 45 (*c*) 70 (*d*) 90

Answers

1382. (*b*) 1383. (*d*) 1384. (*a*) 1385. (*b*) 1386. (*c*) 1387. (*c*) 1388. (*c*) 1389. (*a*) 1390. (*d*)
1391. (*b*) 1392. (*c*) 1393. (*d*) 1394. (*c*) 1395. (*a*) 1396. (*c*) 1397. (*c*)

1398. Composting provides ________ to the soil while keeping organic wastes out of landfills.
(*a*) nutrients (*b*) neutralization
(*c*) moisture (*d*) pest resistance

1399. Decomposing yard waste and fruit and vegetable trimmings can be broken down by bacteria to form ________, which reduces the need for fertilizes.
(*a*) insecticides (*b*) leachate
(*c*) methane (*d*) compost

1400. Biodegradable plastics never completely
(*a*) burn (*b*) melt
(*c*) decompose (*d*) recycled

1401. The goal of waste management is
(*a*) acquiring a labor capable of disposing of garbage.
(*b*) disposing of waste safely and effectively and reducing the amount of waste generated
(*c*) developing management training for supervisor of disposal companies.
(*d*) developing technology for more efficient incineration.

1402. Incineration of waste can be a problem because ________.
(*a*) it takes large areas for storage of waste.
(*b*) most countries have made the practice illegal.
(*c*) hazardous chemicals can be released in the smoke that is generated.
(*d*) leaching to the groundwater table can occur.

1403. The best solution to the solid waste problem is ________.
(*a*) incineration (*b*) recycling
(*c*) sources reduction
(*d*) landfills

1404. A surface impoundment is ________.
(*a*) any artificial lake created by a dam
(*b*) used for nuclear waste storage
(*c*) a permanent disposal option for liquid hazardous waste
(*d*) used to store and concentrated dilute liquid hazardous waste before it is moved to a permanent disposal option

1405. Substances that decompose the metal that make up storage tanks or equipment are called
(*a*) alkaline caustic agents
(*b*) ignitable (*c*) corrosive
(*d*) toxic

1406. Which of the following represents the largest waste disposal practice in the United states:
(*a*) landfills (*b*) incineration
(*c*) composting (*d*) recycling

1407. If the recycling loop is to work and become economically sustainable, then ________.
(*a*) the number of recycled products must increase
(*b*) consumers and businesses must purchase the products made from recycled materials
(*c*) programs must require recycling in all cities
(*d*) genetically modified bacteria must be found to consume toxic components

1408. An abandoned waste site that poses a direct threat to health and the environment and is designated to have U.S. government funds applied to its cleanup is a (n)
(*a*) hazardous waste site
(*b*) Superfund site
(*c*) landfill leachate-producing site
(*d*) surface impoundment site

1409. Consumers can decrease waste generation by
(*a*) purchasing good with the Energy Star label
(*b*) Choosing to buy locally made goods
(*c*) choosing the purchase only imported goods
(*d*) choosing minimally packaged goods

Answers

1398. (*a*) 1399. (*b*) 1400. (*c*) 1401. (*b*) 1402. (*c*) 1403. (*c*) 1404. (*d*) 1405. (*c*) 1406. (*a*)
1407. (*b*) 1408. (*b*) 1409. (*d*)

1410. Which of the following statement about the "pay-as-you-throw" approach to garbage collection is NOT correct?

(*a*) It will not likely be used dye to lake of public support.

(*b*) It reduces the waste stream.

(*c*) "Pay s you throw" encourages consumers to reduce the volume of goods purchased.

(*d*) It uses a financial incentive for influencing consumer behavior.

1411. The conversion of organic waste to mulch or humus usable for enriching soil by encouraging natural biological processes of decomposition in a controlled manner, is called

(*a*) combustion (*b*) composition

(*c*) compromising (*d*) composting

1412. Which of the following conditions results from the operation and subsequent use of landfills?

(*a*) Decommissioned landfills cannot be used as locations for schools, industries, parks, and golf courses for 300 years after the closing of the landfills.

(*b*) Modern sanitary landfills are regulated by the president and lined with a porous materials.

(*c*) Hydrogen gas is generated by the organic decomposition and can be sold as a by-product.

(*d*) Leachate from old landfills can contaminate groundwater.

1413. Which of the following is an example of a product found in a "throwaways society"?

(*a*) reusable cloth diapers

(*b*) fine silverware (*c*) paper napkin

(*d*) reusable razor blade

1414. What is an important factor that determines whether the average citizen of a developed country will produce more or less trash.

(*a*) the cost of disposing of trash

(*b*) how close the country is to the equator

(*c*) how many people live in the country

(*d*) the age structure of the country

1415. How can "final cover" on landfills reduce decomposition of the trush?

(*a*) The cap keeps the landfills moist on the inside, preventing decompositon.

(*b*) The cap prevents hawks and blue jays from decomposing the trash.

(*c*) The cap causes paper to bond to plastic in the landfill.

(*d*) The cap keeps the landfill dry, which decreases the ability of bacteria to decompose trash.

1416. Why do industries tend to produce so much waste even though it is not the most efficient way do business

(*a*) It is more profitable to be efficient with waste.

(*b*) In most cases there is no financial incentive for becoming efficient.

(*c*) Industry ecologists do not recommend streamlining industrial waste.

(*d*) Most of industry is run by disorganized people.

1417. "E-waste" a growing concern, because

(*a*) Only 60% of waste is recycled.

(*b*) Electronic devices die easily so few of the devices thrown out could be re-used.

(*c*) There is no financial incentive to recycle E-waste.

(*d*) E-Waste contains toxic heavy metals and flame retardants.

1418. What do chromium, mercury, copper, and tin have in common?

(*a*) They are all heavy metals.

(*b*) They are spices.

(*c*) They are all corrosive materials.

(*d*) They are all gases.

1419. Our obsession with having only the most current electronic devices and throwing away

Answers

1410. (*a*) 1411. (*d*) 1412. (*d*) 1413. (*c*) 1414. (*a*) 1415. (*d*) 1416. (*b*) 1417. (*d*) 1418. (*a*)
1419. (*d*)

outdated version despite the fact that they work is an example of

(*a*) reduce, reuse, recycle
(*b*) sound science (*c*) living sustainably
(*d*) our throwaway society.

1420. Which of the following statements best describes the concept of phytoremediation?

(*a*) The use of plants to absorb and accumulate toxic materials from the soil.
(*b*) The concentration of toxic material in the tissue of organisms as they move through is successive tropic levels.
(*c*) The breaking down of. toxic materials in the natural environment by sunlight.
(*d*) The remediation of an ecosystem degraded by activities back to its original structure and composition.

1421. An advantage of recycling aluminium rather than disposing of it in landfills is that aluminium can be

(*a*) Produced from recycled metal using much less energy than is required for its production from aluminum one
(*b*) Produced from one that is chemically reactive and dangerous to transport, store and process
(*c*) Produced from ore that is scarce and found primarily in remote, inhospitable region at high latitudes
(*d*) Absorbed by plants and then biomagnified in both terrestrial and aquatic food chains

1422. Critics of incineration of municipal solid waste suggest that it may not be best solution for the future, for which of the following reasons?

(*a*) Incineration produces ash that increases landfill volume.
(*b*) incineration generates methane.
(*c*) incineration involves advanced technology.
(*d*) incineration contributes to air pollution.

1423. Integrates waste management employs all of the following EXCEPT

(*a*) Using refillable soft-drink bottle.
(*b*) Using disposable diapers instead of cloth diapers.
(*c*) Using reusable canvas bags instead of plastic or paper bags.
(*d*) Using tires for the construction of artificial reefs.

1424. Which method reduces the volume of waste but could release toxic emissions into the atmosphere?

(*a*) Sanitary landfill (*b*) Incineration
(*c*) Discharge to sewers, streams, and rivers
(*d*) Chemical treatment

1425. A flammable gas produced by landfills is

(*a*) Carbon dioxide (*b*) Carbon monoxide
(*c*) Methane (*d*) Radon

1426. All of the following are considered toxic metal pollutants, EXCEPT

(*a*) Cadmium (*b*) Potassium
(*c*) Lead (*d*) Mercury

1427. Of the following, which is the most serious immediate problem associated with sanitary landfills?

(*a*) Generation of CO_2 gas
(*b*) Leachate contamination of groundwater
(*c*) Release of disease organisms
(*d*) Incomplete degradation of wastes

1428. Which of the following correctly orders the methods of solid-waste management in term of energy required?

Least Energy → Most Energy

(*a*) Reuse → Recycle → Reduce
(*b*) Recycle → Reuse → Reduce
(*c*) Recycle → Reduce → Reuse
(*d*) Reduce → Reuse → Recycle

1429. Nuclear wastes comprises paper, rags, tools, clothing, filters etc which contain small

Answers

1420. (*a*) 1421. (*a*) 1422. (*d*) 1423. (*b*) 1424. (*b*) 1425. (*c*) 1426. (*b*) 1427. (*b*) 1428. (*d*)
1429. (*b*)

amounts of mostly short-lived radioactivity is called as
(*a*) Medium level waste
(*b*) Low level waste (*c*) High Level waste
(*d*) Transuranic wastes

1430. Nuclear waste which requires 100% shielding during disposal is
(*a*) Medium level waste
(*b*) Low level waste (*c*) High Level waste
(*d*) Transuranic wastes

1431. The waste which contain elements having an atomic number greater than Uranium is
(*a*) Medium level waste
(*b*) Low level waste (*c*) High Level waste
(*d*) Transuranic wastes

1432. Radioactive pollution that is spread through the earth's atmosphere is called ________
(*a*) Transuranic (*b*) Radioactive fallout
(*c*) Radionulides (*d*) Low level waste

1433. The disaster at the Chernobyl Nuclear fallout contaminated approximately ______ Square miles of farmland and villages in Ukrain, Soviet Union.
(*a*) 200 (*b*) 400 (*c*) 800 (*d*) 1,000

1434. The part of human body which is first affected by nuclear radiation is
(*a*) Lungs (*b*) Brain
(*c*) Bone marrow (*d*) Liver.

1435. Verification is disposing nuclear waste by
(*a*) Burying below ground.
(*b*) Solidifying in the form of glass
(*c*) Ice sheet Disposal
(*d*) Deep geological disposal

1436. Converting part of the hazardous nuclear waste into a more stable material that decays quickly is called
(*a*) Transmutation (*b*) Vitrification
(*c*) Ice sheet Disposal
(*d*) Deep geological disposal

1437. International Agreement is necessary
(*a*) Ice sheet Disposal
(*b*) Deep geological disposal
(*c*) Buying below ground
(*d*) Space Disposal

1438. Richness of nutrients in industrial waste water leads to
(*a*) eutrophic condition
(*b*) oligotrophic (*c*) mesotrophic
(*d*) none of the above

1439. Increase I the concentration of a recalcitrant compounds along the food chain is called
(*a*) biomagnifications
(*b*) bioaccumulation
(*c*) biosorption (*d*) none of these

1440. The main problem associated with managing Indian solid waste is
(*a*) segregation
(*b*) high moisture content
(*c*) season variation (*d*) all of these

1441. Sludge is treated prior to ultimate disposal for
(*a*) stabilization of organics
(*b*) volume reduction
(*c*) pathogen reduction
(*d*) all the above

1442. Integrated waste management is
(*a*) co composting (*b*) composting
(*c*) vermicomposting (*d*) all of these

1443. Destructive distillation is called
(*a*) pyrolysis (*b*) incineration
(*c*) controlled tripping
(*d*) none of the above

1444. Match List I items with those in List II using codes given below

	List I		List II
A.	Photo-synthesis	I.	The rapid rate of this process in tropical forests results in low-nutrient soils.

Answers

1430. (*c*) 1431. (*d*) 1432. (*b*) 1433. (*d*) 1434. (*c*) 1435. (*b*) 1436. (*a*) 1437. (*d*) 1438. (*a*)
1439. (*a*) 1440. (*d*) 1441. (*d*) 1442. (*a*) 1443. (*a*) 1444. (*a*)

B.	Eutrophication	II.	The process I which glucose is synthesized by plants.
C.	Denitrification	III.	The process by which a soil nutrient is reduced and released to the atmosphere as a gas
D.	Decomposition		

Code:

(*a*) I-D, II-A, III-C (*b*) I-B, II-A, III-C
(*c*) I-A, II-B, III-C (*d*) I-D, II-B, III-A

1445. Match items in List I with those in List II using codes given below:

	List I		List II
A.	Oxygen	I.	The most abundant element in Earth's crust
B.	Aluminium	II.	The most abundant element in Earth's atmosphere
C.	Iron	III.	The most abundant element in Earth's core
D.	Nitrogen	IV.	The element commercially extracted from bauxite

Codes:

(*a*) I-A, II-B, III-D, IV-C
(*b*) I-A, II-D, III-C, IV-B
(*c*) I-D, II-B, III-C, IV-A
(*d*) I-C, II-D, III-A, IV-B

1446. Match the items in List I with that in List II using the codes given below:

	List I		List II
A.	Nitrification	I.	Ammonia is converted to nitrite, then to nitrite
B.	Denitrification	II.	Plant roots absorb ammonium ions and nitrate ions for use in making molecules such as DNA, amino acids, and proteins.
C.	Assimilation	III.	Nitrate ions and nitrite ions are converted into nitrous oxide gas and nitrogen gas (N_2)
D.	Nitrogen fixation		

Codes

(*a*) I-C, II-A, III-D (*b*) I-A, II-B, III-C
(*c*) I-A, II-C, III-B (*d*) I-B, II-C, III-A

1447. Match items in List I with those in List II using the codes given below.

	List I		List II
A.	Sulfur dioxide	I.	Most often cites as causative factor for acid deposition
B.	Lead	II.	Implicated in human neurological damage
C.	Ozone	III.	Considered harmful in the troposphere but beneficial in the stratosphere
D.	Particulates	IV.	Is the major pollutant that electrostatic precipitators are designed to remove from power-plant smokestack emissions.

Codes:

(*a*) I-A, II-B, III-C, IV-D
(*b*) I-D, II-C, III-B, IV-
(*c*) I-B, II-C, III-D, IV-A
(*d*) I-A, II-D, III-B, IV-C

1448. The factor that likely poses the greatest threat of extinction of species worldwide is

(*a*) weakening of environmental legislation.
(*b*) increase in hunting and fishing.
(*c*) aggressive collecting of specimens by museum and zoo.
(*d*) increasing loss of habitat.

1449. Which of the following energy sources is not derived directly or indirectly from solar energy?

Answers

1445. (*b*) 1446. (*c*) 1447. (*a*) 1448. (*d*) 1449. (*a*)

(a) Geothermal (b) Photovoltaic
(c) Hydroelectric (d) Biomass

1450. Which of the following is an effective alternative to chlorine for disinfecting waste water in a municipal water
(a) Freon (b) Phosphate
(c) Ammonia (d) Ozone

1451. The presence of fiscal coliform bacteria in a sample of river water suggest which of the following?
(a) The pH of the river is very high.
(b) The water is contaminated with animal waste.
(c) The river is devoid of plant life.
(d) The dissolved oxygen level of the river is high

1452. Which of the following best help to explain why phosphorus is often a limiting factor in many ecosystems?
(a) There is usually a gaseous phase in the phosphorus cycle.
(b) Under many conditions, phosphorus forms stable insoluble compounds.
(c) Phosphate is not readily taken up by plants.
(d) There are no anthropogenic sources phosphorus.

1453. Which of the following statement about generic diversity is true?
(a) Genetic uniformity of a crop increases the crops overall resistance pests and disease.
(c) Genetic engineering technology is used to increase genetic diversity by creating new species with synthetic genes.
(d) Genetic diversity is usually high in endangered species.

1454. In India most municipal solid waste is disposed of by
(a) composting (b) recycling
(c) incineration (d) landfilling

1455. Which of the following factors is fundamentally responsible for seasons on Earth?
(a) The varying distance of Earth from the Sun.
(b) The tilt of Earth's axis of rotation
(c) The tidal pull of the Sun and Moon
(d) The anthropogenic greenhouse effect

1456. If waste water treatment plant effluent that contains nitrates and phosphates is allowed to flow into a body of water, which of the following may result?
(a) Chlorination (b) Decomposition
(c) Eutrophication (d) Oxygenation

1457. The water-holding capacity of soil is least likely to be affected by the addition of which of the following?
(a) Clay (b) humus
(c) Manure (d) Pesticide

1458. Incineration of municipal solid waste may not be the best solution for which of the following reasons?
(a) Incineration produces ash that increases landfill volume.
(b) Incineration generation methane.
(c) Incineration involves advanced technology.
(d) No-till agriculture

1459. Which of the following is likely minimize soil erosion?
(a) High-yield crops
(b) Deforestation (c) Herbicide use
(d) No-till agriculture

1460. The polar region radiate away more heat energy than they receive from the Sun I course of a year. However, they are prevented from becoming progressively colder each year primarily by the
(a) absorption of ultraviolet radiation by snow.
(b) transport of heat through the atmosphere and oceans.

Answers

1450. (d) 1451. (b) 1452. (b) 1453. (b) 1454. (d) 1455. (b) 1456. (c) 1457. (d) 1458. (d) 1459. (d) 1460. (b)

(*c*) release of latent het to the atmosphere when the polar ice caps melt.
(*d*) generation of heat by glacial movement.

1461. An advantage of recycling aluminium rather than disposing of it in landfills is that aluminium can be
(*a*) produced from recycled metal using much less energy than is required for its production from aluminum ore.
(*b*) produced from ore that is chemically reactive and dangerous to transport, store, and process.
(*c*) absorbed by plants and then biomagnified in both terrestrial and aquatic food chains
(*d*) leached from landfills in the form of Al^{3+} ions that could increase the pH of lakes and streams

1462. As urbanization increases and natural soil surfaces are covered, the groundwater supply is reduced due to
(*a*) increased evaporation and transpiration
(*b*) decreased surface runoff
(*c*) loss of recharge area
(*d*) confinement of aquifers

1463. Which of the following components of a waste water treatment plant is designed to facilitate the decomposition of organic material by aerobic microorganisms?
(*a*) Bar screen (*b*) Grit-settling tank
(*c*) Activated-sludge tank
(*d*) Chlorination tank

1464. Ozone in the stratosphere is most important to life at Earth's surface because it absorbs
(*a*) gamma rays (*b*) microwaves
(*c*) ultraviolet light (*d*) X-rays

1465. The major source of radon in houses
(*a*) furniture and carpets
(*b*) the underlying bedrock
(*c*) nuclear power plants
(*d*) fossil-fuel combustion

1466. The annual fluctuation in carbon dioxide concentration can best be explained by the
(*a*) regularity of volcanic activity
(*b*) deforestation in the tropics
(*c*) reusable canvas bags instead of plastic or paper bags
(*d*) tires for the construction of artificial reefs

1467. Integrated waste management employs all of the following except using
(*a*) refillable soft-drink bottles.
(*b*) disposable diapers instead of cloth diapers
(*c*) reusable canvas bags instead of plastic or paper bags
(*d*) tires for the construction of artificial reefs

1468. Of the following sources, which supplies the most commercial energy I the world today?
(*a*) Solar (*b*) Oil
(*c*) Biomass (*d*) Hydroelectric

1469. Which of the following is a greenhouse gas that is also a by-product of anaerobic respiration?
(*a*) Methane (CH_4) (*b*) Nitrogen (N_2)
(*c*) Oxygen (O_2)
(*d*) Nitrogen dioxide (NO_2)

1470. If the population of a country grows at a rate of approximately 5 percent per year the number of year required for the population to double is closest to
(*a*) 5 years (*b*) 10 years
(*c*) 15 years (*d*) 25 years

1471. Which of the following will occur, if the trend of global temperature increase continues?
(*a*) Night temperatures will decrease as day temperature increase.
(*b*) Sea levels will drop due to increased evaporation.
(*c*) The incidence of insect-borne diseases will decrease.
(*d*) The troposphere will contain more water vapor.

Answers

1461. (*a*) 1462. (*c*) 1463. (*c*) 1464. (*c*) 1465. (*b*) 1466. (*d*) 1467. (*b*) 1468. (*b*) 1469. (*a*)
1470. (*c*) 1471. (*d*)

1472. Which of the following is the primary environmental problem associated with the use of nuclear power to generate electricity?
(*a*) Radon leaking into buildings
(*b*) Production of greenhouse gases such as carbon dioxide
(*d*) Depletion of the ozone layer

1473. Compared to a coal-fired power plant that produces the same amount energy, a nuclear power plant generates more
(*a*) CO_2 (*b*) fly ash
(*c*) particulate (*d*) thermal pollution

1474. Which of the following identifies the two factors thought to be the most harmful to biodiversity?
(*a*) Acid deposition and increased use of fertilizers for agriculture.
(*b*) Depletion of the ozone layer and oil drilling.
(*c*) Destruction of habitat and invasion by nonnative species.
(*d*) Changes in climate and proliferation of endemic species

1475. For a certain insecticide, the LD-50 dosage level for rats is determined to be 250 milligrams per kilogram of body mass. On the basis of this information, which of the following is the best prediction regarding the consequences of receiving this dosage of the insecticide?
(*a*) Fifty percent of any rat population would be sickened.
(*b*) Fifty percent of the population of any warm-blooded animal would die.
(*c*) Five hundred out of every one thousand people would experience acute effects.
(*d*) Five hundred out of every one thousand rats would die.

1476. The major human health problem related to radon accumulation is
(*a*) lung cancer (*b*) heart disease
(*c*) cataracts
(*d*) malignant melanoma

1477. Acid rain is associated with which of the following?
(*a*) Formation of the Antarctic ozone hole
(*b*) Release of PCB's into the atmosphere
(*c*) Damage to tropical rain forests
(*d*) The burning of fossil fuels

1478. Characteristics that tend to increase the risk of a species becoming endangered include which of the following?
I. Having a very limited distribution.
II. Being a specialist at the end of a long food chain.
III. Having a small population size.
(*a*) I only (*b*) III only
(*c*) II and III only (*d*) I, II, and III

1479. Stratospheric ozone depletion is most likely to result in which of the following?
(*a*) Increased growth of food crops due to increasing amounts of ultraviolet radiation.
(*b*) Extended grazing season for cattle.
(*c*) Disruption of photosynthesis in plants.
(*d*) A higher percentage of cloudless day.

1480. Which of the following is a process that indirectly removes carbon from Earth's atmosphere?
(*a*) Formation of carbonate deposits
(*b*) Out gassing by volcanoes
(*c*) Respiration by mammals
(*d*) Photo dissociation by ultraviolet light

1481. Many scientist believe that a rise in sea level has occurred in the last hundred years as a result of global warming. If this is true, which of the following factors best explains such a rise?
(*a*) Increased precipitation
(*b*) Increased cloud cover
(*c*) Increased evapotranspiration
(*d*) Thermal expansion of the ocean

Answers

1472. (*c*) 1473. (*d*) 1474. (*c*) 1475. (*d*) 1476. (*a*) 1477. (*d*) 1478. (*d*) 1479. (*c*) 1480. (*a*) 1481. (*d*)

1482. Which of the following natural events would be most likely to contribute to the cooling of Earth's atmosphere?

(*a*) Earthquake (*b*) Volcanism
(*c*) Hurricane (*d*) Monsoon

1483. Which of the following best describes the significance of the United nations Conference on the Human Environment, which was held in 1972 in Stockholm?

(*a*) Developing countries and developed countries were largely in agreement about the most important environment problems.
(*b*) For the first time since the end of the Second World war, nation set aside their political difference to achieve a common objective.
(*c*) The conference provided the first global forum for dialogue on environmental problems.
(*d*) The delegates assembled at the conference voted in favor of controlling population growth.

1484. One solution for reducing the amount of atmospheric carbon dioxide would be to

(*a*) increase oceanic temperatures to enhance carbon dioxide uptake
(*b*) decrease the total area of rice paddies
(*c*) decrease the use of fossil fuels
(*d*) decrease the production of chlorofluorocarbons

1485. The drop in stratospheric ozone levels in the Southern Hemisphere (the "ozone hole") is must evident during which season?

(*a*) Antarctic spring (October)
(*b*) Antarctic autumn (April)
(*c*) Antarctic summer only (January)
(*d*) Antarctic winter only (July)

1486. Uranium-235 has a half of 710 million years. If it is determined that a certain amount of stored U-235 will be considered safe only when its radioactivity has dropped to 0.10 percent of the original level, approximately how much time must the U-235 be stored securely to be safe?

1487. Which of the following would be the strongest evidence I support of the contention that a local area was experiencing acid deposition?

(*a*) A sudden die-of all the fish in a local stream
(*b*) A gradual increase in the temperature of a local lake
(*c*) An increase in the rate of photosynthesis of aquatic plants in a local lake
(*d*) An increase in a concentration of soluble heavy metal in a local pond

1488. Which of the following is the usual cause of cultural eutrophication in surface water of both developed and developing countries?

(*a*) Introduction of cyanobacteria to streams and river
(*b*) Runoff of metal ions in bodies of water
(*c*) Runoff of nitrate compounds into bodies of water
(*d*) Runoff of herbicides into bodies of water

1489. If mean global temperature keeps on increasing which of the following is most likely?

(*a*) These will be fewer insect, pests and disease-carrying organisms.
(*b*) Specialists species living I fragile ecosystems will risk extinction.
(*c*) There will be an increase in sulfate concentration ion wetlands.
(*d*) There will be an increase in the reproductive rate of mammals.

1490. Photosynthesis is the major source of which of the following gases in Earth's atmosphere?

(*a*) CO_2 (*b*) H_2O (water vapor)
(*c*) O_2 (*d*) N_2

1491. Which material accounts for the greatest percentage of the weight of municipal solid waste?

Answers

1482. (*b*) 1483. (*c*) 1484. (*c*) 1485. (*a*) 1486. (*d*) 1487. (*b*) 1488. (*c*) 1489. (*b*) 1490. (*c*)
1491. (*b*)

(*a*) plastic (*b*) paper
(*c*) yard waste (*d*) food waste

1492. Which of the following is not a method of sources reduction of solid waste?
(*a*) composting (*b*) incineration
(*c*) recycling
(*d*) making packages that weigh less

1493. Which of the following be not true about incineration of solid wastes?
(*a*) There is a drastic reduction in the volume and weight of the waste.
(*b*) The incinerators of solid wastes?
(*c*) The incinerators can be built to generate electricity.
(*d*) The incinerators ash may contain heavy metals and other toxic substances.

1494. Modern municipal solid waste landfills contain
(*a*) Impermeable barriers and monitoring wells.
(*b*) leachate treatment facilities
(*c*) methane retrieval systems
(*d*) all (*a*), (*b*) and (*c*)

1495. Pick out the wrong statement
(*a*) Essentially five techniques-incineration landfills, source reduction, composting and recycling are used to dispose of solid waste.
(*b*) Per capita production of municipal solid waste in India equal about 10 kg per day.
(*c*) As compared to 1960, net solid waste generation has increased even with recycling programme in place.
(*d*) A European concept that uses solid waste to generate electricity is called mass burn.

1496. Dieldrin is listed by the EPA as a probable human carcinogen. Thus, the presence of dieldrin in the environment is a ________.
(*a*) risk (*b*) disease
(*c*) hazard (*d*) vector

1497. You have lived near a coal-fired power plant for over 20 years. You are now experiencing respiratory illness that your doctor determines was caused by exposure to sulfur dioxide from the power plant. What kind of impact is this?
(*a*) acute (*b*) outdoor
(*c*) chronic (*d*) indoor

1498. There are three categories of impact on respiratory health from airborne pollution. What are they?
(*a*) toxic impacts, persistent impacts, and bioaccumulation impacts
(*b*) positive, negative, and neutral impacts
(*c*) chronic impacts, acute impacts, and carcinogenic impacts
(*d*) nervous system impacts, heart impacts, and respiratory impacts

1499. Suppose less water evaporates from the world's oceans. If all other variables remain the same, what will happen to mean global temperature?
(*a*) Mean global temperature will increase.
(*b*) Mean global temperature will decrease.
(*c*) Evaporation from the oceans does not impact mean global temperature.
(*d*) Mean global temperature will stay the same.

1500. If the number of low clouds and the amount of sulfate aerosols both increase in the atmosphere, and all other variables remain the same, what will happen?
(*a*) Temperatures will increase.
(*b*) Temperatures will decrease.
(*c*) Clouds and sulfate aerosols do not impact temperatures.
(*d*) Temperatures will not be affected.

1501. Which of the following have opposite effects on climate change?
(*a*) increased volcanic activity and increased low clouds
(*b*) increased low clouds and increased fossil fuel burning

Answers

1492. (*b*) 1493. (*b*) 1494. (*d*) 1495. (*b*) 1496. (*c*) 1497. (*c*) 1498. (*c*) 1499. (*b*) 1500. (*b*)
1501. (*b*)

(*c*) increased sulfate aerosols and decreased fossil fuel burning

(*d*) increased volcanic activity and increased sulfate aerosols

1502. If trends continue, which of the following will happen?

(*a*) Global mean sea level will increase, and Arctic sea ice will decrease.

(*b*) Global mean sea level will decrease, and Arctic sea ice will also decrease.

(*c*) Global mean sea level will decrease, and Arctic sea ice will increase.

(*d*) Global mean sea level will increase, and Arctic sea ice will also increase.

1503. If fossil fuel burning continues to increase, what will happen to the acidity of the oceans?

(*a*) Ocean acidity will remain the same.

(*b*) They will become less acidic.

(*c*) They will become more acidic.

(*d*) It is unknown what will happen to the acidity of the oceans.

1504. Which of the following would have the largest impact on reducing the third most important greenhouse gas?

(*a*) eliminating cattle farming

(*b*) banning the use of fertilizers

(*c*) planting trees

(*d*) banning aerosol cans

1505. Which two countries saw the highest percentage of increase in greenhouse gas emissions between 1990 and 2006?

(*a*) Japan and Canada

(*b*) United States and Australia

(*c*) United States and Canada

(*d*) Canada and Australia

1506. If global average temperature were to increase between 2.8°C and 4.0°C, carbon dioxide emissions would have to peak ________.

(*a*) between 2000 and 2020

(*b*) between 2000 and 2030

(*c*) between 2010 and 2060

(*d*) between 2050 and 2090

1507. Which of the following have positive radiative forcing effects?

(*a*) increasing sulfate aerosols and increasing fossil fuel burning

(*b*) decreasing low clouds and decreasing fossil fuel burning

(*c*) increasing sulfate aerosols and decreasing fossil fuel burning

(*d*) decreasing low clouds and increasing fossil fuel burning

1508. Which areas of the world contribute to climate change least by the burning of fossil fuels?

(*a*) Europe and Latin America

(*b*) Africa and Latin America

(*c*) Eurasia and the Middle East

(*d*) Europe and Asia/Pacific

1509. If temperature increases 3°C between 2000 and 2100, what will happen to Antarctic snowfall?

(*a*) It will stop snowing altogether in the Antarctic.

(*b*) Antarctic snowfall will stay the same.

(*c*) Antarctic snowfall will decrease by about 15%.

(*d*) Antarctic snowfall will increase by about 15%.

1510. The city of Arcata, California, is committed to decreasing locally generated greenhouse gases by increasing the use of vehicles that run on alternative fuels and by not venting methane from waste water treatment plant into the atmosphere. This is an example of dealing with climate change through ________.

(*a*) adaptation (*b*) prohibition

(*c*) mitigation (*d*) stabilization

1511. A scientist genetically engineers a variety of corn to grow well in warmer temperatures.

Answers

1502. (*a*) 1503. (*c*) 1504. (*a*) 1505. (*d*) 1506. (*c*) 1507. (*d*) 1508. (*b*) 1509. (*d*) 1510. (*c*) 1511. (*b*)

This is an example of dealing with climate change through ________.

(*a*) mitigation (*b*) adaptation

(*c*) prohibition (*d*) stabilization

1512. Model predictions of global climate change show ________.

(*a*) close agreement on trends but little agreement on values

(*b*) no agreement at all

(*c*) close agreement on trends and values (for instance, predicted carbon dioxide concentrations)

(*d*) general agreement on trends but little agreement on values

1513. Which of the following is true?

(*a*) The greenhouse effect and climate change are the same thing.

(*b*) The greenhouse effect and climate change are not related.

(*c*) The greenhouse effect is a natural effect. Enhancing it is likely resulting in climate change.

(*d*) Climate change is a natural effect. Enhancing it is likely resulting in the greenhouse effect.

1514. Which of the following will increase the planetary albedo?

(*a*) fewer volcanic eruptions and the formation of more low clouds

(*b*) more volcanic eruptions and less formation of low clouds.

(*c*) more volcanic eruptions and the formation of more low clouds

(*d*) fewer volcanic eruptions and less formation of low clouds.

1515. What are some of the differences between the troposphere and the stratosphere?

(*a*) The stratosphere is smaller than the troposphere and also has less vertical mixing.

(*b*) The troposphere is smaller than the stratosphere and also has more vertical mixing.

(*c*) The troposphere is smaller than the stratosphere and also has less vertical mixing.

(*d*) The stratosphere is smaller than the troposphere and also has more vertical mixing.

1516. Which of the following green house gases has been impacting the climate for the least amount of time?

(*a*) carbon dioxide (*b*) water vapor

(*c*) CFCs (*d*) methane

1517. You are a meteorologist. What atmospheric layer do you study?

(*a*) troposphere (*b*) mesosphere

(*c*) stratosphere (*d*) thermosphere

1518. A high-pressure system is moving easterly into a region with low pressure. In which direction is the wind blowing?

(*a*) south (*b*) west (*c*) north (*d*) east

1519. Which of the following statements regarding the relationship between global mean temperature and atmospheric concentrations of carbon dioxide is true?

(*a*) As the concentration of atmospheric carbon dioxide has increased, mean global temperature has decreased.

(*b*) As the concentration of atmospheric carbon dioxide has decreased, mean global temperature has increased.

(*c*) As the concentration of atmospheric carbon dioxide has increased, mean global temperature has also increased.

(*d*) As the concentration of atmospheric carbon dioxide has increased, mean global temperature has stayed the same.

1520. Fluids such as gasoline, paint solvents, and organic cleaning solutions release ________.

Answers

1512. (*a*) 1513. (*a*) 1514. (*c*) 1515. (*b*) 1516. (*c*) 1517. (*a*) 1518. (*d*) 1519. (*c*) 1520. (*b*)

(a) nitrogen oxides
(b) volatile organic compounds
(c) sulfur dioxide (d) carbon monoxide

1521. Which of the following is a pollutant in one part of the atmosphere and valuable filter in another part of the atmosphere?
(a) sulfur dioxide (b) ozone
(c) nitrous oxide (d) oxygen

1522. Which of the following locations is most likely to have a temperature inversion that results in large amounts of trapped air pollution?
(a) a city in the mountains with many large industries
(b) a city in a valley with large numbers of people who use personal vehicles to commute
(c) a city in the plains with large numbers of people who use personal vehicles to commute
(d) a city on a lake with many large industries

1523. Your community is home to a large coal-fired power plant. What pollutants are most likely to be found in the air?
(a) particulate matter, radon, and sulfur dioxide
(b) lead, radon, and particulate matter
(c) particulate matter, nitrous oxides, and sulfur dioxide
(d) lead, nitrogen oxides, and sulfur dioxide

1524. Which of the following air pollutants has shown the greatest decline in recent years?
(a) particulates (b) carbon monoxide
(c) lead (d) sulfur dioxide

1525. If all transportation were suddenly stopped, which three pollutants would decrease the most?
(a) sulfur dioxide, particulate matter, and volatile organic compounds
(b) sulfur dioxide, nitrous oxides, and volatile organic compounds
(c) carbon monoxide, particulate matter, and nitrous oxides
(d) nitrous oxides, volatile organic compounds, and carbon monoxide

1526. Tomato juice has a pH of 4.0. How many times more acidic is lemon juice, which has a pH of 2.0?
(a) 2 times (b) 4 times
(c) 10 times (d) 100 times

1527. Children in your community are suffering from brain damage and respiratory problems. What two pollutants are the likely causes?
(a) carbon monoxide and lead
(b) carbon monoxide and sulfur dioxide
(c) lead and sulfur dioxide
(d) carbon monoxide and particulate matter

1528. If all volatile organic compounds (VOCs) suddenly disappeared, what would happen to ozone concentrations in the troposphere?
(a) It is unknown what would happen to ozone concentrations in the troposphere.
(b) Ozone concentrations in the troposphere would increase.
(c) Ozone concentrations in the troposphere would decrease.
(d) Ozone concentrations in the troposphere would stay the same.

1529. Which of the following can result in the formation of acids in the atmosphere?
(a) volatile organic compounds and radon
(b) ozone and asbestos
(c) carbon monoxide and lead
(d) nitrogen oxides and sulfur oxides

1530. Which of the following crops would be best to grow in areas with high levels of ozone pollution?
(a) soybeans and winter wheat
(b) peanuts and winter wheat
(c) cotton and field corn
(d) sorghum and field corn

Answers

1521. (b) 1522. (b) 1523. (c) 1524. (c) 1525. (d) 1526. (d) 1527. (c) 1528. (a) 1529. (d) 1530. (a)

1531. Which of the following industries would be most impacted if the Environmental Protection Agency mandated the elimination of sulfur dioxide and nitrous oxides from the air?
(*a*) agriculture (*b*) car manufacturers
(*c*) paper manufacturers
(*d*) electrical utilities

1532. Over time, if all primary pollutants were eliminated, what would happen to concentrations of secondary pollutants?
(*a*) Concentrations of secondary pollutants would drop to zero.
(*b*) Concentrations of secondary pollutants would stay the same.
(*c*) Concentrations of secondary pollutants would increase.
(*d*) It is unknown what would happen to concentrations of secondary pollutants.

1533. Photochemical smog is produced when sunlight acts on ________ and ________.
(*a*) ground-level ozone/nitrogen oxides
(*b*) nitrogen oxides/volatile organic carbon compounds
(*c*) airborne particulates/volatile organic carbon compounds
(*d*) sulfur oxides/carbon dioxide

1534. Which of the following is true regarding catalytic converters?
(*a*) They convert an air pollutant into a greenhouse gas.
(*b*) They convert a greenhouse gas into a carcinogen.
(*c*) They convert a primary pollutant into a secondary pollutant.
(*d*) They have increased air pollution.

1535. Which of the following saw the highest increase in fuel economy between 1975 and 2008?
(*a*) Fuel efficiency improved equally for cars and trucks.
(*b*) trucks (*c*) cars
(*d*) Fuel efficiency did not improve for either cars or trucks.

1536. If chlorine were not a catalyst, how would CFCs have affected ozone depletion?
(*a*) It is unknown what would have happened with ozone depletion.
(*b*) Ozone depletion would have been slower.
(*c*) Ozone depletion would have been faster.
(*d*) Ozone depletion would have occurred at the same rate.

1537. Which of the following true?
(*a*) As the concentration of ozone in the stratosphere over the Antarctic increased, the size of the ozone hole also increased.
(*b*) As the concentration of ozone in the stratosphere over the Antarctic increased, the size of the ozone hole remained the same.
(*c*) As the concentration of ozone in the stratosphere over the Antarctic decreased, the size of the ozone hole also decreased.
(*d*) As the concentration of ozone in the stratosphere over the Antarctic increased, the size of the ozone hole decreased.

1538. HCFCs have replaced CFCs for many uses. However, HCFCs still damage ozone. Why?
(*a*) They both contain hydrogen.
(*b*) They both contain oxygen.
(*c*) They both contain chlorine.
(*d*) They both contain carbon.

1539. Which of the following best describes a command-and-control approach to decreasing air pollution?
(*a*) allowing consumers to pay more to ensure that their electricity comes from a renewable resource
(*b*) allowing utility companies to buy emissions from each other
(*c*) fining a utility company for emitting too much sulfur dioxide

Answers

1531. (*d*) 1532. (*a*) 1533. (*a*) 1534. (*a*) 1535. (*c*) 1536. (*a*) 1537. (*c*) 1538. (*c*) 1539. (*c*)

(d) paying a utility company to emit less sulfur dioxide

1540. Eliminating which of the following would help alleviate the problem of acid deposition?

(a) hydrogen ions (b) oxygen gas
(c) hydroxide ions (d) carbon

1541. What would be the best way to eliminate photochemical smog?

(a) banning nuclear power plants
(b) banning HCFCs
(c) banning coal-fired power plants
(d) banning gasoline-powered vehicles

1542. Which of the following is an example of point-source pollution?

(a) acid deposition in Pittsburgh, Pennsylvania
(b) the effluent from a plastics factory in Madison, Wisconsin
(c) runoff from six farms in Hardin County, Iowa
(d) runoff from the streets of Portland, Oregon

1543. Which of the following is an example of nonpoint-source pollution?

(a) the water from the cooling towers of a coal-fired power plant in Atlanta, Georgia
(b) the oil from an oil tanker spill off the coast of Alaska
(c) the effluent from the waste water-treatment plant in Duluth, Minnesota
(d) the water going into the storm drains in Lincoln, Nebraska

1544. Which of the following is an inorganic pollutant?

(a) oil (b) DDT
(c) manure (d) chromium

1545. Which of the following organisms can tolerate low levels of dissolved oxygen in water?

(a) blackfly larvae and mayfly larvae
(b) blackfly larvae and sludge worms
(c) caddis fly and mayfly larvae
(d) sludge worms and caddis fly larvae

1546. You find the following organisms on a stream bottom: mayfly larvae, stonefly larvae, and blackfly larvae. What is the quality of the water?

(a) greenish (b) very cloudy
(c) severely damaged (d) clean water

1547. Which of the following is true?

(a) The sediment bed load impacts only vertebrate life in a stream.
(b) Sediment bed load does not impact the organisms that live in a stream.
(c) Streams with a high sediment bed load support a wider variety of organisms than streams with a low sediment bed load.
(d) Streams with a low sediment bed load support a wider variety of organisms than streams with a high sediment bed load.

1548. Sticks and bottles that wash into sewers fit into which category of materials handled at waste water treatment plants?

(a) dissolved inorganic material
(b) particulate organic material
(c) debris and grit
(d) colloidal and dissolved organic material

1549. The bar screen at a waste water-treatment plant breaks. Which phase of treatment is impacted?

(a) primary treatment
(b) disinfection
(c) secondary treatment
(d) biological nutrient removal

1550. Someone adds oxygen to the sludge digester at a waste water-treatment plant. What will happen?

(a) The digestion of sludge will increase because the bacteria need oxygen.
(b) The digestion of sludge will decrease because the bacteria cannot live in the presence of oxygen.

Answers

1540. (c) 1541. (d) 1542. (b) 1543. (d) 1544. (d) 1545. (b) 1546. (d) 1547. (d) 1548. (c) 1549. (a) 1550. (b)

(c) The digestion of sludge will not change.
(d) The pathogens in the sludge will all die. Clarification will take longer.

1551. Your waste water-treatment plant adds chlorine to the treated water before discharging the effluent to the river. However, many local species of fish are dying because of the presence of this chemical. Which phase of treatment will the plant need to change?
(a) disinfection
(b) secondary treatment
(c) biological nutrient removal
(d) primary treatment

1552. You are a farmer. Which of the following would you most like to have to put on your fields?
(a) water that remains after treatment is finished
(b) untreated waste water
(c) raw sludge (d) dried biosolids

1553. Biochemical oxygen demand (BOD) is an important measure of _______.
(a) oxygen content of water and waste water
(b) the oxygen-using potential of water and waste waters
(c) an organism's natural level of oxygen requirement
(d) a measure of the biological activity of water and waste water

1554. Which of the following is most likely to be oligotrophic?
(a) a pond in the middle of a housing development
(b) a lake surrounded by corn fields
(c) a pond in the middle of a golf course
(d) a lake in Superior National Forest

1555. The ideal modern sewage system is one in which _______.
(a) storm water is collected and treated, but sewer water is not
(b) individual homeowners operate private septic systems on large lots
(c) sanitary sewage water and storm water are collected in a single sewer system and treated in a single treatment plant
(d) sanitary sewage water is collected separately from storm water and is fully treated to remove all pollutants before the water is put back into the environment.

1556. A farmer has a eutrophic pond on her land. Which of the following would work best to reduce the amount of nutrients entering the pond?
(a) tilling the surrounding fields more often
(b) planting a field windbreak around the pond
(c) using more chemical fertilizer on the surrounding fields
(d) putting manure on the surrounding fields

1557. In freshwater systems, _______ is most often the limiting factor, whereas in marine systems the limiting factor is most often _______.
(a) nitrogen, phosphorus
(b) potassium, nitrogen
(c) phosphorus, potassium
(d) phosphorus, nitrogen

1558. Which of the following is true?
(a) Warm water and cold water hold the same amount of dissolved oxygen.
(b) Warm water holds more dissolved oxygen than cold water.
(c) Cold water holds more dissolved oxygen than warm water.
(d) The amount of dissolved oxygen that water can hold does not depend on temperature.

1559. Which of the following includes all of the others?
(a) benthic plants
(b) emergent vegetation

Answers

1551. (a) 1552. (d) 1553. (a) 1554. (d) 1555. (d) 1556. (b) 1557. (d) 1558. (c) 1559. (a)

(*c*) submerged aquatic vegetation
(*d*) pond weeds

1560. Which of the following includes all of the others?
(*a*) diatoms (*b*) cyanobacteria
(*c*) phytoplankton (*d*) green algae

1561. Secondary sewage treatment can best be described as a ________ process.
(*a*) chemical (*b*) geological
(*c*) biological (*da*) mechanical

1562. The largest single component of municipal solid waste in the United States is ________.
(*a*) yard waste (*b*) plastics
(*c*) food waste
(*d*) paper and paper products

1563. You manage the local landfill and wish for materials to degrade faster. What should you mix in with the waste?
(*a*) carbon dioxide (*b*) more waste
(*c*) methane (*d*) water

1564. It has been discovered that a local landfill is contaminating the groundwater. What is the most likely reason that this occurred?
(*a*) There is a problem with the liner.
(*b*) There is a problem with the fill dirt.
(*c*) There is a problem with the compactor.
(*d*) There is a problem with the waste being added.

1565. Which of the following is true?
(*a*) It is usually cheaper to dispose of waste at a waste-to-energy facility.
(*b*) Waste-to-energy facilities combust all of the materials they receive.
(*c*) Waste-to-energy facilities eliminate the need for landfills.
(*d*) One of the benefits of a waste-to-energy facility is that waste is considerably reduced in weight and volume.

1566. Your city recycles 10 tons of newspaper per week. How many trees does this save?
(*a*) about 80 (*b*) about 170
(*c*) about 360 (*d*) about 640

1567. You recycle an aluminium soda can, and it is made into another soda can. What kind of recycling is this?
(*a*) primary recycling
(*b*) quaternary recycling
(*c*) tertiary recycling
(*d*) secondary recycling

1568. You recycle your tires, and they are made into material for a high school track. What kind of recycling is this?
(*a*) primary recycling
(*b*) quaternary recycling
(*c*) tertiary recycling
(*d*) secondary recycling

1569. Eight tons of steel are recycled. About how many tons of coal does this save?
(*a*) 1 (*b*) 4 (*c*) 8 (*d*) 16

1570. If trends since 1960 continue, what will happen in regard to municipal solid waste?
(*a*) The total amount of municipal solid waste recycled will stay the same, but the percentage recycled will increase.
(*b*) The total amount of municipal solid waste recycled and the percentage recycled will both decrease.
(*c*) The total amount of municipal solid waste recycled and the percentage recycled will both increase.
(*d*) The total amount of municipal solid waste recycled will increase, but the percentage recycled will stay the same.

1571. Assuming that people are generating the same amount of trash, if a pay-as-you-throw system is working, what should happen to recycling rates?
(*a*) Recycling rates should stay the same.

Answers

1560. (*a*) 1561. (*c*) 1562. (*d*) 1563. (*d*) 1564. (*a*) 1565. (*d*) 1566. (*b*) 1567. (*a*) 1568. (*d*)
1569. (*c*) 1570. (*c*) 1571. (*b*)

(*b*) Recycling rates should increase.
(*c*) Recycling rates should decrease.
(*d*) There should be no recycling at all.

1572. At the grocery store, you want to make choices that are best for the environment. When asked, "Paper or plastic?" What should you do?
(*a*) Neither. You should bring your own bag to use.
(*b*) It doesn't matter. Both bags use resources and are equally harmful to the environment.
(*c*) plastic (*d*) paper

1573. Which of the following can be categorized as E-waste?
(*a*) yard waste (*b*) clothing
(*c*) an iPod
(*d*) a plastic water bottle

1574. Which of the following includes all of the others?
(*a*) landfilling (*b*) recycling
(*c*) waste management
(*d*) incineration

1575. Ultimately, the fact that the United States has the highest per capita waste production in the world can be tied to ________.
(*a*) the cost of landfilling
(*b*) lack of curbside recycling
(*c*) over consumption
(*d*) inefficient government laws

1576. In the best of all possible worlds, the best strategy for managing MSW is to ________.
(*a*) recycle it
(*b*) reduce it at the source
(*c*) burn it (*d*) landfill it

1577. Which of the following includes all of the others?
(*a*) bottle bill (*b*) recycling
(*c*) production of glassphalt
(*d*) composting

1578. Which of the following is true regarding the Toxics Release Inventory?
(*a*) The amount of toxic waste disposed of on-site has stayed about the same since 1987.
(*b*) The amount of toxic waste disposed of on-site has dropped to zero since 1987.
(*c*) The amount of toxic waste disposed of on-site has decreased by more than half since 1987.
(*d*) The amount of toxic waste disposed of on-site has doubled since 1987.

1579. High-density polyethylene is a type of plastic. Therefore, it is a(n) ________.
(*a*) synthetic organic compound
(*b*) inorganic compound
(*c*) heavy metal (*d*) dioxin

1580. You want to stop contributing to the release of PERC. What should you do?
(*a*) Stop buying plastics.
(*b*) Stop taking clothes to the dry cleaner.
(*c*) Give up your cell phone.
(*d*) Buy only organic food.

1581. You live near an incineration facility. Which pollutants are likely in the air near your home?
(*a*) endrin and toxaphene
(*b*) heptachlor and dioxins
(*c*) chlordane and furans
(*d*) dioxins and furans

1582. Pregnant women are told to avoid consuming shark, swordfish, king mackerel, and tilefish because they often have high concentrations of mercury. This advice is to protect the developing baby from ________?
(*a*) hormone disruption
(*b*) cancer
(*c*) errors in chromosome number
(*d*) birth defects

1583. In deep-well injection, how far down is hazardous waste stored?

Answers

1572. (*a*) 1573. (*c*) 1574. (*c*) 1575. (*c*) 1576. (*b*) 1577. (*b*) 1578. (*c*) 1579. (*a*) 1580. (*b*)
1581. (*d*) 1582. (*d*) 1583. (*a*)

(*a*) It is stored below the drinking-water source.
(*b*) It is stored above the drinking-water source.
(*c*) It is stored at the same level as the drinking-water source.
(*d*) The storage depth varies based on location.

1584. Waste oil easily catches on fire. Therefore, the EPA classifies it as a hazardous material based on its ________.
(*a*) corrosivity (*b*) ignitability
(*c*) toxicity (*d*) reactivity

1585. Lithium-sulfur batteries are unstable under normal conditions. Therefore, the EPA classifies them as a hazardous material based on their ________.
(*a*) corrosivity (*b*) toxicity
(*c*) ignitability (*d*) reactivity

1586. Battery acid can cause metal containers to break down. Therefore, the EPA classifies it as a hazardous material based on its ________.
(*a*) corrosivity (*b*) toxicity
(*c*) reactivity (*d*) ignitability

1587. Bleach is harmful if swallowed. Therefore, the EPA classifies it as a hazardous material based on its ________.
(*a*) toxicity (*b*) corrosivity
(*c*) ignitability (*d*) reactivity

1588. The MCL for nitrates in drinking water is 10 ppm. What does this mean?
(*a*) You must boil your water for 10 minutes before using it if it contains nitrates.
(*b*) Your drinking water should not have more than 10 ppm nitrates.
(*c*) You must let your tap run for 10 minutes before using it if the water contains nitrates.
(*d*) Your drinking water must have at least 10 ppm nitrates.

1589. Many coal-fired power plants have electrostatic precipitators, which remove some of the pollutants. This is an example of ________.
(*a*) pollution prevention
(*b*) reuse (*c*) green chemistry
(*d*) pollution control

1590. Which of the following represents the issue of environmental justice?
(*a*) The United States fails to agree to the principles of the Kyoto Protocol.
(*b*) Carbon dioxide is released from automobile exhaust.
(*c*) General Motors illegally disposes of PCBs near Akwesasne Mohawk lands, on which many people of this Native American tribe live.
(*d*) The EPA limits the use of MTBE because it can pollute groundwater.

1591. A brownfield site is one that ________.
(*a*) is a former industrial or commercial site, now contaminated
(*b*) will not support an agricultural crop
(*c*) has little vegetation remaining as a result of past environmental contamination
(*d*) abandoned, idled, or under-used industrial and commercial facilities where expansion or redevelopment is complicated by real or perceived environmental contamination

1592. In Fort Bragg, California, some people are proposing the use of mushrooms to clean up dioxins in five plots of soil. This is an example of
(*a*) bioremediation (*b*) georemediation
(*c*) ecoremediation (*d*) phytoremediation

1593. You own a coal-fired power plant and wish to legally dispose of the ash in the cheapest way possible. How should you dispose of it?
(*a*) at a secure landfill
(*b*) surface impoundment
(*c*) midnight dumping
(*d*) deep-well injection

Answers

1584. (*b*) 1585. (*d*) 1586. (*a*) 1587. (*a*) 1588. (*b*) 1589. (*d*) 1590. (*c*) 1591. (*d*). 1592. (*a*) 1593. (*b*)

1594. The level below which no ill effects are observed is called ________.
(*a*) absolute limit (*a*) threshold level
(*a*) toxic level (*a*) level of tolerance

1595. Acid Rain is a form of
(*a*) Soil pollution (*b*) Water pollution
(*c*) Solar pollution (*d*) Air pollution

1596. The major contributors to acid rain are known as
(*a*) Precursors (*b*) Processors
(*c*) Protons (*d*) Pollutants

1597. Precursors mainly come from
(*a*) Chemical and physical sources
(*b*) Biogenic and anthropogenic sources
(*c*) Biotic and abiotic sources
(*d*) Arctic and Antarctic sources

1598. Anthropogenic or Man made sources of air pollution are:
(*a*) Spread over large areas.
(*b*) Concentrated around the source
(*c*) Spread all over the Earth
(*d*) None of the above.

1599. Volcanic eruptions and Decay of organic matter are examples of
(*a*) Volcanic source (*b*) Natural resources
(*c*) Biogenic source
(*d*) Anthropogenic source

1600. Particulates falling directly from the atmosphere to the ground are known as
(*a*) Wet forms (*b*) Solid forms
(*c*) Dry deposition (*d*) None of these

1601. Catalytic reduction is a process adopted to
(*a*) Increase the levels of Sulphur dioxide
(*b*) Decrease the levels of Sulphur dioxide
(*c*) Increase the levels of Nitrogen oxides
(*d*) Decrease the levels of Nitrogen oxides

1602. Using less air for combustion can
(*a*) Increase nitrogen oxides
(*b*) Increase Sulphur oxides
(*c*) Decrease Sulphur oxides
(*d*) Decrease Nitrogen oxides.

1603. Lower the temperature of combustion, lower will be the production of
(*a*) Nitrogen (*b*) Hydrogen
(*c*) Nitrogen oxides (*d*) Water

1604. Two stage combustion and flue gas recirculation results in
(*a*) Reduction in temperature
(*b*) Increase in temperature
(*c*) Reduction in emissions
(*d*) None of the above.

1605. Fluidized bed combustion is a complex process of reducing
(*a*) Nitrogen (*b*) Oxygen
(*c*) Hydrogen (*d*) Sulphur dioxide

1606. Flue gas desulphurization is achieved by using
(*a*) Lime stone (*b*) Coal
(*c*) Sand stone (*d*) Diesel

1607. Fluidized bed combustion produces
(*a*) Liquid waste (*b*) Air pollution
(*c*) Solid waste (*d*) None of these

1608. Acid rain on trees results in
(*a*) Brown dead spots (*b*) Photosynthesis
(*c*) White patches (*d*) Plasma

1609. In Black forests of South Western Germany, half of the trees are damaged from
(*a*) Snow (*b*) Water pollution
(*c*) Air pollution (*d*) Acid rain.

1610. Most farm crops are ________ affected by Acid Rain.
(*a*) Most (*b*) Very little
(*c*) Not at all (*d*) Less

1611. Parthenon in Greece and Taj Mahal in India are deteriorating due to
(*a*) Alkali deposition (*b*) Snow fall
(*c*) Acid rain (*d*) All of these

Answers

1594. (*b*) 1595. (*d*) 1596. (*a*) 1597. (*b*) 1598. (*b*) 1599. (*c*) 1600. (*c*) 1601. (*d*) 1602. (*d*)
1603. (*c*) 1604. (*a*) 1605. (*d*) 1606. (*a*) 1607. (*c*) 1608. (*a*) 1609. (*d*) 1610. (*d*) 1611. (*c*)

1612. It is ______ to swim in even the most acidified lakes.
(*a*) Safe (*b*) Unsafe
(*c*) Not advisable (*d*) None of these

1613. Acid rain tends to ______ global warming
(*a*) Increase (*b*) No effect on
(*c*) Slow down (*d*) All the above

1614. Ozone layer was discovered by Charles Fabry in 1913 and properties of ozone layer were explored by Dobson. The equivalent of one Dobson Unit (DU) is
(*a*) 0.1 mm (*b*) 0.01 mm
(*c*) 0.1 m (*d*) 0.01 m

1615. Ozone layer is at a height of ______ above the Earth's Surface
(*a*) 19 to 48 m (*b*) 19 to 480 m
(*c*) 19 to 48 km (*d*) 190 to 148 km

1616. UV radiation is divided into three categories based on its
(*a*) Frequency (*b*) Intensity
(*c*) Wave length (*d*) None of these

1617. The wave length range of UV-A is in the range of
(*a*) 315 to 400 nm (*b*) 315 to 400 nm
(*c*) 315 to 400 cm (*d*) 315 to 400 m.

1618. Excessive exposure to UV-B radiation can cause
(*a*) Kidney failure (*b*) Skin cancer
(*c*) Blood cancer (*d*) None of the above.

1619. UV-C rapidly damages all types of
(*a*) Biota (*b*) Buildings
(*c*) Water bodies (*d*) All the above.

1620. Ozone molecules absorb UV light between
(*a*) 310 to 200 nm (*b*) 310 and 200 m
(*c*) 310 and 200 mm (*d*) 310 and 200 nm

1621. The average thickness of ozone layer in the stratosphere is approximately
(*a*) 3000 DU (*b*) 30 DU
(*c*) 30,000 DU (*d*) 300 DU

1622. Ozone holes is due to decrease in ozone level. Ozone hole is said to occur when the ozone level decreases below
(*a*) 200 DU (*b*) 2000 DU
(*c*) 20 DU (*d*) 2 DU

1623. Hydroxyl, Nitric oxide, atomic chlorine and Bromine can result in ______ of stratospheric ozone.
(*a*) Increase (*b*) Balancing
(*c*) Decreasing (*d*) Neutralizises

1624. UV-C radiation lies in the region
(*a*) < 280 nm (*b*) > 280 nm
(*c*) 315 to 400 nm (*d*) 280 to 315 nm.

1625. Many thousands of ozone molecules are removed from the stratosphere due to atoms of
(*a*) Sulphur (*b*) Oxygen
(*c*) Hydrogen (*d*) Chlorine

1626. CFC molecules can stay in the atmosphere for about
(*a*) 1 year (*b*) 10 yrs
(*c*) 100 yrs (*d*) 1000 yrs

1627. In the Antarctic zone, ozone layer depletion occurs during the period September to December. The decrease in the ozone layer was predicted in the early
(*a*) 1960 (*b*) 1970 (*c*) 1980 (*d*) 1880.

1628. The Antarctic stratosphere is ______ than the Arctic, hence the reason for ozone hole.
(*a*) Colder (*b*) Warmer
(*c*) Brighter (*d*) None of these

1629. One percent decrease in stratospheric ozone would increase the incidence of skin cancer by
(*a*) 0.2% (*b*) 20% (*c*) 2% (*d*) 0.02%

1630. Ozone at ground level is health risky due to its strong ______ property.
(*a*) Acid (*b*) Alkaline
(*c*) Oxidant (*d*) Hygroscopic

1631. World ozone day is celebrated on 18th Sep. The Antarctic ozone hole reached its largest ever size in

Answers

1612. (*a*) 1613. (*c*) 1614. (*b*) 1615. (*c*) 1616. (*c*) 1617. (*a*) 1618. (*b*) 1619. (*a*) 1620. (*a*)
1621. (*d*) 1622. (*a*) 1623. (*c*) 1624. (*a*) 1625. (*d*) 1626. (*c*) 1627. (*c*) 1628. (*a*) 1629. (*c*)
1630. (*c*) 1631. (*a*)

(*a*) Sep 2000 (*b*) Jan 2000
(*c*) Jan 2005 (*d*) Sep 2005

1632. The largest ozone hole size on Antarctic region is reported as
(*a*) 11.5 thousand square miles
(*b*) 11.5 million square miles
(*c*) 11,500 square miles
(*d*) 1,15,000 square miles

1633. Breeding, feeding and managing animals for the production of food, fiber, work and pleasure is termed as Animal husbandry . The purpose of animal husbandry is production of meat. Animal husbandry may result in
(*a*) Acid rain (*b*) Ozone depletion
(*c*) Global warming (*d*) Snow fall

1634. The annual global generation of methane from ruminants accounts for 15% of the total amount released into the atmosphere. Methane produced from cattle is ________ that from the ruminants.
(*a*) Less than (*b*) Equal to
(*c*) More than (*d*) None of the above

1635. Cosmic rays, Gamma rays, X-rays and Ultraviolet radiation are known Ionizing radiation.
Cancer and related diseases are caused due to
(*a*) Primary radiation
(*b*) Non-Ionizing radiation
(*c*) Ionizing radiation
(*d*) Secondary radiation.

1636. The forms of electromagnetic radiation not containing enough energy to form ions are known as
(*a*) Thermal radiation
(*b*) Primary radiation
(*c*) Ionizing radiation
(*d*) Non-Ionizing radiation

1637. 92% of the solar radiation reaching the Earth will be in the range of
(*a*) 315 to 140 m
(*b*) 315 to 1400 m
(*c*) 315 to 1400 nm
(*d*) 315 to 1400 mm

1638. Ionizing radiation waves have
(*a*) Low energy and short wave length
(*b*) Low energy and long wave length
(*c*) High energy and short wave length
(*a*) High energy and long wave length

1639. Electromagnetic radiation travels through space at the speed of
(*a*) Sound (*b*) Wind
(*c*) Light (*d*) None of these

❑❑❑

Answers

1632. (*b*) 1633. (*c*) 1634. (*c*) 1635. (*c*) 1636. (*d*) 1637. (*c*) 1638. (*c*) 1639. (*c*)

IDLH Concentrations of Chemical Substances

Immediately Dangerous to Life/Health (IDLH) Concentrations of Various Substances/Chemicals (as per NIOSH as on 03/01/1995)

S. No.	Substances/chemicals	IDLH Value
1	Acetaldehyde	2,000 ppm
2	Acetic acid	50 ppm
3	Acetic anhydride	200 ppm
4	Acetone	2,500 ppm [LEL]
5	Acetonitrile	500 ppm
6	Acetylene terabromide	8 ppm
7	Acrolein	2 ppm
8	Acrylamide	60 mg/m^3
9	Acrylonitrile	85 ppm
10	Aldrin	25 mg/m^3
11	Allyl alcohol	20 ppm
12	Allyl chloride	250 ppm
13	Allyl glycidyl ether	50 ppm
14	2 Aminopyridine	5 ppm
15	Ammonia	300 ppm
16	Ammonium sulfamate	1,500 mg/m^3
17	n-Amyl acetate	1,000 ppm
18	Sec-Amyl acetate	1,000 ppm
19	Aniline	100 ppm
20	o-Anisidine	50 mg/m^3
21	p-Anisidine	50 mg/m^3
22	Antimony compounds (as Sb)	50 Sb/m^3
23	Arsenic (inorganic compounds, as As)	5 mg As/m^3
24	Arsine	3 ppm
25	Azinphosmethyl	10 mg/m^3
26	Barium (soluble compounds, as Ba)	50 mg Ba/m^3
27	Benzene	500 ppm
28	Benzoyl peroxide	1,500 mg/m^3
29	Benzyl chloride	10 ppm
30	Beryllium compounds (as Be)	4 mg Be/m^3
31	Boron oxide	2,000 mg/m^3
32	Boron trifluoride	25 ppm
33	Bromine	3 ppm
34	Bromoform	850 ppm
35	1,3–Butadiene	2,000 ppm [LEL]
36	2-Butanone	3,000 ppm
37	2-Butoxyethanol	700 ppm
38	n-Butyl acetate	1,700 ppm
39	Sec-Butyl acetate	1,700 ppm
40	Tert-Butyl acetate	1,500 ppm
41	n-Butyl alcohol	1,400 ppm
42	Sec-Butyl alcohol	2,000 ppm
43	Tert-Butyl alcohol	1,600 pm
44	n-Butylamine	300 ppm
45	Tert-Butyl chromate	15 mg Cr(VI)/m^3
46	n-Butyl glycidyl ether	250 ppm

S. No.	Substances/chemicals	IDLH Value
47	n-Butyl mercaptan	500 ppm
48	p-tert-Butyltoluene	100 ppm
49	Cadmium dust (as Cd)	9 mg Cd/m^3
50	Cadmium fume (as Cd)	9 mg Cd/m^3
51	Calcium arsenate (as As)	5 mg As/m^3
52	Calcium oxide	25 mg/m^3
53	Camphor (synthetic)	200 mg/m^3
54	Carbaryl	100 mg/m^3
55	Carbon black	1,750 mg/m^3
56	Carbon dioxide	40,000 ppm
57	Carbon disulfide	500 ppm
58	Carbon monoxide	1,200 ppm
59	Carbon tetrachloride	200 ppm
60	Chlordane	100mg/m^3
61	Chlorinated camphene	200 mg/m^3
62	Chlorinated diphenyl oxide	5 mg/m^3
63	Chlorine	10 ppm
64	Chlorine dioxide	5 ppm
65	Chlorine trifluoride	20 ppm
66	Chloroacetatldehyde	45 ppm
67	Alpha-Chloroacetophenone	15 mg/m^3
68	Chlorobenzene	1,000 ppm
69	o-Chlorobenzylidene malononitrile	2 mg/m^3
70	Chlorobromomethane	2,000 ppm
71	Chlorodiphenyl (42% chlorine)	5 mg/m^3
72	Chlorodiphenyl (54% chlorine)	5 mg/m^3
73	Chloroform	500 ppm
74	1-Chloro-1-nitropropane	100 ppm
75	Chloropicrine	2 ppm
76	Bet-Chloroprene	300 ppm
77	Chromic acid and chromates	15 mg Cr(VI) /m^3
78	Chromium (II) compounds [as Cr(II)]	250 mg Cr(II) /m^3
79	Chromium (III) compounds [as Cr(III)]	25 mg Cr(III) /m^3
80	Chromium metal (as Cr)	250 mg Cr/m^3
81	Coal tar pitchvolatiles	80 mg/m^3
82	Cobalt metal, dust and fume (as Co)	20 mg CO//m^3
83	Copper (dusts and mists, as Cu)	100 mg Cu/m^3
84	Copper fume (as Cu)	100 mg Cu/m^3
85	Cotton dust (raw)	100 mg/m^3
86	Crag (r) herbicide	500 mg/m^3
87	Cresol (o, m, p isomers)	250 ppm
88	Crotonaldehyde	50 ppm
89	Cumene	900 ppm [LEL]
90	Cyanides (as CN)	25 mg/m^3 (as CN)
91	Cyclohexane	1,300 ppm [LEL]
92	Cyclohexanol	400 ppm
93	Cyclohexanone	700 ppm
94	Cyclohexene	2,000 ppm
95	Cyclopentadiene	750 ppm
96	2,4-D	100 mg/m^3
97	DDT	500 mg/m^3
98	Decaborane	15 mg/m^3
99	Demeton	10 mg/m^3
100	Diacetone alcohol	1,800 ppm [LEL]
101	Diazomethane	2 ppm
102	Diborane	15 ppm
103	Dibutyl phosphate	30 ppm
104	Dibutyl phthalate	4,000 mg/m^3
105	o-Dichlorobenzene	200 ppm
106	p-Dichlorobenzene	150 ppm
107	Dichlorodifluoromethane	15,000 ppm
108	1,3-Dichloro 5,5-dimethylhydantoin	5 mg/m^3
109	1,1-Dichloroethane	3,000 ppm
110	1,2-Dichloroethylene	1,000 ppm
111	Dichloroethyl ether	100 ppm
112	Dichloromono-fluoromethane	5,000 ppm
113	1,1-Dichloro 1-nitroethane	25 ppm

S. No.	Substances/chemicals	IDLH Value
114	Dichlorotetra-fluoroethane	15,000 ppm
115	Dichlorvos	100 mg/m^3
116	Dieldrin	50 mg/m^3
117	Diethylamine	200 ppm
118	2-Diethylaminoethanol	100 ppm
119	Difluorodibromo-methane	2,000 ppm
120	Diglycidyl ether	10 ppm
121	Diisobutyl ketone	500 ppm
122	Diisopropylamine	200 ppm
123	Dimethyl acetamide	300 ppm
124	Dimethylamine	500 ppm
125	N,N-Dimethylaniline	100 ppm
126	Dimethyl 1,2-dibromo 2,2-dichlorethyl phosphate	200 mg/m^3
127	Dimethylformamide	500 ppm
128	1,1-Dimethylhydrazine	15 ppm
129	Dimethylphthalate	2,000 mg/m^3
130	Dimethyl sulfate	7 ppm
131	Dinitrobenzene (o, m, p isomers)	50 mg/m^3
132	Dinitroocresol	5 mg/m^3
133	Dinitrotoluene	50 mg/m^3
134	Di sec-octyl phthalate	5,000 mg/m^3
135	Dioxane	500 ppm
136	Diphenyl	100 mg/m^3
137	Dipropylene glycol methyl ether	600 ppm
138	Dndrin	2 mg/m^3
139	Epichlorohydrin	75 ppm
140	EPN	5 mg/m^3
141	Ethanolamine	30 ppm
142	2-Ethoxyethanol	500 ppm
143	2-Ethoxyethyl acetate	500 ppm
144	Ethyl acetate	2,000 ppm [LEL]
145	Ethyl acrylate	300 ppm
146	Ethyl alcohol	3,300 ppm [LEL]
147	Ethylamine	600 ppm
148	Ethyl benzene	800 ppm [LEL]

S. No.	Substances/chemicals	IDLH Value
149	Ethyl bromide	2,000 ppm
150	Ethyl butyl ketone	1,000 ppm
151	Ethyl chloride	3,800 ppm [LEL]
152	Ethylene chlorohydrins	7 ppm
153	Ethylenediamine	1,000 ppm
154	Ethylene dibromide	100 ppm
155	Ethylene dichloride	50 ppm
156	Ethylene glycol dinitrate	75 mg/m^3
157	Ethyleneimine	100 ppm
158	Ethylene oxide	800 ppm
159	Ethyl ether	1,900 ppm [LEL]
160	Ethyl formate	1,500 ppm
161	Ethyl mercaptan	500 ppm
162	N-Ethylmorpholine	100 ppm
163	Ethyl silicate	700 ppm
164	Ferbam	800 mg/m^3
165	Ferrovanadium dust	500 mg/m^3
166	Fluorides (as F)	250 mg F/m^3
167	Fluorine	25 ppm
168	Fluorotrichloromethane	2,000 ppm
169	Formaldehyde	20 ppm
170	Formic acid	30 ppm
171	Furfural	100 ppm
172	Furfuryl alcohol	75 ppm
173	Gycidol	150 ppm
174	Graphite (natural)	1,250 mg/m^3
175	Hafnium compounds (as Hf)	50 mg Hf//m^3
176	Heptachlor	35 mg/m^3
177	n-Heptane	750 ppm
178	Hexachloroethane	300 ppm
179	Hexachloronaphthalene	2 mg/m^3
180	n-Hexane	1,100 ppm [LEL]
181	2-Hexanone	1,600 ppm
182	Hexone	500 ppm
183	sec Hexyl acetate	500 ppm
184	Hydrazine	50 ppm
185	Hydrogen bromide	30 ppm

S. No.	Substances/chemicals	IDLH Value
186	Hydrogen chloride	50 ppm
187	Hydrogen cyanide	50 ppm
188	Hydrogen fluoride (as F)	30 ppm
189	Hydrogen peroxide	75 ppm
190	Hydrogen selenide (as Se)	1 ppm
191	Hydrogen sulfide	100 ppm
192	Hydroquinone	50 mg/m^3
193	Iodine	2 ppm
194	Iron oxide dust and fume (as Fe)	2,500 mg Fe/m^3
195	Isoamyl acetate	1,000 ppm
196	Isoamyl alcohol (primary and secondary)	500 ppm
197	Isobutyl acetate	1,300 ppm [LEL]
198	Isobutyl alcohol	1,600 ppm
199	Isophorone	200 ppm
200	Isopropyl acetate	1,800 ppm
201	Isopropyl alcohol	2,000 ppm [LEL]
202	Isopropylamine	750 ppm
203	Isopropyl ether	1,400 ppm [LEL]
204	Isopropyl glycidyl ether	400 ppm
205	Ketene	5 ppm
206	Lead compounds (as Pb)	100 mg Pb/m^3
207	Lindane	50 mg/m^3
208	Lithium hydride	0.5 mg/m^3
209	L.P.G.	2,000 ppm [LEL]
210	Magnesium oxide fume	750 mg/m^3
211	Malathion	250 mg/m^3
212	Maleic anhydride	10 mg/m^3
213	Mangnese compounds (as Mn)	500 mg Mn/m^3
214	Mercury compounds [except (organo) alkyls, as Hg]	10 mg Hg/m^3
215	Mercury (organo) alkyl compounds (as Hg)	2 mg Hg/m^3
216	Mesityl oxide	1,400 ppm [LEL]
217	Methoxychlor	5,000 mg/m^3
218	Methyl acetate	3,100 ppm [LEL]
219	Methyl acetylene	1,700 ppm [LEL]
220	Methyl acetylenepropadiene mixture	3,400 ppm [LEL]
221	Methyl acrylate	250 ppm
222	Methylal	2,200 ppm [LEL]
223	Methyl alcohol	6,000 ppm
224	Methylamine	100 ppm
225	Methyl (namyl) ketone	800 ppm
226	Methyl bromide	250 ppm
227	Methyl Cellosolve (r) acetate	200 ppm
228	Methyl chloride	200 ppm
229	Methyl chloroform	2,000 ppm
230	Methylcyclohexane	700 ppm
231	Methylcyclonexanol	1,200 ppm [LEL]
232	o-Methylcyclohexanone	500 ppm
233	Methylene bisphenyl isocyanate	600 ppm
234	Methylene chloride	75 mg/m^3
235	Methyl formate	2,300 ppm
236	5-Methyl 3-heptanone	4,500 ppm
237	Methyl hydrazine	100 ppm
238	Methyl hydrazine	20 ppm
239	Methyl iodide	100 ppm
240	Methyl isobutyl carbinol	400 ppm
241	Methyl isocyanate	3 ppm
242	Methyl mercaptan	150 ppm
243	Methyl methacrylate	1,000 ppm
244	Methyl styrene	700 ppm
245	Mica	1,500 mg/m^3
246	Molybdenum (insoluble compounds, as Mo)	5,000 mg Mo/m^3
247	Molybdenum (soluble compounds, as Mo)	1,000 mg Mo/m^3
248	Monomethyl aniline	100 ppm
249	Morpholine	1,400 ppm [LEL]
250	Naphtha (coal tar)	1,000 ppm [LEL]
251	Naphthalene	250 ppm
252	Nickel carbonyl (as Ni)	2 ppm

S. No.	Substances/chemicals	IDLH Value
253	Nickel metal and other compounds (as Ni)	10 mg Ni/m^3
254	Nicotine	5 mg/m^3
255	Nitric acid	25 ppm
256	Nitric oxide	100 ppm
257	p-Nitroaniline	300 mg/m^3
258	Nitrobenzene	200 ppm
259	p-Nitrochlorobenzene	100 mg/m^3
260	Nitroethane	1,000 ppm
261	Nitrogen dioxide	20 ppm
262	Nitrogen trifluoride	1,000 ppm
263	Nitroglycerine	75 mg/m^3
264	Nitromethane	750 ppm
265	1-Nitropropane	1,000 ppm
266	2-Nitropropane	100 ppm
267	Nitrotoluene (o, m, p isomers)	200 ppm
268	Octachloronapthalene	Unknown
269	Octane	1,000 ppm [LEL]
270	Oil mist (mineral)	2,500 mg/m^3
271	Osmium tetroxide (as Os)	1 mg Os/m^3
272	Oxalic acid	500 mg/m^3
273	Oxygen difluoride	0.5 ppm
274	Ozone	5 ppm
275	Paraquat	1 mg/m^3
278	Parathion	10 mg/m^3
279	Pentachlorophenol	1 ppm
280	n-Pentane	Unknown
281	2Pentanone	2.5 mg/m^3
282	Perchloromethyl mercaptan	1,500 ppm [LEL]
283	Perchloryl fluoride	1,500 ppm
284	Petroleum distillates (naphtha)	10 ppm
285	Phenol	100 ppm
286	p-Phenylene diamine	1,100 ppm [LEL]
287	Phenyl ether (vapor)	250 ppm
288	Phenyl etherbiphenyl mixture (vapor)	25 mg/m^3
289	Phenyl glycidyl ether	100 ppm
290	Phenylhydrazine	15 ppm
291	Phosdrin	4 ppm
292	Phosgene	2 ppm
293	Phosphine	50 ppm
294	Phosphoric acid	1,000 mg/m^3
295	Phosphorus (yellow)	5 mg/m^3
296	Phosphrous pentachloride	70 mg/m^3
297	Phosphorus pentasulfide	250 mg/m^3
298	Phosphorus trichloride	25 ppm
299	Phthalic anhydride	60 mg/m^3
300	Picric acid	75 mg/m^3
301	Pindone	100 mg/m^3
302	Platinum (soluble salts, as Pt)	4 mg Pt/m^3
303	Portland cement	5,000 mg/m^3
304	Propane	2,100 ppm [LEL]
305	n-Propyl acetate	1,700 ppm
306	n-Propyl alcohol	800 ppm
307	Propylene dichloride	400 ppm
308	Propylene imine	100 ppm
309	Propylene oxide	400 ppm
310	n-Propyl nitrate	500 ppm
311	Pyrethrum	5,000 mg/m^3
312	Pyridine	1,000 ppm
313	Quinone	100 mg/m^3
314	Rhodium (metal fume and insoluble compounds, as Rh)	100 mg Rh/m^3
315	Rhodium (soluble compounds, as Rh)	2 mg Rh/m^3
316	Ronnel	300 mg/m^3
317	Rotenone	2,500 mg/m^3
318	Selenium compounds (as Se)	1 mg Se/m^3
319	Selenium hexafluoride	2 ppm
320	Silica, amorphous	3,000 mg/m^3
321	Silica, carystalline (respirable dust)	
322	Cristobalite/tridymite:	25 mg/m^3
323	Quartz/tripoli:	50 mg/m^3

S. No.	Substances/chemicals	IDLH Value
324	Silver (metal dust and soluble compounds, as Ag)	10 mg Ag/m^3
325	Soapstone	3,000 mg/m^3
326	Sodium fluoroacetate	2.5 mg/m^3
327	Sodium hydroxide	10 mg/m^3
328	Stibine	5 ppm
329	Stoddard solvent	20,000 mg/m^3
330	Strychnine	3 mg/m^3
331	Styrene	700 ppm
332	Sulfur dioxide	100 ppm
333	Sulfuric acid	15 mg/m^3
334	Sulfur monochloride	5 ppm
335	Sulfur pentafluoride	1 ppm
336	Sulfuryl fluoride	200 ppm
337	2,4,5-T	250 mg/m^3
338	Talc	1,000 mg/m^3
339	Tantalum (metal and oxide dust, as Ta)	2,500 mg Ta/m^3
340	TEDP	10 mg/m^3
341	Tellurium compounds (as Te)	25 mg Te/m^3
342	Tellurium hexafluoride	1 ppm
343	TEPP	5 mg/m^3
344	Terphynyl (o, m, p isomers)	500 mg/m^3
345	1,1,1,2-Terachloro 2,2-difluoroethane	2,000 ppm
346	1,1,2,2,-Tetrachloro 1,2-difluoroethane	2,000 ppm
347	1,1,2,2-Tetrachloroethane	100 ppm
348	Tetrachloroethylene	150 ppm
349	Tetrachloronaphthalene	Unknown
350	Tetraethyl lead (as Pb)	40 mg Pb/m^3
351	Tetrahydrofuran	2,000 ppm [LEL]
352	Tetramethyl lead (as Pb)	40 mg Pb/m^3
353	Tetramethyl succinonitrile	5 ppm
354	Tetranitromethane	4 ppm
355	Tetryl	750 mg/m^3
356	Thallium (soluble compounds, as Tl)	15 mg Tl/m^3
357	Thiram	100 mg/m^3
358	Tin (inorganic compounds, as Sn)	100 mg Sn/m^3
359	Tin (Organic compounds, as Sn)	25 mg Sn/m^3
360	Titanium dioxide	5,000 mg/m^3
361	Toluene	500 ppm
362	Toleuene 2,4-diisocyanate	2.5 ppm
363	o-Toluidine	50 ppm
364	Tributyl phosphate	30 ppm
365	1,1,2-Trichloroethane	100 ppm
366	Trichloroethylene	1,000 ppm
367	Trichloronaphthalene	Unknown
368	1,2,3-Trichloropropane	100 ppm
369	1,1,2-Trichloro 1,2,2-trifluoroethane	2,000 ppm
370	Triethylamine	200 ppm
371	Trifluorobromomethane	40,000 ppm
372	2,4,6-Trinitrotoluene	500 mg/m^3
373	Triorthocresyl phosphate	40 mg/m^3
374	Triphenyl phosphate	1,000 mg/m^3
375	Turpentine	800 ppm
376	Uranium (insoluble compouds, as U)	10 mg U/m^3
377	Uranium (soluble compounds, as U)	10 mg U/m^3
378	Vanadium dustVanadium fume	35 mg V/m^3
379	Vinyl toluene	35 mg V/m^3
380	Warfarin	400 ppm
381	Xylene (o, m, p isomers)	100 mg/m^3
382	Xylidine	900 ppm
383	Xylidine	50 ppm
384	Yttrium compounds (as Y)	500 mg Y/m^3
385	Zinc chloride fume	50 mg/m^3
386	Zinc oxide	500 mg/m^3
387	Zirconium compounds (as Zr)	25 mg Zr/m^3

Note: LEL - Lower Explosion Limit

Note: The current NIOSH definition for an IDLH condition is one that poses a threat of exposure to airborne contaminants when that exposure is likely to cause death or immediate or delayed permanent adverse health effects or prevent escape from such an environment. The purpose of establishing an IDLH value is (1) to ensure that the worker can escape from a given contaminated environment in the event of failure of the respiratory protection equipment and (2) is considered a maximum level above which only a highly reliable breathing apparatus providing maximum worker protection is permitted. In establishing the IDLH value, the following conditions must be assured:

A. The ability to escape without loss of life or immediate or delayed irreversible health effects. (Thirty minutes is considered the maximum time for escape so as to provide some margin of safety in calculating an IDLH value.)

B. The prevention of severe eye or respiratory irritation or other reactions that would hinder escape.

"Highly reliable" respirators (i.e., the most protective respirators) would be selected for emergency situations, fire fighting, exposure to carcinogens, entry into oxygen-deficient atmospheres, entry into atmospheres that contain a substance at a concentration greater an 2,000 times the NIOSH REL or OSHA PEL, and for entry into immediately dangerous to life or health conditions. These "highly reliable" respirators include either a self-contained breathing apparatus that has a full facepiece and is operated in a pressure-demand or other positive-pressure mode, or supplied-air respirator that has a full facepiece and is operated in a pressure-demand or other positive¬pressure mode in combination with an auxiliary SCBA operated in a pressure-demand or other positive pressure mode.

This list includes IDLH values for 85 substances (e.g., benzene and methylene chloride) determined by NIOSH to meet the OSHA definition of "potential occupational carcinogen", For all of these substances, except ethylene oxide and crystalline silica, NIOSH recommends that the "most protective" respirators be worn by workers exposed at concentrations above the NIOSH REL, or at any detectable concentration when there is no REL. For ethylene oxide and crystalline silica, TIOSH recommends that the "most protective" respirators be worn in concentrations exceeding 5 ppm and 25 mg/rng, respectively.

The criteria utilized to determine the adequacy of existing IDLH values were a combination ofthose used during the SCP and a newer methodology developed by NIOSH. These criteria form a tiered approach with acute human toxicity data being used preferentially, followed next by acute animal inhalation toxicity data, and then finally by acute animal oral toxicity data to determine an updated IDLH value. When relevant acute toxicity data were insufficient or unavailable, then the use of chronic toxicity data or an analogy to a chemical with similar toxic effects was considered. In order to facilitate the revision process, secondary toxicological data were primarily used. Once a preliminary IDLH value was developed, it was compared to the existing IDLH value and to several other factors (e.g., existing short-term exposure guidelines and lower explosive limits).

The following "hierarchy" was followed to develop a "preliminary" value for the revised IDLH value:

A. Human acute toxicity data were used if sufficient to determine a concentration that for up to 30 minutes does not cause death, serious or irreversible health effects, or does not impair or impede the ability to escape.

B. Animal acute lethal concentration (LC) data were considered next. The only animal lethal concentration data used involved mammals; the vast majority of the data was from studies of rats, mice, guinea pigs, and hamsters. It was decided to generally use the lowest reliable LC data, with LC_{50} data preferred. If acute LC data determined during a 30-minute period were not available, then the data, based on a study. were "adjusted" to an equivalent 30-minute value using the following relationship:

Adjusted LC_{50} (30 minutes)

$$= LC_{50}(t) * (t/0.5)^{**} (l/n)$$

where:

$LC_{50}(t) = LC_{50}$ determined over t hours

n = constant*

*Note: Ten Berge determined the relationship shown above based on experimental data. The constant "n" was determined by ten Berge to be less than 3.0 for 18 of the 20 substances studied. Although the individual "n" values determined by ten Berge were utilized when applicable during the review and revision of the original IDLHs, as a conservative estimate, an "n" = 3.0 was assumed when "adjusting" the LC data to 30 minutes for all other substances.

This equation with an "n" = 3.0 results in the following correction factors:

t (hours)	Correction factor
0.5	1.0
1	1.25
2	1.6
3	1.8
4	2.0
5	2.15
6	2.3
7	2.4
8	2.5

The LC values (after "adjusting" if necessary to 30 minutes) were divided by a safety factor of 10 to determine a "preliminary" IDLH for comparison purposes.

C. Animal lethal dose (LD) data were considered next. As was the case with the lethal concentration data, the only animal lethal dose data used involved mammals; the vast majority of the data were from studies of rats, mice, guinea pigs, and hamsters. It was decided to generally use the lowest LD data with oral LD_{50} data preferred. The LD data was used to determine the equivalent total dose to a 70-kg worker and, the air concentration containing this dose was determined by dividing by 10 cubic meters. [A worker breathing at a rate of 50 liters per minute for 30 minutes would inhale 1.5 cubic meters of air.] A "preliminary" IDLH for comparison purposes was determined by dividing these air concentrations by a safety factor of 10.

D. Chronic toxicity data were considered if no relevant acute toxicity data existed. However, the fact that chronic exposures may have limited relevance to acute effects was taken into consideration.

E. When relevant toxicity data applying specifically to the chemicals in question were lacking, and if it was determined to be justified, then analogies to substances with similar acute toxic effects were considered.

F. All "preliminary" IDLHs derived during this update were checked against the following factors prior to establishing the final "revised" IDLH:

1. Lower explosive limit (LEL):It was decided to restrict the "routine" entry into a possible explosive atmosphere to concentrations no greater than 10% of the LEL. However, OSHA considers concentrations in excess of 10% of the LEL to be a hazardous atmosphere in confined spaces
2. RD_{50} data: An RD_{50} is defined as the 10-minute exposure concentration producing a 50% respiratory rate decrease in mice or rats and can be used to estimate severe respiratory irritation. Prolonged exposure to an RDso concentration has been shown to produce respiratory tract lesions consistent with irritation.
3. Other short-term exposure guidelines such as the American Industrial Hygiene Association's emergency response planning guidelines (ERPGs) and the National Research Council's emergency exposure guidance levels (EEGLs) and short-term public emergency guidance levels (SPEGLs), and occupational exposure standards or recommendations such as OSHA PELs, NIOSH RELs, or the American Conference of Governmental Industrial Hygienists (ACGIH) TLVs.
4. Based on the NIOSH respirator decision logic, the revised IDLH values could not be greater than 2,000 times the NIOSH REL (or OSHA PEL).

Abbreviations used for IDLH Documentation

LC	lethal concentration
LC_{50}	concentration causing death in 50% of test animals
LD_{50}	dosage causing death in 50%
LEL	lower explosive limit
mg/kg	milligrams per kilogram of body weight
mg/m	milligrams per cubic meter of air
mppcf	millions of particles per cubic foot of air
NIOSH	National Institute for Occupational Safety and Health
OSHA	Occuptional Safety and Health Administration
PEL	permissible exposure limit (OSHA)
Ppm	parts per million parts of air
RD_{50}	concentration producing a 50% decrease in respiratory rate following a 10-minute exposure
REL	recommended exposure limit (NIOSH)
STEL	Short-term exposure limit
TLV	threshold limit value (ACGIH)
TWA	time-weighted average

KHANNA'S MULTI-CHOICE QUESTIONS & ANSWER IN

ENVIRONMENTAL ENGINEERING

O.P. GUPTA

FEEDBACK FORM

1. Which topics from your examination point of view are inadequately or not covered in this book?

2. Is there any misprints/ mistakes/ factual inaccuracy in this book? If so, please specify.

3. What is your assessment of this book as regards the presentation of the subject-matter, expression, precision and price in relation to the other recommended books available on this subject?

4. Which competing books you regard as better than this? Please mention their author and publishers.

5. Any further suggestions/comments you would like to make for the improvement of the book?

PS: If needed, please attach a separate sheet.

Name :

College :

Address :

Email :

Mobile :

Please mail the feedback form at:

KHANNA BOOK PUBLISHING CO. (P) LTD.

4C/4344, Ansari Road, Daryaganj, New Delhi-110002

You can also email this feedback form at contact@khannabooks.com or Whatsapp at +91 – 99109 09320

BEST FEEDBACK WILL BE REWARDED SUITABLY